Unity and Disunity in Evolutionary Biology

Richard G. Delisle • Maurizio Esposito •
David Ceccarelli
Editors

Unity and Disunity in Evolutionary Biology

Deconstructing Darwinism

Springer

Editors
Richard G. Delisle
School of Liberal Education
University of Lethbridge
Lethbridge, AB, Canada

Maurizio Esposito
Department of History and Philosophy of Sciences, Interuniversity Center for the History of Science and Technology
University of Lisbon
Lisbon, Portugal

David Ceccarelli
Department of Philosophy
Communication and Performing Arts
Roma Tre University
Rome, Italy

ISBN 978-3-031-42628-5 ISBN 978-3-031-42629-2 (eBook)
https://doi.org/10.1007/978-3-031-42629-2

This Springer imprint is published by the registered company Springer Nature Switzerland AG
The registered company address is: Gewerbestrasse 11, 6330 Cham, Switzerland

This volume is dedicated to the memory of Mark B. Adams (1944–2024).—The Editors

Contents

Part I

Introductory Essays

Toward a New Historiography

David Ceccarelli, Maurizio Esposito, and Richard G. Delisle

Abstract

The history of evolutionary biology presents well-established categorizations and labels that have significantly influenced the imaginary of evolutionism. The somewhat uncritical understanding and use of such labels demand a thorough reconsideration of traditional narratives, thus paving the way for new research avenues to emerge.

Keywords

History of evolutionary biology · Darwinism · New historiography of evolutionary biology · Presentism · Modern Synthesis · Extended evolutionary synthesis · Darwinian industry

It is not uncommon to see in major areas of research concerned with science and human affairs—the Scientific Revolution, Enlightenment, Materialism, Industrial Revolution, Cold War studies, etc.—that historical studies are accompanied by the rise of complementary or contradictory historiographies. Over time, scholars

D. Ceccarelli (✉)
Department of Philosophy, Communication and Performing Arts, Roma Tre University, Rome, Italy
e-mail: david.ceccarelli@uniroma3.it

M. Esposito
Department of History and Philosophy of Sciences, Interuniversity Center for the History of Science and Technology, University of Lisbon, Lisbon, Portugal
e-mail: mesposito@fc.ul.pt

R. G. Delisle
School of Liberal Education, University of Lethbridge, Lethbridge, AB, Canada
e-mail: richard.delisle@uleth.ca

R. G. Delisle et al. (eds.), *Unity and Disunity in Evolutionary Biology*,
https://doi.org/10.1007/978-3-031-42629-2_1

continually devise new methodologies, leading to the critical reevaluation of established concepts, traditions, periodization, and historical vocabulary. Curiously, the field of evolutionary biology has undergone a somewhat different fate. Here, long-standing historiographic labels such as the "Darwinian Revolution," "Eclipse of Darwinism," and "Modern Synthesis" have persisted largely unchanged for decades. Although scientists, philosophers, and historians of biology have increasingly engaged in debates surrounding these labels (especially since the 1980s), they are still used with ease. Their persistence, largely unaltered despite the shifting landscape of scholarship, invites reflection on the underlying factors influencing their longevity and the potential need for reassessment in light of contemporary insights and perspectives. No doubt, historiographical categories have long performed various important functions in evolutionary biology. Labels such as "Darwinism" and "Lamarckism," for example, were long uncritically believed to hold a heuristic value and exploited by scholars to put some order in the disputes between scientists, creating somewhat archetypical conceptual systems whose borders are, however, highly mutable. At the same time, they have served a dual purpose: delineating distinct research traditions and unifying research and institutional networks. Since the late nineteenth century, evolutionary biology frequently saw scientists grappling with the task of accurately labeling their work and rejecting labels perceived as problematic. These efforts were often instrumental in defining the parameters of research programs, establishing and consolidating schools of thought, as well as efficiently marginalizing inconvenient ideas.

By definition, historiographical categories, and particularly "-isms," are complex objects whose uncritical use could lead to ambiguity. The transformation of a scientific research program into a specific label involves complex conceptual, epistemological, and ideological interplays. Primarily, it consists in abstracting various aspects of an argument into one or more explanatory cores. Additionally, it is a process of "reification," merging diverse layers of significance and potential conceptual ramifications into a singular term. As Lovejoy (1960, pp. 4–6) argued, due to their inherent temporality, historiographical labels rarely designate a singular concept, but "distinct and often conflicting doctrines held by different individuals or groups to whose way of thinking these appellations have been applied."[1]

This volume aims to move beyond the static and often uncritical understanding of historiographic labels in evolutionary biology. Through examining the existing debate in the literature, the authors of this volume started from shared questions developing their investigative lines around profoundly intertwined historiographical, epistemological, and sociological issues:

- How are historiographical labels constructed and deconstructed over time?
- What is the relation between evolutionary biology and its history?
- Has the historiography of evolutionary biology been preponderantly whiggish? If yes, what does that mean specifically?

[1]Lovejoy (1960).

- Are scientists' retrospective interest in the history of evolutionism and historians' call for overcoming old-fashioned dichotomies irreconcilable?
- What is the impact of national styles of thinking on our understanding of evolutionary biology?
- What is the part played by mythical consideration in Darwin studies?
- Have historians and philosophers been co-opted by major evolutionists?
- Was pluralism always maintained since the inception of evolutionary biology?
- What is the impact of looking at evolutionary biology through different lenses such as "theory," "paradigm," "research program," and "research tradition"?
- Have evolutionary biologists been actively engaged in rhetorical arguments?
- Is it useful to rethink the periodization of the development of evolutionary biology by proposing a brand new one?
- Is "Darwinism" a social construct mainly generated under the impetus of the English-speaking world?
- To what extent was the Darwinian revolution an "ideological" one?
- Is it constructive to approach the field of evolutionary biology through a "mechanism-centric" view as largely done thus far?
- What is the impact of positivism and analytical philosophy on our view of the development of evolutionary biology?

The volume comprises four sections. The Introductory Essays section gathers reflections on the current state of the historiography within evolutionary biology, setting the stage for the key themes addressed throughout the book. Here, the co-editors examine the methodological and sociological foundations of conventional historiography, the relevance of "Darwin's paradigmatic story" in shaping the contemporary imaginary of evolutionism, and the multi-layered meanings and ideological ramifications of the term "Darwinism."

In Part I, titled "Deconstructing Darwinism," Antonello La Vergata highlights how the term "Darwinism" conceals a spectrum of interpretations that emerged shortly after 1859. Biology exhibited pluralism both before and after Darwin, and this richness and complexity often fostered diverse approaches and interpretations, sometimes exacerbating the apparent gap between the so-called "two cultures." Mark B. Adams's contribution, "The Evolution of 'Darwinism': Up Close and Personal," reveals an enduring tension between scientists' retrospective interest in history and historians' advocacy for transcending outdated dichotomies. The author examines the multifaceted dimensions that concepts like "evolution," "Darwinism," "natural selection," "micro-evolution," and "macro-evolution" have assumed across various contexts, dismantling the conventional "just-so-story" of Darwinism's history that has solidified within the Modern Synthesis narrative. Delving into narratives that have long lingered in the background of traditional historiography is crucial for unveiling new perspectives in contemporary historical research. In this vein, Nicolaas Rupke presents a thought-provoking and bold analysis, deconstructing the portrayal of Richard Owen in evolutionary biology as propagated by the "Darwinian industry." In his "Darwin, Archaeopteryx lithographica, and the Problem of Intermediate Species," Bogdana Stamenkovic critically analyzes

Darwin's interpretation of species genealogy about paleontological evidence, addressing the inherent challenges of reconciling rigid Darwinian gradualism with the notion of "intermediate form." Richard Delisle's chapter further explores the idea that Darwin's long argument is imbued with the classic nineteenth-century epistemological doctrine of uniformitarianism. Here, the notion of "evolutionary contingency" typically associated with Darwinism is deconstructed by examining the influence of Ernst Mayr's and Stephen Jay Gould's contributions on the subject. The section concludes with Joël Dolbeault's reflection on the epistemological nature of Darwin's theory, probing the ongoing relevance of Karl Popper's characterization of Darwinism as a "metaphysical research program."

The second section of the volume, "Around and Beyond the Synthesis," is dedicated to deconstructing the synthesis discourses that have emerged in the history of evolutionary biology. A multifaceted scenario unfolds, where synthesis attempts manifest as polycentric phenomena spread across various geographical contexts and historical phases. In "Typology/Population Distinction and Its Role in the Marginalization of 19th-Century Non-Darwinian Theories in Modern Historiography," Michał Wagner posits that historical narratives, such as Mayr's "essentialist story," played a crucial role in oversimplifying the species concept within non-Darwinian theories of the nineteenth century, strengthening the connection between Darwinism and the Modern Synthesis as well as fostering unity within the scientific community. The rhetoric of unity conceals a much more intricate history, wherein research traditions, levels, and languages of analysis present themselves in all their complexity. Joël Dolbeault's chapter "Fisher, Wright and Haldane: Three Philosophical Conceptions of Evolution," for example, discusses the different philosophical underpinnings that characterized the works of the advocates of the so-called early synthesis. In their chapter, Jan Badke, Alexander Böhm, Stefan Reiners-Selbach, and Vera Straetmanns challenge the perception of Darwinism as a dominant theory during the interwar period, a time marked by the emergence and diversification of various research pathways. In German theoretical biology, scientists attempted to overcome theoretical and methodological fragmentation by developing a philosophical framework that could unify biological knowledge, a process that marginalized the discussion of Darwinism. Among the various dynamics characterizing the context of the Modern Synthesis, the tendency to overlook figures who were not entirely aligned with its agenda also emerges clearly. Silvia Caianiello delves into this phenomenon in her work, "The Strange Story of Mosaic Evolution," focusing on the case of Gavin de Beer, whose formulation of "mosaic evolution" posed a significant challenge to the predominant selectionist explanation of macroevolutionary trends. The section concludes with Carlos Ocha inviting readers to reconsider the assumptions of the Extended Evolutionary Synthesis, discussing potential alternative theoretical scenarios that could offer a more comprehensive integration among the various multilevel phenomena that characterize biological evolution.

The third section, "Deconstructing the Historiography of Evolutionary Biology," offers a critical analysis of pivotal concepts, themes, and episodes in the history of evolutionary biology. It begins with Maurizio Esposito's reconstruction of the emergence and evolution of crucial philosophical foundations of "evolutionism," i.e., the concepts of materialism and the historicization of nature, which set the conditions of possibility for modern evolutionary biology and, most notably, predate Darwinism. In his chapter, "What if Darwin had published his 1844 Essay?" Derek Partridge challenges the conventional notion that Darwin's evolutionary theory evolved through a strictly gradual and linear process of theoretical refinement and empirical analysis. Partridge examines the main discontinuities within Darwinian theory from the 1840s to the late 1850s through a counterfactual historical lens. Focusing on a wider time frame, David Ceccarelli questions the semantic stability of historiographical labels such as "Darwinism," "Lamarckism," and "non-Darwinian evolution" by examining how scholars assessed Darwin's endorsement of the inheritance of acquired characters during the 1909, 1959, and 2009 celebrations of the publication of the *Origin of Species*. Building upon the analysis of shifting perspectives on evolutionary theory, Juan Manuel Rodríguez Caso delves into the case study of the reception of the Anglo-centric literature on Darwinism in Mexico, showing how such incorporation sometimes led to misinterpretations and the propagation of controversies. In the following chapter, "Historicity, Temporalities and Causality: A Confusion at the Heart of Debates on Darwinism," Mathilde Tahar explores Darwin's multi-layered conceptualization of time in evolution, discussing its philosophical underpinnings and developments in the context of twentieth-century evolutionary biology. The book closes with a contribution by Erica Torrens, Juan Manuel Rodríguez Caso, and Ana Barahona, which critically analyzes the "tree of life" imaginary, discussing its epistemological and ontological commitments as well as its enduring presence despite the growing recognition of its limitations in representing phylogenetic complexity when compared to alternative visualizations.

Through diverse and frequently hybridized methodological approaches, this volume endeavors to challenge and deconstruct the binary oppositions and rigid categorizations prevalent in the historiography of evolutionary biology by favoring a polycentric reading of the history of evolutionary debates; incorporating global and local contexts; highlighting the importance of both professional specificities and contaminations in knowledge production; grappling with the inherent tension between presentist and relativist narratives; and recognizing both the advantages and the burden of embracing pluralism. This list of endeavors can pave the way for innovative avenues, thus laying the foundation for a new historiography of evolutionary biology. More than 40 years after *The Evolutionary Synthesis: Perspectives on the Unification of Biology* (1980), edited by Ernst Mayr and William Provine, it is now obvious that this earlier attempt at presenting the development of evolutionary biology proves too narrow and too biased to make sense of its inherent complexity.

Reference

Lovejoy AO (1960) [1936] The great chain of being: a study of the history of an idea. Harper Torchbooks, New York

"Reformist" and "Radical" Historiographies Behind and Beyond the Unity and Disunity of the Evolutionary Thought

Maurizio Esposito

Abstract

We are in the midst of a revolutionary moment where historians, philosophers, and scientists of different opinions debate how to overcome, update, or replace the received views. This volume is an exemplary instance that records a moment of productive confusion where different alternatives and possibilities are explored and assessed. The fragmentation and disunity of historiography, which, to a certain extent, parallels the disunity of evolutionary biology, should not be a reason to despair. On the contrary, it is a stimulating opportunity to find new insightful ways to understand and rewrite the history of evolutionary thought. Notably, it is an occasion that calls for imagining new historiographies where the traditional place of Darwin and Darwinism in the history of life science is questioned, rediscussed, and reevaluated.

Keywords

Darwinism · History and historiography of biology · Evolutionary biology · Unity and disunity in evolution

History, generally speaking, is the most difficult composition that an author can undertake, or one of the most difficult. It requires a great judgment, a noble, clear and concise style, a good conscience, a perfect probity, many excellent materials, and the art of placing them in

M. Esposito (✉)
Department of History and Philosophy of Sciences, Interuniversity Center for the History of Science and Technology, University of Lisbon, Lisbon, Portugal
e-mail: Mesposito@ciencias.ulisboa.pt

R. G. Delisle et al. (eds.), *Unity and Disunity in Evolutionary Biology*,
https://doi.org/10.1007/978-3-031-42629-2_2

> good order, and above all things, the power of resisting the instinct of religious zeal, which prompts us to cry down what we think to be true.[1]

There is a thriving intellectual industry behind the history of evolutionary thought. Whether directly linked to Darwin or any other significant figures, the feat of reconstructing the pasts and presents of evolutionary biology is as alive as ever. It would be virtually impossible to follow all the yearly publications dedicated to the field, whether in English or any other language. But, for simplicity's sake, we might divide this intellectual industry into two generic historiographic categories, which I tentatively call *reformist* and *radical* ones. These categories do not pretend to pigeonhole anyone—neither in this monograph nor elsewhere-but I see them as "ideal types" that have the sole aim of charting a contentious and complicated territory, which, as it always is the case, supposes many different shades, exceptions, and overlapping. The *reformist* approaches tend to complexify the extant historiography by either adding more details or reinterpreting the received view of one figure, episode, or idea. For most reformists, the existing historiography is far too simple in that it does not consider or include what it should. From the reformist's standpoint, the received historiography either misses some crucial episodes that should not be missed or misinterprets the well-known ones. In short, the reformist scholar wants to expand, and thus reinforce, the extant tradition by making it more sophisticated and complex. On the other hand, the *radical* approach assumes that the existing historiography needs to be entirely rethought from scratch. This does not mean that the extant historiography is useless; it means that the data and interpretations of the extant historiography need to be *radically* refashioned into a brand-new plot. The word "radical" needs to be understood in its etymological sense as *Radix* (root). A change in the historiographical roots means we must go to the very foundations and reconsider them entirely. Radical approaches require audacity and a sort of stubborn self-righteousness, not only because they disrupt ingrained assumptions that have made the careers of other scholars but also because one needs to wander into unexplored territories where the possibility of making mistakes is painfully high.

Thomas Kuhn's famous distinction between normal science and revolutionary science might help clarify the distinction I am introducing here. The reformist option corresponds to what Kuhn calls "normal science." Reformists accept the general paradigm and aim to improve or advance it by adding more confirmations. To bring closer the example to our discussion, the reformist assumes that the historiography of evolutionary thought based on Darwin and Darwinism is largely fine (let's call it *Darwin's paradigmatic story*, DPS onward). The core idea of DPS is that the history of evolutionary biology hinges essentially around Darwin, whether the plot focuses on forerunners, followers, or critics. Those interested in the "pre-Darwinian" theories of evolution, those who believed in the absolute originality of Darwin, and those who study "non-Darwinian" theories of evolution all wittingly or

[1] P. Bayle, Historical Dictionary, quoted and translated in White (2014), p. 49.

unwittingly share the assumption that Darwin is the referential point from which different plots must branch off.

In this context, the sensible task of DPS's reformists is to pile up more people, institutions, countries, ideas, theories, and contexts that confirm and simultaneously expand the basic paradigmatic story. But as it happens in the context of "normal science," we eventually get to an overgrown and chaotic paradigm, which tends to collapse under the burden of those new data, people, and countless conflicting reinterpretations. To make an analogy with the history of astronomy, adding more and more "epicycles" can only work for a while. The paradigm becomes so chaotic and confusing that only a tiny cohort of highly trained people manage to have a relative familiarity with it. The paradigm grows into something so much articulated that it becomes impervious to any genuine critique and reconsideration. Kuhn called "anomalies" those unexpected difficulties and incongruences emerging along the development of paradigm itself. But "anomalies" in the Kuhnian sense can also exist in historiography, and, in particular in the DPS. As it happens with mature paradigms, the "anomalies" that purportedly contradict the DPS can be easily ignored or normalized.

Kuhn recalled that, in revolutionary periods, entrenched paradigms are challenged, and new, *radical* options are explored. Concerning the historiography of evolutionary thought, I think we are in a sort of revolutionary phase where the time for a *radical* approach is ripe. So far as I can see, we are currently enmeshed in a kind of "hundred flowers campaign" where a hundred schools of thought proliferate. I see this monograph as a telling contribution to a historiographical revolution, where reformists and radicals are quarreling on all sorts of issues: on whether we should rethink natural selection and its role in the history of evolutionary biology; on whether and how we should reassess Darwin's role in this history; on whether there is any coherent historical entity called "Darwinism" or whether we have many, often incompatible, Darwinisms; on whether the debates on modern, unmodern, hardened, softened, or extended syntheses make any historical sense or whether the histories written by scientist-historians should be amended or simply transcended. But, even if my Kuhnian diagnosis of the current situation turns out to be spurious, the monograph undoubtedly displays a genuine desire to embrace an open-minded approach to historiography and a sincere interest, for both young and older generations, to reassess DPS seriously. There is a visible eagerness to revise the assumption that sees any history of evolutionary thought as rooted in Darwin. There is growing and legitimate suspicion that, perhaps, historians have exaggerated the uniqueness and exclusivity of the nonetheless great British naturalist (no one would deny that Darwin was indeed a great naturalist). There is an increasing awareness that evolutionary thought is much more than Darwin, Victorian England, or Lamarck in post-revolutionary France. There is a shared feeling that when one opens a book or reads an article dedicated to Darwin and Darwinism, one starts to smell an unmistakable (or perhaps mystic) scent of incense. One perceives that there is something disturbingly religious in the obsession to show that evolutionary thought is essentially Darwin and evolutionary biology is, and must be, rooted in Darwin. My opinion, and I guess the opinion of a growing number of scholars, is that we should get through the incense's smoke and look at the history of evolutionary through more secular, disenchanted eyes. Of course, this monograph is not the first to do the job. Many distinguished scholars-including some who have contributed to this monograph-have campaigned for a radical historiography for decades. And if today we have the feeling that the DPS might be on the verge of collapsing, we owe it to them, including, most notably, Mark Adams (2024), to whom this volume is dedicated.

But if DPS eventually collapses, what would be the task of present and future historians? It is still too early to answer this question and I can only provide a few fragmentary suggestions. Like Kuhn reminded us in his *The Structure of Scientific Revolutions* (1962, 1970)], the revolutionary periods are characterized by many engaging debates over the very conceptual foundations of old and new paradigms. No one has the definitive answers on the paths to follow. When something old is about to die and something new arises, confusion and chaos rule. But disagreement, fragmentation, and disunity are signs of vitality, not hopelessness. One task, for instance, could be to explore the new aims, methods, and ambitions that should characterize a new historical and radical plot. We should ask ourselves what the scope of writing a new history of evolutionary thought could be. After all, every new historiography adopts a philosophy that prescribes what is and is not important to tell. We can undoubtedly be oblivious about this philosophy, but we should be equally aware that such obliviousness gives consent. We can legitimately resist embroiling ourselves in metahistorical debates, yet the price for neutrality is to accept or assume previous historiographies.

Today, it is no longer a mystery that DPS was mainly concocted by professional biologists turned historians who shared, wittingly or unwittingly, a particular agenda. The main rationale for this philosophy was to back up some particular scientific options of the moment. Ernst Mayr, perhaps one of the most committed architects of DPS, was crystal clear when, in his great history's *magnum opus*, *The Growth of Biological Thought*, he wrote that: "...most scientific problems are far better understood by studying their history than their logic" (1982, p. 6). History has a pedagogic and propaedeutic role, and the role is to clarify persistent scientific problems. This is undoubtedly a legitimate prospect for a scientist who wants to figure out where evolutionary biology comes from and where it eventually goes. Yet, pedagogical history would not be a satisfactory answer for someone interested in how, when, and why the idea of "evolution" emerged. Of course, no one should dismiss or belittle the extraordinary service biologists like Mayr, Gould, and others have delivered. After all, we would not be here discussing old and new historiography of evolutionary thought if they did not write commendable pieces of scholarship. But, as scholars, the best service we can deliver to our predecessors is not bowing to their wisdom but criticizing them and, eventually, transcending their perspectives. Elsewhere, I suggested that one first step in the process is recognizing that scientists might have different historical interests from historians and philosophers (Esposito 2021). Scientifically oriented historians hoped that they could get more conceptual transparency through history. What would be the interests and aims underpinning new historiographies? What do we really want or expect from new histories of biology?

As far as I am concerned, a second step could be considering the plot's ambition, scale, and genre. Shall a new history be global or national? General or specific? Shall it include antiquity, early modern, and modern periods? Shall it be mainly intellectual in the sense propounded by historians of ideas? Shall it be more sociological, political, economic, etc.? Shall we use the largely outdated categories of internalism and externalism?[2] If anything, what kind of historical genre would we consider?[3] Although there are many possible answers to these questions, one thing should be clear: a *radical* sort of historiography is, by definition, ambitious. As such, it aims to provide a *novel* understanding of the whole history of evolutionary thought. "Novel" here does not mean there is no connection or continuity whatsoever with the previous "paradigm." It only means we have a *new* historiographic "paradigm," where people, events, theories, and ideas acquire different meanings and relevance. The issue is not to amend one or more details by adding further details and specifications. This is the task of *reformists*, who still cling, implicitly or explicitly, to DPS and intend to it from all threatening "anomalies." The issue here is to entirely rethink a new historiographic framework for the whole history of evolutionary thought, a framework in which all the other specific histories-biographies, national, institutional, intellectual, economic, or political narratives—can be situated.

No doubt, many historians might complain by claiming that we do not need a paradigm. If the old DPS is misleading or uninteresting for us, then we can safely replace it with a proliferation of many different stories. There is not one paradigm, not one history of evolutionary thought, but many overlapping, sometimes contradictory, stories. We have no linear plots but a web of uneven and overlapping narratives covering the globe. In short, we can dispense once and for all from any "master narrative" or "dominant discourses", as a few decades ago some "postmodernists" claimed! We should, therefore, push for pluralism and fragmentation while condemning any form of deceptive unity. For sure, the feat would be politically and morally commendable. Alas, as recent history has repeatedly demonstrated, the postmodern strategy is an excellent medicine for keeping alive the old paradigm, which has become a sort of *zombie* that persists because there are no other more accessible alternatives around. As it might be expected, not everyone has the time and will to read the whole web of all overlapping narratives, which is even difficult, if not impossible, for the insiders. For many, if not the majority, of those who want to learn about the history of evolutionary thought, there is no other option than going back to the old paradigm and, therefore, to the traditional master narratives. DPS is patiently waiting in the doorway to be promptly resurrected once the last version of the postmodern slant goes out of fashion. In short, the appeals for sweeping forms of pluralism are good-intentioned programmatic intents, but they

[2]For the issue, it might be interesting to revisit David Hull's discussion in Hull (2005).

[3]For this, one could reread the classic monograph of Hayden White, *Metahistory*, where the origin and nature of diverse historical genres are introduced and assessed (White 2014).

remain programmatic simply because they are self-defeating in the long term (if not in the short term). The quest for unity might be specious but the celebration of disunity and fragmentation might be intellectually barren. The replacement of a problematic unity with an equally problematic disunity guarantees a frustrating recurrence of the DPS. But if we admit (or at least consider) the inadequacy of postmodern solutions, one question becomes immediately urgent: from where shall we start?

Given the preliminary state of our agenda, I think it is much easier to consider first from where we should not start, i.e., from the widespread intuition, especially popular among scientist-historians, that history itself is a procession of big names and doctrines, all advancing toward the glory of the present. In this volume, La Vergata (2024) puts it straightforwardly: "...many authors still conceive and practise the history of science in the following way: they start with an accepted, familiar idea; they go backwards to its source, minimize oppositions to it, and present us with a coherent history of its triumph. Such an approach can be suitable for clear exposition, but it is essentially a form of fiction or novel" (p. 35). The approach is in fact suitable to clarify particular problems, and this was precisely the kind of history explicitly favored by Mayr and implicitly accepted by Gould. This is a conception of history as a pedagogic tool to understand the present. But I would not define those histories as "fictions" or "novels" because I do not know how a "real" or "realistic" historiography should look like. There are several ways to organize, historically, facts, data, people, and events, and all depend on tour intentions and aims[4] As the Polish American philosopher Alfred Korzybski reminded long ago, "the map is not the territory." We might add that if historical charts represented the past as "it really happened," they would be useless like the 1:1 Borgesian maps. The problem with self-serving narratives is not that they are wrong (or "unreal" like novels)—strictly speaking, all narratives are wrong, or, at least, incomplete like any map—but that they are no longer interesting. They have exhausted their epistemic and heuristic potentials. They do not satisfy the kinds of interests and expectations of the new generations. They refer to a vision of the past that has become what Richard Delisle convincingly shows in the chapter of this volume (2024), an obstacle to move on: a hindrance to think and shape new and more interesting narratives. In short, "presentist" or "whiggish" historical accounts of evolutionary thought have lost their appeal because the historiographical rationales and agendas have changed.[5] A presentist approach is no longer an interesting way to organize historical data. Of course, we should understand the notion of "interesting narratives" as "interesting" for us scholars of the early twenty-first century.

But a careful reader could immediately reply with a reasonable question: how shall we define "interesting" here? After all, what is interesting for someone is

[4]See, for instance, the famous controversy on the nature of historical writing in the 1960s triggered by Carr and Elton (see Carr 1961 and Elton 1967).

[5]On this issue, see Esposito (2021a, b).

uninteresting to someone else. Here, again, Kuhn might help us: to contemporary readers, DPS is as "interesting" as the Aristotelian Physics or Ptolemaic paradigm for a contemporary cosmologist. One might justifiably observe that cosmologists are not interested in Aristotle and Ptolemaic cosmologies simply because the latter were, according to our standards, patently wrong. But suppose we interpret the history of paradigms in the Kuhnian spirit (or, at least, according to my creative interpretation of Kuhn). In that case, we realize that no paradigms can be considered as *absolutely* true or wrong. Those paradigms are held to be true in certain periods because they resonate with a given community's conceptual, social, or political expectations. Paradigms persist as long as they are perceived as "interesting" to most members of a scientific collective. Similarly, I do not think there are *absolutely* true or wrong historiographies but more or less supported, robust, meaningful, coherent, focused, instructive, informative and, especially, interesting ones. And some historiographies are allegedly interesting because they resonate with the curiosities, concerns, and expectations of a specific community. The social mechanisms working behind the preference for certain kinds of historiographies cannot be addressed here. But at least one tentative hypothesis can be tried out and assessed. And my hypothesis hinges on at least three basic elements:

1. The community of scholars focused on the history and philosophy of evolutionary biology is much larger and heterogeneous now than when DPS was concocted.
2. Many members of this community are not primarily biologists and, thus, have interests and aims that differ from those involved in pedagogic historiographies. In other words, if the earlier histories of biology were mainly put forward by biologists who used history as a propaedeutic tool to understand better their discipline, new generations of scholars see history as intrinsically interesting and a self-sufficient discipline.
3. The increasing professionalization of the history of science and philosophy of science has made the disciplines relatively independent and sometimes even impervious to contemporary scientific interests and aims. There are conferences, journals, and departments where specific historical and philosophical issues are hotly debated, independent of what scientists currently do or think.[6]

If these three elements are true (or partially true), we have some element to understand why we are amid a paradigmatic historiographic shift. New people, different training, and readings and, thus, a distinct community with diverse goals and interests make some sort of change expectable. When in the early 1980s Ernst Mayr published his historical magnum opus, *The Growth of Biological Thought*, he could easily expect to monopolize and lead the interests of the majority of those involved in the discussion, whether other scientists or humanists (as M. Adams

[6]Of course, I am not suggesting that scientists are no longer involved. They are and they should be involved. What I am suggesting is that they cannot lead or monopolize the debate anymore.

brillantly shows in Chapter "The Evolution of "Darwinism": Up Close and Personal" (2024) of this volume). This does not mean everyone had to emulate Mayr's scholarship and write like he did. Instead, it meant that anyone interested in the history of evolutionary thought had a limited repertory of topics to address, and those topics hinged around Darwin and Darwinism. An intellectual industry such as Darwin's could only be possible when the interests and careers of many scholars coalesced into one agenda. True enough, one could always be free to explore other topics or other figures, but the risk for mavericks snooping out of the box was to be immediately ostracized or, more plausibly, ignored because they did not fit the real interests and aims of the accepted paradigm. Today, the situation in which the debate is carried on is much more dynamic and confusing because we have no Mayrs or Goulds to push forward a specific intellectual program. As a matter of fact, we do not have one generic Darwinian agenda but many, often incompatible, agendas. Accordingly, we are in a situation where historiographic norms, aims, interests, and goals are constantly challenged and reassessed.

A short consideration of some of the contributions of this monograph is a telling demonstration. When, for instance, Rupke (2024), in this volume, introduces Richard Owen's evolutionary view, which was deeply indebted to Kant and Kantian tradition, he challenges a few fundamental elements of DPS. First, he convincingly shows that Owen was far from being a simple-minded creationist and a staunch opponent of organic transformism. On the contrary, although he rejected Darwin's specific kind of transformism, he was a transformist after all. Perhaps even more importantly, through Owen, Rupke shows us that there were many other alternatives on the nineteenth-century intellectual menu beyond the simplistic dilemma between evolutionism and creationism. When Ceccarelli (2024) persuasively shows in this volume that there have been at least three Darwins and three Darwinisms, each fit for the contextual Darwin's celebrations in 1909, 1959, and 2009, we understand (and are entitled to conclude) that DPS is, in the best case, deeply misleading or highly restrictive. The selective readings of the British naturalist throughout the twentieth and early twenty-first centuries transformed Darwin into a useful flag for various agendas and purposes, most of them extra scientific and eminently political. When Torrens et al. (2024) notice that the very image of the "tree of life" is ambiguous and open to different interpretations, they invite us to question one of the most cherished images of most received historiographic views. When Baedke et al. (2024) argue that in the early twentieth century attempts to unify biology were not customarily subsumed under Darwin and Darwinism, they demonstrate that the plot of Modern Synthesis needs to be dramatically retooled. In short, I venture to say that most, if not all, chapters of this monograph put one or more nails in the coffin containing the corpse of DPS, whether the contributions address historical and historiographic issues or focus on the opportunity (or inconvenience) of keeping some form of "Darwinism" in contemporary evolutionary biology.

So, what do we really expect from new histories of biology? In the fascinating text of Bayle I quoted at the beginning of this short introductory essay, the French skeptic concludes that the historian should resist "...the instinct of religious zeal, which prompts us to cry down what we think to be true." No doubt, few historians of science would have today the "religious zeal" that Bayle meant while proffering these words. But there is a specific sense in which the quote is still actual. Religious zeal does not only apply to traditional religions or divinities. It can also be applied to secular figures. The cult of personality has been a common phenomenon in human history. From Augustus to Napoleon and from Stalin to Mao, the divinization of political figures has been a vast and recurrent social praxis. But the urges of prior or posthumous sanctification not only apply to significant political figures but also affect sportsmen, singers, actors, and to be sure, naturalists and scientists. Even a committed and sophisticated Darwinian like S. J. Gould, in his massive volume, *The Structure of Evolutionary Theory*, noticed the misplaced religious zeal Darwin has often enthused:

> The *Origin*, as a volume of single authorship, maintains a stronger plot line and features fewer inconsistencies than the Bible; but Darwin and the Good Lord do share the common trait of saying something about nearly everything. Wrenched from context and divorced from a crucial assessment by relative frequency, a Darwinian statement can be found to support almost any position, even the most un-Darwinian. . .Since Darwin prevails as the patron saint of our profession, and since everyone wants such a preeminent authority on his side, a lamentable tradition has arisen for appropriating single Darwinian statements as defenses for particular views that either bear no relation to Darwin's own concerns, or that even confute the general tenor of his work. . .(Gould 2002, p. 148)

We can debate whether Gould belongs to the crowd that enthusiastically incensed (and incenses) their Salopian patron saint. I only add that I cannot see any fundamental difference between secular and non-secular patrons. What both have in common, in fact, is a misplaced and committed fetishism that normally blinds most of our critical faculties. Bayle, in this sense, gives a piece of advice that is still meaningful to us: to write good histories, we need, among other things, to refrain from any sort of fetishistic celebration of the figures we are considering in our plot. Unfortunately, the history of evolutionary thought, and evolutionary biology in particular, has been peculiarly affected by an uncompressing religious zeal skirting a cult of personality. I see this monograph as a further attempt to cure the history (and perhaps future historians) of evolutionary thought from such an unnecessary disease. We can continue to enthusiastically study Darwin, Darwinians, and, more generally, different kinds of evolutionism without feeling the urge to canonize anyone. We can lean toward *reformist* or *radical* outlooks and devise more or less convincing plots, but, following Bayle, we can wisely leave the saint patrons to the churches and national heroes to the autocracies.

References

Adams MB (2024) The evolution of 'Darwinism': up close and personal. In: Delisle RG, Esposito M, Ceccarelli D (eds) Unity and disunity in evolutionary biology: deconstructing Darwinism. Springer, Cham, pp 83–121

Baedke J, Böhm A, Reiners-Selbach S, Straetmanns V (2024) A synthesis without Darwin: unification attempts in early theoretical biology. In: Delisle RG, Esposito M, Ceccarelli D (eds) Unity and disunity in evolutionary biology: deconstructing Darwinism. Springer, Cham

Carr EH (1961) What is history? Cambridge University Press, Cambridge

Ceccarelli D (2024) Redrawing the boundaries of Darwinism: addressing Darwin's endorsement of the inheritance of acquired characteristics in Darwin's celebrations, 1909-1959-2009. In: Delisle RG, Esposito M, Ceccarelli D (eds) Unity and disunity in evolutionary biology: deconstructing Darwinism. Springer, Cham

Delisle R (2024) Deconstructing Darwinism with Darwin, Mayr, and Gould: through the lens of evolutionary contingency. In: Delisle RG, Esposito M, Ceccarelli D (eds) Unity and disunity in evolutionary biology: deconstructing Darwinism. Springer, Cham

Elton GE (1967) The practice of history. Thomas Y. Crowell Company, New York

Esposito E (2021a) Cathedrals, corals and mycelia: three analogies for the history of evolutionary biology. In: Delisle RG (ed) Natural selection: revisiting its explanatory role in evolutionary biology. Springer, Cham, pp 11–35

Esposito E (2021b) Lo material, lo ideal y la historia: algunas reflexiones en torno al "materialismo" darwiniano. In: López-Orellana R, Suárez-Ruíz J (eds) Filosofía posdarwiniana. Enfoques actuales sobre la intersección entre análisis epistemológico y naturalismo filosófico. College Publications, London

Gould SJ (2002) The structure of the evolutionary theory. Belknap Press, Cambridge

Hull D (2005) Deconstructing darwin: evolutionary theory in context. J Hist Biol 38:137–152

Kuhn TS (1970) The structure of scientific revolutions. Enlarged, 2nd edn. University of Chicago Press, Chicago

La Vergata A (2024) Constructing, deconstructing and reconstructing. On 'Darwinism' and 'Darwinisms', with some disparate considerations on the history of science. In: Delisle RG, Esposito M, Ceccarelli D (eds) Unity and disunity in evolutionary biology: deconstructing Darwinism. Springer, Cham

Mayr E (1982) The growth of biological thought. Belknap Press, Cambridge

Rupke N (2024) Richard Owen's deconstruction of Darwinian natural selection. In: Delisle RG, Esposito M, Ceccarelli D (eds) Unity and disunity in evolutionary biology: deconstructing Darwinism. Springer, Cham

Torrens E, Caso JMR, Barahona A (2024) Shaking the tree: discussing an evolutionary icon. In: Delisle RG, Esposito M, Ceccarelli D (eds) Unity and disunity in evolutionary biology: deconstructing Darwinism. Springer, Cham

White H (2014) Metahistory: the historical imagination in 19th century Europe. John Hopkins University Press, Baltimore

Darwin as a Unifying Figure in Evolutionary Biology: A Meta-historical Overview

David Ceccarelli

"The choice of unification is not a free one; if science fails to unite, it will fail to exist" Ackoff Churchman, 1946

Abstract

The idea that the unification in evolutionary biology centered around Darwin's theory has been extensively discussed in the literature. "Darwinism" remains an elusive historiographical category due to its heterogeneity and evolution over time. The legacy of Charles Darwin prompts reflection on the continued relevance of invoking his name in contemporary discussions, highlighting the multilayered meanings and ideological implications embedded in the term "Darwinism."

Keywords

Darwinism · Darwinian age · Modern Synthesis · Extended evolutionary synthesis · Charles Darwin · Unified theory

D. Ceccarelli (✉)
Department of Philosophy, Communication and Performing Arts, Roma Tre University, Rome, Italy
e-mail: david.ceccarelli@uniroma3.it

R. G. Delisle et al. (eds.), *Unity and Disunity in Evolutionary Biology*,
https://doi.org/10.1007/978-3-031-42629-2_3

1 Unification as a Regulative Ideal

The pursuit of unification has long served as a regulative ideal in modern science. Over time, this principle has faced growing skepticism from epistemologists and historians of science. Scholars increasingly challenged the notion that sciences can be neatly organized into hierarchically structured systems of explanation based on epistemic monism, and increasingly argued for pluralist approaches to scientific knowledge (Kitcher 1999). This shift is particularly evident in the field of life sciences, where the quest for a "unified theory" capable of comprehensively explaining biological phenomena has captivated biologists over the course of several decades, and yet proved to be extremely complicated (Callebaut 2010; Love 2013; Dupré 2015).

It is not easy to determine when biologists embraced unification as a guiding principle. In hindsight, many would point to the advent of evolutionary theory as a pivotal moment that established a cohesive research agenda in the life sciences (Delbrück 1949; Singh 2023). As early as 1876, American biologist Alpheus Packard (1876, p. 592) argued that it was only after the publication of Charles Darwin's *Origin of Species* that the zoological, paleontological, and embryological data collected by the previous generations of naturalists began to be extensively reexamined based on the study of genealogical relationships among life forms. Almost a century later, Theodosius Dobzhansky famously stated that "nothing in biology makes sense except in the light of evolution" (Dobzhansky 1973). However, well before Charles Darwin published the *Origin* in 1859, the idea that biological science should adhere to the epistemological standards of physics had played a significant role in directing naturalists toward the pursuit of "universal" laws or principles. During the eighteenth and early nineteenth centuries, natural history primarily consisted of the meticulous description and categorization of organisms through a rigorous inductive approach, and yet several naturalists proposed "general laws" aimed at unveiling the dynamics underlying phenomena such as hybridization, development, inheritance, and adaptation (Hoquet 2010; Corsi 2011; see also McLaughlin and Demarest 2022).

The endeavor to establish parallels between biological sciences and physics, along with the pursuit of identifying general laws in biology, gained prominence within the epistemological culture of determinism and methodological uniformitarianism. The concept of nomic spatiotemporal invariance (Haufe 2015), asserting that the laws of nature remain consistent over time and that one could generalize retrodictions about the past based on what we observe in the present, served as a foundational methodology in the nineteenth century. This indeed contributed to emancipating science from non-naturalistic approaches, and deeply influenced various disciplines, including modern evolutionary biology (Gould 1965; Ceccarelli 2018; Delisle 2019).

Scholars have often argued that Newton's unification of terrestrial and celestial mechanics represented the paradigm for evolutionary science during the nineteenth century (Depew and Weber 1995; Weber 2011). Michael Ruse (1975, p. 166) wrote that Charles Darwin regarded Newtonian astronomy as a model for science, aspiring to be the "Newton of biology" by following the methodological ideals of classical mechanics. While Darwin ardently believed that biological evolution adhered to law-like processes, in accordance with the uniformitarian principles advocated, among others, by his mentor Charles Lyell, some scholars have raised doubts about his commitment to emulating Newton's methodology (Pence 2018). Historian Jon Hodge, for instance, argued that Darwin had initially entertained the idea of constructing a theoretical system aligned with Newtonian celestial mechanics. During the nineteenth century, this was perceived as having a threefold pyramidal structure: "at the base were particular astronomical observations [...], in the next level up were lawful generalisations," whereas at the top there were "the lawful causes, the lawful forces of gravitation and inertia" (Hodge 2009, p. 54). To Hodge (Ivi, p. 68), Darwin however gradually moved away from this ideal structure, acknowledging that natural selection lacked a "law of its own."

In spite of this, physics, with its inclination toward generating unifying explanatory frameworks, continued to serve as a reference model in the biological sciences. As Kaufmann and Gare (2015) remarked, many biologists remained "children of Newton" and dreamed of "a grand theory that is epistemologically complete and would allow lawful entailment of the evolution of the biosphere." In line with the nineteenth-century epistemological culture, epistemic monism was often deemed more scientifically reliable and productive than embracing pluralism. This inclination toward a singular perspective was rooted in the belief that a unified framework would offer greater clarity and coherence to scientific discourse. Moreover, the presence of theoretical inhomogeneity was often viewed with skepticism, as it was thought to be potentially detrimental to the construction and preservation of scientific and institutional networks. The lack of theoretical uniformity could indicate weakness, particularly to those on the anti-evolutionary side.

The need to conform to the epistemological standards of physics appears clearly from the works of several evolutionists between the nineteenth and the twentieth centuries. In his brief foreword to neurologist Frederick Tinley's *The Brain from Ape to Man* (1928), American paleontologist Henry Farfield Osborn remarked on the striking contrast between "exact" sciences and biological sciences. In the exact sciences, the "uniformity of laws and principles" allowed for a unified explanatory framework that made physical, chemical, and astronomical phenomena comprehensible. In anatomy, physiology, pathology, and heredity, Osborn highlighted, scientists had not even reached the threshold of exactitude:

> With increasing energy, refinement and ingenuity, we know all the organs revealed in comparative and human anatomy, in both their grosser and their finer structure. We know also the history of the rise of many of these organs in the course of past time and what their functions and relations are, but there is always the Great Beyond of the unknown, and perhaps unknowable, which is summed up in the word life (Osborn 1928, p. xv).

Over time, the attempt to draw parallelisms between the unified explanatory framework of physics and biological sciences has faced criticism for at least two reasons. As early as 1925, philosopher Alfred North Whitehead maintained that modern physics had significantly challenged the traditional deterministic worldview, which appeared to be a no longer tenable reference model. Moreover, biology had long leaned on the epistemological underpinnings of physics and needed to reestablish itself as an independent science, affirming the specificities of its objects of analysis (Dupré and Nicholson 2018, p. 8). During the twentieth century, several biologists would endorse this view. Evolutionary biologist Ernst Mayr, for example, staunchly championed the autonomy of biology in relation to physics. His central argument revolved around the perception of biology as a marginalized science due to its historical association with vitalism and teleology (Mayr 2004). At the same time, he contended that fundamental principles inherent in the physical sciences, such as essentialism, determinism, reductionism, and the postulation of universal laws, found little, if any, applicability in the realm of biology. This discrepancy resulted in a notable methodological and epistemological gap. Unless one advocates for reducing biological phenomena to their underlying physical components, the endeavor to unify all sciences within a singular framework should be relinquished. If a unification process occurred in biological sciences, Mayr remarked, it took place mainly thanks to the principles outlined by Charles Darwin, who acknowledged the specificity of living phenomena without falling into non-naturalistic fallacies (Ivi, Chap. 2).

2 Darwin as a Unifying Figure in Evolutionary Biology

The claim that the unification in evolutionary biology centered around Darwin's theory has been extensively discussed in the literature. When Packard wrote his brief note on the progress of American zoology in 1876, crediting Darwin with initiating a new phase in the study of natural history, Darwin's theory was far from being the only game in town. Packard himself remarked that, in the USA, this new evolutionary era was marked by the emergence of an "original and distinctively American school of evolutionists" he referred to as "neo-Lamarckians" (Packard 1876, p. 597). Darwin had catalyzed a period of scientific ferment rather than a unanimous consensus among his peers. The *Origin of Species* had certainly contributed to inflaming the already rich debate on species transformation, providing the scientific community with invaluable empirical data and a powerful argument. It unequivocally represented a pivotal and irreversible turning point in the history of biological sciences (Bowler 2013). However, as often discussed in historiography, its reception was far from a straightforward process in the 1860s and 1870s. It sparked a vigorous debate that, over time, fragmented into a multitude of critiques, opinions, and reinterpretations (Moore 1979; Glick 1988; Numbers and Stenhouse 2001; Engels and Glick 2008).[1] Since the early reviews of *Origin*, Darwin's ideas have undergone

[1] As shown by Johnson (2020), Darwin's effort to claim originality and emphasize the explanatory superiority of his theory over rival ones indirectly shows how varied was the theoretical panorama of evolutionary studies during the mid-nineteenth century.

a complex reframing process that, in some sense, echoes the words expressed by the protagonist in Robert Musil's *The Man Without Qualities* (Musil 1996):

> As soon as some leading thinker comes up with an idea it is immediately pulled apart by the sympathies and antipathies generated: first its admirers rip large chunks out of it to suit themselves, wrenching their masters' minds out of shape the way a fox savages his kill, and then his opponents destroy the weak links so that soon there's nothing left but a stock of aphorisms from which friend and foe alike help themselves at will. The result is a general ambiguity (Musil 1996, p. 412)

Between the nineteenth and the twentieth centuries, pluralism and theoretical inhomogeneity continued to prevail in evolutionary biology. This stage was later termed as the "eclipse of Darwinism," which, according to the standard historiographical account, was marked by the waning influence of Darwinian evolutionary theory and the rise of alternative theories that downplayed Darwin's key insights, such as random individual variation, gradualism, branching phylogeny, and natural selection. It was only in the mid-twentieth century that evolutionary biologists regained alignment with the principles of Darwinism. This process, better known as "Modern Evolutionary Synthesis," involved the integration of population genetics, zoology, paleontology, and botany within a revitalized Darwinian paradigm, signaling a return to the foundations of evolutionary thought (Huxley 1942). As Betty Smocovitis (1999, p. 279) highlighted, at this stage Darwin was "reinvented" as the founding father of evolutionary biology, establishing order in a field marked by a plurality of methods, level of analysis, and research traditions (Delisle 2009). In a way, the Modern Synthesis substantiated the pursuit of unification that had long characterized biological sciences. This occurred during an era in which explaining complex phenomena through micro-level theories was considered a significant epistemological virtue, aligning with the call for the "unity of science" and the dismissal of metaphysics championed by philosophers associated with the Vienna Circle (Smocovitis 1992). In the words of Stephen Jay Gould (1994, p. 135), such intellectual context stood out for its emphasis on "the abstract, the simplified, the fully universal, the underlying principles that build the unique and complex from the small and general, all fueled the preference within evolutionary biology for a comprehensive micro-level theory that could build all scales and sizes by smooth extrapolation."

The claim that the Modern Synthesis was also, and most importantly, a "Darwinian" synthesis raises various historical and conceptual issues. First of all, it brings us to the even more trap-laden task to determine which aspects of Darwin's theory of evolution must be maintained in order for an evolutionary theory to be considered "Darwinian." Mayr claimed that the evolutionary synthesis had long been erroneously named "neo-Darwinism," a term which originated in the late nineteenth-century debate on biological heredity to distinguish between those who believed that the inheritance of acquired characters was a source of variability and those who dismissed it as an essentially unconfirmed hypothesis. According to Mayr, the more appropriate name for the Modern Synthesis should be simply "Darwinism," as it did not introduce revolutionary changes to Darwin's theory. Instead, it enhanced it with a more robust understanding of speciation and eliminated the concept of soft inheritance, which was included in Darwin's original theory. The Modern Synthesis revolved around the more distinctive component of Darwin's original theory of evolution, namely the "interplay of variation and selection"[2] (Mayr 2004, pp. 129–130; see also Hodge 1977; Ruse 1975, 2013).

It is however worth wondering what it means for "selectionism" to be the core of Darwinism. If by core we mean, precisely, an essential component of the *explanans* of the Darwinian evolutionary theory, it must be noted that Darwin used natural selection as the primary agent responsible for the adaptation of organisms, as a complementary and auxiliary agent capable of reinforcing alternative mechanisms (i.e., use and disuse), and that he dismissed it to make sense of some specific evolutionary patterns (Delisle 2021, pp. 95–97). Darwin zealously considered natural selection as his most precious contribution to the debate on species transformation (Johnson 2020); however, especially in his late works, he consistently relied on pluralistic accounts of organic change, to the extent that scholars would employ his "pluralism" against those attempting to equate Darwinism solely with selectionism.

Presumably, the association between Darwinism and selectionism emerged from regarding natural selection as the most original contribution of Darwin to the nineteenth-century debate on species transformation. This view has however diverted attention from the theoretical and historical complexity of Darwinian theory. Darwin's theory had a complex and multilayered structure. Not only is it trivial to consider it "a unitary entity" (Mayr 1985, pp. 755–772, 755), but it also displays a significant diachronic change. Over the course of the years, Darwin refined his "long argument" by incorporating insights from his peers, assimilating criticisms, and broadening the scope of his analysis. The study of his personal writings, correspondence, and publications reveals several changes of opinion on themes such as the endorsement of gradualism, the role of geographic isolation in speciation, and the use-inheritance hypothesis. Over time, scholars have feasted upon such theoretical and historical complexity, dissecting specific theoretical components, extrapolated from various stages of Darwin's intellectual development, which aligned with contemporary theoretical models. As Barzun (1958, p. 75) provocatively argued, "Darwin's hedging and self-contradiction – enabled an unscrupulous reader to choose his texts from the *Origin of Species* or the *Descent of Man* with almost the same ease of accommodation to his purpose as if he had chosen from the Bible."

From a historical perspective, the heterogeneity inherent in and surrounding Darwinism suggests that seeking rigid and enduring conceptual entities in the history of ideas often leads to dead-end alleys (Esposito 2021). On this matter, David Hull famously proposed to treat Darwinism as an evolving conceptual system. In his work "Darwinism as a Historical Entity," featured in the renowned book *The Darwinian*

[2] As outlined by Richard Delisle in this volume, Mayr seemed to have gradually embraced a broader interpretation of Darwinism, departing from the positions he had initially developed and advocated between the 1940s and the early 1960s.

Heritage (1985), Hull contended that, despite frequent assertions about the "essence" of Darwinism, there is no unanimous agreement among scientists regarding its fundamental tenets. Mayr (1985) famously asserted that Darwin's theory encompassed five distinct sub-theories: evolution itself, common descent, gradualism, multiplication of species, and natural selection. In his early "pan-adaptionist" phase, Stephen Jay Gould stated that the "essence" of Darwinism lay in the assertion that "natural selection creates the fit," and that "variation is ubiquitous and random in direction," providing only "the raw material" (Gould 1977, p. 44). According to Richard Lewontin (1974), if Darwinism does possess an essence, it should be identified at a more general level: the replacement of a metaphysical view of organic change with a naturalistic—if not materialistic—conception. Like any successful conceptual system, Hull remarked, Darwinism was "protean" and demonstrated flexibility over time, rendering the pursuit of any essential nature misleading.[3]

Drawing on Stephen E. Toulmin and evolutionary epistemology, Hull argued that both the network of scientists supporting Darwin and the conceptual system identified as "Darwinism" evolved phylogenetically as a lineage (Hull 1985, pp. 778, 781). These two levels, the Darwinians and Darwinism, are undoubtedly interconnected, yet there seems to be a weak causal nexus between them. In the aftermath of the publication of the *Origin*, scientists actively defended Darwin's ideas without necessarily embracing all its tenets. Conversely, scholars who accepted the major principles of Darwin's proposal did not automatically affiliate themselves with the label "Darwinians." Figures like Thomas H. Huxley and George Mivart shared similar criticisms against Darwin's theory, particularly regarding the belief in gradualism and the creative power of natural selection (Ivi, 797). Despite this, they have gone down in history as two distinctly antithetic figures: Huxley, known as Darwin's bulldog and the foremost advocate of Darwinism in the Victorian Age, and, on the opposing front, Mivart, the scientist who articulated some of the most biting criticisms of natural selection, eventually becoming a prominent anti-Darwinian figure. The history of post-Darwinian debates on the theory of evolution is replete with authors whose embrace or rejection of "Darwinism" was often driven more by ideological and philosophical commitments than purely scientific considerations (Ceccarelli 2021).

3 Past and Present Scenarios

In the contemporary debate on evolution, the dynamics surrounding the unifying role of Darwin's figure and the varied interpretations of Darwin's legacy persist. In his contribution to the volume *Extended Synthesis* (2010), Werner Callebaut pointed

[3]This shift in interpreting Darwin's theory as an evolving historical entity poses the problem whether Darwinism needed to preserve its core identity to exist as a "unified" conceptual lineage or if it could undergo continuous change, as long as it maintained a certain level of internal coherence (Hull 1982, 1988).

out how the continuous calls for new "synthesis," "integrations,"'and "extensions" that characterized the late twentieth-century evolutionary debate, and that had long marked biological sciences, compel us to reflect on the rhetorical functions of such discourses in scientific practice. In Callebaut's view, the history of evolutionary thinking has been marked by coexisting unifying and disunifying tendencies. Like a diaphragm that alternately inflates and contracts, the Modern Synthesis exhibited the same dual nature. On one hand, it demonstrated a proclivity to integrate diverse biological fields, aiming for unification and leveling out inhomogeneity.[4] On the other hand, it displayed hypostatization. This occurred through "vertical extensions," involving a bottom-up expansion of molecular biology models, and "horizontal extensions" toward disciplines such as social sciences, as seen in sociobiology. Amid these poles, developmental phenomena, macroevolution, and epigenetic studies were on the point of being integrated through a major reshuffling of the Modern Synthesis framework (Callebaut 2010, p. 472).

One of the central themes in recent calls for synthesis and new theoretical integrations lies in the shared commitment to non-reductionist approaches, advocated by both scientists and philosophers of biology (cf. Ivi, 460–462; see also Pigliucci and Kaplan 2002; Gould 2002; Doolittle, Bapteste 2007; Danchin et al. 2011; Jablonka and Lamb 2014; Pievani 2016, Araújo Reis 2021). This inclination toward non-reductionism extends beyond evolutionary biology, permeating fields such as environmental sciences and Earth System Science (ESS). As an interdisciplinary domain seeking a unified modeling framework to examine intricate interactions among physical, chemical, and biological processes (Edwards 2011), the maturation of ESS, according to ecologist John Harte, must involve a synthesis of "Newtonian" and "Darwinian" worldviews. This requires steering clear of reductionist temptations and respecting the contingent evolutionary perspective at the core of biology: "particularity and contingency, which characterize the ecological sciences, and generality and simplicity, which characterize the physical sciences, are miscible, and indeed necessary, ingredients in the quest to understand humankind's home in the universe" (Harte 2002).

To what extent is the label "Darwinism"' still applicable to these considerations? Has the term outlived its utility, or does it possess a meaningful degree of conceptual coherence that justifies such associations? Undoubtedly, Darwin's work has permanently altered the landscape of natural sciences. His ideas, principles, and contributions have been incorporated, dissected, and reframed into the body of knowledge and methodology of biology multiple times. Some of these have followed intricate and diverse paths, involving multiple research traditions and leading to varied outcomes (Delisle 2017). If Darwin were alive today, he would easily notice similarities between his "long argument" and the current state of

[4]Talking about Fisher, Haldane, and Wright, the so-called architects of the first phase of the synthesis, William Provine (1992, p. 172) for example observed that they "disagreed, often intensely, about the actual processes of evolution in nature, even when their models were mathematically equivalent," and that their engagement in the synthesis discourse mostly consisted of "exercises in removing barriers, consistency arguments, and forging consensus."

evolutionary biology. At the same time, however, he would likely be astonished by the expansive theoretical and methodological richness characterizing the present discourse on evolution.

Determining whether we are still in a Darwinian age or not—assuming a true "Darwinian era" ever existed—is however an issue that probably conceals a further key question, at least from a meta-historical perspective: why do we feel so compelled to invoke Darwin? The development of scientific theories and ideas follows intricate paths, some branching from the contributions of a few great scientists. The trajectory of theoretical proposals involves debates, reframing, criticism, misconceptions, adjustments, censorship, fanaticism, and various other dynamics that characterize scientific practices within their social and cultural complexities. While exploring the coherence that might qualify a system as "Darwinian," we should probably inquire about the purpose of keeping these labels alive besides the possible continuities and discontinuities between two or more conceptual systems. As historian Jacques Roger (1985) noted, physics has drawn upon the legacies of Newton and, more recently, Einstein, yet we seldom encounter terms like "Newtonism" or "Einsteinism." Paradoxical as it may seem, the history of the term "Darwinism" resembles much more that of eponyms such as "Marxism," due to the philosophical and ideological implications that profoundly marked its resonance in the scientific and general audience. Since the debate that followed the publication of the *Origin*, the term condensed several layers of meaning, turning Darwin's theory into many "Darwinisms."

References

Araújo L, Reis C (2021) Evolutionary pluralism and the ideal of a unified biology. História, Ciências, Saúde 28(2):2–18. https://doi.org/10.1590/S0104-59702021000200004

Barzun J (1958) Darwin, Marx and Wagner: critique of a heritage, 2nd edn. Anchor, Garden City

Bowler PJ (2013) Darwin deleted. Imagining a world without Darwin. The University of Chicago Press, Chicago

Callebaut W (2010) The dialectics of dis/unity in the evolutionary synthesis and its extension. In: Pigliucci M, Müller GB (eds) Evolution. The extended synthesis. MIT Press, Cambridge, pp 443–481

Ceccarelli D (2018) Orthogenetic predictability: orderliness and symmetry in early macroevolutionary explanations. In: Ceccarelli D, Frezza G (eds) Predictability and the unpredictable life, evolution and behaviour. CNR Edizioni, Rome, pp 177–192

Ceccarelli D (2021) Recasting natural selection: Osborn and the pluralistic view of life. In: Delisle RG (ed) Natural selection: revisiting its explanatory role in evolutionary biology. Springer, Cham, pp 171–194

Corsi P (2011) Jean-Baptiste Lamarck: from myth to history. In: Gissis SB, Jablonka E (eds) Transformations of Lamarckism: from subtle fluids to molecular biology. MIT Press, Cambridge, pp 9–18

Danchin EA et al (2011) Beyond DNA: integrating inclusive inheritance into an extended theory of evolution. Nat Rev Gene 12(7):475–486 https://doi.org/10.1038/nrg3028

Delbrück M (1949) A physicist looks at biology. Trans Connecticut Acad Arts Sci 38:173–190

Delisle RG (2009) The uncertain foundation of neo-Darwinism: metaphysical and epistemological pluralism in the evolutionary synthesis. Stud Hist Phil Biol Biomed Sci 40:119–132

Delisle RG (2017) Darwinism or a kaleidoscope of research programs and ideas? In: Delisle R (ed) The Darwinian tradition in context. Springer, Cham
Delisle RG (2019) Charles Darwin's incomplete revolution: the origin of species and the static worldview. Springer, Cham
Delisle RG (2021) Natural selection as a mere auxiliary hypothesis (Sensu Stricto I. Lakatos) in Charles Darwin's origin of species. In: Delisle RG (ed) Natural selection: revisiting its explanatory role in evolutionary biology. Springer, Cham, pp 73–104
Depew D, Weber B (1995) Darwinism evolving: systems dynamics and the genealogy of natural selection. MIT Press, Cambridge
Dobzhansky T (1973) Nothing in biology makes sense except in the light of evolution. Am Biol Teach 35:125–129
Doolittle WF, Bapteste E (2007) Pattern pluralism and the tree of life hypothesis. Proc Natl Acad Sci 104(7):2043–2049
Dupré J (2015) Pluralism and process in understanding human nature. Rivista di Filosofia Neo-Scolastica 107(1-2):15–28
Dupré J, Nicholson DJ (2018) Everything flows. Towards a processual philosophy of biology. Oxford University Press, Oxford
Edwards PN (2011) History of climate modeling. WIREs 2:128–139
Engels E, Glick TF (eds) (2008) The reception of Charles Darwin in Europe, I. Continuum, London
Esposito M (2021) Cathedrals, corals and mycelia: three analogies for the history of evolutionary biology. In: Delisle RG (ed) Natural selection: revisiting its explanatory role in evolutionary biology. Springer, Cham, pp 11–38
Glick TF (ed) (1988) The comparative reception of Darwinism. University of Chicago Press, Chicago
Gould SJ (1965) Is uniformitarianism necessary? Am J Sci 263:223–228
Gould SJ (1977) Ever since Darwin: reflections in natural history. Norton, New York
Gould SJ (1994) Tempo and mode in the macroevolutionary reconstruction of Darwinism. PNAS 91(15):6764–6771. https://doi.org/10.1073/pnas.91.15.6764
Gould SJ (2002) The structure of evolutionary theory. Harvard University Press, Cambridge
Harte J (2002) Toward a synthesis of the Newtonian and Darwinian worldviews. Phys Today 55(10):29–34. https://doi.org/10.1063/1.1522164
Haufe C (2015) Gould Laws. Philos Sci 82:1–20
Hodge MJS (1977) The structure and strategy of Darwin's 'long argument'. Br J Hist Sci 10:237–246
Hodge MJS (2009) The notebook programmes and projects of Darwin's London years. In: Hodge MJS, Radick G (eds) The Cambridge companion to Darwin, 2nd edn. Cambridge University Press, Cambridge, pp 44–72
Hoquet T (2010) History without time: Buffon's natural history as a nonmathematical physique. Isis 101(1):30–61. https://doi.org/10.1086/652688
Hull DL (1982) Exemplars and scientific change. Proc Biennial Meeting Philos Sci Assoc 2:479–503
Hull DL (1985) Darwinism as a historical entity: a historiographic proposal. In: Kohn D (ed) The Darwinian heritage. Princeton University Press, Princeton, pp 773–812
Hull DL (1988) Science as a process: an evolutionary account of the social and conceptual development of science. Chicago University Press, Chicago
Huxley J (1942) Evolution. The modern synthesis. Allen and Unwin, London
Jablonka E, Lamb MJ (2014) Evolution in four dimensions, revised edition: genetic, epigenetic, behavioral, and symbolic variation in the history of life. MIT press, Cambridge
Johnson CN (2020) Darwin's "historical sketch." an examination of the 'preface' to the origin of species. Oxford University Press, New York
Kauffman SA, Gare A (2015) Beyond Descartes and Newton: recovering life and humanity. Prog Biophys Mol Biol 119(3):219–244. https://doi.org/10.1016/j.pbiomolbio.2015.06.003
Kitcher P (1999) Unification as a regulative ideal. Perspect Sci 7(3):337–348

Lewontin RC (1974) The genetic basis of evolutionary change. Columbia University Press, New York

Love AC (2013) Theory is as theory does: scientific practice and theory structure in biology. Biol Theory 7:325–337. https://doi.org/10.1007/s13752-012-0046-2

Mayr E (1985) Darwin's five theories of evolution. In: Kohn D (ed) The Darwin

Mayr E (2004) What makes biology unique? Cambridge University Press, Cambridge

McLaughlin P, Demarest B (2022) Newtonianism and the life sciences. In: Jalobeanu D, Wolfe CT (eds) Encyclopedia of early modern philosophy and the sciences. Springer, Cham. https://doi.org/10.1007/978-3-319-31069-5_98

Moore JR (1979) The post-Darwinian controversies. A study of the Protestant struggle to come to terms with Darwin in Great Britain and America. Cambridge University Press, Cambridge

Musil R (1996) [1952], The man without qualities. Vintage International, New York

Numbers RL, Stenhouse J (eds) (2001) Disseminating Darwinism: the role of place, race, religion and gender. Cambridge University Press, Cambridge

Osborn HF (1928) Forward. In: Tinley F (ed) The brain from ape to man; a contribution to the study of the evolution and development of the human brain. Paul Hoeber, New York, pp xv–xvii

Packard A (1876) A century's progress in American zoology. Am Nat 10(10):592–597

Pence CP (2018) Charles Darwin and Sir John F. W. Herschel: nineteenth-century science and its methodology. HOPOS 8(1):198–140

Pievani T (2016) How to rethink evolutionary theory: a plurality of evolutionary patterns. Evol Biol 43(4):446–455

Pigliucci M, Kaplan J (2002) The fall and rise of Dr. Pangloss: adaptationism and the Spandrels paper 20 years later. Trends Ecol Evolution 15, 2:66–70

Provine W (1992) Progress in evolution and meaning of life. In: Waters CK, Helden AV (eds) Julian Huxley: biologist and statesman of science. Rice University Press, Houston, pp 165–180

Roger J (1985) Darwinism today (commentary). In: Kohn D (ed) The Darwinian heritage. Princeton University Press, Princeton, pp 813–823

Ruse M (1975) Darwin's debt to philosophy: an examination of the influence of the philosophical ideas of John F. W. Herschel and William Whewell on the development of Charles Darwin's theory of evolution. Stud Hist Phil Sci 6(2):159–181

Ruse M (2013) The origin of species. In: Ruse M (ed) The Cambridge encyclopedia of Darwin and evolutionary thought. Cambridge University Press, Cambridge, pp 95–102

Singh BN (2023) Evolution is the unifying theory of biology. Curr Sci 124(3):289–290

Smocovitis VB (1999) The 1959 Darwin centennial celebration in America. Osiris 14:274–323

Weber BH (2011) Extending and expanding the Darwinian synthesis: the role of complex systems dynamics. Stud History Philos Sci Part C Stud Hist Philos Biol Biomed Sci 42(1):75–81. https://doi.org/10.1016/j.shpsc.2010.11.014

Whitehead AN (1925) Science and the modern world. Cambridge University Press, London

Part II

Deconstructing Darwinism

Constructing, Deconstructing, and Reconstructing. On "Darwinism" and "Darwinisms," with Some Disparate Considerations on the History of Science

Antonello La Vergata

Abstract

Recent developments in evolutionary biology rekindled the debate on whether evolutionary theory needs a rethink. Has the path opened by Darwin been abandoned? Or is a new synthesis emerging? What does "Darwinism" mean today? A historian can only say that things were not much different in the past. Biology was pluralistic before and after Darwin. Darwinism was a multifaceted collective movement, and the term meant many things to many people, including Darwin's supporters. This is demonstrated by an overview of the different interpretations of the concepts of "struggle for existence" and "natural selection," particularly by a discussion of the American psychologist James Mark Baldwin's attempt to bring order into a proliferation of struggles and selections and, at the same time, to embrace them, as far as possible, under one label. A linear reconstruction of the events which led to the so-called Modern Synthesis can no longer be accepted, despite the fact that it has inspired fascinating narratives, first and foremost that of Ernst Mayr. The history of evolutionary biology, and of science in general, is only linear in retrospect, from the point of view of those who have an axe to grind, or of scientists-historians who adopt self-serving strategies in order to prove that they are on the right side of science and history. The commitment to concrete research makes many scientists indifferent to history and somewhat refractory to a genuine historical sense. There seems to be a profound difference between the "two cultures," both of which are necessary: the scientific, present-oriented one, and the historical-humanistic one, which shows how the

I am very grateful to Richard Delisle, David Ceccarelli, Maurizio Esposito, and the anonymous reviewers for their comments on a former version of this Chapter.

To my friend and Darwin scholar Mario Di Gregorio, in memoriam

A. La Vergata (✉)
Dipartimento di Studi linguistici e culturali, Università di Modena e Reggio Emilia, Modena, Italy

R. G. Delisle et al. (eds.), *Unity and Disunity in Evolutionary Biology*,
https://doi.org/10.1007/978-3-031-42629-2_4

past conditions the present. Extended outside the realm of scientific research, however, "presentism" can be dangerous. Only a realistic picture of the history of science can help combat that ever-renewing enemy: anti-science.

Keywords

Darwinism · Darwinian revolution · Eclipse of Darwinism · Struggle for existence · Natural selection · Baldwin · Modern Synthesis · Mayr · Historiography · Deconstruction · Two cultures · Presentism

1 Introduction

The history of Darwinism is a history of continuities as well as discontinuities. The word "continuity" tends to suggest a progressive addition of "discoveries" and "facts," and the elimination of "errors" and "false" pathways, whereas "discontinuity" tends to be interpreted in terms of "revolutions," that is, changes of paradigms. However, both attitudes are vitiated by hindsight. History teaches that "confusion" reigns today as it did at the end of the nineteenth century and the early twentieth century. Such confusion is normal, in science as in all human affairs: the more things one sees, the more it increases. Science is the only field of human experience in which it is not only inevitable, but necessary and beneficial. To realize this does not at all mean to diminish science, but to have a realistic image of it as the most, indeed the only, reliable form of knowledge, and to protect it from two opposite dangers, which, paradoxical though it may seem, support and reinforce each other: making science into mythical body of unassailable truths, and making it into an equally mythical body of arrogant and anti-human falsehoods.

The present-day debate on whether "a new evolutionary synthesis is emerging," as S.J. Gould put it as early as 1980, should be accompanied by a sound "*histoire historienne*" (Jacques Roger's term) of what really happened and, more difficult, what is happening. Such a task cannot be entrusted to scientists engaged in research (still less to those who write self-serving narratives), nor to those philosophers who merely analyze the internal *logic* of a theory. The important results achieved by what has come to be generally called "Darwin industry" confirm that the logic, the implications, and the consequences of a theory cannot be fully understood if its genesis and contexts are not investigated: *ordo expositionis non est ordo inventionis*. A vast field lies open for a "post-Darwin industry."

An overview of the debates on Darwin's theory of evolution demonstrates that his key concepts of "struggle for life" and "natural selection" were interpreted and evaluated in many different ways. They were now accepted uncritically, now grossly misunderstood and rejected, now analyzed, discussed, specified, split into sub-concepts, or integrated with supplementary concepts: in a word, each author seems to have read and understood them in his own language. This holds true not only for philosophers, social writers, literary authors, and those who applied those concepts to political issues, but also for professional biologists. It is not overly exaggerated to say that there were as many Darwinisms as readers. This raises some general questions, for instance:

- Was there a Darwinian revolution?
- Was Darwin's theory an open theory? Was it a mere general framework? Has it been completed, integrated, or renovated?
- Is Darwinism a historical construct?
- Are scientists' retrospective interest in the history of evolutionism and historians' call for overcoming simplifications irreconcilable?

I am not going to answer all of these questions directly. I will merely offer some historical food for thought, and a few disparate reflections on some more general issues.

2 Darwinism, Darwinisms, Darwinisticism, etc.

After reading the *Origin of Species* "for the *n*th time," Thomas Henry Huxley said that it was "one of the hardest book to understand thoroughly that I know of."[1] Who is an "absolute and pure Darwinian?" Huxley also asked on another occasion, and he replied: "I doubt if I can ever have seen one alive" (Huxley 1893–1894, p. 474). The very Wallace who coined the term "Darwinism" admitted to being "more Darwinian than Darwin himself." No wonder there were several streams in Darwinism from the outset, and even "Darwinians" disputed among themselves.

The progress of studies on Darwin and Darwinism has now clarified that there were few "true" Darwinians. Both as a scientific theory and as a view of nature, Darwinism meant different things to different people, and many self-styled Darwinians were only superficially so. Not a few of them did not fully accept natural selection as the main mechanism of evolution, or had troubles with, for instance, Darwin's gradualism, or chance variations. We should not wonder at this. The history of scientific ideas is only simple in retrospect, or from the point of view of those who have an axe to grind, or scientists-historians who adopt self-serving narratives to prove that they are on the right side of science and history. There is no via *regia* to truth, whatever you mean by the word "truth."

[1] Letter to Joseph Dalton Hooker, March 9, 1888, in L. Huxley ed. 1900, 2: 192. See also Huxley's letter to Michael Foster, February 14, 1888 (ibid.: 190): "Nothing entertains me more than to hear people call it [the *Origin*] easy reading." These statements are in a curious contrast with what Huxley said he had exclaimed to himself when he "first made [him]self master of the central idea of the 'Origin'": "How extremely stupid not to have thought of that! I suppose that Columbus' companions said much the same when he made the egg stand on end" (ibid.: 170; F. Darwin ed. 1887, p. 197).

Many readers were simply unable to go deep into Darwin's ecological way of thinking. To do this required crossing disciplinary borders and traditions, going beyond specialized fields of research, changing mental approach, and renouncing cherished habits of thought. Not an easy task. In spite of normative philosophers of science, there is no universal recipe for performing it. Scientific theories and conceptual innovation need time not only to be accepted, but to take place. Revolutions only occur through the *revision* of their original content. The Copernican revolution took more than a century to be generally accepted, and, when it was, Copernicus' original view had been transformed. And the Newtonian revolution had to overcome a strong resistance by Cartesians, who had their good reasons to object to such an "occult quality" as "force."

The number of varieties of Darwinism which developed soon after 1859 is impressive. As John Greene (1971) argued, Darwin's theory did not emerge as a response to the crisis of one paradigm and did not replace one paradigm.[2] Biology was pluralistic before and after Darwin. It makes little sense to describe the debates which followed 1859 as a for-or-against fight between two opposing teams. Darwinism was a multifaceted collective movement. Hull (1974, pp. 388, 391, 398) remarked that "Darwinism was many things to many people," and he went as far as to say that "in many instances, Darwin was a comparatively minor figure in the movement termed 'Darwinism' [...] A scientist could be a Darwinist without accepting natural selection as the major evolutionary mechanism." It is, therefore, extremely difficult to capture an "essence" of Darwinism (Hull 1985).[3] In his review of British periodical press from 1859 to 1872, Alvar Ellegård (1958, rept. 1990, p. 334) had written: "Though it was chiefly the Natural Selection theory which made the Descent doctrine scientifically acceptable, popular Evolutionism has by and large remained at the pre-Darwinian, unscientific stage." Historiographically speaking, the term "unscientific" is patronizing, and it would be better not to use it, but the fact remains that Darwin's theory was often presented as merely a "recent modification" of Lamarck's (ibid.: 269–270; see also Bowler 2004). The American paleontologist James Dwight Dana (1871, cit. in Pfeifer 1974, p. 192) said that Darwin's theory of "variation by natural selection" was losing ground in Britain at the same time as evolution was gaining ground. In the same year, the American philosopher Chauncey Wright (1871, p. 65) noticed the following paradox: "It would seem, at

[2]It is obvious that only retrospectively can one speak of a paradigm change. I am inclined to think that Kuhn's term was not particularly well chosen. Despite his own definition of it, it suggests a well-defined and prescriptive model, which is something historians only come across when dealing with fully constituted systems. After all, what Kuhn defined as "normal science" consists of a number of ideas, practices, linguistic uses, etc., which hardly address theoretical issues directly, and are often carried out without having a clear-cut paradigm in mind.

[3]Gould (2002, p. 11) did not agree with Hull: although they are historical entities, "theories of range and power also feature inherent 'essences,' implicit in their logical structure, and operationally definable as minimal sets of propositions so crucial to the basic function of a system that their falsification must undermine the entire structure." On this and related issues see the critical discussion in Esposito (2021).

first sight, that Mr. Darwin has won a victory, not for himself, but for Lamarck. Transmutation, it would seem, has been accepted, but Natural Selection, its explanation, is still rejected by many converts to the general theory, both on scientific and religious grounds." The most vocal Darwinist in Germany, Haeckel, believed he could bring together Goethe, Lamarck, and Darwin, as the subtitle of his *Natural History of Creation* (1868) suggested. In a sense, then, there was also a "non-Darwinian revolution" (Bowler 1988).

Discussing Hull (1985), Roger (1985, p. 820) wrote: "I would prefer to use the label 'Darwinism' only for the thought of Darwin himself." Delisle (2017, p. 157) goes so far as to argue that the time has perhaps come to dispose of the label "Darwinism" altogether.

With respect to Hull's proposal to deal with Darwinism as both a historical entity and a social group, Roger remarked that filiations of ideas, which define historical-intellectual entities, are identified by criteria which are not easily applicable to social groups. Participation in a group or circle, one might add, is based on a form of consensus, and consensus can be due to very disparate reasons. In 1959, Morse Peckham made a distinction between "Darwinism" and "Darwinisticism." The first, he argued, "is a scientific theory about the origin of biological species from pre-existing species, the mechanism of that process being an extraordinarily complex ecology." The second "can be an evolutionary metaphysic about the nature of reality and the universe. It can be a metaphysical and simplistic notion of natural law. It can be an economic theory, or a moral theory, or an aesthetic theory, or a psychological theory. It can be anything which claims to have support from the *Origin*, or conversely anything which claims to have really understood what Darwin inadequately and partially presented" (Peckham 1970, p. 191). This distinction was accepted as "fundamentally sound," but "refined as well," by Moore (1979, pp. 15, 116, 367–368, notes 37–38). Unfortunately, as Moore rightly remarked, "Peckham limits consideration to the *Origin of Species* as a document defining Darwinism and oversimplifies both Darwinism and Darwinisticism, disregarding the various unresolved issues in post-Darwinian evolutionary theory," which led Darwin himself to revisions, and "others, not to a 'perverting and self-deluding acceptance of the work', but to an acceptance conditioned in many cases by their solutions to the unresolved problems" (ibid.: 191, 368). Distinguishing between Darwinian and "Darwinian" (Leeds 1974) is obviously crucial for those who deal with the applications of Darwin's theory to society. The problem of what is to be meant by "Darwinian" and "Darwinism," however, remains. And whatever is meant by these terms, they refer to a cultural and social phenomenon as well as a scientific one. After all, it is not surprising that Darwinism itself changes, like everything else. "More and more studies show that the development of evolutionary biology was never organized around a central and dominant narrative called Darwinism, as currently defined" (Delisle 2021a, p. 6). Has the time perhaps come to dispose of the label "Darwinism" altogether? (Delisle 2017, p. 157)

Camille Limoges' *La sélection naturelle* (1970) marked a turning point in Darwin studies by focusing on the genesis of Darwin's concept of natural selection. Until then, a general history of ideas approach had prevailed in studies on Darwin,

Darwinism, and evolution. The best fruits of that approach were John Greene's important *The Death of Adam* (Greene 1974, 1st edn 1959) and Glass et al. eds. *Forerunners of Darwin* (1959), which latter also contained articles dating back to the beginning of the twentieth century. Many later works, however, were still influenced by a sort of whiggish interpretation of the history of evolutionary ideas. That interpretation, in turn, was strongly influenced by the fact that some leading biologists (first and foremost Ernst Mayr) retrospectively interpreted Darwin's achievement in the light of the so-called Synthetic Theory, or New Synthesis, or Modern Synthesis, that is, as the starting point of a more or less linear process which eventually culminated in the only kind of evolutionary thinking that deserved to triumph.

Limoges', Sandra Herbert's (1971, 1974, 1977), and David Kohn's (1980) pioneering work on Darwin's manuscripts opened a new field of investigations, which proved so fruitful and rewarding as to generate a "Darwin industry" (DI). It focused on Darwin's own development, inevitably neglecting other figures and approaches, especially outside the English-speaking world. Therefore, on the one hand, DI put an end to the historiography of evolutionary ideas in terms of forerunners. On the other hand, it hardly affected a still widespread tendency to read what preceded, and particularly what followed, 1859 as if everything rotated around an allegedly identifiable "Darwinism." It is extremely difficult to identify an essence of Darwinism, and a "populational" approach has not generally been applied to studies on it, not even by those who made "population thinking" their *cheval de bataille*. Although they occasionally turn into historians, scientists seem to prefer "rational reconstructions." They are de facto essentialist.

At the same time, the need to go, in a sense, beyond Darwin has led to a flourishing of contextual studies, most of them focused on the social, religious, and political implications and consequences of evolutionary debates. Important studies have been produced on Weismann's neo-Darwinism, neo-Lamarckism, other alternative evolution theories, and pre-Mendelian and post-Mendelian debates on inheritance, and a vast territory lays open for further historical research. They present us with a much more complex landscape of what happened after 1859 than that which might be suggested by the striking metaphor of an "Eclipse of Darwinism."[4] How could a sun that never fully shone eclipse?

3 "The Elimination of Semantic Confusions"?

It is not just a matter of agreeing on what is meant by words. The confusion surrounding the term "Darwinism" cannot be eliminated by a semantic rectification, or by defining it for the umpteenth time. It is too late. "Darwinism" is not a technical

[4]This is not intended as a criticism of Bowler's (1983) important book, but of those who transformed an impressive title into a historical fact. As a serious scholar, Bowler (2005, 2009, 2013, 2017) revisited and qualified his thesis.

term. But is the case *completely* different with those which were or are considered technical terms in evolutionary biology? Is their success due to their being clearly defined when first introduced? Ernst Mayr (1982, pp. 841–842) wrote:

> "Technical terms, when clearly defined and *well understood*, are a great help in the advance of scientific understanding. Contrariwise, when a term is inadvertently transferred to a different concept (as was done by T. H. Morgan with the term 'mutation') or when the same term is used for different concepts, considerable confusion will ensue until the ambiguity is cleared up. The introduction of new technical terms often helps to disentangle this kind of confusion. Examples are the term 'taxon' (for which previously the term 'category' had also been used), the term 'subspecies' (for which taxonomists had previously used the term 'variety' which was also being used for individual variants),[5] or the term 'isolating mechanisms' (for which previously no term had existed). Examples could be quoted from any branch of biology where the introduction of new terms resulted in clarification in a previously confused area. The evolutionary synthesis of the 1930s and 40s was greatly facilitated by the introduction of the terms 'polytypic' by Huxley and Mayr, 'sympatric' and 'allopatric' by Mayr, 'gene pool' by the Russian school, 'genetic drift' by Sewall Wright, and other terms such as 'founder principle' and 'genetic homeostasis.' When properly defined and clearly delimited against other phenomena with which they had been previously confounded, such terms helped to eliminate controversies. When a term is transferred from concept to concept no matter how much the underlying conceptualization is changing, misunderstandings inevitably result." (my italics)

However, Mayr has to recognize that the reverse is also true:

> "And yet, in most cases the retention of the technical term has been preferable to a continuous introduction of new terms whenever there was a slight or gradual change of the underlying concept. For example, the term 'gene', when proposed by Johannsen, was specifically given for a 'nonmaterial' entity, an 'accounting unit.' In the Morgan school the term was soon applied to a definite, distinctly material locus on the chromosome and in molecular genetics to a certain set of base pairs, likewise a strictly material entity. One could go on and on giving examples of this kind" (ibidem).

This raises a couple of questions. (1). When can we say that a term is "well understood"? As we are going to see in § 4, Darwin's key concept, natural selection, can hardly be said to have been "well understood," if by that phrase we mean that it was generally taken in one sense only. Was this due to its not being "well defined"

[5]Many heterogeneous phenomena, Mayr (1982, pp. 639–640) wrote, were subsumed under the term "variety," and the efforts to bring some order resulted in an "elaborate terminology which [...] did not eliminate the problem, because it did not eliminate the underlying conceptual confusion. Most authors failed to distinguish clearly between (1) genetic and nongenetic variation; (2) continuous and discontinuous variation [...]; and (3) individual and geographical variation. As a consequence, when different authors talked about "varieties," they often had entirely different phenomena in mind. The situation was aggravated by the fact that, beginning with Linnaeus, two different traditions developed which divided botanists and zoologists. When zoologists spoke of varieties, they usually meant geographic races; when botanists did so, they usually meant cultivated varieties or intrapopulation variants. And yet this difference in tradition was the first indication of a sorting out of the different kinds of varieties." Darwin himself sometimes used "varieties" and "variations" indifferently.

by its creator? (2). When do terms become "technical," and how do they come to be accepted as such? In many cases, the processes is not ruled by a "logic," or the decision of a committee, but is influenced by "social" factors as well: prestige, imitation, linguistic inertia, teaching, even suggestions and emotional resonances. And the whole series of such qualifications, restrictions or extension of meanings, subdivisions, etc. may also start again *after* a term is generally accepted. Natural selection is a perfect instance of this. Historians try to reconstruct that process, and, whenever possible, to explain it, without simplifying their task by resorting to "rational reconstructions." It is only by hindsight, that is, after history has done its job, that one can judge whether terms and concepts were "rightly" understood or not. The "elimination of semantic confusion" is not always a matter of definition. Future historians will narrate how *present-day* semantic confusions (see below, § 7) will have been eliminated, and whether they were only due to bad semantics.

These considerations hold true *a fortiori* for metaphors. "Metaphors – Mayr writes (ibidem) – play an important role in the history of science." But, he adds, "there are felicitous metaphors and those that are unfortunate." There is ambiguity hidden in this sentence: whereas "unfortunate" (or "fortunate") is a statement of fact, "felicitous" (or "infelicitous") smells of retrospective evaluation. Mayr mentions the term "genetic drift," which, "introduced by Sewall Wright to designate the stochastic processes of changes in allele frequencies in small populations, was misinterpreted by certain authors as a steady one-directional drift." Another instance is Darwin's natural selection, which, Mayr says, "is on the borderline of the two categories [felicitous and unfortunate] and was *strenuously resisted by the majority of his contemporaries* (my italics). They tended to personify whatever did the selecting and to insist that there was no real difference between selection by Nature and creation by God." This statement is not true. It is based on an unacceptable oversimplification. Resistances to natural selection also came from biologists whose objections Darwin took very seriously. And misunderstandings due to seeing "no real difference between selection by Nature and creation by God" often helped making the notion of natural selection palatable for people who would otherwise have opposed evolution (Moore 1979; Corsi 1988). Those misunderstandings helped spread the notion, and, as a matter of fact, fueled further debates on it and its mechanisms. (Incidentally, Darwin also took religious objections to his theory very seriously. Mayr might consider this irrelevant, but they had important consequences for the fate of natural selection.)

Unfortunately, Mayr goes on to say, Darwin "jumped from the frying pan into the fire" when he adopted Spencer's term "survival of the fittest," because "this new metaphor suggested circular reasoning." Some comments on this statement are in point: (1) Darwin adopted Spencer's term on Wallace's extensively argued advice exactly in order to avoid misunderstandings by demetaphorizing his concept and replace it with a "mere description of the fact." (2) A historian must not merely register the fact, but ask why Wallace made his suggestion and why Darwin accepted it. (3) The circularity of "survival of the fittest" has been amply discussed, and good reasons have been given for rejecting the charge (Sober 1984a, b). (4) One cannot help perceiving in Mayr's sentence a holdover of the old habit of dumping on poor

Spencer the responsibility for social Darwinism (incidentally, another historiographic category which has been, and is being, deconstructed). (5) "Survival of the fittest" was far from "unfortunate" (in the neutral sense of the word), as it was used by many biologists not only when applying Darwinism to social matters, but also when doing biological research (before semantic confusion was eliminated, of course).

Mayr is quite right when he says that "a study of the introduction and subsequent fate of metaphors in biology would be an interesting task for a historian," provided, I add, the historian does his job to the end, that is, investigates why and how metaphors were "misinterpreted," and misinterpreted in different ways. Mayr is not quite right when he uses the word "confusion" in order to suggest, rather surreptitiously, what might (or *should*?) have been clearly seen but was not.

4 Creative Confusion, or, Why So Many "Misunderstandings"?

To give an instance of "confusion," let us consider the fate of Darwin's metaphors of "struggle for existence" (or "struggle for life," as he used both) and "natural selection." We shall see that much "confusion" arose as a consequence of attempts to clear up ambiguity. "Struggle for life" and "struggle for existence" became immensely popular after 1859. They were used by supporters as well as by critics, in an enormous quantity of writings, scientific, philosophical, political, social, and literary. But not everyone meant the same thing, quite the contrary. As Gillian Beer (1985, p. 9) remarked, "the unused, or uncontrolled, elements in metaphors such as 'the struggle for existence' take on a life of their own. They surpass their status in the text and generate further ideas and ideologies." Such was the case not only with laypersons using a fashionable catchphrase whenever it seemed suitable for their purposes. A number of authors, including respected scientists and intellectuals, translated Darwin's phrases into their own conceptual language and built ideas, theories, and doctrines upon their translations.

"Struggle for existence," which is more appropriately described as a bundle of interweaving metaphors, referred to too many things for it to be mastered critically in all its meanings and implications. Darwin's phrase met with a number of different reactions, especially, but not exclusively, with regard to human evolution. They ranged from its acceptance as a commonplace to rejecting it as a gloomy denial of the goodness of nature. Some simply misunderstood what Darwin tried to convey. While some authors, popularizers as well as scientists, took it in a narrowly literal sense (in spite of Darwin's warning), others tried to disentangle it, or to translate it in terms of their own image of nature. As a result, it was easier for many people to reject it as a whole or to accept it as a "magical word" (Haeckel's phrase) than to take advantage of its polysemy.

When taken in the literal (or "gladiatorial," as Huxley put it) sense, the expression generated emotional, as well as scientific and philosophical, reactions.[6] It dismayed some readers, adding to their aversion to evolution, while it delighted others, who welcomed it as a confirmation that nature is a theater of struggle and strife. Misunderstandings and distortions (I use these words for brevity's sake) were due to the most various reasons: scientific, philosophical, ideological, social, and political. Interestingly, and contrary to what might have been expected, they were not limited to Darwin's critics, but were also widespread among many who described themselves, or have been usually considered, as Darwinians. As a result, they did not invariably lead to a rejection of Darwin's theory. On the contrary, in some cases they *favored* its acceptance. In some hands the metaphor withered and dried up. In others the attempt to disentangle the different meanings that Darwin had condensed in one phrase, to revise it or to apply it to new fields generated new investigations. As a term describing the ecological context of evolution, it suffered from objections leveled against natural selection. A metaphor became the basis for further metaphors, and a proliferation of struggles was accompanied by equally amazing proliferations of selections, as we shall see. Confusion was then increased by the introduction of new competing terms into the arena. Faced with it, the historian is offered no consolation by being reminded of the German philosopher Hegel's dictum on "thinking dying in the midst of too many qualifications."

This raises some big questions: Were misunderstandings due to the ambiguity of Darwin's metaphor in a time when mathematics had not yet laid its neutral, objective hands on natural history? Are scientists, as scientists, per se fully able to understand the scientific works they read, or are supposed to read, and upon which they ground their own research? Are they always "scientific" when speaking of science? Like all human beings, they, too, are liable to prejudices, not to mention the influence of disciplinary traditions, different schools of thought, and ideological biases.

As I said, there was a positive side to this mess: disagreements as well as false agreements, restrictions as well as extensions of Darwin's metaphors, even misunderstandings opened the way to further investigations. "Confusion" can be creative. From our vantage point we could go as far as saying that Darwin's "ambiguity" (as some historians delight in calling it), "hesitations" (Delisle and Tierney 2022) and doubts on many points was the price to pay for new approaches to new fields. Departing from Darwin's main road proved fertile in some cases. Studies

[6]Elsewhere (La Vergata 2023) I have tried to emphasize that in many cases *moral* factors, although often implicit, played an important role in the reception of Darwin's theory. *Moral* assumptions lay at the basis of much of what many scientists read into Darwin's work, or derived from it. *Moral* instructions were demanded, or derived from, or read into, nature. I have laid particular stress on the assumption that life is strife, an assumption that runs through Western culture, from the Christian idea of the struggle against the world's temptations to the *Streben* of the Romantics, from the ancient and protean myth of an eternal conflict between Good and Evil to the idea that war is the mother of history, from such commonplaces as "no pain, no gain" to the belief that self-realization is only possible through struggling against something (Boas 1972). It is a great thing to have Nature on one's side.

on the role of geographical isolation, for instance, implied revising Darwin's emphasis on vast and open areas as the main laboratory for struggle and selection to operate. The late nineteenth century and early twentieth century were a period of turmoil in biology, as witnessed by the proliferation of non-Darwinian theories and neologisms, most of them short-lived. It is therefore necessary to analyze linguistic as well as conceptual aspects, and tacit assumptions as well as outspoken opinions and open "ideological" stances. In a word, one needs to investigate what authors implied as well as, if not more than, what they declared.

Not a few historians, first and foremost Young (1969), have argued that Darwin's idea of struggle, indeed his whole view of nature, was influenced by a "common context of biological and social theory." "Darwinism is social," Young (1985) stated. No wonder, then, that the idea of struggle played a role in a debate where what was, so to speak, strictly scientific can hardly be distinguished from what was social, or ideological, although Bellon (2021, p. 40) seems to me to go a little too far when he says: "None of Darwin's contemporaries evaluated natural selection purely on intellectual grounds."[7] To be sure, scientific theories are social constructions, as they are built upon the work of others, and are "true" to the extent that they are accepted by a community of experts. But this does not explain why there were so many differences, indeed so many *different misunderstandings*, not only among people using Darwin's struggle in political and social discourse, but also among scientists sharing his general theory of evolution by natural selection. It does not pay much to stuff it all into politics, as it seems to be the case with a flourishing and much lauded historiography.[8] Interesting as it is, such approach leaves some big questions untouched: Why were there also differences in strictly technical, so to speak, interpretations of Darwin's struggle for life? What became of *science* in this struggle of political interpretations, uses, and adaptations to changing social contexts? I hope I am not going to be branded a piece of positivist junk if I still think that science *is* something different from other discourses and that scientific debates, however influenced by their context, cannot be reduced to political debates. I am aware that

[7]Bellon (ibidem) goes on to say: "Darwin, then, engineered natural selection to aid research. Many allies, and even some erstwhile opponents, approached natural selection on these terms. Others, again both allies and opponents, plugged Darwin's ideas into the rancorous ideological and theological controversies which had rumbled through British life for decades. For the participants in these cultural melees, natural selection's set of interlocking propositions about patterns in the natural world was largely interesting—often exclusively—to the extent that they could be weaponized. Darwin designed natural selection as a scientific scalpel; his contemporaries repeatedly appropriated it as a machete for close-quarter ideological battle."

[8]This approach is brought to extreme consequences in Piers J. Hale's *Political Descent: Malthus, Mutualism, and the Politics of Evolution in Victorian England* (2014). To simplify his job, Hale discusses many authors as if each of them produced only one work containing one thesis, as if people such as Huxley were not prolific and did not hold different positions throughout their life. This "one author-one work-one idea" method consists in extracting the gist of a work and then comparing it with the gists extracted from other works written by other authors. In such narratives authors do not seem to have a life of their own. They become pawns on a chessboard where the historian plays a game with himself.

the term "misunderstanding" smells of "whiggish" historiography, and I use it, as I said, for brevity's sake. But it is impossible not to consider things *also* in the light of what happened later, or, to be more exact, of the different understandings, uses, applications, and reactions a theory generated.

Studies of the reception of Darwin's struggle for life have generally focused on its uses in social and political discourse, that is on that inextricable mess which goes under the label "social Darwinism." I think, however, that we should go deeper and study interpretations of Darwin's struggle, as it were, *before* they were used in social and political discourse. Before applying something to something else, you must think that that thing is applicable, that is you must assume, or be convinced, that there is an analogy or a continuity between two fields, and that they are ruled by the same laws. You must at least be inclined to think that there is a continuity between nature and society. This belief leads you to identify evolution with progress. And this is an inclination or assumption which is at work *while* you read. The very reading of Darwin's metaphor was social, in the sense that it was influenced by the readers' frame of mind, tacit assumptions, disciplinary training, and so on. Spectacles are put on before you start reading, and "ideology," or cultural biases, or mental dispositions, or intellectual leanings—call them whatever you like—work in shaping them.

5 Struggles and Selections[9]

Attempts to revise, qualify, reformulate, restrict, or extend the concept of the struggle for existence inevitably reverberated on that of natural selection, and *vice versa*. Darwin wrote that "several scientific men have thought the term 'Natural Selection' good, because its meaning is not obvious, and each man could not put on it his own interpretation" (as instead was the case with "struggle," one would be tempted to add) (Darwin 1985–, 8, p. 83). He was wrong. In later editions of the *Origin* he devoted a paragraph to dispelling the most common misunderstandings: natural selection does not induce variability; it does not imply conscious choice on the part of animals, and is therefore also applicable to plants[10]; it is not an active power or Deity. "In the literal sense of the word, no doubt, natural selection is a false term," he wrote, but "everyone knows what is meant and is implied by such metaphorical expressions" as chemical affinity or gravity, which were "almost necessary for brevity." It was difficult to avoid personifying the word "nature," but "with a little familiarity such superficial objections will be forgotten" (Darwin 1876, p. 63).

[9]In what follows I use materials and sources I discussed in more detail in La Vergata (2023).

[10]A mistake which was much more common than one could suspect, in spite of more accurate readers warning against it (Carpenter 1860, rept. in Hull ed. 1973; Darwin 1985–, 8, p. 515). The Scottish biologist John Arthur Thomson felt it necessary to point out as late as 1909 that "we might be saved from taking a narrow view of the struggle for existence if we emphasise the fact that the concept must apply to plants as well as to animals" (Thomson 1909, p. 75).

Darwin was being optimistic, as if he did not fully realize what he had set in motion. On the one hand, superficial objections did not cease. On the other hand, a real hunt for definitions, sub-definitions, qualifications, and classifications of the various forms of natural selection was opened. As a result, like I said, a proliferation of struggles was accompanied by a proliferation of selections. Analyzing them leaves the historian with that depressing sense of confusion that assails one in the midst of an animated discussion in which everyone is speaking at the same time, some believe they bring harmony by shouting even louder, and others try to start again by redefining the terms of the debate.[11]

A debate took place in France on whether to use *lutte pour la vie* (or *pour l'existence*) or *concurrence vitale*, and *élection* or *sélection naturelle* (which latter prevailed in the end). In the German-speaking world, many expressions were used at the same time. Even Germany's prophet of Darwinism, Ernst Haeckel, acknowledged that Darwin's phrase "struggle for life" "had not been well chosen in many respects, and its meaning could perhaps be more precisely conveyed by "interaction (*Wechselwirkung*) between organisms, and their necessary competition for the more or less indispensable necessities of existence (*Mitbewerbung der Organismen um die mehr oder weniger unentbehrlichen Lebensbedürfnisse*)." Darwin himself, Haeckel said, occasionally used the term in a metaphorical sense that "pollutes its purity and leads to misunderstandings" (Haeckel 1866, pp. 238–239).[12] In illustrating its applications he mixed proper and improper examples. Improper examples were those which referred to struggle against the environment, in which, as Darwin himself recognized, we should instead speak of "dependence of the organism on specific conditions." In such cases, observed Haeckel, in ordinary language we would say that living beings struggle with hunger, thirst, and so forth, but even if we admit that this is struggle, it acts in these cases by adapting organisms to the state of necessity, not by selecting them. We can therefore speak of a "real struggle for existence only if many organisms struggle among themselves for food, water, etc. So the real struggle for existence can only be a competition (*Wettkampf*) between the many organisms which compete (*ringen*) in order to satisfy their own vital needs."[13] Haeckel's distinctions confirmed Ludwig

[11] I am taking liberal advantage of a striking image in Fleck (1980, p. 124). As Rossi (1986, p. 439) commented, "one could broaden the metaphor by adding that languages are often different and raise no easy problems in translation; that the group is enriched by the presence of more or less legitimate translators; that some dispute over contents while others, in disagreement with each other, announce pretended rules of conversion which, if respected, would make all live harmoniously and happily."

[12] Cf. Haeckel (1876), pp. 160–161: "The 'Struggle for Existence' has rapidly become a watchword of the day. Yet this designation is, perhaps, in many respects not very happily chosen, and the phenomena might probably have been more accurately described as 'Competition for the Means of Subsistence.' For under the name of 'Struggle for Life,' many relations are comprehended which properly and strictly speaking do not belong to it."

[13] Darwin was positively impressed by Haeckel's remarks. In his copy of Haeckel's book he annotated: "good criticism of my theory of struggle for existence—says ought to be confined to struggle between organisms for same end—all other cases are dependence—misseltoe [sic] depends on apple" (Di Gregorio ed. 1990, p. 356b-c).

Büchner's distinction between "active struggle" and "passive struggle" for survival: "The struggle can be both active and passive, as it is fought at times against competing organisms, at times against the ravages of nature." Both competition (*Concurrenz oder Mitbewerbung*) between single individuals and the restrictedness of external conditions of existence and the struggle (*Kampf oder Ringen*) for existence that derives from it intervene to prevent and contain the extraordinary fertility and multiplication of organisms (Büchner 1879, p. 43; see also 52–53). In his unrelenting drive for precision, Haeckel (1866, p. 231) managed to use four different renderings of "struggle for life" in the space of a few lines: "*Der Kampf ums Dasein oder das Ringen um die Existenz oder die Mitbewerbung um das Leben, am passendsten vielleicht als Wettkampf um die Lebensbedürfnisse zu bezeichnen*," which I can awkwardly try to translate as "The struggle for life, or striving for existence, or competition for life, which should perhaps more adequately characterized as battle over vital needs"...

The first translator of the *Origin*, the respected geologist Heinrich Georg Bronn, initially came up with a rather bizarre *Wahl der Lebensweise* (literally: choice of way of life). A more accurate translation, approved by Darwin, appeared in 1867. Its author, the zoologist Julius Victor Carus, used *natürliche Zuchtwahl* (Kelly 1981, pp. 30–31). However, the literature in German teemed with alternative forms: *natürliche Auserwahlung, natürliche Auslese, natürliche Auswahl, Auslese der Natur*. The zoologist Georg Seidlitz carefully distinguished between *Naturzüchtung* and *Naturauslese*, but at a certain point either felt unsatisfied or loosened control and wrote *natürliche Zuchtwahl* (Seidlitz 1871, p. 210, note 28). Here as ever, Haeckel beat all of them, able as he was to use three synonyms in a very short space: "*Diese Auslese der Besten, d.h. die Auswahl der am meisten Begünstigten zur Naturzucht*" (Haeckel 1866, vol. 2, p. 232; approximate translation: "This selection of the best, that is, sorting the most favoured by nature's choice"). Similarly, to him the fittest were variously *die Begünstigsten, die Tüchtigsten, die Fähigsten, die Passendsten* (or *Passenderen*), when not simply and curtly *die Besten,* "the best."

Deconstructing and reconstructing natural selection was not limited to the German-speaking world. Two Scots, the biologist (and pioneering town planner) Patrick Geddes and the naturalist John Arthur Thomson (1911, p. 169), distinguished two basic forms of it: "lethal selection," which physically eliminates the less fit, and "reproductive selection", which acts through the greater reproductive success of the fittest. The statistician Karl Pearson proposed a more articulated distinction: there was "autogeneric selection," consequent on struggle between individuals of the same kind; "heterogenic selection," consequent on struggle between individuals of different but more or less allied kinds; and "inorganic selection," due to the struggle between organisms and their physical environment. Natural selection was the result of the three forms of struggle taken together (Pearson

1892, repr. 1900, p. 380; on p. 358 Pearson used "physical selection" instead of "inorganic selection").[14] Yet Pearson (1896a, b) also pointed out a "reproductive selection," or "selection by relative reproductivity," which he said was what Galton had referred to as "genetic selection" (but also as "proliferal selection"). It acted on the variable fecundity from one individual to another and produced a continuous and gradual change of type. This change could only be checked by a stronger selection, that is, natural selection, which kept the new generation within the limits of the preceding one by maintaining a selective rate of mortality (ibid.: 439; cf. 1897, 1: 63–102; see Gayon 1998, p. 220). From another perspective, however (and to simplify things for historians), Pearson split natural selection into "periodic selection," that is selective mortality during growth (which "repeats itself in each generation and maintains the species in a practically permanent and stable condition"), and "secular selection" (which, "in the long course of time, has brought a species from a less developed to its present form") (Pearson 1897: 159 note).

The Dutch botanist Hugo de Vries, one of the rediscoverers of Mendel and the author of the mutation theory, allowed no creative role for Darwinian natural selection of "individual differences," but only the destructive one of eliminating non-adaptive forms. It is only "a sieve," he said, as "it creates nothing." Species change was due to sudden, well-defined "mutations." In the light of his theory, de Vries reformulated the usual distinction between intraspecific and interspecific struggle. The term "struggle for existence, that is to say the competition for the means of subsistence," might refer to "two entirely different things": the struggle between individuals "of one and the same elementary species" and that between "the various species themselves."

> "The former is a struggle between fluctuations, the latter between mutations. In the former case those that survive are the individuals which find conditions favourable to them—that is to say, as a rule, the strongest individuals. It is by this process that local races arise [. . .] If the new conditions of life are relaxed, the adapted race reverts to the form from which it sprang. The natural selection of newly arisen elementary species in the struggle for existence is an entirely different matter. They arise suddenly and without any obvious cause; they increase and multiply because the new characters are inherited. According as the young or the parent form is better fitted to the environment will the one or the other of them survive. Species no more *arise* as the result of this struggle for existence, than they do as the result of the struggle between the variants of one and the same type—though for different reasons in the two cases. In order that species may engage in competition with one another it is evidently an essential condition that they should already be in existence; the struggle only decides which of them shall survive and which shall disappear." (de Vries 1909–1910, pp. 211–212; cf. 609–610)[15]

[14]These distinctions were also important from an ideological perspective. The socialist and imperialist Pearson used them to attack the "individualists," i.e., defenders of laissez-faire, who saw humans as only subject to autogeneric selection.

[15]De Vries praised his countryman, the theologian and politician (and future Prime Minister of the Netherlands) Abraham Kuyper, for distinguishing "selection," which "aims at the maintenance of species," from "election," which "is the choice (*Auswahl*) of persons" (de Vries 1901–1903, p. 667; Engl. tr. 1909–1910, 1: 609–610; Kuyper 1899, p. 11).

One need only leaf through the meritorious review by Ludwig Plate (1908) to realize how many other authors had their own distinctions and alternatives for the term and/or the concept of natural selection. We find the physiologist and comparative psychologist George John Romanes' "physiological selection,"[16] the botanist Liberty Hyde Bailey's (Bailey 1896a, b) "survival of the unlike," the zoologist and psychologist Conwy Lloyd Morgan's (1888, 1891, 1896a, b, 1900, 1902, 1909) "natural elimination" and "functional selection,"[17] the entomologist Erich Wasmann's (1901, p. 739; 1904, p. 230) "*Amical-Selection*" (in social insects), and the zoologist Wilhelm Haacke's "*constitutionelle Auslese*" (1893, p. 70) and "*dotationelle Selection*"... In 1897, the American psychologist James Mark Baldwin counted no less than 12 different types of selection currently under discussion. They became 14 by 1902 (see Table 1),[18] Baldwin himself having in the meantime introduced "organic (or indirect) selection" (Bellomy 1984, p. 88)[19]—which generated what has come to be known as "Baldwin effect" (Simpson 1953)—and "functional selection" (Baldwin 1896a, b, 1897b, c, 1902a, b).

As was to be expected, interpretations of the struggle for life varied most when it came to human and social evolution. Independent of overtly political discourse by inhabitants of that jungle that goes under the misleading name of social Darwinism, behind the use, and abuse, of that phrase there hides a vast array of differing shades of meaning and a tangle of implications and connections.

Some authors set themselves the task of linking the various meanings of the phrase and the various manifestations of struggle in a continuous chain from the most to the least bloody, that is, from the least to the most metaphorical. Some did so by subsuming all of its manifestations under a more general concept, whether it be "endeavor" or "effort" or the omnipresent "competition" and its synonyms. All of

---

[16]Romanes' "additional suggestion on the origin of species" opened a long debate with Wallace, who defended the all-sufficiency of natural selection: see Romanes 1886a, b, c, 1887; Wallace 1886, 1890, and many other articles, all listed in Richards (1987, p. 408).

[17]Morgan (1891, p. 79) subdivided natural selection under "two heads": "*natural elimination*, of widespread occurrence throughout the animal world; and *selection proper*, involving the element of individual or special [i.e. of the species] choice." Selection properly so called was a "selection of the fit" (ibid. 92–93).

[18]This Table, like Figs. 1 and 2 in this chapter, is reproduced from La Vergata (2023).

[19]Weismann (1909, p. 41) claimed he had anticipated organic selection in 1894, but Baldwin (1909, p. 17, note 2) vindicated his priority, as well as that of Lloyd Morgan and Osborn, who had both independently formulated the same idea in 1896 (Osborn 1896a, b; Morgan 1896a, b; Osborn and Poulton 1897). He added that Morgan and E.B. Poulton preferred to call it "indirect selection." The original papers of Baldwin, Morgan, and Osborn, together with expositions by Poulton and others, were collected in Baldwin (1902c), where Baldwin discussed his differences from Morgan and Osborn. As Ceccarelli (2021a, b) remarks, Osborn lost his interest in organic selection very soon (Osborn 1897). On organic selection in Baldwin, Morgan, and others I can simply refer to the excellent treatment in Richards (1987). Raymond Hovasse (1950, pp. 12–13) proposed an "improved reformulation" of Baldwin's organic selection (the term "organic" being "imprecise") and of the Baldwin effect: better to say "selection of parallel mutations, or, more simply, parallel selection." On the Baldwin effect see (Continenza 1984, 1986; Weber and Depew 2003; Pertile 2018; Ceccarelli 2019).

Table 1 Sorts, means, and results of selection

1, 2. Natural selection	1. Struggle for existence	1. "Survival of the fittest" individuals
	2. Inherent weakness	2. Destruction of the unfit individuals
3. Germinal selection	3. Struggle of germinal elements	3. Survival of fittest germinal elements
4. Intra-selection	4. Struggle of parts	4. Survival of fittest cells or organs
5. Functional selection	5. Overproduction of movements	5. Survival of fittest functions
6. Organic selection	6. Accommodation, modification, growth processes	6. Survival of accommodating and modified individuals
7. Artificial selection	7. Choice for planting and mating together	7. Reproduction of desirable individuals
8. Personal selection	8. Choice	8. Employment and survival of socially available individuals
9. Sexual selection	9. Conscious selection through display, courting, etc.	9. Reproduction of attractive individuals
10. Social selection [group selection]	10. Social competition of individuals and groups with natural selection	10. Survival of socially fittest individuals and groups
11. Social suppression	11. Suppression of socially unfittest (by law, custom, etc.)	11. Survival of the socially fit
12. Imitative selection	12. Imitative propagation from mind to mind with social heredity	12. Survival of ideas, customs, etc.
13. Physiological selection	13. Relative infertility	13. Survival of the divergent
14. Reproductive or genetic selection	14. Enhanced fertility	14. Survival of the most fertile

From Baldwin (1902b), p. 166

these terms, note, were no less anthropomorphic, and had no lesser moral implications, than "struggle." And the whole circus started again. Here is a list of selections, in addition to artificial, natural, and sexual selection, from Plate's (1908) Index: amical, inorganic, cormal, decimating, properly called, fecundative, germinal, histonal, catastrophic, constitutional, organic, personal, physiological, non-qualitative, reproductive, inverse, unconscious (*amicale, anorganische, cellulare, coincidierende, cormale, dezimierende, eigentliche, fekundative, germinale, histonale, katastrophale, konstitutionnelle, organische, personale, physiologische, qualitätslose, reproduktive, umgekehrte, unbewußte*). Note that also "selection properly called" (*eigentliche*) figures among them. On "decimating selection" Plate remarks that it is "often used to refer to the elimination of diseased individuals" and comments: "This expression is unclear, and it would better be avoided, as any selection 'decimates', i.e. diminishes the quantity of individuals" (Plate 1908, p. 419).

When seeing this list, one understands, and to some extent sympathizes with, those scientists-historians who bypassed such complications and invented a via *regia* from Darwin to the Synthesis. For them it was a matter of sheer survival.[20] It is then no wonder all these forms of selections are not mentioned in important books on the history of Darwinism. The fact remains, however, that the concept of natural selection has itself a history, and such a history would correspond to the different uses and meanings that the concept acquired not only in disparate theoretical systems (Schwartz 2021) but in different cultural contexts.

Things get even more confused when it comes to social and political applications. As professional biologists were themselves lost in a mess, we should not wonder that what goes under the name of "social Darwinism" was an intricate phenomenon. This is not the place to dwell on it. However, it is useful to offer at least one instance of the extent to which the *rage des définitions* affected those who wanted to address both biological and social processes. The case of the already mentioned Baldwin is particularly interesting because he himself commented extensively on his own definitions and distinctions, in a desperate effort to generalize and distinguish, split and subsume at the same time. I apologize for the length of the next section, which might seem, but is not, a digression.

6 Subsuming and Splitting

In 1902, Baldwin wrote that the struggle for existence "may be defined as the attempt to remain alive, or technically to 'survive,' on the part of an organism."[21] He distinguished three "quite different forms" of it, even in the animal world.

> "(1) The competition for food, etc., that arises among organic beings from the overproduction of individuals or from a limited supply of food. This is called the 'Malthusian form' of struggle for existence.

[20]Needless to say, the creation of new terms did not come to an end after the birth of genetics. Just think of Schmalhausen's (1949) "stabilizing selection" (which, according to Waddington 1940, 1957, 1960, can be "normalizing" or "canalizing"), "balanced selection" and "diversifying selection" (Dobzhansky 1955, 1968), "organismic selection" (which Sewall Wright used to distinguish his view of selection from Fisher's "genic" view: see Wright 1980), "group selection," "kin selection"... As Sewall Wright (1956) said, selection operates at different levels of complexity: genes, cells, organisms, demes, populations, and species. The debate on the units of selection is still going on, among biologists as well as philosophers of biology. But this is another story, which will hopefully be narrated by genuine historians.

[21]Baldwin quotes Darwin's famous words on the "large and metaphorical sense" of the term "struggle," but, incredible though it may seem, he takes the quotation from the philosopher Bosanquet, not from Darwin himself! Did he not care to check on the original? Did he not have a copy of the *Origin* at hand? Whatever the case, it is strange, to say the least, to find an author discussing one of the key concepts of another author without consulting him. Baldwin also mentions Pearson's (*Grammar of Science*, 2nd ed., p. 364) distinction between "struggle of individual man against individual man, struggle of individual society against individual society, struggle of the totality of humanity with its organic and inorganic environment."

(2) Competition in any form of active contest in which individuals are pitted against one another.

(3) Survival due to greater fitness for life in a given environment, whether combined with direct competition with other organisms or not.

The second case (2) is that in which animals either (*a*) fight with, or (*b*) prey upon, one another; only the former of these (*a*) having any *analogy* to the form of competition due to the overproduction of individuals or to a limited supply of food, etc., and then only in the case in which the strife results from the circumstances of getting a living — not in the case very common in nature of mere combativeness of temper, through which the stronger animal kills the weaker *simply from aggressiveness*. In case (*b*) one animal feeds upon members of another group as his natural prey, as is seen in the eating of insects by birds. This is also extremely widespread, and leads to some of the most beautiful special adaptations — for concealment, warning, etc.—in the species preyed upon. This has *nothing* to do with the overproduction of individuals in the sense given under (1), except in so far as the species preyed upon overproduces in the way of compensation for the constant drain upon it; but this is a very different thing." (Baldwin 1902b, pp. 213–214; cf. entries "Existence (Struggle for)" and "Rivalry" in Baldwin ed. 1902–1905; my italics, here as well as in further quotations from Baldwin, unless otherwise stated).

Baldwin then specifies that:

"A case of (2*a*), of extreme importance in its effects upon the next generation, is that of the struggle of males for the female, occurring often apparently irrespective of the number of available females.

The third case of struggle (3) is that in which *individuals struggle against fate—the inorganic environment—not directly against one another*. This is really a '*struggle to accommodate'*—to reach a state of adjustment or balance under which continued living is possible. As the other forms may perhaps be styled respectively '*struggle to eat'* (*in a large sense*) and '*struggle to win*,' so this may be called 'struggle to accommodate,' or 'struggle to live.' The distinction between cases (2) and (3) disappears in instances in which the animal accommodates actively in order to coping with his enemies; for these then become part of his environment in the sense of case (3).

The relation of large productiveness to this last form (3) of the struggle for existence would seem to be but indirect. It would not matter how many individuals perished provided some lived; and any amount of overproduction would not help matters if none of the individuals could cope with the environment. Yet on the theory of indeterminate or indefinite variation, the chances—under the law of probability—of the occurrence of any required variation is a definite quantity, and these chances are of course increased with large productiveness; for with more variations, more chances of those that are fit, and with increased production, more variations." (ibid.: 215)

Not entirely happy with informing the reader that "struggle to live" is different from "struggle for life," Baldwin goes on to say:

"In recent evolution theory [*note the singular*] the doctrine of natural selection has come to rest more and more upon the second and third sorts of struggle (2 and 3), and less on the Malthusian conception (1). Experimental studies which support the selection view [...] show the eliminative effect of the environment, and the preying of some animals upon others, rather than direct competition *inter se*, among individuals of the same species, for food or other necessities of life. It is these forms of the struggle, too, that we find nature especially providing to meet [...] The result common to all the sorts of struggle for existence, however, is the survival of an adequate number of the fittest individuals; and

> this justifies the use of the term in the theory of evolution to cover so *wide a variety of* instances." (ibid.: 216)

Darwin's and Wallace's original conception "has been broadened with the development of the theory to include the other less Malthusian forms." To sum it up:

> "If we consider the three forms of struggle pointed out above, as together making up the conception, we may for convenience designate it as 'biological' struggle, inasmuch as individuals are directly brought into conflict with one another for life and death, and as moreover the end is not attained through the struggle alone, but requires the further biological function of reproduction to make it effective." (ibid.: 217)

Baldwin was not very tolerant with respect to other biologists's analogies and extensions that were "in greater or less contrast" with his distinctions. He objected to theories "in which there is a shading one way or the other away from this form of competition with its indirect results." There are, he wrote, "certain hypotheses of a biological sort which utilize the conception of struggle without distinguishing it clearly as a process preliminary [sic] to that of survival. In Roux' [1881] conception of 'struggle of the parts' physiological processes accentuate or develop certain cells and organs at the expense of others." The idea of "a preferential supply of nourishment, blood — whatever aids the anabolic processes in these particular directions" is analogous to the struggle for food. But the mechanism of it is entirely unknown. In Weismann's "germinal selection," a similar reason for survival is postulated, and "differences of some sort between germ-cells" result in some of them being "more favourably situated or otherwise conditioned for survival, and this may be called a 'struggle.'" "But here again there is no precise notion of what takes place. In both of these cases the struggle is merely a hypothetical means to the end, which is selection and survival; it is not a clearly described phase in the process." Weismann's "intraselection" also involves struggle, "in an obscure way"; and "the selection of 'motor functions' by what has been above called 'functional selection' involves the survival of movements from among a series of overproduced or excessive discharges; but it again seems to strain the notion of 'struggle for existence' to speak of these movement variations as engaged in a struggle with one another or with the environment." "In all these cases — Baldwin concludes — Spencer's term 'survival of the fittest' is more applicable" (ibid.: 217–218). Therefore, and contrary to what many, starting from Darwin and Wallace themselves, have taken for granted, Spencer's term could *not* replace Darwin's phrase.

Of course, Baldwin admitted, things are more complicated when intelligence comes into play. "It secures the widest possible range of personal adjustments, and by so doing widens the sphere of organic selection, so that *the creature which thinks has a general screen from the action* of *natural selection.* The struggle for existence, depending upon the physical qualities on which the animals rely, is *in some degree done away* with" (ibid.: 145). This is why Baldwin introduced the concepts of "conscious competition and rivalry."

> "If the facts of conscious 'competition' and 'rivalry' are compared with those of biological struggle, the result is that *only under certain conditions* do they even show an analogy with struggle for existence in the sense principally employed by Darwin and Wallace—the struggle for food. So also, when we come to subject the conception of 'selection' to a thorough analysis, we have *distinctions to make which forbid our using the biological conception in the mental and social spheres except under the very restricted limitation*, namely, that results of the selection in question normally fall under the laws of physical reproduction and heredity for their conservation. Yet again, in the matter of conservation of type, with regression, where there is the question of the application of such a principle to mental transmission, we find that the mental products do not, in respect to their effectiveness for the future movement of social evolution or development, follow such a law—that they follow, moreover, a very different and in no wise analogous law." (ibid.: 43–44)

However, in spite of so many similarities-cum-distinctions, and all his "buts" and "yets," Baldwin was unwilling to give up all analogies between nature and society, even at the cost of dialectical contortions. He did not object to "the extensions of the meaning of the concept of survival with struggle in the direction of conscious and social functions" (ibid.: 218). More than that, there were "certain processes which we may *distinguish under the general heading* of 'Rivalry' — a broad term which may be used to designate the entire field, including biological struggle for existence." Then Baldwin embarked again in the splitting-and-subsuming process he had just performed with respect to "struggle." As with biological struggle,

> "[t]here are three great cases of Rivalry which it is essential to distinguish, especially in view of current confusions arising from lack of discrimination: (1) Biological Rivalry, or struggle for existence, of which the forms have been pointed out above; (2) Personal or Conscious Rivalry, or emulation, to which the term rivalry is more generally restricted; and (3) commercial and industrial rivalry, known as Economic Competition." (ibidem)

"Rivalry" enables Baldwin to embrace nature and society under one term, to draw moral lessons from nature to society, and to project his view of society back onto nature, in a circle. Thus, the whole circus starts again, in spite of so many distinctions, indeed thanks to them. Here lies the rationale underlying the table reproduced in § 5, where he lists all "sorts, means, and results of selection."[22]

Baldwin's own comments on his table are interesting. I therefore ask for a little more patience from the reader.[23] Concerning terms 4, 5, and 6, he says that "by a

---

[22] As Baldwin (1902b, p. 165) says in a note, "the terms in the table which relate to social evolution are fully explained in the work, *Social and Ethical Interpretations* [Baldwin 1897a], Index and Appendix B, where acknowledgment is made of suggestions from Professor Lloyd Morgan."

[23] Mayr (1982, pp. 58, 71) only mentions Baldwin twice. He ascribes him two merits: 1. "As far back as 1909 Baldwin specified two reasons why biological events are so often unpredictable: the great complexity of biological systems and the frequency at which unexpected novelties emerge at higher hierarchical levels." 2. "When Baldwin in 1909 pointed out how much the acceptance of Darwinism had changed the thinking of biologists, he concluded that 'the reign of physical science and of mechanical law over the scientific and philosophical mind is over now.'" Mayr does not seem interested in the reasons why Baldwin took this position, but only in what Baldwin contributed to the advancement of Darwinism by objecting to the notion that all science had to be like the

singular coincidence" Delage's phrase "sélection organique" [Delage 1903, p. 732] describes Roux' "struggle of the parts," which is akin to "functional selection." However, as Weismann has directly applied his "intra-selection" (4) to his interpretation of Roux's "struggle," Delage's phrase "is not likely to have currency as a substitute for intra-selection." And, "as 'Functional Selection' (5) is a special means of motor accommodation, it is additional (and in a sense subordinate) to Intraselection, since it has a functional reference" (Baldwin's italics). Concerning 7, 8, 9 Baldwin writes:

> "A separate heading might be given to Professor Lloyd Morgan's phrase 'Conscious Selection', but it will be seen that, as he uses it, i.e., in broad antithesis to 'Natural Selection,' it really includes all those special forms of selection in which a state of consciousness plays the selecting role (7, 8, 9, 11, 12).[24] It would be ambiguous if used for cases where natural selection operates on mental and social variations (5, 6, 10), since it might then mean he survival of the conscious; and even when applicable, as in sexual selection (9)[25] with respect to the 'means' of the selection, it may be ambiguous with respect to the 'result' of the selection. This last ambiguity, which is brought out in the table (8, 9),[26] makes it desirable to confine the phrase 'Conscious Selection' (if used at all) to cases which result in continuance of what is desirable for consciousness or thought. 'Personal Selection' is suggested (8) for selection by human personal choice, analogous to Sexual Selection (9) and to Romanes' 'Physiological Selection' (13). Furthermore, Darwin's 'Artificial Selection' should be used, as he used it, with reference only to securing results by induced mating (his 'Methodical' as opposed to his 'Unconscious' Selection)."

Concerning 10, 11, 12, Baldwin says:

> "In all the different sorts of 'selection,' *considered as factors in progress from generation to generation in which the laws* of *natural selection and physical reproduction do not operate together*, it seems extremely desirable that we qualify the word 'selection' carefully, giving each case a name which shall apply to it alone. The cases of the preservation of individuals and groups by reason of their social endowments do illustrate natural selection with physical reproduction, and 'Social Selection' (10) is proposed for that. In the instances in which either physical heredity is not operative (12), or in which it is not the only means of transmission (11), we cannot secure clearness without new terms; for these two cases 'Social Suppression' (11) and 'Social Generalization' (12) are suggested. The phrase 'Imitative Selection' is given

---

physical sciences or to be based on mathematics. After distilling this 'Mayrian' gist from Baldwin's convolutions, Mayr concludes, rather inconsistently: "In his optimism [Baldwin] was mistaken; there are still many *philosophers* who write as if Darwin had never existed and as if evolutionary biology were not part of science" (ibid.: 71; my italics).

[24] "This, indeed, is still liable to the question as to *whose is the state of consciousness*, giving the difference (both in means and result) seen between 'Artificial' (7) and 'Sexual' (9) selection. [Lester Frank] Ward's suggestion of the phrase 'subjective selection' (i.e. by consciousness) in antithesis to natural selection (Encyclopedia Britannica, 9th ed., Art 'Psychology') was earlier" (Baldwin's note and italics).

[25] Baldwin is referring to Morgan (1896b, pp. 219, 271).

[26] "The bird 'selects' (sexually) for the sake of the experience, and it is a secondary result that she is also thus 'selected' for mating with the male and so for continuing his attractive characters with her own characters in the offspring" (Baldwin's note).

> in the table alternately for the latter (12), seeing that the discussions of the topic usually employ the term 'Selection' and use (wrongly) the 'Natural Selection' analogy. Selection may be used also when there is no reference to race-progress (and so no danger of the misuse of the biological analogy), since it then means presumably the 'conscious choice' of psychology and of pre-Darwinian theory." (ibid.: 167–168)

This was Baldwin in 1902. In *Darwin and the Humanities* (1909), he seemed to have left behind his reservations about embracing nature and society under *one* law. He wrote that "natural selection is in principle the universal law of genetic organization and progress — in human nature no less than physical nature," and that "the theory of natural selection is to be accepted not merely as a law of biology as such, but as a principle of the natural world, which finds appropriate application in all the sciences of life and mind" (Baldwin 1909, p. ix, 89; cf. La Vergata 2023, pp. 563–567).

Before going on, a few short considerations on Baldwin's table are in point.

1. It documents at the same time a disorienting proliferation of terms and concepts, and an attempt to bring order by embracing them, as far as possible, under *one* label.
2. It shows that, despite "confusion," a lively and even creative debate was taking place.
3. It is a striking instance of the pervasive attempt to make biological reality and social normativity coincide.
4. It shows an astounding overlapping of evolutionary levels. The question inevitably arises: How could so many people apply struggle and selection to biological *and* social phenomena without realizing that they act on different, indeed incommensurable, time scales and tempos, and that their effects in society are insignificant when compared with the extremely long times of biological evolution, especially if the latter are conceived, in Darwin's view, as extremely gradual and slow? In other words, how could people base sermons concerning a fast-moving present on a very slow past? I see only one possible answer: those people felt an irresistible urge to draw *moral*, and ideological, teachings. Such urge was widespread among both supporters and opposers of Darwinism(s).

As we saw, Baldwin was not alone in his exercise in qualifications. Just one more example, before we move on. The Hawaiian-born Reverend John Thomas Gulick, a missionary and a specialist of land snails who has a place in the history of Darwinism for his insistence on the importance of geographical isolation (Lesch 1975; Finnegan 2021), went even further. He did not intend as much to bring together other people's terms as to offer terms of his own coinage for mechanisms that others had not seen or had neglected. He further enriched the catalogue of selections (see Figs. 1 and 2) by pointing out a variety of forms of what Baldwin and Morgan (1902, in Baldwin

Fig. 1 "Classified Table of Forms of Segregation." From John Thomas Gulick, Divergent Evolution through Cumulative Segregation (Communicated by Alfred Russel Wallace, read 15th December, 1887), *Journal of the Linnean Society of London. Zoology,* 20, 1890 (120): 273–274. Courtesy of the Linnean Society of London

Classified Table of Forms of Segregation

A

Environal Segregation
- (a) Industrial Segregation
  - Sustentational
  - Defensive
  - Nidificational
- (b) Chronal Segregation
  - Cyclical
  - Seasonal
- (c) Spatial Segregation
  - Geographical (Migrational, Transportational, Geological)
  - Local
- (d) Fertilizational Segregation
- (e) Artificial Segregation

B

Reflexive Segregation
- (a) Conjugational Segregation
  - Social
  - Sexual
  - Germinal
  - Floral
- (b) Impregnational Segregation
  - Segregate Size
  - Segregate Structure
  - Prepotential Segregation
  - Segregate Fecundity
  - Segregate Vigour
- (c) Institutional Segregation

C

Intensive Segregation
- (a) Assimilational Intension
- (b) Stimulational Intension
- (c) Suetudinal Intension
- (d) Correlated Intension
- (e) Integrational Intension
- (f) Selectional Intension
- (g) Fecundal Intension
- (h) Eliminational Intension

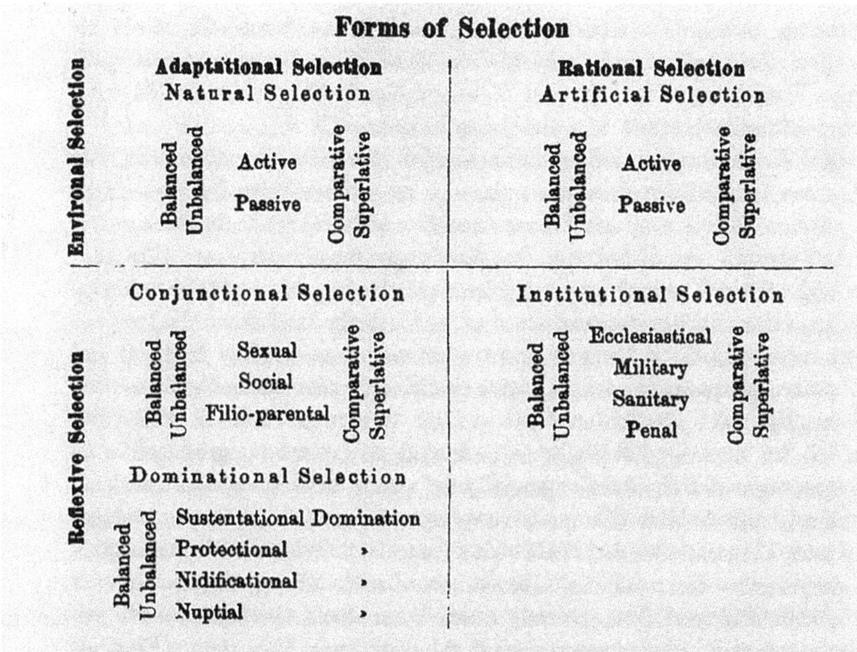

Fig. 2 Forms of selection. From John Thomas Gulick, Intensive Segregation, or Divergence through Independent Transformation (Communicated by W. Percy Sladen, read 19th December, 1889), *Journal of the Linnean Society of London. Zoology*, 23 (145) 1891: 329. Courtesy of the Linnean Society of London

1901–1905, 2: 506) were apparently happy to refer to simply as "social selection."[27] He introduced "balanced" and "unbalanced" selection, "comparative" and "superlative" selection, "active" and "passive" selection," then "environal [*sic*] selection," "denominational selection," and, along with Darwin's "artificial selection" (which was obviously a form of "rational selection"), an "institutional selection," which could be "ecclesiastical," "military," "sanitary," "penal" (Gulick 1888, 1890, 1905).[28]

After dwelling so long on Baldwin, it would be too much abuse of the reader's patience to comment on Gulick's classifications. Let me just say that Mayr (1982) mentions Gulick (1872) as the first supporter of "random (stochastic) differentiation," that is, the view according to which "neither the environment (directly through natural selection) nor internal factors" influence the direction of variation and

[27] However, Baldwin later could not resist pointing out an "intra-group" and an "extra-group" selection, and mentioning a "racial selection" (Baldwin 1909, pp. 44, 71).

[28] Starting with Haeckel (of course) there had, and were to, come up "economic selection," "matrimonial selection," "medical selection," "military selection," "juridical selection," "psychic selection," "religious selection," and so forth (see, for example, Vacher de Lapouge 1896; Ferri 1900; Kirchhoff 1910).

evolution," and that "much evolutionary change is simply a result of chance variation" (Mayr 1982, pp. 361, 555). He praises Gulick for making a clear distinction between the two "very different components" or "modes" of evolution: "transformation," that is, "the 'vertical' (usually adaptive) component of change in time)" and "diversification," that is, "processes that occur simultaneously, like the multiplication of species," which "can also be called the 'horizontal' component of change manifested by different populations and incipient species." Gulick (1888) used "monotypic evolution for transformation and *polytypic evolution* for diversification." Mayr adds that Romanes (1887), who adopted Gulick's terminology, also "referred to transformation as 'transformation in time' and diversification as 'transformation in space.'" Unfortunately, he comments, this was "an insight that was largely forgotten after 1897, until Mayr (1942) and others revised it during the evolutionary synthesis" (ibid.: 400–401; Mayr's italics).[29]

It is time to conclude this long section.

Darwin's multifaceted metaphors branched out in a variety of struggles and selections, which were often contrasted to analogous classifications by other authors. Confusion was fed by attempts to bring order by substituting one term with another or by articulating concepts in sub-concepts and sub-sub-concepts—which had a snowballing effect. Yet this very confusion demonstrates that the complexity of the Darwinian concepts of struggle and selection had set something important in motion. It was a time of creative turmoil in the history of biology. The churning out of neologisms — generally ephemeral — is evidence of the ferment of ideas. Darwin had undoubtedly opened the door, which led to unpredictable developments, including what a posteriori could be described as dead ends.

7 Where Are We Now?

Have things changed since then? Thanks to mathematics, there certainly is less semantic confusion, but standardization of terms and procedures is not necessarily accompanied by universal agreement on theoretical issues. Daily research activity is not based on a continuous reflection on "philosophical" problems. I can hardly imagine a biologist going to a laboratory in the morning and asking "How is

[29]Mayr (ibid.: 564–565) criticized his past colleagues in his usual way: "One of the major difficulties was that most of those who joined the controversy in the ensuing years — Romanes, Gulick, and even Wallace (Lesch 1975) — made no clear separation of geographical and reproductive isolation, nor of individual and geographical variation, and often dealt with speciation as if it was the same as natural selection. The confusion is particularly painful in the writings of Romanes, who invented the misleading term 'physiological selection' for reproductive isolation. There still is no completely critical analysis of this literature, but one can, on the whole, recognize two camps, those who followed Darwin in not clearly distinguishing the two kinds of isolation (among them, Weismann, Semper, Romanes, Gulick, and Wallace) and those who, following Wagner, considered geographical isolation as a factor *sui generis* and indispensable for speciation (for example, Seebohm, K. Jordan, D. S. Jordan, Grinnell, a good many entomologists such as H. W. Bates and perhaps Meldola and Poulton, as well as Kerner and Wettstein among the botanists)."

Darwinism today?" Paradigms work tacitly, through normal routine. Still less am I able to imagine that biologist engaged in an effort to make sense of obsolete terms and concepts he came across when reading the works of past biologists.

Present-day research in evolutionary biology does not consist in a continuous re-examination of what Darwin wrote. Those who study neuronal Darwinism, invertebrate paleontology, or the behavior of mosquitos do not need to refer to his works. They are under no academic, social, or intellectual pressure to do that. Whereas no student of philosophy or literature could think of getting a degree without reading Kant or Shakespeare, a student of science is not requested to have any acquaintance with the "classics" of his/her field. They approach a form of knowledge which is much more structured. Students in biology departments are not trained to read *On the Origin of Species*, but to assimilate the form of evolutionary biology that their teachers consider as "true," or prevailing, or deserving attention. They are taught where we are now, not how things came to their present stand. Science also progresses, indeed mainly progresses, through an obscure work which does not necessarily translate into brilliant books or revolutionary papers, and which only too often is an end to itself, or proves to be a blind alley. Only those who conceive the history of science as an acrobatic jumping from one revolutionary peak to another can forget that it is normality that "normal science" normally consists of.

One needs only to look at the 2014 debate in *Nature* (Laland et al. 2014) on "Does evolutionary theory need a rethink?" to get an idea of the contraposition between those who answer "Yes, urgently" and those who answer "No, all is well." Was it a match between orthodox Darwinian loyalists and reformers? As Minelli (2018, p. 126; see also Minelli 2017) remarked, "it would be reductive, to say the least, to define reformist rather than revolutionary the attitude of [such evolutionary biologists as] Laland, Uller, Feldman, Sterelny, Müller, Moczek, Jablonka, and Odling-Smee." They say that the confrontation "is no storm in an academic tearoom, it is a struggle for the very soul of the discipline." They argue that "rather than selection being free to traverse across any physical possibility, it is guided along specific routes opened up by the process of development [...] Another kind of developmental bias occurs when individuals respond to their environment by changing their form – a phenomenon called plasticity [...] In other words, often it is the trait that comes first; genes that cement it follow, sometimes several generations later."[30] Likewise, Minelli went on to say, it would be reductive to describe as conservative, rather than reformist, the position of Wray, Hoekstra, Futuyma, Lenski, Mackay, Schluter, and Strassmann, according to whom all the phenomena that Laland and colleagues promote "are 'add-ons' to the basic process that produce evolutionary process" and "none of these additions are necessary for evolution." Therefore, an "extended evolutionary synthesis" (Pigliucci and Müller 2010), as opposed to the "standard evolutionary synthesis" is not necessary. Minelli is right in

[30] One could wonder if what is meant here is a mere revision, or an updated version, or a revolutionary reformulation of what was formerly, though somehow improperly, called "Baldwin effect."

commenting that it is a matter of words whether these perspectives just imply routine updating of neo-Darwinism or revolutionary changes, or to what extent integrations are alterations. As a matter of fact, variation is subject to constraints, selection operates in a context of structural limitations, and it is necessary to take new empirical domains — such as phenotypic and developmental plasticity, niche construction, epigenetic variation and inheritance, horizontal gene transfer, exaptive phenomena, etc. — as relevant. Focusing on natural selection alone is not enough to explain the routes in which evolutionary change can be channeled. An organism must be built before examining whether and how it works. It is, therefore, necessary to open the black box which lies between genes and the phenotype. Evolutionary developmental biology is a "trading zone," as Rasmus Winther called it, where negotiating on new facts and interpretations may lead to unpredictable results. Darwin himself pointed out many facts that would falsify his theory: "If this were true, adios theory," he wrote in one of his Notebooks.

Who decides where the border is between "updating," "reforming," "transforming," "revolutionizing," and "disrupting"? Here again, defending one's own image of Darwin and "Darwinism" amounts to defending one's own stance in ongoing debates, and, after all, one's own view of what evolutionary biology is and should do. I allow myself to make an easy prediction: future historians will show that no two members of the the above teams exactly agreed on the reasons for ranging in the same team.

A genuine historical history of what is happening is necessary. The problem is: Who is going to attempt it? Who is equally endowed with historical sense and information, philosophical open-mindedness, and updated scientific information? Who can really go beyond a mere registration of what happens in "laboratory life" (Latour and Woolgar 1979)? The ideal is that scientists are the historians of themselves and of their context. But…

8 "Two Cultures"

Maurizio Esposito (2021) rightly distinguishes between "scientist" and "humanist" historiographies of evolutionary biology. Scientists and historians see history differently. Scientists "often conceive history of science as a strategic tool for clarifying the conceptual issues haunting contemporary science." Scientist historiography has frequently used history for underpinning particular scientific options (i.e., in Mayr's or Gould's historiographies, different though they are) and is based on a linear narrative emphasizing great figures, core concepts, and foundations. On the contrary, in the humanist historiography Darwinism seems to have no essence: the cognitive system we call Darwinism would be whatever "Darwinians" think it is, or was, within a given timeframe. According to Esposito, we should not expect an obvious and consistent overlapping between the two historiographies. Richard Delisle agrees with Esposito: the two agendas are "conflicting" but also "complementary." Indeed, they should cooperate, for the sake of both history and science, although Delisle recognizes that "things only get worse when historians and philosophers become co-opted" by the two camps.

> "As strange as it seems, the future of evolutionary biology may in part belong to a re-discovery of its past [...] The inspiration sought by contemporary biologists in pursuit of greater explanatory pluralism can be accelerated, I would argue, by the rediscovery of older ideas capable of reformulation. Just as no one can legitimately claim that solutions to all current and future problems are to be found in past proposals, neither can it be denied that re-engaging with them may lead to some fruitful inspirations." (Delisle 2021a, b, p. 5)

Delisle hopes that "a new alliance will be forged that will generate an overall narrative that is, at least, consistent throughout" (ibid.: 6). I am not optimistic about that. I will explain why by pushing Esposito's and Delisle's distinction between the two approaches to its extreme consequences, which they probably do not accept. I believe that there is an unbridgeable gap between the "two cultures," as the physicist and crystallographer Charles P. Snow described them in 1959. They are rooted in two essentially different attitudes (or *formae mentis*), aims, ways of operating, and effects on society. However, Snow underestimated the aspect that I think makes the difference between scientific and technological culture and what he called, rather generically, "humanistic" or "traditional" culture (even a scientist can be a humanist in the traditional sense of the term: Snow himself wrote novels). The difference is made by the sense of history and the use which is made of history (if any).

A biologist or a physicist can be awarded the Nobel Prize even if she/he has never read articles or books published earlier than, say, 15 years. All scientists obviously know that science has a past and feel veneration for the founding fathers of their field of research. Some also read them, but a direct knowledge of the pages of Galileo, Newton, or Darwin is not required for a scientist to be accepted or esteemed as such. Those who have read those pages are clearly more cultivated persons than those who have not. But reading Galileo, Newton, or Darwin or not is in a sense like listening or not listening to music, or going or not going to painting exhibitions. Many scientists almost fall from the clouds when it is explained to them that in Aristotelian physics the term "motion" encompassed both the motions of bodies in space and the processes of generation and corruption, that Copernicus believed in the existence of solid spheres on which the planets were embedded, that for Kepler the universe was finite, that for Galilei comets were merely "vapors" of the upper atmosphere, and that for Newton comets served to "recharge" the world machine. When one tells the members of a physics department about the trunk that was opened 200 years after Newton's death, and the content of which led John Maynard Keynes in the early 1940s to refer to Newton as "the last of the magicians" (Keynes 2010), the great majority of them manifest a surprise verging on disbelief. The history of science is full of statements on objects which have long disappeared from textbooks and what the practitioners of science generally consider the "real world." It is full of theories which were later abandoned: that a body needs a force to be kept in motion, that the Moon has nothing to do with tides, that fossils are "freaks of nature," that continents and oceans are stable entities, that there is a real correlation between the shape of the skull and mental powers, that cosmical rays are identical with gamma rays, that to each character there corresponds one single gene... It is also full of theories that dealt with objects designated by terms that are now considered to be without a real referent, and with entities which were thought to be real, and later discarded as non-existent: the "Angelic Powers" which according to Kepler's moved the spheres, phlogiston, caloric, luminiferous ether, germs boxed within germs since the creation, and so on. Those objects have simply disappeared from the physical world and only preserve a meaning in libraries. Many terms have also disappeared, whereas others have changed their meaning (for instance, "mutation"); others have been given a new name (for instance, "melancholy").

In order not to write triumphalist stories (which are useless and, in the end, uninteresting), historians must pay attention not only to the discoveries that have fruitfully become part of the growth of knowledge, not only to the theories that still form the basis of our current knowledge of the world. They should also take a serious interest in those objects, experiments, and theories that have been excluded from present scientific discourse, set aside or thrown away, and, so to speak, consigned to oblivion. They take into consideration not only the history written by the winners, but also that of the defeated.

Just as antique dealers routinely rummage through cellars and attics, and sometimes find interesting things, historians rummage through the past and often come across the unexpected. They rummage through garbage as well, in spite of many philosophers of science teaching them what they should and should not do. In an article of 1945 bearing the strange title "The vindication of rubbish," the great historian of medicine Walter Pagel said a simple thing that has too often been forgotten: if we want to make history, we have to deal not with how we think, but above all with how *they* thought. Often, they considered things to be self-evident that to us are not at all so. To understand them, we must also go through discourses that would appear outdated or even nonsensical today. In a sense, the history of science ends up making the history of precisely that which is not science, or, in other words, it must study what science forgets in order to be science.

We should never forget that disciplines or fields of research have not always been so distinct from one another as they are now; that the truths we find presented as self-evident in science handbooks are always results; and that those results have long and complicated processes behind them, including contrasts, difficulties, doubts, enthusiasms, attempts to identify crisis situations and to emerge from them, changes of opinions, and stubborn insistence in spite of recalcitrant evidence.

It is quite understandable that many active scientists do not show much interest for the strange ways in which their field of activity became what it is. They simply take it as their starting point and go on. They have little interest in "outdated" theories or positions. These appear to them either as outright errors or as partial truths or as necessary steps to the truth. Scientists become members of a community by adopting its language and accepting its running rules, or those of the particular group to which they have come to belong. Inevitably, as Kuhn remarked, handbooks introducing to a particular field contain an element of "dogmatism." They do not mainly address problems currently under discussions in scientific periodicals. They mostly deal with problems which either have been already solved or can be taken as typical or paradigmatic. Although originality may be encouraged, students are not expected to offer new and unexpected solutions, but to solve them again in the same ways they have been solved. Humanists should not smile condescendingly at this, as they sometimes do, least of all feel a sense of superiority. That dogmatism is necessary, and the kind of knowledge it generates is capable of rapid transformations which call into question its very assumptions with a frequency which is unknown to the majority of intellectuals.[31]

Scientific journals go into storage every 5 years. Science textbooks age the quicker the more a science grows. Their authors delete "obsolete" statements and replace them with "truer" ones. They are aware that there will be deletions and corrections (and if they have at heart the progress of science they sincerely wish that there will be many of them) in subsequent editions.[32] Textbooks do not only hide the role, but the very existence of most past events that helped generate current ideas. Information on how such a collective enterprise has been carried out, and on past debates and relations between innovations and their contexts is not relevant in them, and can be ignored, or confined to a few footnotes, or a few lines in the introduction, if ever. Only fragmentary information about the "tradition" of the science they expound is given, and reference is made only to "anticipations" and "precursors" of the "truths" that are expounded in the text.

> Partly by selection and partly by distortion, the scientists of earlier ages are implicitly represented as having worked upon the same set of fixed problems and in accordance with the same set of fixed canons that the most recent evolution in scientific theory and method has made seem scientific [...] Why dignify what science's best and most persistent efforts have made it possible to discard? The depreciation of historical fact is deeply, and probably functionally, ingrained in the ideology of the scientific profession, the same profession that places the highest of all values upon factual details of other sorts (Kuhn 1996, p. 138).

Why place "errors" among the things worth remembering? In the "ideology" of the scientific profession a devaluation of history is rooted. Even "revolutions" are sooner or later institutionalized in science, and prizes and accolades are awarded to those who held opinions which were earlier thought to be unorthodox or unacceptable. The result is an indissoluble interweaving of traditionalism and critical spirit, a "generalized skepticism" (as Michael Polanyi described it in 1946) in which, on the one hand, "plausibility" and "scientific value" tend to reinforce professional standards and, in a sense, conformism, while, on the other hand, "originality" tends to encourage dissension: rebellion in the context of discipline, as Polanyi put it: "the scientist is a traditionalist who likes to play difficult games according to

[31] I am tempted to go as far as saying that in many great scientists there is a sort of *daimon*. Otherwise, so many of them would never have tenaciously insisted on their path in spite of the contrariety of their colleagues. In some cases, such 'stubbornness' led to positive results; in others it did not. In this they are like those artists who keep loyal to their inspiration at the cost of living in poverty. The history of science is full of examples of both tenacity and mental flexibility, and of coexistence of the two: Darwin himself is a wonderful instance of the latter attitude. Why should the whole field of "Darwinists" present a different scenario?

[32] And, if they belong to the very rare species of militant Popperians, they try hard to refute their own ideas as soon as they conceive them.

well-established rules, in order to be an innovator able to discover new rules and new pieces with which to play."

> "Both the criteria of plausibility and of scientific value tend to enforce conformity, while the value attached to originality encourages dissent. This internal tension is essential in guiding and motivating scientific work. The professional standards of science must impose a framework of discipline and at the same time encourage rebellion against it. They must demand that, in order to be taken seriously, an investigation should largely conform to the currently predominant beliefs about the nature of things, while allowing that in order to be original it may to some extent go against these. Thus, the authority of scientific opinion enforces the teachings of science in general, for the very purpose of fostering their subversion in particular points." (Polanyi 1969, pp. 54–55)

Scientists inevitably tend to place their activities under a linear conception of progress. Just as they are constantly rewriting their textbooks, they are constantly rewriting history backward. As Kuhn (1996, p. 139) wrote, "the result is a persistent tendency to make the history of science look linear or cumulative, a tendency that even affects scientists looking back at their own research."

Too many historians of science still conceive and practice the history of science in the following way: they start with an accepted, familiar idea, or prevailing theory; they go backward to its "discovery," minimize oppositions to it, and present us with a coherent history of its triumph. Such an approach can be suitable for clear exposition, but it is essentially a form of fiction or novel. It makes any and all forms of creativity disappear from the story. In the word "discovery" and similar concepts the danger of equating facts with theories waits in ambush. But facts must be interpreted in the light of theories, and theories are not found, but constructed, in many cases by tortuous subterranean processes. For this reason, I cannot accept those forms of a "continuist" history of science which fuel the idea of a "triumphal path" of science and the mythical image of science as embodied reason.

Nor can I completely endorse that more sophisticated "epistemological history" approach whose "unique aspect [. . .] is that it takes into consideration the difficult question of the validity of different theories" (Gayon 1998, p. 6). Gayon states that "a rational reconstruction in no way excludes the social reconstruction of the history of science." He writes that "in order for the history of science not to loose itself in a morass of relativist descriptions of customs and beliefs, [. . .] we have to allow for a certain retrospectivity and avoid thinking that the ignorance of the current state of science in some ways helps the historian. Relatively few episodes of the history of what we call science can resist resolution by a strong dose of hindsight" (ibid.: 7–8). A *certain* "dose of hindsight" is inevitable in all reconstructions of the past, and might even be useful, provided the historian has a clear idea of what "the current state of science" is, which is not so simple a thing as Gayon, influenced by Canguilhem's epistemology, seems to take for granted. His excellent book on

Darwinism's Struggle for Survival is influenced by the "Modern Synthesis" retrospective view of the empirical and theoretical discussions which led to population genetics. As a result, "neo-Lamarckian" theories and authors are virtually neglected, and so are the debates on isolation, the fossil record, and classification. Authors such as Romanes, Wagner, Roux, Gulick, Volterra, Gause, Pearl, Simpson, and many others are not discussed, if they are mentioned at all.

To sum up and conclude this section: When reflecting or writing about the past, the easiest thing to do, and that which responds to the most obvious temptation, is to project backward the ways of thinking, feelings, categories, and beliefs of today. Then we are inclined to think that the same clarity and distinctions were something obvious in the past too, and were operating with the same force and vigor with which they operate today. Today we know, or believe we know, where right and wrong, facts and prejudices, were, or where the boundary between acceptable beliefs and superstitious beliefs, between free open thinking and intolerant and sectarian fanaticism, was. We know how things turned out and what the fate of the main characters in the drama was. It is hard, very hard, to keep constantly in mind that uncertainties and "errors" which characterized the past can also characterize our attitude today in many ways. We do not know what is to come. Neither did our predecessors. Pluralism and "confusion" characterized the debates on "Darwinism" in the past, just as they characterize present-day debates on what is meant by it, what is being done in evolutionary biology, and what one should do. If it is so hard to bring order today, and to write a linear history of what is happening, why should this be easier when dealing with the past? Most historians (with the exception of a substantial number of historians of philosophy) know that writing about history means one thing above all: trying to reconstruct, as far as possible, the past as as "another present,"[33] with its load of expectations, uncertainties, and "errors." This brings me to my last point.

9 Presentism

Are scientists *always* scientific? Certainly not, as they are human beings. Are they scientific whenever they talk about science, particularly that of their fellows? Not necessarily. The reactions to Darwinism show that many scientists (including some of Darwin's best supporters) did not even care to analyze the bases of his argument completely.[34] Have things changed since then?

[33]Or "another now", a phrase coined by the Italian philosopher Giulio Preti ("*un altro presente*") and adopted by the historian of science and philosophy Paolo Rossi (1999).

[34]This was even the case with one of Darwin's best friends, the botanist Joseph Dalton Hooker. After the end of American Civil War, he wrote to his colleague Asa Gray: "Did you ever read that painful book, Malthus on Population? I did the other day, and was painfully impressed by it. I had supposed he was a sort of materialist, who advised the checking of the population by restrictive means, and was surprised to find nothing of the sort, and a rather fine exordium at the end on a future state and the benefits of Christianity! His arguments seem incontrovertible to me" (Huxley 1918,

Scientists *can* dispense with history, I said. Nor are they requested to read other people's works, including those of their colleagues, with a philologist's eye. To keep pace with an enormous, ever-growing literature is virtually impossible. But what about a *minimum* of critical and self-critical sense when taking position on delicate issues or discussing matters of general interest? Cases of scientists rushing headlong into problems they do not fully master or making astonishingly rash statements are known to everybody, from the well-known statement attributed to Lord Rutherford "all science is either physics or stamp collecting"[35] to E.O. Wilson's prediction that ethics will be "cannibalized" by biology, and the "unsubstantiate and reckless opinions Dr James D. Watson [co-discoverer of the structure of DNA and a founder of the Human Genome Project] expressed on the subject of ethnicity and genetics", which led the Cold Spring Harbor Laboratory to revoke all titles and honors conferred on him [...] who led the lab for many years" (*The Guardian*, 13 January 2019). Not to mention scientists offering their sperm for eugenic experiments, on the evident assumption of their mental superiority.[36]

As historians keep documenting, a large number of biologists embraced some form of eugenics. It is too easy to say that in doing so they were being "unscientific" and to dismiss their behavior as a deviation, due to political interests or bad faith, from scientific caution. Eugenics was not a mere application of biology. It was all of a piece with the study of inheritance. To make just one important, and well-known, example, R.A. Fisher himself got interested in mathematical genetics because he was a eugenicist, and he devoted one-third of his 1930 foundational book to eugenics.[37]

p. 2: 43; no exact date given). His friend's book had turned the world upside down, but Hooker either was not bothered to read it carefully or had cavalierly skipped over what Darwin had said about the decisive inspiration of his main idea! This is all the more astonishing as Hooker was not only interested in political and social matters, but did not refrain from crudely applying "scientific principles" to politics and society (La Vergata 2023).

[35] I am aware that there is no direct evidence that Rutherford pronounced exactly these words. He is also reported to have uttered other perplexing sentences, such as "If your experiment needs statistics, you ought to have done a better experiment." By an irony of fate, in 1908, he was awarded the Nobel Prize for Chemistry, not Physics. He expressed surprise, but, after all, chemistry was nearer to physics than to stamp collecting.

[36] Here is a quote from Cavalli-Sforza and Cavalli Sforza (1995, pp. 376–377), who remind us of the interesting case of the great geneticist Herman J. Muller: "He suggested using the sperm of exceptionally able and intelligent men for the artificial insemination of women who volunteered (a process known as 'eutelegenesis'). In politics, he was a communist and spent a lot of time in Russia between the wars. It is said that he tried, unsuccessfully, to interest Stalin in his positive eugenics programme. Dissatisfied with his Soviet experience, he deleted the great communists from the list of possible elected representatives." The authors' comment is: "There is something snooty and even ridiculous about the noble gesture of the important person who donates his sperm for the betterment of humanity. More modest donors would be preferred, in the event that modesty is hereditary."

[37] After all, the criminal Mengele, a pupil of the geneticist Otmar von Verschuer (who was held in esteem by many American colleagues even after the war), must have been a reliable doctor, for otherwise the Nazis, who were very efficient in their job, would not have entrusted him with so important a task. Verschuer corresponded with Fisher, who had emigrated to Australia in 1957 and was senior research fellow at the Australian Commonwealth Scientific and Industrial Research

Eugenicists were discredited by the Holocaust, the war, and the Nuremberg trial, not by their colleagues or by scientific evidence. And eugenic societies survived the war. The factors which led to such "unscientific" practices are still with us, even if underground. What biologist can honestly say that being a scientist is per se an insurance against all forms of "unscientific" reasoning on the part of her/his colleagues?[38]

Fisher is also interesting for another reason. He was a conservative in politics and religion. So far so good. But he believed that social differences and racial differences had a biological basis. He was a convinced opponent of "miscegenation." He used statistics to support his views on these issues. He also used it in his attempt to reject the causal association between lung cancer and smoking (Stolley 1991; Christopher 2016; Proctor 2022; Gelman 2022). In 1957, he ridiculed the idea of cigarettes causing cancer as "a catastrophic and conspicuous howler," and, confounding cause and effect, argued that cancer gives its victims an "itch" that only smoking can scratch. Fisher disparaged experts and fellow statisticians arguing to the contrary, and he became a sort of recruiter for the tobacco industry, which funded thousands of scholars, including at least 25 who went on to win the Nobel Prize (interestingly, many eugenicists were willing to shill for that industry). If Fisher was not purposely spreading misinformation, was he being just naive? But how could such a brilliant statistician be so naive? Was he simply, inadvertently wrong? Were his conclusions, which he repeated over and again (Fisher 1957, 1958a, b, c, d and 1959) merely prematurely drawn? Was he unconsciously biased? And by what? Was he flawed by an unwillingness to examine the entire body of available data? May his views also have been influenced by personal and professional conflicts, by his work as a consultant to the tobacco industry, and by the fact that he was himself a smoker?[39]

Organisation in Adelaide. Fisher denied tobacco harm (see below) and enlisted von Verschuer to his cause (Keane 2022).

[38] See the "Facing Our History" Report of the American Society of Human Genetics (January 1923). From its foundation in 1948 to the early 1970s, many presidents of the ASHG also covered important positions in the American Eugenics Society, founded in 1922. The AES changed its name to the Society for the Study of Social Biology (1973) and then to the Society for Biodemography and Social Biology (2008), until it was disbanded in 2019.

[39] I apologize for allowing myself a personal reminiscence. At a conference at the Accademia Nazionale dei Lincei (Rome) I gave a talk in which I happened to mention (I do not remember why) the Fisher-tobacco affair. I wish I had never done it! One of the most important population geneticists of the late twentieth century, and Fisher's collaborator, Luigi Luca Cavalli Sforza, was in the audience. He suddenly interrupted, and then kept interrupting, me—much to the chairperson's embarrassment—protesting that the issue was trivial, that Fisher's contribution to population genetics could not be called into question (which I had no intention to do at all), and that what I was saying had nothing to do with science. He was simply unable to admit, indeed, he could simply not realize, that the same statistician used the same statistic methods and approach both when founding population genetics and when attacking other statisticians on the tobacco–cancer correlation. This is an interesting case, among many, of selective blindness, to say the least, in an otherwise wonderfully open-minded and interdisciplinary scientist. One can wonder whether Cavalli-Sforza thought he was speaking as a scientist or, shedding his professional robes, as a free citizen concerned about the fate of humanity when he subscribed to the one-child policy

Not a few scientists feel and speak as if they believe they are more scientific than their colleagues. Both they and their adversaries appeal to scientific data and facts. Exactly as it happened during the debate on Darwin and Darwinism. Coming to more recent events, during the COVID-19 pandemic we heard all kinds of things, even from respected scientists. The Nobel Prize winner for medicine Luc Montagnier defined vaccination "a crime," and even participated in no-vax meetings. A well-known Italian doctor, Alberto Zangrillo, declared on May 31, 2020, that the virus was "clinically dead" (there followed tens of thousands of deaths), except that he later said he had been "misunderstood."[40] Asked by a TV interviewer to comment on a microbiologist's opinions about what to do, an immunologist replied condescendingly that "Professor X is speaking for himself" (as if that immunologist had been commissioned by all the world's biologists to speak on behalf of Science...).[41] Conversely, many "humanists" demonstrated no better critical attitude and sense of proportion. The pandemic unleashed a flood of uncontrolled verbiage in the form of epochal diagnoses and predictions, including attacks on science as such (for an impressive and well-documented overview, see Rossi Monti 2020).

---

followed (and imposed) by the Chinese government as a remedy to overpopulation. The Chinese government—he wrote—encourages abortion for birth control purposes, "a very severe and undoubtedly bitter but unavoidable restriction, which deserves the gratitude of the rest of the world," because otherwise China's already huge population would grow at an explosive rate and "would soon have to drive out the rest of the world to occupy its living space, sharing it at most with India and other nations that have so far been unable, despite some attempts, to control births" (Cavalli-Sforza and Cavalli Sforza 1995, p. 367).

[40]Zangrillo also accused an important infectiologist who recommended extreme caution and was in favor of lockdown for being a "catastrophist" and "a survived left-wing protester from 1968." On June 8, 2021, he said: "Now the virus is hibernating." On March 31, 2023, Zangrillo was asked: "Is the virus really dead now?" He replied: "According to me, yes." Having contracted the virus two times (June 2023 and November 2024) I can witness that there is at least one case in which Zangrillo was wrong and that my disease was not "little more than a cold," as the infectiologist Matteo Bassetti said on December 4, 2021, about the effects of the Omicron variant, and repeated on October 7, 2023 (sources: https://osf.io/f7txy/; *Il fatto quotidiano*, December 4, 2021; *Il Sole 24 Ore*, December 4, 2021, *Corriere della sera*, June 8, 2021, March 31, 2023, October 7, 2023).

[41]Complaining that "Italians are a reluctant population [to vaccinate], and polls say that only 78% believe in vaccines [*not too bad, after all*]," even an important and balanced pharmacologist let himself go to a simplistic explanation: "This depends to a great extent on the fact that science has not entered schools: Italian education is literary-philosophical-artistic. Science as knowledge is not present in our culture" *La Nazione*, December 29, 2021. A little exaggeration, after all, since the Italian system produced so many authoritative scientists able to catechize the whole nation. One could turn the criticism and say that there is no taint of "humanism" at all in most science curricula, which might be one of the reasons why they churn out specialists refractory to a critical, and self-critical, attitude. Pronouncements on these issues sometimes degenerate into sheer arrogance. For instance, the zoologist Carlo Alberto Redi (2011, p. 31) said in anger that he had heard "many times" his students complain about the fact that, in order to carry on their studies, "they have to give up their evenings, Sundays, and parties while their humanist colleagues fuck around with the theses of this or that author!" Lord knows how many social events Mayr and Gould missed and how much serious research they neglected in their wish to ape humanists by "fucking around with the theses of this or that past author"!

Faced with the unavoidable differences in approach, and the avoidable controversies, between immunologists, infectious pathologists, epidemiologists, virologists, microbiologists, parasitologists, clinicians, pneumologists, pharmacologists, and statistical biologists on this or that aspect of the pandemic, and on the measures needed to deal with it, the mass media raised the absurd demand that "science should speak with one voice." The barbarity of those who reduced everything to a conspiracy of occult "strong powers" was sometimes countered by an equally mythical scientistic confidence. The historian perceives a family air in such easy security and complacency. Something recalls to his/her mind what often happened in the debate on Darwin and Darwinism, when technically precise objections and "ideological" resistances, disciplinary biases and sweeping generalizations, intellectual constraints and superficial enthusiasms mixed up in what at first glance looks as a mess. The story of Darwinism, whatever one means by this term, is an example of the necessity of historical sense, yesterday as today.

Of course, during the pandemic the public wanted a black-or-white certainty. However, from scientists one would expect caution, critical sense, problematism, and the ability to inform accordingly, at the cost of being unpopular. But a critical and self-critical attitude is not a gift of nature. It must be taught and practiced in a sort of recycling education and self-education. Scientific "dogmatism" (in the sense used above, § 8) can be dangerous when transplanted into public discourse on science. Truly historical sense is the best vaccine against it. History and historical education make scientists more human, and *more scientific*. Unfortunately, an uncritical, naive, ideological conception of scientific objectivity reigns in many Science Faculties. The past is only taken into consideration in the light of its immediate usefulness in the present. The virus of "presentism"[42] has virtually no effect on scientists as long as they are engaged in their research, but serious effects on them as citizens, especially when they intervene in public discourse. The same applies to those philosophers who, despite the ever-growing cooperation between history and philosophy of science, still debate over *the* scientific method, thereby de facto turning the analysis of what "is" into an implicit and devious "it is good that it is so." I say it once again: the only vaccine against that virus is history. All sorts of whiggish historiography weaken the immune system, and so do all sorts of self-serving historiographical strategies and of "rational reconstructions" describing science as a linear progress.

A form of presentism is a psychological and social bias affecting all cultures (Hartog 2003). It conditions and threatens to disfigure the image of the past. However, today it seems to be stronger than ever, fostered as it is by more powerful factors, mainly the tremendous successes and all-pervading power of technology, and the grandiose delusion that everything is ready at hand. We have achieved so many results which were unthinkable until a few years ago that we are inclined to

[42] I am using this word in a different sense from Esposito (2021), referring not to presentist historiography but, more generally, to a sort of syndrome which affects much of our culture, including science. In what follows I use information provided by my colleague (and, despite this, friend) Alberto Burgio, to whom I am indebted.

believe that the way to the best has already been mapped out. It's like we are deceiving ourselves into believing that we are living in an epoch that has had no history. "[The principle of the present] rages against the principle of hope and against all the illusions of the past" (Kluge 1990, p. 13). It is "a sort of permanent present lacking any organic relation to the public past of the times they live in [. . .] The destruction of the past, or rather of the social mechanisms that link one's contemporary experience to that of earlier generations, is one of the most characteristic and eerie phenomena of the late twentieth century" (Hobsbawm 1994, p. 3). In his inquiry into the metamorphosis of contemporary social memory, Paul Connerton (2009, p. 87) spoke of a "hyper-present," "an intensified immediacy which, by training the viewer's attention on a rapid succession of micro-events, makes it ever more difficult to envisage even the short-term past as 'real' [and] to conceive of any long-term historical coherence."

Science is presentist by its very nature. It *must* be so. No wonder so many scientists simply ignore history, or do not care for it. No wonder scientists-turned-historians write presentist histories (in Esposito's sense). From a scientist's perspective, historians may appear, quite understandably, as relativists for whom there is no difference between "truths" and "errors." It is a price historians have to pay. There is no escape. But in spite of some intellectuals proclaiming "the end of history" (Fukuyama 1992), we have not yet reached that point. On the contrary, the more humanity progresses, the more there is history to reckon with, and the more we are, so to say, "lived by history," as the Italian philosopher Antonio Labriola put it in 1897 (Labriola 1968, p. 174). In their never-ending job of deconstructing and reconstructing, historians do have something to teach: they warn scientists that their science is always not only progressing in a cumulative way, but being deconstructed and reconstructed by scientists themselves, and that, in spite of their increasing prestige and their dealing with "facts," they are more or less influenced and conditioned by their intellectual, social, and moral context, like all their fellow citizens (including historians, of course), whether they are aware of it or not.

Presentism also characterizes an increasingly spreading way of conceiving the role and task of philosophy. As the following quotation shows, the "need to enlighten the present" seem to many analytic philosophers (fortunately not all of them) "to be frustrated by historical-philosophical research."

> "Analytic philosophers deal with *philosophical problems* – not with authors or periods in the history of philosophy. And they deal with philosophical problems not in order to describe them or to grasp the cultural-historical sense of their occurrence, but to try to *solve* them [...] Analytic philosophers are interested in philosophical problems in their current formulation, that is, in the formulation they have received in philosophical research in recent decades, in the light of the events of the contemporary world and in particular the current order of the natural sciences and mathematics. They are not interested in the past history of problems, unless ways of dealing with a problem can be found in the past of philosophy that are different from those circulating in today's philosophical literature, and at least as promising." (Marconi 2014, p. 23)

Maybe I exaggerate, but I perceive in such statements something more than merely registering the separation between two different jobs. To me they sound like a polite (?) way of saying: "We, the true and original thinkers, have more important things to do than to encumber ourselves with historical junk." There seems to me to be an analogy between the attitude of these philosophers and that of scientists. Indeed, many of these philosophers claim they are "scientific." However, they go much further than scientists themselves, who do not even think that their scientific work can be "frustrated by historical research," either because they easily get rid of the complexity of history by simply ignoring it or because they protect themselves from the danger by building a fortress in the form of a self-serving reconstruction of the past. After all, no historian aims at preventing scientists or philosophers to go their own way, just as a paleontologist does not even think of interfering with those in search of a vaccine. Then why be so (politely...) aggressive? My answer is: inter-group competition—an animal struggle—for predominance (academic, professional, cultural, etc.) in an environment which favors what seems to be most current, or immediately expendable on the market. All very much like applied scientists patronizing "pure" scientists, with all competing against all for funding, and all pushing useless "humanists" away.

There is a never-ending debate on the relation between philosophy, the history of philosophy, and in general the history of ideas. I think my position is clear enough from what I have been saying. Therefore, I will restrict myself to a quotation from the important historian of philosophy Richard H. Popkin (1992: 325–326):

> "In the thirty–nine and one–half years that I have been teaching in philosophy departments, I have heard and participated in dreary discussions concerning whether study of the history of philosophy is part of what a philosophy department should engage in; whether historians of philosophy should be in philosophy departments, and whether advanced students should be expected to have some knowledge of the history of philosophy. Over the years I have been challenged about whether what I do is *philosophy*. By now I am not sure what *philosophy* is, but I am aware that people who discuss the issue of philosophy versus the history of philosophy are part of the history of philosophy."

Which amounts to saying that even the most presentist of present-day philosophers will be a subject for future historians, who will have the last word. This is even truer with respect to the relation between science and the history of science. Richard Delisle raises this point in an incisive way: "How many practicing scientists will still continue to read the work of Mayr in 30 years from now? Very few. Scientists of the future will read the work of historians to learn more about Mayr [...] Who is working for whom, then? From my viewpoint, biologists are working for me [as a historically-inclined philosopher], and not the other way around."[43] As to the question "what the history of philosophy (or science, or ideas, or the forms of

[43]Personal communication. I am grateful to Richard Delisle for allowing me to quote from our exchange, here and later on.

culture) is for?" I will simply quote a parable, which I found in the anthropologist Clifford Geertz' *After the Fact* (1995, p. 167).[44]

> A sage is squatted before a real elephant that is standing right in front of him. The sage is saying, "This is not an elephant." Only later, as the elephant turns and begins to lumber away, does a doubt begin to arise in the sage's mind about whether there might not be an elephant around after all. Finally, when the elephant has altogether disappeared from view, the sage looks down at the footprints the beast has left behind and declares with certainty, "An elephant *was* here."

By reconstructing even "elusive, rather ethereal, and by now wholly departed elephants," history helps us who are in the present also to *see* things which were not seen, or noticed, or seriously taken into account, or rightly evaluated when they happened, because agents and bystanders took them for granted or considered them irrelevant for what those people were doing in *that* moment. In a sense, their vision of the present was blurred by too "presentist" an attitude. To how many things can presentism blind us!

The wonderful successes of science and the increasing, inevitable, need for specialization contribute to a weakening of the historical sense, to screwing ourselves into the present, to placing more and more trust in operational techniques, to reflecting less and less on the how and why of the present state. It is as if we feel we have finally taken the right path; more than that: as if we were always destined to take it. Hardships, disagreements, alternatives closed and reopened, revisions and controversies, uncertainties, and misplaced security seem to belong to a past which has been forgotten and deserves to be forgotten, so as not to upset the security achieved (at least until the next "scientific revolution" makes it advisable to write a new linear history of the events which led to it. . .). Science, I repeat, is, and must be, presentist. The society in which it operates cannot and must not.

10 By Way of Conclusion

I opened this essay by mentioning Jacques Roger (1984) on the necessity of a "historical history." The philosophy of science cannot dispense with the history of science. If I am not mistaken, it was Gaston Bachelard who said that philosophers are always "one mutation behind" with respect to science. This is even truer with historians. They have to wait. History must happen before they attempt to write it. We are always running the risk of replicating what happened in the 1960s and 1970s, when the so-called "Modern Synthesis" was prevailing, and two generations of historians just took this for granted, and therefore wrote a "Synthesis Historiography."

[44]Geertz says that he came across it in a footnote of Jerome Bruner's *Acts of Meaning* (Harvard University Press, Cambridge, MA, 1990). The parable is from Kalidasa's *Sakuntala*, "the most famous drama in Sanskrit literature."

I am not optimistic about the chance of "a new alliance" between the two historiographic approaches distinguished by Delisle and Esposito, and I am inclined to push their distinction to its extreme consequences. I think that there is an irreconcilable gap between the "two cultures." The ideal of a field bound in harmony—biologists, philosophers, historians—has been a fabrication useful for scientists writing history for a self-serving strategy. What we need, I think, is a continuous re-foundation along the fault lines of the two cultures: historians and historically inclined philosophers on the one side, with normative philosophers and biologists on the other. The two sides should *hopefully* be close enough for historians to keep an eye on the development of evolutionary biology, but far enough for them to work on their own agenda (i.e., "reinvent" their area of study by cutting and recutting the boundaries and divisions of issues from as many different angles as possible), and for biologists to acquire a more nuanced and realistic image of the history of their field. *Expectedly*, historians' discourses will leave biologists (and normative philosophers) with their own separate narratives. However, historians and historically inclined philosophers will be at least prevented from offering for "free" a discourse supporting textbook history.

At the end of the introduction to this chapter I raised some general questions. Now I must give at least a tentative answer to them, in order for the reader to recover after going through so convoluted a text as mine.

Question 1: Was There a Darwinian Revolution?

Yes, *in the long run*, and it was both a scientific and an epistemological revolution. I cannot but repeat what I said elsewhere (La Vergata 2023, p. 590):

> "It is obvious that, if by 'revolution' a sudden and unanimous acceptance of Darwin's theory is meant, there was obviously no such thing. By this standard, however, neither a Copernican nor a Newtonian revolution ever occurred. Yet, it does make sense to use the word 'revolution' in their cases, as it is undeniable that things were no longer the same after their scientific and intellectual achievements, although Copernicus still believed in circular orbits, and electrons do not behave in a Newtonian way. For the historian the question is not 'Was Darwin right?', but "Did things change? And how? We do live in a Darwinian world, as we live in a Copernican and a Newtonian world (and an Einsteinian, and a quantum-physical world, and a world which will be altered by all the 'revolutions' to come). It is quite legitimate to use the names of Copernicus, Newton and Darwin to sum up in a name very complex processes, although this does not mean making those scientists into icons or patron saints in a Comtean-style positivist calendar. Like Copernicus and Newton, Darwin set in motion a process, or, rather, a ramification of unpredictable processes which is still unrolling. His was a 'rolling revolution' (Herbert 2005, p. 53). Where it will lead us, no one can say."

Question 2: Was Darwin's Theory an Open Theory? Was It a Mere General Framework? Has It Been Completed, Integrated, or Renovated?

Yes, Darwin's theory was an open theory. As far as I can say, every scientific theory is an open one. It can be enriched by further research, completed, confirmed, and revised. Darwin himself revised his theory, before and after publishing it. Further revisions followed when, for instance, other biologists stressed the role of isolation (geographic and/or reproductive and/or behavioral), and challenged his uniformitarian gradualism. After a long debate, an end was put to the belief in the inheritance of acquired characters (or "soft inheritance") and in "blending" inheritance. Far from endangering the role of natural selection, this led to emphasizing its power (its "omnipotence" in Weismann's language), whereas Darwin regretted in subsequent editions of the *Origin* that he had overemphasized it. The meaning of "chance variations" was revised, and the search for the mechanisms producing them is still going on. The more one gets into the box, the more the problem of the relation between chance and necessity, to use Jacques Monod's famous phrase, is pushed forward. The problem is open whether and to what extent there are constraints on the role of selection, whether it acts at different levels, and what lies between genes and fully developed organisms. Darwin must not made into an icon or patron saint,[45] as those do who, whatever happens, just keep repeating that "he was right." To reassure them, the historian says that what that saint set in motion has been, and will go on, producing miracles, until other, by now unpredictable, developments and events will change the whole scenery. Darwin's theory has also been a general framework—but not a *merely* general framework—within which to work as long as natural selection, however qualified and reformulated, remains its core notion.

Question 3: Is Darwinism a Historical Construct?
Yes, "Darwinism" is a historical construct. I go further: since the historian is faced with so many different interpretations of what Darwin said, it would better to talk about "Darwinisms," and, as Jacques Roger's (1985, p. 820) suggested, "use the label 'Darwinism' only for the thought of Darwin himself." Delisle (2017, p. 157) argues that the time has perhaps come to dispose of the label "Darwinism"

[45]Gould (2002, p. 12) calls Darwin "our profession's founder," and "the patron saint of our profession." He also says that "everyone wants such a preeminent authority on his side..." (ibid.: 148). As Richards (1988, p. 146) rightly remarked, "Among contemporary evolutionary theorists Darwin functions as an icon, an image against which theories receive approbation or reprobation. To select from the historical Darwin those features that best comport with one's own predilections in the contemporary scientific debate is to have those predilections sanctioned by the master." Shanahan (2004, pp. 293–294) agrees: "One way of coming to terms with Darwin's influence is to appropriate him in support of one's own views [...] But Darwin's view must be assessed in its own terms, not simply as a reflection of current beliefs." All the more so, I would add, because current beliefs are far from unanimous. These considerations apply not only to those who, being, like Gould, "extremely weary of the notion of biological 'progress' [...], are predisposed to find in Darwin's writings the validation they seek," but also to Modern Synthesis historians. Needless to say, the tendency to pull Darwin to one's side is even more evident in attempts to preserve his purity from the slightest taint of social Darwinism. Only historical misinformation and disinterest in history can explain statements like the following: "Several religions and many other bodies that play a fundamental role in human society show an over-reliance on a form of 'social Darwinism', such as was fashionable *in the last century*. *Never* accepted by Darwin [...], this misinterpreted form of Darwinism explained social relations by a natural selection understood *exclusively* as a 'red in tooth and claw' fight to the death for existence (Cavalli-Sforza 1996, p. 322; my italics). Only in the nineteenth century? Never? Exclusively? As if a number of historians had not been engaged in deconstructing social Darwinism and showing that it did not always consist in a crude application of a 'red in tooth and claw' fight to for existence to society... That sentence does not appear in the English translation of the work (Cavalli-Sforza 2000).

altogether, but I am afraid this is not a realizable ideal, not even with many historians. I think that "evolution theory" (or "theories," depending on contexts) is preferable to "Darwinism," as more respectful of past and present pluralism. But it is difficult to persuade those biologists who do not care for historians' sophistry to give up a comfortable icon. That icon has a powerful social function. It unites different opinions under one label. It strengthens the individual biologist's confidence that he/she is working in mainstream science and is on the "right side," with history supporting her/his position, however heterodox it may look to his/her colleagues. Therefore, Darwinism is also a social construct. I am not using "social" here to mean that the Victorian social and ideological context helped shape Darwin's theory and left an "indelible stamp" on it (as some historians have done, quite legitimately), but in the sense of "collective" (provided, of course, the term is not taken as referring to a sort of consensus by a well-defined collectivity). After all, "historical" is the best possible word: it includes both "social" and "collective," and embraces scientific as well as ideological positions, and all varieties of Darwinism, whether real and documented or fictitious.

Question 4: Are Scientists' Retrospective Interest in the History of Evolutionism and Historians' Call for Overcoming Simplifications Irreconcilable?

I am afraid they basically are irreconcilable, as I argued in §§ 7–9. Scientists and historians have different agendas. They deal with different aspects of reality and see the same things with different eyes. They belong to different "cultures" (both of which are indispensable). Mutual help and cooperation is desirable, and often actually possible. But is the way things are now favoring cooperation? In the correspondence which followed my sending him the first version of the present Chapter, Richard Delisle said that "perhaps the problem as of now in the area of evolutionary biology is not so much the risk of a schism between the two cultures as, on the contrary, the current lack of complete rupture between them." We seem, he argued, to be living "in an ill-defined and confused gray zone." According to him, a two-step process is necessary, or desirable: "(1) normative knowledge followed by (2) a delayed discourse generated 20 or 30 years later by historians, thus reestablishing the diversity and complexity of things." Over the long run, Richard asked, "who gets to control the discourse about the development of evolutionary biology?" Historians, he answered, "because biologists-philosophers control the short term, and historians the long term." I agree with him when he says that "there is no need for historians to desperately run behind biologists in order to pick up the bits and pieces dropped by them. They just need to wait for time to pass by and engage the literature with a delay of at least 20 or 30 years (or more, of course). And who gets the chance to generate a consistent discourse throughout (excluding the last 20 or 30 years)? Historians and historically inclined philosophers."

Delisle's forecast sounds convincing, even for a pessimist. The forecast *can* prove true, provided presentism does not overwhelm the humanities themselves, that "disinterested" and "pure" science is not swallowed by applied science as a consequence of the tremendous, ever-increasing, and ever more funded progress of technology, and that too much of philosophy is not replaced by the most aggressive and presentist sub-species of analytic philosophy.

References

Bailey LH (1896a) The survival of the unlike. Proc Am Philos Soc 35:88–111
Bailey LH (1896b) The survival of the unlike. A collection of evolution essays suggested by the study of domestic plants. Macmillan, London
Baldwin JM (1896a) On criticisms of organic selection. Science 4:727
Baldwin JM (1896b) A new factor in evolution. Am Nat 30(441–451):536–553
Baldwin JM (1897a) Social and ethical interpretations in mental development: a study in social psychology. Macmillan, New York
Baldwin JM (1897b) Organic selection. Nature 55:558. (also Science 5: 634–636)
Baldwin JM (1897c) Organic selection. Science 5:634–636. (same as 1897b)
Baldwin JM (ed) (1901–1905) Dictionary of philosophy and psychology. Macmillan, New York
Baldwin JM (1902a) Addendum to C. Lloyd Morgan, “Selection”. In: Baldwin ed. 1901–1905, 2(1902):506
Baldwin JM (1902b) Development and evolution. Macmillan, New York
Baldwin JM (1902c) Fragments in philosophy and science. Scribner’s Sons, New York
Baldwin JM (1909) Darwin and the humanities. Review Publishing Company, Baltimore
Baldwin JM, Morgan CL (1902) Social selection. In: Baldwin ed. 1901–1905 2(1902):213–218
Baldwin JM et al. [Morgan CL, Poulton EB, Stout GF, Groos K] (1902) Organic (or indirect) selection. In: Baldwin ed. 1901–1905 2:213–218
Beer G (1985) Darwin’s reading and the fictions of development. In: Kohn D (ed) The Darwinian heritage. Princeton University Press, Princeton, in association with Nova Pacifica, Wellington, NZ, pp 543–588
Bellomy DC (1984) Social Darwinism revisited. Perspect Am Hist New Series 1:1–130
Bellon R (2021) Guiding a train of discoveries: Charles Darwin, Charles Daubeny, and the reception of natural selection, 1859–1865. In: Delisle RG (ed) Natural selection. Revisiting its explanatory role in evolutionary biology. Springer, Cham, pp 39–72
Boas G (1972) Welfare in the cosmos. Diogenes 78:38–51
Bowler PJ (1983) The eclipse of Darwinism. Anti-Darwinian evolution theories in the decades around 1900. The Johns Hopkins University Press, Baltimore
Bowler PJ (1988) The non-Darwinian revolution. The Johns Hopkins University Press, Baltimore
Bowler PJ (2004) The specter of Darwinism: the popular image of Darwinism in early twentieth-century Britain. In: Lustig A, Richards RJ, Ruse M (eds) Darwinian heresies. Cambridge University Press, Cambridge, pp 46–68
Bowler PJ (2005) Revisiting the eclipse of Darwinism. J Hist Biol 38(1):19–32
Bowler PJ (2009) The eclipse of pseudo-Darwinism? Reflections on some recent developments in Darwin studies. Hist Sci 47(4):431–443
Bowler PJ (2013) Darwin deleted. University of Chicago Press, Chicago
Bowler PJ (2017) Alternatives to Darwinism in the early twentieth century. In: Delisle RG (ed) The Darwinian tradition in context. Springer, Cham, pp 195–217
Büchner L (1879) Sechs Vorlesungen über die Darwin’sche Theorie. T. Thomas, Leipzig
Carpenter WB (1860) Darwin on the origin of species. In: Hull DL (ed) Darwin and his critics. The reception of Darwin’s theory of evolution by the scientific community. Harvard University Press, Cambridge, pp 88–114
Cavalli-Sforza LL (1996) Gènes, peuples et langues. Odile Jacob, Paris
Cavalli-Sforza LL (2000) Genes, peoples and languages. Allen Lane, London. (Engl. tr. of Cavalli-Sforza 1996)

Cavalli-Sforza LL, Cavalli Sforza F (1995) Chi siamo. La storia della diversità umana. Mondadori, Milan

Ceccarelli D (2019) Between social and biological heredity: Cope and Baldwin on evolution, inheritance, and mind. J Hist Biol 52:161–194

Ceccarelli D (2021a) Recasting natural selection: Osborn and the pluralistic view of life. In: Delisle RG (ed) Natural selection. Evolutionary biology – New perspectives on its development. Springer, Cham, pp 171–191

Ceccarelli D (2021b) Theistic evolution and evolutionary ethics: Henry Fairfield Osborn and Huxley's legacy. Hist Philos Life Sci 43:114. https://doi.org/10.1007/s40656-021-00467-7

Christopher B (2016) Why the father of modern statistics didn't believe smoking caused cancer. *Priceonomics--In Data We Trust.* Published September 21, 2016 by Ben Christopher. https://priceonomics.com

Connerton P (1989) How societies remember. Cambridge University Press, Cambridge

Connerton P (2009) How modernity forgets. Cambridge University Press, Cambridge

Continenza B (1984) Tra lamarckismo e darwinismo. L'effetto Baldwin. In: Somenzi V (ed) Evoluzione e modelli. Il concetto di adattamento nelle teorie dei sistemi biologici, culturali e artificiali. Editori Riuniti, Rome, pp 107–191

Continenza B (1986) Simulation of Lamarckism or stimulation of Darwinism. Rivista di storia della scienza 3(2):155–190

Corsi P (1988) Science and religion: Baden Powell and the Anglican debate, 1800–1860. Cambridge University Press, Cambridge

Dana JD (1871) The Darwinian theory. The Independent 23 (1164, March 23), p 1

Darwin C (1876) On the origin of species, or the preservation of favoured races in the struggle for life. 6th edn. with additions and corrections to 1872. Murray, London

Darwin F (ed) (1887) The life and letters of Charles Darwin. Murray, London

Darwin C (1985) In: Burkhardt F et al (eds) The correspondence of Charles Darwin. Cambridge University Press, Cambridge

de Vries H (1901–1903) Die Mutationstheorie. Versuche und Beobachtungen über die Entstehung der Arten im Pflanzenreich. Veit, Leipzig

de Vries H (1909–1910) The mutation theory: experiments and observations on the origin of species in the vegetable kingdom. Open Court, Chicago (Engl. tr. of de Vries 1901–1903)

Delage Y (1903) La structure du protoplasme et les théories de l'hérédité et les grands problèmes de la biologie générale, 2nd edn. Schleicher Frères, Paris. (1st edn. 1894)

Delisle RG (2017) From Charles Darwin to the evolutionary synthesis: weak and diffused connections only. In: Delisle RG (ed) The Darwinian tradition in context: research programs in evolutionary biology. Springer, Cham, pp 133–168

Delisle RG (2021a) Introduction: In search of a new paradigm for the development of evolutionary biology. In: Delisle RG (ed) The Darwinian tradition in context: research programs in evolutionary biology. Springer, Cham, pp 1–10

Delisle RG (ed) (2021b) Natural selection. Revisiting its explanatory role in evolutionary biology. Springer, Cham

Delisle RG, Tierney J (2022) Rereading Darwin's Origin of Species: the hesitations of an evolutionist. Bloomsbury Academic, London

Di Gregorio M. (with the assistance of N.W. Gill)(1990) Charles Darwin's Marginalia. Part I. Books. Garland, New York

Dobzhansky T (1955) Evolution, genetics and man. Wiley/Chapman and Hall, New York/London

Dobzhansky T (1968) On some fundamental concepts of Darwinian biology. Evol Biol 2:1–34

Ellegård A (1990) Darwin and the general reader. The reception of Darwin's theory of evolution in the British periodical press, 1859–1872. With a new foreword by David L. Hull. The University of Chicago Press, Chicago. (orig. 1956. Gothenburg: Gothenburg Studies in English, vol. VIII – Acta Universitatis Gothoburgensis, vol. XLIV)

Esposito M (2021) Cathedrals, corals and mycelia: three analogies for the history of evolutionary biology. In: Delisle RG (ed) Natural selection. Revisiting its explanatory role in evolutionary biology. Springer, Cham, pp 11–38

Ferri E (1900) Socialism and positive science (Darwin-Spencer-Marx). Independent Labour Party, London (Engl. tr. of Socialismo e scienza positiva (Darwin, Spencer, Marx). Casa editrice italiana, Rome, 1894)

Finnegan DA (2021) Of snails and salvation: the theological construction of John Thomas Gulick's theory of evolution. Hist Stud Nat Sci 51(5):567–604

Fisher RN (1957) Alleged dangers of cigarette-smoking. Letters to the Editor (6 July, 3 August). British Medical J 2:43., 297–298

Fisher RN (1958a) Cigarettes, cancer, and statistics. Centennial Review 2(2):151–166

Fisher RN (1958b) The nature of probability. Centennial Rev 2(3):261–274

Fisher RA (1958c) Lung cancer and cigarettes? Letter to the Editor. Nature 182:108

Fisher RN (1958d) Cancer and smoking. Letter to the editor. Nature 182:596

Fisher RN (1959) Inhaling. In: Id (ed) Smoking. The cancer controversy: some attempts to assess the evidence. Oliver and Boyd, Edinburgh, pp 45–47

Fleck L (1980) Genesis and structure of a scientific fact. University of Chicago Press, Chicago (Engl. tr. of Entstehung und Enwicklung einer wissenschaftlichen Tatsache. Einführung in die Lehre vom Denkstil und Denkkollektiv, 1935)

Fukuyama F (1992) The end of history and the last man. Free Press, New York

Gayon J (1998) Darwinism's struggle for survival: heredity and the hypothesis of natural selection. Cambridge University Press, Cambridge (Engl. tr. of Darwin et l'après-Darwin. Une histoire de l'hypothèse de la sélection naturelle. Éditions Kimé, Paris, 1992)

Geddes P, Thomson JA (1911) Evolution. Williams & Norgate, London

Geertz C (1995) After the fact. Two countries, four decades, one anthropologist. Harvard University Press, Cambridge

Gelman A (2022) How can a top scientist be so confidently wrong? (R. A. Fisher and smoking example). Posted on June 8, 2022 9:24 AM by Andrew Gelman, professor of statistics and political science at Columbia University. https://statmodeling.stat.columbia.edu/2022/06/08/how-can-a-top-scientist-be-so-confidently-wrong-r-a-fisher-and-smoking-example/

Glass B, Temkin O, Straus WL Jr (eds) (1959) Forerunners of Darwin, 1745–1859. The Johns Hopkins Press, Baltimore, pp 415–437

Gould SJ (1980) Is a new and general theory of evolution emerging? Paleobiology 6:119–130

Gould SJ (2002) The structure of evolutionary theory. The Belknap Press of Harvard University Press, Cambridge, Mass.

Greene JC (1971) The Kuhnian paradigm and the Darwinian revolution. In: Roller DHD (ed) Perspectives in the history of science and technology. University of Oklahoma Press, Norman, pp 3–23

Greene JC (1974) The death of Adam. Evolution and its impact on western thought, 1st edn. The Iowa State University Press, Ames. 1959

Gulick JT (1872) Diversity of evolution under one set of external conditions. J Linn Soc London, Zoology 11:496–505

Gulick JT (1888) Divergent evolution through cumulative segregation. J Linn Soc London 20:189–274

Gulick JT (1890) Intensive segregation, or divergence through independent transformation. J Linn Soc London (Zool) 23(145):312–380

Gulick JT (1905) Evolution, racial and habitudinal. Carnegie Institution, Washington

Haacke W (1893) Gestaltung und Vererbung. Weigel Nachfolger, Leipzig

Haeckel E (1866) Generelle Morphologie der Organismen. Allgemeine Grundzüge der organischen Formen-Wissenschaft, mechanisch begründet durch die vom Charles Darwin reformirte Descendenz-Theorie. Georg Reimer, Berlin

Haeckel E (1876) History of creation; or the development of the Earth and its inhabitants by the action of natural causes, 2nd edn. Henry S. King, London (Engl. tr. by ER Lankester of

Natürliche Schöpfungsgeschichte. Gemeinverständliche wissenschaftliche Vorträge über die Entwicklungslehre im Allgemeinen und diejenige von Darwin. In, Von Darwin, Goethe und Lamarck im Besonderen, über die Anwendung derselben auf den Ursprung des Menschen und andere damit zusammenhängenden Grundfragen der Naturwissenschaft. Georg Reimer, Berlin 1868)

Hale PJ (2014) Political descent. Malthus, mutualism, and the politics of evolution in Victorian England. The University of Chicago Press, Chicago

Hartog F (2003) Régimes d'historicité. Présentisme et expérience du passé. Seuil, Paris. (Engl. tr. Regimes of historicity: presentism and experiences of time. Columbia University Press, New York, 2015)

Herbert S (1971) Darwin, Malthus, and selection. J Hist Biol 4:209–217

Herbert S (1974) The place of man in the development of Darwin's theory of transmutation. Part I. J Hist Biol 7:217–258

Herbert S (1977) The place of man in the development of Darwin's theory of transmutation. Part II. J Hist Biol 10:155–227

Herbert S (2005) The Darwinian revolution revisited. J Hist Biol 38:51–66

Hobsbawm EJ (1994) The age of extremes. The short twentieth century, 1914–1991. Abacus, London

Hovasse R (1950) Adaptation et évolution. Hermann, Paris.

Hull DL (ed) (1973) Darwin and his critics. The reception of Darwin's theory of evolution by the scientific community. Harvard University Press, Cambridge

Hull DL (1974) Darwinism and historiography. In: Glick TF (ed) The comparative reception of Darwinism. University of Texas Press, Austin, pp 388–402

Hull DL (1985) Darwinism as a historical entity: a historiographic proposal. In: Kohn D (ed) The Darwinian heritage. Princeton University Press, Princeton, in association with Nova Pacifica, Wellington, NZ, pp 773–812

Huxley TH (1893–1894) Collected essays. Macmillan, London

Huxley L (ed) (1918) The life and letters of Sir Joseph Dalton Hooker. Murray, London

Keane D (2022) Nazi scientist Otmar von Verschuer's correspondence with British biologist illuminates corruption of medicine ABC News, 31 August 2022. Australian Broadcasting Corporation

Kelly A (1981) The descent of Darwin. The popularisation of Darwinism in Germany, 1860–1914. University of North Carolina Press, Chapel Hill

Keynes JM (2010) Newton, the Man. In: Id. Essays in biography. Palgrave Macmillan, Basingstoke, pp 363–374

Kirchhoff A (1910) Darwinismus angewandt auf Völker und Staaten. Gebauer-Schwetschke, Halle

Kluge A (1990) The assault of the present on the rest of time. New German Critique, Winter, 1990, n. 49 (Special Issue on Alexander Kluge), pp. 11–22 (Engl. tr. of Der Angriff der Gegenwart auf die übrige Zeit. Das Drehbuch zum Film. Syndikat, Frankfurt a.M, 1985)

Kohn D (1980) Theories to work by: rejected theories, reproduction, and Darwin's path to natural selection. In: Coleman W, Limoges C (eds) Studies in history of biology, vol 4. The Johns Hopkins University Press, Baltimore, pp 67–70

Kuhn T (1996) The structure of scientific revolutions, 3rd edn. The University of Chicago Press, Chicago. (1st edn. 1962)

Kuyper A (1899) Evolutie. Rede bij de overdracht van het rektoraat aan de Vrije Universiteit of October 1899 gehouden. Höveker & Wormser, Amsterdam

La Vergata A (2023) Images of the economy of nature, 1650–1930. From "nature's war" to Darwin's "struggle for life". Springer, Cham

Labriola A (1968) Discorrendo di socialismo e di filosofia (1897). In: Id. Saggi sul materialismo storico, a cura di V. Gerratana e A. Guerra. Editori Riuniti, Rome

Laland K, Uller T, Feldman M et al (2014) Does evolutionary theory need a rethink? Nature 514: 161–164

Latour B, Woolgar S (1979) Laboratory life. The social construction of scientific facts. Sage Publications, Beverly Hills
Leeds A (1974) Darwinian and "Darwinian" evolutionism in the study of society and culture. In: Glick TF (ed) The comparative reception of Darwinism. University of Texas Press, Austin, pp 437–485
Lesch JE (1975) The role of isolation in evolution: George J. Romanes and John T. Gulick. Isis 66: 483–503
Limoges C (1970) La sélection naturelle: étude sur la première constitution d'un concept. Presses Universitaires de France, Paris
Marconi D (2014) Il mestiere di pensare. La filosofia nell'epoca del professionismo. Einaudi, Turin
Mayr E (1942) Systematics and the origin of species. Columbia University Press, New York
Mayr E (1982) The growth of biological thought. Harvard University Press, Cambridge
Minelli A (2017) Sviluppo ed evoluzione. Rivista di filosofia 108(1):94–99
Minelli A (2018) Risposta a Telmo Pievani. Rivista di filosofia 109(1):124–128
Moore JR (1979) The post-Darwinian controversies. A study of the Protestant struggle to come to terms with Darwin in Great Britain and America, 1870–1900. Cambridge University Press, Cambridge
Morgan CL (1888) Natural selection and elimination. Nature 38:370
Morgan CL (1891) Animal life and intelligence. Ginn & Company, Boston
Morgan CL (1896a) On modification and variation. Science 4(99):733–740
Morgan CL (1896b) Habit and instinct. Edward Arnold, London
Morgan CL (1900) Animal behaviour. Edward Arnold, London
Morgan CL (1902) Selection (in biology). In: Baldwin ed. 1901–1905, 2(1902):504–506
Morgan CL (1909) Mental factors in evolution. In: Seward AC (ed) Darwin and modern science: essays in commemoration of the birth of Charles Darwin and of the fiftieth anniversary of the publication of the 'Origin of Species'. Cambridge University Press, Cambridge, pp 424–445
Osborn HF (1896a) A mode of evolution requiring neither natural selection nor the inheritance of acquired characters. Trans N Y Acad Sci 15:141–148
Osborn HF (1896b) Ontogenic and phylogenic variation. Science 4(100):786–789
Osborn HF (1897) The limits of organic selection. Am Nat 31:944–951
Osborn HF, Poulton EB (1897) Organic selection (A discussion introduced by Professor Henry F. Osborn and Professor Edward B. Poulton at the Detroit meeting of the American Association, Wednesday, August 11). Science 6(146):583–587
Pagel W (1945) The vindication of rubbish. Middlesex Hospital Journal 45:42–45. (repr. in Id. Religion and Neoplatonism in Renaissance medicine, Variorum Reprints, London, 1985, pp 1–14)
Pearson K (1892) The grammar of science. Adam and Charles Black, London
Pearson K (1896a) Mathematical contributions to the theory of evolution. III. Regression, heredity and panmixia. Philos Trans R Soc London A 187:253–318
Pearson K (1896b) Mathematical contributions to the theory of evolution. Note on reproductive selection. Proc R Soc London 59:301–305
Pearson K (1897) The chances of death, and other studies in evolution. Edward Arnold, London
Pearson K (1900) The grammar of science. Adam and Charles Black, London (1st edn. 1892)
Peckham M (1970) Darwin and Darwinisticism. In: Id. The triumph of romanticism: collected essays. University of South Carolina Press, Columbia, pp 176–201. (orig. in Victorian Studies 3: 19–40)
Pertile C (2018) Selezione organica ed eredità sociale. Sguardo sul pensiero evoluzionistico di James Mark Baldwin. Noema 9:24–39
Pfeifer EJ (1974) United States. In: Glick TF (ed) The comparative reception of Darwinism. University of Texas Press, Austin, pp 168–206
Pigliucci M, Müller GB (2010) Evolution. The extended synthesis. MIT Press, Cambridge
Plate L (1908) Selectionsprinzip und Probleme der Artbildung. Ein Handbuch des Darwinismus. Dritte, sehr vermehrte Auflage. Engelmann, Leipzig

Polanyi M (1969) Knowing and being. Ed. by M. Grene. The University of Chicago Press, Chicago
Popkin RH (1992) The third force in seventeenth-century thought. Brill, Leiden
Proctor R (2022) Why did Big Tobacco love (and fund) eugenicists like R.A. Fisher? Lecture delivered on 22nd April, 2022, at "Fisher in the 21st Century. Academic Conference", 21–22 April, 2022, Gonville and Caius College, Cambridge. https://www.cai.cam.ac.uk/discover/fisher-21st-century
Redi CA (2011) Il biologo furioso. Provocazioni d'autore tra scienza e politica. Sironi, Milan
Richards RJ (1987) Darwin and the emergence of evolutionary theories of mind and behavior. University of Chicago Press, Chicago
Richards RJ (1988) The moral foundations of the idea of evolutionary progress: Darwin, Spencer, and the neo-Darwinists. In: Nitecki M (ed) Evolutionary progress. Univeristy of Chicago Press, Chicago, pp 129–148
Roger J (1984) Per una storia storica delle scienze. Giornale critico della filosofia italiana 63(3): 285–314
Roger J (1985) Darwinism today (commentary). In: Kohn D (ed) The Darwinian heritage. Princeton University Press, Princeton, in association with Nova Pacifica, Wellington, NZ, pp 813–823
Romanes GJ (1886a) Physiological selection: an additional suggestion on the origin of species. J Linn Soc Zool 19:337–411
Romanes GJ (1886b) Physiological selection. Nature 34:314–316, 336–340, 362–365
Romanes GJ (1886c) Physiological selection and the origin of species. Nature 34:407–408, 439
Romanes GJ (1887) Physiological selection. Nineteenth Cent 21:59–80
Rossi P (1986) I ragni e le formiche. Un'apologia della storia della scienza. Il Mulino, Bologna
Rossi P (1999) Un altro presente. Saggi sulla storia della filosofia. Il Mulino, Bologna
Rossi Monti M (2020) In the name of the virus. Intellectuals and the pandemic. Lares Quadrimestrale di studi demoantropologici 86(3):407–439
Roux W (1881) Der Kampf der Theile im Organismus. Engelmann, Leipzig
Schmalhausen II (1949) Factors of evolution. Blakiston, Philadelphia
Schwartz JH (2021) What's natural about natural selection? In: Delisle RG (ed) Natural selection: revisiting its explanatory role in evolutionary biology. Springer, Cham, pp 329–390
Seidlitz GS (1871) Die Darwinische Theorie. Elf Vorlesungen über die Entstehung der Thiere und Pflanzen durch Naturzüchtung. Mattiessen, Dorpat
Shanahan T (2004) The evolution of Darwinism. Selection, adaptation, and progress in evolution biology. Cambridge University Press, Cambridge
Simpson GG (1953) The Baldwin effect. Evolution 7 (2, Jun. 1953), pp. 110–117
Snow CP (1961) The two cultures: the Rede Lectures 1959. Cambridge University Press, Cambridge
Sober E (ed) (1984a) Conceptual issues in evolutionary biology. An anthology. MIT Press, Cambridge
Sober E (1984b) The nature of selection: evolutionary theory in philosophical focus. MIT Press, Cambridge
Stolley D (1991) When genius errs: RA Fisher and the lung cancer controversy. Am J Epidemiol 133(5):416–425. https://doi.org/10.1093/oxfordjournals.aje.a115904. PMID 2000852
Thomson JA (1909) Darwinism and human life. The South African Lectures for 1909. Andrew Melrose, London
Vacher de Lapouge G (1896) Les sélections sociales. A. Fontemoing, Paris
Waddington CH (1940) Organisers and genes. Cambridge University Press, Cambridge
Waddington CH (1957) The strategy of the genes. Allen & Unwin, London
Waddington CH (1960) Experiments on canalizing selection. Genet Res 1:140–150
Wallace AR (1886) Romanes versus Darwin. Fortnight Rev 46:300–316
Wallace AR (1890) Dr. Romanes on physiological selection. Nature 43:79–150
Wasmann E (1901) Gibt es tatsächlich Arten, die heute noch in der Stammesentwicklung begriffen sind? Biologisches Zentralblatt 21(689–711):737–752

Wasmann E (1904) Die moderne Biologie und die Entwicklungstheorie. Herder, Freiburg im Breisgau

Weber BH, Depew DJ (eds) (2003) Evolution and learning: the Baldwin effect reconsidered. MIT Press, Boston

Weismann A (1909) The selection theory. In: Seward AC (ed) Darwin and modern science: essays in commemoration of the birth of Charles Darwin and of the fiftieth anniversary of the publication of the 'Origin of Species'. Cambridge University Press, Cambridge, pp 18–65

Wright C (1871) The genesis of species. North Am Rev 113(232, July):63–103 (Repr. in Id. Philosophical discussions. Henry Holt & Co., New York, 1878, pp 127–167)

Wright S (1980) Genic and organismic selection. Evolution 34:825–843

Young RM (1969) Malthus and the evolutionists: the common context of biological and social theory. Past Present 43:109–145. (Repr. in Id. Darwin's metaphor. Cambridge University Press, Cambridge)

Young RM (1985) Darwinism is social. In: Kohn D (ed) The Darwinian heritage. A centennial appraisal. Princeton University Press, Princeton, in association with Nova Pacifica, Wellington, NZ, pp 609–639

The Evolution of "Darwinism": Up Close and Personal

Mark B. Adams

This chapter is dedicated to the memory and legacy of Garland Allen (1936–2023): scientist, historian, teacher, mentor, confidant, and close friend for the past 60 years; to the memory and legacy of Everett Mendelsohn (1931–2023), my mentor who pioneered the creation of the history of biology and its social context and taught me how to be an historian; and to the memory and legacy of Ernst Mayr (1904–2005), who schooled me in evolutionary biology, helped launch my career, and was always patient, supportive, and helpful at the beginning and over the years, until the end of his remarkable, century-long life.

"When I use a word,' Humpty Dumpty said in rather a scornful tone, 'it means just what I choose it to mean — neither more nor less.' 'The question is,' said Alice, 'whether you can make words mean so many different things.''The question is,' said Humpty Dumpty, 'which is to be master — that's all."
— Lewis Carroll, Through the Looking Glass

Abstract

This is a sequel to a recent article arguing that the standard "just-so-story" of the history of Darwinism through the so-called Modern Synthesis has obfuscated the central persistent problem Darwin's natural selection theory has faced ever since 1859, namely its failure to explain or account for macroevolution and the origin of higher taxa. It also "deconstructed" the standard story and explored the actual

Author Mark B. Adams has died before the publication of this book.

R. G. Delisle et al. (eds.), *Unity and Disunity in Evolutionary Biology*,
https://doi.org/10.1007/978-3-031-42629-2_5

disunity underlying the apparent "unity" of synthesis supporters. Here, I explore the various international meanings of "evolution," "Darwinism," "Dismal Darwinism," "Visionary Biology," "Modern Synthesis," "natural selection," "microevolution," and "macroevolution" and then focus on the contextual origins, creation, and development of the modern "evolutionary synthesis," how and why it was so widely and quickly endorsed, how it became the standard "just-so-story" taught by evolutionary biologists and historians alike for so long, and who, how, why, and when the disunity underlying the apparent "unity" re-emerged and the so-called "synthesis" unraveled, as I observed firsthand and experienced it over the past 50 years.

Keywords

Darwin · T.H. Huxley · Evolution · Darwinism · Neo-Darwinism · Creative Darwinism · Modern Synthesis · Evolutionary biology · Molecular biology · Mutual aid · International reception of the *Origin* · J.B.S. Haldane · H.J. Muller · Julian Huxley · Ernst Mayr · Th. Dobzhansky · G.G. Simpson · Population genetics · Microevolution · Macroevolution · Stephen Jay Gould · William Provine · Garland Allen · Everett Mendelsohn

1 "Darwinism"

Darwin's *Origin* was published in London in late 1859, under the title *On the Origin of Species by Means of Natural Selection, Or the Preservation of Favoured Races in the Struggle for Life*. It was widely reviewed in Britain. Richard Owen's review, and those of other experts older than Darwin, tended to be critical; the positive reviews generally came from Darwin's friends, as well as predominantly younger professionals. And it was in the book's review by T. H. Huxley, in 1860, that the word "Darwinism" entered the Anglophone lexicon.

Karl Marx is said to have declared, "I am not a Marxist!" and Darwin might well have denied being a "Darwinist," because the word "Darwinism" very quickly acquired different meanings that had lives of their own. Darwin's book was what he wrote, but "Darwinism" was an entirely different matter, and quickly went "public." There was a popular fascination with dinosaurs from "digs" in the nineteenth century, as the exhibits at the Crystal Palace demonstrated. Meanwhile, rumors about the exchange (not a "debate") between Huxley and Bishop Wilberforce at the Oxford meeting soon became the fascination of *Vanity Fair*, and the British public became alert and engaged.

Understandably, the interest quickly spread beyond the Anglophone world. As we all know, different cultures and languages parse reality differently, and the immediate problem was how to translate Darwin's work into other European languages, such as French, German, and Russian. Some translated editions actually changed the title, adding "*in the animal and plant kingdoms*"; others characterized it as "Progress in the Animal and Plant Kingdom." Many were "edited" or heavily footnoted, sometimes disagreeing with Darwin's text, the editor or translator "interpreting" its relevance to national or cultural issues.

Especially troublesome was a key word in Darwin's title: "selection." No other languages had a word that meant what "selection" did to English speakers. Struggling through various editions of Darwin's book titles in French, German, and Russian, words were initially chosen that meant "choice," "breeding," "election," "matching," "matching of kin," "selectivity," and "electivity." (A Japanese student told me that in the first two Japanese translations, one meant "breeding," and the other was "a military term.") Finally, in the end, many chose to adopt the "colorful foreign neologism," *selection* (*sélection*, *selektsiia*).(Krementsov 2018).

Another problem in the nineteenth century was the birth of new words that soon spread, and whose meaning and references evolved and changed over time in different settings, including such terms as "biology," "evolution," "science," and "scientist." Both the German word *Wissenschaft* and the Russian word "*nauka*" meant "scholarship," not "science." Meanwhile, in English, "science" became separate from the humanities and social studies as a different, privileged kind of knowledge.

In that process, Huxley played a key role. From his own middle-class background, his broader agenda was to remake science from a gentleman's pastime into a *profession*, and to eliminate the British clergy from their dominant position supporting "natural theology." (Interestingly, Darwin was often supported by the younger, liberal clergy, at home and abroad, who argued that "miracles" are incompatible with God's orderly, lawful Creation.) For Huxley (who coined the term "agnostic"), a "scientist" should have no conflicting commitments, following reasoning and evidence wherever it leads, regardless of one's religious beliefs. When he attacked the conservative Bishop Wilberforce (who had been prepped by Owen), ostensibly saying he would rather be descendent from a monkey than from a man who used his skill to defame science, he was laying down the gauntlet.

The word "Darwinism" also began to spread abroad (e.g., the "Darwinismus" of Ernst Haeckel, the "neo-Darwinismus" of August Weismann), and was often seen as a challenge to religious and traditional beliefs, and orthodoxies of various nations and cultures. Indeed, whatever the word Darwinism "meant" in various cultures, and whomsoever its different and varied opponents were, the use of the word became at least as often negative as positive. In early 1871, Darwin brought home the human implications of his theories with the publication of *The Descent of Man, and Selection in Relation to Sex,* which caused explosive public controversy. In the 1870s, the phrase "social Darwinism" came into broad use, first coined (and most often used) as a pejorative.

In Huxley's case (as was the case with other "Darwinists" as well), praise for Darwin's book did not necessarily signify agreement with Darwin about the nature of evolution, or the efficacy of natural selection. (After first reading Darwin's *Origin*, Huxley remembered, his first reaction was "How extremely stupid not to have thought of that!") (Huxley 1887, p. 197). Although we allude to him as "Darwin's bulldog," that term only came into use beginning in the twentieth century (see

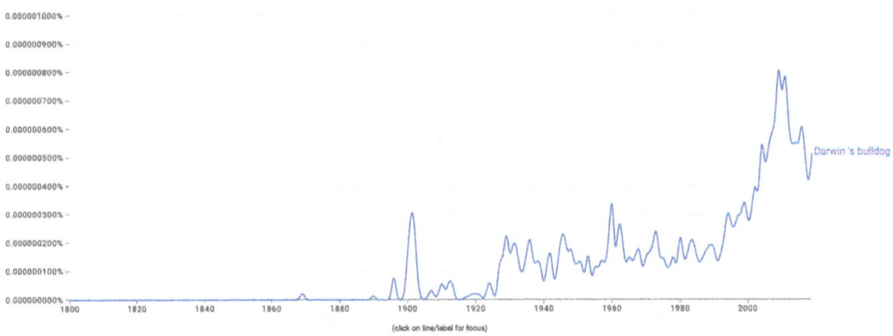

Fig. 1 The gradual increase of the use of the expression "Darwin's bulldog" in the literature over the years

Fig. 1). Huxley and Darwin had strong disagreements (van Wyhe 2019). Indeed, although Huxley welcomed a rational alternative to creationism and "natural theology," he declared in 1887 that he was "indifferent" as to whether Darwin's ideas turned out to be right or wrong.

In *Origin,* Darwin made two central claims: evolution (in his term, "descent with modification")[1] and its cause and mechanism ("natural selection"). In general, in Darwin's time and the decades following his death, "evolution" became the central issue about "Darwinism," and, for many, being a "Darwinist" meant a belief in evolution. By the so-called Eclipse period (1890–1920), however, being a "Darwinist" now meant a belief in "natural selection." Most biologists at the time had come to believe in evolution, but many doubted that Darwin's "natural selection" was its sufficient cause. Ironically, during this period, those who agreed with Darwin's claims in the final edition of the *Origin* were branded as "anti-Darwinians." (For many, the problem was not the "*survival* of the fittest," but rather "the *arrival* of the fittest," and many who believed in evolution considered its "cause" as yet unknown.) (DeVries 1906; Cock and Forsdyke 2008, p. 2022).

In general publications, the word "Darwinism" has been in wide perpetual use from 1860 through 2019, hundreds of times more frequently than "Modern Synthesis" or "evolutionary synthesis" or "synthetic theory of evolution," and more than twice as often as the generic term, "evolutionary biology" (see Fig. 2). It meant all sorts of different and contradictory things, including a "worldview," a "religion" (Ruse 2016), an atheist campaign against Christianity, an "ideology" (Adams 1987), or an arrogant, immoral "scientism" (C.S. Lewis). But, ironically, it *never* referred

[1]When Darwin wrote, the English word "evolution" had several different meanings (Richards 1992), and in his first edition he never used the word, although in the final paragraph he did use "evolving."

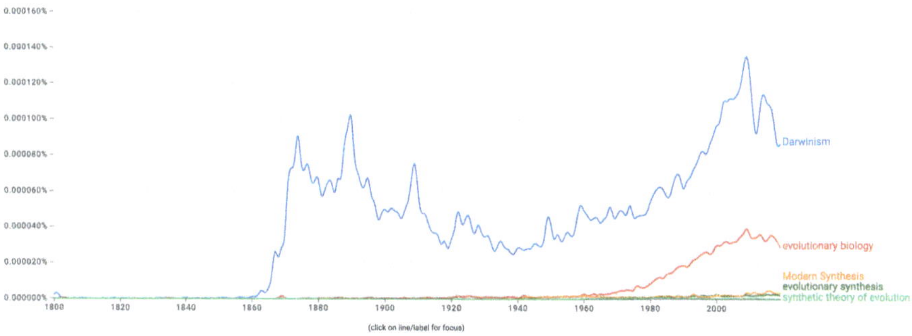

Fig. 2 Comparison of the use of different expressions in the literature over the years. The term "Darwinism" is significantly more popular than other comparable expressions

precisely to what Darwin believed and asserted in the *Origin* in any of its various editions, or what Darwin actually wrote in his other books.[2]

2 Dismal Darwinism

In the course of my various researches spanning science and literature, I noticed a generally overlooked (but valid and consequential) reaction to "Darwinism" by both scientists and the broader public alike, a "Dismal Darwinism," which in turn generated a kind of "Visionary Biology" among concerned evolutionists during the so-called *fin de siècle* and the "Eclipse" (Adams 2000, 2004).

As the end of the nineteenth century approached, many publications appeared celebrating its unparalleled achievements. The triumphs of the "Century of Progress"—the Industrial Revolution, the rise of scientific medicine, the heroic age of invention, and the creation and efflorescence of a host of new sciences—were everywhere evident. No wonder, then, that leading intellectuals had been preoccupied with explaining progress itself: such disparate thinkers as Auguste Comte, Herbert Spencer, and Karl Marx shared in common the effort to understand its causes and nature. For Spencer, for example, the "law of progress" was everywhere evident in the universe, and always, inevitably, over time, the "homogeneous" increasingly became more "heterogeneous" and "complexified" (because every cause had more than one effect), producing universal development and progress—in the solar system, the Earth, the embryo, diversified living creatures (leading to humans), and of human society (from families, to hunter-gatherer tribes, to modern civilization). Many saw Spencer as a founder of the new discipline of "Sociology."

Yet, many such works, even those of a celebratory tone, contained troubling worries based on those many discoveries of nineteenth century science. By the end of the "century of progress," there was a widespread sense that progress was over. At

[2]In recent days, I came upon an article about "A Darwinian Approach to Investing," and another about "A Darwinian Summer Cuisine."

the *fin de siècle,* in the words of one prominent witness, "the prevalent feeling is one of perdition and extinction" (Bergonzi 1961, p. 5). Most troubling was not what science was saying about the past, but rather what science seemed to be saying about the future. Charles Darwin, probably more than any other, both provided the progress theorists with best scientific evidence and, at the same time, threw cold water on the human future.

It was not "evolution" per se, but his "natural selection" that proved most troubling. If the marvelous adaptations everywhere evident in animate nature were not the product of God's design, but simply the result of overpopulation, variation, and natural selection—Spencer's *survival of the fittest,* reminiscent of Tennyson's *nature red in tooth and claw*—then humanity's progress from apes was the result of the extinction of "less fit" forms in the struggle for life. Darwin had made it clear that there was nothing inevitable about "progress" in Spencer's sense: cave fish whose ancestors had eyes could lose them and become blind through "natural selection" if doing so helped them survive in their lightless environment. Almost by definition, the "fittest" were not the most admirable or advanced, but whatever happened to survive, for whatever reasons.

Others drew stronger lessons. If humanity was "higher" than their primate precursors, it was because the fitter had triumphed over the weaker through fierce competition—the fiercer the competition, the faster the "progress." (For example, at the time, the common assumption was that Neanderthals had been wiped out by the Cro-Magnon race, which had more advanced, long-range weapons.) Future human progress of the sort everyone was celebrating was by no means inevitable, quite the contrary: the outcome of a perpetual "struggle for life" had led to mass extinctions of the "losers." Most agreed that altruism, civility, freedom, immunization, and charity toward the less fortunate were characteristics of a more advanced, higher, more socially evolved form of civilization. Yet, drawing on Darwin, many had come to believe that, by sparing humans the struggles that weeded out the "unfit," advanced civilized behavior would almost certainly lead to human degeneration and the undoing of civilization itself. At the turn of the century, in England, France, and Germany (often comparing their populations to ancient Greek and Roman statues) many feared their own national "degeneration" had been increasing over many centuries, and would weaken them against their potential rivals and enemies. Like that of any other species, our future—even our continued existence—was subject to natural laws beyond our control.

This view was embodied, appropriately, in the world-renowned works of Huxley's most famous pupil, H. G. Wells (1866–1946). Launched to international fame by *The Time Machine* (1895), his multiple "stories" through 1905 were translated into many languages, often dwelling on biological themes to make the same point: the human species will inevitably soon degenerate beyond salvation, and, in the longer term, we will almost certainly go extinct, sooner rather than later. (A leading British physicist, Lord Kelvin, assured us that the sun would die and the Earth would become uninhabitable relatively soon.)

The broader public, as well as many scientists, had grasped the human "gist" of Darwin's theory perfectly well: the history of life on Earth has experienced vast extinctions of countless species, and if we can't take control of our own destiny, we'll soon become one of them. The "scientific laws of nature" would soon lead humans to rapid degeneration and extinction. (Recall that even Alfred Russel Wallace, in some sense a co-discoverer of Darwin's "natural selection," opposed vaccination and other practices that would keep the "unfit" alive, for fear of further degeneration and ultimate extinction.)

Historians of biology have noted that, at the turn of the century, the waxing of "morphology," which had so dominated the nineteenth century, now gave way to the rise of "experimental biology," and the move to the laboratory (Nyhart 1994, 2009; Rainger et al. 1988). As Jacques Loeb put it, we can and must master "the control of life".

Among the most progressive thinkers, one of the ways they thought we might be able to overcome degeneration and ultimate extinction was "eugenics" or "racial hygiene." After WWII, "eugenics" became a "dirty word" (Kevles 1985), and the "just-so-story" that "everyone knew to be true" was that it was a radical, rightwing, Nazi abomination. At one meeting in the early 1970s, where I'd given a talk urging the comparative study of the eugenics societies and movements in more than 30 different countries, the next speaker began by shaking his head, and declared, "A bad idea... We already know far too much about the history of eugenics!" The post-war world apparently had conveniently forgotten that "eugenics" was originally regarded as a "progressive," leftist policy based on Darwinian science, allied with conservation, abortion, and feminism (Indeed, progressive feminist Margaret Sanger was a devoted advocate of eugenics) (Adams 1990a, b, c, 2008).[3]

Along with eugenics came other approaches, such as a new and hopeful science called "rejuvenation." Among the variants were surgical "rejuvenation" from vasectomies or removing female parts ("getting Steinached"), thought to not only improve the behavior of young hooligans, but also to renew the youthfulness of the aged. Another "rejuvenation" strategy, in Paris, was Voronoff's transplants of monkey testicles into the mesothelium of old men. Also in Paris, at the turn of the century, Nobelist Elie Metchnikoff believed that the "natural" human lifespan was 120 years, and could be achieved by eating the "yogurt" he brought back to Europe from Central Asia, which would kill the harmful bacteria in our guts that shorten our lives. Rejuvenation was one of the most exciting "new sciences" in the first three decades of the twentieth century, heavily funded by the elder Nobel, and undergone by Sigmund Freud and many other renowned people. (In the mid-1920s, the most popular show on Broadway was a plot involving three old ladies, and a familiar fourth "young girl" ...an old friend who had gotten "rejuvenated.")

This is a dimension of "Darwinism" that scientific specialists studying evolution have often ignored. But it was not ignored by the "big three" in the Anglophone world who ostensibly "saved Darwinism": Ronald Fisher, J.B.S. Haldane, and Sewall Wright. What guided and inspired their statistical work is evident from

[3] A master's degree I supervised found that in the early twentieth century, a much higher percentage of women were engaged in eugenics research than in any other science (Bettes 1975).

their classic books and articles. Fisher's book *The Genetical Theory of Natural Selection* (Fisher 1930) assumed an infinitely large population and random breeding, but its last five chapters were devoted to his passionate analysis and advocacy of "eugenics."(In scientific circles, the first half of his book is often lauded for "saving Darwinism," while the second half of the book goes unmentioned.)

Haldane went even further. In his famous "Daedalus: or, Science and the Future" (1924), he imagined that humans would have degenerated and gone extinct in a few hundred years, were it not for "ectogenesis," human reproduction controlled by the state through in vitro fertilization, as a successful eugenic tool for creating a stable society and better individuals. His "ectogenesis" idea would be satirized by fellow Etonian, Julian's younger brother, Aldous Huxley, in *Brave New World* (1932). (This was only the first of five novels, satirizing every hopeful scientific way to save humanity from degeneration and extinction, each proving merely to advance the inevitable more quickly!) Haldane's sequel to *Daedalus,* "Last Judgment" (1927), went further into the future, arguing that, to avoid extinction, we must deliberately experiment with humankind to produce future, altered human "species" engineered to live on other planets in order to survive the inevitable, eventual forthcoming destruction of the Earth in a few million years. Ultimately, he argued, the central issue was not the extended longevity of our own individual lives, but the survival of the human race into the far distant future (Adams 2000). (Later, during WWII, he experimented on himself.)

Meanwhile, in the final chapter of his classic *The Causes of Evolution* (1932), Haldane expressed similar views of the human future, and explicitly praised, as the most realistic, Olaf Stapledon's *Last and First Men: A Story of the Near and Far Future* (1930), which was a vastly expanded version of Haldane's "Last Judgment." Stapledon's lifelong philosophical effort to create a plausible myth for the scientific, Darwinian age continued with *Last Man in London* (1932), *Odd John* (1935), and *Starmaker* (1937), soon followed by similar novels by Robert Heinlein (*Beyond This Horizon,* 1942) and Arthur C. Clarke (*Against the Fall of Night,* 1948; and *The City and the Stars,* 1956). By contrast, C.S. Lewis responded to Haldane's futuristic ideas with his "Ransom Trilogy" of novels: *Out of the Silent Planet* (1938), *Perelandra* (1943), and finally, *That Hideous Strength* (1945), in which a character explicitly modeled on Haldane is portrayed as Satan incarnate. Lewis also attacked Haldane's "scientistic" immoral visions elsewhere. In response, Haldane "reviewed" the trilogy, criticizing Lewis for his "complete mischaracterisation of science, and his disparagement of the human race" (Adams 2004).

Envisioning the eventual artificial breeding of superior individuals, Haldane was also one of the first (with Julian Huxley) to describe the genetic basis for human cloning, introducing the terms "clone" and "cloning" in his speech on "Biological Possibilities for the Human Species of the Next Ten Thousand Years" at the *Symposium on Man and his Future* (1963).

The fear of degeneration and extinction was not limited to the Brits: Consider the pre-eminent geneticist H.J. Muller, discoverer of the "arrival of the fittest" by mutations, caused by radiation and chemical mutagenesis. A strong advocate of eugenics as an undergraduate at Columbia, he turned in the 1930s to human "artificial insemination" (as developed for animals). While directing a laboratory in Vavilov's Institute of Genetics in Russia, Muller sent a letter to Stalin, urging him to establish breeding camps to inseminate women with the sperm of great thinkers to accelerate the "Soviet experiment." In his book, *Out of the Night* (Muller 1935), he urged that all societies do the same. (He made his own seminal contribution to such an organization in the early 1960s).

Julian Huxley was another "Darwinian" troubled about its implications for the human future. In *Daedalus*, Haldane had chided H. G. Wells for being behind the times, and not up-to-date with the developments in biology. Quick to rectify his futurist reputation, Wells turned to his son "Gip," and Julian Huxley, to research and write an "up-to-date" work titled *The Science of Life* (1930). Huxley did most of the research and writing for almost 3 years (Huxley 1970), and the resulting hefty three volumes expressed many of the ideas for which he would be famous later in life, including his ideas on "scientific humanism," "religion without revelation," "eugenics," "conservation," and "birth control." The final chapter concluded with the hope that mankind could avoid extinction by leaving the earth and scattering among the stars. In the late 1920s, Huxley also published a science fiction story about human "cloning" (a word he also has been given credit for coining), keeping embryonic cells alive in the laboratory for long periods of time. Throughout his life, he was a staunch supporter of eugenics, serving as Vice-President (1937–1944) and President (1959–1962) of the British Eugenics Society. Fearing human self-destruction resulting from overpopulation, he was a strong advocate of birth control, abortion, and Planned Parenthood, and supported the sterilization of the genetically defective. His concerns and policies were evident in his advocacy as the first head of UNESCO, 1946–1948.

These "visionary biologists" all believed in both evolution and natural selection, but, concerned about the human future, they all advocated seizing control of our destiny through "engineering life" by some kind of "experimental biology." All were leaders in their own fields, and were engaged writing popular science, giving talks and public lectures, and publishing essays and stories addressed to the broad public. All were strongly motivated and obsessed with keeping humanity from degeneration and extinction and avoiding our own likely destiny. The new emphasis on "laboratory biology" in the early twentieth century, then, in many respects was motivated by scientists who accepted Darwin's view and sought consciously to overcome its implications for the human future. Indeed, Muller's famous warnings about the human "genetic load" resulting from radiation from atomic bombs in the late 1940s (which Mayr and Dobzhansky strongly critiqued) were part of this lifelong campaign. Later, Kevles (1985) would term these visionaries part of "reformed eugenics," because most of them were acknowledged atheists and politically leftist. (Haldane was active in the British Communist Party.)

Historians of biology have noted the "abandonment" of nineteenth-century "morphology" and the rise of laboratory-based "experimental biology" in the early twentieth century (Nyhart 1994, 2009). I believe the concerns above contributed to that "shift" of focus. Rather than being passively subject to natural "laws," it was time to create and control our own destiny through "bio-engineering," and this view was spread in popular essays and science fictional tales written through the first half of the twentieth century by "visionary biologists" and their popularizers.

Perhaps diseased people could be "frozen," to be defrosted when a cure had been discovered. When the body had broken down, perhaps "human brains" could be kept alive in electrified vats of nutritious fluid. Perhaps some human variants could be bred or taught "precognition," or "telepathy" with other human minds. Perhaps if we created a language that all humans could easily learn, that "embodied" the way the universe works, we could think faster and eliminate war. Perhaps the elderly could be "rejuvenated" communally by replacing their blood with that of youths (Adams 1989a, b). This was not "unscientific nonsense," nor only the central ideas of science fictional movies and stories—some were written by serious scientists, and every single thing I described above was actually under scientific and laboratory investigation by scientists in Russia and elsewhere (Krementsov 2014).

Looking back at biologists' professional obsessions relating to Darwinism from more disciplinary, narrower points of view, three major issues were debated among "biologists" in the late nineteenth century: evolution, heredity, and development. Specialists like Haeckel, Weismann, or Spencer, studying various topics, explored how and whether Darwin's *Origin* applied to their own work. But by the so-called "eclipse" period and into the early 1930s, the next generation of "sociologists" had turned hostile to Spencer's view, and embraced "nurture" over "nature." Experimental embryologists related similarly to Haeckel, correcting his "faulty" claims ("ontogeny recapitulates phylogeny"), and utterly rejecting his approach, becoming more interested in a variant of "vitalism" similar to "orthogenesis" (Rainger 1991), or to the *Élan vital* ("vital impetus") introduced in French philosopher Henri Bergson's famous, widely read book *L'Évolution créatrice* (1907, English ed. *Creative Evolution,* 1911). Meanwhile, paleontologists were preoccupied with notions of orthogenesis and inner-driven tendencies to explain Huxley's observed "evolutionary progress," which they noted as the increase of fossilized "progressive advanced" forms of life emerging, as evidenced in the fossil record. In temporally successive, newer strata, some types had survived unchanged, others had died out and disappeared, and some seemed to be "new" or more "advanced" forms of earlier types. Many paleontologists (like Cope and Marsh) still believed in a variant of neo-Lamarckism. Not until the 1930s did specialists in various fields "update" recent discoveries in their own domains, and their relationship to their understanding of "Darwinism" and evolution (many re-enacting the age-old academic pattern of "killing the fathers.")

3 Russian Darwinisms

I have concentrated above on the Anglophone world of Britain and the USA, but, of course, the various issues related to "Darwinism" played out differently in different countries and cultures. The developments in Russia and the USSR illustrate this diversity of "Darwinisms" in differing contexts and settings (Adams 1991).

Prior to the 1917 revolution, the Russian reaction to Darwin's *Origin* in Russia was much the same as in other countries: Some were excited by it and accepted it; others felt it incomplete or in need of "fixing"; and still others were critical, denying evolution, and/or the efficacy of "natural selection".

But even proponents gave it a special Russian twist. V.O. Kovalevsky was a Darwinian first who became a paleontologist (using material from, not digs, but museum basements), but Russia's paleontologists were not impressed (Todes 1978). Karl Kessler, the then Rector of St. Petersburg University in the Russian capitol, was aware of the influence of Malthus on Darwin's theory, but was also aware that, although Malthus's argument may apply to the industrial, capitalist, "crowded" British Isles, Malthus himself had admitted his argument did *not* apply to Russia's vast unpopulated territory, where intraspecific competition seemed relatively unimportant. Russian naturalists noted that over its territory, the main dangers to "survival" were predators, and the harsh climate, and that "mutual aid" in "herds" and "flocks" of the animals there provided best protection of the young against both predators and harsh climate by keeping warm (Todes 1989). There emerged a standard "Darwinian" revision among Russian naturalists. When (Prince) Peter Kropotkin was exiled in Britain, he published that standard Russian view in his volume, *Mutual Aid: A Factor in Evolution* (1902), but, although a commonplace to Russians, it was apparently "news" to Brits, and many there classified him and his book as "anti-Darwinian."

Following the 1917 Russian Revolution, however, Russian Marxists adopted "Darwinism" as part of their worldview, in part precisely because it seemed to offer scientific proof against religion, the "opiate of the masses." Indeed, in the late 1920s, the Communist Academy funded secret attempts to hybridize African women with apes using artificial insemination. (When I informed a leading American medical geneticist of this, his first question was, "Were there any offspring?"!) In the 1920s, eugenics (or *evgenika*) was actively studied in Petersburg, Moscow, and elsewhere. But Leo Berg's *Nomogenesis* (1926)—a candidly anti-Darwinian volume (whose English translation Haldane declared "the best anti-Darwinian book in the 20th century")—had difficulty getting published in Russia. There, "Darwinism" was part of the new Communist national ideology.

In 1930 at the beginning of Stalinism, however, "eugenics" became unacceptable, suggestive of Nazi "race hygiene," and eugenics journals and researches essentially "ceased." (Actually, *evgenika* was simply rebranded as "medical genetics," and opened in a large institute just 4 years later.) (Adams 1989a). Still, "Darwinism" was part of official doctrine. But a curious thing happened as the 1930s wore on: a debate began about what Darwinism actually *was,* and which of its variant forms best fitted Marxist "dialectical materialism."

Among traditional biologists, Russian work paralleled Western, and they argued that their approach embodied "dialectical materialism" and was a scientific vindication of Russian Marxism. But, with the rise of Lysenko, a new "form" of "Darwinism" (called "Creative Darwinism," mimicking Henri Bergson's *L'Évolution créatrice* 1907), was forged with Lysenko's philosophical ally, I. Prezent. Their "Creative Darwinism" denied not only "intraspecific competition," but also the existence of genes, and claimed that environmental factors could create new species and genera in one generation, and that "plants" should be clustered tightly together to take advantage of "mutual aid." In the West in the 1930s, "Darwinism" denied the inheritance of acquired characteristics (though Darwin believed in it, especially the heritability of "use and disuse"); in the USSR, "Creative Darwinism" agreed with Darwin, and endorsed it. In the West, "natural selection" was intraspecific; in the USSR, Darwin was wrong about that (his error understandable, since he had lived in capitalist, crowded Britain). In Russian "Creative Darwinism," "mutual aid" held sway. In the West, "genes" now governed what is inherited; in "Creative Darwinism," there are no such things as genes, the very idea critiqued (pejoratively) as the "formalist" "idealism" of "Mendelism-Weissmanism-Morganism." (Russian philosophers knew their history of science.)

By the late 1930s, "Darwinism" was made a scientific discipline in the USSR, and universities opened new departments to teach and study it. In Moscow University, Shmalgauzen (Severtsov's student) became chair, and taught the standard scientific form; in St. Petersburg University, the Lysenkoist Prezent became chair, and taught "Creative Darwinism." The arguments and discussions continued until 1948, when Lysenko's variant was endorsed (and edited by Stalin personally, who had believed in neo-Lamarckian inheritance throughout his life), and "genetics" was officially forbidden until Khrushchev's ouster in 1964 (Adams 2001). Consider: throughout the period, "Darwinism" was accepted as part of Marxist ideology. The only dispute was which of the two completely opposite "Darwinisms" was it?(Roll-Hansen 2015, Adams 1991).

4 The "Modern Synthesis"

In many respects, it was Julian Huxley who initiated the scientific disuse of the word "Darwinism" in scientific, Anglophone culture, in *The Science of Life.* In many respects, he was the ideal choice for Wells, not only because he was the grandson of Wells's teacher, but because he had been engaged in many different scientific approaches (including studies of animal behavior, and ecology), and after writing the book over almost 3 years of study about all branches of "life," he knew even more (Huxley 1970; Waters and Van Helden 1991).

In a single volume edition of *The Science of Life* published in New York in 1934 (a hefty volume of 1514 pages in a tiny font) there are only two references to "Darwinism," covering only seven pages. There, Huxley lists the myriad critics of "Darwinism," noting "To this day, in common talk, Evolution, Darwinism, and Natural Selection are hopelessly mixed and muddled" (p. 429). Elsewhere, he insists that both evolution and natural selection are *facts*, but the notion that natural selection can explain all of evolution is *not* a fact, but a "theory," Darwin's theory, which remains neither proven nor disproven.

The origins of Huxley's (1942) classic *Evolution: The Modern Synthesis* is evident in the volume's preface. Here he mentions that the book derived from an address he gave in 1936 titled "Natural Selection and Evolutionary Progress," and explains:

> Even among professional zoologists the modern conception of natural selection and its mode of operation is quite different from that in Darwin's day, but much of the research on which the changed outlook is based is so recent that the new ideas have not spread far. The idea of evolutionary progress, on the other hand, has been undeservedly neglected. Thus it seemed to me valuable to attempt to give a broad account of the two concepts and their relation to each other.... The result is the present volume. (Huxley 1942, p. 7)

On page 389 of both editions (1942, 1963, 1964), in a chapter on "Speciation, Evolution, and Taxonomy," appears the following text:

> The formation of many geographically isolated and most genetically isolated species is thus without any bearing upon the main processes of evolution.... These major processes in evolution thus consist essentially in ... a progressive increase of life's control over and independence of the environment. Superimposed upon these processes, and having little or no bearing upon them, are the processes of species-formation which are the consequences of accidents in the environment or in the genetic machinery of life. Much of the minor systematic diversity to be observed in nature is irrelevant to the main course of evolution, a mere frill of variety superimposed upon its broad pattern. Species-formation constitutes one aspect of evolution; but a large fraction of it is in a sense an accident, a biological luxury, without bearing upon the major and continuing trends of the evolutionary process.

What does Huxley mean by "synthesis," and by the "reborn Darwinism" he celebrates? The "synthesis" he welcomes is a gradual sharing of discoveries and information from previously isolated branches of biology, previously unaware of what others are finding or doing. He welcomes the contributions of Fisher, Haldane, Dobzhansky, and Goldschmidt:

> The time is ripe for a rapid advance in our understanding of evolution.... Genetics, developmental physiology, ecology, systematics, paleontology, cytology, mathematical analysis, have all provided new facts or new tools of research; the need to-day is for concerted attack and synthesis. (Huxley 1942, p. 8)

Unlike other specialists, who claimed their own fields as *the* evolutionary discipline (Dobzhansky, population genetics; Severtsov, morphology; Mayr, systematics; Simpson, paleontology), Huxley welcomes all who can contribute to advance our understanding of evolution.

Praising this "sharing," he celebrates a "reborn Darwinism," but what he means by it is quite specific. In the first chapter, he writes:

> The Darwinism thus reborn is a modified Darwinism, since it must operate with facts unknown to Darwin; but it is still Darwinism in the sense that it aims at giving a naturalistic interpretation of evolution, and that its upholders, while constantly striving for more facts and more experimental results, do not, like some cautious spirits, reject the method of deduction. . . . By Darwinism I imply that blend of induction and deduction which Darwin was the first to apply to the study of evolution. (Huxley 1942, pp. 13, 28)

In like manner, as a would-be "moderator" for evolution, he had welcomed both Dobzhansky's *Genetics and the Origin of Species* (1937) and Richard Goldschmidt's "critique" of it, *The Material Basis of Evolution* (1940) (which strongly disagreed with Dobzhansky and the so-called "synthesis"). Huxley would also welcome the forthcoming classic by Ernst Mayr, *Systematics and the Origin of Species* (Mayr 1942), and G.G. Simpson, *Tempo and Mode in Evolution* (1944), which he also disagreed with, and would delight in the Darwin Centenary celebration in Chicago in 1959.

Yet, despite coining the key terms they would use, Huxley was not part of the core group that would formulate the so-called "Evolutionary Synthesis" that was being organized and formulated by Dobzhansky, Mayr, and Simpson in the post-WWII years. Their correspondence (archived at the American Philosophical Society) indicates that they wanted to use Huxley's "title," while keeping him as uninvolved as possible in the "Society for the Study of Evolution," as well as their new journal, "Evolution" (Ruse, personal communication). They immediately recognized that his approach and beliefs were not theirs.

5 "The Evolutionary Synthesis"

I have detailed my becoming an advocate and teacher of the "evolutionary synthesis," in a paper largely written around 1988 (based on my personal knowledge of James Watson, Matthew Meselson, Ed Wilson, Ernst Mayr, G.G. Simpson, Th. Dobzhansky, and Stephen Jay Gould) (Adams 2021). There I describe the agreements and disagreements among those publicly endorsing the new, "hardened" synthesis, and I showed that Darwin's major problem ever since, in every period, was the failure to address the "macroevolution" of higher taxa, while the "speciation" question remained contested territory throughout. And in another piece (Adams 2021), after teaching to my students the "evolutionary synthesis" "just-so-story," I described what led to my "epiphany" in the 1980s, realizing that what I had believed and taught for decades was wrong.

There is no need to detail or reiterate either paper here. Instead, now, some 30 years later, I think I have a fuller understanding of the "evolutionary synthesis," "just-so-story" phenomenon. I now regard it an enormously productive (but ultimately unsuccessful) "history," and have pondered how, when, and why it unraveled, and where we are now.

The "Intellectual Migration" The inter-war period (1919–1939) was a troubled time. The Russian Revolution, the Depression, and the rise of Stalin and Hitler led many to flee imminent dangers. The period witnessed the so-called "intellectual migration" of leading European thinkers in many fields to the UK or the USA. Scientists often argue and discuss their work as simply rational "ideas," but scientists are human, have careers, worry about financial support for their research, seek professional advancement, and are subject to public, political, and institutional demands of differing cultures. Some "follow the money," some "go along to get along," some stay and fight, others leave for somewhere else that they hope will be safe and best suit their wishes or needs. When they arrived, they brought with them knowledge and traditions from their homelands that differed from the particular Anglophone world they were moving to, and they had to accommodate to new cultures.

Among those coming to America were not only Albert Einstein and George Gamow, but also Ernst Mayr and Richard Goldschmidt (from Germany), and Theodosius Dobzhansky (from the USSR). Upon arrival, they must have realized that the USA was not Europe, and that it was a welcoming but alien setting. The Depression posed no obstacles: Mayr found a job as a curator at New York's American Museum of Natural History; Goldschmidt, a professorship in genetics at Berkeley; and Dobzhansky, membership in the Morgan group in California, and a professorship at Columbia. Before World War II had ended, all had published important books on evolution. All had come from countries largely devoid of fundamentalist religious leaders who fought against the teaching of evolution, but their adopted country was different.

The aftermath of the Scopes Trial in Tennessee was resonant, undermining the teaching of evolution in biology high school classes (due to textbook publishers, who avoided the topic in all their high school biology texts, rather than having to create different editions for each state).These émigré biologists also knew that in this unusually religious and democratic country, with large fundamentalist and evangelical populations, many Anglophone evolutionists were self-identified agnostics or atheists, and eugenicists. They also were aware that genetics was "in," but "Darwinism" was not, and the most "scientific" science at the time was "physics" (a "model science" whose primacy was reinforced by the subsequent development of the atomic bomb).

Dobzhansky Theodosius Dobzhansky (who insisted his friends, colleagues, and students call him "Doby") was the first to publish, and his classic *Genetics and the Origin of Species* (1937) was in retrospect a masterful accommodation with his new setting (Adams 1994a, b). First, his time with the Morgan group gave him "genetics" authority. Yet (as Mayr soon discovered), Doby was a fellow "naturalist" and systematist, and he also, independently, agreed with Mayr, endorsing the "biological species" concept (Krementsov 1994, Konashev 2023). In 1927–1930, in his weekly letters to his mentor, Iurii Filipchenko, Doby was highly negative and suspicious about the Moscow "Chetverikov" group as opposed to Filipchenko's group (Adams

1994b, Konashev 2023). But, after Chetverikov was arrested and exiled to Siberia, Doby, now in the USA, began to praise his work, especially for his group's testing their ideas by studying natural populations (Adams 1990c). Doby introduced, to the West and to Mayr, Chetverikov and his 1926 paper by translating it (a somewhat Bowdlerized, edited, and updated version) into English. As it happened, Chetverikov was not a statistician (that was his brother), but he was also a "naturalist" and "systematist" who discovered that wild populations hide considerable genetic recessive variability (Adams 1968a). Understandably, he became Mayr's "favorite" population geneticist (aside from Doby): Mayr attacked "bean-bag genetics" throughout his life, pointing out, correctly, that natural selection acts not on individual "genes," but on "phenotypes."

Actually, Doby was also a strong supporter of "*evgenika*" in his youth. During a weeklong interview I had with "Doby" (where he insisted I call him that) at the Mather Camp in Yellowstone, Doby greatly praised his Leningrad mentor. At that time (1925–1927), Filipchenko was head of the Soviet "Eugenics Bureau," and Doby was his assistant director (Adams 1990c). After coming to work in the Morgan lab, he attended various US "eugenics" meetings, and in his weekly letters to Filipchenko, he indicated that American eugenics seemed to him exceedingly racist, not at all like his mentor's voluntaristic, more Galtonian "*evgenika.*" For him *evgenika* (Filipchenko's Russian eugenics) was desirable, but *eugenics* (the American variant) was not. Later, in his book *Mankind Evolving*, his take was that human adaptive advancements were accelerated as a result of language, which could quickly spread change through science and collective social sharing of information and advancement, rather than traditional evolution through natural selection, which took a far longer time.(Curiously, his analysis of speciation in other animals did not consider that they may also have ways of communicating, teaching, and learning, which allowed much more rapid "adaptation" than the "genes" he focused on. In fact, not only was "macroevolution" not explained, but even "speciation" remained "disputed territory.")

Nor was "agnosticism" or "atheism" a problem. He was publicly a follower and believer of Russian Orthodoxy, often addressing American Greek Orthodox churches, and would later state, "I'm both a creationist, and an evolutionist," explaining that the evolution was the Creator's method. From the time of his 1936 Jessup lectures, and his 1937 book, he consulted with Wright, whose calculations he checked against the realities in natural populations, paying no attention to his statistical calculations, which he admitted he didn't understand (Provine 1986).

In his introduction, and his first chapter, he brings up the "issue" of microevolution vs. macroevolution:

> Experience seems to show, however, that there is no way toward an understanding of the mechanisms of macro-evolutionary changes, which require time on a geological scale, other than through a full comprehension of the micro-evolutionary processes observable within the span of a human lifetime and often controlled by man's will. For this reason we are compelled at the present level of knowledge reluctantly to put a sign of equality between the

> mechanisms of macro- and micro-evolution, and, proceeding on this assumption, to push our investigations as far ahead as this working hypothesis will permit. (Dobzhansky 1937, p. 12)

But, even more revolutionary was Dobzhansky's "redefinition" of "evolution." In his first chapter, he declares categorically: "Since evolution is a change in the genetic composition of populations, the mechanisms of evolution constitute problems of population genetics" (Dobzhansky 1937, p. 11). Thus, in one great rhetorical finesse, he defined "evolution" as microevolutionary change, specifically a change in the genetic composition of populations; and then he made a daring claim of territoriality for a new field he called "population genetics" by declaring it to have primacy in dealing with the mechanism of evolution. On this basis, the book subsequently asserted that the problem of the origin of species is resolved.

In retrospect, it proved to be a brilliant and consequential move. Consider: When it is uncertain whether population genetics can illuminate what most people mean by "evolution," and evolution itself is under fundamentalist attack for being just a "theory," what to do? Dobzhansky's answer: kill two birds with one stone—redefine "evolution," and do so in a way that makes it a proven fact!

But Goldschmidt sought to counter Doby's book with his own, *The Material Basis of Evolution* (1940), divided into two sections: "Microevolution" (where he deals with intraspecific natural selection, which he endorses) and "Macroevolution," where he argues that "natural selection" *cannot* explain major changes in species which would be selected *against* until they are fully formed (a long-standing argument against "natural selection").

In his own book (1942), Julian Huxley praised both Doby's book and Goldschmidt's, as "contributions" to the gradual "synthesis process" he was encouraging. By contrast, the growing number identifying as supporting the "evolutionary synthesis" attacked Goldschmidt as an "outlier" and, following the "hardened" (Gould) or "constricted"(Provine) synthesis in the early 1950s, he was ignored and treated as a "crackpot." (A popular book then was *Fads and Falacies in the Name of Science*: Lysenko and Velikovsky were included, but it was said that Goldschmidt could also have gotten in.) The "hardened" synthesis of Dobzhanky no longer emphasized "trying to see how far his working hypothesis could go"; it couldn't be disproved, so Doby's "gene pools" became a core idea of the synthesis, and "micro/macro" suddenly became useless language.

In addition, Doby was a person most colleagues liked, and was a close friend of L.C. Dunn (from Columbia), botanist G. Ledyard Stebbins, Sewall Wright, and many others studying other phenomena in many diverse forms of life. As a professor at Columbia, he welcomed teaching both undergrads and grad students, building his own "school." It was in a co-authored paperback arguing against racism and eugenics with Dunn and Dobzhansky (1946) that he first introduced the word "gene pool," translated from the term "genofund" ("gene fund") Serebrovsky had coined in the late 1920s, in arguing for eugenics, and to trace the human migrations in central Asia by analyzing the genetics of their domesticated chickens (Adams 1979). By the early 1950s, Doby had generalized the word into a core element of the newly "hardened" "evolutionary synthesis."

Mayr An avid bird collector from childhood, Mayr's family aim for him was medicine, but Gustav Stresemann convinced him to switch to "biology," and helped him arrange collecting trips to the wild, especially where Wallace had worked in New Guinea. In the early 1930s, Mayr was funded by the American Museum of Natural History to collect there, and later was hired as the museum's ornithology curator. Based in New York, he became friends with the other curators, among them G.G. Simpson. Mayr told me that, after WWII, he shared lunch with Simpson every day. When I asked him whether they were discussing evolution, Mayr said, "No. We never ever discussed science at lunch." So, what did they talk about? "Baseball. Only baseball!" Mayr kept up with the literature, including Julian Huxley's edited volume, *The New Systematics* (1940); believing that Huxley made mistakes, he wrote his own classic, *Systematics and the Origin of Species: From the Viewpoint of a Zoologist* (1942), to "correct" it.

Getting a sense of the USA from New York, Mayr was aware that museums were low down among the "sciences," but he believed that naturalists' study of diverse living forms was indeed science, albeit a different kind of science from physics, or other "lab" subjects. His book drew attention from Harvard, which hired him as a professor in biology in 1953 (the year Stalin died, Watson and Crick announced their "double helix" structure of DNA, and Jonas Salk announced his polio vaccine in Ann Arbor, then my home town). As was Harvard's habit, the university usually hired those who had achieved greatness somewhere else; it hired James Watson as professor of biology in 1955. His arrival proved an enormous threat to field biologists and other biological disciplines. As Wikipedia accurately described Watson's arrival:

> Watson championed a switch in focus for the school from classical biology to molecular biology, stating that disciplines such as ecology, developmental biology, taxonomy, physiology, etc. had stagnated and could progress only once the underlying disciplines of molecular biology and biochemistry had elucidated their underpinnings, *going so far as to discourage their study by students.*[Italics added.]

My freshman year at Harvard in 1962, Watson won the Nobel Prize, and his campaign against his fellow biologists intensified.

At one point, a small E. O. Wilson seminar I was auditing was constantly riddled with students breaking into the room, interrupting, and asking where Watson's seminar was being held. No wonder Wilson once described Watson as "the most unpleasant human being I had ever met," but in a later TV interview said that he considered them friends and *their rivalry at Harvard "old history" (when they had competed for funding in their respective fields)* [Italics added.](Wikipedia). Watson's campaign had enormous energy behind it, and, in many biology departments, when "classic" biologists retired or died, they were being replaced by "molecular biologists." Right through the night at Harvard, the only buildings lit then were the Watson labs trying to break "the genetic code." Gar Allen told me that at a Harvard faculty meeting he attended, Watson had declared that the only biological "science" was molecular biology, and all the other biologists (looking at Mayr) were just "stamp collectors."

Immediately following Watson's hiring, Mayr acted, defending "natural history" and "evolutionary biology" as equally scientific, but of a different kind, publishing many articles and books, including philosophical articles, a long history of biology, and scientific books expanding on the new, hardened "evolutionary synthesis." In 1959, he was able to hire G.G. Simpson as professor and curator of Harvard's Museum of Comparative Zoology from 1959 to 1970 (although Simpson moved to Arizona in 1968). Mayr had hoped that he would train undergrads and grads and create a "school" of paleontologists at Harvard, but (as he learned) that was not Simpson. From Mayr's point of view, Simpson had no students.

Despite his stern, demanding treatment of museum grad students, Mayr was always charming and welcoming to historians of biology. Later, he told me that, when he died, his legacy would be shaped not by his own students, but by the historians of science he taught. And another time, when I asked him how he wished to be remembered and what his "legacy" would be, he answered (surprisingly), "That is up to historians like you!" In fact, upon his retirement, his Festschrift dinner and celebratory publication was organized by more than a dozen historians of biology he had taught. (His "museum" students were few, and weren't interested.)

Simpson Perhaps the most perplexing of the three core advocates of "the synthetic theory of evolution" was George Gaylord Simpson. He originally wanted to be a writer, but he became fascinated by digs and fossil finds where he grew up. His personal life was troubled, and he was often sick or wounded, and hospitalized (e.g., having a leg mashed by a falling tree while on a South American "dig," from which he almost became lame). He wrote many books, and won many awards. He also had various distinguished academic positions, but was rarely there, always anxious to go digging in the USA, South America, Russia, wherever. What he did do was write excellent, readable books, and dig. Although he was a "professor" at Columbia, Harvard, and Arizona, he disliked teaching, and had few if any students. Early on, he had a passion to "correct" American paleontology, from the Marsh and Cope "competitive" days and traditions, and especially from orthogenecist Henry Fairfield Osborn, the head of the American Museum of Natural History 1890–1935 where Simpson had been a curator of geology and paleontology (Rainger 1991). His "mission" was to "update" paleontology, and obliterate its neo-Lamarckism and especially orthogenist obsessions (perhaps a strong case of "killing the fathers."). He could be stubborn, opposing Wegener's "continent drift" theory until the data on "plate tectonics" became overwhelming.

Simpson's classic book *Tempo and Mode in Evolution* (1944) was begun in 1938, and published when he was still in WWII military service (in intelligence), thanks to the help of his wife, Dr. Ann Roe. Nonetheless, it was open, candid, fresh, and interesting, full of visuals and ideas. The third of seven chapters was entitled "Micro-Evolution, Macro-Evolution, and Mega-Evolution." There, he declares paleontology the core evolutionary subject, but attacks Goldschmidt, suggesting that the difference between micro- and macro- is trivial, with macro dealing largely with species and genera, compared to "mega-evolution" (a word he coined for the occasion), stating candidly that the "mega" (the sudden appearance in the fossil record of entirely new *Baupläne* without plausible prior ancestors) is what is not explained by "natural selection," however well it works on the "micro" or "macro" level. (It was a trick, of course, since Goldschmidt's "macro" included Simpson's "mega.") But it is a great "contribution" to the gradual synthesis process of sharing information and expertise among formally isolated specialties, which Julian Huxley welcomed, whether he agreed with it or didn't. (I don't believe he did, since he needed an explanation of "progress.")

It is in the second edition of this work, published in 1953 as *The Major Features of Evolution,* where he turns on a dime. Now, he claims that the terms that "micro," "macro," and "mega" should all be dropped, as they had outlived their usefulness. Now, although he admits there is no possible way to prove it, he "believes" that natural selection is the cause of all evolution . . . while also warning the readers that not a thing in his book is proven or provable, so readers may choose to believe whatever they want to! (Simpson had been unhappy at the American Museum of Natural History since the late 1940s, feeling abused and unappreciated by his boss and associates, and rumors had it he was "shopping" for a more welcoming situation. Was Simpson's 1953 "turn-about" (the same year Mayr went the Harvard) a signal to his old "baseball friend" that he might be willing to join him? I don't know.)

It is this "switch," in both Doby's and Simpson's publications in the early 1950s (Dobzhansky 1950, 1951a, b; Simpson 1953) that constitute what Gould later called the "hardened" version of the "evolutionary synthesis," or what, later still, Will Provine called the "constricted" version. It was that "hardened" version that I learned in Ed Wilson's introductory course, "Evolution" (as Gar had, a year or two before me), and elaborated in the graduate seminar on the "evolutionary synthesis," taught jointly by Mayr and Simpson (and a year or two after Gar took it).

6 The "Hardened" Evolutionary Synthesis

I can still remember the "hardened synthesis" I learned in the 1960s, in Ed Wilson's "Evolution" course. It goes like this: A "species" is a group that can hybridize and interbreed (the "*biological species*" concept). The unit that "evolves" is a *population* of a species. The population contains a *"gene pool"* that governs the transmissible traits in the population. New genes can be added to the "pool" by *mutation* and *migration*; they can be lost by *random drift* and *natural selection.* If parts of the population become *isolated* from one another (physiologically, geographically, ecologically, behaviorally, or by niche) the different effects of *drift, natural selection, mutation,* and/or *migration* on each can evolve each differently over time, producing two groups that can no longer interbreed and produce fertile offspring, yielding two distinct species. Over long periods of time, this is the *cause* of all evolution, with "natural selection" producing both increased diversity and much extinction.

Mayr, Dobzhansky, and Simpson all realized what some popular writers ignored, and still ignore: natural selection does not act on "genes," but on "phenotypes." Eliminating all non-genetic transmissible features (neo-Lamarckism, Orthogenesis, etc.), and unable to test in the laboratory what happens over millions and billions of years, the only remaining possible or acceptable cause of evolution, they claimed, is natural selection.

But the "downside" is that all three core advocates (Mayr, Doby, and Simpson, as well as Ed Wilson and Julian Huxley) were all field *naturalists* and *zoologists*. Plants, fungi, bacteria, and viruses don't behave or reproduce the way sexual animals do. A second problem: the way "genes" relate to "phenotypes" is *expressed through developmental biology* (formerly called "embryology"), but, as Mayr once remarked, "They weren't particularly anxious to join us, and we weren't particularly interested in having them join us." Unlike the late nineteenth-century "Darwinists" such as Haeckel and Weismann, whose approaches encompassed heritable variations, evolution, and embryology, the "synthetic theory of evolution" left a key piece out of the puzzle (Adams 1980b, 1988).

Many "went along" with "the synthetic theory of evolution," but at its core were Doby, Mayr, and Simpson. Why? Because the hardened "evolutionary synthesis" had become not a gradual ongoing "process," but a fixed "doctrine" that purported to explain microevolution (population genetics, covered by Doby), speciation (covered by Mayr), and macro- and megaevolution (covered by Simpson). All were leading figures in their fields, so the new "big three" constituted solid pillars that could sustain and legitimate the new "doctrine."

So, why did so many accept and teach it for so long? When did it "unravel," and why? As I wrote in a previous article (Adams 2021):

> Throughout its history, the word "Darwinism" has been used in radically different ways, in different countries, by different people, for different purposes, to mean different things.... Like other named "isms," the term's use is always problematic, because its meaning and utility are always granularly context dependent.

So, what constituted the "granular context dependence" that allowed the "synthetic theory of evolution" to flourish for so long, and when and why did it unravel? The context, I believe, was both a time and a place: post-WWII America, and Harvard, which became, almost simultaneously, the central core of the synthetic theory, and the place where the history of biology first appeared as a new discipline, thanks to a timely arrangement between Ernst Mayr and Everett Mendelson.

7 The Hybrid Nexus

The results of the war brought "Big Science" to the USA, and new funding opportunities for the sciences. (In letters to L.C. Dunn, Dobzhansky urged him to see if his evolutionary studies could be funded by the Atomic Energy Commission (AEC), since radioactive mutations were an opening.) Physics was the immediate concern (given the war's—and Cold War's—jets, rockets, computers, and atomic bombs), but most believed that the USSR's developments were the result of Russian spies. But in 1957, with the successful launching of Sputnik, things quickly changed. I was in a basketball game in junior high when, suddenly, we heard over the speakers that the Russians had launched Sputnik, which was traveling over the USA at that moment. We kids were shocked, and angry. The "Russkies" were literally looking down on us. How could that be? I was never very good at basketball—that was my older brother—but this striking episode reinforced my fascination with Russian history and science.

The Eisenhower administration moved quickly. Funds poured out of the federal government to support science; new paperback science and history of science books by leading experts were published for use in primary and secondary schools; the National Defense Education Act (NDEA) provided grants to college and grad students studying science, engineering, and Russia; "Advanced Placement" (AP) high school classes in college English, Math, French opened (I took all 3). And the possibilities and competitions for science funding intensified. The history of science experienced new growth, and the "history of biology" was born in the 1960s as a new discipline. And Harvard was at the center of it.

In earlier times, the "history of science" concentrated on "intellectual history" and the "history of ideas." The journal *ISIS* was founded in 1913 by George Sarton, who once turned down a paper on technology dealing with the history of "tinkers," explaining: "We're not interested in tinkers, we're interested in thinkers!" Prior to the late 1950s, almost all the accounts of the history of science were formulated by the scientists themselves, from the "presentist" point of view, focusing less on "true history" than on the "genealogy" of how their current progress had come to be. The scattered few "historians of science" before then were incidental members of history departments, and most focused on the history of the more distant past, usually concentrating on the so-called "scientific revolution" or on physics (then, the "model science"). Those who studied and taught the history of biology were few and far between, often focusing on medicine or physiology. (Note: there is no Nobel Prize for "Biology.")

It was only the late 1950s and early 1960s that saw the major expansion of "the history of science" as a professional discipline in the USA. As president of Harvard from 1933 to 1953, James B. Conant emphasized the importance of education on science and its history. A protégé of Sarton, I. Bernard Cohen (1914–2003) was the first American to earn a Ph.D. in the "History of Science"; he taught at Harvard 1942–2003, inviting Alexandre Koyré for lectures and courses; he himself studied the "Scientific Revolution," especially Newton. In 1960, Cohen founded "History of Science" as a Harvard department, hired Everett Mendelsohn that year, and created an undergraduate major. Charles Gillespie (1918–2015) earned a Harvard Ph.D. in 1949, and was hired by the History Department at Princeton. In the 1950s, he published two influential books (*Genesis and Geology,* 1951; and a textbook, *The Edge of Objectivity*). Thomas Kuhn received his Ph.D. in physics at Harvard the same year, 1949, and taught a "history of science" course as a Harvard Junior Fellow at the suggestion of President Conant, which was when Kuhn decided to become an historian/philosopher of science. Hired by the Berkeley History Department, he published his most famous and influential book, *The Structure of Scientific Revolutions* (1962), and introduced the term "paradigm shift" into our lexicon. He was hired back to the Princeton History Department and its newly organized "Program on History of Science."

The earliest tranche of historians of biology—Everett Mendelsohn (1931–2023, Harvard), Robert Olby (1933–2020, Leeds), Bill Coleman (1934–1988, Johns Hopkins), and Rob Richards (1942–, Chicago)—were first hired by universities soon after the "Sputnik" debacle, and it was a time when whole departments and "programs" dedicated to the "history of science" were created, along with graduate training and undergraduate majors. By 1970, five major universities had searches open for assistant professors in the history of biology: Berkeley, Chicago, Penn, Princeton, and Toronto.

I was on the younger side of that "history of biology" group (1944-, Ph.D. Harvard 1973), along with Garland Allen (1936–2023, Ph.D. Harvard 1967), Fred Churchill (1932–2017, Ph.D. Harvard 1967), Jon Hodge (Ph.D. Harvard 1970), Will Provine (1942–2015, Ph.D. Chicago 1970), Polly Winsor (1943, Ph.D. Yale 1972), and other grads of those first institutions. Frank Sulloway (1947, Ph.D. Harvard 1978) was a bit younger, as were Sharon Kingsland, Jane Maienschein, and many others.

As a freshman in 1962, I was fascinated by Everett Mendelsohn's history of science course, "Science and Society," and selected as my major the newly created "History and Science," which required coursework in a history specialty and a science: I chose Russian history, and Biology, and took advanced courses in Russian and Russian history, as well as "Life" (George Wald), "Evolution" (Ed Wilson), and "Genetics" (Mathew Meselson), and later, the graduate seminar co-taught by Mayr and Simpson on "The Evolutionary Synthesis."[4]

---

[4]Of course, at the time, I knew nothing of the history and background of the special "nexus." Raised in Ann Arbor (a classic 50:50 "town and gown" city), the public schools were excellent, many teachers were spouses of University of Michigan professors, and most of my friends were their sons and daughters (from philosophy, anthropology, computers, physics, chemistry, psychology, astronomy, French, English, etc.). I had an early membership in both the Science Book Club and the Science Fiction Book Club, and had read the paperbacks published in the 1950s by Huxley, Doby, and Simpson. The Russian history professor lived next door with his young wife and baby (Prince Lobanov-Rostovsky), who once served me Russian tea and summarized Russian history for me (from Peter the Great to the 1917 putsch). Jonas Salk had worked there, and as a kid I knew Michigan had a great football team (the stadium was just down the street), and every Saturday home game us kids were allowed in for free. I knew I wanted to become a professor somewhere, and assumed after high school, I'd enroll in the University of Michigan, as did many of my high school friends, but three older friends had gone to Harvard, and I visited them to have a look. When I got into Harvard, I assumed that most universities had "history of science" programs and "biology" departments focused on evolution. And so, at Harvard, when I became aware of the programs at Hopkins, Princeton, Yale, and Berkeley, I was not surprised. So, I thought no matter what university I went to, I'd probably take the same kind of courses I ended up taking. Only later, as a new "historian of biology," did I become aware of the uniqueness of the evolution/history of biology nexus I was lucky enough to have had access to, or became aware of the prior events that had

Curiously, throughout the 13 weeks of 3-h sessions, Simpson sat at the head table, but Mayr taught the course. At the end of each session, he would turn to Simpson respectfully and ask, "Professor Simpson, is there anything you would care to add?" Every time, Simpson would shake his head no, without uttering a word. I attended every session, early, and Simpson uttered only 20 words the entire semester (though it was supposedly a "joint" seminar!). How strange (I thought at the time): the seminar was on "the evolutionary synthesis," and I got to attend a course co-taught by two of the three core people who created it, yet Simpson had nothing to add, and Mayr never mentioned fossils! Surely, Simpson had much to add; why did he have nothing to say or contribute to that seminar?

Later, reading his autobiography (Simpson 1978), I was curious about what he wrote about his years as a professor (especially at Harvard). Of 24 chapters, only one—the 14th—covered his life as a professor. In it, he never mentioned any courses, teaching, or students. He once mentioned how he hid in his professor's office during days and evenings writing his books, and once about getting locked into the building late one night. The rest of the chapter is devoted to the dozens of digs he remembers fondly from those years! So, for the first time, I understood Simpson's behavior in that seminar.[5]

In that seminar, for our final required research paper, it was Mayr who suggested I do my paper on Sergei Chetverikov (a Russian population geneticist on a par with Fisher, Haldane, and Wright, and, in some ways, ahead of them, but largely unknown in the West). Simpson suggested V.O. Kovalevsky (a Russian Darwinist paleontologist from the nineteenth century). I took Mayr's advice, but years later I suggested Kovalevsky as a topic to my first "major" and graduate student at Penn, Daniel Todes.

In the 1960s, Mendelsohn was working to create the first *Journal of the History of Biology*, but Mayr wanted to be the editor. Apparently (as Gar Allen explained to me then), they came to a mutual "understanding." Mendelsohn remained in control of the journal, but any history of science graduate student interested in evolution would be supervised by Mayr, leaving any grad students interested in other areas of biology to Mendelsohn. Mayr gave me an A on my Chetverikov paper, and it was published in the first issue of *Journal of the History of Biology* (Adams 1968a), followed by an article about his contribution to the synthesis (Adams 1970).

happened before I was born, or during my childhood. Before college, I learned almost everything from "books"; once at Harvard, it became courses, and mentors, and I was a witness and minor "participant" as to what was happening, and can contribute to the "historical" story from then on. Hence, from Darwinism's "history" above to the "up close and personal" of whom I knew, what I witnessed, what I heard and saw, and what I thought and did to contribute to the history henceforth.

[5]During a trip to Moscow many years later, my hotel concierge told me to go the room of a new American visitor, who needed my help translating. To my surprise, it was Simpson, planning a "dig" in Siberia. He understood spoken Russian, but he couldn't speak it. As always, he was abrupt, didn't remember me, and neither thanked me nor said goodbye. I took no offense, and wasn't surprised: as I had learned many years before, that was Simpson.

From Gar Allen, I learned that, as a supervisor of dissertations, Mayr would not approve any chapter that he regarded as "inaccurate" or "wrong." Although Mayr took an active interest in my research, my dissertation dealt with Russian genetics and molecular biology during the Lysenko years (Adams 1978), so Mendelsohn was my official supervisor.

In retrospect, the Mayr/Mendelsohn arrangement seemed a unique, successful "marriage." In a sense, Mayr's career was dedicated to establishing the scientific legitimacy and nature of evolutionary biology, field natural history, and "classic biology," and its continuous funding and support, despite the "molecular biology" threat posed by Watson. And, despite its omissions and failings, many "classical" biologists were happy to become associated with the "hardened synthesis" for that reason, despite their disagreements among one another. In that sense, Mayr's focused campaign was successful: "classic biologists" remained and continued to be funded, and were still there when Watson left Harvard for Cold Spring Harbor.

For many years, Mayr had the opportunity to train and support the work of historians of biology, whether from Mendelsohn's department or visitors from elsewhere. Throughout the subsequent years, he supported my work, introduced me to Doby, and helped get me access to Russian sources.[6] Meanwhile, he succeeded making Harvard's biology department the "center" of evolutionary biology, including the appointments of Ed Wilson, G.G. Simpson, Richard Lewontin, and Stephen Jay Gould. The many history of biology students he mentored were hired by other departments elsewhere, spreading the word.

Both Mayr and Dobzhansky stayed focused, keeping clear of incidental academic "politics" that might have taken "their eyes off the prize." When Wilson's *Sociobiology* was attacked by both Gould and Lewontin (and Gar), both Doby and Mayr took no side in that public squabble. Although Mayr was an atheist (as were Simpson and so many others), he never advertised his beliefs, nor did he ever debate religious opponents, as did many others. They remained almost totally unengaged with eugenics (and debates about it). And Doby regarded himself as religious. Mayr went out of his way to encourage and help younger historians of science who visited Harvard during a sabbatical, or a visit for a week, semester, or year. Even those who disagreed with his "evolutionary synthesis," or were researching other subjects, testify that he helped them enormously, which they greatly appreciated. He showed interest in their work, made helpful suggestions, led them to sources, and could be quite humble, charming, and engaging. Throughout the period from 1960 through roughly 1990, he encouraged and welcomed the "Darwin Industry," the gradual complete publications of Darwiniana, all the while helping shape the story that many biologists still teach to their classes about evolution.

[6] He did so by "legitimating" me as his student, and urging his Russian colleagues to help, which was crucial to my success: my "Soviet handler" had been spreading the rumor that I was an Israeli spy from Mossad (because of my "Jewish name," Mark Adams!), and my hotel room was "bugged" in the "fire alarm"!

8 The Unraveling

Some of the failings of the "hardened" synthesis were already evident in 1974. Preparing with Provine for the papers that would lead to *The Evolutionary Synthesis: Perspectives on the Unification of Biology* (Mayr and Provine 1980), Mayr organized two meetings (in the Spring and Fall).[7] At both meetings, interesting doubts surfaced. Coleman asked, "What is the 'evolutionary synthesis', and exactly what did it synthesize?"(At the time, the remark reminded the audience of what it *hadn't* synthesized, for example, embryology, botany, ecology, bacteriology, virology, ethology. . . .) Gould critiqued the "hardened version" of the synthesis; he had already republished Goldschmidt's contrary book with a semi-supportive introduction (Goldschmidt 1982), and had joined with Niles Eldredge in arguing for the "punctuated equilibrium" theory and the "unfinished synthesis" (which Mayr strongly opposed).

Also indicative were those who declined invitations to the meetings: everyone was invited, but not everyone came. Simpson didn't attend, nor (if I remember correctly) did Stebbins, nor did any developmental biologist. Apparently, they felt "unappreciated" (Smocovitis 1996), and no longer willing to play along with things they disagreed with (perhaps because they no longer needed to).

In subsequent years, the pendulum had swung: former "advocates" came to strongly disagree with what they had once vigorously endorsed. My own "epiphany" came around 1988 (Adams 1990a, b, c, 2021). Mayr remarked to me in a phone call that "we may have gone too far" with the evolutionary synthesis. After reading my 1990 paper, developmental biologist (and friend) Scott Gilbert co-authored a call for "evo-devo" (Gilbert et al. 1996), part of which came almost verbatim from my "samizdat" (which is what he called it; he couldn't read the French version published in 1990, although he cites it). Since then, Gilbert has constantly updated his widely used textbooks on "evo-devo" and the new "eco-evo-devo," trying to put "developmental biology" back in the "synthesis" in the hope that it might result in an explanation for both "macroevolution" as well as Simpson's "megaevolution."

By the early 2000s, even Mayr's co-editor, Will Provine, had vigorously renounced his former views. In the 2001 edition of his famous first book, *The Origins of Theoretical Population Genetics* (1971), he added a 14-page "Afterword" renouncing "most of what he believed" when he first wrote that book. He no longer believed that speciation and macroevolution have been satisfactorily explained, and mentioned a dozen other issues that rendered the "evolutionary synthesis" completely suspect. His last book was *The "Random Genetic Drift" Fallacy,* which attacks the work of Fisher, Haldane, and Sewall Wright, the man whose biography he had written (Provine 1986). To quote his own words (Provine 2005: vii):

[7]I was at both sessions in 1974, and contributed two articles to the book (Adams 1980a, b). It was at the first meeting I met Doby, and the next summer I interviewed him for a week at Mather Camp, Yellowstone.

> Much of my life has been devoted to the history of population genetics. My early book was my Ph.D. thesis still in print: *The Origins of Theoretical Populations Genetics* (1971, 2nd edition, 1991). I stated in the 2nd edition in the Afterword that "random genetic drift" was giving me pause, as does the evolutionary synthesis. My later book was *Sewall Wright and Evolutionary Biology* (1986) and is also still in print. Now I am writing this book against "random genetic drift," invented by R. A. Fisher and followed by Sewall Wright and J. B. S. Haldane. "Random genetic drift" is the core of population genetics. Any person who believes in "random genetic drift" should read this book.

I suspect what Provine had problems with was Wright's famous 1932 "diagram" of variously high "adaptation peaks." At a meeting, during a break, he suddenly asked *me* to explain it to *him.* I suggested that, as Wright's biographer, *he* was the expert on the subject. Agitated and almost yelling, he responded: "Explain it to me, try! You can't! I can't! Nobody can!... Its nonsense!" Alas, as for me, Wright's "random drift" was one of my favorites.

As I often insist, I am not a biologist nor a philosopher, but an historian, so I am neither qualified nor interested in declaring what evolution is or how it works. But, I do like to follow the unfolding "history," and I can (like Provine) share my views, for whatever they are worth. As to my view of the status of Huxley's "Modern Synthesis" as a gradual, continuing process, I have hope for "evo-devo" and "eco-evo-devo" developments, but during the last decade, researches concerning nutrition and COVID-19 have, I think, "killed" the "gene" fixation and obliterated the "hardened," "constricted" "synthetic theory of evolution." How?

We now know that the operation of genes is regulated, not only by other genes, and by the immune system, but also (predominantly?) by chemicals produced in the digestive track, coming from the "microbiome" of many different kinds of bacteria and viruses, which, from "digesting" what we eat in our guts, exude off the active chemicals our bodies can use, and which can also "regulate" our genes. Bacteria and viruses can move and hybridize new combinations rapidly in a gut and in body cells. We also know that identical twins can have radically different health conditions, weights, and appearances, because of their radically different microbiomes. I presume, then, that natural selection can act differently on genetically identical genotypes, but utterly different "phenotypes," partially formed by their microbiomes, and what they eat! All of which may well depend on what is available to eat, when and where.

The complexity of such systems includes heritable genes, to be sure, but genes and "gene pools" are a much smaller part of the process than "the evolutionary synthesis" ever imagined, and development, ecology, and ethology evidently play a much bigger role than previously considered. In a sense, Julian Huxley's image of his gradual "synthesis" process over time, sharing discoveries from many different, semi-isolated fields, is still active, but the "hardened" (or, in Provine's word, "constricted") synthesis is no longer viable. The *process* is still continuing, but the *doctrine* seems dead.

9 Who Creates the History?

When and how did the historical information about "Darwinism" come about, from 1859 until the present? Almost all we now reliably know has been published since the 1960s, most written by historians of biology. Why not earlier? Because "the history of biology" as a discipline was not in existence before the 1960s! But there were histories written about Darwinism throughout. Who wrote those?

The "historical information" available prior to then were usually accounts made not by historians but by scientists themselves. Each successive generation looked back on the development of their field from their own point of view, to serve their own agendas, summarizing their received "story" about the genealogy of their field as it related to their current work and interests, and taught them by their scientific mentors. Often, such accounts were "just-so-stories" used to introduce their subjects to their students, passed down (and updated) generation after generation.

Alongside this, with the accelerating transformation of science beginning in the nineteenth century, and exploding throughout the twentieth, "science" has diversified into "sciences," each of which produced new "disciplines," which in turn yielded new "specialties," and newly named "fields," "topics," and "problems." In doing so, working scientists became increasingly "silo-ed" (e.g., from "astronomy" to "cosmology" to "astrophysics" to "astrobiology"; from "alchemy" to "chemistry" to "organic chemistry" to "biochemistry" to "molecular biology"; from "natural history" to "morphology" to "biology," to "experimental biology," to "cell theory," to "genetics," etc.). And, in almost every case, the protocol for "scientific publishable papers" in top journals rejected the earlier patterns of casual discussion and argument, and instead demanded beginnings with a short history "situating" the current paper by what led to it. That "history," in turn, is passed on, edited, simplified, creating over time the scientists' version of their own history, taught by their mentors, and passed on to their students through the standard "historical introductions" common to textbooks, training future scientists with what they "needed to know" about the triumphal advances that led to present advanced frontiers of the field. In almost all cases, what was presented was not "history," but the "relevant past" of the current work, more genealogy than "history."

That approach is sometimes termed "Whiggish history" or "presentism."Consider a cat, who climbed a tree to get away from the dog that had chased it, who looked up from the ground. Looking down, the cat sees a single route it took from the ground to the branch it sits on, but the dog, looking up, sees endless and diverse branches, and many plausible paths. Metaphorically, the cat's view is "Whiggish" or "presentist," repeating as "history" the route it chose, but the dog, looking up, sees a tangle of splitting branches, any of which the cat could have taken. Put another way, the "cat" saw history from the viewpoint of "here, now," whereas the dog saw the potential situations unfolding from the "there, then."

A "history" written by a scientist is usually not based on a study of the history. Their mission is to do research, and publish, in an attempt to answer the question, "how does the universe work"? They rarely have the time or inclination to become a careful "historian," especially since they already think they know—from their reading and training and mentoring—the standard "just-so-story" which they, in turn, teach their students.

There were and are, of course, a few who are really interested in history, and have written excellent, intriguing "historical" analyses. In the nineteenth century, Merz (a German trained Scottish engineer) wrote a fascinating four-volume study of the different scientific and philosophical problems and national "styles" of science in various European countries (Merz 1904). There were "philosophers" who analyzed and compared various scientists in search of the most reliable "method" to gain knowledge. Just recently, a lab scientist did a "deep dive" based on a trunk full of papers (including personal and professional letters, notebooks, and a treasure trove) of William Bateson papers that were thought lost, and "discovered" by him and his co-author in the attic of a Massachusetts barn, the basis for a beautifully documented, long, year by year biography of Bateson (Cock and Forsdyke 2008, 2nd ed. 2022). If they care, and are interested and motivated, they can produce very good history.

However, the "norm" of professional scientists is usually much more absorbed by the scientific research they are engaged in, understandably. Teaching courses to graduate students generally focuses on the understandably "presentist" things they are mastering; teaching to undergrads is more likely to be "satisfied" by "just-so-stories" that suffice.

The new breed of "historians of biology" was busy exploring in detail Darwin and other historical figures. Yet, those same historians often did not challenge the "evolutionary synthesis." And, as a result, for many years, many (along with me) accepted our own "just-so-story" of the "evolutionary synthesis." At the time, it was understandable: we were newly minted "historians," but the "evolutionary synthesis" work was not "history"—it was living, current science, and its authors were teaching us, and *making history*, often literarily, writing their own rich "histories of the evolutionary synthesis" themselves, as well as creating their own "philosophies" (and "philosophers") of evolutionary science. (Looking back, shortly before his death, Mayr said he thought his most important contributions were not his "historical," but his "philosophical" works.)

As time passed, new generations of historians of biology emerged, and the early ones were already well established in their careers... and what they had earlier learned and experienced had *become* history. Perhaps that is why it was in the mid-1980s that the worm began to turn, and many of those who had most embraced it in the 1950s, 1960s, and 1970s were ready to join in the unraveling.

A frequently asked question I get is: why did so many historians of biology who should have known better keep teaching and endorsing the "just-so-story" for so many years? Smocovitis (1996) has suggested it came from a broader mission to unify science, and to unify biology. Others have blamed the historians: how can scientists raise doubts about the "just-so-story" if they think that leading biologists endorse it, and so do most historians of biology? Perhaps we were too easily coopted. Did we desire to get "recognition" from biologists, "go along to get along" as a safe way to promote our careers, reciprocally endorsing the views of our mentors and, in return, being endorsed by them? Those ideas are plausible. But I think they are too "presentist," too "Whiggish."

It is easy to spread the blame when a building collapses, but when it seems solid, there is no blame to spread! It is like Will Provine's remark that if two past people had collaborated instead of fighting and arguing, Darwinism could have been saved decades earlier (Provine 1971). Maybe, maybe not, but. . . they didn't, and it wasn't. If the "hardened" evolutionary synthesis had collapsed earlier. . . what then? No different story to replace it?

When pondering those questions, I was one of those historians who taught the "corrected" "just-so-story" for 20 years; so did many of my colleagues. So I rehearsed in my mind how my own career and research unfolded, as well as how the research and teaching of my history of biology colleagues developed.

Why did it take me and my cohort of historians of biology so long to drop the just-so-story of the "evolutionary synthesis"? For years, young historians of biology and graduate students were taught the "hardened synthesis" by its creators. The "evolutionary synthesis" was a complete story of all the developments (post-Darwin), and there was nobody who was part of that history that Mayr, Doby, and/or Simpson didn't know about, in detail. It was the "just-so-story" the first "historians of biology" had learned when young.

But, as their researches and careers proceeded, most historians of biology focused on investigating *individual* scientists or specific subjects that interested them, all of which were only a small part of the "complete" story. Gar went after Morgan; Churchill went after Weismann; Richards went after the Romantics, Goethe, and Haeckel; I went after Chetverikov, then eugenics, then "sci & lit," then Doby; Provine went after Fisher, Haldane, and Wright; Sulloway chose Darwin, Freud, and the Galapagos; Maienschein picked embryology; Winsor, classification.

If you read their writings, you will notice that they end up "correcting" the relevant "mistakes" about their person or subject the just-so-story assumed or promoted. Their work needs no apology: much of it is excellent! But a mistake about one scientist or another was NOT evidence to scrap the whole "myth," only to "edit" their special "piece" of it. Consider: "What the evolutionary synthesis assumes about Adolph von Schnekt is mistaken... so let's throw out the whole damn thing!" Nope: Where is the evidence that "the whole story" is wrong? Are you a qualified "counter" to Mayr, Doby, and Simpson, and can know what is wrong about everything? Do deep dives, and share with us what you learned. If the garage door is broken, you have no justification to tear down the whole house.

A good example is Gar Allen. Mayr almost did not pass his dissertation, disagreeing with only *one chapter* on Morgan! Mendelsohn, Gar, and Mayr talked, and it got passed. (That is why Gar usually said his dissertation supervisors were both Mayr and Mendelsohn.) And, in his teaching, and the new books he has been preparing for many years (much of which I've read), Gar ended up "changing" lots of what the "evolutionary synthesis" claimed, with interesting and surprising "dives" into Mendel, Gulick, Richard Wagner, Karl Marx, neo-Lamarckism, Goldschmidt, and even Lysenko. But he still stuck with the "story," because it was the only coherent story around, and his many "corrections" did not change the whole, complete standard "just-so-story."

Only after reading my "Little Evolution, BIG Evolution" paper (Adams 2021) was he ready to convert and toss out the story. And why? Three things: I proved that those scientists who ostensibly "supported" the story, in fact, didn't (some of which he had already uncovered). I suggested a plausible substitute for the "complete" just-so-story of "ever since Darwin" with a new one to replace it! And my "revision" made sense to him because it synchronized with what he had already learned about the people and topics he had studied.

It is one thing to share your historical findings with scientists interested in their "true" history; most of us would welcome the opportunity. It is also great to share your findings with other historians of biology and evolution. When I shared my 2021 paper with leading history of biology colleagues who, like me, once accepted, endorsed, and taught the "evolutionary synthesis" (Gar Allen, Michael Ruse, Lynn Nyhart, Frank Sulloway, Robert Richards, Jon Hodge, and many others), I was surprised and gratified by their responses. All told me it was fair and compelling, and agreed that I was right (with occasional minor quibbles on the side), both about the "evolutionary synthesis" and the replacement of the standard "just-so-story" with a more correct historical account. (In a few cases, it led some to "redraft" things they were about to publish.). As with Gar, when they had read my article, it was consistent, and resonated, with their own researches.

10 Does It Matter?

Many historians of biology feel it is vital and important that current biologists learn the "true" history of biology, replacing their "just-so-stories" with the actual facts. It is an understandable wish. Having worked to come up with "actual history," there is a natural yearning to share the information with scientists who understand the scientific details. And some believe that knowing the "true" history of their field will somehow make them better scientists. Yet, many of those historians who seek to "share" their knowledge with scientists are often frustrated by their disinterest (e.g., Smocovitis, private communication). Which raises an intriguing question: Does it matter?

Two things seem clear to me. First, I am happy to share my findings with anyone interested (and I imagine that is true of most historians). I am also open and happy to engage fellow historians in discussions and disagreements where our views differ. But trying to thrust my work down the throats of people who are not interested strikes me as foolish, uncivil, counter-productive, and impolite.

Second, sometimes this effort is predicated on the belief that it is important that scientists know their "true" history because it makes them better scientists. Some scientists claim that; some historians do. I do not. Let me explain why.

Let's start with the delightfully amusing article in *Science* by Stephen J. Brush many years ago, entitled "Should the History of Science Be X-Rated?" (Brush 1974). It details the many physics textbooks' opening chapters that emphasize Isaac Newton and his "clockwork universe." When he asked them why their physics texts so often open with a history chapter highlighting Newton for his "clockwork universe," they told him it was vital to know the field's past and traditions leading to the current achievements, so their students can become better physicists! Of course, little did they know that a recent biography of Newton (funded by the NSF) had gained access to previously unavailable Newtonian papers and archives. What they showed was that the prior 100+ biographies of Newton had been wrong. The "clockwork universe" was "French Newtonianism," a creation of the French Enlightenment, and historically false about the man and his actual beliefs. In fact, Newton spent much more time studying alchemy than gravity, and 10 times more on his unorthodox religious concerns, studying Biblical and ancient religious texts to determine the date of Creation (4004 BC), and its Ending (we're currently into it!). He was fascinated by Solomon's Temple. Thus Brush's title: If it is vital for the future of physics that students be taught Newton's clockwork universe, "Should the History of Science Be X-Rated?"

When I was a grad student, a popular book on Darwin was by Penn anthropologist Loren Eiseley, *Darwin's Century* (Eiseley 1961). I hadn't read it when, in 1968, Frank Sulloway and I spent the summer in South America to document and film what Darwin reported discovering, seeing, describing, and learning in his *Voyage of the Beagle.* We carried with us (and almost memorized!) many passages from the *Voyage,* and also had copies of every page and letter Darwin had written in those voyage years.

After returning to Cambridge, I read Eiseley's compelling book and immediately realized it had so many misunderstandings, falsehoods, and mistakes about the voyage that it wasn't worth reading; he was apparently simply stating a "just-so-story" about Darwin, no doubt passed down from his own mentor. When I first came to Penn, 1 year Eiseley was the "acting" chair of my department, and, as undergraduate chairman, I met with him: a charming, helpful man with a large book collection. He was an anthropologist who wrote a popular book, "spreading the word" in readable prose for the broader public. Unfortunately, not being an "historian," the "word" he was spreading was wrong. Did it make him a worse anthropologist? Perhaps, but I don't think so; it just made him a bad, careless historian.

When teaching the history of biology, I teach what I know. Indeed, one time I gave my class an assigned competition: during the spring break, they were to read *Darwin's Century* and count up all the things he claimed that were wrong! (It was a good way to see how many students were absorbing and understanding the course.) The student that could correctly count the most actual mistakes won. (They lost points if they thought that something he wrote was false when it wasn't). When teaching the history of science, teach it!

When I came to Penn in the early 1970s and was chatting with a scientist who taught that kind of history to introduce his subject (he was a geologist), I reminded him that Darwin believed in the inheritance of acquired characteristics, and the results of "use" and "disuse," and he quickly corrected me: "No that was Lamarck. You're confusing Darwin with Lamarck!" Which led me to ask (perhaps impolitely) "Have you ever read anything that Darwin ever wrote?" He fell silent, and admitted that he hadn't... but claimed he didn't need to, to believe in "Darwinism"! (Or to teach it?) That was the early 1970s. Thirty-five years later, I co-taught a course with the biologist who taught the "evolution" course for the Penn biology department. To my surprise, I got much the same response: they were still teaching the false "just-so-story," with a few molecular biology updates.

Scientists and philosophers frequently have much to say about "history," most of which (in my experience) is simplified, concocted, or false. Usually, such stories are Whiggish and triumphalist (e.g., the "Modern Synthesis"), and almost universally are accepted as what "everybody knows to be true." Looking at the past from such a silo-ed present often distorts the past, imposing modern "divisions" and "words" on periods that lacked both, and hiding the past under misleading simplified "just-so-stories." Much of my life's work has been focused on "just-so-stories" that almost everyone accepted, which, after digging, turned out to be false.

Doing good history is hard, and demanding, and time-consuming. Why on earth would a scientist, who is seeking to understand how *nature* works, spend much time thinking about their own history, surely by now (in their view) advanced from and improved upon? Consider the hypothesis that knowing their true history helps them become better scientists. For the last 200 years, physicists were taught a false history about Newton. Have physicists over that time been stymied? Would they have been better, brighter, more productive had they known the "true" Newton? (Would it have made them more religious?). I see no causation. Indeed, I don't even see a correlation! If I thought I was on the trail to figure out how nature works, I couldn't care less whether my "mythic," gerrymandered "just-so-story" was right or wrong. I'd have a much more important mission than those "pedantic historians" do. Physicists have been doing important, brilliant work, all the while concocting, advocating, and believing triumphalist "just-so-stories," routinely.

As an historian, over time I learned many things I did not know when I first began, and was lucky enough to learn some useful lessons. Among them, "just-so-stories" are almost always wrong, and especially when "everybody knows they are true" (because if everybody knows them to be true, nobody considers studying or challenging them). Don't believe or accept secondary sources without checking the primaries. (I got a copy from the H. J. Muller archives; when "quoted" by Sonneborn at Muller's memorial, he lied, omitting all reference to "eugenics" in the original manuscript.) Don't assume words in the past mean what they mean today (Fleming 1967, Adams 1979). (Any guesses what Darwin meant when he wrote in his notebook, "*solutions of continuous structure*"?) And (following the admonitions of both Bacon and Descartes), "Doubt everything!" So, my motto: when you next encounter any "just-so-story" that everybody ostensibly knows to be true, remember: "It ain't necessarily so!" (And often, perhaps usually, it is *not.*) That is my advice, to historians.

A colleague, Richard Delisle, asked me a question about this: "What will happen if within the next 20 years the narrative of historians and the narrative of biologists continue to diverge away, to the point of incompatibility?" I thought about it, and here is my answer. The future is not determined. But, on the basis of my experience, I don't think things would change very much. The "Evolution" course taught by the biology department was never an "evolution" course: it was a course in "population genetics," updated by relevant molecular biology discoveries. The biology professor's contributions and lectures were about breeding and speciation. He explained what "natural selection" meant, and worked it out dealing with breeding and speciation, and never mentioned or taught anything about "macro-" or "mega" evolution. He stuck with Doby's 1937 approach: "evolution" is any change in gene frequency in a population, so evolution is not a "theory" but a proven "fact." And in one lecture, he briefly attacked "eugenics," and then hypothesized about how humankind had evolved, and might evolve in the near future.

I wouldn't claim his "history" is bad or false; I would say the name of the course ("Evolution") was wrong. Its subject stayed close to what biology departments and laboratories deal with. And, if criticized, they would probably say: "Macroevolution? Megaevolution? Paleontology? Fossils? You're talking to the wrong department. Go talk to the Geology Department!" (Which is now named the "Environmental Studies" department.)

I also asked a close friend who is in that department, a paleontologist who specializes in dinosaurs, what did he think of Simpson? ("For my generation, he was our God!"). How has the "Modern Synthesis" affected paleontology? He answered: "I wasn't originally aware of how much variability existed in wild natural populations. So, I ended up reducing the number of "species" from 12 to 7, and the number of genera from 9 to 5." How did he explain "megaevolution," or the sudden appearance in the fossil record of new *Baupläne* that lack any plausible precursors, or the evolution of the higher taxa? ("Nothing. We just don't know yet what causes it. Simpson was right: we may never know. With all the extinctions over billions of years, we haven't got sufficient evidence, and may never. So much extinction, so much forever lost in the past. It might be natural selection, and it might not.") What struck me about those two courses (one in the biology department and the other in geology) is that there wasn't much to complain about either one of them.

A real evolution course? The best "Evolution" course? Perhaps it would be overflowing with guest lecturers who come from different places and disagree with each other. That way, students will learn of the "different ways of seeing and understanding." No harm of "compromise pro-mingling" in such a course: bring in paleontologists, geneticists, botanists, bacteriologists, virologists, ornithologists, a couple philosophers, a couple of historians, and a broad sample of the different "kinds" of biologists, including molecular biologists, population geneticists, microbiome researchers, anthropologists, etc. (That would be a course I'd love to audit!)

In some ways, I suppose I'm selfish. Both a distinguished historian of technology and another of medicine admitted they were attracted to their fields "to change the world." But that's not me. From what I've studied, efforts in the past to use the most up-to-date science to "change the world" have often been harmful, and usually counter-productive. I am simply curious. I want to *know* what really happened in the past, and I want to *know* "how the universe works," for myself. Frankly, historians who want to "change the world" raise doubts for me about their historical reliability, because they may have allowed their political "activism" to pollute their history. If they approach it knowing what they want and expect to find, to "verify" or "justify" their politics, I fear they are likely to exhibit "confirmation bias."

I conclude this section with a question asked of me. One Sunday, I got a notice from UPenn that the coming Thursday, there would be a big conference on "Evolution" for undergrads, grad students, the broader public, and all those who are interested. There would be three principle speakers: a molecular geneticist, an evolutionist, and a young philosopher of biology, who I knew and respected—the fellow is very knowledgeable and informed about science and evolution. And it suddenly occurred to me that I hadn't sent him my 2021 paper! So I emailed him, sent him a copy, and jokingly urged him to read it before the Thursday meeting, because it might change his mind! Apparently, he took my advice, because on Tuesday, I got an email from him.

There, he told me he had read my paper carefully, and he found it compelling. Despite his teaching the "just-so-story" in his classes, he said he had always been uneasy about Doby's "rhetorical finesse" (the name he used) redefining "evolution" as any change in the gene frequencies in a population's "gene pool." That wasn't what he regarded as "evolution." So, I thought, he is "converted," but then, what would he say on Thursday? A clever fellow, he anticipated my question with another.

He concluded by asking me this: *"If the old story is untrue, what are we supposed to teach our students now?* (A good, relevant question!) How would you answer him?

11 Final Thoughts

Judging from Google Ngrams, "evolutionary biology" seems alive and well, but "Modern Synthesis," "evolutionary synthesis," and "synthetic theory of evolution" seem to be fading, and "Darwinism" remains a rare word in "evolutionary biology," a generic term which has largely replaced it. By contrast, in the broader public, "Darwinism" remains everywhere, completely overwhelming all competitors, and has done so for more than a century and a half (see Fig. 3).

Meanwhile, today, many biologists still teach the "just-so-story" that the historians and some biologists have successfully discredited. To a historian and observer, that is not something to challenge or correct; that is something interesting! Is it not our job to observe and study the way science works and operates, and not to bias or judge it? As to the future of evolutionary biology and Darwinism, we don't know. Life is full of unexpected surprises, and the future is unpredictable.

But I've always been curious: *Can we use evolutionary thinking to understand how evolutionary thinking "evolves"?*

Another favorite of mine was Wright's view that evolution proceeded most rapidly in small, semi-isolated populations. Does that not fit how science has evolved, with small semi-isolated groups making unexpected discoveries? I wrote a paper applying those ideas to the history of science (Adams 1968b), and philosopher Steven Toulmin invited me to co-author a book about it. Toulmin followed up

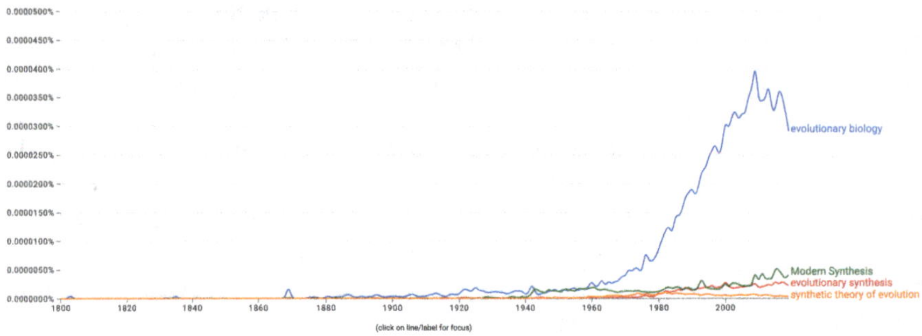

Fig. 3 The expression "evolutionary biology" is doing well in the literature, whereas expressions such as "Modern Synthesis," "evolutionary synthesis," and "synthetic theory of evolution" seem to be fading

emphasizing the "evolution" of science in his own publications, challenging Kuhn's analysis of "scientific revolutions" and "paradigm shifts."

At Harvard, Toulmin even publicly debated Kuhn on the topic, "Evolution, or Revolution?" in front of a large audience of several hundred people. I was at that debate. In their opening statements, Kuhn went first, and explained what a revolutionary "paradigm shift" was. Toulmin's opening statement responded by giving a quick history of the gradual "evolution" of the meaning of "paradigm," from 300 AD to Kuhn's use of the term. Kuhn was furious, and accused Toulmin of "cheating." The audience, however, was much amused.

I turned down Toulmin's invitation, but I was not the only one trying to apply evolutionary theory to the history of science. Others explored their own versions of the "evolution" of science or biology, including Nash (1967), Gillespie (1968), Richards (1987), Hull (1988), Basalla (1988), and Sapp (2003), among many others.

I wonder: Were we right in applying evolutionary models to evolutionary science itself? Will the outcome be determined by "natural selection" and the "survival of the fittest"? Will science evolve more quickly in small, semi-isolated settings? As selection acts not on genes, but on phenotypes, will the coming history of science depend not only on the surviving "ideas," but also on the surviving scientific "phenotypes"—the disciplines and scientists themselves? Which will advance and prosper, gaining interest, participants, and funding? And which biologists and disciplines will wane, shrink in numbers, vitality, viability, and relevance, losing out in the scientific-social competition, possibly getting reorganized or split up, and simply disappear?

A century ago, "rejuvenation" was an exciting new scientific "discipline" throughout Europe and America. Today it no longer exists. What surprises will the future of "evolutionary biology" and "Darwinism" hold? If "history of science" and "historians of biology" survive and are selected for and prosper, they will have interesting things to watch and study as time passes and the future unfolds. I envy them!

Acknowledgments Many thanks are due to many friends, notably among them: Daniel Todes, Nikolai Krementsov, Garland Allen, Polly Winsor, Joe Maline, and Donald Forsdyke. Thanks also to Richard Delisle and his fellow editors, for their support and patience.

References

Adams MB (1968a) The founding of population genetics: contributions of the Chetverikov school, 1924–1934. J Hist Biol 1(1):23–39

Adams MB (1968b) Towards an evolutionary model of the history of science. 67 pp, manuscript

Adams MB (1970) Towards a synthesis: population concepts in Russian evolutionary thought,1925–1935. J Hist Biol 3(1):107–129

Adams MB (1978) Genetics and the soviet scientific community, 1948–1965. University Microfilms, Ann Arbor

Adams MB (1979) From 'gene fund' to 'gene pool': on the evolution of evolutionary language. Stud Hist Biol 3:241–285

Adams MB (1980a) Sergei Chetverikov, the Kol'tsov Institute, and the evolutionary synthesis. In: Mayr E, Provine W (eds) The evolutionary synthesis: perspectives on the unification of biology. Harvard University Press, Cambridge, pp 242–278

Adams MB (1980b) Severtsov and Schmalhausen: Russian morphology and the evolutionary synthesis. In: Mayr E, Provine W (eds) The evolutionary synthesis. Harvard University Press, Cambridge, pp 193–225

Adams MB (1987) Darwinism as science and ideology. In: Burke JG, Adams MB et al (eds) Science and culture the western tradition. Gorsuch, Scarisbrick, Scottsdale, pp 199–234

Adams MB (1988) A missing link in the evolutionary synthesis (essay review of I. I. Schmalhausen, factors of evolution). Isis 79:281–284

Adams MB (1989a) The politics of human heredity in the USSR, 1920–1940. Genome 31(2): 879–884

Adams MB (ed) (1989b) The wellborn science: eugenics in Germany, France, Brazil, and Russia. Oxford University Press, New York

Adams MB (1990a) La génétique des populations était-elle une génétique évolutive? In: Fischer JL, Schneider WH (eds) Histoire de la génétique. A.R.P.E.M, Paris, pp 153–171

Adams MB (1990b) Aleksandr Sergeevich Serebrovskii. In: Dictionary of scientific biography, supplement II. Charles Scribner's Sons, New York

Adams MB (1990c) Iurii Aleksandrovich Filipchenko. In: Dictionary of scientific biography, supplement II. Charles Scribner's Sons, New York

Adams MB (1991) Through the looking glass: the evolution of soviet Darwinism. In: Warren L, Koprowski H (eds) New perspectives on evolution. Alan R. Liss, New York, pp 37–63

Adams MB (1994a) Theodosius Dobzhansky in Russia and America. In: Adams MB (ed) The evolution of Theodosius Dobzhansky. Princeton University Press, Princeton, pp 3–11

Adams MB (ed) (1994b) The evolution of Theodosius Dobzhansky: essays on his life and thought in Russia and America. Princeton University Press, Princeton

Adams MB (2000) Last judgment: the visionary biology of J. B. S. Haldane. J Hist Biol 33(3): 457–491

Adams MB (2001) Networks in action. In: Allen GE, MacLeod RM (eds) Science, history and social activism: a tribute to Everett Mendelsohn (Boston studies in the philosophy of science), vol 228. Kluwer, Dordrecht, pp 255–276

Adams MB (2004) The quest for immortality: visions and presentiments in science and literature. In: Post SG, Binstock RH (eds) The fountain of youth: cultural, scientific, and ethical perspectives on a biomedical goal. Oxford University Press, New York, pp 38–71

Adams MB (2008) Eugenics. In: Ravitsky V, Fiester A, Caplan A (eds) The Penn Center guide to bioethics. Springer, pp 371–382

Adams MB (2021) Little evolution big evolution: rethinking the history of Darwinism, population genetics, and the 'synthesis'. In: Delisle R (ed) Natural selection. Springer, Cham. chap 8, pp 195–229
Basalla G (1988) The evolution of technology. Cambridge University Press, Cambridge
Berg LS (1926) Nomogenesis, or evolution determined by law. Constable, London. reprint, MITPress, Cambridge, MA, reprint 1969
Bergonzi B (1961) The early H.G.Wells. Manchester University Press, Manchester
Bergson H (1911) Creative evolution (trans: Arthur Mitchell). Henry Holt, New York
Bettes M (1975) Women and the eugenics movement. Masters Thesis, History and sociology of science, University of Pennsylvania
Brush SG (1974) Should the history of science be X-rated? Science 183(413022):1154–1172
Chetverikov SS (1926) O nekotorykh momentakh evoliutsionnogo protsessa s tochki zreniiasovremennoi genetiki. Zhurnal eksperimental'noi biologii ser A 2(1):3–54
Cock AG, Forsdyke DR (2008) Treasure your exceptions: the science and life of William Bateson, 2nd edn. Springer, New York. 2022
DeVries H (1906) Species and varieties: their origin by mutation. In: MacDougal DT (ed) Lectures delivered at the University of California, 2nd edn. Open Court/Kegan Paul, London/Chicago
Dobzhansky T (1937) Genetics and the origin of species. Columbia University Press, New York
Dobzhansky T (1950) Mendelian populations and their evolution. Am Nat 84:401–418. reprinted in Dunn 1951
Dobzhansky T (1951a) Genetics and the origin of species, 3rd revised edn. Columbia University Press, New York
Dobzhansky T (1951b) Mendelian populations and their evolution. In: Dunn LC (ed) Genetics in the 20th century: essays on the progress of genetics during its first fifty years. Macmillan, New York, pp 573–589
Dunn LC, Dobzhansky T (1946) Heredity, race and society. Mentor (New American library), New York
Eiseley L (1961) Darwin's century : evolution and the men who discovered it. Anchor
Fisher RA (1930) The genetical theory of natural selection. Oxford University Press, Oxford
Fleming D (1967) Attitude: the history of a concept. Perspect Am Hist 1:285–365
Gilbert S, Opitz J, Raff R (1996) Review: synthesizing evolutionary and developmental biology. Dev Biol 173:0032
Gillespie CC (1968) Remarks on social selection as a factor in the progressivism of science. Am Sci 56(4):439–450. (first presented at the XIIth international congress of the history of science held in Paris, 25 August–1 September 1968)
Goldschmidt R (1940) The material basis of evolution. Yale University Press, New Haven. CT; reprint 1960 Pageant, New Jersey
Goldschmidt R (1982) The material basis of evolution, introduction by Stephen Jay Gould. Yale University Press, London
Haldane JBS (1924) Daedalus, or science and the future. Kegan Paul, Trench, Trubner, London
Haldane JBS (1932) The causes of evolution. Longmans, Green, London. facsimile reprint 1966, Cornell University Press, Ithaca, NY
Hull DL (1988) Science as a process: an evolutionary account of the social and conceptual development of science. Chicago University Press, Chicago
Huxley TH (1887) On the reception of the origin of species. In Francis Darwin. In: The life and letters of Charles Darwin, including an autobiographical chapter, vol 2. John Murray, London, pp 179–204. p. 197
Huxley JS (ed) (1940) The new systematics. Clarendon, Oxford
Huxley JS (1942) Evolution: the modern synthesis. George Allen & Unwin, London
Huxley JS (1963) Evolution: the modern synthesis, 2nd edn. George Allen & Unwin, London
Huxley JS (1964) Evolution: the modern synthesis, 2nd edn. Wiley, New York
Huxley JS (1970) Memories, vol 1. Harper & Row, New York
Kevles D (1985) In the name of eugenics. Knopf, New York

Konashev M (2023) The Russian backdrop to Dobzhansky's Genetics and the Origin of Species, J Hist Biol, accepted 7 February 2023, on-line
Krementsov N (1994) Th. Dobzhansky and Russian entomology: the origin of his ideas on species and speciation. In: Adams MB (ed) The evolution of Theodosius Dobzhansky: his life and thought in Russia and America. Princeton University Press, Princeton, pp 31–48
Krementsov N (2014) Revolutionary experiments: the quest for immortality in Bolshevik science and fiction. Oxford University Press, New York
Krementsov N (2018) With and without Galton: Vasilii Florinskii and the fate of eugenics in Russia. Open Book Publishers, Cambridge. xxv+666 pp
Kropotkin P (1902) Mutual aid: a factor of evolution. McClure Phillips, New York
Kuhn TJ (1962) The structure of scientific revolutions. Chicago University Press, Chicago
Mayr E (1942) Systematics and the origin of species. Columbia University Press, New York
Mayr E, Provine W (eds) (1980) The evolutionary synthesis: perspectives on the unification of biology. Harvard University Press, Cambridge
Merz JT (1904) A history of European thought in the 19th century. 4 vols. Blackwood, Edinburgh and London
Muller HJ (1935) Out of the night: a biologist's view of the future. Vanguard Press, New York
Nash LK (1967) The nature of the natural sciences. Harvard University Press, Cambridge., especially "The Evolution of Scientific Theories", 254–297
Nyhart L (1994) Biology takes form: animal morphology and the German universities, 1800–1900. University of Chicago Press, Chicago
Nyhart L (2009) Modern nature: the rise of the biological perspective in Germany. University of Chicago Press, Chicago
Provine WB (1971) The origins of theoretical population genetics, 2nd edn. Chicago University Press, Chicago. 1991; 3rd ed. 2001
Provine WB (1986) Sewall Wright and evolutionary biology. Chicago University Press, Chicago
Provine WB (2005) The "random genetic drift" fallacy. CreateSpace Independent Publishing, Scotts Valley
Rainger R (1991) An agenda for antiquity: Henry Fairfield Osborn and vertebrate paleontology at the American Museum of Natural History, 1890–1935. University of Alabama Press, Alabama
Rainger R, Benson KR, Maienschein J (eds) (1988) The American development of biology. University of Pennsylvania Press, Philadelphia
Richards RJ (1987) Darwin and the emergence of evolutionary theories of mind and behavior. University of Chicago Press, Chicago. (see the conclusion, entitled "Darwinism is Evolutionary")
Richards RJ (1992) The meaning of evolution. Chicago University Press, Chicago
Roll-Hansen N (2015) On the philosophical roots of science policy: any lessons from the Lysenko affair? Stud East Eur Thought 67:91–109. Springer
Sapp J (2003) Genesis: the evolution of biology. Oxford University Press, New York
Simpson GG (1944) Tempo and mode in evolution. Columbia University Press, New York
Simpson GG (1953) The major features of evolution. Columbia University Press, New York
Simpson GG (1978) Concession to the improbable: an unconventional autobiography. Yale University Press, New Haven
Smocovitis VB (1996) Unifying biology: the evolutionary synthesis and evolutionary biology. Princeton University Press, Princeton
Todes D (1978) V. 0. Kovalevskii: the genesis, content, and reception of his paleontological work. Studies in History of Biology 2:99–165
Todes D (1989) Darwin without Malthus: the struggle for existence in Russian evolutionary thought. Oxford University Press, New York
van Wyhe J (2019) Why there was no 'Darwin's bulldog': Thomas Henry Huxley's famous nickname. The Linnean 35(1):26–30
Waters CK, Van Helden A (eds) (1991) Julian Huxley: biologist and statesman of science. Rice University Press, Houston Texas
Wright S (1932) Mutation, inbreeding, crossbreeding and selection in evolution. Proc Sixth Int Congr Genet 1:355–366

Richard Owen's Deconstruction of Darwinian Natural Selection

Nicolaas Rupke

Abstract

From Darwin's *Origin of Species* to Richard Dawkins' *The Greatest Show on Earth*, the efficacy of natural selection in evolution has been overrated. The central strategy of the two men and their allies, in coping with the inadequacy of their evolutionary views, has been to typecast their critics as creationists. Misrepresentation, even vilification, has been repeatedly used. The most prominent victim of this argumentation strategy has been Darwin's contemporary Richard Owen who was a public proponent of evolution theory well before Darwin. As a scientist, Owen was Darwin's superior. He cautiously and tactfully promoted a structuralist approach to evolution, as developed by several German contemporaries and predecessors. When Darwin, in his *Origin of Species*, misrepresented Owen as a creationist, the latter defended himself in a(n) (in)-famous critical essay, published in *Edinburgh Review*. Subsequently, Owen clarified his views in a booklet entitled *The Derivative Hypothesis*, comparing and contrasting his belief in "natural rejection" to/with Darwin's "natural selection," and using the Malagasy lemurs as examples to demonstrate pre-adaptational logic in evolutionary development. All the while, Owen managed to establish the greatest cathedral of natural history ever, the BM(NH) in South Kensington, a project that Darwin and his allies tried to subvert. In this chapter, Owen's critique of the Darwinian arguments is presented and contextualized.

N. Rupke (✉)
Georg-August Universität Göttingen, Göttingen, Germany

Washington and Lee University, Lexington, VA, USA
e-mail: nrupke@gwdg.de

R. G. Delisle et al. (eds.), *Unity and Disunity in Evolutionary Biology*,
https://doi.org/10.1007/978-3-031-42629-2_6

Keywords

Owen · Darwin · Natural rejection vs natural selection · Owenian structuralist evolution · Darwin's mendacity · Deconstructing "the struggle for life"

1 Introduction: Doubts About Darwin

The many commemorations of Darwin, his life, the *Origin of Species*, and a medley of related events have been accompanied by panegyrics, witness Richard Dawkins' (in)famous "only game in town."

> We are surrounded by endless forms, most beautiful and wonderful, and it is no accident, but the direct consequence of evolution by natural selection—the only game in town, the greatest show on earth (Dawkins 2009, p. 426).

This hyped praise of Darwin did and does obscure the fact that many serious scientists and scholars, from the time of Darwin's own mid-Victorian days till today, have expressed doubt about natural selection as the major factor of evolution.

Robert M. Young's teachings and writing at Cambridge during the late 1960s and early 1970s, and his later compilation of articles under the title *Darwin's Metaphor. Nature's Place in Victorian Culture* (1985) helped many historians of science think of Darwinism as shaped by a Malthusian common context of biological and social theory. Darwin's evolutionism, Young contended, was irreducibly anthropomorphic, social, political, and ideological. Others, not least the leading Darwin scholars Adrian Desmond, James Moore, and Janet Browne, have fleshed out Young's work and turned his common context, which was defined primarily in terms of ideas, into a more embodied sociopolitical one, extensively taking account of place, time, institutions, career, family, vested interests, rivalries, strategies, and more. An inventory of much of this work, for the most part dealing with Britain, was presented in the collected volume *Victorian Science in Context* (1997). Further studies, by David N. Livingstone, Ronald L. Numbers, and others, expanded the geographical scope of this work to include other parts of the English-speaking world, delineating spaces of science, religion, and politics in which Darwinism was reconstituted to suit the local or regional needs of different constituencies.

As Moore commented, biography has proved to be a powerful contextualizing tool. In several new, substantive biographical studies of both Darwin and his followers, evolutionary biology has been dissected down to the bare bones of its varied ideologies, showing how basic notions in biological science were molded by the cultural, ideological, and socioeconomic forces that shape the world in which the biologists live. This contextualist work represented a fundamental revision of the mostly essentialist Darwin hagiography, spawned by the 1959 centenary celebrations, and has had a deconstructionist or, in a later fashion to enter the history of science, a constructivist effect on our perception of Darwinism. In other words, we see Darwin's theory as constitutively affected by social conditions and, what's more, we understand the literature on Darwin, too, as reflecting, to a significant extent, the worlds of its authors, each with a distinct set of concerns and interests (Rupke 2010, p. 140).

A remarkably perceptive as well as slightly sardonic quotation is from the historian of science John C. Greene, whose *The Death of Adam: Evolution and its Impact on Western Thought* (1959) earned him widespread commendation. Greene observed:

> It is a curious fact that all, or nearly, all, of the men who propounded some idea of natural selection in the first half of the nineteenth century were British. Given the international character of science, it seems strange that nature should divulge one of her profoundest secrets only to inhabitants of Great Britain. Yet she did. The fact seems explicable only by assuming that British political economy, based on the idea of the survival of the fittest in the marketplace, and the British competitive ethos generally predisposed Britons to think in terms of competitive struggle in theorizing about plants and animals as well as man (Greene 1981, p. 7).

A genre of critique of Darwin and Darwinism that perhaps has received insufficient attention is parody / satire. An early author was Charles Kingsley, who in his *Waterbabies* (1862–1863), famously lampooned Owen and T.H. Huxley over their hippocampus controversy (Rupke 2009, pp. 221–224). Much later, the perhaps most brilliant parody ever of Darwin's *Origin of Species* was produced by the German zoologist Gerolf Steiner, pseudonymously under the title *Professor Dr. Harald Stümpke, Bau und Leben der Rhinogradentia* (1961). It introduced what in English became known as *The Snouters* (1967; 1982), whose bizarre, comical anatomy and morphology mocked Darwinian functionalism. Greene, too, followed up on his critique of Darwinism with an extended spoof on evolution and natural selection, *A Nonimous, The Wonderful Adventures of Nat Selleck and Eva Lou Shinn in Sci Fi Land* (2007). It is delightful to read with a cast of characters that includes Chuck Loyall (Charles Lyell), Tom Huxtable (Thomas Henry Huxley), Frank Gallstone (Francis Galton), Ronnie the Fish (Ronald Fisher), Doby Russky (Theodosius Dobzhansky), Dick Dockins (Richard Dawkins), and Mike Beehive (Michael Behe). Even the 45th President of America wouldn't be able to come up with a better set of deprecating nicknames.[1]

This genre of critique highlights weak points in Darwinism. At the same time, it has a weakness of its own, namely that it does not offer a substantive alternative theory of evolution. And yet, serious alternatives did exist, and not merely in French

[1] See also the delightful spoof by Phillip Sloan, in which he personifies Owen in a commemorative conversation with Darwin. "Darwin and Owen at Notre Dame: a conversation after 150 years," Nov. 17, 2009. Sloan, personal communication.

and German, but also in English. Darwin, whose knowledge of foreign languages was poor or non-existent, had at least access to the sources in his mother tongue. Inexcusable is that Darwin refused to engage with an earlier alternative, the Kant–Humboldtian theory of evolution, which attributed evolutionary change to a cosmic and holistic process of logical complexification. This structuralist approach to evolution was familiar to Darwin, because it was made known to the British public, including Darwin, by Richard Owen, the greatest British paleobiologist of the Victorian era. It exposed fatal flaws in Darwin's theory of evolution. For one, it spotlighted the fact that natural selection cannot cope with the fundamental question of the origin of life. For another, natural selection by means of small steps of functional adaptation fails to account for large, logical, predetermined evolutionary changes, as documented by Owen in several brilliant monographs (see below), making use of his magisterial two-volume *Odontography* (1840–1845) on the comparative anatomy and physiology of the teeth in vertebrate animals.

Darwin and several of his followers responded with malice and misrepresentation. They lacked the ability to counter Owen's arguments, but managed to outmaneuver him in the public arena by making his awkward personality and tendency to produce overly complex and contorted phrases the issue. Still, the Darwin bandwagon has rolled on till our own day. To quote Greene's Dick Dockins: "The creationist Owen, . . ., failed to draw the obvious conclusion. Any intelligent designer would have hived off the laryngeal nerve on its way down . . ." (Dawkins 2009, p. 427). The false implication is that Owen was blinded by his alleged "creationism" and did not see obvious facts in a giraffe's elongated neck.

Here I present a slimmed down version of the Owen vs Darwin affair as dealt with in my scientific biography of Owen and other publications (see also Sloan 2003).

2 Owen's Turn Toward Evolutionism

Admittedly, throughout the 1830s, Owen had been an outspoken and well-known advocate of the doctrine of the creation of species, but so were a majority of the fellows of the Royal Society, the Geological Society, and the Linnean Society, including Darwin for much of the time. Having followed the famous debate between Georges Cuvier and Étienne Geoffroy Saint-Hilaire in 1830, he initially had been left unconvinced by Geoffroy's transmutationist ideas, siding instead with Cuvier.

Yet Owen did become a proponent of evolution theory. An early indication that he had abandoned creationist-Cuvierian species fixism was his tolerant, partly appreciative reaction to Robert Chambers' anonymously published *Vestiges* (1844). Owen's friends and patrons, especially the Cantabrigians among them, strongly condemned the evolutionary ideas of the book and were deeply worried about its popularity. The redoubtable geologist Roderick Murchison urged Owen to consider writing a damning piece for the *Quarterly Review*; "a real man *in armour* is required, and if you would undertake the concern you would do infinite service to *true* science and sincerely oblige your friends." Adding to the flattery of his appeal, Murchison continued: "I cannot say how much you would gratify your friends and admirers by this effort, which would entitle you to another niche in the temple of good works in which you already occupy so high a place" (Murchison to Owen, in Rev. Owen vol. 1, 1894, p. 254). But Murchison failed to move Owen who left the public condemnation of *Vestiges* to the Cambridge "stars" Adam Sedgwick and William Whewell. The latter issued a selection from both his history and philosophy of the inductive sciences under the title *Indications of the Creator* (1845), reaffirming the validity of the creationist design argument. Sedgwick went further and wrote a deeply emotive attack on *Vestiges* for the *Edinburgh Review* (Sedgwick 1845).

Owen did not back down. In a series of prestigious public lectures, he reiterated and expanded upon his evolutionism, at considerable risk to his career ambitions. As president of the British Association in 1858, Owen had the opportunity to speak to a national audience of colleagues on issues of his choosing. His address to the plenary session of the Leeds meeting, on September 22, is a fascinating document. Between the lines one can read that Owen had been led away by his bio-paleontological/zoogeographical studies from the doctrine of special creation toward a belief in natural causes, whatever these may have been, and in front of the gathered multitude he declared that the appearance of reptiles, birds, and beasts had to be explained on the basis of "[t]he axiom of the continuous operation of Creative power, or of the ordained becoming of living things" (Rupke 2009, p. 159). Although his language was ornately traditional, his message was indubitably non-creationist. In more recent terminology, Owen put forward a theory of theistic evolution. Judiciously, he declared himself ignorant of precisely what laws of nature had been at work in the "creating" of new species.

More explicitly, he referred favorably to the ideas which Darwin and the multi-talented naturalist Alfred Russel Wallace had only just put forward during the historic meeting of the Linnean Society on July 1 of that year, less than 3 months before the British Association meeting. Owen had cited environmental changes as the cause of extinction; now Darwin and Wallace put the same changes forward as a mechanism for the origin of varieties. Owen quoted Darwin's example of an island on which a canine animal preys on rabbits and to a lesser extent on hares. A slight "plasticity" of the dogs or foxes, combined with a decrease in the number of rabbits and an increase in the hare population, would favor the canines "with the lightest forms, longest limbs, and best eyesight." In the course of a thousand generations, the dog or fox would become adapted to the catching of hares instead of rabbits, just as greyhounds can be improved by selection and breeding. Owen cautioned that the degree of plasticity or variability needed to be established by observation of animals in the wild and that the fossil record should be scrutinized to discover the "ante-types," the "more generalized structures," from which varieties might have originated, "so as to give rise ultimately to such extreme forms as the Giraffe for example" (Rupke 2009, p. 159).

Owen was one of the few naturalists, if not the only one who some time before the publication of the *Origin of Species* publicly alerted the scientific community to the importance of the novel ideas of Darwin and Wallace, and, while not losing sight of empirical obstacles to a theory of descent, he hinted that his own studies of the fossil record might reveal the steps that had led from a more general form to such extreme adaptations as the giraffe's long neck.

This Fullerian lecture "On the extinction and transmutation of species" was printed as an appendix to Owen's Rede lecture *On the Classification and Geographical Distribution of the Mammalia* (1859). With its publication, a period of 15 years had elapsed during which Owen had repeatedly drawn attention to the difficulties which the existence of past and present biogeological/zoogeographical provinces poses, and had highlighted their importance in speculating about a naturalistic origin of species, cautiously drawing attention to Continental theories of autogenesis and self-organization. He had played a major role in the discovery and delineation of the provinces to which some Late Tertiary and Pleistocene mammalian fossils belonged. He had been the leading authority on the issue of a progressive succession of fossil types in opposition to the Lyellian doctrine of non-progression (Owen 1851). He had advanced a non-catastrophist, ecological theory of extinction. And although, with respect to the mechanism of the origin of species, he had consistently preached caution, it was no secret that Owen had nailed his colors firmly to the mast of an origin of species by natural law, holding out against opposition from those who controlled his career advancement.

3 Deconstructing Darwin's Natural Selection

Imagine Owen's feelings when in November 1859 the *Origin of Species* appeared in which Darwin cast Owen in the role of a leading advocate of the immutability of species, lumping him together with outspoken anti-evolution creationists such as Harvard's Louis Agassiz and Sedgwick at Cambridge. Owen was outraged, and whereas he had passed up the opportunity to review *Vestiges* all those 15 years ago, this time he forsook caution and wrote an essay on the *Origin of Species* for the *Edinburgh Review*. It added up to a demolition job. Some have cited Owen's critique as evidence that he opposed evolution (Beer 1963, p. 164), but it shows nothing of the sort. In fact, the review was most of all a forceful reiteration of Owen's own evolutionary stance, to which was added a critique of natural selection. In addition to Darwin's book, nine other publications were included in this review, among them the French translation of Owen's own work on the archetype, his presidential address to the British Association, and his *Palaeontology* which had just come off the press.

Apart from some well-known evolutionists/transmutationists, zoologists had kept aloof from any hypothesis of the origin of species.

> One only [Owen wrote in reference to himself] . . . has pronounced in favour of the view of the origin of species by a continuously operative creational law; but he, at the same time, has set forth some of the strongest objections or exceptions to the hypothesis of the nature of that law as a progressively and gradually transmutational one ([Owen] 1860, pp. 503–504).

In Owen's view, Darwin made the fundamental mistake of confusing the issue of the origin of species by natural law with that of the nature of such a law. He quoted from his *Palaeontology* "that perhaps the most important and significant result of palaeontological research has been the establishment of the axiom of *the continuous operation of the ordained becoming of living things*." With respect to the mode of operation, various ideas had been put forward by the French evolutionists and by the author of *Vestiges*. As these and other possibilities had by no means been exhaustively explored, why consent to Darwin's additional speculation?

> The natural phenomena already possessed by science are far from being exhausted on which hypotheses, other than transmutative, of the production of species by law might be based, and on a foundation at least as broad as that which Mr. Darwin has exposed in this Essay ([Owen] 1860, p. 503).

Not that Owen was committed to any rival hypothesis to that of natural selection; "we merely affirm that this at present rests on as purely a conjectural basis." If the struggle for life indeed represented a mechanism of natural selection, it would have to be demonstrable in such organs of struggle as the antlers of deer. Even though antlers are the most variable appendages of the quadruped, millennia of combat have not produced changes beyond the specific limit. Individual variability obviously exists, but

> [w]e have searched in vain, from Demaillet [Benoît Demaillet, French diplomat and author of the proto-evolutionary *Telliamed*] to Darwin, for the evidence or the proof, that it is only necessary for one individual to vary, be it ever so little, in order to [validate] the conclusion that the variability is progressive and unlimited, so as, in the course of generations, to change the species, the genus, the order, or the class. We have no objection to this result of "natural selection" in the abstract; but we desire to have reason for our faith. What we do object to is, that science should be compromised through the assumption of its true character by mere hypotheses, the logical consequences of which are of such deep importance ([Owen] 1860, p. 521).

In his presidential address to the British Association, Owen maintained that the question of the origin of life, in its simplest forms, "cannot be answered until every possible care and pains have been applied to its solution" (Owen 1858, p. lxxvi). This view was in line with his Kant–Humboldtian belief in a naturalistic origin through self-organization. Darwin, by contrast, left the problem of the primordial origin of life unsolved, wrapping it in a biblical phrase when he stated "that probably all the organic beings which have ever lived on this earth, have descended from some one primordial form, into which life was first breathed" (Darwin 1859, p. 484). This gave Owen a chance to "out-naturalize Darwin," as Neal Gillespie puts it (Gillespie 1979, p. 93).

Owen's self-portrayal as the misrepresented and ignored evolutionist went on till the very last page of his review. The assumption that Owen wrote this critique of the *Origin of Species* in subservience to the powers-that-be, or to his own or his patrons' religious belief in Genesis (Himmelfard 1962, p. 277), is incorrect. Why then had Owen changed his reaction to natural selection since the British Association meeting in Leeds from "could be" to "can't be"? Was it just jealousy, as Darwin himself suggested, and with him many historians since (Beer (ed.) 1983, p. 61)?

Perhaps Owen, who lacked magnanimity, did feel pangs of jealousy when he read Darwin's *magnum opus*. Professional jealousy was by no means the whole story, however. More substantively, Owen fundamentally objected to the concept that a process of chance phenomena is the cause of the origin of life and its history, even though he was willing to assign a limited, subsidiary role to natural selection, just as he did to the Lamarckian inheritance of acquired characteristics. When Darwin lumped Owen together in the *Origin of Species* with a number of vehemently outspoken opponents of evolution, this must have seemed an intentional misrepresentation, part of an anti-Owen strategy. Darwin's correspondence with Hooker, Huxley, Lyell, and a few other friends indeed shows a conspiratorial obsession with Owen, and a tendency to suspect Owen's hand in more attacks on the Darwinians than those for which he was actually responsible. A critique of Lyell's *Antiquity of Man*, for example, published in the *Quarterly Review*, was ascribed to Owen (Huxley 1918, vol. 2, 50), although it was written by John Phillips, professor of geology at Oxford.

Darwin later attributed his "misunderstanding" of Owen's position to the difficulty "to understand and to reconcile with each other" Owen's "controversial writings" (Darwin 1872, p. xxviii), but this was a feeble excuse. The *Manchester Spectator* had experienced no difficulty in perceiving the implications of Owen's advocacy of the origin of species by "natural laws or secondary causes," nor had several creationist fellows of the Geological Society. The influential clergyman-geologist William Conybeare, for one, disagreed with Owen as to the extent of the operation of natural laws in producing organic diversity.[2] Sedgwick, too, was worried about Owen's non-creationist recourse to nature's laws (Sedgwick to Owen, March 28, 1860, in Clark and Hughes 1890, vol. 2, p. 361).

Conversely, the pro-evolution Baden Powell fully understood the naturalistic implications of Owen's *Nature of Limbs* and paid ample tribute to it in his "Essay on the Philosophy of Creation" (Powell 1855, p. 370). Across the Atlantic, the pro-Darwin Asa Gray admitted, albeit with a well-targeted sarcasm, that he had appreciated the drift of Owen's position which had prepared him—Gray—to a certain extent, for Darwin's book (Gray 1973: 63). Moreover, the people who attended Owen's Hunterian lecture courses, the last of which was delivered in 1855, had not come away with the impression that Owen was a creationist. The statesman and polymath George Douglas Campbell, 8th Duke of Argyll, and author of *The Reign of Law*, recalled: "I had heard many of his lectures, and all these had left on my mind an impression that he himself believed that somehow or another different races of animals had descended from each other by ordinary generation (Duchess of Argyll (ed.) 1906, vol. 1, 581)." The Owen pupil and Catholic biologist St. George Jackson Mivart also confirmed that his revered teacher had in no way been averse to the idea of evolution; "on the contrary, he was its decided advocate" (Mivart 1895).

[2]William Daniel Conybeare to Richard Owen, n.d., BM(NH), L, OC 8 (fols 374-75).

Owen's reviled review of 1860 was not what in Darwinian mythology it has become, namely an obscurantist attack on the honest and persecuted prophet of truth Darwin. The self-portrayal by Owen as a misrepresented evolutionist was a true likeness, and his discussion of natural selection correctly exposed a weak link in the chain of Darwin's reasoning, namely the nature of organic variability.

4 Natural Rejection vs Natural Selection

Owen's textbook *Palaeontology* appeared very shortly after Darwin had set the cat among the pigeons, but it did little to restore calm. In the introduction, Owen merely re-stated his axiom "of the continuous operation of the ordained becoming of living things," and the book's conclusion of exactly eleven pages of text was no more than a collage of paragraphs from his presidential address of 1858, his concluding Fullerian lecture of 1859, and some remarks he was about to repeat in his essay for the *Edinburgh Review* of 1860.

Owen's approach no longer worked. He had been content to add to the observational basis of the fact of evolution, keeping speculations about the mode of evolution at arm's length. More clarity about a non-Darwinian alternative was required. In 1862, Owen addressed the British Association "On the characters of the aye-aye, as a test of the Lamarckian and Darwinian hypothesis of the transmutation and origin of species" (Owen 1862a, pp. 114–116). He included an expanded version of this talk in his superb *Monograph on the Aye-aye* (Owen 1862b), having first lectured on the subject to the Zoological Society (Owen 1863a, b). Via the colonial secretary for Madagascar, Owen had received an adult male *Chiromys madagascariensis*. It feeds on, among other things, wood-boring grubs by gnawing down to the burrows with its strong incisor teeth and extracting the larvae with the claw of its attenuated middle finger.

This most remarkable of the Malagasy lemurs (some now regard it as separate from lemurs) provided Owen with a unique opportunity to elucidate his views. He began by clarifying that the succession of species by a continuously operating law was not a "blind operation," but a process of preconceived progress, guided by divine volition. In a footnote he recalled the 1849 fracas with the *Manchester Spectator.* Owen explicitly and lucidly stated that he subscribed to the "derivative hypothesis" of the origin of life by the spontaneous generation of a variety of primordial forms, and the origin of species by organic descent. Such a belief did not imply the adoption of either the Lamarckian or the Darwinian mode of evolution. The following is the first comprehensive summary of Owen's evolutionary beliefs:

> What I have termed the "derivative hypothesis" of organisms [.....] holds that these are coming into being, by aggregation of organic atoms, at all times and in all places, under their simplest unicellular condition; with differences of character as many as are the various circumstances, conditions, and combinations of the causes educing them, – one form appearing in mud at the bottom of the ocean, another in the pond on the heath, a third in the sawdust of the cellar, a fourth on the surface of the mountain rock, etc., but all by combination and arrangement of organic atoms through forces and conditions acting

> according to predetermined law. The disposition to vary in form and structure, according to variation of surrounding conditions, is greatest in these first-formed beings; and from them, or such as them, are and have been derived all other and higher forms of organisms on this planet. And thus it is that we now find energizing in fair proportions every grade of organization from Man to the Monad. Each organism, as such, also propagates its own form for a time under such similitude as to be called its kind.

Furthermore, the "derivative hypothesis" is merely "a kind of vantage-ground artificially raised to expand the view of the outlooker for the road to truth." This road did not run via the transmutational hypothesis, if only because it failed the aye-aye test. Lamarck's notion that the increased use of an organ increases its development through a succession of generations cannot explain the incisors of the *Chiromys.* These teeth already possess their specific size and shape before protruding from the gum. "The great scalpriform front teeth thus appear to be structures fore-ordained – to be predetermined characters of the grub-extracting Lemur; and one can as little conceive the development of these teeth to be the result of external stimulus or effort, as the development of the tail, or as the atrophy of the *digitus medius* of both hands" (Owen 1863a, b, pp. 64–65). As far as the Darwinian "preservation of favoured races in the struggle for life" was concerned, there was no evidence of a shortage of food for, and therefore no selective pressures on, the various lemurs (n.b.: this statement was made before Darwin introduced the notion of sexual selection). "We know of no changes in progress in the Island of Madagascar, necessitating a special quest of wood-boring larvae by small quadrupeds of the Lemurine or Sciurine types of organization" (Owen 1863a, b, p. 66).

Yet Owen did believe that the various *Lemuridae* were related by descent from a common ancestral form. The evidence for the phenomenon of descent existed, in general, in the fossil succession and "the progressive departure from a general to a special type"; in the analogy of embryonal stages in higher species to the mature forms of lower species; and in the "unity of type" which Owen had demonstrated in his homology work. Non-adaptive organs, too, negated an origin of species by primary creation. Owen ended by stating that further work to determine the way in which species originate would prove beneficial to the progress of zoology.

In a notice of the first two volumes of Owen's three-volume *On the Anatomy of Vertebrates* (1866–68), the *London Review* pointed to "a significant though partial admission which he makes of the truth of the principles of Natural Selection" (Anon 1866a, p. 483). The *Popular Science Review* also commented that Owen's book "contains a partial concurrence, on the part of the author, in the theory of *natural selection*" (Anon 1866b, p. 212). In the introduction to the first volume, Owen had repeated his long-standing theory of the extinction of large mammals, enunciated as long ago as 1850 in a paper on the dinornis, namely that they are less well equipped than small species to cope with environmental changes "in the contest... against the surrounding agencies." This time he added the expression that the smaller species "have fared better in the 'battle of life.'" In a "courteous and temperate letter" to the editor of the *London Review*, Owen objected to the imputation of his concurrence with the Darwinian view; if he was thought to have followed Darwin's *magnum opus* of 1859, then the fact that he—Owen—had actually expressed his theory in 1850 would place Darwin in the position of "adopter" (Anon 1866a, p. 516).

Darwin immediately seized this opportunity, in the fourth edition of the *Origin of Species* (1866), to maneuver his opponent into the embarrassing position of having claimed that he—Owen—had given to the world the theory of natural selection long before the appearance of the *Origin of Species* (Darwin 1866, p. xviii). Some historians, uncritically taking the Darwinian line, have accused Owen of shameless inconsistency, anonymously attacking Darwin's ideas and publicly claiming priority in having developed them. Even Roy MacLeod maintained in his pioneering 1965 paper on Owen "that Owen had become more convinced of the probable truth of Darwin's concept, and was trying desperately to regain his fallen prestige by accepting and claiming as his own as much of the Darwinian synthesis as he could" (MacLeod 1965, p. 278). A number of leading Darwin scholars, including Moore (1979, p. 88) and Ruse (1979, p. 228), have put forward the same interpretation of what MacLeod termed "this blot upon his honor."

There was nothing disingenuous, however, in Owen's claims, and the allegation that Owen was a turncoat with respect to the idea of natural selection was a further instance of misrepresentation by Darwin and his allies. Darwin's complaint has often been cited, namely that Owen in his review of the *Origin of Species* "scandalously misrepresents many parts" (Darwin to Lyell, April 10, 1860, in K.M. Lyell (ed) 1881, vol. 2: 94). "I have just read the Edinburgh, which without doubt is by Owen. It is extremely malignant, clever & I fear will be very damaging. He is atrociously severe on Huxley's lecture, & very bitter against Hooker. So we three enjoyed it together: not that I really enjoyed it, for it made me uncomfortable for one night; but I have got quite over it today. It requires much study to appreciate all the bitter spite of many of the remarks against me; indeed I did not discover all myself.— It scandalously misrepresents many parts. He misquotes some passages altering words within inverted commas" (Darwin to Lyell, in Lyell 1881, vol. 2, p. 462).

Owen's review, as we have seen, was calculated to present an accurate picture, not so much of Darwin but of Owen himself. Darwin's accusation has never been critically examined. No Darwin scholar has considered the possibility that Darwin was perfectly capable of misrepresenting Owen. The latter had publicly enunciated his notion of the "contest with surrounding agencies" nearly a decade before Darwin came out of the closet, and in the course of the 1850s, Owen had applied it on several occasions to the extinction of species. During the early 1860s, unwisely perhaps, he changed his original, awkward phrase of "contest against the surrounding agencies" to the Darwinian-sounding "battle of life" and "contest for existence." But this was merely a terminological adoption; at no time did Owen suggest that competition between animals contributes to the origin of species. On the contrary, he repeatedly maintained that it explained extinctions only (Owen 1866–8, vol. 3, p. 798).

For example, Owen believed that extinction may have been a function of the size of an animal. The larger a species is, the less adaptable it will be to a major change in environmental conditions. A drought, a diminished water and food supply, or the introduction of new enemies will affect larger species more than smaller ones. Also, the latter tend to be more numerous. It is the "contest... against the surrounding agencies" which explains the extinction of species. This view bears more than a superficial resemblance to Darwin's "the struggle for life," and in order to solve a priority dispute between the two men that flared up in the early 1860s we need to quote Owen's actual words:

> In proportion to its bulk is the difficulty of the contest which, as a living organized whole, the individual of such species has to maintain against the surrounding agencies that are ever tending to dissolve the vital bond, and subjugate the living matter to the ordinary chemical and physical forces. Any changes, therefore, in such external agencies as a species may have been originally adapted to exist in, will militate against that existence in a degree proportionate, perhaps in a geometrical ratio, to the bulk of the species. If a dry season be gradually prolonged, the large Mammal will suffer from the drought sooner than the small one: if such alteration of climate affect the quantity of vegetable food, the bulky Herbivore will first feel the effects of stinted nourishment: if new enemies are introduced, the large and conspicuous quadruped or bird will fall a prey, whilst the smaller species conceal themselves and escape. Smaller animals are usually, also, more prolific than larger ones (Owen 1866–8, vol. 3, p. 798).

The *Quarterly Review,* known for its traditional Anglican and Tory values, nevertheless regarded Owen's case as just, reiterating that he had put forward "natural selection" as "the cause of *extinction* of species," whereas Darwin's theory "applies not only to the extinction, but also to the *origin* of species" ([Mozley] 1869, p. 163). The accusation by the *London Review* and by Darwin irritated Owen, and he returned to the issue in a comparatively long entry on "species" for the new edition of Brande and Cox's *Dictionary of Science, Literature, and Art* (1867). It was Owen's most lucid, concise, and well-written contribution to the controversy on the origin of species, and the one most readily available to a large readership. He summarized the main evidence for, and the main theories of, a natural origin of species, adding that changes in external conditions are the cause of a "*natural rejection*" of species, not of a "*natural selection*," as Darwin maintained. External stimuli, whether those of French or of English transmutation theories, do not add up to an adequate cause of new species, Owen argued.

5 The Teeth of Owen's Critique

The fullest refutation of the Darwinian accusation that Owen had claimed priority in formulating the theory of natural selection appeared as a long footnote to the concluding chapter of the third volume of Owen's *Anatomy of Vertebrates* (1868). This chapter formed the longest and most detailed statement of where he stood with respect to the theory of evolution. It was reprinted in its entirety in the *American Journal of Science* and published as a separate, forty-page booklet, entitled *Derivative Hypothesis of Life and Species* (1868). Owen reminded his readers of Cuvier's opinion that, if evolution had occurred, one should be able to find intermediary forms in the fossil record, connecting, for example, the palaeotherium with today's hoofed quadrupeds.

This condition had now amply been met—Owen maintained—in particular by the discovery of *Hipparion*, a fossil genus of the *Equidae*, from the Miocene at Eppelsheim and the Sewalik Hills in India. Their molar teeth differed from those of existing horses mainly by the greater extent of separation of the inner column, and the second and fourth digits of their feet, suppressed in the form of splint bones in the existing horse, were retained in a functional state.

In his lectures on fossil mammals at the School of Mines in 1857, Owen had earlier illustrated the consecutive modifications of teeth and feet in the morphological series from palaeotherium to hipparion and eventually to today's horse. He rejected the possibility that these forms were successive, separate creations: "If the alternative – species by miracle or by law? – be applied to *Palaeotherium*, *Paloplotherium*, *Anchitherium*, *Hipparion*, *Equus*, I accept the latter, without misgiving, and recognise such law as continuously operative throughout Tertiary time" (Owen 1866–8, vol. 3, p. 793). If no direct, creative acts had taken place, how then had *Palaeotherium* ultimately become *Equus?* One could try to explain the development from a three-hoofed to a one-hoofed foot, the latter with splint bones, on the basis of "Lamarckian volition" or "Darwinian selection," but these hypotheses "are less applicable, less intelligible, in connection with the changes in the structure and proportion of the molar series of teeth." Owen envisaged no gradual transmutation but sudden "mutations." He drew attention to the phenomenon of reverse evolution when modern horses are born with supplementary hoofs in which the second and fourth metacarpals, normally present in the form of splint bones, carry phalanges and a terminal hoof. The feet of such polydactyl horses resemble those of *Hipparion* and the pairing of a tridactyl colt with a tridactyl filly "might restore the race of hipparions" (Owen 1866–8, vol. 3, p. 795).

> Now the fact suggesting such possibility teaches that the change would be sudden and considerable: it opposes the idea that species are transmuted by minute and slow degrees. It also shows that a species might originate independently of the operation of any external influence; that change of structure would precede that of use and habit; that appetency, impulse, ambient medium, fortuitous fitness of surrounding circumstances, or a personified "selecting Nature," would have had no share in the transmutative act (Owen 1866–8, vol. 3, p. 796).

Thus, Owen continued to prod the Darwinians about the nature of organic variability. Objecting to a chance process of variation, he believed that evolution was a teleological process, a movement toward a pre-ordained goal. Mutations are not randomly useful or useless but a logical embroidering on the archetype.

From among the invertebrates, Owen selected the reef-building corals as a further test of evolutionary hypotheses. The calcareous cups of certain corals show internal, longitudinal partitions or septa which in some types are arranged in a pattern of four or its multiples, and in others show a six-part symmetry. In the course of earth history, the hexapartite corals superseded the tetrapartite ones. Considering the very many species of Anthozoa which have come and gone in the course of geological time, the miracle of special creation becomes hard to believe, "inconsistent with any worthy conception of an all-seeing, all-provident Omnipotence!" A Lamarckian mode of evolution would not work either, "from the impotency of a coral-polyp to exercise volition." Nor would a Darwinian mode—Owen asserted. Just as in the case of the whales, the required external influences do not exist in the oceanic environment. In the Red Sea, for example, a minimum of 100 types of coral exist under more or less the same conditions:

> So, being unable to accept the volitional hypothesis, or that of impulse from within, or the selective force exerted by outward circumstances, I deem an innate tendency to deviate from parental type, operating through periods of adequate duration, to be the most probable nature, or way of operation, of the secondary law, whereby species have been derived one from the other (Owen 1843, pp. 32, 79).

Whereas Darwin's theory left the origin of species to the chance occurrence of individual variations and milieu conditions, Owen's "derivation" of species recognized that the evolution of life follows a preordained course, determined by an organism's "innate capacity or power of change."

Darwin's hypothesis of the evolution of all life from a primordial form "into which life was first breathed" amounted to a belief in "thaumatogeny," a miraculous origin of life, whereas Owen advocated "nomogeny," a coming into existence caused by natural law, possibly at multiple locations and times. Over the years Owen had paid little attention to the issue of spontaneous generation, although in his *Lectures on the Comparative Anatomy and Physiology of the Invertebrate Animals* he had explicitly rejected the hypothesis of "equivocal generation," calling it "gratuitous" (Owen: 1866–8, vol. 3, pp. 817–818). But now, in the heat of the Darwinian debate, Owen took the opposite stance, speculating in his *Anatomy of Vertebrates* that life in the form of "protogenal" jelly-specks originated as a result of a favorable concurrence of physicochemical conditions that produced "organic crystallization,"[3] a prescient anticipation of the importance of crystallography in deciphering DNA.

In a manuscript note on "spontaneous generation," Owen even more explicitly than in his *Anatomy of Vertebrates* contrasted himself with Darwin by means of this issue. Natural selection could only work once life exists, and to get over the hurdle of abiogenesis a different mechanism is required:

> The doctrines of the *generatio spontanea* and of the transmutation of species are intimately connected. Who believes in the one, ought to take the other for granted, both being founded on the faith in the immutability of the laws of nature.[4]

A fairly lengthy sketch of Owen's views *in casu* had already appeared in 1863 in the form of a review in the *Athenaeum* of William B. Carpenter's magnificent *Introduction to the Study of the Foraminifera.* The anonymous reviewer, recognized by the Darwinians as Owen (Huxley 1918: vol. [. . .], 51), noted with disapproval Darwin's "primordial form, into which life was first breathed." This—he charged—

---

[3]Owen, "Spontaneous generation," BM(NH), L, OC 90, vol. 3, 30, 33.

[4]Ibid.

was invoking an "occult, miraculous, creative cause operating once for all." "The inevitable corollary of the Darwinian hypothesis [i.e., the generation of the first form or forms of life by means of a divine act of creation] seems to us to demonstrate its weakness" ([Owen] 1862b, p. 419).

Darwin, seeing the weakest part of his position prodded, was furious: "Who would ever have thought of the old stupid *Athenaeum* taking to Oken-like transcendental philosophy written in Owenian style!" he exclaimed to Hooker, and a public exchange with Owen in the correspondence pages of the *Athenaeum* followed (Darwin 1863, pp. 554–555, 617; [Owen] 1863b, pp. 586–587).

6 Concluding Observations

The falling out between Owen and Darwin was not merely about Owen's (1860) critique of Darwin's *Origin of Species*. Two major issues separated Owen and Darwin. The first concerned the British Museum of Natural History in South Kensington, perhaps the greatest architectural embodiment of Victorian science anywhere. Shortly after Owen had initiated in his presidential address of 1858 the drive to establish this museum, and before the appearance of his essay on the *Origin of Species* in the *Edinburgh Review*, the Darwinians attempted to sabotage Owen's plans for what was to become his most monumental accomplishment—the Natural History Museum in South Kensington. Darwin was an active part of the attempt at a political assassination of this initiative. "[T]he Darwinian camp acted with the same jealous duplicity of which Owen habitually is accused." Owen's attack on Darwin's crowning glory appeared in April 1860, nearly a year and a half after Huxley, Darwin, and others had begun sabotaging Owen's museum movement (Rupke 1988, p. 85). Darwin and his cronies lost; Owen won, with the help of Gladstone and other Liberal science supporters, and in 1881 was ultimately enthroned as the first director of the new and separate British Museum (Natural History) in South Kensington.

All Darwin had left in his effort to defeat Owen was misrepresentation, dishonestly characterizing him as a creationist. This brings us to the second source of separation between the two men. Owen preceded his scheming adversary as a scientifically better informed paleobiologist and a supporter of the Kantian unified theory of evolution. Disappointingly, even Curtis N. Johnson's comprehensive *Darwin's Historical Sketch. An Examination of the 'Preface' to the Origin of Species* (2020) misses the political context of the Owen–Darwin dispute and perpetuates a conventional Darwin idolatry. Darwin's theory of evolution was a product of British—more particularly English—culture. In Scotland (Edinburgh) and Ireland (both Belfast and Dublin), German structuralism had strong scientific representation and held its ground till well after the appearance of *The Origin of Species*. His theory of evolution by natural selection, with its emphasis on adaptation, was an anastomosis of Malthusian and Paleyan thought and, although Darwin inverted Paley's argument of design, one can argue that his theory remained an integral part of homegrown functionalist preoccupations. As a result, Owen did not win the evolution debate. An indication of this was a uniquely humiliating blot on Owen's memory. In 1894, Owen's grandson published the two-volume *Life and Letters* of his famous relative. However, none other than Huxley, Owen's worst enemy and Darwinian fiend, was invited to contribute an essay on "Owen's position in the history of anatomical science" (Rev. Owen 1894). It will have made Owen turn in his grave.

Still, the main reason for Darwin's win in this fight was nationalism (Rupke 2020). Owen's leadership in the British Museum movement attracted a following and was admired. The cathedral-like edifice outshone the national museums of natural history in Paris, Vienna, Madrid, and other capital cities, while the collections that Owen managed to amass from around the world, exhibit, and store inside were a proud part of the work of empire. By stark contrast, Owen's contributions to evolution theory were in line with Continental thought as developed by Kant, Blumenbach, Humboldt, and other non-British thinkers. Darwin, who was a relatively simple gentleman-naturalist, could not cope with "foreign" learning and stuck to a mixture of insular, perhaps parochial Malthusian and Paleyan, notions. This "Brexit" attitude was adopted by others, disciples who treated Darwin as a latter-day Moses and thought of the *Origin of Species* as the new *Book of Genesis*. This Darwinian lack of appetite for German evolutionary thought turned into disapproval and, at times, excoriation and nationalistic ridicule, especially after WWI and even more so WWII. Still, Owen's legacy in the English-speaking world remained visible in writings from contemporaries such as Argyll and Mivart to Simon Conway Morris (Conway Morris 2022) today. In spite of the excesses of nationalism that have influenced evolution theory, a trickle of Owenian insight and knowledge has kept flowing through the twentieth century. Owen's belief in "inner tendencies" as the mechanism of evolutionary change and teleological progress in evolution is present in Conway Morris' work on convergencies and, for example, *Andreas Wagner's Arrival of the Fittest,* in which he cites the classic saying by Hugo de Vries that natural selection may explain the survival of the fittest, but it cannot explain the arrival of the fittest (Wagner 2014; see also Ball 2009).

References

Anon (1866a) Professor Owen's comparative anatomy. London Rev Politics Society Literature Art Sci:12

Anon (1866b) The comparative anatomy of vertebrate animals. Popular Sci Rev 5:211–213

Ball P (2009) Shapes. Nature's patterns. A tapestry in three parts. Oxford University Press, Oxford

Clark JD, Hughes TM (1890) The life and letters of the Reverend Adam Sedgwick. Cambridge University Press, Cambridge. 2 vols

Conway Morris S (2022) From extraterrestrials to animal minds. Six myths of evolution

Darwin C (1859) On the origin of species by means of natural selection, or the preservation of favoured races in the struggle for life (1st edn)

Darwin C (1863) Doctrine of heterogeny and modification of species. Athenaeum 25:554–555

Darwin C (1866) Historical sketch. On the origin of species, 4th edn. John Murray, London

Darwin C (1872) An historical sketch. In: The origin of species, 6th edn. John Murray, London

Dawkins R (2009) The greatest show on earth. The evidence for evolution. Free Press, New York

de Beer G (1963) Charles Darwin: a scientific biography. Doubleday, New York

de Beer G (ed) (1983) Charles Darwin and T.H. Huxley. Autobiographies. Oxford University Press, Oxford

Duches AD (ed) (1906) George Douglas, Eighth Duke of Argyll, K.G, K.T., (1823–1900), autobiography and memoirs. John Murray, London. 2[?] vols

Gillespie NC (1979) Charles Darwin and the problem of creation. University of Chicago Press, Chicago

Greene JC (1981) Science, ideology and world view. University of California Press, Berkeley

Himmelfarb G (1962) Darwin and the Darwinian revolution. Doubleday, New York

Huxley L (1918) Life and letters of Sir Joseph Dalton Hooker. London, 2 vols

Johnson CN (2020) Darwin's Historical sketch. An examination of the 'Preface' to the Origin of Species. Oxford University Press, New York

Lyell KM (ed) (1881) Life, Letters and Journals of Sir Charles Lyell, Bart. John Murray, London. 2 vols

MacLeod RM (1965) Evolutionism and Richard Owen, 1830-1868: an episode in Darwin's century. Isis 56:259–280

Mivart SGJ (1895) A century of science. Living Age 205:771–787

Moore JR (1979) The Post-Darwinian controversies: A study of the Protestant struggle to come to terms with Darwin in Great Britain and America. Cambridge University Press, Cambridge

Mozley JB (1869) The argument of design. Q Rev 127:134–176

Owen R (1843) Lectures on the comparative anatomy and physiology of the invertebrate animals, delivered at the Royal College of Surgeons, in 1843. Longman, London

Owen R (1851) Lyell—on life and its successive development. Q Rev 89:412–451

Owen R (1858) Address. *BAAS, Report*: [pages]

Owen R (1860) Darwin on the origin of species. Edinburgh Rev 3(111):487–532

Owen R (1862a) On the characters of the Aye-aye, as a test of the Lamarckian and Darwinian hypothesis of the transmutation and origin of species. BAAS, Report, Transactions of the Sections: 114–116

Owen R (1862b) Monograph on the Aye-aye (Chiromys madagascariensis, Cuvier). Taylor and Francis, London

Owen R (1863a) On the Aye-aye (Chiromys, Cuvier; Chiromys madagascariensis, Desm.; Sciurus madagascariensis, Gmel., Sonnerat; Lemur psilodactylus, Schreber, Shaw). Trans Zool Soc 5(2):33–101

Owen R (1863b) Origin of species," Athenaeum, May 2, 586–87, 617

Owen R (1866–8) On the anatomy of vertebrates. Longmans, Green, London., 3 vols

Owen Rev R (1894) The life of Richard Owen by his grandson. With the scientific portions revised by C. Davies Sherborn. 2 vols. John Murray, London

Powell B (1855) Essays on the spirit of the inductive philosophy, the unity of worlds, and the philosophy of creation. Longman, Brown, Green and Longmans, London

Rupke NA (1988) The road to Albertopolis: Richard Owen (1804-1892) and the founding of the British Museum of Natural History. In: Rupke NA (ed) Science, politics and the public good. Macmillan Press, London, pp 63–89

Rupke NA (2009) Richard Owen: biology without Darwin. Chicago University Press, Chicago

Rupke NA (2010) Darwin's choice. In: Alexander DR, Numbers RL (eds) Biology and ideology from Descartes to Dawkins. University of Chicago Press, Chicago

Rupke NA (2020) Putting the structuralist theory of evolution in its place. In: Mayhew RJ, Withers CWJ (eds) Geographies of knowledge. Science, scale and spatiality in the nineteenth century. Johns Hopkins University Press, Baltimore Templeton Press, West Conshohocken

Ruse M (1979) The Darwinian revolution: science red in tooth and claw. Chicago University Press, Chicago
Sedgwick A (1845) Natural history of creation. Edinburgh Rev 82:1–85
Sloan P (2003) Whewell's philosophy of discovery and the archetype of the vertebrate skeleton: the role of German philosophy of science in Richard Owen's biology. Ann Sci 60:39–61
Wagner A (2014) Arrival of the fittest. How nature innovates. Penguin, New York

Darwin, *Archaeopteryx lithographica* and the Problem of Intermediate Species

Bogdana Stamenković Jajčević

Abstract

This chapter explores Darwin's conception of transitional forms and challenges his view on the problem of intermediate species. Starting with a brief historical introduction to the epistemological status of fossils before Darwin's time, I proceed to the analysis of Darwin's theory of evolution and identify three fundamental hypotheses that substantially influence his investigation of the fossil record. Although Darwin's theory implies the necessary existence of transitional forms, he believes that the paleontological record remains incomplete due to lacking discovery of intermediate forms. I find several reasons for such a radical conclusion. I deconstruct his specific yet ambiguous conception of intermediate forms and show that, in addition to the fundamental assumptions of his theory, this conception is at the root of the emergence of the intermediate species problem. Further, I explicate how this problem arises at two distinct but connected levels of Darwin's theory. Whilst at the first level it appears in the form of a vague interpretation of the intermediate form concept, at the second level this problem takes a form of a stringent requirement for classification of an organism as an intermediate form: the discovery of a complete chain of transitional forms that connects such an organism to at least one known species. *Archaeopteryx lithographica* could not satisfy such a requirement; actually, no organic form can. The problem of the intermediate species thus remains one of 'the gravest difficulties' for Darwin's theory of evolution.

B. Stamenković Jajčević (✉)
Faculty of Philosophy, University of Belgrade, Belgrade, Republic of Serbia

Department of Philosophy, Institute for Philosophy, University of Belgrade, Belgrade, Republic of Serbia

R. G. Delisle et al. (eds.), *Unity and Disunity in Evolutionary Biology*,
https://doi.org/10.1007/978-3-031-42629-2_7

Keywords

Darwinism · Theory of evolution · Fossil record · Intermediate form · *Archaeopteryx lithographica*

In 1973, Theodosius Dobzhansky (1900–1975) wrote that 'nothing in biology makes sense except in the light of evolution' (Dobzhansky 1973). Contemporary Darwin scholarship often emphasises the importance of *Darwinism*—Darwin's achievement, suggesting that his theory of evolution represents the starting point for the development of modern biology. The reasons for such a stance seem apparent. Darwin's theory dethroned the influential creationist tradition which advocated immutability of species. Contra Creationism, Darwin argued for mutability of natural species and claimed that members of distinct species are interconnected by hereditary relations. In other words, Darwin suggested that organisms of different species and genera share a common ancestor. Further, he discovered the mechanism guiding the evolution of species, i.e. the principle of natural selection. Thus it seems that Darwin's theory represents a turning point in the historical development of biology. Scholars often refer to this turning point as 'Darwin's scientific revolution' or 'Darwinian revolution'.[1] The term 'Darwinism' can be said to generally refer to Darwin's theory of evolution, his essential theoretical assumptions about the evolutionary process and its mechanism as well as the implications of his theory. Whilst this represents a theoretical aspect of Darwinism (and Darwinian revolution), its ontological aspect can also be observed where 'Darwinism' refers to the natural system that behaves according to the principles of Darwin's theory of evolution.[2] So, Darwin's revolution, or *Darwinian revolution*, implies scientific changes in both theoretical and ontological understanding of the natural system. However, I believe we can detect an additional meaning of 'Darwinism' that remains overlooked by contemporary Darwin scholarship. Contemporary scholars usually omit the fact that Darwin was a scientist of the nineteenth century and that he wrote in the time of intersection of two conceptually different worlds: (1) the world of Romanticism, i.e. the world of a harmonious, static nature shaped by actions of the 'Author of Nature', and (2) the world of a dynamic nature governed by natural laws, filled with perturbations and multiplicity of variable species. Darwin's theory certainly embraces the elements of both those worlds, which has led many authors to question the influence of Darwin's forerunners—scholars whose research influenced the formulation of his theory of evolution.[3] This suggests a historical aspect of 'Darwinism', pertaining to the very history of Darwin's theoretical (and ontological) conception. Given these different meanings, the *deconstruction of Darwinism* refers to

[1] For example, Bowler (2003).

[2] As Gould notes, "although "Darwinism has often been equated with evolution itself in popular literature, the term should be restricted to the body of thought allied with Darwin's own theory of mechanism" (Gould 1982: 380).

[3] For example, Gale (1972); Bowler (1976); Sloan (2001) and Richards (2002, 2009).

(1) the identification and analysis of the essential aspects of Darwin's theory (e.g., fundamental assumptions, method, etc.), followed by the examination of its implications and possible problems of the theory, and (2) the philosophical and historical inquiry of the formulation of Darwin's theory of evolution, focused on the research of pre-Darwinian beliefs and theories about transmutation of species. Each of these 'modes' of deconstruction provides additional insight and understanding of Darwin's theory and its problems.

In this chapter, I will focus on the first 'mode' of the deconstruction of Darwinism, i.e. on the epistemological and methodological difficulty relating to the fossil evidence in his theory, and explore why 'the missing link' problem remains an unsolvable conundrum for Darwin. I start with a brief historical review of the status of the fossil record in the scientific period before Darwin's theory. Next, I analyse the crucial aspects of Darwin's theory in order to identify the fundamental hypothesis of his theory of evolution. Moving forward, I turn to Darwin's investigation of the fossil record and deconstruct the *function* of fossils and, specifically, the *notion* and *function* of intermediate forms in his theory of evolution. Following his claims about the creation and preservation of fossils, intermediate species, incompleteness of geological data and evolution by natural selection, I show that the famous problem of the absence of intermediate species resurfaces at two distinct yet related levels of the theory of evolution. Firstly, this problem arises at the *conceptual level* in the form of a vague concept of intermediate species; secondly, the problem of intermediate forms appears at the *ontological level* where classification of a form as intermediate requires additional discovery of a *complete chain* of related transitional forms linking it to at least one known species. It seems that *Archaeopteryx lithographica* could not satisfy such rigorous requirement of Darwin's theory and could not be recognised as an intermediate form. However, it appears that no form, extinct or living, can meet this requirement. The problem of the absence of intermediate species thus remains one of 'the gravest difficulties' for his theory of evolution.

1 Fossils Through History

Not one attempt to provide a historical insight about the epistemological status of fossil record can avoid discussing the Creationistic interpretation of fossils. As is well-known, the nineteenth century study of the fossil record was strongly influenced by the paradigm of special creation of species (Archibald 2017: 02). It was believed that God created each species separately, and since the moment of their creation, these species have not undergone any morphological changes. In other words, the paradigm of special creation of species implies the thesis of *immutability of natural species* meaning that organic beings, in the moment of their creation, looked morphologically the same as they do now. Therefore, the succession of species in the fossil record is not more than a series of replacement species created by the Author of Nature (Archibald 2017: 07).

Creationist assumptions about the origin of species were not weakened by the geological record that indicated (1) a significantly greater diversity of species and

(2) an older age of the Earth than presumed by the Creationist paradigm. As Flannelly observes, fossils represented nothing more than the pieces of a puzzle whose value was not recognised by many naturalists of the nineteenth century (Flannelly 2017: 29–30). It is therefore not surprising that most naturalists of that time were not convinced in the accuracy of the evolutionary hypothesis. For example, Thomas Huxley believed that the evidence for the transmutation of species (at the time) was not sufficient. Further, there were no adequate assumptions about the causes of transmutation of species that could scientifically explain such a phenomenon (Darwin 1892: 169). Accordingly, fossils were interpreted in two ways:

1. Fossils are nothing more than 'curiously shaped rocks, perhaps formed by some mysterious "plastic virtue" that could mimic living structures' (Bowler 2003: 35). Since the fossils represent mineral entities, their unexpected appearance in areas such as mountain tops ceases to be an anomaly.
2. Fossils represent the consequence of the biblical Great Flood that caused mass extinction of all species; their displacement, elevation by waves and aggressive flood currents are the causes of fossil appearance on mountain tops (Gould 2011: 26–27).

It is worth noting that both interpretations were challenged by Leonardo Da Vinci (1452–1519) in the fifteenth century.

Da Vinci's study of fossils was not motivated by the questions of the origin and evolution of natural species. His investigation of the fossil record was motivated by another research. In fact, Da Vinci aimed to empirically prove his own theory stating that the same processes occur on Earth (macro-cosmos) as in the human organism (micro-cosmos). He believed that the Earth and the human organism are created from the same four ancient elements: earth, water, air and fire. Apparently, the circulation of these elements, especially water (i.e. blood) that rises from the legs up to the brain, enables the preservation of the human body. The analogy between the micro- and macro-cosmos is sustainable only if it is successfully shown that the Earth also possesses some kind of maintenance mechanism through the circulation of the ancient elements. Accordingly, if Da Vinci's theory should be true, a similar process *must* occur in both, the micro- and macro-cosmos (Gould 2011: 33).

It seems that the crucial explanation needed to clarify the rise and fall of two elements—earth and water—that tend to stabilise on lower levels. Stabilization of earth and water prevents their circulation, implying that the maintenance mechanism of the macro-cosmos cannot be found in their behaviour. It was therefore necessary to explain the rise and descendance of earth and water. Although he failed to provide the necessary explanation for the movement of water, Da Vinci succeeded in explaining the movement of the Earth by studying the fossil remains of shells in various mountain strata discovered several thousand feet above the sea level (Gould 2011: 34). At the same time, Da Vinci dismissed the two aforementioned interpretations of fossils. Da Vinci's investigation led him to conclude that the Great Flood could not have caused deposition of shells in different mountain layers in one burst because their presence in these layers is a result of long-term deposition during different geological periods of time. Further, fossils do not represent the so-called 'game of nature'; they are not a consequence of plastic forces within rocks or emanations of stars. If that were the case, their positions on mountain tops would not be surprising. So, Da Vinci concluded that deposition of fossils on mountain tops resulted from the Earth's frequent uplifting (Gould 2011: 34).

Despite Da Vinci's research, the Creationistic explanation of the existence of fossils retained its influence throughout the first half of the nineteenth century. Darwin challenged the Creationists' paradigm by (1) investigating the fossil record in different parts of the world with distinct paleontological histories, and by (2) emphasizing the similarities between species of South America and Galapagos Archipelago (Archibald 2017: 03–04). His work *On the Origin of Species* convincingly showed that fossils are neither anomalies of the natural system nor 'curiously shaped rocks'; fossils are not remains of natural species that were wiped out by the biblical Great Flood. Fossils are remains of extinct organisms related to the present species by hereditary relations. Evidently, Darwin recognised fossils as the evidence for his main hypothesis—that the evolution of natural species occurs in nature.

2 Fundamental Assumptions of Darwin's Theory of Evolution

The history of natural science endured a major change in the years following the first publication of the *Origin of Species*. Yet, one should not overlook the fact that Darwin did not *discover* evolution. Scholars of the eighteenth and nineteenth centuries were acquainted with the idea of transmutation of organic beings.[4] However, it seems that academic community began to accept the hypothesis as scientifically true in the years after the publication of Darwin's famous work. The reasons for such a change in the epistemological status of the evolutionary hypothesis are different. Darwin discovered the missing piece of the evolutionary puzzle—the very mechanism of evolution, i.e. natural selection. Further, he recognised the fossil record as evidence for the occurrence of evolution in nature. In other words:

> In considering the Origin of Species, it is quite conceivable that a naturalist, reflecting on the mutual affinities of organic beings, on their embryological relations, their geographical distribution, geological succession, and other such facts, might come to the conclusion that species had not been independently created, but had descended, like varieties, from

[4]For example, Leibniz (1646–1716) notices the structural similarity between fossils and living species and emphasises in *Protogaea* (2008) that a closer look at their body parts eliminates doubts about their origin: 'Therefore, there is no reason not to assume the same origin where the earth turned to stone' (Leibniz 2008: 65). Carl Linnaeus (1707–1778) creates a taxonomic system of plants and animals based on a number and arrangement of their reproductive organs and draws attention to the (at the time) newly discovered phenomenon of sexual reproduction in plants. This idea was the subject of research of Comte de Buffon (1707–1788), Lamarck (1744–1829) and Erasmus Darwin (1731–1802).

> other species. Nevertheless, such a conclusion, even if well founded, would be unsatisfactory, until it could be shown how the innumerable species inhabiting this world have been modified, so as to acquire that perfection of structure and coadaptation which justly excites our admiration. (Darwin 2009: 02)

Darwin realises that the scientific acceptance of his theory rests on providing the explanation of *how* natural species are modified. Recognizing that environmental changes can gradually lead to the emergence of new species, Darwin concludes that the survival of such species depends on the individual characteristics—*variations*—of an organism. Preservation of useful and rejection of harmful variations is called natural selection. As we know, variations that are neither beneficial nor harmful will not be affected by natural selection:

> Variations, however slight, and from whatever cause proceeding, if they be in any degree profitable to the individuals of a species, in their infinitely complex relations to other organic beings and to their physical conditions of life, will tend to the preservation of such individuals, and will generally be inherited by the offspring. The offspring, also, will thus have a better chance of surviving, for, of the many individuals of any species which are periodically born, but a small number can survive. I have called this principle, by which each slight variation, if useful, is preserved, by the term Natural Selection, in order to mark its relation to man's power of selection. (Darwin 2009: 49)

However, Darwin argues that we cannot observe the changes in variations of organisms or the operation of natural selection. 'So imperfect is our view into long-past geological ages, that we see only that the forms of life are now different from what they formerly were' (Darwin 2009: 66).

These parts of the *Origin* allow us to identify three fundamental epistemological assumptions of Darwin's theory:

1. The evolution of natural species occurs in nature; it represents a *slow*, *gradual*, *continuous* and, for human mind, *imperceivable* process of modification of organisms that finally causes the emergence of new species—a final product observable by our mind.
2. *All* natural species are mutually connected by hereditary relations due to an identical origin in a *common ancestor*.
3. The *mechanism* of evolution of organisms is *natural selection*.

In order to provide adequate understanding of Darwin's theory and its implications on his investigation of the fossil record, I will briefly explain each of these assumptions.[5]

[5]Given that Darwin's theory of evolution is a *scientific* theory, these assumptions are not to be regarded as *scientific hypotheses*. However, their epistemological and methodological aspects should not be overlooked. On the one hand, the epistemological aspect reflects Darwin's belief that distinguishes him from his predecessors—that knowledge about the evolution of species is possible. On the other, the methodological aspect shows how Darwin used and connected those assumptions in order to confirm the reality of evolution and avoid potential criticism. And as I will demonstrate in due course, Darwin's assumption about the incompleteness of geological data takes

Contemporary Darwin scholarship acknowledges the scientific influence of several scholars on Darwin's theory of evolution.[6] Charles Lyell (1797–1875) and Thomas Malthus (1766–1834) represent probably the most prominent figures whose research highly influenced Darwin's theory. Malthus's *Essay on the Principle of Population* (1789) was based on the assumption that conflicts between organisms are due to limited amounts of natural resources (such as food). This hypothesis acquires a biological aspect in the *Origin of Species* and transforms into the idea of the *cause* of the struggle for survival. Accordingly, Darwin argues these causes can be found precisely in the limited resources of a particular ecosystem. Although Malthus's impact on Darwin is indisputable, it seems that Lyell's research had a more profound impact on Darwin's theory. Lyell is known as the father of uniformism—a conception of nature founded on the hypothesis that the Earth was shaped in the past by *slow* and *gradual* action of natural forces whose activity can be detected in the present as well. Lyell's influence can be observed in the first hypothesis of Darwin's theory. Darwin adopts Lyell's fundamental assumption and applies it to the problem of natural species transmutation. In parallel to the Earth having been shaped by slow and gradual action of geological forces that remain active in the present, natural species are *slowly* and *gradually* created in the process of evolution:

> The species of the same genus, and the genera of the same family, can increase only slowly and progressively; the process of modification and the production of a number of allied forms necessarily being a slow and gradual process—one species first giving rise to two or three varieties, these being slowly converted into species, which in their turn produce by equally slow steps other varieties and species, and so on, like the branching of a great tree from a single stem, till the group becomes large. (Darwin 2009: 293)

If the transmutation of organisms and the creation of new species occur excessively slowly and gradually, then the evolution of species stretches over several long geological periods of time. This 'stretch' of geological time represents a consequence of Lyellian-Darwinian conception of nature. As Delisle notices, Darwin's theory is grounded on two major principles. The first principle—*principle of uniformism*—establishes that the *same* natural laws and processes operate in the natural system during all geological periods (Delisle 2019: 208). The second principle—*principle of gradualism*—ensures the connection of *all geological* and *evolutionary* periods *without any disruption* (Delisle 2019: 135).

Notice how Darwin's first fundamental hypothesis implies an additional epistemological-methodological assumption: gradual, evolutionary changes of organisms *cannot* be observed:

> We see nothing of these slow changes in progress, until the hand of time has marked the lapse of ages, and then *so imperfect is our view* into long-past geological ages, that we see

up precisely this role and could be seen as a methodological decision that should eliminate one of the problems compromising his theory.

[6]For example, Egerton (1970); Bowler (1976); Sloan (2001) and Richards (2002, 2009).

> only that the forms of life are now different from what they formerly were. (Darwin 2009: 66, my emphasis)

In other words, cognitive limitations of our epistemological apparatus disable our perception of slow, gradual and delicate modifications that arise during the evolutionary process of an organism or a whole species. All we can observe is the final product—an individual with new variations, or a new species. As I will show in due course, the intermediate forms should testify precisely the manifestation of these delicate, imperceivable morphological changes of organisms.

Darwin's second fundamental hypothesis establishes *monophyletism* and states that all living and extinct species have a common origin, i.e. represent the descendants of the same ancestor:

> By the theory of natural selection all living species have been connected with the parent-species of each genus, by differences not greater than we see between the natural and domestic varieties of the same species at the present day; and these parent species, now generally extinct, have in their turn been similarly connected with more ancient forms; and so on backwards, always converging to the common ancestor of each great class. (Darwin 2009: 266)

Along with the evidence provided in the form of fossils, the second hypothesis should emphasise the materialistic and historical reality of evolution of natural species (De Ricqles 2010: 88). As I will show in the next section, Darwin believes that the fossils of intermediate forms *should* assert the reality of evolution in nature and confirm the hypothesis of the common ancestry of organic beings. However, due to the problem of absence of intermediate forms, Darwin turns to the investigation of geographical distribution of species in his further argumentation.[7]

Finally, the third hypothesis of Darwin's theory postulates the mechanism of evolutionary process—natural selection, 'the most important, but not the exclusive, means of modification' (Darwin 2009: 04). Darwin defines the principle of natural selection as the preservation of useful and rejection of harmful variations (Darwin 2009: 63). As noted, the mechanism of evolutionary process represented one of the main naturalistic problems before Darwin's time. So, it is no surprise that Darwin's discovery of natural selection influenced the scientific acceptance of the theory of evolution. As Delisle notices, at the first glance, natural selection is assigned with the exceptional power to cause evolutionary change and separate closely related organic forms from their parental group (Delisle 2019: 86). In the *Origin*, Darwin writes that natural selection represents a force 'incessantly ready for action' (Darwin 2009: 49). Yet, it appears that Darwin's gradualism confines the action of such a force. Darwin writes:

> Natural selection acts only by the preservation and accumulation of small inherited modifications, each profitable to the preserved being; and as modern geology has almost banished such views as the excavation of a great valley by a single diluvial wave, so will

[7]Due to the scope of this paper and the main research, I will not discuss Darwin's study of geographical distribution of species.

> natural selection banish the belief of the continued creation of new organic beings, or of *any great and sudden modification* in their structure. (Darwin 2009: 75–76, my emphasis)

Specifically, Darwin's fundamental assumption about evolution as a slow and gradual process affects the mode and activity of natural selection. Such a force *necessarily* operates slowly and gradually, creating morphological modifications to organisms that remain imperceivable for the human mind. Evolution can be viewed as a gradual process of modification of species that diverge towards present forms. But, as Delisle states, natural species evolve, but not solely in Darwinian manner. Whilst it is possible that individual organisms diverge from each other (*divergence*), it is also possible for those organisms to maintain the equal phenotypic distance (*parallelism*). Finally, there is a possibility of mutual convergence of the same individuals up to the point of their crossbreeding (*convergence*) (Delisle 2019: 134). Further, Darwin's gradualist interpretation of evolution and activity of natural selection eliminates any possibility for the appearance of organic forms which, although not related by a series of intermediate forms, exhibit certain morphological similarities. It seems that Darwin does not acknowledge the possibility of *evolutionary leaps* (Delisle 2019: 135) and fails to notice that natural selection can cause sudden and rapid changes to organisms in just a few generations of a certain species.[8]

Darwin's gradualist conception implies a specific, rigid interpretation of the main natural phenomena discussed in his theory: evolution and natural selection. He argues that evolution is a slow and gradual process guided by the slow and gradual operation of natural selection. As noted, the discovery of natural selection represents one of the reasons for the acceptance of the theory of evolution after the publication of the *Origin*. However, the second reason is found in fossil remains of extinct species—Darwin's crucial evidence for the fundamental hypothesis that evolution of organisms occurs in nature.

3 Darwin's Investigation of Fossils: Incompleteness of the Geological Record

In the previous section, I identified three fundamental epistemological-methodological assumptions of Darwin's theory of evolution:

1. Evolution of natural species occurs in nature; it represents a *slow*, *gradual*, *continuous* and, for human mind, *imperceivable* process of modification of organisms that finally causes the emergence of new species—the final product observable by our mind.

[8]In 1959, the Russian geneticist Dmitry Belyaev (1917–1985) showed that such a situation was possible. Belyaev conducted a long-term experiment on silver fox domestication in order to confirm his hypothesis about the domestication syndrome—all domesticated species acquire the same characteristics over time. During the experiment, Belyaev observed that the selection for domestication, *in just six generations*, created silver fox offspring that behaved like domestic dogs (Dugatkin 2018: 1–2).

2. *All* natural species are mutually connected by hereditary relations due to the identical origin in a *common ancestor*.
3. The *mechanism* of evolution of organisms is *natural selection*.

Whilst the discovery of natural selection represents the first reason for the acceptance of Darwin's theory in the nineteenth and twentieth centuries, the second reason is found in what is for Darwin the key evidence of his theory—the fossil record. Specifically, fossils were supposed to confirm Darwin's hypothesis that nature is characterised by a slow, gradual and continual modification of organisms, which finally causes emergence of new species. As Buckland remarked, 'extinct species can all be classed either in still existing groups, or between them. That the extinct forms of life help to fill up the intervals between existing genera, families, and orders, is certainly true' (Darwin 2009: 301). That being the case, the fossil evidence should be abundant with intermediate forms confirming Darwin's assumptions about the evolution of organisms. However, contrary to Darwin's expectations, the fossil evidence of the nineteenth century indicated evolutionary leaps between species, whilst the transitional, i.e. intermediate forms apparently remained extremely rare. In other words, it seems that the fossil record provided little support to Darwin's hypothesis of the slow and gradual evolution of natural species (Sepkoski 2008: 42–43).

The discovery of intermediate forms was certainly important to Darwin. Such discovery would confirm the fundamental hypotheses of his theory, namely, the (1) occurrence of a slow, gradual and continuous evolution of organisms that are (2) mutually connected by hereditary relations because of their common ancestry. Additionally, the discovery of intermediate forms would also confirm that (3) the evolutionary process is guided by slow and gradual operation of natural selection. Yet, it seems that palaeontologists did not encounter many (if any) transitional forms. In order to preserve the compatibility of the fossil record with his theory, Darwin formulates an additional assumption about *the incompleteness of geological data* (Sepkoski 2008: 43). This additional hypothesis is supported by identification of three crucial problems in the nineteenth century palaeontology (Darwin 2009: 264–289):

1. The absence of intermediate varieties at present and in any geological formation
2. The sudden appearance of groups of species in the youngest and oldest fossil strata we know
3. The scarcity of palaeontological collections

As previously noted, this chapter explores (1) the Darwinian problem of the absence of intermediate species. However, I should highlight here that Darwin attempts to provide answers to the remaining two issues of palaeontological data. In his study of the sudden appearance of groups of species, Darwin emphasises that we must consider the long intervals of time which are necessary for the creation of each formation: 'These intervals will have given time for the multiplication of species from a single parent-form; and in the succeeding formation, such groups or species will appear as if suddenly created' (Darwin 2009: 283). Turning to the third problem of the palaeontological data scarcity, Darwin accentuates the difficulties of the process of formation and preservation of fossils: no soft organism can be adequately preserved so that its remains turn into a fossil; insufficient deposition of sediments on the seabed (if such occurs at all) prevents the adequate protection of organic bodies from decomposition and, therefore, prevents the formation of fossils, etc. (Darwin 2009: 270–275).[9] Thus, the palaeontological data scarcity should be expected, and we should not hope that geology will supply us with a complete collection of intermediate forms between natural species (Darwin 2009: 380). Expressly:

> Why does not every collection of fossil remains afford plain evidence of the gradation and mutation of the forms of life? Although geological research has undoubtedly revealed the former existence of many links, bringing numerous forms of life much closer together, it does not yield the infinitely many fine gradations between past and present species required on the theory; and this is the most obvious of the many objections which may be urged against it. Why, again, do whole groups of allied species appear, though this appearance is often false, to have come in suddenly on the successive geological stages? Although we now know that organic beings appeared on this globe, at a period incalculably remote, long before the lowest bed of the Cambrian system was deposited, why do we not find beneath this system great piles of strata stored with the remains of the progenitors of the Cambrian fossils? For on the theory, such strata must somewhere have been deposited at these ancient and utterly unknown epochs of the world's history. I can answer these questions and objections only on the supposition that the geological record is far more imperfect than most geologists believe. (Darwin 2009: 408)

Each of the three introduced problems of paleontological data can be said to have interesting implications for Darwin's theory of evolution. However, it seems that the problem of the absence of intermediate forms was most engaging for the scientists of Darwin's time. And Darwin himself recognises the significance of this problem:

> Why then is not every geological formation and every stratum full of such intermediate links? Geology assuredly does not reveal any such finely-graduated organic chain; and this, perhaps, is the most obvious and serious objection which can be urged against the theory. The explanation lies, as I believe, in the extreme imperfection of the geological record. (Darwin 2009: 264–265)

Apparently, the problem of the absence of intermediate forms remains the 'most obvious and serious objection' for Darwin; this objection could not be clarified even with the discovery of the famous *Archaeopteryx lithographica*. This fossil was discovered 2 years after the first publication of the *Origin of Species*, and it was secured by the British Museum on the recommendation given by Richard Owen (Rupke 2009: 45). Today, we know that *Archaeopteryx* lived approximately 60 million years ago in the late Jurassic period. Its reptile-like skeleton displayed

[9] Darwin's contemporary, Hutton provides a detailed analysis that supports Darwin's insights about the difficulties of the formation and preservation of fossils (Hutton 1861).

morphological similarities with both avians and reptiles. To be specific, *Archaeopteryx* had feathers, beak and the skull of a bird, whilst its skeleton was reptilian (Ayala 2011: 592–593). In the nineteenth century, Owen examined the specimen and concluded that 'the best-determinable parts of its [*Arcaheopteryx's*] preserved structure declare it unequivocally to be a Bird, with rare peculiarities indicative of a distinct order in that class' (Owen 1863: 46). As is known, Thomas Huxley disagreed with Owen, igniting the famous dispute about the origin of the birds. Huxley believed that *Archaeopteryx* represents an intermediate form between reptiles and birds. Otherwise put, he opted for a bold hypothesis that birds originated from reptiles. As Switek notes, Huxley believed that *Archaeopteryx* 'was not a direct ancestor of birds, . . ., but an illustration that birds evolved from reptiles' (Switek 2010: 256).

Contrary to Huxley, who recognises *Archaeopteryx lithographica* as a specimen of an intermediate form (Gawne 2015: 1078),[10] Darwin refers to the discovery of *Archaeopteryx* as evidence for the hypothesis that birds suddenly appeared in the Eocene period (Darwin 2009: 284).[11] Such a methodological decision indicates that Darwin does not recognise *Archaeopteryx lithographica* as an intermediate form specimen. Why is that the case? As noted, Darwin scholarship often regards *Archaeopteryx* as a transitional form specimen that confirms Darwin's theory of evolution.[12] And Darwin himself, however briefly, discusses some *potential* transitional forms:

> As we here and there see a thin straggling branch springing from a fork low down in a tree, and which by some chance has been favoured and is still alive on its summit, so we occasionally see an animal like the Ornithorhynchus or Lepidosiren, *which in some small degree connects by its affinities two large branches of life*, and which has apparently been saved from fatal competition by having inhabited a protected station. (Darwin 2009: 104–105, my emphasis)

A few pages later, Darwin questions 'if the Ornithorhynchus had been covered with feathers instead of hair, this external and trifling character would have been considered by naturalists as an important aid *in determining the degree of affinity of this strange creature to birds*' (Darwin 2009: 367, my emphasis). Yet, it seems that Darwin remains uncertain about the classification of these and similar forms as the intermediate ones. In the next section, my aim is to explore reasons for Darwin's hesitation. If my analysis is successful, Darwin's apparently sceptical stance regarding *Archaeopteryx* (and other possible transitional forms) represents a methodological decision grounded in a specific conception of intermediate forms. As I will show, this conception is highly influenced by Darwin's fundamental gradualist assumptions in his conceptual framework.

---

[10] Yet, Huxley's insights do not indicate that he accepted Darwin's theory of evolution. As Darwin himself writes in his *Autobiography*, Huxley believed that there was not enough of convincing evidence for such theory at the time (Darwin 1892: 169).

[11] Gawne notes that *Compsognathus* receives even less attention as another potential intermediate form, whilst amphibian *Archegosaurus* and the dinosaur *Hypsilophodon* receive no mention whatsoever (Gawne 2015: 1079).

[12] For example, McGowran (2014); Ayala (2011), Bowler (2020).

4 *Archaeopteryx lithographica* and the Problem of Intermediate Species

In the *Origin of Species*, Darwin often uses the term 'intermediate form'. Yet Darwin provides the explanation of this term only in Chap. 9:

> In the first place, it should always be borne in mind what sort of intermediate forms must, on the theory, have formerly existed. I have found it difficult, when looking at any two species, to avoid picturing to myself forms directly intermediate between them. But this is a wholly false view; we should always look for forms intermediate between each species and a common but unknown progenitor; and the progenitor will generally have differed in some respects from all its modified descendants. To give a simple illustration: the fantail and pouter pigeons are both descended from the rock-pigeon; if we possessed all the intermediate varieties which have ever existed, we should have an extremely close series between both and the rock-pigeon; but we should have no varieties directly intermediate between the fantail and pouter; none, for instance, combining a tail somewhat expanded with a crop somewhat enlarged, the characteristic features of these two breeds. (Darwin 2009: 265)

Hence, an intermediate form represents an organic form (organism) related to both its parental and descendant species. Evidently, the progenitor and descendant species will be morphologically distinct to a certain degree. Therefore, intermediate forms have a crucial function to *clearly exhibit the developmental chain of modifications between the parental and descendant species*. An intermediate or transitional form can be construed as representing an organism or species in which a quantitative trait of interest is an approximate morphological median between two other species (Gawne, 2015: 1080). Additionally, Darwin seems to adopt the assumption of *blending inheritance* and argues that intermediate form should manifest the qualities of the progenitor as well as the descendant species:

> It is worthwhile to reflect for a moment on the character of the new species F^{14}, which is supposed not to have diverged much in character, but to have retained the form of (F), either unaltered or altered only in a slight degree. In this case, its affinities to the other fourteen new species will be of a curious and circuitous nature. Being descended from a form which stood between the parent-species (A) and (I), now supposed to be extinct and unknown, it will be *in some degree intermediate in character* between the two groups descended from these two species. (Darwin 2009: 95, my emphasis)

As Delisle argues, Darwin apparently believes that offspring manifests intermediate characteristics of the parent organisms. For example, if the first organism of the parental pair is tall, whilst the other is short, the offspring should be of a medium height (Delisle 2019: 87). Similarly, the transitional form should exhibit intermediate characteristics of both the progenitor and the descendant species:

> A considerable catalogue, also, could be given of forms intermediate between two other forms, which themselves can only doubtfully be ranked as species; and this shows, unless all these closely allied forms be considered as independently created species, that *they have in varying assumed some of the characters of the others*. (Darwin 2009: 127, my emphasis)

Evidently, Darwin assigns a specific meaning to the term of intermediate form. Intermediate or transitional form represents an organism or species which is

(1) positioned between the progenitor and descendant species, (2) formed by the process of modification of the parental species, and (3) exhibits morphological characteristics of remote, but hereditary species, i.e. the progenitor and descendant species. Or, to put it otherwise, an intermediate form is *a result of the slow, gradual and continuous evolution of the progenitor species into later varieties which, by accumulation of delicate modifications, finally becomes a new—descendant—species*. Darwin's intermediate form conception is therefore *gradualist*. It represents a methodological consequence of the fundamental hypothesis of his theory of evolution: (1) evolution of natural species is a *slow*, *gradual*, *continuous*, but for human mind imperceivable process of modification of organisms that are (2) mutually connected by hereditary relations due to their common ancestry. And obviously, this process is guided by (3) the mechanism of evolution—natural selection. That being the case, the discovery of intermediate forms would confirm Darwin's fundamental assumptions about evolution, common ancestry and natural selection at the *ontological level*. In particular, the discovery of intermediate forms would affirm the reality of Darwinian slow, gradual and continuous evolution of natural species mutually connected by hereditary relations. And the main mechanism of such evolutionary process in nature would be natural selection.

In the *Origin of Species*, Darwin states: 'In the first place, it should always be borne in mind what sort of intermediate forms *must*, on the theory, have formerly existed' (Darwin 2009: 265, my emphasis). His theory of evolution thus necessarily implies the existence of intermediate forms. If evolution occurs in a Darwinian manner, then such a process produces specific organic forms—transitional links—that, during distinct periods of time, ontologically connect parental and descendant species:

> Lastly, looking not to any one time, but to all time, if my theory be true, numberless intermediate varieties, linking closely together all the species of the same group, must assuredly have existed; but the very process of natural selection constantly tends, as has been so often remarked, to exterminate the parent-forms and the intermediate links. Consequently, evidence of their former existence could be found only amongst fossil remains, which are preserved, as we shall attempt to show in a future chapter, in an extremely imperfect and intermittent record. (Darwin 2009: 138)

As noted, Darwin believes that palaeontological data do not show many, if any at all, of transitional forms, leading him to formulate the famous problem of the absence of intermediate forms, and declares that the geological evidence is 'extremely imperfect and intermittent record'. Yet, one should not overlook that, beside the difficulties of successful formation and preservation of fossils, Darwin identifies another common cause for the absence of transitional forms and geological data scarcity—the activity of natural selection which tends to 'exterminate the parent-forms and the intermediate links'. A similar observation is found in the following extract from the *Origin*:

> We have seen that species at any one period are not indefinitely variable and are not linked together by a multitude of intermediate gradations, partly because the process of natural selection is always very slow, and at any one time acts only on a few forms; and partly

> because the very process of natural selection implies the continual supplanting and extinction of preceding and intermediate gradations. (Darwin 2009: 164)

Natural selection favours individuals with useful adaptive variations that could help them triumph in the struggle for survival. As intermediate forms represent organisms of average strength and ability, natural selection will not favour them. Due to the lower chances for triumph in the struggle for survival, intermediate forms will eventually disappear, whilst their place in the natural system will be taken by individuals with better chances of survival. So, natural selection is a force that causes modification of organisms and increases their chances of survival. However, Darwin apparently believes that the *passivity* of natural selection, in the case of average organisms, simultaneously leads towards their extinction. In fact, it seems that Darwin attempts to establish the absence of intermediate species as an *expected* ontological consequence of the overall activity of natural selection (De Ricqles 2010: 90). Yet, he does not discuss why the assumption about the slow and gradual extinction of intermediate species causes the geological data scarcity. Instead, Darwin returns to his explanation of the difficulties of formation and preservation of fossils (soft morphological structure, insufficient deposition of sediments on the seabed, etc. (Darwin 2009: 270–275)).

It is time now to return to the question why Darwin does not recognise *Archaeopteryx lithographica* as an intermediate form specimen. As already explained, an intermediate or transitional form represents an organism which is (1) positioned between the progenitor and descendant species, (2) formed by the process of modification of the parental species, and (3) exhibits morphological characteristics of remote, but hereditary species, i.e. the progenitor and descendant species. In fact, it seems that an intermediate form represents a *variety* of the progenitor species—a link that temporally, geologically and evolutionary connects the parental and descendant species. However, if we suppose the existence of a whole chain of such varieties that eventually become a new species, the question that can be asked is *when* a variety actually becomes a new species. Or *what* is the difference between varieties and species? Darwin writes:

> It is all-important to remember that naturalists *have no golden rule by which to distinguish species and varieties*; they grant some little variability to each species, but when they meet with a somewhat greater amount of difference between any two forms, they rank both as species, unless they are enabled to connect them together by the closest intermediate gradations; and this, from the reasons just assigned, we can seldom hope to effect in any one geological section. Supposing B and C to be two species, and a third, A, to be found in an older and underlying bed; even if A were strictly intermediate between B and C, *it would simply be ranked as a third and distinct species, unless at the same time it could be closely connected by intermediate varieties with either one or both forms*. (Darwin 2009: 278, my emphasis)

There is obviously no rule that distinguishes species from varieties. Further, it cannot be known when exactly a variety becomes a new species (Gawne 2015: 1079–1080). Overall, we can say that classification of the form A as a variety or as a member of a new species rests on the amount of morphological similarities between A and the known species B. If those similarities are substantial, A will be classified as a variety of B. However, if similarities between A and B are minimal or non-existent, A will be recognised as a member of a new species. Yet, one should not overlook the following: even though morphological characteristics of A clearly show it represents a transitional form between two remote species B and C, the form A will *not* be recognised as such without discovery of additional intermediate links that connect A to B or C. It therefore seems that *classification of a specific organic form A as intermediate requires additional discovery of a complete chain of transitional forms that connect A with at least one remote, known species*. As Darwin states: 'Even if A were strictly intermediate between B and C, *it would simply be ranked as a third and distinct species, unless at the same time it could be closely connected by intermediate varieties with either one or both forms*' (Darwin 2009: 278, my emphasis). And here we finally find the reason for Darwin's neglect of *Archaeopteryx lithographica* as an intermediate form specimen. Despite the important morphological similarities between *Archaeopteryx* and avians and reptiles, its discovery was not followed by further findings of additional transitional forms that would establish an undeniable link between *Archaeopteryx* and avians and reptiles and demonstrate *a gradualist change of one species into another*.

The problem of the absence of intermediate forms can be said to manifest itself at two distinct but related levels of Darwin's theory. At the first—*conceptual*—level, the problem of intermediate species appears as a difficulty in distinguishing between the concepts (and terms) of species and varieties. In his discussion of such difficulties, Darwin explains that he uses them arbitrarily:

> I look at the term species as one arbitrarily given, for the sake of convenience, to a set of individuals closely resembling each other, and that it does not essentially differ from the term variety, which is given to less distinct and more fluctuating forms. The term variety, again, in comparison with mere individual differences, is also applied arbitrarily, for convenience sake. (Darwin 2009: 42)

Since an intermediate form represents a variety connecting two remote species, the impossibility to distinguish the notion of species from variety makes us unable to determine the conceptual range of the concept of an intermediate form. Further, the problem of the absence of intermediate species appears at the second—*ontological*—level where the classification of a form A as intermediate requires discovery of additional transitional forms that will link A with *at least* one remote known species B or C. Notice how these difficulties arise as a consequence of the Lyellian-Darwinian conception of nature. As explained in the previous sections, the fundamental hypothesis of Lyell's uniformism states that the Earth was shaped by the slow and gradual operation of natural forces whose activity can be detected in the present as well. Darwin welcomes Lyell's uniformism and applies his fundamental assumption to the problem of the natural species transmutation: just as the Earth was shaped by slow and gradual action of geological forces that remain active in the present, so are the natural species slowly and gradually created in the process of evolution. Intermediate forms should assert this uniform and gradualist progress of the evolution of natural species that are mutually connected by hereditary relations due to their common ancestry. A dual function of intermediate forms in Darwin's theory can here be identified, each aspect reflecting one of the three fundamental hypotheses of the theory. Firstly, intermediate forms should connect distinct species, i.e. ensure the existence of their mutual hereditary relations and thus confirm Darwinian assumption about the common ancestry. Secondly, intermediate forms should affirm the hypothesis about evolution as *slow*, *gradual* and *continuous process*. Additionally, since evolutionary modifications are too delicate to be observed by human mind, intermediate forms should *show us precisely those slight morphological modifications of organisms*. And here we find the reason why the classification of a specific form as intermediate requires discovery of a complete chain of transitional forms that would link it to at least one known species: discovery of additional intermediate forms would show the gradual and continuous modification of organisms and ensure the *temporal*, *geological* and *evolutionary* connection of all natural species. *Archaeopteryx lithographica* could not satisfy such a rigorous methodological requirement. I believe this is the reason Darwin does not recognise it as an intermediate form. However, it seems that no form, living or extinct, can meet such a strict criterion. If that is the case, then the problem of the absence of intermediate species remains 'the gravest objection' to Darwin's theory of evolution.

As Sepkoski observes, Darwin's theory revolutionised palaeontology; the fossil record became the crucial (if not only) potential evidence for the reality of Darwinian evolution (Sepkoski 2008: 43). Darwin's study of the fossil record emphasised the importance of the morphology of organic beings, making it one of the crucial aspects of palaeontological investigations. As Bowler remarks, 'morphology thus became the first centre of evolutionary biology' (Bowler 1996: 41). And whilst the problem of intermediate species remains,[13] the contemporary Darwinists attempt to reduce its significance. Some authors particularly emphasise Darwin's aim in the *Origin*—to convince the reader of species mutability and to demonstrate the fallacy of Creationism (Warren 2015: 17–18). Others insist on the importance of the necessary conditions for formation and preservation of fossils; the absence of such conditions imply the absence of certain fossil remains and appearance of specific 'holes', i.e. missing links, in the chain of deceased beings. To clarify, the problem of intermediate species ceases to be the problem once we realise how complex the process of formation and preservation of the fossil record is; in fact, the absence of some deceased forms should be expected (Thanukos 2009: 84–85). Finally, some Darwinists highlight the overall state of the nineteenth century palaeontology and the development the discipline achieved in the period after the publication of the *Origin of Species*. Darwin's conclusions about imperfection of the geological data and the

[13] As Raup notices, 'the geological record did not then and still does not yield a finely graduated chain of slow and progressive evolution. In other words, there are not enough intermediates' (Raup 1979: 23).

problem of intermediate species are understandable given the scarcity of available paleontological data at the time (Warren 2015: 19).[14]

Sepkoski notes that, without Darwin's theory of evolution, palaeontology seems to be nothing more than a discipline focused on providing descriptive observations about the morphological form and distribution of extinct creatures (Sepkoski 2008: 43). That being the case, Darwin's conclusion about the paleontological data scarcity and incompleteness of the geological record on the whole certainly had an impact on the development of the discipline after the publication of the *Origin*. Warren notices that modern commentators 'have bemoaned this emphasis on "imperfection" because they think that it left a negative legacy for palaeontology that took more than a century to overcome' (Warren 2015: 17). Apparently, Darwin provided palaeontologists with three research alternatives. Firstly, they could solely focus on the descriptive studies of fossil morphology and their stratigraphic positioning. Secondly, palaeontologists could embrace Darwinism and direct their research towards the discovery of intermediate forms. Finally, they could deny Darwinism and explore alternative evolutionary theories that explained the actual fossil record (Sepkoski 2008: 43). One such alternative was developed by Niles Eldredge (1943–) and Stephen J. Gould (1941–2002) whose controversial theory of *punctuated equilibrium* holds that 'the great majority of species, as evidenced by their anatomical and geographical histories in the fossil record, originate in geological moments (punctuations) and then persist in stasis throughout their long durations' (Gould 2007: 40). Realising the impact of Darwinism on the contemporary palaeontological research and the general aim towards discovery of intermediate forms, Eldredge and Gould developed an alternative approach that will complement the palaeontological data. This approach was founded on the three essential hypotheses: (1) speciation occurs by splitting the lineages, and (2) new species arise *rapidly* (3) in an isolated group of organisms on the periphery of ecosystems. Due to limitations of this paper's scope and its main subject, I shall not endeavour a deeper analysis of Eldredge and Gould's interesting theory. However, it is worth noting that the theory of punctuated equilibrium, whilst remaining controversial, seems to be successful in resolving Darwin's problem of the fossil record, challenging his hypothesis of the geological data incompleteness. Namely, Eldredge and Gould emphasise that new species develop rapidly in isolated, peripheral parts of an area occupied by their parental species. That being the case, 'in any local section containing the ancestral species, the fossil record for the descendant's origin *should consist of a sharp morphological break between the two forms*. This break marks the migration of the descendant, from the peripherally isolated area in which it developed, into its ancestral range… *Many breaks in the fossil record are real; they express the way in which evolution occurs, not the fragments of an imperfect record*' (Eldredge and Gould 1971: 96, my

[14]Whilst the text refers to some interesting approaches towards the problem of intermediate species, I should note that there are scholars (e.g. Bowler (2020), Ayala (2011)) who believe this problem is resolved through discovery of several specimens such as *Archaeopteryx lithographica* and *Tiktaalik*.

emphasis). Otherwise put, the theory of punctuated equilibrium posits that we *should* expect sharp morphological differences between parental and descendant forms in the fossil record. However, sharpness as a feature of these differences *does not* indicate absence of some additional, intermediate forms that temporally, geologically and evolutionary connect the parental and descendant species. Rather, it reveals the steps of evolutionary process of those two species.

5 Concluding Remarks

The general aim of this chapter was to examine the role of the fossil record in Darwin's theory of evolution. Specifically, I intended to deconstruct one of the central concepts of Darwinism, the concept of intermediate species, and the function of transitional forms in Darwin's theory of evolution. After a brief discourse on the historical changes in the epistemological status of the fossil record, I identified three fundamental hypotheses of Darwin's theory:

1. Evolution of natural species occurs in nature; it represents a *slow*, *gradual*, *continuous* and, for human mind, *imperceivable* process of modification of organisms that finally causes the emergence of new species—the final product observable by our mind.
2. *All* natural species are mutually connected by hereditary relations due to the identical origin in a *common ancestor*.
3. The *mechanism* of evolution of organisms is *natural selection*.

In the subsequent sections of the chapter, I examined Darwin's study of the fossil record, asking the following questions: What is an intermediate form for Darwin? What is the Darwinian problem of the absence of intermediate forms? Why does Darwin neglect the discovery of *Archaeopteryx lithographica* in his argumentation about intermediate species? *What is the function of fossils and intermediate forms in Darwin's theory of evolution?* As explained, Darwin believes that an intermediate form represents an organism which is (1) positioned between the progenitor and descendant species, (2) formed by the process of modification of the parental species, and (3) exhibits morphological characteristics of remote, known species, i.e. the progenitor and descendant species. Although his theory implies the necessary existence of transitional forms, Darwin believes that the palaeontological record remains incomplete due to failure in discovery of intermediate forms. My analysis, however, shows that the famous problem of intermediate species arises at two distinct but connected levels of Darwin's theory. The first level is conceptual; here the problem of intermediate species appears as the difficulty to provide a precise and adequate interpretation of the concept of intermediate forms. As I have explained, this *conceptual* difficulty represents a consequence of Darwin's ambiguous use of some of the key concepts of his theory, the concepts of species and variety. Further, it seems that the problem of intermediate species also arises at the ontological level; here the existence of intermediate forms should ensure temporal, geological and evolutionary connection of all organic beings and confirm the reality of Darwinian evolution in the natural system. However, it appears that the Darwinian way of classifying a specific organism as an intermediate form requires additional discovery of a complete chain of transitional forms. No form, extinct or living, can satisfy such a strict requirement. Hence, according to Darwin's theory, no organic form can be recognised and classified as an intermediate form. That being the case, I concur with Delisle stating that Darwin offers us to *imagine* transitional forms, and not to *observe* the actual ones (Delisle 2019: 123). Nevertheless, one should not overlook that the problem of intermediate forms is tied with Darwinian gradualism. Thus, abandoning the gradualist view of nature eliminates Darwin's problems of geological and palaeontological data. Geology and palaeontology do not represent incomplete scientific disciplines if we opt for alternatives that explain the actual fossil record. As noted, the punctuated equilibrium theory is one such alternative. However, as long as we pertain to the gradualist conception of evolution, it seems that the problem of intermediate species remains 'the gravest objection' to Darwin's theory.

We can say that Darwin's investigation of the fossil record and the role of fossils in the *Origin of Species* is shaped by the fundamental hypothesis of his theory (Archibald 2017: 38). As De Ricqles observes, palaeontological data contribute to the explanatory value of Darwin's hypotheses and allow them to ascend to the level of a theory. Nonetheless, 'in Darwin's mind, the explanatory mechanism of evolution (the process of natural selection) takes precedence over the "fact" or historical reality of evolution (pattern)' (De Ricqles 2010: 89). To put it otherwise, Darwin overlooks what the fossil record *clearly* showed: evolutionary divergence, convergence and parallelism as well as the possibility of evolutionary leaps. The reasons for this neglect, and for the emergence of various problems in his investigation of geological data, can be found in Darwin's uniformist and gradualist conception of nature. As I have explained, this conception is based on the fundamental hypothesis of Lyell's uniformism. However, one should not overlook that Darwin formulated his theory in the second half of the nineteenth century, at the time when two conceptually different worlds were intersecting: (1) the world of Romanticism, i.e. the world of a harmonious, static nature shaped by actions of the 'Author of Nature' and (2) the world of a dynamic nature governed by natural laws, filled with perturbations and multiplicity of variable species. Darwin's theory embraces certain aspects of both those worlds. Thus, in his theory we encounter a uniform, steady-state system of nature characterised by delicate, dynamic change in the form of the evolution of natural species. As Delisle states, in addition to the revolutionary Darwin, we should oppose another Darwin:

> Holding to a rigid pan-divergent view negating the evolutionary contingencies, complexities, and novelties, a view organised around fixed independent classes/phyla looked at from the present time only, in addition to having failed at proposing a theory of evolution even remotely unified. This other Darwin ultimately based his view on a commitment toward a steady-state view of biology, wherein evolutionary motion (change) constitutes a superficial feature of a world fundamentally static. (Delisle 2019: 43–44)

It is that other image of Darwin that the contemporary Darwin scholarship tends to forget. That image does not compromise his scientific accomplishment but reminds us that, after all, Darwin was a scholar of the nineteenth century, who viewed nature as a uniform, mostly static system with occasional but limited manifestations of change in the form of the evolution of natural species. This, however, can be the subject of future research.

Acknowledgements I thank Richard Delisle for the invitation to contribute to this special issue and expand my research on the deconstruction of Darwinism. I also thank Richard for his support, advices and inspiring discussions during the process of writing this chapter. Finally, I thank two anonymous referees for useful suggestions that greatly improved the quality of the chapter.

References

Archibald JD (2017) Origins of Darwin's evolution: solving the species puzzle through time and place. Columbia University Press, New York

Ayala FF (2011) Darwin in the twenty-first century: natural selection, molecular biology, and species concepts. In: de Bruijn FJ (ed) Handbook of molecular microbial ecology, Volume II: metagenomics in different habitats. John Wiley & Sons Inc, pp 589–596

Bowler PJ (1976) Malthus, Darwin, and the concept of struggle. J Hist Ideas 37:631–650

Bowler PJ (1996) Life's splendid drama: evolutionary biology and the reconstruction of life's ancestry: 1860–1940. University of Chicago Press, Chicago

Bowler PJ (2003) Evolution: the history of an idea. University of California Press, Berkeley

Bowler PJ (2020) Making modern science: a historical survey. The University of Chicago Press, Chicago

Darwin C (1892) Charles Darwin: his life told in an autobiographical chapter, and in a selected series of his published letters. John Murray, London

Darwin C (2009) The origin of species: by means of natural selection, or the preservation of favoured races in the struggle for life, 6th edn. Cambridge University Press, Cambridge

De Ricqles A (2010) On Darwin's palaeontology in the origin of species. C R Biol 333:87–94

Delisle RG (2019) Charles Darwin's incomplete revolution: the origin of species and the static worldview. Springer Nature Switzerland AG, Cham

Dobzhansky T (1973) Nothing in biology makes sense except in the light of evolution. Am Biol Teach 35:125–129

Dugatkin LA (2018) The silver fox domestication experiment. Evol Educ Outreach 11:10–15

Egerton FN (1970) Humboldt, Darwin, and the concept of struggle. J Hist Biol 3:325–360

Eldredge N, Gould SJ (1971) Punctuated equilibria: an alternative to phyletic gradualism. In: Thomas J, Schopf M (eds) Models in paleobiology. Freeman Cooper, pp 82–115

Flannelly KJ (2017) 19th century evolutionary thought before Charles Darwin. In: Flannelly KJ (ed) Religious beliefs, evolutionary psychiatry, and mental health in America: evolutionary threat assessment systems theory, vol 1, pp 29–37

Gale BG (1972) Darwin and the concept of a struggle for existence: a study in the extrascientific origins in scientific ideas. Isis 63:321–344

Gawne R (2015) Fossil evidence in the origin of species. Bioscience 65:1077–1083

Gould SJ (1982) Darwinism and the expansion of evolutionary theory. Science 216:380–387

Gould SJ (2007) Punctuated equilibrium. Harvard University Press, Cambridge

Gould SJ (2011) Leonardo's mountain of claims and the diet of worms. Harvard University Press, Cambridge

Hutton FW (1861) Some remarks on Mr. Darwin's theory. The Geologist 4(5):183–188

Leibniz GW (2008) Protogaea. The University of Chicago Press, Chicago

McGowran B (2014) Organic evolution in deep time: Charles Darwin and the Fossil record. Trans R Soc S Aust 137:102–148

Owen R (1863) On the Archeopteryx of Von Meyer, with a description of the fossil remains of a long-tailed species, from the lithographic stone of Solenhofen. Philos Trans R Soc Lond 153: 33–47

Raup DM (1979) Conflicts between Darwin and palaeontology. Field Mus Nat Hist Bull 50:22–29

Richards RJ (2002) The romantic conception of life: science and philosophy in the age of Goethe. The University of Chicago Press, Chicago

Richards RJ (2009) Darwin's place in the history of thought: a re-evaluation. Proc Natl Acad Sci USA 106:10056–10060

Rupke N (2009) Richard Owen: biology without Darwin, a revised edition. The University of Chicago Press, Chicago

Sepkoski D (2008) The unfinished synthesis?: palaeontology and evolutionary biology in the 20th century. J Hist Biol 52:687–703

Sloan PR (2001) 'The sense of sublimity': Darwin on nature and divinity. Osiris 16:251–269

Switek B (2010) Thomas Henry Huxley and the reptile to bird transition. Geol Soc Lond 343:251–263

Thanukos A (2009) Darwin's "extreme" imperfection? Evol: Educ Outreach 2:84–89

Warren AD (2015) Darwin and palaeontology: a re-evaluation of his interpretation of the fossil record. Hist Biol Int J Paleobiol 28(5):1–27

Deconstructing Darwinism with Darwin, Mayr, and Gould: Through the Lens of Evolutionary Contingency

Richard G. Delisle

Abstract

The label "Darwinism" has been with us for so long that it seems totally unreasonable to propose to dispose of it. Yet, we maintain it has outlived its utility in so far as it acts more like a distorting lens than as a magnifying one. This will be illustrated by investigating the notion of "evolutionary contingency" in Charles Darwin, Ernst Mayr, and Stephen Jay Gould. It is a common assumption of the received view that Darwinism is consubstantial with the idea of an open-ended evolutionary process in time. However, Darwin's many commitments toward uniformitarianism seriously limited his understanding of evolutionary contingency, while the early Mayr (roughly between 1940 and 1970) stayed too close from this forgotten Darwin to fully engage with it (notwithstanding his rhetoric to the contrary). Indeed, the way Darwin and Mayr exploited the principle of divergence closed off several open avenues to evolutionary change, thus channeling the expression of contingency. For his part, Gould made a significant contribution toward a more robust appreciation of evolutionary contingency (a view partly supported by the late Mayr (1990–2005) who by that time had retreated significantly from his earlier view), although Gould's justification has its intellectual roots in a mythical Darwin that never existed. Evolutionary contingency has thus far played the role of a blanket notion artificially uniting "Darwinism" and obscuring important features part of Darwin's, Mayr's, and Gould's brands of biology. Our analysis points to the usefulness of distinguishing between competing ontologies of evolutionary time.

This chapter contains a very partial discussion of topics to be presented in full in a monograph in progress and tentatively entitled *The Changing Landscape of Evolutionary Biology: Darwin, Mayr and Gould*.

R. G. Delisle (✉)
School of Liberal Education, University of Lethbridge, Lethbridge, AB, Canada
e-mail: richard.delisle@uleth.ca

R. G. Delisle et al. (eds.), *Unity and Disunity in Evolutionary Biology*,
https://doi.org/10.1007/978-3-031-42629-2_8

Keywords

Darwinism · Rhetoric · Evolutionary contingency · Creative time · Ontology of time · Cone of increasing diversity · Decimation model · Charles Darwin · Ernst Mayr · Stephen Jay Gould

1 Introduction

It is hard to understand the nature of evolutionary biology if one pays too much attention to what practicing biologists explicitly say about it. In constructing their own working narratives, it is fair game to select bits of theory here and there from its past as suits their needs. At the same time, historians and philosophers have no obligation to follow that self-interested road, especially considering that alternative narratives are available by engaging directly with the writings of these evolutionists, provided enough precautions are taken. The label "Darwinism" constitutes, in our view, such a problematic narrative. We hold that Darwin could not have subscribed to evolutionary contingency except in a relatively modest way. His central commitment toward the principle of divergence implies constraining the action of a creative process. For his part, Mayr started his career by following Darwin on many of the same choices that thwart the rise of evolutionary novelties. It is only when Mayr had to defend his key commitments—in the face of competing approaches and theories in the 1970s and after—that he began to loosen-up some of them. By so doing, Mayr slowly moved away from the historical Darwin we describe in this chapter, simultaneously making his overall view more open to evolutionary contingency. To give our analysis more epistemological leverage, it will be useful to introduce a third figure in our story: Stephen Jay Gould. The critical approach adopted here should not be taken to be part of a normative enterprise: it is merely a tool allowing us to probe beyond the facades of discourses.

2 Beyond the Apparent Unity

The explanatory core of "Darwinism" is often conceived in terms of an adaptive evolutionary process, a truly creative one driven by natural selection. As summarized by Peter J. Bowler (1988: 9):

> Darwin devised the theory of natural selection as an explanation of how populations adapt to changes in their environment...The variations from which the environment selects are essentially random and undirected...The fact that variations are not directed along fixed lines emphasizes the haphazard or open-ended nature of a process governed only by adaptation.

Darwin's commitment to an open-ended view of evolution is thought to rest on his classic exposition of this idea in *On the Origin of Species* (1859: 314):

> These several facts accord well with my theory. I believe in no fixed law of development, causing all the inhabitants of a country to change abruptly, or simultaneously, or to an equal degree. The process of modification must be extremely slow. The variability of each species is quite independent of that of all others. Whether such variability be taken advantage of by natural selection, and whether the variations be accumulated to a greater or lesser amount, thus causing a greater or lesser amount of modification in the varying species, depends on many complex contingencies—on the variability being of a beneficial nature, on the power of intercrossing, on the rate of breeding, on the slowly changing physical conditions of the country, and more especially on the nature of the other inhabitants with which the varying species comes into competition. Hence it is by no means surprising that one species should retain the same identical form much longer than others; or, if changing, that it should change less.

As Darwinism developed over time—so the argument goes—this key explanatory feature was kept as one of its essential features. As argued by Ernst Mayr in *The Growth of Biological Thought* (1982a: 27):

> Interestingly, Darwin's approach was completely in line with modern theory. He realized he could never demonstrate evolutionary conclusions with the certainty of a mathematical proof. Instead, in about twenty different parts of the *Origin* he asks: 'Is this particular finding—whether a pattern of distribution or an anatomical structure—more easily explained by special creation or by *evolutionary opportunism*?' Invariably, he insists that the second alternative is the more probable one. Darwin anticipated many of the most important tenets of the current philosophy of science. [my emphasis]

With the publication of the *Origin of Species*, Mayr (1982a: 21–82) continues, a new conception of science was developed, one devised to confront unique and unpredictable evolutionary events in the argumentative style of historical narrative. Indeed, as Stephen Jay Gould also seems to be saying (1986: 60), the "long argument" found in Darwin's *Origin* sees history being the organizing principle of the relationship among organisms, evolution's order reflecting a historical pathway. Stating this differently in *Wonderful Life* (1989: 289–290), Gould writes:

> Replay the tape [of life] a million times from the Burgess beginning, and I doubt that anything like *Homo sapiens* would ever evolve again...When we set our focus upon the level of detail that regulates most common questions about the history of life, contingency dominates and the predictability of general form recedes to an irrelevant background. Charles Darwin recognized this central distinction between laws in the background and contingency in the details. [italics original]

For the moment, let us stress the uniformity of a narrative which takes for granted that Darwin had already been fully committed to the notion of evolutionary contingency. This is precisely what is being contested here. The matter is complex since it involves these scholars' self-perception: how they see the nature of their own scientific contribution and how they see their relation to ideas of other scholars (for a different approach on evolutionary contingency, see Tahar (2024)).

3 Questioning the Label "Darwinism"

Before tackling the positions of Darwin, Mayr, and Gould, it will be useful to first briefly summarize aspects brought to the forefront in recent analyses concerning "Darwinism." Since readers are likely to find some of this surprising, we invite them to suspend judgement pending the full expositions of the arguments. It is now an ongoing reality that a historiographic shift is in the making, with traditional labels like the *Darwinian Revolution*, the *Eclipse of Darwinism*, and the *Modern Synthesis* coming under greater scrutiny (e.g., Ceccarelli 2024a; Esposito 2021, 2024; La Vergata 2023: 591–600, 2024; Ceccarelli et al. 2024). Although space prevents us from providing a fuller exposition, three complementary features of this historiographical shift should be noted:

1. Darwinism, broadly conceived, does not represent a dominant intellectual thread running throughout the twentieth century, considering that it interacted and fused in significant ways with other research entities running in parallel, thus blurring its separateness in intellectual space and time and making its isolation from other such entities impossible (Caso 2024; Ceccarelli 2021, 2024b; Delisle 2008, 2009a, b, 2011, 2017a, b; Dolbeault 2024; La Vergata 2024; Levit and Hossfeld 2006, 2011, 2017; Wagner 2024).
2. The notion that natural selection is at the causal and explanatory core of Darwinism is not supported by recent analyses, on the grounds that many "Darwinians" like Darwin himself, Julian S. Huxley, Theodosius Dobzhansky, Ernst Mayr, and Bernhard Rensch do not fully subscribe to the "mechanism-centric" view claimed by the traditional historiography (Delisle 2019, 2021b; Delisle and Tierney 2022; Van der Meer 2021; see also La Vergata 2024 for other evolutionists).
3. Neo-Darwinism (in the guise of the Modern Synthesis) in the 1930–1960 period constitutes a confused reality for two related reasons: (a) several attempts at a synthesis were made before, during, and after; and (b) the so-called Modern Synthesis largely failed at the conceptual level under the weight of its inherent pluralism, confining it to a sociological attempt (Adams 2021, 2024; Baedke et al. 2024; Cain 2009; Ceccarelli 2021; Delisle 2008, 2009a, 2017a; Largent 2009; Levit et al. 2008; Ochoa 2021, 2024; Smocovitis 1999).

This chapter further expands upon this ongoing historiographic shift by focusing, this time, on the notion of periodization: epistemological/ontological discontinuities that were introduced in evolutionary thinking since the nineteenth century.

4 Charles Darwin's Ontology of Evolutionary Time

An obstacle to a better understanding of Darwin's contribution to evolutionism comes from reading his work through the paradigm of the Modern Synthesis. As summarized by Mayr (1980a) and Gould (1980a), that synthesis is presumably characterized by two main explanatory components: (1) gradual evolution is explained by small variations oriented by an adaptive process (natural selection); (2) evolutionary trends and manifestations in general are macroevolutionary phenomena explainable solely in terms of a process occurring at the microevolutionary level. This takes us back to evolutionary contingency considering that evolution is seen as an open-ended process that generates a whole range of evolutionary patterns driven by natural selection (adaptation), with selection holding a central explanatory role in evolutionary theory.

As argued at length elsewhere (Delisle 2014, 2017a, 2019, 2021b; Delisle and Tierney 2022), the Darwin of the Modern Synthesis is a recent fabrication. In contradistinction to the common assumption that *process* (natural selection) explains *pattern* (the history of life), the two sides of this equation are now known to be connected in a more complex way in Darwin. Indeed, the philosopher Elliott Sober offers us an important clue which is reflected in his book title, *Did Darwin Write the Origin Backwards?* (2011). While the first few chapters of the *Origin* focus on natural selection, the last few chapters are devoted to the ideas of common ancestry and divergence in fields such as biogeography, systematics, morphology, embryology, and comparative anatomy. As suggested by its title, Sober wonders about the conceptual relationship established by Darwin between common ancestry and natural selection. He writes:

> The first chapter [of Sober's book] examines a simple but fundamental question about how Darwin organized his argument in the *Origin of Species*. The two big ideas in Darwin's theory are common ancestry and natural selection. Why did Darwin put selection first and foremost in the *Origin* and allow his views about common ancestry to emerge only gradually and as a secondary theme? This is a question about Darwin's *rhetoric*—the strategy he used to lay out his argument. The question becomes pressing when we look at Darwin's *logic*—at how the elements of his theory are related to each other and to the evidence he assembles. (Sober 2011: 9; italics original)

Sober argues that although common ancestry and natural selection are entangled in Darwin's theory, they hold a particular relationship to one another:

> Darwin used the fact of common ancestry to reason about natural selection...However, Darwin's Principle says that you don't need to look at traits that evolved by natural selection to argue for common ancestry. In fact, to defend hypotheses of common ancestry, Darwin thinks you need traits whose presence was *not* due to natural selection. This is the evidential asymmetry that separates common ancestry from natural selection in his theory. So, did Darwin write the *Origin* backwards? The book is in the right causal order, but evidentially, the book is backwards. (Sober 2011: 44; italics original)

Sober's full reasoning is as follows: if natural selection is envisioned as an all-powerful force, its action (adaptation) will delete traces connecting affiliated forms, thus making the task of identifying common ancestors impossible. Darwin's theory, then, required a selective force weak enough to leave in place affiliative traces (Sober 2011: 35). This explains why Sober (2011: 21) holds that instead of describing Darwin's theory as "evolution *by* natural selection" it would be better to describe it as "common ancestry *plus* natural selection." As much as Sober (2011:

43) insists on the epistemological distinction at the heart of Darwin's theory—for "the inference of common ancestry does not depend on the occurrence of selection"—he does not go as far as to propose a complete disjuncture between natural selection and common ancestry/divergence. That step, however, is more clearly taken by the historian Mary P. Winsor, who argues that the pattern of divergence (common ancestry) had an autonomous status for Darwin, independently of natural selection. Here, we follow Winsor (2009: 47) in holding that pattern has explanatory precedence over process in Darwin:

> The leaders of the Modern Synthesis and their epigones have convinced us that this means that belief in evolution could not be scientific unless it included a cause, which they have achieved by overlooking Darwin's claim that a reasonable person ought to accept branching evolution 'even if it were unsupported by other facts or arguments' (Darwin 1859: 458)... Yet after the Modern Synthesis, many writers treat natural selection and evolution as interdependent and inextricable parts of one unified theory. This flies in the face of the historical facts.

Moving beyond the traditional narrative of the "Synthesis Historiography" (Amundson 2005: 11) and taking Winsor's reasoning one step further, we argue that behind evolutionary pattern, Darwin's "tree of life" (see Fig. 1) hides a forest of decisive commitments relegating process (natural selection) to an auxiliary role (Lakatos 1970: 132–138; Delisle 2021b). Although a modern mind sees in the notion of "divergence" an evolutionary pattern among a whole range of other possible evolutionary manifestations—reticulate evolution, parallelism, convergent evolution, etc.—Darwin, for his part, elected to recognize in it a true *principle*, one that predominates and structures the entire evolutionary process. Indeed, the pattern of divergence is at the small scale what the cone of increasing diversity is at the large scale (see Fig. 2). To reverse the perspective, if one assumes as Darwin does that the overall view of life is organized as a cone of increasing diversity, the pattern of divergence becomes of necessity a small-scale justification for the large-scale reality. And if the pattern of divergence predominates in Darwin's evolutionary biology, we hold, it is because he made the crucial assumption that life's history as seen in the *present time* is at its highest level of biological diversity ever, implying that this degree of diversity is ever more reduced as one travels back in geological time toward a limited number of common ancestors (and, perhaps, only a single one). Darwin expresses this key idea in these words:

> Let us now look to the mutual affinities of extinct and living species. They all fall into one grand natural system; and this fact is at once explained on the principle of descent. The more ancient any form is, the more, as a general rule, it differs from living forms. *But...all fossils can be classed either in still existing groups, or between them.* That the extinct forms of life help to fill up the wide intervals between existing genera, families, and orders, cannot be disputed. For if we confine our attention either to the living or to the extinct alone, the series is far less perfect than if we combine both into one general system. With respect to the Vertebrata, whole pages could be filled with striking illustrations... showing how extinct animals fall in between existing groups.... (Darwin 1859: 329; see also 1872: 301) (my emphasis)

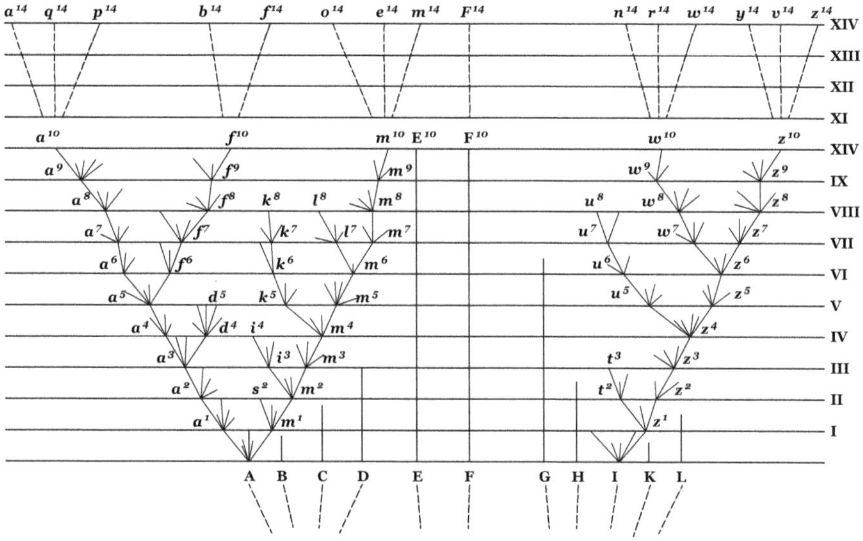

Fig. 1 Darwin's Diagram in the *Origin of Species* represents a genealogy of extant forms and not an actual phylogenetic tree as we might imagine it today. In addition, past forms therein are not genuine "common ancestors" but imagined forms created out of features selected from among extant ones and projected backward in geological time. In a sense, Darwin's Diagram constitutes a "preformationist machine"

It has been overlooked among Darwin scholars that what he expresses through the cone of increasing diversity is more than a dominant pattern called "divergence" since it constitutes, in itself, an "ontology" in the full sense of the expression: *a claim about the nature and fundamental properties of life's history.* As such, branching evolution is, in Darwin's *Origin of Species* (1859–1872), the tip of a conceptual iceberg, whose several properties allow us to penetrate his ontology. Among such properties one finds the following:

1. The principle of divergence—whose ultimate function is to ensure that extinct forms fall in-between extant ones—prevents the rise of variations among extinct forms that are not already recorded among extant forms.
2. The principle of gradation—which levels out discontinuities by tightly connecting past forms to recent ones through intermediate links—further prevents the rise of such novelties.
3. A tree of life organized around the notion of "fractal geometry" wherein change is straightjacketed under a rigid, geometrical, and predictable pattern—divergence—endlessly repeating itself and ruling at all taxonomic levels.
4. Fixed taxonomic positions since the beginning of life.

Fig. 2 The cone of increasing diversity. This evolutionary model promoted by Charles Darwin and applied within separate classes assumes that variability observed among extant forms entirely captures that of the past, considering that all fossils are either aligned with or fall in-between extant forms. Such a model allows for little evolutionary room as far as the rise of true novelties is concerned

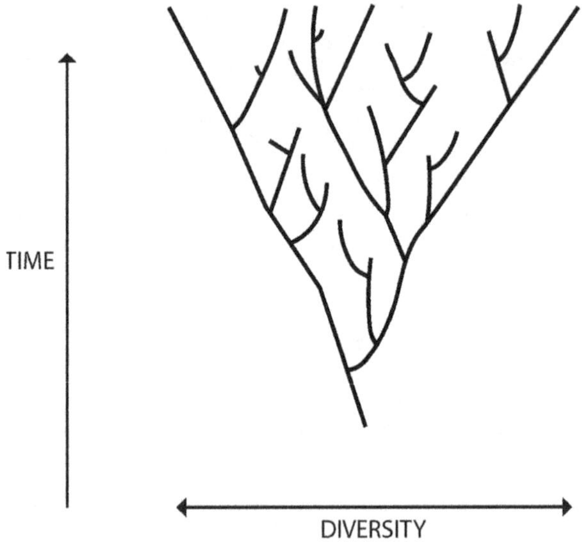

5. A segmented tree of life organized around separate and independent classes/phyla, with change restricted to the rise and fall of entities at lower taxonomic levels only (at and under the order level).
6. Permanent and stable high taxonomic groups (classes and phyla) deeply entrenched in nature, none of which ever went extinct.
7. A biological modernity acquired in its main features as early as the Cambrian period.
8. A concealed yet pervasive support for a notion akin to preformationism under many guises. Darwin's Diagram (see Fig. 1) in the *Origin* constitutes a kind of preformationist machine wherein so-called common ancestors are imagined creatures constructed out of selected features seen among extant forms and projected backward in geological time.

We are aware that what we said thus far about Charles Darwin flies in the face of orthodoxy (see also the original proposal of Partridge (2018, 2024)), but the strength of our thesis will be enhanced, we think, when seen against the backdrop of a range of similar commitments in Mayr. In a nutshell, the *Origin of Species* is less about "natural selection" than a "view of life," which owes much to the uniformitarian doctrine of the eighteenth century.[1] Darwin's view of life is *structured* in such a way as to powerfully restrict the free action of natural selection by channeling its force along a pre-established pattern—divergence—a direct logical consequence of his

[1] Space is lacking here to expand upon how Darwin managed to generate evolutionary thinking by adjusting the eighteenth-century uniformitarian conceptual envelope to nineteenth-century evolutionism (see Delisle 2014, 2019: 1–64; Delisle and Tierney 2022: 63–120).

presentist ontology which assumes that all past and extinct forms *must* fall in-between extant ones, with the notion of "common ancestry" being another natural companion to the cone of increasing diversity. What follows summarizes a long, robust, and consistent list of mutually reinforcing commitments to be found in Darwin's *Origin*.

5 Neontology Versus Paleontology

The *Origin* devotes only two chapters to the fossil record (1859: 279–345; 1872: 264–289). In addition, these provide little direct support for Darwin's theory based on common ancestors and divergent evolutionary lines, for instance, and serve more as a negative argument explaining why the fossil record should not be expected to reveal such manifestations (e.g., 1859: 280–306). The *Origin* is, therefore, focused on neontology much more than on paleontology. While this is a *methodological* approach, it naturally emerges from his *ontological* view of the living world.

While it is well known that for Darwin the principle of divergence constitutes a cornerstone of his theory—accompanied by its sister notion, "common ancestry"—this principle also serves the overarching explanatory function of bracketing or shoehorning all past forms under the biological variability of currently living ones (neontology), as explicitly stated in the *Origin of Species* (1859: 329; 1872: 301–302). In other words, the present time is believed to be always more revealing than the past, not because the present exposes these forms to direct observation—in contradistinction to a concealed past—but because it is only in the present time that one sees the full expression of biological diversity. Differently stated, it is only with evolutionary time that one sees all the potential diversity being expressed during life's history. In Darwin's ontology, then, the biological diversity of the past falls wholly within that of the present, the latter including everything that existed in that past, *plus* more (see also Stamenkovic (2024)). If this is so, why even bother studying the past at all? In fact, this is not far removed from Darwin's attitude.

The implications of Darwin's ontology are significant: they prevent him from dealing properly with evolutionary contingency considering that novelties of the past are denied by being shoehorned under categories designed for currently existing (extant) forms. Interestingly, William Whewell (1794–1866) foresaw the inherent limitations accompanying Darwin's brand of uniformitarianism, the same that was promoted in Charles Lyell's *Principles of Geology* (1830–1833) and directly criticized by Whewell:

> It seems to us somewhat rash to suppose, as the uniformitarian does, that the information which we at present possess concerning the course of physical occurrences, affecting the earth and its inhabitants, is sufficient to enable us to *construct classifications, which shall include all that is past under the categories of the present*. Limited as our knowledge is in time, in space, in kind, it would be very wonderful if it should have suggested to us all the laws and causes by which the natural history of the globe. . .is influenced—it would be strange, if it should not even have left us ignorant of some of the most important of the agents

which, since the beginning of time, have been in action.... (Whewell 1832: 126; my emphasis)

The ontological meaning of Darwin's choice is best revealed when contrasted against an alternative ontology, as expressed under the "decimation model" (compare Figs. 2 and 3). The implication is that Darwin's discourse about an overpopulated world putting adaptive pressure on organisms and species—by creating the condition for evolution through divergent pathways—is a *post hoc* justification of a foregone conclusion: divergence can only be the main path open in a world organized around the cone of increasing diversity.

In addition, Darwin enshrined within his view a complementary means to keep the past under the epistemic control of the present: the principle of gradation/gradualism, *Natura non facit saltum* (Darwin 1859: 194, 206, 210, 243; 1872: 156, 166, 234). This motto has the ultimate explanatory function of keeping the *past* tightly riveted to the *present* through presumed intermediate links—leaving no empty space between the two by preventing the rise of discontinuities and *unknown novelties* (e.g., 1859: 432; 1872: 380)—thus ensuring that nothing from the past will escape the purview of the present by thwarting the potential rise of unrecorded diversity from the past.

6 Unpacking Darwin's View of Life

With the dual principles of divergence and gradation at Darwin's theory's core, he was free to impose additional interpretative constraints within that explanatory matrix. First, Darwin's Diagram (see Fig. 1) does not apply to the entire tree of life; rather, evolution is confined to separate classes or phyla (e.g., 1859: 420–422; 1872: 369–370). His biology is actively constructed around these segmented and independent entities. Second, Darwin holds that the taxonomic position of life forms one relative to another within each class/phylum remained unchanged throughout the entire evolutionary history—as seen in Darwin's Diagram for lineages A, F, and I—thus preventing the blurring of genealogical connections over time and thus keeping all *major* events of the past within view (e.g., 1859: 420–422; 1872: 369–370). Third, Darwin argued that the biological modernity of each class/phylum was acquired as early as the Silurian (1859: 302–310) or Cambrian (1872: 282–289). This explains why the *Origin of Species* imposes upon the reader the strange sight of a deep past inhabited by fairly modern-looking forms (e.g., Darwin 1859: 303–306)—the past largely being a mirror image of currently living representatives—albeit in a diminished state. In reality, Darwin's "tree of life" is a *genealogy of living forms*: biological variations observed in extant forms are projected backward in time in order to create an *imagined* past, instead of a discovered one. So-called common ancestors or progenitors are thus fictitious, being mere shadows or pale reflections built out of parts selected among extant forms (e.g., 1859: 437; 1872: 384–385). As one would expect, Darwin's approach is thus assimilationist to the extreme (e.g., 1859: 303–306). Fourth, Darwin holds that

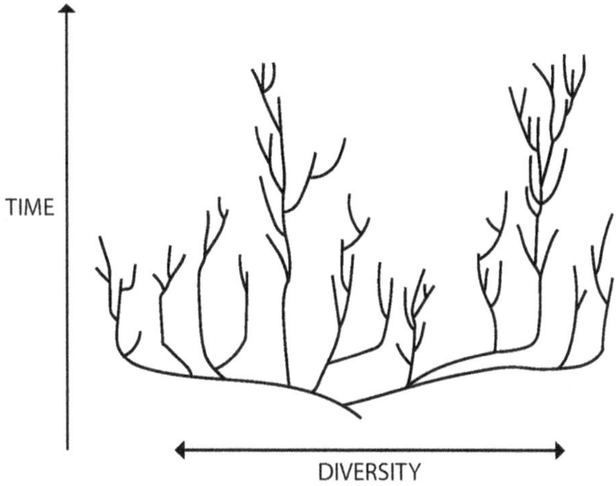

Fig. 3 Decimation model. This evolutionary model promoted by Stephen Jay Gould is based on a rapid multiplication of major biological types followed by a significant amount of decimation (extinction), especially at high taxonomic levels. This model assumes that the present poorly represents the evolutionary past since many body plans that once existed are now extinct

high taxonomic entities like classes or phyla are permanent and stable groups deeply entrenched in nature. It is no surprise, then, that 9 out of 11 lineages represented in his Diagram are shown in a perfect state of evolutionary stagnation, a topic discussed by Darwin under different guises (1859: 52, 73, 313, 331–332, 351, 368, 390, 400–402, 421, 429, 488; 1872: 57, 290–292, 304, 332, 319, 332, 348, 355–356, 370, 428). For Darwin, then, the only two evolutionary options open inside his cone of increasing diversity are "divergence" or "stability," as far as surviving lineages are concerned. Fifth, Darwin (1859: 446, 449; 1872: 396) reinforces the idea of permanence and stability by embracing an embryological theory that anchors past forms in fixed types dated back to the earliest moment of the rise of each class/phylum.

7 Darwin and Contingency

The most robust contribution the reader encounters in the *Origin of Species*, we argue, is a theory that allows for a limited amount of evolutionary change, one which is channeled along pre-established divergent paths, when change occurs at all (see also Torrens et al. 2024). It cannot be otherwise when one holds that all extinct forms fall in-between extant ones, a product of Darwin's presentist ontology and expressed in the cone of increasing diversity. In this interpretative context, it is obvious why the action of natural selection is severally restricted, serving the epistemological function of justifying *after the fact* what has already been established in his view of life. That is why it is argued here that Darwin's *Origin* cannot be the primary source for thinking about evolutionary creativity and contingency.

Does this mean that Darwin's contribution to evolutionary contingency is entirely lacking? Of course not since he creates evolutionary room by allowing for a greater or lesser amount of divergence to take place among related lineages, although it should be noticed that about 2/3 of the speciation events that are successful in his Diagram (see Fig. 1)—those giving rise to lineages that survive into the next geological time horizon—are the most divergent ones. Such a feature indicates a level of predictability and rigidity not obviously congenial to the notion of evolutionary contingency. Darwin's frequent appeal to contingency, we think, can be understood in this limited sense only, considering that a truly open-ended process should, by definition, allow for the full expression of all evolutionary alternatives such as convergent evolution, reticulate evolution, polyphyletism, mass extinctions, and major evolutionary discontinuities. But clearly, Darwin's theoretical core, founded on the dual notion of divergence-gradation, thwarts such alternatives.

The exploration of other components part of Darwin's thinking will be delayed until these can be discussed in direct association with the views of Mayr and Gould.

8 The Early and the Late Mayr

When someone publishes a paper entitled "How to Carry Out the Adaptationist Program?" as Mayr (1983) did, one might expect it will contain a reflection about an evolutionary process driven by adaptation, one that opens up the road to a whole range of phenomena falling under the purview of evolutionary contingency. Unfortunately, things are not that straightforward. Mayr certainly thought his work served the purpose of methodologically reinforcing Darwin's approach to evolution, explaining why he claimed that 1859, the year *On the Origin of Species* was published, should rightly be considered "the year in which the modern science of evolution was born. . .this zero year of history" (1959a: 1). By so doing, however, he may not have realized that he also took on board Darwin's ill-conceived ontology to account for life's historicity.

In what follows, it will be useful to introduce a distinction between the early Mayr (1940–1970) and the late Mayr (after 1970) as a *rough device* to register major shifts during his long career. In no way should this device, however, be taken to imply a sharp break occurred at a particular moment in his life. Rather, the transformation of his ideas followed a complex process—occurring at different paces, about various topics, and at several distinct explanatory levels—a story made more opaque by Mayr's use of rhetoric to cover over his earlier positions, thus dissimulating complete or partial retreats from such positions. It seems fair to say that while Mayr established his original core position between 1940 and the early 1960s, subtle changes started to appear in the 1960s and early 1970s, with an obvious acceleration in the 1980s and after. Around the year 1990, Mayr had given up on the ontology accompanying the cone of increasing diversity—Darwin's ontology—which had been his since around 1940. This being said, full consistency in Mayr's theoretical matrix at each moment should not be sought, considering that his system of thought was modified in a "mosaic" fashion (to resort to that biological analogy). This explains why the following analysis of Mayr's work cannot strictly respect the historical periods introduced here, although these are of real heuristic value. This also explains why we will often venture beyond the 1970 demarcation line, continually moving back and forth in time to create the necessary dialectical perspective to better see his shifting positions.

9 The Early Mayr (1940–1970): Putting the Theoretical Framework in Place

A fundamental idea unites Darwin and Mayr: the association of the concepts of "common ancestry" and "divergence."[2] Mayr's way of supporting this link is revealing of many of his key choices. Notice how, in the following passage, Mayr presents his view as an obvious fact accepted by *all* evolutionists, whereas, in reality, this commitment is part of a specific theoretical agenda:

> It must be emphasized that the terms *common descent* and *branching* describe exactly the same phenomenon for an evolutionist. Common descent reflects a backward-looking view and branching a forward-looking view. . .Branching, by necessity, means divergence: and every modern student of phylogeny, regardless of which school he belongs to, takes it for granted that evolutionary lines, once they have become completely separated from each other, will steadily diverge to a greater or lesser degree. And the evidence indicates that Darwin made the same assumption in his earlier branching diagrams. . .Soon it was demonstrated that even animals and plants, seemingly so different from each other, could be derived from a common, one-celled ancestor. (Mayr 1985a: 759–760; also reproduced in Mayr 1988: 200–202) (italics original)

The reader will recognize in this quotation Darwin's "cone of increasing diversity" (see Fig. 2), wherein the notions of "common ancestry" and "divergence" are conceptually fused under the cone ontology. But here, Mayr is more ambitious than Darwin. For the latter, the field of evolution is actively constructed around separate and well-defined classes and phyla, restricting his comment about life's single common ancestry to vague and hypothetical considerations only (Darwin 1859: 484, 488, 490; 1872: 424, 429). For his part, Mayr embraced that view at a time when the molecular support for such a common ancestry had considerably firmed up. The fact that Mayr straightjacketed the entire evolutionary process within a single cone of increasing diversity has direct implications for the rise of evolutionary novelties. Is an ongoing pan-divergent process really the only road open to novelties? Are lineages truly prevented from acquiring true novelties through other evolutionary patterns than divergence? What is the role of the adaptationist doctrine in Mayr's cone of increasing diversity? Let us keep these questions in mind while conducting our investigation.

[2]Here, I would like to correct a mistake introduced by myself concerning the view of Mayr (1985a). In the annotated bibliography of a recent monograph (Delisle and Tierney 2022: 142), I wrote: "Ernst Mayr has argued that in Darwin the notions of 'divergence' and 'common ancestry' are two separate theories." My memory served me badly, as I thought that these two notions were listed by Mayr under two distinct theories. Mayr is totally right: these two notions are, indeed, entirely fused in Darwin.

Mayr concluded *Systematics and the Origin of Species* (1942) with a chapter on higher taxonomic categories, one with an implicit yet pervasive discourse promoting a specific view of the epistemological relation between *classification* (taxonomy) of extant forms and *phylogeny*. He states: "The organisms of a 'natural' systematic category agree with one another in so many characteristics because they are descendants of one common ancestor!. . .The natural system is based on similarity, the phylogenetic system on the degree of relationship" (Mayr 1942: 276). Then, to make his meaning explicit he directly quotes Dobzhansky's *Genetics and the Origin of Species*. Again, the reader should pay attention to his insistence that the taxonomy of extant forms already captures most of the relevant information, with paleontology (phylogeny) contributing nothing essential:

> The fact is that the classification of organisms that existed before the advent of evolutionary theories has undergone surprisingly little change in the times following it, and such changes as have been made have depended only to a trifling extent on the elucidation of the actual phylogenetic relationships through paleontological evidence. *The phylogenetic interpretation has been simply superimposed on the existing classification; a rejection of the former fails to do any violence to the latter*. (quoted from Dobzhansky in Mayr 1942: 276; reproduced in Mayr 1999: 276; my emphasis)[3]

Driving home his point, Mayr (1969: 61) later adds: "It is not surprising in the least that the adoption of the evolutionary theory had virtually no impact on the established classifications." Bringing further precision concerning the exact relation between taxonomy and phylogeny, he continues: "Since we do not know the actual phylogeny, we can never base classification on phylogeny, but from our classification we can draw inferences on the relative probability of various possible phylogenies" (Mayr 1964a: 21). Mayr's approach is nicely summarized by Richard A. Richards (2016: 105): "[F]or Mayr a *monophyletic* grouping must contain *only* the descendants of a single species, but need not contain *all* the descendants of that species. . .In this way classification, according to Mayr, was only 'based on' phylogeny, but did not 'express' it" (italics original). By giving precedence to the taxonomy of extant forms in phylogenetic analyses as done by Darwin before him—a commitment Mayr will eventually give up around 1990—he shoehorned the past into the categories of the present. On that specific issue, Mayr (as Darwin before him) exposed himself to the same sort of epistemological criticism levelled by William Whewell (1832: 126) against Lyell's *Principles of Geology*, and already quoted above. The nature of that criticism is captured by Marjorie Grene and David Depew (2004: 171) when they state: "In light of mass extinctions spread across quite different geological eras, Whewell thought it decidedly unscientific for Lyell to shoehorn the history of life into categories cobbled up to fit contemporary life forms."

[3]To get the full texture of this quotation, see Dobzhansky (1982[1937]: 303–307). As much as Mayr and Dobzhansky agree with respect to this notion, the fact remains that they do not subscribe to the same ontology or worldview (see Delisle 2009a, 2011).

We reach here the crux of our argument concerning Mayr's commitment to the cone of increasing diversity: for him, this device is part of a tightly knit conceptual bundle of mutually reinforcing components (whose explanatory organization will be unpacked below). Mayr (1969: 61) adds this revealing comment: "Darwin realized that two processes occur during phylogeny—branching and subsequent divergence." And elsewhere he states: "Evolution always includes two sets of phenomena: the splitting (branching) of lineages, and the greater and lesser divergence of these lineages from each other" (Mayr 1964a: 21; see also Mayr 1969: 61). Is that all there is, speciation and divergence? If so, are we supposed to assume that this conceptual matrix is flexible enough to explain all manifestations expressed under evolutionary contingency?

Let us postpone any firm conclusions at this point. It will be sufficient to stress how Mayr's considerable efforts at proposing an adequate speciation theory constitutes *an extension* of a much more general and ambitious theoretical agenda: "Without speciation there would be no diversification of the organic world, no adaptive radiation, and very little evolutionary progress" (Mayr 1963a: 621). A most revealing aspect of Mayr's approach, we hold, is the statement found in the very last paragraph of *Systematics and the Origin of Species* (1942: 298): "In conclusion we may say that all the available evidence indicates that the origin of the higher categories is a process which is nothing but an *extrapolation* of speciation" (my emphasis). Mayr's speciation theory finds full justification in the cone of increasing diversity, one being the expression of the other, with "extrapolation" from low taxonomic levels to higher ones serving as a binding concept between the two, as well as the reason why the present time is, allegedly, epistemologically more revealing than the past.

Let us inspect the Diagram proposed by Mayr in *Systematics and the Origin of Species* (1942: 282; also reproduced in Mayr 1999: 282) under the heading "Diagram of part of a phylogenetic tree" (see Fig. 4). There, the evolutionary branches and twigs are represented in a perfect state of divergence, leaving the strange impression of geometrical perfection (and rigidity). Surely, such a symmetry cannot be a representation of how things happen in nature when driven by an adaptive process. Is Mayr, then, merely trying to make a more abstract and theoretical point? If so, why call this Diagram a "part of a phylogenetic tree"? In fact, even when Mayr (1964a: 23–24; also reproduced in Mayr, 1969: 227) illustrates graphically the reality of other patterns like parallel evolution and convergent evolution, one is offered mere variations upon the theme of "divergence": these other patterns are represented as an extension of a general evolutionary view organized around divergence (see Fig. 5).

When it is assumed that Mayr sees in adaptation a key notion at the heart of evolutionary biology, this idea certainly requires qualifications. Apparently, for Mayr, the action of natural selection is constrained—as for Darwin before him—irrespective of his rhetoric to the contrary. One thing is certain: his approach does not adequately address alternatives to divergence, although Mayr does recognize them (Mayr et al. 1953: 41–43). For instance, in the case of *polyphyletism*, he writes: "there is no doubt that even among recent animals the number of polyphyletic

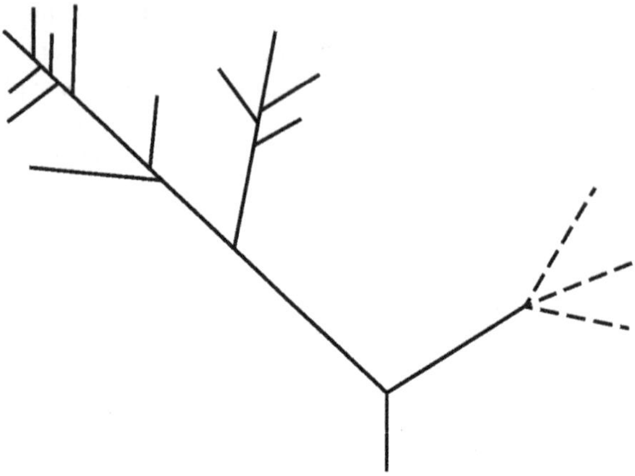

Fig. 4 Reconstitution of a figure appearing in Ernst Mayr's *Systematics and the Origin of Species* (1942), p. 282, entitled "Diagram of part of a phylogenetic tree." It should be noticed how this representation leaves one with a strange sense of perfect symmetry not obviously congenial with evolutionary contingency

categories which the taxonomist has not yet been able to eliminate is considerable" (Mayr 1942: 279). Furthermore, as far as *convergence* is concerned, one reads: "Similar selection pressures may produce phenotypic similarity of independent and sometimes not even closely related lines" (Mayr 1964a: 22). As will be unpacked below, the issue is clearly not the formal recognition of alternatives to "divergence," but rather the role given to them by Mayr in his overall explanatory matrix.

It is our thesis that Mayr frames these alternatives inside a conceptual continuum that takes us from the pattern of speciation to the cone of increasing diversity (the reverse also being true). Indeed, it seems that Mayr (and Darwin before him) busily reconstructs a past that merely reflects a pan-divergent theory neontologically based and projected backward in time. Not only is Mayr fully conscious of that risk, he is also confident that it poses no difficulty:

> Since classifications are not based on established phylogenies, the objection that phylogeny is still largely unknown is irrelevant. What we do base our classification on is the universal fact that all organisms are the product of evolution... Evolutionism, as a philosophical basis for classification, is a valid approach only if and when natural groups of organisms are the result of divergent evolution. If reticulate evolution were common owing to the frequent fusion of previously separated evolutionary lines, or if convergence were frequently so complete as to lead to groupings that could not be unmasked as having a polyphyletic origin, then the claims of evolutionism as a proper theoretical basis for biological classification would indeed be questionable. No one has so far presented convincing evidence that these two potential difficulties are at all frequent in the evolution of animals. (Mayr 1969: 78–79)

Mayr takes his own approach to classification—called evolutionary taxonomy—to be the only way to make sense of the fact of evolution. Yet, other evolutionary alternatives could be conceived that are more careful about the risk of restraining evolution within a pan-divergent theory, by taking care to respect evolution's complexities. It will be seen that debates between evolutionary taxonomists and other schools of classification in the 1970s and after will force a much more carefully articulated and moderate response from Mayr.

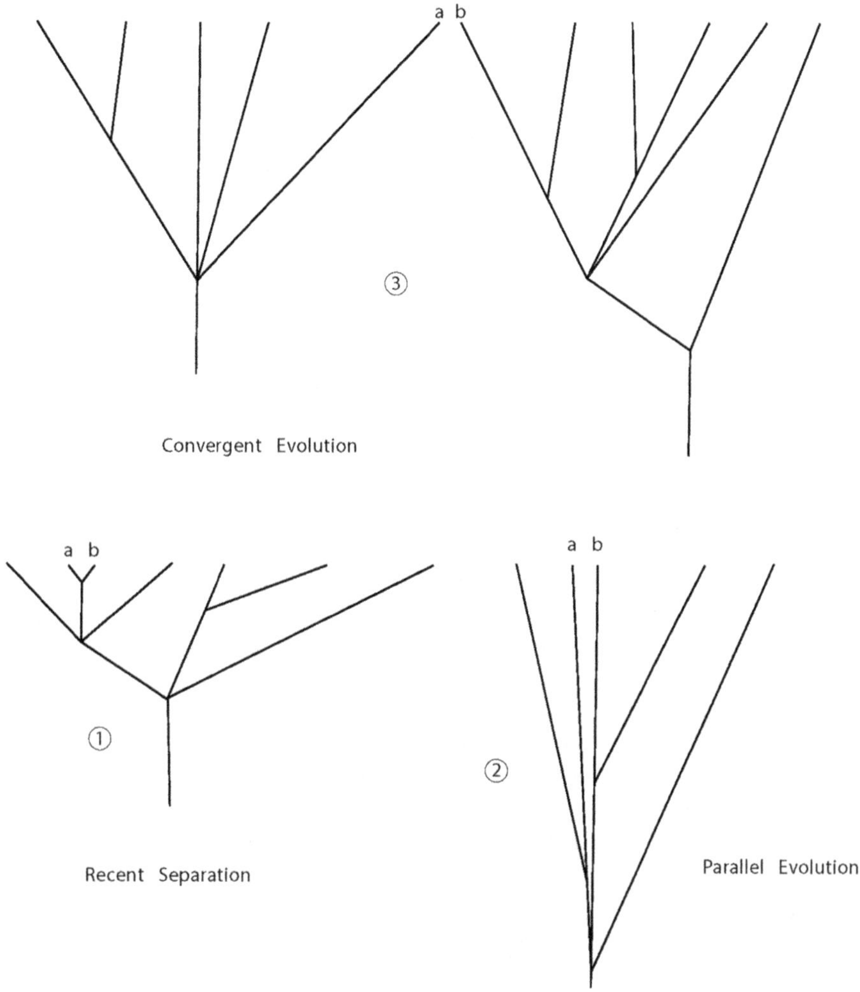

Fig. 5 Reconstitution of a figure appearing in Ernst Mayr's *Principles of Systematic Zoology* (1969), p. 227, illustrating three different reasons why two distinct taxa (*a* and *b*) are showing similarities: (1) because of a recent separation; (2) because of parallel evolution; and (3) because of convergent evolution. It should be noticed how such separate explanations are, in reality, by-products of an evolutionary view organized around the principle of divergence. Again, the impression is given that this rigidity leaves little room for the full expression of evolutionary contingency

The nature of the relation Mayr establishes between "divergence" and "selection" is worth considering here. Precisely because his main theoretical assumption rests on a *monophyletic conceptual complex*—which includes the notions of common ancestry, divergence as a dominant pattern, and the cone of increasing diversity—it is with full consistency that he applies this agenda throughout. It will be seen that Mayr's commitment to *adaptationism* is not properly speaking at the core of his evolutionary view, because it is superimposed upon a conceptual structure for which adaptationism serves the explanatory function of justifying an already established cone of increasing diversity. Whereas "pattern" and "process" are apparently fused in Mayr's biology, the former has explanatory precedence over the latter, except when obvious facts force him to think otherwise. And when this is the case, Mayr turns to a bag of conceptual tricks to keep exceptions under control: (1) parallel evolution is *narrowly defined* to comprise forms exclusively arising from common ancestors; (2) convergent evolution is *restricted* to a limited number of features; and (3) reticulate evolution is said to occur only *rarely*. If adaptive pressure was truly believed to be at the core of evolutionary biology, Mayr's theoretical apparatus would be differently designed, allowing more explanatory room to face a whole range of evolutionary manifestations driven by adaptation. Let us delve into Mayr's bag of conceptual tricks.

It is interesting to see the preemptive strikes Mayr delivers to limit the impact of blurring (adaptive) evolutionary phenomena on his view of taxonomy. In the case of reticulate evolution, Mayr assures us that *divergence* prevails (Mayr 1985a: 759). He reassures us that *reticulate evolution* occurs only on occasion, because it is limited to animal species, genera, and families that can hybridize successfully (Mayr 1942: 280; but see O'Malley 2010 for obvious problems). Returning to the *Origin of Species*, it has been overlooked that not only was Darwin well-aware of threatening options for his divergence theory—reticulate evolution, polyphyletism, parallelism, convergence, erased evolutionary links owing to adaptation—but also that he forced his theoretical agenda on them (Delisle 2019: 83–105, 115–129, 161–194). Mayr is here walking in the footsteps of Darwin. Indeed, Mayr took some precautions to prevent the pattern of divergence from being disloged from its central explanatory position. Let us review other cases. When speaking of evolutionary *convergence*, he writes:

> Convergence is believed by many taxonomists to be the greatest obstacle against achieving a truly phyletic classification. Actually the difficulties caused by convergence have been exaggerated. . .[C]onvergence between entire phenotypes of unrelated organisms is rare, if not absent. . .No flying mammal, reptile, or arthropod could ever be mistaken for a bird. Nor can a whale, a seal, or a manatee be mistaken for a fish. *Convergence is a problem only among fairly close relatives in rather uniform groups*. . . Most cases of convergence are restricted to a single character or functional character complex. The utilization of new characters *invariably* leads to the unmasking of unnatural assemblages. (Mayr 1969: 226, 228; see also Mayr and Ashlock 1991: 146; my emphasis)

Let us set aside Mayr's utilization of the uncompromising word "invariably" and focus on the statement that convergence could be a problem only among fairly closely related entities of uniform conformation. If this is the case, then, can one really rule out that such groups are uniform precisely because of convergence in the first place?

Turning now to the case of *parallel evolution*, Mayr (1969: 202) chooses to rely on a narrow definition that explicitly incorporates the notion of common ancestry:

"Similarities resulting from joint possession of independently acquired phenotypic characteristics produced by a shared genotype inherited from a common ancestor (similarity through parallel evolution)" (see also Mayr et al. 1953; Mayr and Ashlock 1991: 147). Again, the reader cannot help but notice how Mayr's phylogenetic tree manages to ensconce parallel evolution within an overall divergent pattern that combines with the notion of common ancestry (Mayr 1969: 71, 74, 227; see Fig. 5).

And when the reader is presented with a possible case of *polyphyletism* in the rise of mammals, he/she is reassured that the monophyletic explanation prevails on the grounds that all mammals were endowed with an internal condition—a common genetic program—that predisposed them to evolve in the same direction:

> The diagnostic mammalian structure of the jaw-ear region, for instance, is believed to have evolved several times from ancestral therapsid reptiles, which had the needed genetic program to predispose them toward evolving the mammalian grade when exposed to the same selection pressures. This is not polyphyly, because the genotype permitting these parallel evolutionary changes goes back to the same ancestral program...The class Mammalia is monophyletic because all mammalian lines were derived from the immediately ancestral taxon of the therapsid reptiles. Most cases of alleged polyphyly reported in the literature do not stand up under critical analysis. (Mayr 1969: 75; see also 1964a: 22; 1969: 243–244)

This explanation is rather convenient. We shall not go so far as to hold here that Mayr treads on the thin ice of internalistic explanations acting inside the biological matter itself, but it is surprising to see how far he is willing to go in advancing the monophyletic agenda, especially for someone presumably embracing "adaptationism." Suddenly, it seems, genetics is preferred to organismic biology in settling this question; an interesting position for a self-proclaimed anti-reductionist. This issue will be discussed below.

Finally, Mayr subtly attempts to restrict the blurring impact of an adaptive process that could potentially erase the traces of affiliation by insisting on several convenient "facts." Among his claims are the following (Mayr 1942: 276–277): First, it is argued that monophyletism regroups forms that have the greatest number of features in common, "and there are *only a few exceptions* to this" (Mayr 1942: 276; my emphasis). It is hard to miss the potential circularity of Mayr's reasoning: by definition, monophyletic groups are those that are united by sharing a great number of morphological features owing to common ancestry, and it just so happens that when one encounters forms sharing numerous common features these must be explained by monophyletism. Apparently, little room is left for other possible explanations, such as evolutionary analogies. Second, Mayr stresses the alleged superficiality of some adaptations by identifying more significant and stable features solidly anchored in phylogeny:

> Special adaptations, for example for living in deserts, in the sea, in mud, for flying, and so forth, produce ecological types which are suitable for a classification based not on relationship, but on the *superficial similarity of adaptive structures*. That such a classification is universally rejected...[is] simply because such ecological adaptations generally affect *only*

> *one or a few characters* and encounter, therefore, many more practical difficulties than the phylogenetic systems. (Mayr 1942: 277; my emphasis)

In such cases, it is worth noting how traits associated with either special adaptations or ancestry can presumably be easily distinguished from one another; key discriminatory factors being the number of features involved and the opposition between superficial/stable features.

We are not concerned here with contesting Mayr's claims cited above. From case to case, these may even be quite reasonable. Rather, our point is epistemological: taken collectively, they show that Mayr is trying to reduce the impact of blurring adaptive factors on his view of evolution by justifying why divergence should be the cornerstone of evolutionary biology. Apparently, Mayr doesn't see reticulate evolution, convergence, and parallelism to be significant creative factors, thus confining such manifestations to minor roles within an overarching cone of increasing diversity.

An explanatory tension therefore runs throughout Mayr's biology when confronted with the notion of *creative evolutionary time*: inside his cone of increasing diversity, the evolutionary past carries little weight when it comes to revealing potential manifestations of novelties: "it has become clear that there is nothing in the past history of the earth that cannot be interpreted in terms of the processes that are known to occur in the Recent fauna" (Mayr 1963a: 617). Brave words, indeed, which run directly counter to Whewell's warnings already quoted above. Mayr did not suspect, at that time, how much he would have to compromise the very foundation of his biology in the 1970s and after.

Thus far we have been busy examining the impact of Mayr's view of evolutionary contingency as interpreted through his conceptual toolkit, which is composed of the following mutually reinforcing elements: the cone of increasing diversity, the principle of divergence, monophyletism (common ancestry), and a neontological agenda. While this toolkit as unpacked so far reveals the early Mayr's broad interpretative framework—with a focus on a pre-established or dominant divergent *pattern* in evolution—its application also extends to issues that are more properly classified under the *process* of evolution. Exploring this complementary set of explanatory components will allow us to reflect upon the nature of the relation between pattern and process in the early Mayr, keeping in mind our interest on the role played by the doctrine of adaptationism within this explanatory apparatus. Here, too, we will see that this complementary set of concepts also acts to restrict the expression of creative evolutionary time.

Let us begin by considering how Mayr deals with evolutionary phenomena that occur at several distinct biological levels. Levels constituted of molecules, genes, organisms, species, and higher categories are, for Mayr, compressed, so to speak, around the levels of organisms and species. Molecules and genes are put in a position of explanatory dependence on the level of organisms-species, whereas higher taxonomic entities owe their existence to a mere extrapolation of a pattern-process that deploys itself at the organism-species level. In short, organismic biology is viewed as being the center of evolutionary biology, but one that can only be investigated by surveying currently existing populations distributed in geographical space. Suddenly, it seems, the temporal dimension of evolution is flattened and stretched out horizontally in geographical space. For Mayr, as for Darwin before him, to travel in geographical space is to travel in geological time.

Of course, this peculiar method of investigation of evolutionary time—irrespective of its virtues—comes with implications for the recognition of the rise of novelties. Mayr (1963a: 617) has no other epistemological option but to assume that what can be observed today is representative enough of all past evolutionary phenomena. This is uniformitarianism in its most uncompromising form, reflected in a gradualism that he applies throughout: Mayr tries to catch evolution on the march, a fleeting reality if there is one. Speaking of *Systematics and the Origin of Species* (1942), he retrospectively writes: "Speciation, being ordinarily a slow historical process, can not be observed directly but must be *inferred*. In order to be convincing I piled up a very large number of indirect proofs for the *universal* occurrence of geographic speciation" (Mayr 1992a: 10; my emphasis). This is an ambitious claim, one mostly founded on the ornithology literature focused on the geographical area of the Pacific Islands (Mayr 1942: xiii), a bias later openly acknowledged by Mayr (1992a: 10, 29). Then, Mayr perceptively described the nature of his method (including his tacit assumptions):

> One of Darwin's major contributions was the theory of gradualism. He insisted that there are no jumps, no discontinuities, in evolution...Geneticists, who worked exclusively in the vertical dimension, had been quite defeated in their effort to explain the origin of new species without saltations. They were able to demonstrate the change of a gene pool over time—that is, evolution—but not the origin of a new species. One of my major contributions in *Systematics and the Origin of Species* was to show that by supplementing vertical evolutionism with horizontal evolutionism—that is, by studying geographical variation—one could explain how discontinuities could gradually arise... If an isolated population gradually becomes a different species in a period of thousands or millions of years, one should be able to illustrate all stages of this process of ongoing speciation by looking at scores if not hundreds of cases of incipient species. Several chapters of *Systematics and the Origin of Species* are devoted to the presentation of these successive stages. No case was found that did not fit the scenario of geographic speciation. Darwin's theory of gradualism was abundantly confirmed by this approach. (Mayr 1999: xxiii)

This collection of statements intertwined in a rich conceptual bundle is worth being untangled. First, Mayr's commitment toward gradualism—presumably supported by cases in nature documenting all stages accompanying the speciation process (see also Mayr 1942: 159)—cannot entirely escape the charge of theory-ladenness: the risk exists that facts have been gathered selectively, leaving aside counterfactual evidence. It is because Mayr believed in gradualism that he thought he could document all key stages part of speciation. Reijer Hooykaas (1963: 102) explains the epistemological implications accompanying Darwin's gradualism, an argument that can be transferred to Mayr:

> [Darwin] is of opinion, however, that the intermediate types must exist, and, when speaking about the evolution from the most simple eye to that of the highest vertebrates, he remarks

> that it is indispensable that the 'reason should conquer the imagination'. The fundamental assumption is 'natura non facit saltum'. We see leaps in nature, but according to the doctrine of natural selection they cannot exist, consequently they do not exist.

Unsurprisingly, then, Mayr (1960: 351) collects his data in accordance with his speciation theory: "As long as we believe in gradual evolution, we must be prepared to encounter immediate evolutionary stages." No one can blame him for collecting facts in accordance with his theory. But this is not quite the same as claiming that counterexamples to gradualism *cannot* exist. In our view, Mayr crossed that line, thus foreclosing the debate altogether:

> No special evolutionary processes need to be postulated, even in groups where such missing links have not yet been found and where the primitive roots of the various stems always seem to be missing. . .The fact that there are many gaps in our knowledge of the phylogenetic lines does not indicate. . .that evolution of the higher categories proceeds by big steps, by veritable leaps. (Mayr 1942: 297)

In fact, he later added: "The evidence, whether genetic, morphological, or functional, is so uniformly opposed to a saltationist origin of new structures that no choice is left but to search for explanations in terms of a gradual origin" (Mayr 1960: 357). To drive home the point, but even more clearly: "The students of speciation, and more broadly the neontologists as a whole, accepted as an article of faith not only that all macroevolutionary phenomena were consistent with the laws of genetics but also that they could be explained in terms of the phenomena of geographic variation and speciation" (Mayr 1982b: 1120). Mayr's inclusion of *all* neontologists under his speciation theory was not without rhetorical dimension! Be that as it may, if Mayr is wrong about the gradualistic nature of evolution—and here we are not taking a position in the scientific debate—then his speciation theory would be significantly weakened. The debate around the theory of punctuated equilibrium in the 1970s and 1980s shows that Mayr's confidence was somewhat premature.

For Mayr, then, organismic biology essentially means speciation (and subsequent divergence), the moment during which the most significant evolutionary events allegedly occur (Mayr 1963a: 555, 621). To support his case, he deploys an articulated and fleshed-out conceptual matrix by turning to such notions as population structure, polytypic species, population thinking, biological species concept, allopatric and peripatric speciation, isolating mechanisms, gene flow, founder effect, size of effective breeding population, genetic drift, unity of the genotype, and genetic revolution. Although the last six notions are, properly speaking, genetic concepts, the fact remains that Mayr put them to work in the context of organismic biology: "What is exposed to natural selection is not the individual gene nor the genotype but rather the phenotype, the product of the interactions of all genes with each other and with the environment" (Mayr 1959b: 7–8). This quotation is drawn from the famous "beanbag genetics" paper (1959b: 2) designed to capture the intellectual trend of moving away from an atomistic view of genetics based on abstract individualistic selective values of separate genes and toward a collection of genes or gene-complexes, themselves only meaningfully studied at the population level. Of course, it is argued these populations of genes confront the environment at the level of organisms, races, subspecies, and species, hence Mayr's promotion of populational concepts listed above (see also Mayr 1954, 1955, 1963a: 588). By so doing, Mayr manages to incorporate some aspects of the lower levels of living matter into organismic biology, including recent developments in molecular biology. Delivering a preemptive strike against the molecular reductionism that was gaining momentum in the wake of the revelation of the structure of the DNA molecule by James Watson and Francis Crick in 1953, he writes: "No case is known to me in which a change in body chemistry initiated a new evolutionary trend. Invariably it was a change in habits or habitat which created a selection pressure in favor of chemical adjustments" (Mayr 1964b: 1233). But there is more to this question.

Mayr's approach consists in envisioning biological populations (species) not as uniform and cohesive entities but rather as ones variably organized around smaller subgroups of organisms that can eventually be irreversibly separated (Mayr 1942: 225, 297). In his view, it is under such conditions that speciation (and ongoing divergence) can be instituted and sustained successfully, especially when groups are already quite isolated from a main population (Mayr 1963a: 521). Mayr is here attempting to counter the arguments of mathematical population geneticists like Ronald Fisher and J.B.S. Haldane, whose theoretical approach focus on large panmictic populations in which gene frequencies are shifted and circulated openly, under an adaptive impetus (Mayr 1992a: 3; 1999: xii-xiv). In this context, Mayr argues genuine breakthroughs and innovations are prevented from arising under the inherent stabilizing or equalizing factors associated with the free circulation of small genetic changes inside large and self-contained populations, with the result being restricted to vertical evolution (anagenesis or phyletic adaptation) (Mayr 1949).

It is hard to avoid the conclusion that Mayr's system of thought is far from flexible enough when it comes to explaining the rise of evolutionary novelties. He was probably aware of this challenge, given that he took the trouble to write a paper entitled "The Emergence of Evolutionary Novelties" (1960). In it, the reader is surprised to find him falling back on classic ideas already exploited long ago by Darwin (1859: 171–206)—mainly declinations of the general idea of "preadaptation"—especially considering that Mayr (1954: 170) had already opened to door to what he called a "genetic revolution." The chasm between the notions of "genetic revolution" and "preadaptation" is rather deep! Mayr (1960: 377–378) thus summarized his conservative view:

> The emergence of new structures is normally due to the acquisition of a new function by an existing structure. In both cases the resulting 'new' structure is merely a modification of a preceding structure. The selection pressure in favor of the structural modification is greatly increased by a shift into a new ecological niche, by the acquisition of a new habit, or by both. A shift in function exposes the fully formed 'preadapted' structure to the new selection pressure. This, in most cases, explains how an incipient structure could be favored by natural selection before reaching a size and elaboration where it would be advantageous in a new role. (see also Mayr 1963a: 602–604; 1992a: 20)

Considering the simplicity of early life forms on Earth, it is not easy to see how the complexity level observed in a multitude of extant classes can be explained exclusively in terms of preadaptation, unless one is prepared to embrace a sort of preformationism wherein all future features are thought to have already existed in a diminished state. In fact, this puts Mayr (1963a: 621; 1974a; 1982a: 531–532; 1997a: 197–198) in a difficult position relative to the idea of biological progress he so openly embraces, recognizing it in the rise of manifestations such as photosynthesis, nuclear cells, multicellular structures, division of labor among specialized organs, central nervous systems, homeothermy, predation, and parental care (Mayr 1974a: 8; 1982a: 532; 1991: 62; 2001a: 215).

10 Evolutionary Contingency (1959–2005): Exposition and Meaning

The theoretical matrix of the early Mayr is remarkably well fleshed out, coherent, tightly knit, well-articulated, and composed of mutually reinforcing elements. This rather abstract construct, however, was ill-suited to confront the complexity of the biological world, which explains why it gradually fell apart in the post-1970 period. Epistemologically speaking, this early matrix contains two overlapping subsets, one incorporated into a more inclusive subset, each filling different yet complementary explanatory functions: while the *overall evolutionary framework* provides the broad theoretical context (including specific explanatory constraints), the *conceptual support* puts the latter to work. The various components of each subset are the following:

A. **Overall evolutionary framework**: cone of increasing diversity; the principle of divergence; monophyletism (common ancestry); a neontological agenda wherein extinct forms entirely fall in-between extant ones; and gradualism-uniformitarianism
B. **Conceptual support**: horizontal view of evolution (to travel in geographical space is to travel in geological time); organismic biology focused on the organism-species level; speciation theory; evolutionary contingency; evolutionary taxonomy; and adaptationism/natural selection

Differently presented, subset **A** constitutes an *ontological* representation of the living world (its overall structure and texture), whereas subset **B** is a *conceptual/methodological* extension of that ontology, playing a justificatory role as well as allowing for the investigation of that world. Here, we are drawing on the kind of epistemology described by Thomas S. Kuhn (1970) and Larry Laudan (1977).

Considering the regulatory role played by the overall evolutionary framework upon its conceptual/methodological support, it is difficult to see how the theoretical matrix of the early Mayr can be envisioned as a significant contribution to evolutionary contingency. Although Mayr maintained his official discourse when talking about evolutionary contingency from the late 1950s onward, the fact remains that major changes were brought to his overall evolutionary framework in the 1970s and after. What Mayr explicitly said about evolutionary contingency, then, inevitably changed *meaning* throughout his career, irrespective of the apparent stability of his discourse on such a topic. This being the case, taking a piecemeal approach by studying Mayr's explanatory components in isolation from one another can only lead to a partial and distorted understanding of his life's work. Let us delve deeper into the topic of evolutionary contingency.

Already by the early 1960s, Mayr's narrative on evolutionary contingency was in place (Mayr 1959b, 1961, 1962; see also 1982a: 21–82, 1996). He held that twentieth-century Darwinism (the synthetic theory of evolution) is characterized by two postulates: (1) the production of new genotypes under non-finalistic and random events (mutation, recombination, fertilization) and (2) the ordering of that variability as displayed in numerous adaptations to biotic and abiotic environments due to natural selection. Selection is said to be a genuinely "creative process" wherein the statistical/populational nature of living entities manifests itself in differential reproduction. Interestingly, Mayr turned to a similar heuristic device used later by Stephen Jay Gould (1989: 289)—"replaying the tape of life" conjectural approach—as early as 1959:

> All evolution is a sequence of unique events. This is true not only for selective values in different genotypes and different environments, it is likewise true for segregation, crossing over, and whatever other events determine evolutionary success and evolutionary change. If we set up 10 parallel selection experiments, starting each from the same, apparently identical foundation stock, we are liable to get in the end 10 somewhat different answers. The more I study evolution the more I am impressed by the uniqueness, by the unpredictability, and by the unrepeatability of evolutionary events. (Mayr 1959b: 6)

Distinguishing between functional biology and evolutionary biology (Mayr 1961: 1502), he argued that individual organisms or species as seen from within evolutionary biology are "the product of a long history" which dates back more than 2000 million years. In this space-bound and time-bound context, permanent validity is denied, since life forms are links in a long chain of changing forms: structures and functions can only be fully understood when studied against changing historical backgrounds. Hence, the diversity of the organic world is explained by the numerous adaptive pathways taken by them. Reflecting upon the issue of prediction in evolution, Mayr writes:

> Probably nothing in biology is less predictable than the future course of evolution. Looking at the Permian reptiles, who would have predicted that most of the more flourishing groups would become extinct (many rather rapidly), and that one of the most undistinguished branches would give rise to the mammals? Which student of the Cambrian fauna would have predicted the revolutionary changes in the marine life of the subsequent geological eras? Unpredictability also characterizes small-scale evolution. Breeders and students of natural selection have discovered again and again that independent parallel lines exposed to the same selection pressure will respond at different rates and with different correlated effects, none of them predictable. (Mayr 1961: 1504)

Again, this statement is not very different from what Gould will later declare in *Wonderful Life* (1989). Not only did the early Mayr (1959b: 6) turn to the heuristic device of conjectural history to support his narrative about evolutionary contingency, he also appealed to the unpredictability of evolutionary events such as the extinction of Permian reptiles, an event that would have opened to door to an "undistinguished" group of mammals (Mayr 1961: 1504). Gould apparently followed in Mayr's footsteps by using the same examples: the replay of the tape device and the rise of mammals owing to the unpredictable extinction of dinosaurs (Gould 1989: 289, 318, 320). Here too, Mayr seems to be running ahead of Gould by reviewing the improbability of the rise of higher intelligence on Earth. Among such unpredictable events, one encounters (Mayr 1985b) (1) a symbiosis between at least two kinds of prokaryotes (one carrying cytoplasmic organelles, the other a nucleus); (2) the potential rise of intelligence only in the animal kingdom among all other eukaryotes; and (3) a potentiality kept open only among the chordates (out of 40 animal phyla), then in the class of mammals (among the chordates), and again only in the order of primates (among the mammals) and, finally, only in the human line (among the primates). As summarized by Mayr (1985b: 27): "[T]he origin of new taxa is largely a chance event. Ninety-nine of 100 newly arising species probably become extinct without giving rise to descendant taxa." For his part, Gould (1989: 289) writes: "Replay the tape a million times from a Burgess beginning, and I doubt that anything like *Homo sapiens* would ever evolve again."

However, the key point we are making in this chapter is that the apparent rapprochement between the two scholars is deceptive: apparently similar declarations by evolutionists take on quite different *meanings* when interpreted against their competing theoretical frameworks. Indeed, regardless of similarities in the discourses of these two scholars, they are separated by a fundamental difference in ontology: whereas Mayr subscribes to the "cone of increasing diversity," Gould embraces the "decimation model" (compare Figs. 2 and 3). No matter how chancy evolution might be for Mayr, the fact remains that he channels the process under a pan-divergent outlook: "I have shown that at each level of this pathway there were scores, if not hundreds, of *branching points* and separately evolving phyletic lines, with only a single one in each case forming the ancestral lineage that ultimately gave rise to Man" (Mayr 1985b: 27; my emphasis). It will be reminded that Mayr (1985a: 759) claimed that same year: "It must be emphasized that the terms *common descent* and *branching* describe exactly the same phenomenon for an evolutionist.... Branching, by necessity, means divergence... [Evolutionary lines, once] they have become completely separated from each other, will steadily diverge to a greater or lesser degree" (italics original). Mayr (1988: 200–202) reproduced these statements a few years later. The major theoretical shift in Mayr's *overall evolutionary framework* announced by us was not yet conspicuous in the years 1985–1988. Yet, Mayr will soon be forced to profoundly reconceptualize many of his cherished beliefs. Contesting the cone ontology, for instance, Gould (1989: 46–47) explains:

> I know no greater challenge to the iconography of the cone—and hence no more important case for a fundamentally revised view of life—than the radical reconstructions of the Burgess anatomy presented by Whittington and his colleagues... By recognizing so many unique anatomies in the Burgess, and by showing that familiar groups were then experimenting with designs so far beyond the modern range, they have inverted the cone... The maximum range of anatomical possibilities arises with the first rush of diversification. Later history is a tale of restriction, as most of these experiments succumb and life settles down to generating endless variants upon a few surviving models.

With the decimation ontology playing the role of an *overall evolutionary framework*, Gould had to rely on other processes than those Mayr usually appealed to:

> With so many Burgess possibilities of apparently equivalent anatomical promise—over twenty arthropod designs later decimated to four survivors, perhaps fifteen or more unique anatomies available for recruitment as major branches, or phyla, of life's tree—our modern pattern of anatomical disparity is thrown into the lap of contingency. The modern order was not guaranteed by basic laws (natural selection, mechanical superiority in anatomical design), or even by lower-level generalities of ecology or evolutionary theory. (Gould 1989: 288)

By contrast, Mayr attempted to justify his commitment to the cone of increasing diversity a few years prior to Gould's *Wonderful Life* (1989) by relying on a definition of "chance" that appealed to the gradual action of creative selective forces playing on random variations:

> [T]he origin of new taxa is largely a chance event. Ninety-nine of 100 newly arising species probably become extinct without giving rise to descendant taxa. And the characteristic of any new taxon is to a large extent determined by such chance factors as the genetic composition of the founding population, the special internal structure of its genotype, and the physical as well as biotic environment that supplies the selection forces of the new species population. (Mayr 1985b: 27)

Eventually, Mayr realized that the mounting challenges to the cone of increasing diversity pointed to a different understanding of his discourse about evolutionary contingency. Whereas the discourse itself, taken in isolation, was still believed to be valid by Mayr, its meaning had changed:

> Let us look at the bizarre types in the Burgess shale fauna. One cannot escape the suspicion that many of them were due to mutational accidents that were not eliminated by selection. Indeed, I sometimes wonder whether the elimination process is not sometimes a good deal more permissive than is usually assumed. Furthermore, one must not forget that chance always plays a considerable role even at the second step of evolution [the first step being the production of variation], that of survival and reproduction. And not all aspects of adaptedness are tested in every generation. Or let us look at the 35 or so living phyla of animals. They are the survivors of the 60 or more body plans that existed in the early Cambrian. When one studies their differences, one does not get the impression that they are necessities. Many or even most of their unique characteristics may have had their origin in a developmental accident that was tolerated by selection, while the seeming failure of those that became extinct may have been the result of a chance event (like the Alvarez asteroid extinction

> event). S.J. Gould (1989) made such contingencies a major theme in *Wonderful Life*, and I have come to the conclusion that here he may be largely right. (Mayr 2001a: 229)

Here, Mayr is implicitly recognizing that a new ontology of life requires a new *overall evolutionary framework*, considering that the decimation model might require doing away with a *creative* role for selection, the latter being confined to a mere *elimination* process. In the following sections, it will be seen how Mayr's system of thought gradually disintegrated during the post-1970 period, thus creating more explanatory room for the expression of evolutionary contingency.

11 The Late Mayr (Post-1970): Facing Challenges

Mayr's overall evolutionary framework came under stress, especially when confronted to competing theories in the 1970s and 1980s such as the neutral theory of evolution, alternatives to evolutionary taxonomy (phenetics and cladistics), and the rise of paleobiology. He responded to them in phases: first by combating them and then by trying to adjust to them and finally, when forced, by giving up on key commitments. To run ahead of our story, the more Mayr moved away from Darwin (as depicted earlier in this chapter), the better equipped he became to face evolutionary contingency, implying that, at the end of his life, not much was left of Mayr's "Darwinism."

The fate of Mayr's cone of increasing diversity can be assessed differently than has been so far. In 1985, Mayr published a paper entitled "Darwin's Five Theories of Evolution." We earlier quoted a passage from this text starting with: "Branching, by necessity, means divergence…" (Mayr 1985a: 759), thoughts later literally reproduced in Mayr (1988: 200–201). In that passage, it will be remembered that the notions of "divergence" and "common ancestry" were treated as the expression of the same general idea, the cone of increasing diversity. However, when "Darwin's Five Theories of Evolution" was republished in 2004 that passage was deleted (Mayr 2004: 100–102), implying that the notions of "common ancestry" and "divergence" were now uncoupled in Mayr's thinking. In fact, that coupling had disappeared in the 1990s in discussions concerning Darwin's "five theories" (Mayr 1991: 36; 1997a: 179–182; see also 2001a: 86). On this basis, it is safe to assume that the cone of increasing diversity lost its ontological status for Mayr, no longer serving as a general organizing principle for the living world. In fact, he would even go so far as to say that "it is now evident that the principle of divergence is invalid" (Mayr 1992b: 358).[4]

[4]Having himself realized by the late 1980s or early 1990s that the concepts of "speciation" and "divergence" had to be uncoupled, it is no surprise that Mayr suddenly looked back critically at Darwin, who had fused the two notions. He wrote: "Too many problems were as yet unsolved in Darwin's day to allow a resolution of the problem of speciation…It is now no longer necessary to invoke a principle of divergence for this particular process of evolution" (Mayr 1992b: 358). By that time not only had Mayr given up on the ontology of the cone, he also concomitantly realized that

12 Threats to Evolutionary Taxonomy

Mayr's commitment to "evolutionary taxonomy" is an especially useful tool for investigating evolutionary contingency. It will be reminded that Mayr summarizes the principles at the heart of evolutionary classification/taxonomy in these terms: "The theory of evolutionary classification...delimits taxa on the basis of two considerations—common ancestry and subsequent divergence" (Mayr 1968: 548; see also 1964a: 21). While life forms should be organized under genealogical relationships, it is argued that classification should also reflect that related forms are rarely evolving away from common ancestors in the same proportions (or rate), requiring a formal distinction between various taxonomic levels among related taxa as they evolve away from each other. Mayr expressed these ideas more clearly later:

> [Darwin] used two criteria in the construction of a classification: common descent and amount of evolutionary divergence. A classification...must reflect the common ancestry of the members of each taxon. However, Darwin also stressed that the unequal amounts of subsequent divergence of descendant groups must be duly considered in the ranking of taxa...Although birds and crocodilians are sister groups, one would not give the widely divergent birds the same categorical rank as the crocodilians, which in most respects are still reptiles. (Mayr and Ashlock 1991: 122)

When Mayr's support in favor of evolutionary taxonomy is taken in isolation from his other commitments, the impression is given that he strongly embraces *adaptationism* (evolutionary contingency) in his taxonomic practice—as presumably seen in how evolutionary changes (differential rates) are acknowledged and taxonomically registered. Again, we are contesting this interpretation on the grounds that the exact *meaning* of explanatory components like "evolutionary contingency," "adaptationism," and "evolutionary taxonomy," among others, can only be revealed by the role they play in Mayr's overall evolutionary framework. These components are neutral enough to have remained fairly stable throughout Mayr's career and discourse, going through a change of meaning only when the general framework supporting them was modified.

The Mayr of the 1960s, 1970s, and 1980s did not make use of a full flexibility range when it came to accounting for adaptive change in taxonomic practice, for the obvious reason that he was still committed to the cone of increasing diversity (Mayr 1974a: 3, 6, 8; 1985a: 759–760; 1988: 200–202). This explains why Mayr's use of notions like "common ancestry" and "divergence" are framed in the service of a particular view of the monophyletic agenda, until their decoupling around 1990. To see the implications accompanying this commitment more clearly, it is useful to

---

although "speciation" is about diversification or multiplication, that process need not be followed by prolonged and sustained "divergence." This is reflected in his recognition that anagenesis (phyletic evolution) can, after all, be accompanied by a substantial amount of evolutionary change instigating novelty. In other words, cladogenesis (speciation) lost the privileged status the early Mayr awarded to it in evolutionary biology.

contrast it with what Stephen Jay Gould had to say about such notions as they appear in his own "decimation model" (see Fig. 3):

> Yet, within these constraints of *monophyly* and *divergence*, the geometric possibilities for evolutionary trees are nearly endless. A bush may quickly expand to maximal width and then taper continuously, like a Christmas tree. Or it may diversity rapidly, but then maintain its full width by a continuing balance of innovation and death. Or it may, like a tumbleweed, branch helter-skelter in a confusing jumble of shapes and sizes. Ignoring these multifarious possibilities, conventional iconography has fastened upon a primary model, the 'cone of increasing diversity.' (Gould 1989: 38; italics original)

The two notions of "common ancestry" and "divergence" are, therefore, also explanatory neutral, with their functions being given by the distinct ontologies in which they are enshrined.

Mayr's discourse concerning evolutionary taxonomy remained constant throughout his career when viewed from the perspective of his unfailing support for it. Yet, at the same time, it is possible to detect subtle inflections in his discourse reflecting challenges to his view. Whereas in the 1960s Mayr presents things as if his discourse was a fair representation of the real world—through the intermediary of what he believed was a well-supported evolutionary theory—by the early 1980s, he had become much more aware of the specific choices, strengths, biases, and limitations accompanying evolutionary taxonomy. It could not have been otherwise, since new voices began challenging the orthodoxy and articulating competing approaches while forcefully opposing Mayr's own taxonomic school (Hull 1988). The implication is that Mayr had not yet realized in the 1950s and 1960s the extent to which he looked at evolution through the distorting lens of a *theory*, that is, a guess about what the world is about. He writes: "a healthy advance of taxonomy depends on a sound theoretical foundation" (Mayr 1968: 548, see also 1969: 79–80; 1974b: 94). Presumably, he believed at that time that his own theory constituted such a sound foundation. This illusion arose out of the context of the theoretical domination created by the "triumphant march" (Hull 2001: 228) of evolutionary taxonomy (or the "new systematics") within the systematics community. In these early years, Mayr apparently thought that the two key principles of his taxonomic school—common ancestry and subsequent divergence—were an objective reality.

However, the changed epistemological status of evolutionary taxonomy in Mayr's mind in the following decades constitutes one more reason he decided to move beyond his system of thought erected in the 1940–1970 period. In a perceptive study ahead of its time, especially considering the dominant position of evolutionary taxonomy in the English-speaking countries, Hull (1964) pinpointed the inherent limitations of an evolutionary biology organized around the Linnaean hierarchy, that is, nested taxonomic levels of ever more inclusive categories:

> The Linnaean hierarchy is the schema tacitly assumed in taxonomy. Its only modes of representation are rank assignment and class inclusion or exclusion. Because all taxa at a particular level must be mutually exclusive and because two taxa once separated cannot be classed together again at a lower level, the *Linnaean hierarchy is a constantly diverging*

> *affair... The form of the Linnaean hierarchy actually prohibits the consistent classification of polyphyletic taxa.* (Hull 1964: 10; my emphasis)

This criticism is damning. Enshrined within the Linnaean hierarchy itself are biases that nicely support Mayr's overall view. With hindsight—and enlightened by a massive literature that later arose to support competing schools of systematics—Hull became even more skeptical about what could be represented through the Linnaean hierarchy:

> Taxonomists began by thinking that the Linnaean hierarchy is the ideal way of representing phylogenetic trees, but as they attempted to make the relationship between hierarchical classifications and phylogeny clearer and more objective, they were forced to realize that representing even the most straightforward phylogenetic relationship (e.g., sister groups) is very difficult, if not impossible. One solution is to abandon the Linnaean hierarchy. (Hull 2001: 236)

Originally designed in the eighteenth century to deal with clusters of extant forms, the Linnaean hierarchy served Mayr's neontological agenda well, but at the expense of a fictitious past!

To see more clearly why Mayr eventually had to make concessions in the face of mounting challenges, it is useful to go back to the 1960s, at the time when he thought his position was self-evident: "Is it really necessary to repeat all this once more in the middle of the 1960's?" (Mayr 1965a: 167). His confidence in the explanatory power of a neontologically based evolutionary classification (taxonomy) extends not only to competing schools of systematics and molecular biology but also to paleontology: "That so many phylogenetic trees and phylogenetic diagrams were brilliantly confirmed when more adequate fossil records became available proves clearly that at least some taxonomists are able to draw the right inferences." And he added: "I know of only a few cases in which the conclusions of the sound morphologist were not confirmed by subsequent fossil discoveries" (Mayr 1965a: 169, 173; see also 1965b: 94; 1974b: 98; 1982a: 231–232). More brave words!

Although it would be possible to treat Mayr's responses to competing schools of systematics and molecular biology separately, their intrusion in the field of evolutionary biology at about the same time elicited a unified (and fairly predictable) reply from him. At first, Mayr hoped to steer an intellectual road around obstacles. Sooner or later, however, he had to engage in a series of strategic retreats that left his ontology of the living world largely disproved. It will therefore be more fruitful not to adopt a strict disciplinary breakdown of the issues ("morphology versus molecules") but to focus, instead, on a thematic approach.

12.1 The Challenge to the Monophyletic Agenda

The early Mayr was not seriously bothered by blurring processes like parallel evolution and convergent evolution, arguing against their significance and focusing on divergence instead. But after a confident silence concerning these potential threats, he could no longer ignore the issue as it imposed itself with a new urgency in the guise of the school of phenetics (Mayr 1965b: 86). The ideal of pheneticists consists in refusing to weigh taxonomic traits in the elaboration of their classifications, identifying clusters of similarities by treating these traits equally, each being given the same value (Mayr 1965b: 90). If left unanswered, this approach would jeopardize Mayr's monophyletic agenda: since selective forces impact features and life forms unequally, reasoned Mayr, their effect can blur true genealogical connections, such as when selection may produce phenotypic similarities in independent lines, "and sometimes not even closely related lines" (Mayr 1964a: 22; see also 1965b: 86; 1969: 68–69; Richards 2016: 119–120). Carefully selecting or weighting the right traits becomes, for Mayr, a real concern in order to establish clear genealogical lines diverging away from presumed common ancestors. The following quotation is interesting for what it does not say about Mayr's monophyletism:

> Sokal and Sneath [ask]: 'Why should we wish to make taxonomies based on monophyletic groups?' They continue: 'Suppose the convergence had become so absurdly extreme that the two forms are almost indistinguishable and can readily and successfully hybridize: what is the purpose of separating them on grounds of ancestry when in all other attributes they are virtually the same.' This, to me [Mayr], is an altogether unrealistic argument. The possibility of such an 'absurd' convergence is nil and hence this is not a legitimate argument. Even if a whale and a shark should become so similar by convergence that they could no longer be separable by morphological-phenetic means, no one in his right mind would expect them to become cross-fertile or biochemically identical. Not a single case of cross-fertility is known to have resulted from convergent evolution. Convergence is bothersome at a level well above that of the species, well above that of reproductive compatibility. (Mayr 1965b: 83)

Is the issue here really about the possibility of long separated lineages being able to hybridize? Granted, Sokal's and Sneath's objection was not couched under the most favorable terms. But to compare sharks and whales is not entirely fair, perhaps even insulting. Mayr's admission that convergence is "bothersome" well above the species level is a nice way to avoid the hard questions, considering that one can think of cases for which related lines could have become blurred under convergent evolution; for instance, when closely related genera and families are involved. Actually, he admits this in a rather ambiguous formula: "Convergence. This is a more theoretical than practical objection, particularly in groups in which numerous characters can be scored. Convergence can be a real problem in *fairly closely related groups* with rather few characters, where, however, it poses a problem to all known taxonomic methods until additional characters are discovered" (Mayr 1965b: 86; my emphasis). We again see Mayr's confidence in his theoretical choice at work, backed up by the ontology of the cone of increasing diversity. Because "choices" and "biases" can be two sides of the same coin, the following statement only makes sense within an already accepted theoretical framework: "Monophyly in taxonomy is a retrospective, not prospective principle" (Mayr 1965b: 84). Even Mayr's classic "On the Evolution of Photoreceptors and Eyes" (Salvini-Plawen and Mayr 1977)—in which the independent development of eyes across a wide taxonomic spectrum are described—would not diminish his confidence in the cone of increasing diversity.

Having responded to the pheneticists, Mayr now confronted the threat coming from scholars making use of other features than organismic ones, in this case molecular evidence. Raising again the specter of evolutionary analogies such as under convergent evolution, he warned against the blurring of related filiations by appealing to the cardinal pillars of evolutionary taxonomy: common ancestry and subsequent divergence. Molecular biologists are targeted for not paying enough attention to the differential values of characters:

> Unfortunately some of those who have recently entered taxonomy from other fields, like biochemistry, cytology, serology, or medicine, have not yet burned their fingers. Some of them are sincerely convinced that the new characters which they are finding are—for various theoretical reasons—more basic, and hence more important and more informative, than the characters which have been used hitherto. The greatest weakness of this assumption is that it ignores several well-known evolutionary or phylogenetic phenomena. It ignores convergence, it ignores the fact that taxonomic distance is determined by two independent variables (the chronology of splitting and the rate of evolutionary divergence, mentioned previously), and it ignores the phenomenon of multiple pathways. (Mayr 1964a: 26–27; see also Mayr, 1965a: 171; 1974b: 103; Salvini-Plawen and Mayr, 1977)

Raising the issue of convergence and parallelism so openly might have made for a decent argument against phenetics and molecular biology in the 1960s—and cladistics in the 1970s (Mayr 1974b: 116–118)—but sooner or later these objections would come back to haunt Mayr's own taxonomic school, owing to inherent limitations common to all approaches. On this view, "On the Evolution of Photoreceptors and Eyes" (Salvini-Plawen and Mayr 1977) appealing so openly to the significance of the independent rise of eyes in life's history is a piece of bravery that would eventually contribute to undermining Mayr's cone of increasing diversity.

Let us project ourselves forward in time with Mayr for a moment. He writes: "When one uses taxonomic characters to draw inferences about classification, one must always balance the potentially conflicting information derived from different character domains... While molecular characters are subject to the same problems of convergence, parallelism and reversal that any character set may have, it is highly unlikely that morphological and molecular characters will be affected in the same manner" (Mayr and Ashlock 1991: 173; see also Mayr 1981: 511, 512; 1982a: 228–229). If this is so, which features should be privileged under Mayr's monophyletic agenda? We will see shortly that a major chasm is about to open between genotypic evolution and phenotypic evolution in evolutionary biology. For the time being, Mayr felt confident enough in the 1960s and early 1970s in the power of his own taxonomic practice to admonish pheneticists with molecular evidence: "The pheneticist forgets that the visible phenotype is like the visible portion of an iceberg. It reflects only part of the genotype" (Mayr 1965a: 167; see also 1965b: 85).

12.2 The Challenge to the Varying Rates of Divergence

Under the cone of increasing diversity, Mayr needed just the right application of the doctrine of adaptationism. A view based on pan-selectionism would blur the divergent pattern owing to overwhelming adaptive changes, either giving rise to analogies or significantly obliterating the genealogical connections. A view based on too weak a selective force would be impotent in driving evolution forward along divergent paths. This dilemma was also Darwin's (Sober 2011: 35), since both scholars wanted to support their pre-established pattern of divergence by turning to a *justificatory* use of the selective process. In other words, Mayr had to be cautious not to provide arms to his adversaries, since they might want to use them against his main assumption: the cone of increasing diversity. The 1960s, 1970s, and 1980s were years when Mayr modulated his use of adaptationism, stressing selectionism against some opponents and easing up on it against others. That apparent explanatory flexibility accompanying the doctrine of adaptationism reveals that Mayr was not committed to it in the sense usually anticipated, thus offering him genuine rhetorical leeway to use against his opponents. Let us keep that rhetorical dimension of Mayr's discourse in the back of our mind while we pursue our investigation.

Cladistics and molecular biology posed two different challenges for Mayr in this context. Against cladistics, Mayr criticizes the principle of dichotomy which consists in proposing classifications exclusively based on sister groups, allegedly representing branching lineages. He claims that this practice introduces a distortion by not recognizing that related lineages evolve at different rates: "It is altogether wrong to base classifications only on the location of the branching points (forks) and to make the silent assumption that the rate of change is the same on all lineages" (Mayr 1964a: 22; see also 1965a: 167; 1965b: 78; 1968: 547–548; 1969: 70–71; 1974b: 106). As much as Mayr wants to push the principle of divergence forward, he also needs some evolutionary flexibility to allow for the diversification process to also express itself during phyletic evolution (Mayr 1974b: 102).[5] It is here that evolutionary contingency fills its explanatory role, under Mayr's own specific application as discussed above. The cladists's assumption, on his view, is just too rigid to be realistic, process-wise. In fact, he continues, different characters within a lineage change at different rates and in different directions, resulting in mosaic

[5]Of course, this is not to deny that earlier on, Mayr (see 1963a, b: 621) had excessively undervalued the role of phyletic evolution under his notion of "genetic revolution" wherein most key evolutionary changes allegedly occur at the precise moment of speciation. The rise of cladistics and its focus on "dichotomies" in classification forced him to proceed with a *re-evaluation*, stressing more than previously phyletic evolution. This move reconnected Mayr with the two main components of "evolutionary taxonomy": common ancestry (monophyletism) and subsequent evolution (phyletic evolution). In Mayr (1968: 547) we read: "branching is only one of several processes during phylogeny, the rate of the subsequent divergence of the branches, for instance, being at least as important as the branching." See also Mayr (1974b: 106–107). A case could be made—as we are making that case here—that this small conceptual shift in Mayr's thinking in the 1960s and 1970s contributed to his slow transition toward a fuller appreciation of "evolutionary contingency." Many more intellectual steps, however, awaited Mayr on this long road.

evolution (Mayr 1974b: 106; see also Caianiello 2024). In addition, two taxa may be very similar but for various reasons: (a) because they have branched very recently from a common ancestor; (b) because they have branched some time ago, but have since remained stable; or (c) because they have been subject to parallel evolution or convergent evolution (Mayr 1964a: 22–23; see also Salvini-Plawen and Mayr, 1977). About the latter case, he added several years later: "The frequency of nonhomologous 'apomorphies' [uniquely derived characters] has been underestimated by many cladists. How often a given, even seemingly rather improbable, adaptation can be acquired independently is well illustrated by the evolution of eyes... The independent loss of a character in separate lineages is a particularly frequent form of convergence" (Mayr 1982a: 228). But as already alluded to, Mayr is playing with fire here, as his use of convergent evolution will eventually undermine his cone of increasing diversity.

Mayr criticizes cladists on the more general grounds that their dichotomous agenda introduces an overall distortion about evolutionary divergence, seeing it everywhere, even when it is not necessarily there to be found (Mayr 1965a: 167; 1965b: 78; 1974b: 106). For instance, the amount of change during phyletic evolution of a single lineage can easily be misrepresented under the cladistics' formalistic methodology as divergence involving two or more lineages (Mayr 1974b: 109–111, 124): He writes: "This means that the branching point [dichotomy] fixes irrevocably the categorical rank of subsequently evolving taxa without any consideration of evolutionary events that happen later (except branching)" (Mayr 1974b: 124). Again, lurking in the background is the issue of the minimal amount of adaptive flexibility required for the evolutionary process to operate properly under Mayr's theory. In the following quotation, it should be noted how Mayr even relies on the genotypic level—and not the phenotypic one—to defend his case:

> If one of the [evolutionary] lines is exposed to severe selection pressure and as a result diverges dramatically from its genealogically nearest relatives, it may become genetically so different that it would be a biological absurdity to continue calling them near relatives. Even though the crocodilians are cladistically nearest to the birds (both having descended from pseudosuchians), the crocodilians are still closer to many of the other reptiles, as far as the total gene composition is concerned, than they are to the birds, which have so drastically altered their genetic composition. (Mayr 1965b: 79)

Soon enough, however, Mayr will become fairly disenchanted with molecular biology.

12.3 The Challenge to Speciation as a Major Organismic Event

Mayr uses the biological species concept—founded on organisms bound in potential reproduction (1942: 120)—as the gravitational center for his taxonomic reasoning, a choice directly related to his organismic-based biology. Whereas taxonomic categories cannot unambiguously be assigned, given that the taxonomic scale (genus, family, order, etc.) is subject to the personal appreciation of each taxonomist

(Mayr 1968: 548; 1969: 77), the only exception is at the "species" level, since it constitutes both a taxonomic category and a nonarbitrary entity operationally defined under the biological species concept (Mayr 1964a: 19). We have already encountered the extensive conceptual matrix deployed by Mayr to ensure that the species level would play such a central role in evolutionary biology.[6]

Faced with potential threats also arising from molecular biology, Mayr delivered a preemptive strike by inviting organismic biologists not to rely too extensively on genetics: "Molecules are far simpler structures than anatomical structures or behavior patterns, and they may permit convergent evolution in nonrelated groups up to the point of essential identity" (Mayr 1969: 81). Organismic biologists are therefore invited to stay on top of their field. As Maureen O'Malley (2010: 533) stresses: "And because real diversity was morphological, not molecular, for Mayr, then any major classification based on molecular evidence was not sufficient for him. Mayr certainly recognized the value of molecular data. . . [but it must] always be understood against the background of existing classifications, and used to 'strengthen soft spots' in such systems." This is a good characterization of the *official* position of Mayr in the 1960s and 1970s, but problems are lurking behind the scene. When speaking of the *mutual interest* of taxonomists and molecular biologists, Mayr's formulation is rather one-sided: while taxonomists need genetics to fill voids in classifications that are already largely established, molecular biologists, for their part, require taxonomists to make sense of their own data: "The findings of the biochemist often make little sense unless placed against the background of a well established classification; the taxonomist, in turn, is vitally interested in new characters in order to strengthen soft spots in his system" (Mayr 1965a: 165). This early attitude toward molecular biology is quite similar to his approach to the field of paleontology.

In fact, Mayr's relationship with molecular biology is a complex one—often based on a double epistemological standard—using it when it serves his research program and rejecting it when it does not. He even writes: "The amount of genetic similarity now becomes the dominant classifying criterion for the biologist" (Mayr 1968: 548), or the "phenotype is susceptible to all sorts of irrelevant similarities, and it is only the analysis of the inferred genotype which permits us to determine what similarities in the phenotype are due to convergence and what others are an expression of the ancestral genotype" (Mayr 1969: 77).

Ultimately, Mayr aims to avoid a rupture or disjunction between studies focused on the phenotype and those on the genotype. During the process of speciation, for instance, he uses his own proposed concept of "genetic revolution" to hold that small populations bud away from larger parental populations through a "drastic" event that "may affect all loci at once," exposing such genetic changes to positive selection (Mayr 1954: 170; 1963a: 533). Profound phenotypic changes are thus closely linked

[6]This conceptual matrix is constituted of the following components: population structure, polytypic species, population thinking, biological species concept, allopatric and peripatric speciation, isolating mechanisms, gene flow, founder effect, size of effective breeding population, genetic drift, unity of the genotype, and genetic revolution. Ultimately, such a matrix is put at the service of Mayr's ontology of the living world encapsulated in his cone of increasing diversity.

to underlying genotypic changes, on the assumption that life is organized around an "overall unity of the genotype" (Mayr 1965b: 89). Mayr's rationale holds that natural selection acts on the reproductive success of individual organisms, with genes contributing to the selective valence of phenotypes, and thus being indirectly affected by natural selection. Going further in his explanatory connection between biological levels, Mayr argues that since evolution is gradual at the phenotypic level, the same must apply to lower levels: "The new system evolved very gradually, requiring undoubtedly many biochemical adjustments but not requiring any sudden biochemical revolutions" (Mayr 1964b: 1232).

To drive home his point about the unidirectionality of the information flow during the selective process, let us present that quotation again: "No case is known to me in which a change in body chemistry initiated a new evolutionary trend. Invariably it was a change in habits or habitat which created a selection pressure in favor of chemical adjustments" (Mayr 1964b: 1233). Under this construal, one might expect Mayr not to be an avid supporter of the notion of "neutral mutations" just about to emerge in evolutionary biology: "As an evolutionist I am furthermore impressed by the power of natural selection, and I doubt for this reason that any particular amino acid configuration was ever replaced by a different one, unless this replacement was favored by natural selection" (Mayr 1964b: 1234).

Mayr's gradual retreat from this original position is nicely captured by his changing view on hominoid phylogeny. Implicit in this retreat are significant implications for his overall view, considering that key notions inherent to "evolutionary taxonomy"—such as common ancestry (monophyletism), rates of evolutionary change (divergence), and adaptation—become blurred and literally duplicated under competing discourses in both organismic biology and molecular biology. The shifts in his position can be divided into four main phases.

Phase 1 Mayr entered the debate in 1963 by relying on molecular biology. He was faced with a choice as to whether the human evolutionary line diverged away from the great apes stem at an early moment before the three of them (Orang, Gorilla, Chimpanzee) became separated from one another or, alternatively, later on in geological time when the Orang had already become isolated on its own evolutionary road, implying that humans, chimps, and gorillas are more closely related with the Orang being more distant. Mayr chose the second alternative. The challenge consisted in disentangling the impact of evolutionary parallelism, since great apes are all adapted to a brachiating lifestyle. In the absence of fossil evidence, Mayr claimed that hemoglobin structure and serum proteins conclusively proved that the second alternative was correct (Mayr 1963a: 627–628).

Phase 2 In the same publication year of 1963, but relying on slightly more recent references, Mayr added the technique of chromosomal morphology to his list of molecular arguments. He now also mentions that Pliocene fossil remains like *Ramapithecus* and *Kenyapithecus* may well belong to the hominid (human) line exclusively. More importantly, Mayr had to explain the apparent disjunction between, on the one hand, molecular similarities uniting humans and African great apes and, on the other hand, their morphological differences. He did so by pressing the notion of "mosaic evolution," explicitly using the expression in his text:

> Perhaps the simplest interpretation would be to assume that Man's shift into the niche of the bipedal, tool-making, and speech-using hominid necessitated a drastic reconstruction of his morphology, but that this reconstruction did not, in turn, require a complete revamping of his biochemical system. Different characters and character complexes thus diverged at very different rates. (Mayr 1963b: 344)

Mayr later made clear that he supported a splitting time for the human line at 15, if not 20, million years ago, relying on the discovery of *Ramapithecus punjabicus* (Mayr 1970: 378–379).

Let us pause to reflect about the disjuncture so far accepted by Mayr between genotypic and phenotypic levels. Indeed, it is no longer obvious that he is busy merely filling in the "soft spots" of a morphologically-based taxonomy already largely established, as seen in the quotation above (Mayr 1965a: 165). It should be noted how he reversed the priority order of his organismic agenda: molecular biology suddenly becomes the key criterion for supporting the monophyletic origin of the *Homo-Pan-Gorilla* group, with *Pongo* being more distant. This implies that the issue of evolutionary parallelism is believed to be disentangled by molecular evidence, considering that morphology points in the opposite direction, as seen in numerous adaptations explained by brachiation in *Pan*, *Gorilla*, and *Pongo*. At that point, Mayr chose molecular biology over organismic biology to push forward his monophyletic agenda. But was his view still compatible with how he originally described the speciation process?

Looking at Mayr's (1970: 379) phylogenetic tree of hominoid evolution, one gathers the following: (1) the human line departs from the common ancestor shared with *Pan* and *Gorilla* around 18 million years ago; (2) between 18 and 2.5 million years ago, the human line gradually diverges away from the African great apes at a fairly constant rate; (3) around 2.5 million years ago, a drastic morphological shift occurs taking the human line down the path of unique adaptations (such as an increase in brain size and tool use), some of them being acquired nearly instantly, geologically speaking. This entire process, spanning nearly 20 million years, deployed against a fairly uniform molecular background common to *Homo*, *Pan*, and *Gorilla*, as far as the analyzed proteins and chromosome structures are concerned. Apparently, Mayr believes that "mosaic evolution" is endowed with a remarkable (suspiciously so) conceptual plasticity!

At this juncture, a portion of Mayr's theoretical apparatus—monophyletism, divergence driven by selection, speciation simultaneously involving genotypic and phenotypic events, the preeminence of organismic biology over molecular biology—fell under significant explanatory pressure. In other words, the two main components of Mayr's "evolutionary taxonomy" (common ancestry and subsequent divergence) faced challenges at the level of organismic biology.

Phase 3 By 1975, Mayr, who had thus far tried to deal with the rise of molecular biology as best he could by containing it within the purview of his own early system of thought—despite long displayed epistemological and institutional resistance (Dietrich 1998; Milam 2010)—could no longer follow along its latest developments. In fact, Mayr was an early skeptic of the possibility of recognizing constant rates for molecular change (Mayr 1965a: 171–172). But a wider recognition among scholars of notions such as "neutral mutations" and "molecular clock" pushed him to the breaking point. Whereas he previously exploited the notion of mosaic evolution to cover a growing explanatory tension between phenotypic and genotypic evolutions, he now retreated to what was for him a more standard position. This is nicely captured in a long quotation that must be reproduced in full:

> There is little doubt, however, that those who enter biology from the physical sciences make rather unrealistic assumptions on the evolutionary rate of changes in macromolecules. From the fact that some molecules such as the histones evolve very slowly, while others have intermediate, or very rapid evolution, some authors have drawn the conclusion that such molecules have a built-in, so to speak, orthogenetic rate of evolution. Depending on the particular macromolecule they assume that there will be an amino-acid replacement every 2 million, 5 million, or 10 million years. All the known facts contradict this naïve assumption. Certain biochemists, on the basis of the assumption of standard rates in the evolution of molecules, postulated that the splitting of the hominid line from that of the African apes occurred as recently as 3 to 5 million years ago. The fossil record, however, demonstrates conclusively that the split happened at least 14 million years ago.[7] It seems to me that it leads to a far more realistic interpretation of evolutionary rates if we assume that evolutionary change of a given type of macromolecule is not due to a built-in rate but rather due to the need for co-adaptation with other molecules which it has to interact. As the other molecules change in the course of time in response to *ad hoc* selection pressures, every molecule occasionally has to adjust to its changed molecular environment; it must adjust to the cohesion of the genotype. This undoubtedly, is the reason for the continuous evolutionary change of molecules even of those in which the active site has not changed since the days of the most primitive eukaryotes or even prokaryotes... The genes are not the units of evolution nor are they, as such, the targets of natural selection. Rather, genes are tied together into balanced adaptive complexes, the integrity of which is favored by natural selection. (Mayr 1975: 386)

Yet, no matter how much Mayr was willing to restate—or shall we say "reinstate"—his speciation model, it was no longer possible to reestablish genotypic/phenotypic ties that were as strong as he originally conceived them to be. Recall that whereas he once described a genetic revolution (speciation process) as involving a "drastic genetic change" that "may affect all loci at once," something that "may well have the character of a chain reaction" (Mayr 1954: 170; 1963a: 533), he now had to concede that molecular evidence did not support the involvement of

[7]Mayr's view about human evolution in 1975 reflects what many paleoanthropologists believed at the time, seeing in *Ramapithecus* the earliest known hominid (Gribbin and Cherfas 1982; Lewin 1987: 85–127). For Mayr, however, it was most convenient to rely on this apparent consensus in order to shut the door on threatening ideas coming from the molecular clock, as this allowed him to cling to his early system of thought. Yet, it was shown in 1982 that *Ramapithecus* was not a hominid at all, allowing molecular biologists to successfully push their agenda in the field of human evolution (Lewin 1997). By then, Mayr had lost the argumentative advantage he thought he had gained from the fossil record through *Ramapithecus*.

such a massive amount of genetic material during speciation. As noted by William B. Provine (1989: 70–71), Mayr acknowledged in a letter to Alan R. Templeton dated March 24, 1980, "that only a relatively few genes needed to be involved in the initial speciation process," to borrow here Provine's own summary. This left Mayr with a wider genotypic/phenotypic gap than he originally cared to recognize, implying that all sorts of genetic events might occur during speciation, phyletic evolution, or adaptation, many of them being concealed from an organismic viewpoint.

Phase 4 By the early 1980s, Mayr had to comply with what has become since the standard view of hominoid phylogeny, as largely supported by molecular biology (Mayr 1982a: 621–622; 1997a: 229–230; 2001a: 236–237; Mayr and Ashlock 1991: 173–74): (1) the fossil evidence from *Ramapithecus* was discarded; (2) the split between the human line and the African great apes was put between 5 and 8 million years ago, an idea that was eventually restricted to the relationship between humans and chimps exclusively, with a common ancestor around 5 or 6 million years ago; (3) the notion of mosaic evolution was standardized against this illustrated case, with different morphological features evolving at different rates, just as do different molecules. Of course, Mayr could not accept the proposal, which consisted in suggesting two distinct classifications, one based on molecular evidence and one on morphological characters (Mayr 1982a: 237–238); after all, organisms are somehow integrated wholes. But the suggestion certainly points out that the apparently straightforward approach to a monophyletic agenda is more torturous than originally believed by Mayr: "Molecular methods are needed desperately to discover the connecting points of the higher taxa, let us say the orders of birds or the phyla of invertebrates. Here morphological analysis has so far failed, because it has been impossible to find a sufficient number of clearly homologous characters and because the polarity of evolutionary trends is often uncertain" (Mayr 1982a: 237).

The rupture between genotypic and phenotypic evolutions was confirmed by Mayr in unequivocal words: "Morphologically exceedingly similar species of frogs and toads may display remarkable molecular divergence, while in groups like birds and mammals pronounced divergence in morphology and color pattern may not reflect any major molecular divergence" (Mayr 1982a: 241). Clearly, Mayr no longer holds that molecular biology is restricted to strengthening "soft spots" of a largely established morphologically based taxonomy (Mayr 1965a: 165). He finally expressed his recognition of the metronomic component of the molecular clock in these words: "the amount of difference between two species has been shown to be roughly correlated with the time at which the evolutionary lines leading to the two species separated" (Mayr 1982a: 237; see also 1988: 532).

13 Evolutionary Taxonomy: The Decline of a Domination

Mayr's encounters with other alternatives forced him to confront a complex reality by adjusting to it. Working with the principle of the Linnean hierarchy of clusters nested into ever more inclusive clusters, Mayr knew that different classifications could be generated by analyzing the same data (Mayr 1969: 229; Mayr and Ashlock 1991: 125). Although "species" are self-operationally defined through the testing of isolating mechanisms, clusters at higher taxonomic levels are subjective categories. The exercise of classifying them under different genera, families, orders, and so on leaves much room for appreciation and disagreement. Still, Mayr held that something "objective" (Mayr et al. 1953: 59; Mayr 1969: 91; Mayr and Ashlock 1911: 134) could be found in the subjective evaluation of higher categories, considering that properly identified natural clusters are, in his view, descendants of common ancestors. The difficulty, however, consisted in finding such an objective genealogical trail among contradictory signals generated by adaptation (convergence) and neutral mutations, in light of varying assessments from different schools of taxonomy and different disciplinary perspectives (morphology versus molecular biology). The early Mayr was in favor of evolutionary contingency provided it would not jeopardize the foundation of his cone of increasing diversity. Unfortunately, the revelation of evolution's complexies under alternative views proved anything but reassuring for Mayr's cone ontology.

Mayr's eclectic ideal for evolutionary taxonomy consisted in integrating as much information as possible in a classification: branching patterns, inferences about selective pressure, shifts of adaptive zones, passages to new evolutionary grades, and rates of evolutionary change (Mayr 1974b, 1981). It is this broad information agenda that eventually led to the realization that a widening conceptual gap was opening between a "classification" and its "representation" of the world. Pheneticists, for instance, quickly realized the inherent difficulty (or impossibility) of attempting to pack too much information in classification (Richards 2016: 120). At first, Mayr assumed that a classification could contain all that information by merely keeping in mind that the endeavor comprised two goals that are fused in a unified practice:

> The complex relationship of species with each other and the varying rates of branching and of evolutionary divergence must be translated into a system of taxa, conveniently ranked in the appropriate categories, in such a way that it forms as efficient as possible a system of information storage. It is the purpose of the ensuing discussions to make the taxonomist aware of various practical and theoretical considerations that will permit him to produce a classification meeting two objectives: grouping nearest relatives near each other and facilitating information retrieval. It must be the principal concern of the classifier to delimit and rank the taxa in such a way as to meet both objectives. (Mayr 1969: 229)

Then, as the debate over taxonomic practices and their representation matured, the complexity of the task became obvious to Mayr: "It is now realized that classification is not a one-step procedure, and that therefore simplistic methods rarely lead to satisfactory results. Classifying consists of a whole series of steps...

and at each of these steps different procedures are needed and are most useful" (Mayr 1982a: 244–245). As much as Mayr targeted opponents for their so-called simplistic practices (Mayr 1981: 511), he himself had moved well beyond his original assessment. Now, although the two main objectives of classifying were said to be "largely compatible" (Mayr 1981: 511), the fact remains that they could no longer be presented together: "[S]ince evolutionary taxonomists incorporate a great deal more information in their classifications than do the cladists, they cannot express all of it directly in the names and ranking of taxa in their classifications. Therefore, they consider a classification *simply to be an ordered index* that refers them to the information that is stored elsewhere (in detailed taxonomic treatments)" (Mayr 1981: 511; my emphasis). David L. Hull assessed the meaning of this epistemological partitioning in the following way:

> Advocates of the new systematics wanted to make biological classification more scientific. Toward this end they urged the classification of organisms according to their phylogenetic descent. Because they wanted to reflect so many different relationships in their classification, they admitted that no precise relationship could exist between a biological classification and all the relations that went into its construction. Mayr (1981) for one argued that an evolutionary classification is not itself a repository of information but serves only as an ordered index to this information stored elsewhere. (Hull 2001: 235)

We are far from the "triumphant march" (Hull 2001: 228) of evolutionary taxonomy as seen in the 1960s. The Linnaean hierarchy used for so long to mediate the discourse between evolutionists and its representation of the world may well be, after all, fundamentally flawed. Marc Ereshefsky (2001: 199–237) explains that the system's rigid application—the limited number of hierarchical levels, the naming and ranking rules, the requirement of a hierarchical classification—thwarts the evaluation of life's diversity as seen, for instance, in the rich branching of life forms, ancestor-descendant relationships, the integration of fossil taxa, and the reality of hybrid taxa or reticulate evolution. He writes: "The Linnaean system has served biology well. Few would deny that. But biological theory has changed drastically in the last two hundred years. Those changes have rendered the Linnaean system both theoretically outdated and pragmatically flawed" (Ereshefky 2001: 11). With such a reassessment in mind, it is difficult not to ask obvious questions: what is there to be represented with evolutionary taxonomy? Are many of evolution's complexities evading its representation capacity—convergent evolution, neutral genetic change, rates of morphological transformation, etc.—thus making Mayr's early commitment to evolutionary contingency superficial?

Mayr never claimed that competing schools of systematics were entirely without interest. For instance, he valued phenetics for its attempt to evaluate traits in an unbiased way (Mayr 1965b: 74; see also 1982a: 221–226; Mayr and Ashlock 1991: 197–198). He praised cladistics for having fully developed the principles of cladistic analysis (not to be confused with cladistic classification) (Mayr 1974b: 98; see also Mayr 1982a: 226–235; Mayr and Ashlock 1991: 207–208). The following tribute is particularly significant as it concerns the impact of cladism on his monophyletic agenda:

> The greatest virtue of cladistic analysis is that it is an effective method for testing the 'naturalness' (that is, the monophyly) of groups originally delimited by phenetic methods [including evolutionary taxonomy]... In order to appreciate how fundamental the impact of Hennig's methodology has been, one needs only to look at recent taxonomic revisions, particularly those relating to fishes and certain groups of insects... The new cladograms it has produced have shown that many previously accepted taxa were *polyphyletic*. (Mayr 1982a: 231; my emphasis)

Of course, it was Mayr's hope that these components would be employed to improve evolutionary taxonomy itself, the latter remaining, in his view, the unifying banner. By 1981–1982, the conquest rhetoric of Mayr was toned down—as reflected both in a title like "Biological Classification: Toward a Synthesis of Opposing Methodologies" and in the relevant pages of *The Growth of Biological Thought* (1982: 221–250)—but certainly not to the point of giving up on the alleged superiority of evolutionary taxonomy:

> I have no doubt that moderates will be able to develop an eclectic methodology, one that contains a proper balance of phenetics and cladistics that will produce far more 'natural classifications' than any one-sided approach that relies exclusively on a single criterion, whether it be overall similarity, parsimony of branching pattern, or what not. Evolutionary taxonomy, from Darwin on, has been characterized by the adoption of an eclectic approach that makes use of similarity, branching pattern, and degree of evolutionary divergence. (Mayr 1981: 515)

By the early 1990s, Mayr recognized more openly still the legitimacy of concerns raised under other approaches. The following quotation is remarkable for it encapsulates, in a single breath, the threats from competing schools of systematics, molecular biology, and adaptationism (blurred affiliation erased under adaptive changes):

> Current differences of opinion concern both the methodology of taxonomy and the most basic principles of the philosophy of classification. What should a classification reflect, genealogy only or the total aspect of the included taxa? ... What components of the phenotype should one choose as the basis of one's classification, considering the frequency of mosaic evolution and homoplasy?[8] ... Should one give equal weight to the quasi-neutral replacement of base pairs in DNA and to evolutionary significant changes?... Perhaps never before in the history of systematics have there been so many uncertainties and controversies. The representatives of each of the three major current schools of classification—phenetics, cladistics, and evolutionary classification—are trying to convince us that their philosophy and methodology are superior to those of the other two. All three schools have numerous followers, indicating that the methods or principles of each school have considerable appeal. It would appear that none of them could be all wrong. (Mayr and Ashlock 1991: xviii-xix)

[8]Mayr defines homoplasy in these terms: "Homoplasy is similarity in a character independently acquired in two taxa. The term is particularly useful for similar taxa in an intermediate degree of relationship that makes it impossible to determine whether the similarity is due to a common intrinsic potential (parallelism) or to entirely unconnected evolutionary developments (convergence)" (Mayr and Ashlock 1991: 145).

By then, Mayr was less optimistic about the possibility of achieving a synthesis between taxonomic schools owing to wide conceptual or methodological gaps, a synthesis that would require all proponents to abandon some of their cherished principles. As far as evolutionary taxonomy is concerned, he lists the following concessions that could be made: "evolutionary taxonomists would have to adopt cladistic analysis (some already have), work out more concrete methods for the delimitation and ranking of taxa, and come to a consensus on the definition of monophyly" (Mayr and Ashlock 1991: 124). It is unclear whether Mayr was willing to concede on all counts.

Something had changed in evolutionary biology and Mayr could not deny it. He never formally relinquished his support for evolutionary taxonomy but his reasons for using it changed. For the early Mayr (1940–1970), evolutionary taxonomy filled the function of conceptually supporting or implementing his ontology of the living world, the cone of increasing diversity. As he gave up on that ontology, first by essentially pointing at challenging facts (Mayr 1991: 160; Mayr and Ashlock 1991: 13–14) and then by explicitly doing so (Mayr 2001a: 56, 59–61, 203, 209–210, 228–229), the role of evolutionary taxonomy was no longer part of a highly coherent theoretical matrix imposing upon the world a strongly preconceived view. *Mayr's initial view slowly dissolved and was replaced by a much looser theoretical framework, something closer to a general toolkit rather than a coherent overarching theory.* By criticizing evolutionary taxonomy, its opponents (unconsciously?) contributed to undermining Mayr's ontology of the cone, while simultaneously showing the inherent limitations of his approach to classification. This forced him to give up a number of tacit assumptions about evolutionary taxonomy: its *meaning* was transposed to apply to a more modest and ambiguous reality. In that sense, we hold that Mayr's commitment to evolutionary taxonomy was neutral enough to survive the downfall of his original ontology; it finally became a tool in his hands that offered him the *possibility* of exploring more fully the manifestations of "evolutionary contingency," although this came quite late in his career.

To summarize, the early Mayr exploited evolutionary taxonomy as a conceptual/methodological tool allowing him to investigate a world he assumed to be ontologically structured around two complementary features: (a) at the large scale, that world was believed to be organized around a cone of increasing diversity, and (b) at the small scale, that world was mainly organized around divergence. By so doing, Mayr was following in Darwin's footsteps. Unfortunately for Mayr, persistent criticisms seriously undermined evolutionary taxonomy, simultaneously revealing a world too complex to fall under his simplified assumptions:

1. Empirically speaking, the confrontation between competing schools of systematics made Mayr realize that evolution was characterized by blurring phenomena—convergence, polyphyletism, reticulate evolution—that were not easily accommodated under the tandem of divergence/cone of increasing diversity.
2. Disciplinary speaking, Mayr's organismic agenda faced a frontal assault by the rise of molecular biology, as seen in the disjuncture between morphological

evolution and molecular evolution, thus making notions like monophyletism and divergence less legible.

3. Epistemologically speaking, the original assumption of a tight match between "evolutionary taxonomy" and the world it was meant to "represent" was dissolved, disclosing the biases of a taxonomic school that imposed upon the world a pre-established order that did not obviously exist.

14 Behind an Equation: Microevolution = Macroevolution

The demise of Mayr's ontology not only impacted his postulated *structure* of the world—the divergence/cone of increasing diversity tandem—but also its accompanying *texture* whose manifestation is seen in the dual notion of gradualism/gradation. In fact, such a demise instigated in Mayr's mind a cascade of changes in related notions, including adaptationism and uniformitarianism. In order to unpack this conceptual bundle, let us examine this topic through the lens of the microevolution-macroevolution equation.

Here, again, a permanent difficulty for the exegete arises, that of detecting conceptual shifts taking place beneath an apparent terminological stability. Just as it is expected that the cognitive content of concepts can be modified as scientific knowledge progresses, one should also be aware that the use of the same terms throughout a career may also disguise an attempt at rewriting one's personal history. This is the case when one considers Mayr's apparent unwavering commitment toward the microevolution-macroevolution equation. Our thesis can thus be summarized thusly: whereas the weakening of Mayr's commitment toward such an equation is not readily detectable in his choice of language, this weakening becomes more apparent when attention is paid to the reality that is described by these words. Our investigation will therefore focus both on terminology and its interpreted reality. Early in his career, Mayr expressed in these words his commitment toward the microevolution-macroevolution equation:

> By the term macroevolution we understand the development of major evolutionary trends, the origin of higher categories, the development of new organic systems—in short, evolutionary processes that require long periods of time and concern the higher systematic categories... There is only a difference of degree, not one of kind, between the two classes of phenomena. They gradually merge into each other and it is only for practical reasons that they are kept separate... In conclusion we may say that all the available evidence indicates that the origin of the higher categories is a process which is nothing but an extrapolation of speciation. All the processes and phenomena of macroevolution and of the origin of the higher categories can be traced back to intraspecific variation, even though the first steps of such processes are usually very minute. (Mayr 1942: 291, 298; see also Mayr 1963: 586–587)

These words are used to describe what reality, exactly? For the early Mayr, life's history is a chalice-like structure (the cone of increasing diversity) whose texture or fabric is characterized by tightly knit entities conceptually bound together under the doctrine of gradualism/gradation:

> Hitherto we have spoken only of the delimitation of contemporary (synchronic) species. The delimitation of species which do not belong to the same time level (allochronic species) is difficult. In fact, it would be completely impossible if the fossil record were complete. The species of each period are the descendants of the species of the previous period and the ancestors of those of the next period. The change is slight and gradual and should, at least theoretically, not permit the delimitation of definite species. In practice, the fossil record is fragmentary, and the gaps in our knowledge make convenient gaps between the 'species.' In the few cases, in which an almost complete record of a continuous line is already available, the paleontologist follows the reasoning of the taxonomist who is confronted with an unbroken intergrading series of geographic populations. He breaks them up for convenience. (Mayr 1942: 153–154)

In this world filled to capacity, the evolutionary drive is said to be propelled by a selective process continually feeding upon readily available variations. Differently stated, if the texture of the world had been characterized by discontinuous "blocks of matter" (biological types) rather than continuous and contiguous atomistic variations, natural selection could not have played a creative adaptive role, for it would have been confined to a mere eliminative process:

> In the day when the validity of 'Darwinism' was still vigorously disputed, the crucial question was whether or not selection is 'creative'. If evolutionary change were due to the sudden appearance of entirely new types, the only possible function of selection could be the a posteriori one of acceptance or rejection. Likewise, thinking in early Mendelian terms, if one were to consider mutations the principal actors on the evolutionary stage, one would have little use for selection. As long as selection is defined typologically, as something that eliminates all deviations from the type. . .we must assign to it a purely destructive role. . . A different interpretation of the process of selection leads to the opposite answer. Is not a sculptor creative, even though he discards chips of marble? As soon as selection is defined as differential reproduction, its creative aspects become evident. Characters are the developmental product of an intricate interaction of genes and since it is selection that 'supervise' the bringing together of these genes, one is justified in asserting that selection creates superior new gene combinations. . . Natural selection. . .by always being strongly opportunistic and by selecting simultaneously for the preservation of genetic variability as such, is ready in each generation to jump off in new directions. (Mayr 1963: 201–203)

In order to establish a sharp contrast between the early and the late Mayr, let us skip ahead to the year 2001. In *What Evolution Is* (2001a: 188–207), the reader encounters Mayr's rhetoric about gradualism and the microevolution-macroevolution equation. A cursory reading of these pages suggests nothing has changed. A closer look, however, reveals that Mayr now presents issues as part of an open and unresolved debate, insisting that although numerous *observations* run counter to his choices, the *theory* nonetheless supports them:

> The reason why this controversy has not been fully settled is because there seems to be an astonishing conflict between theory and observation. According to Darwinian theory, evolution is a populational phenomenon and should therefore be gradual and continuous. This should be true not only for microevolution but also for macroevolution and for the transition between the two. Alas, this seems to be in conflict with observation. Wherever we look at the living biota, whether at the level of the higher taxa or even at that of the species, discontinuities are overwhelmingly frequent. Among living taxa there is no intermediacy

> between whales and terrestrial mammals, nor between reptiles and either birds or mammals. All 30 phyla of animals are separated from each other by a gap. There seems to be a large gap between the flowering plants (angiosperms) and their nearest relatives. The discontinuities are even more striking in the fossil record. New species usually appear in the fossil record suddenly, not connected with their ancestors by a series of intermediates. Indeed there are rather few cases of continuous series of gradually evolving species. (Mayr 2001a: 189)

This description contrasts with the one presented by the early Mayr (1942: 153–154) and for which a quotation was provided just above. The very late Mayr, we argue, lived in a different ontological reality than the early Mayr. In fact, the quotation just given is an understatement of Mayr's view in *What Evolution Is* (2001). Although he formally maintained his original commitments, their explanatory functions had been modified. Before moving forward with substantive issues, let us first consider the terminological ones. It seems possible to detect in Mayr's choice of words the moment when his thinking about the microevolution-macroevolution question began to be impacted. Speaking in 1982 of connections between phenomena studied by mathematical population geneticists and those studied by paleontologists/comparative anatomists, Mayr insists that these connections only become visible when analyzed through the findings of population systematics, "because they provide a perfect bridge between micro- and macroevolution" (Mayr 1982b: 1126). However, only few years later, in 1988, his choice of terminology when discussing the same topic is presented in a rather convoluted way. Here's a mixed bag of his claims:

1. Macroevolution is an autonomous area of study that should develop and test its own theories, since these are not reducible to microevolutionary theories, most likely in principle but at least given the current state of our knowledge (Mayr 1988: 403).
2. Since the individual is the target of selection and the population the locus of evolutionary change, macroevolutionary processes are automatically connected with the microevolutionary level (Mayr 1988: 403), thus reducing all macroevolutionary processes to the microevolutionary level (Mayr 1988: 479).
3. Although mutation, stochastic processes, and selection are as much involved in microevolution as in macroevolution, these three factors fall short of fully explaining macroevolutionary phenomena (Mayr 1988: 404).
4. The genetics of microevolutionary processes has thus far been unable to offer a full explanation of macroevolution (Mayr 1988: 405).
5. In 1988 Mayr reiterates his previous claim originally made in 1982 that findings in population systematics provide a perfect bridge between microevolution and macroevolution (Mayr 1982b: 1126; 1988: 450).

This sample of claims is enough, we think, to show that Mayr's position had been challenged to such a degree that he felt the need to formulate things differently. Now, because semantic issues are no substitute for substantive ones, let us contrast the ontologies of the early and the very late Mayr.

The early Mayr inhabited a world in which the fabric of life is closely knitted together, with living matter being largely continuous and contiguous as assumed under gradualism/gradation. Of course, gaps or voids in this fabric are created when life forms become ever more separated by climbing the taxonomic scale along the multiple divergent paths, as postulated under the cone of increasing diversity. Yet, such gaps or voids are conceptually filled by the extrapolation principle: Mayr's tree of life is generated in the atemporality of the present time through speciation, with currently observable processes being extrapolated. Within that theoretical matrix, natural selection is believed to be a creative force feeding upon this fabric (variation), never running out of material in the continuity and contiguity of the present time. The world of the early Mayr is, in essence, the same as Darwin's world (Delisle 2019: 85–96, 135; Delisle and Tierney: 85, 111–112, 124–125): the cone of increasing diversity plays the role of a container—a chalice filled to capacity—one without significant voids.

In sharp contrast, the very late Mayr lived in a world wherein the fabric of life is significantly "torn," so to speak, with the rise of contiguous and continuous atomistic variations being interrupted or disrupted. It will be sufficient here to summarize his view:

1. At the high taxonomic level of animal phyla, the overall history of life is a depletion or impoverishment process, with 70 or 80 different structural types in the late Precambrian and early Cambrian, and about 35 remaining today (Mayr 2001a: 59–60, 209, 229).
2. Life's history had been largely disrupted by at least nine mass extinction events (Mayr 2001a: 202).
3. Resistance to atomistic change is seen in the significance of a tight genotypic integration constraining the capacity to produce new structural innovations, also impacting divergent evolution to some degree (Mayr 2001a: 60–61, 209–210, 218).
4. Drastic change is possible through "instantaneous speciations"—the doubling of existing parts like genes, whole chromosomes, or even entire genomes—thus instigating novelty, at least in plants, in lower animals, and even in higher animals like fish, amphibians, and reptiles (Mayr 2001a: 38–39, 181–182).

As much as Mayr continues to officially promote the microevolution-macroevolution equation, as well as notions like divergence and gradualism, his ontological shift away from the cone of increasing diversity obliged him to give new meanings to old concepts. Indeed, the unknitted fabric of his new world presents more gaps and discontinuities, creating maneuvering room for major stochastic factors and chance events. At this point, Mayr downplays and contextualizes the "creative" role of natural selection in favor of an "eliminative" one:

> Natural selection, of course, is no protection against mass extinction. Indeed, there is a considerable probability that the successful survival through such an extinction event includes a considerable chance factor… No amount of natural selection succeeded in

> producing genotypes enabling them to survive [dinosaurs, nautiloids, ammonites]... Biological causes and natural selection are dominant in background extinction [the normal extinction of outcompeted forms], whereas physical factors and chance are dominant in mass extinction. Species are involved in background extinction, and entire higher taxa in mass extinction. (Mayr 2001a: 203; see also p. 229)

To put it differently, in a world with more gaps (voids) and discontinuities than expected, selection as a directional force is deprived of continuous material to "feed" upon, hence the necessity to accept the function of an eliminative role. Once again, Mayr's use of the *adaptationist doctrine* was modulated to fit his latest conceptions, showing that this doctrine had always been envisioned by him as quite flexible in terms of its explanatory purpose. The following quotation is even more revealing of Mayr's shift of perspective, being very explicit about the benefit for the expression of "evolutionary contingency." At this point, Mayr supports what we have been anticipating all along: his commitment to evolutionary contingency could not have been fully expressed when conceived under the cone of increasing diversity. That ontology, which appeared with Darwin, inherently included too many constraining factors:

> I feel strongly that if natural selection is primarily selecting for the best, you would never have got a peacock's tail. So selecting for the best means eliminating most anything you see, eliminating variation altogether. If you eliminate only the worst, however, you still have a lot of variation left... [S.J.] Gould made this point in his book *Wonderful Life*. He's quite right that if you have a lot of variation left after eliminating the worst, it's just a matter of chance which of all these variants will be picked up. That's why the *unpredictability of evolution is much more easily explained by the eliminating concept rather than under a strict election concept that gives primacy to selecting for the best*. The minute you realise that, you will agree that there *is* a difference between selecting for the best and eliminating the worst. (Mayr 2002: 964–965; my emphasis)

Having moved away from the cone ontology, Mayr is now more at liberty to think about unleashing the creative potentialities of evolution.

15 Mayr, Adaptationism, and "Darwinism"

We have intentionally delayed tackling head-on the topic of "adaptationism," a notion often believed to be consubstantial with "Darwinism" in the traditional historiography. Mayr's name is often believed to be synonymous with both labels and he worked hard to ensure that this remained so. Our thesis in this section is the following: Mayr's connection with the adaptationist doctrine needs to be seriously qualified in light of his many other commitments. Historical and philosophical analyses are often focused on mechanisms in evolutionary biology, by resorting to such labels as: Darwinism, neo-Darwinism, Lamarckism, neo-Lamarckism, Mutation theories, Orthogenetic theories, Modern Synthesis, Neutral Theory of

Evolution, and Expanded Evolutionary Synthesis. This tendency, however, obscures the fact that evolutionists are often separated by more significant commitments than mere evolutionary mechanisms. In addition, reference to natural selection does not constitute, in itself, a demarcation criterion when discriminating between evolutionists from the nineteenth century onward since *most* of them exploited the concept (Delisle 2009a, 2011, 2017a, 2021a; La Vergata 2024). To expand upon our thesis, the essential commitment of Mayr's career is not to *adaptationism* but to the *cone of increasing diversity*. It is this ontology which allowed him to articulate all his other commitments—actualism, principle of divergence, gradualism, geographic speciation, and extrapolationism. Nowhere are these other commitments more powerfully coherent and self-reinforcing than when applied in the context of the cone ontology. When that ontology was abandoned by Mayr around the year 1990, the idea of a dominant creative role for natural selection lost its foundation. Differently stated, it is the ontology of the living world that defines the explanatory role played by mechanisms, and not the other way around. Logically speaking, ontology precedes mechanisms. Knowing what we already know of Mayr, it is difficult not to feel some degree of cognitive dissonance when reading the following statement of 1991:

> Where evolutionists today differ from Darwin is almost entirely on matters of emphasis. While Darwin was fully aware of the probabilistic nature of selection, the modern evolutionist emphasizes this even more. The modern evolutionist realizes how great a role chance plays in evolution. Darwin never said, 'Selection can do anything'; neither do we. On the contrary, there are powerful constraints on selection. And selection is, for various reasons, appallingly often unable to prevent extinction. One hundred and thirty years of unsuccessful refutations have resulted in an immense strengthening of Darwinism. Controversies within evolutionary biology about such matters as the occurrence of sympatric speciation, the relative frequency of complete stasis in species, the rate of speciation, the significance of neutral allele replacement, and whatever else all take place within the framework of Darwinism. The basic Darwinian principles are more firmly established than ever. (Mayr 1991: 164)

Thus presented, "Darwinism" is so *inclusive* that it seems to be nearly irrefutable since it excludes little. Because two precautions are better than one, Mayr reverses his perspective by definitively *excluding* any possible threatening developments. "Darwinism" has thus truly become irrefutable:

> [T]he Darwinian paradigm that was adopted during the evolutionary synthesis is essentially valid. The basic Darwinian formula—evolution is the result of genetic variation and its ordering through elimination and selection—is sufficiently comprehensive to cope with all natural eventualities. To search for a new evolutionary theory (paradigm) seems now a futile enterprise. For the last fifty years or longer almost every year a new paper or even book was published in which a serious error or omission in Darwinism was claimed. The author proposed a new theory, or new theories, which he or she claimed would correct this error and fill the gap. Alas, not a single one of these proposals turned out to be constructive. Invariably the now classical Darwinism was confirmed and it was possible to refute the putative improvements or corrections. This suggests to me that Darwinism is nearing full maturity. There are, of course, still numerous unsolved puzzles like the function of much of the

> noncoding DNA, but I do not see how the solution of any of the remaining puzzles can have any noticeable effect on the basic Darwinian paradigm. (Mayr 2004: 128)

Mayr had the supreme ability to largely disguise the demise of his ontological world—including his own specific conceptual shifts over the years—by covering it up under the Darwinian blanket, a unifying discourse held to be as valid today as it was for Darwin. Indeed, the Darwinian Revolution, Mayr continues, is a journey that ultimately triumphed through unshakable principles:

> Let me now try to summarize Darwin's contributions to the thinking of modern men. He was responsible for the replacement of a world view based on Christian dogma by a strictly secular world view. Furthermore, his writings led to the rejection of several previously dominant world views such as essentialism, finalism, determinism, and the suitability of Newtonian laws for the explanation of evolution. He replaced these refuted concepts with a number of new ones of wide-reaching importance, also outside of biology, such as biopopulation, natural selection, the importance of chance and contingency, the explanatory importance of the time factor (historical narrative), and the importance of the social group for the origin of ethics. Almost every component in modern man's belief system is somehow affected by one or another of Darwin's conceptual contributions. (Mayr 2001b: 494–495)

One gets the impression that Mayr is here echoing the Book of Genesis: In the beginning Darwin created the world... (see also Esposito 2024). The more Mayr retreated from his earlier commitments, the more he tried to compensate with a definition of Darwinism enshrined in a cloud of broad, vague, and nebulous ideas; an argumentative strategy that makes "Darwinism" a moving target. Not only does Mayr hide the Darwinian core inside that cloud but, in addition, by confining its cognitive content to a much simplified or "softer" knowledge, he ensures that this explanatory core cannot easily be criticized:

> What name should we apply to the version of Darwinism developed in the 1940s? ... The new evolutionary theory, the product of the synthesis of the theories of the students of anagenesis and cladogenesis, has been called the synthetic theory of evolution. Actually the best solution would be to call it again simply Darwinism. Indeed it is essentially Darwin's original theory with a valid theory of speciation and without soft inheritance. But because such inheritance was refuted more than 100 years ago, no mistake can occur if we go back to the simple term Darwinism, because it encompasses the essentials of Darwin's original concept. In particular it refers to the *interplay of variation and selection, the gist of Darwin's paradigm.* It certifies that the evolutionary paradigm adopted by the modern evolutionists, after a long period of maturation, is best simply referred to as Darwinism. (Mayr 2004: 129–130; my emphasis)

Is Mayr here implying that any theory appealing to the interplay of variation and selection should, by definition, be included within "Darwinism"? Apparently, Mayr means what he says: "An apt summary of Darwin's paradigm is the phrase 'variation and selection'" (Mayr 2004: 136). Here, Mayr attempts to protect the "Modern Synthesis" from competing views by incorporating it under an ever more diffuse "Darwinism." This is in sharp contrast with his view of 1959:

> The question is sometimes asked whether or not it is justified to apply the term "Darwinism" to the modern evolutionary theory... When we compare modern evolutionary theory with the original conceptual structure of Darwin, we find that many of Darwin's ideas have been eliminated and many new ones added. It avoids a good deal of confusion not to designate the currently held interpretation of evolution as Darwinism or Neodarwinism. Simpson...and others have therefore rightly insisted that the term "Darwinism" be replaced by 'the synthetic theory' to indicate the multiple roots of the new theory. (Mayr 1959a: 4–5)

Setting aside some of the more grandiloquent declarations made by Mayr in the above quotations, we will now investigate some of the more specific commitments allegedly binding Darwin and Mayr. In contradiction to Mayr's self-serving narrative, several key commitments supported by these two scholars do not fit well with Mayr's new definition of "Darwinism."

Our analysis in this chapter focuses on "evolutionary contingency" precisely because it is often believed to be a key feature of an evolutionary biology centered around natural selection. In a truly open-ended world—the argument goes—natural selection constitutes the organizing principle of the encounter between the variations of organisms and the changing environment (biotic and abiotic). The expected overall picture thus consists in a score of independent evolutionary lines, which find all sorts of unique adaptive solutions in accordance with available variations in different spaciotemporal contexts. On this understanding, one allegedly finds at the explanatory core of "Darwinism" key concepts such as adaptationism, natural selection, and unpredictability.

So far, so good for the *abstract reasoning*. Yet, in order to be realistically applied, an evolutionary mechanism like natural selection must be put back in an incarnated world. This is what was done by Darwin and Mayr, for instance, each using the cone ontology as their main interpretative matrix, thus setting the explanatory role for natural selection. Thus interpreted, "Darwinism" is quite unlike what Mayr said about it.

15.1 Essentialism and Determinism

Let us begin by two features mentioned in Mayr's (2001b: 494–495) quotation above presumably jettisoned under Darwinism: essentialism and determinism. Mayr's thesis is weak on both counts. His repeated claim that Darwin had moved away from "essentialism" is somewhat superficial, being confined to the idea that natural selection picks, individually, traits or organisms inside "species," under a populational process. Elsewhere he writes:

> Darwin's variational evolution was based on entirely new philosophical concepts... [T]he essentialist, as was emphasized by Lyell, Sedgwick, Herschel, and others, believes that there is a definite limit to the amount of variation possible within a species or, to put it differently, permissible to a single inherent essence. For the population thinker, variation is unlimited. Hence, there is a definite possibility of going beyond the confines of an existing species. The nature and extent of variability was the crucial difference between the population thinker Darwin and his essentialistic opponents. (Mayr 1991: 98)

This interpretation overlooks the fact that there is more than one way to continue to support essentialism. Our point is that entities like "species" should not be considered independently of the theoretical matrix ("a view of life") giving meaning to them.[9] Darwin's biology is actively designed to apply within a segmented tree of life, wherein each class/phylum belongs to its own separate world. Darwin's vague appeal to a unified tree of life rooted in few (or a single) hypothetical common ancestors adds nothing to the story, since this conjecture has no impact on his biology. In addition, although each class/phylum is assumed to have arisen from its own single common ancestor, that ancestor has no real evolutionary existence considering that Darwin reconstructed its "history" by selecting traits among extant forms if only to project them backward in geological time. For Darwin, so-called common ancestors are nothing more than downsized forms only seen fully blown among extant forms. This gives his biology a definite preformationist flavor. In a real sense, Darwin constructed a biology around *separate biological types called extant classes or phyla*. The populational process described by Mayr is, indeed, applied by Darwin but in the very limited context of classes and phyla that are closed-in on themselves, each recycling the same biological variations during evolution. Under this interpretation, it is difficult to significantly get beyond essentialism.

Seen against Darwin's own brand of biology, how can one assess Mayr's own populational approach? As already alluded to, the early Mayr lived at a time when the unity of the tree of life had firmed up considerably, especially after the discovery of a single DNA code for all of life. Furthermore, the early Mayr could also easily avoid the pitfall of Darwin's preformationist inclination, since the fossil record was, by the 1940s onward, dense enough to teach everyone that presumed common ancestors are not just downsized versions of more fully developed extant forms. Some sense of historicity could no longer escape evolutionists. Mayr's populational approach applied in a truly unified world, thus providing a firmer footing to the cone of increasing diversity than Darwin's own application of it.

It is precisely here that we encounter another major limitation to the implementation of "population thinking," this time characterizing both Darwin's and Mayr's approach: the cone of increasing diversity. The cone ontology assumes that biological variability is at its highest level among extant forms, implying that variability is ever more reduced as one travels back deeper in geological time toward presumed common ancestors. In light of the decimation ontology, however, we now know that this assumption is more than debatable. Darwin and Mayr (the former to a greater extent) restricted the explanatory power of population thinking—thwarting the rise of potential variations—by applying this concept inside the cone of increasing diversity. This approach was further strengthened by Darwin's and Mayr's

[9] It was to Mayr's advantage to pretend that his general view directly emerged from considerations about low level taxonomic entities like "species," through notions such as population thinking, polytypic species, the new systematics, the biological species concept, allopatric and peripatric speciation, etc. But again, that conspicuous series of notions make sense because collectively reunited under a less conspicuous ontological foundation underpinning the whole edifice: the cone of increasing diversity.

neontological methodology, which consisted in probing the evolutionary past by merely traveling in geographical space equipped with extant categories, both overlooking (each to different degrees) the full extent of the historicity of the world. Unfortunately for them, it seems that life's history exhibited greater ingenuity in generating novelties than they could ever have imagined through the prism of the cone ontology.

Turning now to the notion of "determinism," Mayr takes for granted that Darwin had also definitively moved away from it. Again, there is more than one way to be a determinist, just as there are different brands of determinism. As much as Mayr argues that Darwin did not subscribe to an internally driven determinism *à la* Bergson's *élan vital* or other kinds of entelechy (Mayr 1961: 1503, 1506), Darwin nonetheless did embrace determinism to a degree, but one emanating from *outside*, so to speak. For him, evolutionary change only occurs under the form of "divergence" to various degrees. In fact, there is something truly predictable about the application of his evolutionary process considering that his pan-divergent outlook concedes little room to other evolutionary alternatives such as reticulate evolution, parallelism, and convergence when contemplated in their full potentialities. Whereas Mayr is busy combatting determinism in the context of internal mechanisms to life forms, he underestimated Darwin's externally imposed determinism accompanying his commitment toward the cone of increasing diversity. The early Mayr was blind to this commitment, we think, for the obvious reason that he, too, was blind to the implication accompanying his own biology focused on speciation-divergence. Mayr's biology, therefore, cannot entirely escape the charge of determinism: it is only that his was weaker than Darwin's, owing to a unified tree of life and a greater sense of historical depth. One must wait for Mayr's abandonment of the cone ontology around 1990 to see him turn his back on the brand of determinism just described, although he had started to loosen up his grip on the supposed prevalence of "divergence" in the 1970s and 1980s when faced with challenges raised by molecular biology, competing schools of taxonomy, and paleobiology.

15.2 Divergence and Common Ancestry

Darwin and Mayr further locked-up the world system by treating the notions of "common ancestry" and "ongoing divergence" as two expressions of the same general idea. This can be seen clearly throughout the *Origin of Species* (as summarized in Darwin's Diagram) and Mayr's (1985: 759; 1988: 200–201) statement already quoted above. Why should this be so? This assumption only makes sense when one embraces the cone of increasing diversity. Again, through another explanatory device, the expression of evolutionary contingency is thwarted by preventing the evolutionary process from expressing itself in full through a whole range of evolutionary manifestations—divergence, parallelism, and convergence—once two related forms evolve away from a common ancestor.

Committed to the cone ontology, Darwin and Mayr applied *adaptationism* only under very specific conditions, perhaps not realizing that their approach was ultimately self-defeating. Unless, of course, adaptationism merely constituted for them an auxiliary component part of a much larger theoretical apparatus, one which did not have natural selection as the explanatory core. This is precisely the thesis we are defending here.[10] Eventually, under the pressure of alternative views, Mayr uncoupled the notions of "common ancestry" and "divergence" around 1990—something Darwin never relinquished—thus creating more room for the expression of evolutionary contingency. This was done at the price of giving up on the cone of increasing diversity.

It is difficult for those who were brought up with the traditional view to accept the idea that *adaptationism* does not necessarily require placing natural selection at the explanatory core of evolutionary biology. But it should be remembered that Mayr had already established his research program based on the dual association of *speciation* and *ontology of the cone* before he became more openly committed to "adaptationism," whatever its meaning. Mayr's recollections of his becoming a "Darwinian" should not distract us from how he viewed evolutionary mechanisms in the early 1940s. Speaking in 1980 of his early support of Lamarckism, he wrote: "In a paper on the origin of bird migration (written in joint authorship with W. Meise mostly in 1927 but not published until 1930), I still maintained that some phenomena, like regular annual transoceanic migrations, could not be explained by natural selection. I think this is the last antiselectionist statement found in my publications" (Mayr 1980b: 416–417). Fair enough, but in *Systematics and the Origin of Species* (1942), Mayr shows considerable openness toward some kind of soft inheritance. It is worth quoting him at some length:

> The opinion was formerly widespread among taxonomists (including even Darwin) that the germ plasm can react to the needs of the body. A change in the body would thus result in an appropriate reaction of the germ plasm. The extremists among the geneticists, on the other hand, insisted that the germ plasm lived a happy life of complete independence in the body, completely unaffected by the physiological processes and the physical states of the body itself. Most modern geneticists are somewhat more cautious. They know that changes in the body may cause reactions in the germ plasm, that the mutation rate may be changed as a consequence of a change in temperature, and so forth. It is still believed, however, that a gene never mutates in response to a need. The connection between the gene and the body is,

[10] In terms of explanation, it is the case that C. Darwin, E. Mayr, J.S. Huxley, Th. Dobzhansky, and B. Rensch gave explanatory priority to their separate ontologies, only inserting evolutionary mechanisms inside them after the fact (see Delisle 2009a, 2009b, 2011, 2019, 2021b; Delisle and Tierney 2022). Of the scholars listed above, only Darwin and Mayr embraced the ontology of the cone. For their part, Huxley and Dobzhansky subscribed to an open/linear ontology (from inert matter, to life, to humans, to a future state yet undefined), whereas Rensch supported a closed/cyclical ontology (a growing complexification of matter, followed by its collapse, followed by another rise, etc.). It is unsurprising, therefore, that the use of natural selection was differently instrumentalized under these three distinct ontologies. The doctrine of adaptationism, as traditionally defined, does not apply to these authors. With S.J. Gould, one encounters a fourth ontology (the decimation model) with yet a different understanding of the role of natural selection. One wonders how many different ontologies have been proposed since the rise of evolutionism in the late eighteenth or early ninteenth centuries?

> for most geneticists, strictly like a one-way street. This may be true, but if so it would be strangely different from most other biological processes, which are based on biological equilibria. A gene controls certain physiological processes, and it is therefore thinkable that the mutation rate might be raised at those loci which, owing to certain environmental conditions, have to be exceptionally active in the production of the particular chemical (enzyme?) which is produced by that gene (locus). I have mentioned this possibility merely to indicate that, in view of our present ignorance of the physiology of the gene and of the chemistry of the mutation process, it seems premature to assert with too much positiveness that all gene mutation is strictly random. (Mayr 1942: 67–68)

This openness toward soft inheritance was also expressed two years earlier in his paper "Speciation Phenomena in Birds" (Mayr 1940: 263). Undoubtedly, *Systematics and the Origin of Species* (1942) is filled with allusions to selective processes. Yet, this is also the place where one finds the first outline of Mayr's support for an evolutionary biology centered around speciation—"In conclusion we may say that all the available evidence indicates that the origin of the higher categories is a process which is nothing but an extrapolation of speciation" (Mayr 1942: 298)—as well as his support for the cone of increasing diversity (Mayr 1942: 275–280).

Coming back to the question of adaptationism, our point is the following: Mayr had no need of natural selection to succeed in putting his overall theoretical framework in place. Presumably, having accepted that "divergence" was the prevalent pattern in evolution in the 1930s, it mattered little to him whether natural selection was the only acceptable evolutionary mechanism. To reverse the perspective, Mayr's pan-divergent outlook is not the product of a reflection upon the action of a process called selection. With pattern coming first, process becomes a mere *justification* of a foregone conclusion. This is also the intellectual road taken by Darwin (Delisle 2021b) and other major evolutionists (see footnote 10).

While Mayr fiddled with the role of natural selection inside his theory between the 1930s and about 1990—eventually committing himself to a fairly strong selectionism (Mayr 1947: 285–285; 1949: 284; 1963a: 60, 201–203; see also 1992a: 25–26; 1993: 33), and later reverting to a softer position (Mayr 1982c, 1983)—the fact remains during those years that he never wavered in his support for a theoretical matrix based on the dual notion of divergence/cone of increasing diversity. The thesis that there was a loss of pluralism in mid-twentieth-century evolutionary biology—the so-called hardened phase which favored selectionism at the expense of stochasticism (Gould 1983)—only makes sense when it is recognized that it may have impacted only one of several other explanatory levels in evolutionists' systems of thought (Delisle 2017a: 160–161). In no way did this shift with respect to evolutionary mechanisms jeopardize Mayr's ontological foundation. His intellectual edifice only crumbled when ultimately confronted with an alternative ontology: the decimation model.

Mayr's overall position on natural selection in the span between 1940 and 1990 can be presented as promoting a "creative agent or selection." His general framework (cone ontology) was *consistent* with the many alterations to his understanding of natural selection during those years, but consistency is not a very high bar, as many proposals can be made to fit a given explanatory framework, especially in the case of evolutionary mechanisms. Mayr could easily have tinkered the role of natural selection even more than he actually did in the period between 1940 and 1990, had he decided to. By the time he had abandoned the cone ontology, however, it was no longer possible for him to continue to support a creative selection acting in all circumstances (Mayr 1997b: 2091; 2001a: 140, 149–150, 228–229; 2002: 964–965; 2004: 134–138). Ontological shifts are fundamental epistemological changes often requiring the reinterpretation of accompanying notions. In this new context, we hold that Mayr was forced to create explanatory room for eliminative selection, come what may. It is our thesis that "adaptationism" played a rather secondary role in Mayr's general theoretical framework, irrespective of his (and others') claims to the contrary. The fact that he continually used adaptationism to combat his adversaries has, in our view, less to do with Mayr's actual use of the notion and more to do with rhetorical and sociological strategies. Such considerations are beyond the scope of this chapter.

15.3 Gradualism and Natural Selection

Another conceptual development Mayr assumes to be part of eternal "Darwinism" is the notion of gradualism, one closely tied to natural selection:

> Darwin's rejection of essentialism and the general emergence of population thinking strengthened his adherence to gradualism and led him to totally reject saltations. As soon as one adopts the concept that species evolve as populations are transformed, owing to the differential reproductive success of unique individuals over generation, and this is what Darwin increasingly believed, one is automatically forced also to believe that evolution must be gradual. (Mayr 1991: 45–46)

Mayr uncompromisingly adds: "The result was that gradualism was able to celebrate a complete victory during the evolutionary synthesis" (Mayr 2004: 104–105; see also 1985a: 763). These brave words pronounced late in Mayr's career also need to be qualified, as these are not devoid of rhetorical intentions. Whereas Darwin and the early Mayr exploited a similar approach to gradualism—confidently combining empirical and theoretical claims—the late Mayr significantly reduced the explanatory scope of his original theoretical claim by finessing both the definition of gradualism and the empirical facts not obviously subsumable under it.

Let us turn first to Darwin's own words. Although it is not possible to entirely decouple empirical claims from theoretical ones in the case of gradualism, the following two quotations are empirical in nature:

> Let us now look to the mutual affinities of extinct and living species. They all fall into one grand natural system; and this fact is at once explained on the principle of descent. The more ancient any form is, the more, as a general rule, it differs from living forms. But, as Buckland long ago remarked, all fossils can be classed either in still existing groups, or between them. That the extinct forms of life help fill up the wide intervals between existing genera, families, and orders, cannot be disputed. For if we confine our attention either to the living or to the

> extinct alone, the series is far less perfect than if we combine both into one general system. (Darwin 1859: 329)

Here, Darwin argues that his cone of increasing diversity is filled to capacity, being a tightly knit fabric of continuous and contiguous life forms. He goes on to claim that although intermediate connections associated with a gradual process of evolution are not always known, the high proportion of those that are actually known is impressive by itself:

> Although in many cases it is most difficult to conjecture by what transitions an organ could have arrived at its present state; yet, considering that the proportion of living and known forms to the extinct and unknown is very small, I have been astonished how rarely an organ can be named, towards which no transitional grade is known to lead. The truth of this remark is indeed shown by that old canon in natural history of 'Natura non facit saltum.' (Darwin 1859: 194)

Nonetheless, the strength of Darwin's empirical assertions was dampened by holes in the tightly knit fabric of life he envisioned. This obliged him to carry out preemptive strikes of a theoretical sort. First, he claimed that gradual intermediate links are missing owing to the nature of the evolutionary process itself, for while natural selection requires slight variations, it simultaneously contributes to the extinction of these intermediate forms:

> I endeavoured, also, to show that intermediate varieties, from existing in lesser numbers than the forms which they connect, will generally be beaten out and exterminated during the course of further modification and improvement. The main cause, however, of innumerable intermediate links not now occurring everywhere throughout nature depends on the very process of natural selection, through which new varieties continually take the places of and exterminate their parent-forms. But just in proportion as this process of extermination has acted on an enormous scale, so must the number of intermediate varieties, which have formerly existed on the earth, be truly enormous. Why then is not every geological formation and every stratum full of such intermediate links? Geology assuredly does not reveal any such finely graduated organic chain; and this, perhaps, is the most obvious and gravest objection which can be urged against my theory. The explanation lies, as I believe, in the extreme imperfection of the geological record. (Darwin 1859: 279–280)

Then, Darwin proceeds to expose his theoretical agenda, explaining that the nature of the evolutionary process can only make use of small and continuous variations. To put it another way, only a world filled to capacity and contiguously organized will allow natural selection to function. Here, Darwin argues that natural selection constitutes a *creative process* that feeds upon small variations that must imperatively be contiguously distributed in time and space: were life characterized by matter unevenly distributed—owing to gaps or discontinuities—natural selection would stall. Natural selection is therefore ineffective as an *eliminative process*. Darwin writes: "Why should not Nature have taken a leap from structure to structure? On the theory of natural selection, we can clearly understand why she should not; for natural selection can act only by taking advantage of slight successive variations; she can never take a leap, but must advance by the shortest and slowest steps" (Darwin 1859: 194). Committed to the idea that natural selection cannot be a major evolutionary force in a non-gradualistic world, Darwin was prepared to raise the stakes by forcing the theoretical issue, casting it as an either-or between *full victory* or *complete defeat*: "If it could be demonstrated that any complex organ existed, which could not possibly have been formed by numerous, successive, slight modifications, my theory would absolutely break down. But I can find out no such case" (Darwin 1859: 189). Of course, such a victory would be secured if Darwin could convince his readers that the world should be looked at through his own theoretical lens:

> He who will go thus far, if he find on finishing this treatise that large bodies of facts, otherwise inexplicable, can be explained by the theory of descent, ought not to hesitate to go further, and to admit that a structure even as perfect as the eye of an eagle might be formed by natural selection, although in this case he does not know any of the transitional grades. His reason ought to conquer his imagination.... (Darwin 1859: 188)

Darwin's understanding of the limitations of natural selection in the context of a non-gradualistic world puts Mayr in an awkward position. Indeed, the very late Mayr openly recognized the significance of eliminative selection in evolution (Mayr 1997b: 2091; 2001a: 140, 149–150, 228–229; 2002: 964–965; 2004: 134–138), while simultaneously claiming that "Darwinism" had never been as firmly established as it was at the time of his writing. The obvious questions to raise here are: do Darwin and Mayr truly belong under the same Darwinian umbrella, and should one of them be excluded? Another way out of this strange dilemma—and this is our proposal in this chapter—is to dispose of the label "Darwinism" entirely, on the grounds that, far from acting as a magnifying glass that allows us to understand the issues, it has become, rather, a distorting one. Mayr's shifting views over several key issues in the 1970s and after, including gradualism and natural selection, makes this point more obvious still. Indeed, and interestingly, the early Mayr once held that if evolution were to be based on an eliminative process, that process would be "very inferior" at generating changes or novelties. For him, gradualism was a precondition for creative selection, affirming Darwin's gamble at putting his own theory in the balance of refutation:

> The evidence, whether genetic, morphological, or functional, is so uniformly opposed to a saltationist origin of new structures that no choice is left but to search for explanations in terms of a gradual origin. The role of natural selection in evolution would indeed be a *very inferior one* if, as was believed by the saltationists, it did nothing but weed out 'hopeless monsters' in favor of 'hopeful monsters.' Darwin was fully aware of this situation: 'If it could be demonstrated that any complex organ existed, which could not possibly have been formed by numerous, successive, slight modifications, my theory would absolutely break down. But I can find out no such case.' (Mayr 1960: 357; my emphasis)

It is difficult to imagine a more radical turnabout for a self-proclaimed "Darwinian" than the following claim:

> Curiously, it has never been remarked that the consequences of an elimination process may be quite different from those of a selection process. A selection process results in the survival of the truly best and there will be only relatively few individuals that qualify for such a designation. In the process of actual selection a bird with a cumbersome tail such as that of a peacock would never emerge as 'the best.' By contrast, elimination would leave in an average year a much larger percentage of survivors than would the selection of only the best. This large pool of survivors provides ample material for sexual selection and for chance. It provides an explanation for the haphazardness of much of evolutionary change. Evolution by elimination provides a far better explanation for the actual course of events during evolution than the 'selection of the best' of the classical evolutionary literature. The elimination of the inferior takes place, of course, simultaneously with the selection of the best, but it differs in strength in different situations. The unpredictability of much of evolution, described so graphically by Gould in his *Wonderful Life* (1989), is well explained by the process of elimination but could not be accounted for by a restriction to the selection of the best. (Mayr 2004: 135)

On Mayr's (1960: 357) original account, Darwin's theory has indeed broken down. Of course, the late Mayr had no other choice but to try to escape of the corner he put himself in. He therefore cushions the blow with the rather timid word, "overlooked": "Darwin borrowed the term selection from the vocabulary of the animal breeders and plant cultivators. But he *overlooked* that the breeders actually utilized two very different approaches to improve their stocks and so does nature" (Mayr 2004: 134; my emphasis). Here, Mayr contrasts "selecting the best" and "eliminating the inferior." Several years before, his attempt to salvage his understanding of Darwin over gradualism read this way:

> When one proposes a new theory, it is good strategy to make it as drastically different from the existing theory as possible. This is the best way to draw attention to it and to emphasize its novelty. What Darwin opposed in his theory of gradualness was the saltationism implicit in two ideologies dominant in his time, creationism and essentialism. To insist on gradualness, thus meant for him to reject both creationism and essentialism. This is forgotten by some of those who object to even the slightest deviation from perfect gradualism. (Mayr 1988: 411)

Does this mean that the very late Mayr had actually given up on the notion of "gradualism" in exchange for more creativity (novelty) generated under eliminative selection? His answer was rather convoluted. Mayr originally started from a similar approach to Darwin, fusing empirical and theoretical claims about the gradualistic nature of life in geographical space and geological time (Mayr 1942: 153–154; 1960). Despite persistent claims to the effect that gradual evolution is encountered everywhere in the form of geographic variation (Mayr 1985a: 763; 1991: 46–47; 2004: 104), the late Mayr made use of a rather flexible definition as far as gradual manifestations are concerned. In 1982, for instance, he speaks of genetic revolutions, major genotypic reconstructions, and chromosomal reorganizations, in these words: "They are in all cases populational phenomena and therefore by necessity gradual, regardless of the size of the building stones of the genetic reconstruction and no matter how drastic the end product may differ from the initial condition" (Mayr 1982b: 1130). In 2001, when speaking of bacteria and protists, Mayr explains that evolutionary diversification is often generated by doubling: “The doubling sometimes affects not merely a single gene, but a whole chromosome set or even an entire genome” (Mayr 2001a: 39). And under the appellation of “instantaneous speciation,” he now concedes that plants and more animals sometimes undergo rather drastic events:

> Through various chromosomal processes an individual may be produced that is instantaneously reproductively isolated from individuals of the parent species. For instance, it occurs quite frequently in plants that a sterile species hybrid...experiences a doubling of its chromosomes... The new polyploid is now a viable species... By further hybridization and chromosome doubling entire series of polyploids can be produced. What occurs instead in some animals (but has not yet been found in either mammals or birds) is that a sterile species hybrid shifts to parthenogenesis and asexual reproduction. Such cases are known from fishes, amphibians, and reptiles. Again, as in the case of polyploidy, it seems as if such cases of nongeographical speciation are rather rare and likely to be evolutionary dead-ends. Too little is known about reproduction and speciation in the lower animals to say how widespread nongeographical speciation is in those groups. (Mayr 2001a: 181–182)

The very late Mayr hoped to keep his gradualistic agenda alive by transferring its explanatory weight to a more general theoretical definition—gradualism being minimally defined as a populational process—an argumentative strategy protecting his claims from empirical refutation. From there, he could freely acknowledge that numerous gaps or discontinuities are observed in nature. To reproduce that quotation again:

> The reason why this controversy has not been fully settled is because there seems to be an astonishing conflict between theory and observation. According to Darwinian theory, evolution is a populational phenomenon and should therefore be gradual and continuous. This should be true not only for microevolution but also for macroevolution and for the transition between the two. Alas, this seems to be in conflict with observation. Wherever we look at the living biota, whether at the level of the higher taxa or even at that of the species, discontinuities are overwhelmingly frequent. Among living taxa there is no intermediacy between whales and terrestrial mammals, nor between reptiles and either birds or mammals. All 30 phyla of animals are separated from each other by a gap. There seems to be a large gap between the flowering plants (angiosperms) and their nearest relatives. The discontinuities are even more striking in the fossil record. New species usually appear in the fossil record suddenly, not connected with their ancestors by a series of intermediates. Indeed there are rather few cases of continuous series of gradually evolving species. (Mayr 2001a: 189)

As already stated, this description contrasts with the one presented by the early Mayr (1942: 153–154). In addition, the late Mayr (2001a: 190–207) replied with a rather outdated list of examples documenting gradual evolution, declinations of the idea of preadaptation. In the context of a rising decimation ontology, was it possible for Mayr to respond differently? Under this different ontology, it will be reminded, the world is characterized by immovable blocks (biological types) with stochastic processes functioning as a privileged way for life forms to break away from these types, events associated with mass extinctions and the rise of profound novelties. It is now obvious why the doctrine of gradualism, as originally defined by Darwin and the early Mayr, was far from being universally valid. To our question "did the very late Mayr still support Darwin's gradualism?", we reply that this issue constitutes yet another epistemological rupture within so-called Darwinism.

16 Stephen Jay Gould and "Darwinism"

A brief incursion into Gould's late work reveals that he, too, could not resist the temptation of attaching his name to the "Darwin bandwagon." Several scholars have noticed that among more recent evolutionists, Darwin has come to play a specific role: an icon against which theories may receive approbation or reprobation (Richards 1988: 146); a name to be exploited to endow a more recent theory with his authority (Jablonka and Lamb 2020: 8) and thus gather support for it (Shanahan 2004: 293). Darwin's shadow seems to be "hovering" over evolutionary biology (see Gould 2002: 96), a "burden," to use John Horgan's (1996: 114) terminology: "No other field of science is as burdened by its past as is evolutionary biology... [It] can be defined to a large degree as the ongoing attempt of Darwin's intellectual descendants to come to terms with his overwhelming influence." In light of our analysis provided thus far, the main thesis of this end section will likely seem anticlimactic: Gould's decimation ontology is inherently incompatible with Darwin's brand of biology (as defined in this chapter), irrespective of Gould's claims to the contrary. Indeed, one reads in *The Structure of Evolutionary Theory* (2002) declarations such as "My allegiance to Darwinian theory, and my willingness to call myself a Darwinian biologist..." (p. 9–10). But trouble lurks and Gould expresses it in the following convoluted formula:

> When I ask myself how all these disparate thoughts and items fell together into the one long argument of this book, I can only cite...my love of Darwin and the power of his genius. Only he could have presented such a fecund framework of a fully consistent theory, so radical in form, so complete in logic, and so expansive in implication. No other early evolutionary thinker ever developed such a rich and comprehensive starting point. From this inception, I only had to explicate the full original version, tease out the central elements and commitments, and discuss the subsequent history of debate and revision for these essential features, culminating in a consistent reformulation of the full corpus in a helpful way that leaves Darwin's foundation intact while constructing a larger edifice of interestingly different form thereupon. (Gould 2002: 47)

It is difficult to do better at claiming in the same breath both *identity* and *difference*. The first part of Gould's (2002: 1–48) *magnum opus* is a very long balancing act, in which effusive praise can hardly disguise a certain uneasiness: he is forced to walk a fine line separating old ideas and recent innovations. The attempt consists in claiming that Darwinism constitutes a historical entity that has changed tremendously since inception, although it retained its essence. Even more explanatory tensions reveal themselves in the background of the following statement:

> [M]y general love of history in the broadest sense spilled over into my empirical work as I began to explore the role of history's great theoretical theme in my empirical work as well—contingency, or the tendency of complex systems with substantial stochastic components, and intricate nonlinear interactions among components, to be unpredictable in principle from full knowledge of antecedent conditions, but fully explainable after time's actual unfolding. This work led to two books on the pageant of life's history [*Wonderful Life* (1989) and *Full House* (1996)]... Although this book [*The Structure of Evolutionary Theory* (2002)], by contrast, treats general theory and its broad results (pattern *vs.* pageant in the terms of this text), rather than contingency and the explanation of life's particulars, *the science of contingency must ultimately be integrated with the more conventional science of general theory as explored in this book*—for we shall thus attain our best possible understanding of both pattern and pageant, and their different attributes and predictabilities. The closing section of the book... offers some suggestions for these future efforts. (Gould 2002: 46–47; my emphasis)

Actually, the explanatory tension in Gould's late work is not just binary—evolutionary contingency versus expanded/modified Darwinism—but ternary. As a founder of the theory of punctuated equilibrium and a strong proponent of the decimation ontology, which implies that biological types often go extinct instead of reinventing themselves, Gould was caught in the triangular conceptual web composed of "evolutionary stability," "evolutionary contingency," and "expanded/modified" Darwinism. Considering that his hands were already full with the first two notions, would it not have been easier for him just to jettison the third? Why bother trying to save the label "Darwinism," an operation that apparently requires forceps? Was Gould looking for a genuine conceptual payoff, or was there something less tangible, more emotional, about saving "our profession's founder" (Gould 2002: 12). If the latter, we are definitively entering mythical territory, a step already taken by the very late Mayr. Let us keep in mind that Gould (2002: 148) wrote: "Since Darwin prevails as the patron saint of our profession, and since everyone wants such a preeminent authority on his side...." Oh boy!

17 A Mechanism-Centric View?

Ontological choices are so only in a weak sense of the expression. Often, these are not explicitly articulated by evolutionists although they constitute the very foundation of competing systems of thought. Darwin undoubtedly thought he was trading in revolutionary territory when investigating evolutionary issues in the 1830s, 1840s, and 1850s, which partly explains his cautiousness in letting people know about it. From his viewpoint, he was likely struck by the contrast he was busy building between evolutionary and non-evolutionary explanations. Yet, Darwin was not fully aware of the extent to which he took on board old uniformitarian tools originally designed in the context of a non-evolutionary worldview (Delisle 2014, 2019; Delisle and Tierney 2022). Only hindsight could reveal that. For his part, Mayr was committed early on to the cone of increasing diversity, something he took to be a fact and not a theoretical gamble. Apparently, only the rise of the decimation model in the 1980s revealed to him the implications of his original "choices."

Similarly, we hold that Gould was also blind to some implications flowing from his own view. Apparently, few scholars were better equipped than he to see the difference between the cone of increasing diversity and the decimation model. Yet, Gould failed to see that what he (and some other paleobiologists) was promoting in the 1980s and after was epistemologically incompatible with a smooth conceptual transition from one ontology (cone) to another (decimation). Gould's lack of awareness is reflected in the way he explains the development of evolutionary biology in *The Structure of Evolutionary Theory*—the iconography of the fossilized coral of Agostino Scilla—a kind of genealogical tree of ideas (Gould 2002: 18). In explanations therein, he grafted more recent innovative developments upon what he took to be Darwin's original ideas. Most significantly, he organizes the Darwinian tree of ideas around a central trunk—the process of natural selection—taking for granted that the original foundation stone of modern evolutionary biology was rooted in such a process. To give this up, continues Gould, is to destroy Darwinism: "The central trunk (the theory of natural selection) cannot be severed, or the creature (the theory) dies" (Gould 2002: 18).

In light of Gould's double criticism toward a so-called hardened synthesis (Gould 1983) and a so-called selectionism (Gould and Lewontin 1979), that central trunk is presumably envisioned as a pluralistic approach to natural selection, wherein "creative selection" and "eliminative selection" are believed to be compatible with Darwinism. Reproducing and completing this last quotation, Gould writes: "The central trunk (the theory of natural selection) cannot be severed, or the creature (the theory) dies. (The roots, if you will, represent sources of evidence; any one may be excised, if recognized as incorrect by later study, *so long as enough remain to anchor the structure*)" (2002: 18; my emphasis). This formulation does little to instill clarity. On Gould's account, then, one can stress eliminative selection over creative selection while remaining within Darwin's original thinking. This is expressed more clearly in the following words:

> Any reasonably honest and intelligent biologist could easily understand that Darwin had identified a *vera causa* (or true cause) in natural selection. Thus, the debate in his time (and, to some extent, in ours as well) never centered upon the existence of natural selection as a genuine causal force in nature. Virtually all anti-Darwinian biologists accepted the reality and action of natural selection, but branded Darwin's force as a minor and negative mechanism, capable only of the headsman's or executioner's role of removing the unfit, once the fit had arisen by some other route, as yet unidentified. This other route, they believed, would provide the centerpiece of a 'real' evolutionary theory, capable of explaining the origin of novelties. Darwin insisted that his admittedly weak and negative force of natural selection could, nonetheless, under certain assumptions (later proved valid) about the nature of variation, act as the positive mechanism of evolutionary novelty—that is, could 'create the fit' as well as eliminate the unfit—by slowly accumulating the positive effects of favorable variations through innumerable generations. (Gould 2002: 14; my emphasis)

We are not concerned here with the putative legitimacy of Gould's association between Darwin and eliminative selection—even in the name of a so-called Darwinian logic (see Gould 2002: 140–169)—although we have seen that Darwin claimed that natural selection was a creative force and not an eliminative one. To quote that passage again: "Why should not Nature have taken a leap from structure to structure? On the theory of natural selection, we can clearly understand why she should not; for natural selection can act only by taking advantage of slight successive variations; she can never take a leap, but must advance by the shortest and slowest steps" (Darwin 1859: 194). Or even more damning for Gould, to also quote again the following: "If it could be demonstrated that any complex organ existed, which could not possibly have been formed by numerous, successive, slight modifications, my theory would absolutely break down" (Darwin 1859: 189).

For the sake of our analysis, however, let us just assume that Gould is right about his association between Darwin and eliminative selection. This being accepted, it should be noted that Gould's argument above makes sense only if one accepts from the start a "mechanism-centric" view of evolutionary biology. In contradistinction to Gould, however, we have argued that Darwin's brand of biology was structured around a *pattern*—a pan-divergent outlook enshrined in the cone of increasing diversity—and not around a *process*. We have taken pains to show that the core of Mayr's biology between 1940 and 1990 also rested on the same foundation. With ontology coming first in the logic of argumentation, processes or mechanisms can only come afterward, serving a justificatory function. This explains why Mayr could so easily tinker with the doctrine of adaptationism during those years. In contradistinction to the common assumption that *process* (natural selection) explains *pattern* (divergence/cone of increasing diversity), we have seen that the two ends of this equation are now known to be connected in a more complex way for Darwin (Sober 2011; Winsor 2009) and for Mayr. For his part, Gould elected to fuse "natural selection" and "divergence," placing them at the core of Darwin's theory. Although Gould keeps them separate conceptually speaking, they are reunited by him in the service of a process-oriented view:

> Darwin's principle of divergence has puzzled many biologists: why did Darwin rank the concept so highly, and as a principle separate from natural selection? May we not view divergence as a logical consequence or simple spin-off from natural selection? . . . After all, the earthly stage of evolution provides ecological and biogeographical prerequisites for diversification. . .Insofar as Darwin considered the issue at all between 1838 and the early 1850's, his thinking followed this general line. . .But Darwin grew dissatisfied with a theory that featured a general principle to explain adaptation. . .He decided that a fully adequate theory of evolution required an equally strong principle of diversity, one that acted intrinsically and predictably. If adaptation and diversification specify the central phenomena of evolution, each must have its principle, and their union would then define his complete theory. (Gould 2002: 225–226)

Here, Gould reads Darwin through the distorting lens of the "Modern Synthesis" (Winsor 2009: 47) wherein it is taken for granted that pattern (divergence) and process (natural selection) are intertwined at the core of Darwin's theory. From there, it is Gould's hope that Darwin's alleged process-oriented biology would be absorbed inside his own pattern-oriented theory—which appeals to notions such as punctuated equilibrium, eliminative selection, and stochastic processes—itself buttressed against his decimation ontology. It is our thesis that Gould's theoretical agenda is unrealizable for the following epistemological reason: Darwin's so-called process-oriented biology is, in reality, a pattern-oriented biology buttressed against its own ontology. Differently stated, for Darwin, "divergence" does not constitute a neutral notion that can be indifferently inserted into different ontologies of life.[11] On the contrary, the principle of divergence is, for him, an integral part of a conceptual bundle connected with the principle of gradation, the cone of increasing diversity, and a neontological agenda. What we encounter here, we argue, is a clash between the cone ontology and the decimation ontology, explaining why the very late Mayr struggled to put together a coherent evolutionary theory incorporating components from both ontologies. The cone of increasing diversity and the decimation model are much more than mere competing "iconographies."

18 Darwin: Disposing of Process

To see more clearly how Darwin treats "convergence" and "natural selection" as two independent components playing quite different explanatory roles in his theory, let us review two of his strategies (among others not reviewed here).[12] In his first argumentative strategy, Darwin *dismisses natural selection in order to protect the core of his argument*: the divergent pattern. When forced to choose between "divergence" or "natural selection," Darwin selects the former. This is seen in the classic case of the Galapagos Islands (1859: 397–403), in which Darwin argues that a group of islands in the Pacific Ocean 500 or 600 miles off the shores of Ecuador (South America) have received populations of faunas and floras from the nearby continent. This migration is presented as a series of divergences:

> Thus the several islands of the Galapagos Archipelago are tenanted. . .in a quite marvellous manner, by very closely related species; so that the inhabitants of each separate island, though mostly distinct, are related in an incomparably closer degree to each other than to the inhabitants of any other part of the world. And this is just what might have been expected on my view, for the islands are situated so near each other that they would almost certainly receive immigrants from the same original source, or from each other. (Darwin 1859: 400)

[11] Of course, modern evolutionists are free to use the notion of "divergence" as a neutral component to be enshrined in a whole range of competing theories. But by doing so, they move out of Darwin's own brand of biology by severing it from the theoretical matrix in which it was originally created. In our view, Gould wants to have it both ways: keeping all of Darwin but inside Gould's own brand of biology. Unfortunately for him, the result is unstable, explanatorily speaking.

[12] In his attempted to pull Darwin over to his side, Gould (2002: 240–246) notices that Darwin, in his unpublished book *Natural Selection* (1856–1858), turned to what Gould himself calls an "ad hoc hypothesis" in order to support the predicted principle of divergence with an unconvincing utilization of natural selection. While Gould hoped to take advantage of this allegedly unconvincing passage to argue that Darwin hinted therein at the notion of "species selection" ahead of time, we, in contradistinction, holds that this is yet another confirmation that the role of natural selection is confined, in Darwin, to a mere justification for an already pre-established pattern (divergence).

This description seems to suggest that divergence and natural selection are intertwined, with life forms being differentially adapted to different islands. But one should not be deceived by the way Darwin presents things. Having explained how populations first came to the Galapagos, he now had to explain why specific forms residing on one island or another did not, in general, invade other islands which are within sight of each other. Darwin's scenario contains, therefore, two main phases: (1) the initial phase of arrival on the Galapagos and (2) the lack of subsequent migration within the Galapagos themselves, once forms have been established on specific islands. The reader should notice Darwin's way of *disactivating* the adaptive drive or process during this second phase:

> But we often take, I think, an erroneous view of the probability of closely allied species invading each other's territory, when put into free intercommunication. Undoubtedly if one species has any advantage whatever over another, it will in a very brief time wholly or in part supplant it; but if both are equally well fitted for their own places in nature, both probably will hold their own places and keep separate for almost any length of time...In the Galapagos Archipelago, many even the birds, though so well adapted for flying from island to island, are distinct on each; thus there are three closely-allied species of mocking-thrush, each confined to its own island...I think we need not greatly marvel at the endemic and representative species, which inhabit the several islands of the Galapagos Archipelago, *not having universally spread from island to island.* In many other instances, as in the several districts of the same continent, *pre-occupation* has probably played an important part in checking the commingling of species under the same conditions of life. (Darwin 1859: 402–403; my emphasis)

Whereas the force of natural selection is appealed to during the first phase of the scenario (adaptation to separate islands), that force is conveniently disactivated during the second phase. In fact, natural selection becomes a kind of embarrassment for Darwin: after all, how can so potent a force during the first phase become so ineffective in making subsequent migrations possible, especially for flying creatures residing on separate islands within sight of each other? In order to save the core of his argument—an overall pattern of divergence accompanying the original colonization of the Galapagos—Darwin was willing to venture into speculation, by turning to the idea of a "pre-occupation" as a way to justify why islands were not colonized by forms coming from nearby islands, thus keeping closely akin populations geographically separate and isolated.[13]

But is not natural selection supposed to be a force much more powerful than presented in the Galapagos case, especially considering that Darwin (1859: 351, 487, 1872: 84) holds that evolutionary change (adaptation) is more susceptible to occur when forms are confronted with other forms (biotic) than to mere physical conditions (abiotic)? Elsewhere, Darwin writes: "But Natural Selection...is a power incessantly

---

[13]Our point here is not to claim that Darwin's suggestion concerning "pre-occupation" is entirely implausible. Rather, we are making an epistemological point: Darwin modulated his explanation of the evolutionary process in accordance with his real priority: the pattern of evolutionary divergence in the Galapagos.

ready for action, and is as immeasurably superior to man's feeble efforts [as seen under artificial selection]" (Darwin 1859: 61; 1872: 49). The way natural selection works to perform evolutionary breakthroughs, Darwin continues, is by favoring ever more divergent forms finding ways to occupy different places in the economy of nature:

> I am far from thinking that the most divergent varieties will invariably prevail and multiply: a medium form may often long endure, and may or may not produce more than one modified descendant; for natural selection will always act according to the nature of the places which are either unoccupied or not perfectly occupied by other beings; and this will depend on infinitely complex relations. But as a general rule, the more diversified in structure the descendants from any one species can be rendered, the more places they will be enabled to seize on, and the more their modified progeny will be increased. (Darwin 1859: 119; see also 1872: 91–92)

Darwin then goes on to specify that a tiny detail may just be what it takes to win the day, although the breakthrough may take time to come: "Battle within battle must ever be recurring with varying success; and yet in the long-run the forces are so nicely balanced, that the face of nature remains uniform for long periods of time, though assuredly the merest trifle would often give victory to one organic being over another" (Darwin 1859: 73; see also 1872: 57).

Our next example showing the epistemological predominance of *pattern* over *process* in Darwin's mind is revealed by his general justification for the cone of increasing diversity, as seen in an overall pan-divergent pattern. Darwin's second argumentative strategy consists in *overlooking all the implications flowing from his own acknowledged reasoning*. The following quotation is interesting for it provides, implicitly, the inherent limitation accompanying Darwin's proposed process:

> But the struggle almost invariably will be most severe between the individuals of the same species, for they frequent the same districts, require the same food, and are exposed to the same dangers. In the case of varieties of the same species, the struggle will generally be almost equally severe... As species of the same genus have usually, though by no means invariably, some similarity in habits and constitution, and always in structure, *the struggle will generally be more severe between species of the same genus, when they come into competition with each other, than between species of distinct genera*...We can dimly see why the competition should be most severe between allied forms, which fill nearly the same place in the economy of nature. (Darwin 1859: 75, 76; see also 1872: 58, 59; my emphasis)

Darwin refers here to what was later termed the "competitive exclusion principle." What seems to be another description of an intertwined process/pattern is, actually, being uncoupled before our eyes. Darwin's argument breaks down for a reason he himself provides, revealing to us the ultimate underpinning of his reasoning, i.e., that life's history is organized around a cone of increasing diversity (for each class/phylum), come what may. Indeed, Darwin is powerless to make sense of the ongoing divergence between entities separated at higher taxonomic levels (families and orders) for the simple reason that the exclusion principle he postulates should gradually be disactivated as these related entities continue to move away (diverge) during their respective taxonomic ascents. Related entities separated at high taxonomic levels have become so distinct from one another that there is little reason to believe that the exclusion principle will still be active between such entities.

Under Darwin's proposed process, then, the overall picture of evolution at middle and high taxonomic levels should not be an "ongoing divergence" but something closer to a "random evolutionary walk," or perhaps even an "evolutionary stagnation." Yet, Darwin's Diagram (see Fig. 1) when read at fairly high taxonomic levels (Darwin 1859: 331–332, 412–413; 1872: 303–304, 364) shows that related families and orders continue to diverge away (see also 1859: 126; 1872: 96). This is easily explained: Darwin was already committed toward the pattern of divergence, a corollary of the cone ontology wherein all extinct forms must, by definition, fall in-between extant ones. *Considering Darwin's foundational commitment, the game was up even before it started, with process (competitive principle) being a poor justification for phenomena occurring at high taxonomic levels!*

The above examples show that the role of natural selection was being instrumentalized by Darwin: where divergence is established, natural selection must somehow justify it, even if one has to get rid of it or overstretch its justificatory role. Could it have been otherwise in Darwin's profoundly structured or patterned view of life? Again, if natural selection had reigned in Darwin's biology—by being at its explanatory core—there would have been no reason for him to restrict evolutionary change to divergence. In a truly process-oriented biology, Darwin would have had no other choice but to interpret his empirical facts as documenting a whole range of phenomena in their *full expression*: divergence, parallel evolution, convergence, reticulate evolution, slow evolution, rapid evolution, gradualism, punctuationism, continuities, discontinuities, etc. This is what evolutionary opportunism is all about, as there is little reason for an adaptively driven evolution not to take advantage of all available opportunities. Instead, Darwin elected to negate or limit the impact of several such listed phenomena on his theory in order to push forward its ultimate agenda: a pan-divergent outlook.

19 Evolutionary Contingency

The table is set to complete our analysis of the fundamental differences separating Darwin and Gould. Two key issues intersect: (1) opposing patterns of evolution and (2) evolutionary change within competing ontological spaces:

Opposing Patterns Darwin conceived overall change as organized around "divergence" throughout, a fractal-like evolutionary geometry that repeats itself at all taxonomic scales, with a pattern that deploys itself over time by going from the lower to the higher taxonomic levels. For his part, Gould envisioned overall change as characterized by an inverse pattern (compare Figs. 2 and 3), wherein high taxonomic entities (*Baupläne*) arose quite suddenly in life's history with many of them rapidly going extinct, confining subsequent evolution to an impoverished range of possibilities and diversity (disparity) when compared to the Cambrian explosion. Such a decimation model is also akin to "fractal geometry," since Gould recognizes that the same pattern applies at lower taxonomic levels within each surviving *Bauplan*, with a loss or depletion of forms through evolutionary time.

Competing Ontological Spaces The differences between Darwin and Gould are already significant when one compares their overall evolutionary patterns. But the depth of the conceptual gap separating them is captured only when one also takes into account the *ontological space* within which each scholar deploys his own pattern. This essential issue has largely been neglected in the historiography, on the presumption that Darwin had successfully managed to put the topic of biological evolution on a firm modern footing, allowing for the field of evolutionary biology to develop within the same epistemological horizon ever since. We have argued above that this was far from being the case. Darwin's pan-divergent outlook operates within a presentist ontology, with extinct common ancestors being reduced to the status of pale or diminished reflections of extant forms, the former falling in-between the biological diversity of the latter. For his part, Gould was prepared to confront an evolutionary past quite unlike what the modern living world offers, opening up much more ontological space when it comes to thinking in terms of evolutionary change. It is difficult to imagine two conceptions of historical/evolutionary time more different than these.

To get a sense of the conceptual gap that separates Darwin and Gould, one only need to think about Mayr: although Mayr was not as deeply committed to the presentist ontology as Darwin was (some of their conceptual differences have already been outlined above), Mayr was nonetheless sufficiently ensconced in it to eventually realize around 1990 that his theory based on the cone of increasing diversity had entirely broken down under the empirical pressure created by the rise of the decimation ontology. Moving from one ontology to another proved perilous to Mayr—as seen, for instance, in his "interesting" convoluted positions concerning evolutionary taxonomy, gradualism, and adaptationism—requiring crossing a kind of intellectual no man's land. Gould's assumption that he could somehow significantly connect his view to Darwin's is based on superficial considerations only. Let us unpack the key features of Gould's brand of biology erected after the mid-1980s. Speaking of the evolutionary pattern he thought had emerged from the Burgess Shale discoveries dated to the Cambrian period,[14] he writes:

> We need no more than the *descriptive pattern* of Burgess disparity and later decimation to impose a major reform upon our traditional view of life. For the new iconography not only alters but thoroughly inverts the conventional cone of increasing diversity. Instead of a narrow beginning and a constantly expanding upward range, multicellular life reaches its maximal scope at the start, while later decimation leaves only a few surviving designs. (Gould 1989: 233; my emphasis)

[14] On Gould's view of the Cambrian explosion, see also Gould (2002: 1155–1161).

The model Gould extracted from the Cambrian explosion was also applied to groups at lower taxonomic levels, thus suggesting a general pattern among multicellular life. Under the expression "bottom-heavy asymmetry," it is argued that newly arising groups show that the highest level of diversity (disparity) is reached in their earlier phase of existence, followed by a loss of this diversity. Again, the issue is evolutionary *pattern*:

> In my recent studies I concluded that the pattern of maximal early breadth is a general characteristic of lineages at several scales and times, not only of major groups at the Cambrian explosion. In fact, we have proposed that this 'bottom-heavy' asymmetry may rank among the few natural phenomena imparting a direction to time, thus serving as a rare example of 'time's arrow'...The bottom-heavy signature is not an oddity of Cambrian invertebrate life, but a general statement about the nature of evolutionary diversification. For example, mammalian lineages that arose during the Paleocene epoch, the initial period of explosive diversification following the demise of dinosaurs, tend to be bottom-heavy, while lineages arising later are symmetrical. We may interpret this bottom-heavy pattern in several ways. I like to think of it as 'early experimentation and later standardization.' Major lineages seem able to generate remarkable disparity of anatomical design at the outset of their history—early experimentation. Few of these designs survive an initial decimation, and later diversification occurs only within the restricted anatomical boundaries of these survivors—later standardization. The number of species may continue to increase, and may reach maximal values late in the history of lineages, but these profound diversifications occur within the restricted anatomies—nearly a million described species of modern insects, but only three basic arthropod designs today, compared with more than twenty in the Burgess. (Gould 1989: 302, 304)

Gould did more than just propose to put the cone of increasing diversity on its head. A rising theme of evolutionary biology in the 1970s concerned evolutionary stability (Eldredge and Gould 1972; Gould and Eldredge 1977; Stanley 1979). What was originally mainly focused on individual species/lineages rapidly spread to larger entities, like biological types and *Baupläne* (Gould 1977; Mayr 1975). The reinterpretation of the Burgess Shale discoveries only made the challenge more acute still, with life's history being characterized by a general loss of evolutionary plasticity. While accepting this phenomenal reality, Gould proposed a way around the inherent limitations accompanying morphological stability. What originally seemed a serious constraint imposed upon evolutionary change itself became, for Gould (and others), an opportunity for widening it. This allowed him to create much more *ontological space* than Darwin and Mayr had done in terms of the expression of evolutionary contingency, propelling the field of evolutionary biology into new epistemological territory.

What Gould proposed was to multiply the number of ontological spaces for evolution to deploy itself in, although only one such space actually materialized (that is, our world) if only because all others remained as possible states. Gould's thought experiment of replaying the tape of life's history fills the epistemological function of opening a great number of potential ontological spaces, any of which might have been actualized. Gould's brand of biology thus contains two epistemological spheres: the sphere dealing with the world that actually materialized (our own world)—and for which he applies conceptual tools that mitigate constraints associated with evolutionary stability: species selection, mass extinctions, and stochastic processes (Stanley 1975, 1979; Gould 1980a, b, 1985, 1991; Raup 1991; see also Sepkoski 2012; Turner 2011)—and the sphere of potential worlds that did not materialize but might have. Although these two epistemological spheres are kept distinct from one another—under the polarity "potential state/realized state"—Gould maintains a connection between them under the notion of a potential "replay." Indeed, in Gould's universe a new range of evolutionary possibilities would have opened up had another world succeeded in establishing itself. In fact, points of contact between the realized and the potential worlds have remained open ever since life arose on Earth, considering, for instance, that a mass extinction event today would profoundly modify the future of life's history. Theoretically speaking, then, a new world can spring up at any moment, taking evolution down quite a different pathway. Before turning to Gould's explication, it is worth keeping in mind that Ernst Mayr (1959b: 6) also resorted to the conjectural approach consisting in "replaying the tape of life." Yet, Gould's own approach to it is of a different nature, as revealed in the different implications accompanying the cone ontology and the decimation ontology on the evolutionary outcome. Gould writes:

> Under the traditional model of evolutionary history as a 'cone of increasing diversity,' life moves ever upward to greater progress, and outward to a larger number of species—from simple Cambrian beginnings for multicellular animals to our modern levels of progress and range of diversity. *Under this iconography, pathways actually followed run along predictable courses that would be at least roughly repeated in any replay.* But a radically different view, suggested by a thorough restudy of soft-bodied fossils in the Burgess Shale and other Cambrian faunas, indicates that an inverted iconography may be more appropriate—with maximal anatomical range of disparity reached early in life's history, followed by the extinction of most initial experiments and the 'settling down' of life's diversity to just a few of the original possibilities. *Moreover, we have strong reason to suspect that the loss of most, and survival of just a few, occurred more through a distribution of lottery tickets than by victories for predictable cause based on higher levels of progress among winners. In the 'pure' lottery model, 'tickets' are distributed at random and few initial lineages receive such a blessing. Any replay distributes the tickets to another random set, and leads to a radically different group of survivors.* (Gould 2011[1996]: 215; my emphasis)

Differently stated, inside the cone of increasing diversity, a replay would generate a limited amount of evolutionary novelties; inside a decimation model, a replay would induce much more in the way of novelties. Gould is right, as even Mayr himself realized the limitations accompanying his own conjectural approach. Let us reintroduce an earlier quotation here, this time drawing the reader's attention to Mayr's use of the word "somewhat":

> All evolution is a sequence of unique events. This is true not only for selective values in different genotypes and different environments, it is likewise true for segregation, crossing over, and whatever other events determine evolutionary success and evolutionary change. If we set up 10 parallel selection experiments, starting each from the same, apparently identical foundation stock, we are liable to get in the end 10 *somewhat* different answers. The more I

> study evolution the more I am impressed by the uniqueness, by the unpredictability, and by the unrepeatability of evolutionary events. (Mayr 1959b: 6; my emphasis)

What is Gould's rationale, here, albeit expressed above in a somewhat implicit manner? It should be noticed that Gould (unconsciously?) introduced an asymmetrical argument that can be made explicit:

1. On the one hand, the cone of increasing diversity is seen as a by-product of an evolutionary *process* driven forward by competition in which more complex forms usually win out over simpler ones. According to Gould, it is the process (adaptation) that generates the pattern (cone).
2. On the other hand, the decimation model is presented as an evolutionary *pattern* which sees the survival of a limited number of forms out of a plethora of perfectly capable initial biological types, indicating that chance more than adaptive performance rules the game. According to Gould, it is the pattern (decimation) that decide on the process (chance).

We have arrived at the core of Gould's understanding of the nature of evolutionary biology, we think. To repeat our main thesis in this section, we hold that Gould did not realize that the cone of increasing diversity as expressed, for instance, by Darwin and Mayr is, in itself, an ontology (and a pattern) of equal epistemological standing to his own decimation ontology. This being the case, evolution inside the cone of increasing diversity is "predictable," to use Gould's own term (2011[1996]: 215; see also quotation just above), not only because the process is directed by the addition of ever more complex (adaptively progressive) and diversified forms in time, as Gould claims, but also because "divergence" constitutes a straitjacket that channels evolution. In our view, the issue is less a process than a pre-established pattern that closes off other creative alternatives when expressed in full, such as under reticulate evolution, parallel evolution, and convergent evolution. In addition, and even more importantly, as long as Mayr overvalued the present time relative to the geological past—through his neontological agenda according to which all extinct forms fall in-between extant ones—the potential "morphospace" created under his conjectural approach ("replaying the tape of life") considerably curbed his imagination when thinking about alternative worlds. Indeed, it is doubtful that Mayr could ever had conceived life forms significantly different from those already indexed among currently living ones. The Burgess Shale discoveries and, eventually, their specific interpretations in the 1970s and after certainly came as a shock to Mayr (Conway Morris 1998; Gould 1989; see also Baron 2011; Brysse 2008; Erwin 2016; Sterelny 2007: 115–141; Turner 2011: 180–196).

The two "iconographies" presented by Gould, therefore, are more than mere "visual representations"; they are fundamental ontologies involving whole sets of accompanying commitments. Gould thought he could select from Darwin's ontology whatever elements would serve to complete his own ontology. He was apparently prevented from seeing this incompatibility because of his asymmetrical argument—process (cone ontology) versus pattern (decimation ontology), as discussed above—thus confusing the respective roles these two components played for Darwin and Mayr. Differently stated, the issue is not the adaptive process of natural selection, since this *detachable* explanatory component can be inserted in competing ontologies. Rather, the issue consists in realizing that, when inserted in the cone ontology or the decimation ontology, natural selection (as well as many other explanatory components) is consigned to a different explanatory role. In the end, perhaps Gould was prevented from seeing that the cone of increasing diversity was, in itself, an ontology of evolutionary time for the simple reason that he also overlooked the ontological underpinnings of his own decimation model.

20 Gould and Darwin: Contingency Versus Steady-Statism

Faced with the seemingly irresistible temptation of seeing himself as an inheritor of Darwin's legacy, Gould finds in him a true inspiration: "I regard Charles Darwin as the greatest of all historical scientists. Not only did he develop convincing evidence for evolution as the coordinating principle of life's history, but he also chose as a conscious and central theme for all his writings. . .the development of a different but equally rigorous methodology for historical science" (Gould 1989: 282). More specifically, Gould credits the *Origin of Species* for having established a line of demarcation between the epistemology of the natural sciences and the epistemology of the historical sciences. Turning to the very last phrase of the *Origin*—the metaphor appealing to the "grandeur in this view of life" (Darwin 1859: 490; 1872: 429)—Gould writes:

> Note how Darwin contrasts the dull repetitiveness of planetary cycling. . .with the gutsy glory of rich diversity on life's ever rising and expanding tree. Darwin even gives his metaphor a geometric flavor, as he contrasts the horizontal solar system, its planet cycling around a central sun to nowhere, with the vertical tree of life, starting in utmost simplicity at the bottom, and rising right through the horizontality of this repetitive physical setting towards the heavenly heights of magnificent and ever expanding diversity, into a contingent and unpredictable future of still greater possibility: 'There is grandeur in this view of life. . . [and] whilst this planet has gone cycling on according to the fixed law of gravity, from so simple a beginning endless forms most beautiful and most wonderful have been, and are being, evolved.' (1859, p. 490) (Gould 2002: 1334)

Here, we contest Gould's exegesis entirely, holding that he neglects the originality of his own contribution, by conflating his brand of biology with Darwin's. We will examine each of the statements in the above quotation, following the order of Gould's exposition:

Statement 1 Gould opposes the dull repetitiveness of planetary cycling around a central sun with an expanding and diversifying tree of life. What shall we make, then, of Darwin's rather surprising description that follows:

> On these principles, I believe, the nature of the affinities of all organic beings may be explained...[A]ll animals and all plants throughout all time and space [are]...related to each other in group subordinate to group...namely, varieties of the same species most closely related together, species of the same genus less closely and unequally related together, forming sections and sub-genera, species of distinct genera much less closely related, and genera related in different degrees, forming sub-families, families, orders, sub-classes, and classes. The several subordinate groups in any class cannot be ranked in a single file, but seem rather to be *clustered round points, and these round other points, and so on in almost endless cycles*... [This] is explained through inheritance and the complex action of natural selection, entailing extinction and divergence of character, as we have seen illustrated in the diagram. (Darwin 1859: 128, 129; see also 1872: 103, 104; my emphasis)

Endless cycles? Is Darwin here merely talking about the evermore inclusive categories up the taxonomic scale? Darwin's choice of words is more than a slip of the tongue, we think, whose meaning is given in other of his commitments. Interestingly, Gould (2011[1996]: 39) writes elsewhere: "But classifications are not passive ordering devices in a world objectively divided into obvious categories. Taxonomies are human decisions imposed upon nature—theories about the causes of nature's order." This is precisely our point: Darwin's use of the Linnaean taxonomy does not necessarily rest on the same theoretical foundation as our modern theories do. Knowing that classes/phyla are, for Darwin, nearly immovable entities by being deeply entrenched in nature's economy, evolutionary motion is largely confined to the perpetual rise and fall (extinction)—endless cycles—of entities at and below the order level.

Statement 2 Gould talks about Darwin's recognition of a rich biological diversity. For Darwin, however, extinct forms are reduced to pale reflections of extant ones: imagined and hypothetical creatures built out of repetitive extant parts.

Statement 3 Gould holds that Darwin lends a geometric flavor to astronomy only. However, Darwin's Diagram in the *Origin* is of a profoundly geometrical configuration: (1) evolutionary change is restricted to the pattern of divergence, with about 2/3 of the surviving lineages in each geological time period being the most divergent ones; and (2) as in fractal geometry, the same pattern (divergence) recurs at all taxonomic levels. There is something profoundly rigid, geometrical, and predictable about Darwin's view of life, a view largely inimical to evolutionary contingency and historical epistemology, as we understand them today.

Statement 4 Gould opposes a vertical tree of life to the horizontality of repetitive physical settings. Still, horizontality is precisely how Darwin envisions evolutionary time given that his presentist ontology projects extant forms as a single homogenous block—horizontally—backward in geological time.

Statement 5 Gould stresses a Darwinian tree of life starting in utmost simplicity. Yet, despite Darwin's limited rhetoric about truly simple life forms, the *Origin* is actively constructed around full-fledged classes/phyla profoundly entrenched in nature.

Statement 6 Gould insists on Darwin's promotion of an ever-expanding diversity of life forms. While Darwin did envision such an expansion moving forward in time, that process was busy going nowhere, merely recycling the same biological variations recorded in extant forms.

Statement 7 Finally, Gould finds in Darwin a contingent and unpredictable approach to evolution. However, in the cone of increasing diversity, no room exists for another evolutionary pattern than divergence, a fact that downplays the full expression of other legitimate patterns/processes.

Gould (2002: 1335) proceeds to quote a passage of Darwin already reproduced in full in the early part of this chapter, the one beginning with "I believe in no fixed law of development, causing all the inhabitants of a country to change abruptly, or simultaneously, or to an equal degree..." (Darwin 1859: 314). Therein, Darwin explicitly uses the expression "complex contingencies" to describe the various circumstances instigating change, such as variability, physical conditions, and biotic competition. Gould (2002: 1336) thus concludes by "Darwin's own commitment to contingency has been underemphasized, or even unrecognized, by his later followers." According to him, Darwin's commitment to contingency is revealed in the richness of a tree of life, as expressed in "the very last line of the *Origin of Species*, where he rejected the traditional claims of quantitative physical science to represent the apotheosis of sophistication, and awarded higher honor to his own discipline of natural history and evolutionary biology, as embodied in the gnarly and meandering icon of the luxuriantly, but contingently, branching tree of life" (Gould 2002: 1334). Unfortunately for Gould, such a conclusion is not warranted. The expression "complex contingencies" under Darwin's pen carries a much weaker meaning, as "contingency" for him meant only *lesser or greater amount of divergence*. Beyond that, Darwin eschewed other alternatives to evolutionary change considering that his "tree of life" is anything but profoundly haphazard.

Let us revisit the very last sentence of the *Origin*, a rather intriguing combination of astrophysics, biology, and theology:

> There is grandeur in this view of life, with its several powers, having been originally breathed into a few forms or into one; and that, whilst this planet has gone cycling on according to the fixed law of gravity, from so simple a beginning endless forms most beautiful and most wonderful have been, and are being, evolved. (Darwin 1859: 490)

In the last edition of the *Origin of Species*, the reader finds attached to it the expression "by the Creator" (1872: 429). Is this apparently out-of-context phrase placed at the very end of a long biological treatise only a matter of style or rhetoric, echoing similar astro-biologico-theological comments of the seventeenth and eighteenth centuries? More than two decades prior to the publication of *On the*

Origin of Species, it caught the attention of William Whewell (1837: vol. 3, pp. 615–620) that direct comparison of celestial and terrestrial phenomena was, after all, perhaps not entirely without consequence. As Whewell explains (1837: vol. 3, p. 618), "The analysis of the heavenly motions, it has been said, supplies no trace of a beginning, no promise of an end." This was the motto popularized by James Hutton in the late eighteenth century for geological investigations. In the context of a Newtonian view of our solar system, *eternally cycling* on but static, the use of an analogy binding terrestrial phenomena to celestial ones may have been the way of transferring this stability from one area of knowledge to another. The idea behind Darwin's final line in the *Origin* appears repeatedly in his *Sketch* (1842), his *Essay* (1844), and in the six editions of the *Origin of Species* (1859–1972). The formulation of 1842 is even more interesting, for it contains a reference to a steady-state geology, echoing a steady-state astronomy:

> There is a simple grandeur in the view of life with its powers of growth, assimilation and reproduction, being originally breathed into matter under one or a few forms, and that whilst this our planet has gone circling on according to fixed laws, *and land and water, in a cycle of change*, have gone on replacing each other, that from so simple an origin, through the process of gradual selection of infinitesimal changes, endless forms most beautiful and most wonderful have been evolved. (reproduced in F. Darwin 1909: 52; my emphasis)

An isomorphism was thus established by Darwin between a steady-state astronomy and a steady-state geology. Gould himself repeatedly claims throughout *The Structure of Evolutionary Theory* (2002) that Darwin's view owed much to Charles Lyell's uniformitarian geology. Gould's mistake, in our view, comes from his assumption that uniformitarianism impacted Darwin's thinking mainly through processes. But we have seen that there is more for Darwin, as uniformitarianism is ultimately ontologically based. For instance, *gradualism* is merely the process side of an ontological reality: *gradation*. The notion of *intermediate forms* pervading Darwin's fabric of life binds related forms under the pattern of divergence, itself a small-scale manifestation of the cone of increasing diversity. Furthermore, the cone ontology (as its name implies) is merely an expression of an even more inclusive epistemological system: the presentist ontology.

In contradistinction to Gould's thesis, we cannot see how Darwin could have managed to entirely extricate steady-statism from his biology maintaining, instead, consistency throughout. Indeed, we have presented in this chapter numerous features of Darwin's biology more coherently explained by his connection with universal uniformitarianism than with modern concepts of biology (see also Delisle 2014, 2019; Delisle and Tierney 2022). We hold that an intellectual space existed in the nineteenth century for a brand of evolutionism developed within universal uniformitarianism. Once it is accepted that Darwin could not have been a modern evolutionist—contrary to today's mythology—it becomes possible to see how he managed to expand the conceptual envelope of the uniformitarian doctrine in order to forge his own specific brand of evolutionism.

In what is perhaps the last summary given by Gould about his own theory, one can now see more clearly the ways in which it is profoundly un-Darwinian in logic, despite his repeated claims to the contrary:

> In most general terms, and in order to form a more perfect union among evolution's hierarchy of structural levels and tiers of time, this revised theory rests upon an expansion and substantial reformation of all three central principles that build the tripod of support for Darwinian logic: (1) the expansion of Darwin's reliance upon organismal selection into a hierarchical model of simultaneous selection at several levels of Darwinian individuality (gene, cell lineage, organism, deme, species and clade); (2) the construction of an interactive model to explain the sources of creative evolutionary change by fusing the positive constraints of structural and historical pathways internal to the anatomy and development of organisms (the formalist approach) with the external guidance of natural selection (the functionalist approach); and (3) the generation of theories appropriate to the characteristic rates and modalities of time's higher tiers to explain the extensive range of macroevolutionary phenomena (particularly the restructuring of global biotas in episodes of mass extinction) that cannot be rendered as simple extrapolated consequences of microevolutionary principles. (Gould 2002: 1340)

Must we really take Gould's Darwinian rhetoric seriously? After all, he was honest enough to raise doubts about his own attempt to connect his biology to Darwin's: "In the text of this book, I speak most often of a 'tripod' since central Darwinian logic embodies three major propositions that I have always visualized as supports—perhaps because I have never been utterly confident about this entire project, and I needed some pictorial encouragement to keep me going for twenty years" (Gould 2002: 15). Although our view of the development of evolutionary biology differs significantly from the one proposed by the historian Betty Smocovitis, we share her assessment when she states: "But when it came to understanding the history and structure of evolutionary theory, however, Gould struggled with finding a conceptual core and tracing its history; it took him some 1500 pages to attempt that (Gould 2002)" (Smocovitis 2023: 44).

21 Conclusion

The Synthesis Historiography continues to weigh heavily on our understanding of the development of evolutionary biology, making it difficult to see some of the deepest intellectual forces playing out on its epistemic landscape. The competing ontologies of evolutionary time reviewed in this chapter belong, in our view, to these deep forces. Ernst Mayr, for one, used labels like "Evolutionary Synthesis" and "Darwinism" to camouflage major shifts in his thinking on issues like evolutionary taxonomy, gradualism, evolutionary contingency, and adaptationism, simultaneously obscuring the collapse of the ontology supporting the cone of increasing diversity. For his part, Stephen Jay Gould erected a kind of mirror image of the same historiography—a symmetrical yet inverted representation—through critiques of the "Adaptationist Programme" and the "Hardening of the Modern Synthesis," eventually trying to squeeze in the underlying ontology supporting his decimation model.

These interested and rhetorical discourses make it more difficult for historians and philosophers to steer a clear and independent road through this conceptual minefield. As stated by Antonello La Vergata: "There is *one* way of doing science only in the mind of some philosophers and (unfortunately still too many) scientists who are convinced that they possess the key to it. The role of the history of science, and of ideas, can be said to consist in overthrowing their intolerable simplifications" (La Vergata 2023: 591; for a full discussion see Esposito 2021, 2024; La Vergata 2024). Ultimately, perhaps the most important victim of the Synthesis Historiography (and its twin yet reverse image) is Charles Darwin himself, whose name is exploited to claim the same and its opposite. Disposing of the label "Darwinism" altogether, we suggest, will help remove one obstacle on our way to better understanding the nature of evolutionary biology.

Acknowledgments I am grateful to Maurizio Esposito and Antonello La Vergata for their thoughtful suggestions, although not all of them could be implemented here. Once more I am grateful to my colleague and collaborator, James Tierney (Yale University), for having improved the quality of my English in this chapter.

References

Adams MB (2021) Little evolution, big evolution: rethinking the history of Darwinism, population genetics, and the 'Synthesis'. In: Delisle RG (ed) Natural selection: revisiting its explanatory role in evolutionary biology. Springer, Cham, pp 195–229

Adams MB (2024) The evolution of 'Darwinism': up close and personal. In: Delisle RG, Esposito M, Ceccarelli D (eds) Unity and disunity in evolutionary biology: deconstructing Darwinism. Springer, Cham, pp 83–121

Amundson R (2005) The changing role of the embryo in evolutionary thought: roots of evo-devo. Cambridge University Press, Cambridge

Baedke J, Böhm A, Reiners-Selbach S, Straetmanns V (2024) A synthesis without Darwin: unification attempts in early theoretical biology. In: Delisle RG, Esposito M, Ceccarelli D (eds) Unity and disunity in evolutionary biology: deconstructing Darwinism. Springer, Cham, pp 333–356

Baron C (2011) A web of controversies: complexity in the Burgess Shale debate. J Hist Biol 44: 745–780

Bowler PJ (1988) The non-Darwinian revolution. Johns Hopkins University Press, Baltimore

Brysse K (2008) From weird wonders to stem lineages: the second reclassification of the Burgess Shale fauna. Stud Hist Philos Biol Biomed Sci 39:298–313

Caianiello S (2024) The strange story of mosaic evolution. In: Delisle RG, Esposito M, Ceccarelli D (eds) Unity and disunity in evolutionary biology: deconstructing Darwinism. Springer, Cham, pp 357–401

Cain J (2009) Rethinking the synthesis period in evolutionary studies. J Hist Biol 42:621–648

Caso JMR (2024) The 'Darwinian revolution' as a presentist discourse: ideological implications beyond the Anglo-Saxon context. In: Delisle RG, Esposito M, Ceccarelli D (eds) Unity and disunity in evolutionary biology: deconstructing Darwinism. Springer, Cham, pp 539–549

Ceccarelli D (2021) Recasting natural selection: osborn and the pluralistic view of life. In: Delisle RG (ed) Natural selection: revisiting its explanatory role in evolutionary biology. Springer, Cham, pp 171–191

Ceccarelli D (2024a) Darwin as a unifying figure in evolutionary biology: a meta-historical overview. In: Delisle RG, Esposito M, Ceccarelli D (eds) Unity and disunity in evolutionary biology: deconstructing Darwinism. Springer, Cham, pp 19–29

Ceccarelli D (2024b) Redrawing the boundaries of Darwinism: addressing Darwin's endorsement of the inheritance of acquired characteristics in Darwin's celebrations, 1909-1959-2009. In: Delisle RG, Esposito M, Ceccarelli D (eds) Unity and disunity in evolutionary biology: deconstructing Darwinism. Springer, Cham, pp 509–537

Ceccarelli D, Esposito M, Delisle RG (2024) Toward a new historiography. In: Delisle RG, Esposito M, Ceccarelli D (eds) Unity and disunity in evolutionary biology: deconstructing Darwinism. Springer, Cham, pp 3–7

Conway Morris S (1998) The crucible of creation: the Burgess Shale and the rise of animals. Oxford University Press, Oxford

Darwin C (1859) On the origin of species. John Murray, London

Darwin C (1872) The origin of species, 6th edn. John Murray, London

Darwin F (1909) The foundation of the origin of species: two essays written in 1842 and 1844 by Charles Darwin. Cambridge University Press, Cambridge

Delisle RG (2008) Expanding the framework of the holism/reductionism debate in neo-Darwinism: the case of Theodosius Dobzhansky and Bernhard Rensch. Hist Philos Life Sci 30:207–226

Delisle RG (2009a) The uncertain foundation of neo-Darwinism: metaphysical and epistemological pluralism in the evolutionary synthesis. Stud Hist Philos Biol Biomed Sci 40:119–132

Delisle RG (2009b) Les philosophies du néo-darwinisme. Presses Universitaires de France, Paris

Delisle RG (2011) What was really synthesized during the evolutionary synthesis? A historiographic proposal. Stud Hist Philos Biol Biomed Sci 42:50–59

Delisle RG (2014) Evolution in a fully constituted world: Charles Darwin's debts towards a static world in the *Origin of Species* (1859). Endeavour 38:197–210

Delisle RG (2017a) From Charles Darwin to the evolutionary synthesis: weak and diffused connections only. In: Delisle RG (ed) The Darwinian tradition in context: research programs in evolutionary biology. Springer, Cham, pp 133–168

Delisle RG (2017b) Introduction: Darwinism or a kaleidoscope of research programs and ideas? In: Delisle RG (ed) The Darwinian tradition in context: research programs in evolutionary biology. Springer, Cham, pp 1–8

Delisle RG (2019) Charles Darwin's incomplete revolution: the origin of species and the static worldview. Springer, Cham

Delisle RG (2021a) Introduction: in search of a new paradigm for the development of evolutionary biology. In: Delisle RG (ed) Natural selection: revisiting its explanatory role in evolutionary biology. Springer, Cham, pp 1–8

Delisle RG (2021b) Natural selection as a mere auxiliary hypothesis (sensu stricto I. Lakatos) in Charles Darwin's *Origin of Species*. In: Delisle RG (ed) Natural selection: revisiting its explanatory role in evolutionary biology. Springer, Cham pp, pp 73–104

Delisle RG, Tierney J (2022) Rereading Darwin's Origin of Species: the hesitations of an evolutionist. Bloomsbury Academic, London

Dietrich M (1998) Paradox and persuasion: negotiating the place of molecular evolution within evolutionary biology. J Hist Biol 31:85–111

Dobzhansky T (1982 [1937]) Genetics and the origin of species. Columbia University Press, New York

Dolbeault J (2024) Fisher, Wright and Haldane: three philosophical conceptions of evolution. In: Delisle RG, Esposito M, Ceccarelli D (eds) Unity and disunity in evolutionary biology: deconstructing Darwinism. Springer, Cham, pp 301–331

Eldredge N, Gould SJ (1972) Punctuated equilibria: an alternative to phyletic gradualism. In: Schopf TJM (ed) Models of paleobiology. Freeman, Cooper & Co, San Francisco, pp 82–115

Ereshefsky M (2001) The poverty of the Linnaean hierarchy. Cambridge University Press, Cambridge

Erwin DH (2016) Wonderful life revisited: chance and contingency in the Ediacaran-Cambrian radiation. In: Ramsey G, Pence C (eds) Chance in evolution. Chicago University Press, Chicago, pp 277–298
Esposito M (2021) Cathedrals, corals and mycelia: three analogies for the history of evolutionary biology. In: Delisle RG (ed) Natural selection: revisiting its explanatory role in evolutionary biology. Springer, Cham, pp 11–35
Esposito M (2024) "Reformist" and "radical" historiographies behind and beyond the unity and disunity of the evolutionary thought. In: Delisle RG, Esposito M, Ceccarelli D (eds) Unity and disunity in evolutionary biology: deconstructing Darwinism. Springer, Cham, pp 9–18
Gould SJ (1977) Ontogeny and phylogeny. Belknap Press, Cambridge, MA
Gould SJ (1980a) Is a new and general theory of evolution emerging? Paleobiology 6:119–130
Gould SJ (1980b) The promise of paleobiology as a nomothetic, evolutionary discipline. Paleobiology 6:96–118
Gould SJ (1983) The hardening of the modern synthesis. In: Grene M (ed) Dimensions of Darwinism. Cambridge University Press, Cambridge, pp 71–93
Gould SJ (1985) The flamingo's smile: reflections in natural history. W.W. Norton, New York
Gould SJ (1986) Evolution and the triumph of homology, or why history matters. Am Sci 74:60–69
Gould SJ (1989) Wonderful life: the Burgess Shale and the nature of history. W.W. Norton, New York
Gould SJ (1991) Introduction to D Raup, extinction: bad genes or bad luck? W.W. Norton, New York, pp xiii–xvii
Gould SJ (2002) The structure of the evolutionary theory. Belknap Press, Cambridge, MA
Gould SJ (2011 [1996]) Full house: the spread of excellence from Plato to Darwin. Belknap Press, Cambridge, MA
Gould SJ, Eldredge N (1977) Punctuated equilibria: the tempo and mode of evolution reconsidered. Paleobiology 3:115–151
Gould SJ, Lewontin RC (1979) The spandrels of San Marco and the panglossian paradigm: a critique of the adaptationist programme. Proc R Soc Lond B205:581–598
Grene M, Depew D (2004) The philosophy of biology: an episodic history. Cambridge University Press, Cambridge
Gribbin J, Cherfas J (1982) The monkey puzzle: reshaping the evolutionary tree. Pantheon Books, New York
Hooykaas R (1963) The principle of uniformity in geology, biology and theology. E J Brill, Leiden
Horgan J (1996) The end of science: facing the limits of knowledge in the twilight of the scientific age. Addison-Wesley, Reading
Hull DL (1964) Consistency and monophyly. Syst Zool 13:1–11
Hull DL (1988) Science as a process. University of Chicago Press, Chicago
Hull DL (2001) The role of theories in biological systematics. Stud Hist Philos Biol Biomed Sci 32: 221–238
Jablonka E, Lamb MJ (2020) Inheritance systems and the extended evolutionary synthesis. Cambridge University Press, Cambridge
Kuhn TS (1970) The structure of scientific revolutions, 2nd edn. University of Chicago Press, Chicago
La Vergata L (2023) Images of the economy of nature, 1650–1930: from "Nature's War" to Darwin's "Struggle for Life". Springer, Cham
La Vergata A (2024) Constructing, deconstructing and reconstructing. On "Darwinism" and "Darwinisms", with some disparate considerations on the history of science. In: Delisle RG, Esposito M, Ceccarelli D (eds) Unity and disunity in evolutionary biology: deconstructing Darwinism. Springer, Cham, pp 33–82
Lakatos I (1970) Falsification and the methodology of scientific research programmes. In: Lakatos I, Musgrave A (eds) Criticism and the growth of knowledge. Cambridge University Press, Cambridge, pp 91–196

Largent M (2009) The so-called eclipse of Darwinism. Trans Am Philos Soc (New Series) 99(1): 3–21
Laudan L (1977) Progress and its problems. University of California Press, Berkeley
Levit GS, Hossfeld U (2006) The forgotten 'Old-Darwinian' synthesis: the evolutionary theory of Ludwig H Plate (1862–1937). Zeitschrift für Geschichte der Wissenschaften, Technik und Medizin 14:9–25
Levit GS, Hossfeld U (2011) Darwin without borders? Looking at 'generalised Darwinism' through the prism of the 'hourglass model'. Theory Biosci 130:299–312
Levit GS, Hossfeld U (2017) Major research traditions in 20th century evolutionary biology: the relations of Germany's Darwinism with them. In: Delisle RG (ed) The Darwinian tradition in context: research programs in evolutionary biology. Springer, Cham, pp 169–194
Levit GS, Simunek M, Hossfeld U (2008) Psychoontogeny and psychophylogeny: Bernhard Rensch's (1900–1990) selectionist turn through the prism of panpsychistic identism. Theory Biosci 127:297–322
Lewin R (1987) Bones of contention: controversies in the search for human origins. Simon and Schuster, New York
Lewin R (1997) Patterns in evolution: the new molecular view. Freeman, New York
Lyell C (1830–1833) Principles of geology, vol 1–3. John Murray, London
Mayr E (1940) Speciation phenomena in birds. Am Nat 74:249–278
Mayr E (1942) Systematics and the origin of species. Columbia University Press, New York
Mayr E (1947) Ecological factors in speciation. Evolution 1:263–288
Mayr E (1949) Speciation and systematics. In: Jepsen GL, Mayr E, Simpson GG (eds) Genetics, paleontology, and evolution. Princeton University Press, Princeton, pp 281–298
Mayr E (1954) Change of genetic environment and evolution. In: Huxley JS, Hardy AC, Ford EB (eds) Evolution as a process. George Allen & Unwin, London, pp 157–180
Mayr E (1955) Integration of genotypes: synthesis. Cold Spring Harb Symp Quant Biol 20:327–333
Mayr E (1959a) Darwin and the evolutionary theory in biology. In: Evolution and anthropology: a centennial appraisal. Anthropological Society, Washington, DC, pp 1–10
Mayr E (1959b) Where are we. Cold Spring Harb Symp Quant Biol 24:1–14
Mayr E (1960) The emergence of evolutionary novelties. In: Tax E (ed) Evolution after Darwin, vol 1. University of Chicago Press, Chicago, pp 349–380
Mayr E (1961) Cause and effect in biology. Science 134:1501–1506
Mayr E (1962) Accident or design, the paradox of evolution. In: Leeper GW (ed) The evolution of living organisms. Melbourne University Press, Melbourne, pp 1–14
Mayr E (1963a) Animal species and evolution. Belknap Press, Cambridge, MA
Mayr E (1963b) The taxonomic evaluation of fossil hominids. In: Washburn SL (ed) Classification and human evolution. Aldine, Chicago, pp 332–346
Mayr E (1964a) The new systematics. In: Leone C (ed) Taxonomic biochemistry and serology. Ronald Press, New York, pp 13–32
Mayr E (1964b) From molecules to organic diversity. Fed Proc 23:1231–1235
Mayr E (1965a) Classification and phylogeny. Am Zool 5:165–174
Mayr E (1965b) Numerical phenetics and taxonomic theory. Syst Zool 14:73–97
Mayr E (1968) Theory of biological classification. Nature 220:545–548
Mayr E (1969) Principles of systematic zoology. McGraw-Hill, New York
Mayr E (1970) Populations, species, and evolution. Belknap Press, Cambridge, MA
Mayr E (1974a) The challenge of diversity. Taxon 23:3–9
Mayr E (1974b) Cladistic analysis or cladistic classification? Zeitschrift für Zoologische Systematik und Evolutionsforschung 12:94–128
Mayr E (1975) The unity of the genotype. Biol Zentralbl 94:377–388
Mayr E (1980a) Prologue: some thoughts on the history of the evolutionary synthesis. In: Mayr E, Provine WB (eds) The evolutionary synthesis: perspectives on the unification of biology. Harvard University Press, Cambridge, MA, pp 1–48

Mayr E (1980b) How I became a Darwinian. In: Mayr E, Provine WB (eds) The evolutionary synthesis: perspectives on the unification of biology. Harvard University Press, Cambridge, MA, pp 413–423
Mayr E (1981) Biological classification: toward a synthesis of opposing methodologies. Science 214:510–516
Mayr E (1982a) The growth of biological thought. Belknap Press, Cambridge
Mayr E (1982b) Speciation and macroevolution. Evolution 36:1119–1132
Mayr E (1982c) Adaptation and selection. Biologisches Zentralblatt 101:161–174
Mayr E (1983) How to carry out the Adaptationist program? Am Nat 121:324–334
Mayr E (1985a) Darwin's five theories of evolution. In: Kohn D (ed) The Darwinian heritage. Princeton University Press, Princeton, pp 755–772
Mayr E (1985b) The probability of extraterrestrial intelligent life. In: Regis E (ed) Extraterrestrials: science and alien intelligence. Cambridge University Press, Cambridge, pp 23–30
Mayr E (1988) Toward a new philosophy of biology. Harvard University Press, Cambridge, MA
Mayr E (1991) One long argument: Charles Darwin and the genesis of modern evolutionary thought. Harvard University Press, Cambridge, MA
Mayr E (1992a) Controversies in retrospect. In: Futuyma D, Antonovics J (eds) Oxford surveys in evolutionary biology, vol 8. Oxford University Press, Oxford, pp 1–34
Mayr E (1992b) Darwin's principle of divergence. J Hist Biol 25:343–359
Mayr E (1993) What was the evolutionary synthesis? Trends Ecol Evol 8:31–34
Mayr E (1996) The autonomy of biology: the position of biology among the sciences. Q Rev Biol 71:97–106
Mayr E (1997a) This is biology: the science of the living world. Belknap Press, Cambridge, MA
Mayr E (1997b) The objects of selection. Proc Natl Acad Sci U S A 94:2091–2094
Mayr E (1999) Systematics and the origin of species from the viewpoint of a zoologist, with a new introduction. Harvard University Press, Cambridge, MA
Mayr E (2001a) What evolution is. Basic Books, New York
Mayr E (2001b) The philosophical foundations of Darwinism. Proc Am Philos Soc 145(4):488–495
Mayr E (2002) Interview with Ernst Mayr. Bioessays 24:960–973
Mayr E (2004) What makes biology unique? Cambridge University Press, Cambridge
Mayr E, Ashlock PD (1991) Principles of systematic zoology, 2nd edn. McGraw-Hill, New York
Mayr E, Linsley EG, Usinger RL (1953) Methods and principles of systematic zoology. McGraw-Hill, New York
Milam EL (2010) The equally wonderful field: Ernst Mayr and organismic biology. Hist Stud Nat Sci 40:279–317
O'Malley MA (2010) Ernst Mayr, the tree of life, and philosophy of biology. Biol Philos 25:529–552
Ochoa C (2021) Inertia, trend, and momentum reconsidered: G.G. Simpson—an orthogeneticist? In: Delisle RG (ed) Natural selection: revisiting its explanatory role in evolutionary biology. Springer, Cham, pp 261–290
Ochoa C (2024) Deconstructing the extended evolutionary synthesis: do we need a new theory of evolution? In: Delisle RG, Esposito M, Ceccarelli D (eds) Unity and disunity in evolutionary biology: deconstructing Darwinism. Springer, Cham, pp 403–439
Partridge D (2018) Darwin's two theories, 1844 and 1859. J Hist Biol 51:563–592
Partridge D (2024) What if Darwin had published his 1844 essay? In: Delisle RG, Esposito M, Ceccarelli D (eds) Unity and disunity in evolutionary biology: deconstructing Darwinism. Springer, Cham, pp 475–507
Provine WB (1989) Founder effects and genetic revolutions in microevolution and speciation: an historical perspective. In: Giddings L, Kaneshiro K, Anderson W (eds) Genetics, speciation, and the founder principle. Oxford University Press, Oxford, pp 43–76
Raup DM (1991) Extinction: bad genes or bad luck? W.W. Norton, New York
Richards RA (2016) Biological classification: a philosophical introduction. Cambridge University Press, Cambridge

Salvini-Plawen L, Mayr E (1977) On the evolution of photoreceptors and eyes. Evol Biol 10:207–263
Sepkoski D (2012) Rereading the fossil record: the growth of paleobiology as an evolutionary discipline. University of Chicago Press, Chicago
Shanahan T (2004) The evolution of Darwinism: selection, adaptation, and progress in evolutionary biology. Cambridge University Press, Cambridge
Smocovitis VB (1999) The 1959 Darwin centennial celebration in America. Osiris 14:274–323
Smocovitis VB (2023) Every evolutionist their own historian: the importance of history, context, and the *extended* evolutionary Synthesis. In: Dickins TE, Dickins BJA (eds) Evolutionary biology: contemporary and historical reflections upon core theory. Springer, Cham, pp 25–54
Sober E (2011) Did Darwin write the origin backward? Philosophical essays on Darwin's theory. Prometheus, Amherst
Stamenkovic B (2024) Darwin, Archaeopteryx lithographica and the problem of intermediate species. In: Delisle RG, Esposito M, Ceccarelli D (eds) Unity and disunity in evolutionary biology: deconstructing Darwinism. Springer, Cham, pp 141–162
Stanley SM (1975) A theory of evolution above the species level. Proc Natl Acad Sci U S A 72: 646–650
Stanley SM (1979) Macroevolution: pattern and process. Johns Hopkins University Press, Baltimore
Sterelny K (2007) Dawkins vs. Gould, new edition. Icon Books, Cambridge
Tahar M (2024) Historicity, temporalities and causality: a confusion at the heart of debates on Darwinism. In: Delisle RG, Esposito M, Ceccarelli D (eds) Unity and disunity in evolutionary biology: deconstructing Darwinism. Springer, Cham, pp 551–573
Torrens E, Caso JMR, Barahona A (2024) Shaking the tree: discussing an evolutionary icon. In: Delisle RG, Esposito M, Ceccarelli D (eds) Unity and disunity in evolutionary biology: deconstructing Darwinism. Springer, Cham, pp 575–599
Turner D (2011) Paleontology: a philosophical introduction. Cambridge University Press, Cambridge
Van der Meer J (2021) The concept of natural selection in Theodosius Dobzhansky. In: Delisle RG (ed) Natural selection: revisiting its explanatory role in evolutionary biology. Springer, Cham, pp 291–326
Wagner M (2024) Typology/population distinction and its role in the marginalization of 19th-Century non-Darwinian theories in modern historiography. In: Delisle RG, Esposito M, Ceccarelli D (eds) Unity and disunity in evolutionary biology: deconstructing Darwinism. Springer, Cham, pp 277–300
Whewell W (1832) [Review of volume 2 of Lyell's] principles of geology. Q Rev 47:103–132
Whewell W (1837) History of the inductive sciences, vol 3. John Parker, London
Winsor MP (2009) Taxonomy was the foundation of Darwin's evolution. Taxon 58:43–49

Is Darwinism a Metaphysical Research Program?: Analysis and Discussion of Karl Popper's Position

Joël Dolbeault

Abstract

Whether defended or criticized, Darwinism is generally considered a scientific theory. However, Popper contests this point. According to him, Darwinism (defined as the explanation of evolution by natural selection) is less a scientific theory than a metaphysical research program: a theoretical framework that has a heuristic function for science, but that does not fulfill the necessary condition for any science, namely, testability. In terms of deconstructing Darwinism, Popper's position is therefore interesting. It raises the question of the scientificity of Darwinism. More generally, it raises the question of the scientificity of the debate on the mechanisms of evolution: what type(s) of argument(s) does this debate bring into play? Can this debate be reduced to an opposition between a scientifically founded theory (Darwinism) and other theories of evolution that are purely philosophical, or even ideological? First, I analyze Popper's position. According to him, Darwinism is rationally defensible, but difficult to test because it is not capable of making precise and sufficiently specific predictions. Darwinism is therefore more metaphysical than scientific, even if it has a heuristic function for science. Moreover, Popper proposes an improvement of Darwinism by asserting that it must give a more important place to the activity of organisms in evolution. In a second step, I discuss Popper's position, taking into account the criticisms that this position has generated. As regards the non-testability of Darwinism, it appears that the tests proposed in the literature do not resolve the problems pointed out by Popper. On the metaphysical character of Darwinism, I discuss Popper's assertion on two levels: on the possibility of a clear demarcation between science and nonscience, and on the value of the demarcation criterion proposed by Popper. Finally, on the improvement of Darwinism, I attempt to underline the originality of the Popperian proposal.

J. Dolbeault (✉)
Department of Philosophy, University of Lille, Lille, France

R. G. Delisle et al. (eds.), *Unity and Disunity in Evolutionary Biology*,
https://doi.org/10.1007/978-3-031-42629-2_9

Keywords

Darwinism · Popper · Demarcation problem · Testability · Falsificationism · Baldwin effect · Teleology

1 Introduction

In 1974, in his "Intellectual Autobiography," Popper asserts that Darwinism is not a scientific theory, but a metaphysical research program for science (Popper 1974a: 133–143). The aim of this chapter is to analyze and discuss this position. More precisely, it is to answer the following two questions: by claiming that Darwinism is less scientific than metaphysical, what exactly does Popper mean? And considering the discussion that this position has generated in the literature, what should we think of it today?

Popper's position has been commented on by many philosophers of biology. But the debate does not seem to be settled. Proof of this is that this position has never stopped being discussed since the 1970s (Ruse 1977, Watkins 1995, Settle 1996, Stamos 1996, Hull 1999, Stamos 2007, Vecchi and Baravalle 2015, Bradie 2016). Another element is that there is no real consensus on this position. Many commentators have criticized Popper, but others have instead defended his position (Platnick and Rosen 1987, Settle 1996, Vecchi and Baravalle 2015). A reexamination of the issue is thus warranted.

In this new examination, I will insist on the most misunderstood points of Popper's position. Thus, we will see that, for him, there are degrees of testability in science and that Darwinism is only *hardly testable*. We will also see that, for Popper, science and metaphysics are always linked and that the notion of metaphysical research program is not at all pejorative. Finally, we shall see that, contrary to a widespread idea, Popper never changed his mind about the metaphysical, i.e., speculative, character of Darwinism.

Regarding the deconstruction of Darwinism, this new examination of Popper's position allows us to address three questions. The first question concerns *the nature of Darwinism*: does Darwinism satisfy the criterion(s) that would make it a scientific theory? Otherwise, is Darwinism just a philosophical theory of evolution[1] serving as a research framework for science? The second question relates more generally to *the nature of evolutionary biology*: as raised by Popper, doesn't the problem of the scientificity of Darwinism concern the whole of evolutionary biology, at least if we define the latter as the study of the mechanisms of evolution (and not just as the historical study of evolution)? In other words, doesn't Darwinism's difficulty in making testable predictions apply to any theory of evolutionary mechanisms? And if so, should we continue to think that the study of evolutionary mechanisms is scientific, in the sense that genetics, molecular biology, developmental biology,

[1] In the sense of macroevolution

etc., are scientific? In short, in terms of scientificity, isn't the study of evolutionary mechanisms a special case in biology? Finally, the third question concerns *the history of evolutionary biology*. Some thinkers describe this history as the progressive development of a scientific theory of evolution in struggle against nonscientific views, in the sense of approaches essentially motivated by ideology (see in particular Mayr 2002: chap. 4; 2004: 39–43). However, if Darwinism fails to meet the criterion (s) that would make it a scientific theory, isn't this description partly wrong? In other words, shouldn't we rather see the history of evolutionary biology as a place where different philosophical conceptions of evolution oppose each other, including Darwinism, motivated by a concern to explain evolution in a mechanistic way? From Popper's point of view, such an opposition could have the form of a rational debate, linked to facts, even if this debate does not concern scientific theories strictly speaking.

For the sake of clarity, this chapter will be divided into two distinct parts. The first part will be essentially devoted to the analysis of the position defended by Popper. The second part will be devoted to the discussion of this position. In the chapter's conclusion, I will attempt to draw the consequences of these analyses for the deconstruction of Darwinism.

2 Analysis of Popper's Position

The idea that Darwinism is an MRP (metaphysical research program) for science is developed in one writing (Popper 1974a). But Popper addresses the question of Darwinism in other writings, before and after 1974. Here are the main ones:

- *The Poverty of Historicism*, chap. IV, section 27 (Popper 1957)
- *Objective Knowledge*, chap. VI and VII (Popper 1972)
- "Natural Selection and the Emergence of Mind," in *Dialectica* (Popper 1978)
- "Evolution," a letter to *New Scientist* (Popper 1980)
- "The Place of Mind in Nature," in *Mind in Nature* (Popper 1982a)
- "A New Interpretation of Darwinism," Medawar lecture 1986 (Popper 2014)

When Popper asserts that Darwinism is an MRP, he is speaking about modern Darwinism, i.e., the Modern Synthesis (1974a: 135). Moreover, he is speaking only of the theory of natural selection (1974a: 120, 136–137), that is, of the Darwinian theory of the mechanism of evolution. In other words, Popper does not question the scientificity of all the ideas contained in what is commonly called "Darwinism," in particular the idea that living species are the result of evolution. This is clear in the 1974 writing and repeated in the years that follow (1978: 344; 1980: 611; 2014: 117–118). Thus, in his letter to the *New Scientist*, Popper explains that he in no way denies that the "historical sciences" have a scientific character, notably paleontology and the science of evolutionary history. For him, indeed, the latter satisfy the criterion of testability that he defends (Popper 1980).

In summary, with regard to Darwinism in the sense of *the theory of natural selection*, Popper defends the following three theses:

1. Darwinism is rationally defensible, but difficult to test.
2. Therefore, it is not a scientific theory, but an MRP for science.
3. Moreover, as an MRP, Darwinism needs to be improved.

In the following sections, I analyze each of these theses in turn.

2.1 Darwinism Is Rationally Defensible, but Hardly Testable

In most of his writings on the subject, Popper emphasizes the interest of Darwinism in understanding evolution. In his view, the Darwinian approach to evolution is clearly more interesting than all other approaches: theistic (Popper 1972: 267; 1974a: 137), animist and vitalist (Popper 1972: 270; 1974a: 143), or Lamarckian (Popper 1972: 268; 1974a: 143). Certainly, these non-Darwinian approaches assert different things. To explain evolution, the theistic approach invokes a purposeful cause external to the living; the animist approach, a purposeful cause internal to the living; the vitalist approach, a special cause internal to the living; and the Lamarckian approach, a physiological cause internal to the living (the principle of use and nonuse, and the heredity of acquired characters). However, all these non-Darwinian approaches have in common that they formulate an ad hoc hypothesis to explain evolution. This is their main weakness. In contrast, Darwinism proposes to explain evolution without invoking an ad hoc force or mechanism. This is its main strength (Popper 1972: 267–270; 1974a: 137).

In other words, for Popper, the main value of Darwinism is to propose a reductionist explanation of evolution, because from a methodological point of view, reductionism has a great interest:

> In the course of this discussion, I will defend three theses. First, I will suggest that scientists have to be reductionists in the sense that nothing is as great a success in science as a successful reduction… A successful reduction is, perhaps, the most successful form conceivable of all scientific explanations, since it achieves what Meyerson (1908, 1930) stressed: an identification of the unknown with the known. (Popper 1974b: 259–260)

From this point of view, for Popper, Darwinism is rationally defensible. However, in his view, this is not enough to make it a scientific theory. For Popper, indeed, the domain of rationality is broader than that of science because the latter implies a precise condition: *testability*.

Let us first recall some fundamental elements concerning what Popper calls testability.

First, for Popper, the testability of a theory lies essentially in its falsifiability, or refutability. This means that the theory must be formulated in such a way as to define the empirical conditions that could show that it is false. Without this, it is impossible to use experience to criticize the theory, which is to say that it is impossible to estimate the empirical value of the theory.

Secondly, for Popper, testability is *a necessary condition* for the scientificity of a discourse. This is true for theoretical sciences (which formulate universal statements) but also for historical sciences (which formulate singular statements) (Popper 1957: 143–144; 1972: 354). The question of whether, for Popper, testability is also a sufficient condition for scientificity is discussed. Popper is unclear on the issue (Hansson 2021: 4.2). We will have to talk about it in the discussion part.

Thirdly, for Popper, the testability of a discourse admits of degrees. Popper underlines this point as early as 1959, in *The Logic of Scientific Discovery* (1959/1992: chap. VI), and he comes back to it several times, as shown by this text from 1963:

> There are, moreover (as I found later), *degrees of testability*: some theories expose themselves to possible refutations more boldly than others. ... This indicates that the criterion of demarcation cannot be an absolutely sharp one but will itself have degrees. There will be well-testable theories, hardly testable theories, and non-testable theories. Those which are non-testable are of no interest to empirical scientists. They may be described as metaphysical. (Popper 1963: 346)

In particular, Popper emphasizes the fact that, all other things being equal, the degree of testability of a theory depends largely on its capacity to make precise numerical predictions (Popper 1959/1992: 108–110; 1963: 346).

Let us now turn to the question of the testability of Darwinism. On this point, Popper sometimes claims that Darwinism is not testable (1974a: 134, 136). But in other passages, he only asserts that Darwinism is *hardly* testable (1974a: 137; 1978: 344; 1994: 90). In fact, the arguments given by Popper allow us to understand that, for him, Darwinism is only *hardly* testable. This means that its testability is not zero, but very weak.

First of all, it is obvious that Darwinism cannot make precise predictions about the formation of a new species, in a given place and at a given time. Its confrontation with experience must therefore depend on more global predictions. Popper mentions two possibilities.

The first concerns evolutionary divergences. According to Popper, Darwinism suggests that life must evolve in a *strongly* divergent manner, at least in a relatively large and varied environment. But this suggestion is not a true prediction in the sense that Darwinism remains logically compatible with a *weakly* divergent evolution:

> For assume that we find life on Mars consisting of exactly three species of bacteria with a genetic outfit similar to that of terrestrial species. Is Darwinism refuted? By no means. We shall say that these three species were the only forms among the many mutants which were sufficiently well adjusted to survive. And we shall say the same if there is only one species (or none). Thus Darwinism does not really *predict* the evolution of variety. It therefore cannot really *explain* it. At best, it can predict the evolution of variety under "favourable conditions." But it is hardly possible to describe in general terms what favourable conditions are—except that, in their presence, a variety of forms will emerge. (Popper 1974a: 136)

Since Darwinism is logically compatible with extremely different degrees of divergence, it cannot be said to make any predictions on this subject. Popper thus writes that the theory "almost predicts," or "suggests," a great variety of forms of life (1974a: 136, 137). This suggestion can be presented as a test for the theory. But this test is not severe enough, in the sense that a negative result cannot be enough to refute the theory.

The second possibility mentioned by Popper concerns the speed of evolution. In this case, Popper claims that Darwinism makes a true prediction. But he explains that this prediction cannot stand as a true test:

> Gradualness is thus, from a logical point of view, the central prediction of the theory. (It seems to me that it is its only prediction.) Moreover, as long as changes in the genetic base of the living forms are gradual, they are—at least "in principle"—explained by the theory; for the theory does predict the occurrence of small changes, each due to mutation. However, "explanation in principle" is something very different from the type of explanation which we demand in physics. While we can explain a particular eclipse by predicting it, we cannot predict or explain any particular evolutionary change (except perhaps certain changes in the gene population *within* one species). (Popper 1974a: 137–138)

The prediction of Darwinism is that evolution is gradual, and observations in genetics are consistent with this prediction. However, this prediction is very general, in the sense that Darwinism does not predict the gradual formation of a particular species. Therefore, the gradual formation of a particular species is just compatible with Darwinism. This compatibility is what Popper calls "explanation in principle." The problem is that other theories than Darwinism can predict that evolution is gradual: certain forms of animism, vitalism or Lamarckism (knowing that the notion of gradualness, as such, is not very precise, and therefore easily adaptable to many different theories). Therefore, the gradual formation of a particular species is also compatible with various non-Darwinian theories. In other words, these theories can also explain "in principle" the gradual formation of any species. Conclusion, the prediction of gradualness is not a sufficiently discriminating test for Darwinism.[2]

Without talking about prediction, we can think that the capacity of Darwinism to explain the survival of a species by its adaptation to an environment (interpreted as an effect of natural selection) plays as a real test in favor of the theory. According to Popper, this is not the case. Indeed, for this to be a true test, it would be necessary to determine adaptation *independently* of survival. But this appears to be very difficult, so Darwinism is tempted to define adaptation *by* survival. This leads to the following tautology: the survival of a species in a given environment can be explained by the adaptation of this species to this environment, defined as the survival of this species in this environment. This is why Popper writes that, on the question of adaptation,

[2]Some may think that other tests can falsify animism, vitalism, and Lamarckism, so that these theories are off the table, and that a test of gradualness is sufficiently severe for Darwinism. But the situation is not so simple because these theories are very general, which means that they can take various forms (incorporating various empirical data). Extending Popper's analysis, we can say that these theories are also difficult to test.

Darwinism is "almost tautological" (1972: 241–242; 1974a: 135, 137). As some commentators have noted, the formula is problematic: a theory is either tautological or it is not (Hull 1999: 486). However, it is easy to understand that, by this formula, Popper means this: Darwinism is not fundamentally tautological; but for lack of a true test on adaptation, it is tempted to function as a tautology.

To conclude this analysis, let us briefly point out what Popper adds in 1978 about the testability of Darwinism. Some commentators suggest that Popper changed his mind on this topic in favor of Darwinism (Bradie 2016: 152–153, 164; Hansson: 4.2). A detailed analysis shows that this is not the case.

In the 1978 article, Popper argues that Darwinism (understood as the theory of natural selection) can be defined in a restricted way, *without including sexual selection*. From this point of view, according to him, Darwinism is empirically testable but also empirically refuted: refuted precisely by the known cases of sexual selection that force one to think that, at best, Darwinism only partially explains evolution (1978: 343–346). On the one hand, therefore, it is true that in 1978 Popper proposes something new about the testability of Darwinism, which he himself emphasizes (1978: 345). On the other hand, this change is very relative because it consists essentially in playing on the definition of Darwinism—with a consequence on its testability. When it comes to Darwinism in the broad sense (including sexual selection), Popper's position remains unchanged: this theory is hardly testable. Moreover, Popper's new proposal is not favorable to Darwinism since it places the latter before the following dilemma: to be based on a broad definition of natural selection (including sexual selection) and to be an untestable theory, or to be based on a narrow definition of natural selection (excluding sexual selection) and to be a testable but refuted theory (1978: 346).

Popper knows that his narrow definition of Darwinism is questionable (1978: 346; 2014: 127–128). Moreover, this definition is not accepted by biologists today, who include sexual selection in natural selection. Therefore, for the remainder of the chapter, the most interesting point of view is to consider the "broad definition" of Darwinism. And from this point of view, Popper's position is that Darwinism is hardly testable.

2.2 Darwinism Is a Metaphysical Research Program for Science

For Popper, the idea that Darwinism is an MRP (1974a: 120, 133–134, 137) is the consequence of three things. The first is that Darwinism is difficult to test. The second is that it is inspired by some metaphysical conception of reality. The third is that it serves as a guide for scientific research. I will analyze these three points in turn.

The first point refers to the previous part. However, something needs to be clarified. As we have seen, Popper thinks that Darwinism is *hardly* testable. But in 1963, concerning the degrees of testability, he asserts that only *non*-testable theories should be considered metaphysical (see the text quoted above). Why assert then that Darwinism is metaphysical? The most likely explanation is that there is a certain wavering of Popper's position on the demarcation between science and nonscience. On the one hand, in 1963, Popper affirms that there are degrees of testability, but without affirming that there are degrees of scientificity. And he adds that a hardly testable theory is still scientific. On the other hand, if there are degrees of testability, it is logical to think that there are degrees of scientificity, and then to underline the rather scientific, or rather metaphysical, character of a theory. In 1974, Popper seems to adopt this point of view: because he considers that Darwinism is very difficult to test, he emphasizes its metaphysical character.[3]

In Sect. 3.2, we will return to the question of the criterion of demarcation between science and nonscience. We will see that Popper's position has been discussed by many philosophers.

The second point concerns the inspiration of Darwinism, bearing in mind that we are not talking here about Darwin's Darwinism, but about modern Darwinism, developed in the first part of the twentieth century.[4] For Popper, an MRP is inspired by a metaphysical conception of reality, defined as an attempt to explain the fundamental structure of reality (Popper 1982b: 161). A metaphysical conception answers questions like: Is reality only made of ideas (idealism) or not (realism)? Of determined processes (determinism) or not (indeterminism)? etc. Therefore, if Darwinism is an MRP, it must be inspired by a certain conception of the fundamental structure of reality. Popper does not explicitly specify this conception. But given its reductionist approach to evolution, it is easy to understand that Darwinism is inspired by *mechanism*. Metaphysical mechanism considers that all the processes in nature, including those which are apparently intentional, are in fact mechanical processes (i.e., obeying blind laws). And indeed, Darwinism seeks to explain evolution without any recourse to intentions, external or internal to the living.

Obviously, this does not mean that Darwinism, as an MRP, is dictated by professional philosophers. For Popper, the scientific world itself determines its own MRPs. Rather, it means that, since the beginning of the twentieth century, evolutionary biology has presupposed a certain metaphysical conception of reality—a conception considered taken for granted, and therefore rarely discussed.

The third point concerns the usefulness of Darwinism for science. Regarding the usefulness of MRPs in general, Popper writes:

> I call them "research programmes" because they incorporate, together with a view of what the most pressing problems are, a general idea of what a satisfactory solution of these problems would look like. (Popper 1982b: 161. See also 1974a: 120)

An MRP serves as a guide for a science. It helps determine which problems are relevant to a science and what kinds of solutions are acceptable. Thus, Popper believes that Darwinism serves as a guide for biology.

To illustrate this idea, Popper uses the following example:

[3]Note that this amounts to adopting a position similar to that of Lakatos on the degrees of scientificity (Lakatos 1999: 20).

[4]For the place of the theory of natural selection in Darwin, see in particular Delisle (2021).

> In trying to explain experiments with bacteria which become adapted to, say, penicillin, it is quite clear that we are greatly helped by the theory of natural selection. Although it is metaphysical, it sheds much light upon very concrete and very practical researches. It allows us to study adaptation to a new environment (such as a penicillin-infested environment) in a rational way: it suggests the existence of a mechanism of adaptation, and it allows us even to study in detail the mechanism at work. And it is the only theory so far which does all that. (Popper 1974a: 137)

The experiment shows that some bacteria adapt to penicillin. The scientist wants to understand why. Darwinism suggests an explanatory scheme: a random genetic variation in these bacteria, with possible consequences on the ability to live in a penicillin-infested environment, and a natural selection mechanism of the individuals best adapted to this environment. In this sense, it directs the research toward the discovery of precise empirical elements (corresponding to this explanatory scheme).

According to Popper, the possible success of this kind of research reinforces the interest of Darwinism as an MRP, but it does not constitute a real test for the theory. First, because this success concerns a limited number of cases: it says nothing about evolution as a whole, especially macroevolution. Secondly, because this success does not constitute a sufficiently discriminative test: it does not allow us to exclude a non-Darwinian explanation of adaptation.

For Popper, Darwinism thus fulfills the three conditions that make it an MRP: weak testability, metaphysical inspiration, and the role of guide for scientific research. But before ending this section, let us emphasize that Popper has a positive conception of metaphysics, as something rational and useful to science. Concerning the metaphysical theories of matter,[5] here is what he writes:

> Such research programmes are, generally speaking, indispensable for science, although their character is that of metaphysical or speculative physics rather than of scientific physics. Originally they were all metaphysical, in nearly every sense of the word (although some of them became scientific in time); they were vast generalizations, based upon various intuitive ideas, most of which now strike us as mistaken. They were unifying pictures of the world—the real world. They were highly speculative; and they were, originally, non-testable. Indeed they may all be said to have been more of the nature of myths, or of dreams, than of science. But they helped to give science its problems, its purposes, and its inspiration. (Popper 1982b: 165)

This point must be emphasized because certain criticisms addressed to Popper about Darwinism can no doubt be explained by the fact that many philosophers and scientists have a very negative conception of metaphysics, as arbitrary discourse,

[5]Popper mentions in particular the Parmenidian conception of matter, opposed to the existence of the vacuum; ancient atomism (which admits the existence of the vacuum); the Pythagorean and Platonic idea that the physical world is geometrizable; Aristotle's potentialism; the theory of the clock-world in the seventeenth century (Hobbes, Descartes, Boyle); the dynamical conception of matter (Newton, Kant, Boscovich); the idea that matter could be reduced to fields (Faraday, Schrödinger, Einstein) (Popper 1982b: 162–163).

even devoid of meaning, and in any case as useless for science. This very negative conception is largely due to philosophy itself, in particular to the positivist critique of metaphysics—in Auguste Comte first, then in logical positivism. In several writings, Popper sought to show that this positivist view of metaphysics is erroneous (Popper 1963: chap. VIII and XI; 1983: chap. II and III).[6] According to him, metaphysics is not an arbitrary discourse: it aims at *rationally interpreting experience*, even if it is not able to predict it. Nor is metaphysics a discourse devoid of meaning: despite all their efforts, logical positivists have never managed to demonstrate this. Finally, metaphysics is not totally separated from science: it suggests ideas to be developed in science and empirical research to be undertaken.

To avoid the negative ideas associated with the word "metaphysics," one could also say that, for Popper, Darwinism is a *philosophical research program* for science. This would also express quite well the view he defends.

As Popper points out in this text, let us add that a metaphysical (or philosophical) idea can eventually become testable, and therefore scientific.[7] In the case of the theory of matter, think, for example, of atomism, that is to say the idea that matter is fundamentally made up of clearly distinct particles. This idea was initially metaphysical, in antiquity and in the seventeenth century, before becoming scientific. By integrating science, it has also become more complex[8] than what the philosophers had first conceived.

Thus, in Popper's mind, the fact that Darwinism is *today* hardly testable does not imply that it will always be the case. Human inventiveness and/or the accumulation of new empirical data can always change the situation.

2.3 Darwinism Is a Research Program that Needs to Be Improved

As I explained at the beginning of Sect. 2.1, Popper defends Darwinism against all other approaches to evolution: theistic, animist, vitalist, and Lamarckian. However, according to him, Darwinism needs to be improved. Indeed, it does not explain some things well:

- *Orthogenetic trends*, that is, the "sequences of evolutionary changes in the same 'direction' (nonrandom walks)" (Popper 1974a: 138). Popper does not give an explanation. But one can assume that he is referring to the classical problem of the appearance of a complex organ in multiple separate steps.

[6] According to Popper, positivism's error about metaphysics is linked to its error about science. In both cases, positivism fails to see that the demarcation between science and nonscience (of which metaphysics is only a part) is based on falsifiability, not on verifiability.

[7] For this point, see also Popper (1963: 347; 1994: 88–89).

[8] And partially refuted, because of the importance taken by the notion of field in physics, which corresponds rather to the idea of a fundamental continuity of the material world.

- *Evolution toward "higher" forms of life*, such as the human species, for example, as opposed to bacteria (Popper 1974a: 141). According to Popper, from the point of view of Darwinism, the appearance of "higher" forms of life is less likely than the improvement of the reproduction rate of "lower" forms of life.[9]
- The *separation of species during speciation*. According to Popper, Darwinism must necessarily appeal to the geographical separation of two populations to explain speciation, whereas another type of separation is possible (Popper 1974a: 141).

Note, however, that for Popper, these criticisms of Darwinism are not refutations. Otherwise, that would imply that Darwinism is testable. For him, these criticisms point out weaknesses of Darwinism, but without refuting it.

Similarly, Popper does not claim to make Darwinism more testable. His proposal to improve Darwinism only aims at eliminating certain weaknesses of the theory. Contrary to what some commentators say (Bradie 2016: 153, 158), there is no ambiguity on this point, and Popper emphasizes it again in 1982 (1982a: 44).

Popper's proposal can be summarized as follows: to Darwinism as it exists, we must add the idea that *the preferences of living organisms* have a decisive influence on the natural selection that they undergo, and thus on their evolution. But before going into detail about this idea, let us clarify a few things right away. The first is that Popper conceives this idea as an improvement (1974a: 120) or an enrichment (1974a: 138) of Darwinism, not as something opposed to it. The second is that Popper does not claim that, from a fundamental point of view, his proposal is totally new. In all the writings in which he presents it, he explicitly refers to certain thinkers who had already made a similar proposal, notably Baldwin[10] (1972: 268; 1982a: 42; 1992: 209; 2014: 119), Schrödinger (1972: 268; 1974a: 138), Waddington (1972: 268; 1974a: 138), and Hardy[11] (1982a: 42; 1992: 209). The third is that, according to him, Darwinism has already more or less integrated the idea that the activity of living organisms plays a determining role in evolution. The problem is that some Darwinians minimize this role:

> My problem is exactly the same as that of my forerunners, such as Baldwin, who felt that the activities, the idiosyncrasies, and the preferences of individual organisms have played a far more important role in the history of evolution than Darwinists as a rule have admitted. My problem is that some Darwinists are inclined to attribute creative powers to what they call natural selection, forgetting for the moment that "natural selection" is no more than a highly suggestive and very useful metaphor. (Popper 2014: 119)

[9] Popper suggests this point in 1972 (Popper 1972: 271).

[10] James Baldwin (1861–1934) was an American psychologist. For his contribution to evolutionary thought, see in particular Baldwin (1896, 1897).

[11] Alister Hardy (1896–1985) was an English biologist. For his contribution to evolutionary thought, see in particular Hardy (1965).

Popper develops his proposal in two writings: *Objective Knowledge* (1972) and "Intellectual Autobiography" (1974a). As the 1974 writing is the more accomplished of the two, it is the one that we will analyze.

Popper first postulates that there are different classes of genes: p-genes that control preferences (or goals), s-genes that control skills, and a-genes that control anatomy. Moreover, he postulates that, because of their genes, some organisms may have a certain variability of behavior. On this basis, he proposes a four-step mechanism:

1. Due to certain changes in the environment, some individuals in a given population may adopt *new preferences*. This is an attempt to adapt to the environment.
2. If this attempt is successful, then natural selection will tend to favor individuals carrying the p-genes that induce the new preferences.
3. Following this, natural selection will tend to favor individuals carrying the s-genes that improve the skills related to the new preferences (which will reinforce the selective value of p-genes).
4. Following this, natural selection will tend to favor individuals carrying the a-genes that improve the organs useful for the skills related to the new preferences (which will reinforce the selective value of p-genes and s-genes) (Popper 1974a: 138–139).

In summary, according to this mechanism, the change of preferences in a population determines the genetic change of that population (of p-genes, s-genes and a-genes), via natural selection. Even more strikingly, this is equivalent to saying that *the aims of organisms* determine their biological evolution, via natural selection.

Of course, this mechanism involves feedbacks. The change in p-genes also determines the change in preferences, in the sense that it reinforces the adoption of the new preferences within the population. Similarly, the change in s-genes also determines the change in p-genes, in the sense that it reinforces the selective value of the latter. However, the impulse comes from the behavior of organisms.

Let us now see how Popper uses this mechanism to answer the different points raised against Darwinism as MRP.

On *orthogenetic trends*: according to Popper, the proposed mechanism helps to understand how natural selection can act continuously in a certain "direction." The idea is that when a population adopts new preferences, it creates a kind of need (in instinct, skill, and anatomy) that makes it possible for natural selection to act for a long time in the same direction (Popper 1974a: 139).

On *the evolution toward "higher" forms of life*: as this mechanism affirms that evolution is determined by the preferences of organisms, it helps to understand that evolution can go toward "higher" forms of life because the preferences of organisms are not necessarily to maximize the rate of reproduction. They can be, for example, to build a shelter, to find a place for this shelter, to exploit this or that food source, etc. (Popper 1974a: 141–142).

On *the separation of species during speciation*: as this mechanism states that the change of preferences in a population precedes and determines the genetic change of this population, it makes it possible to understand that the separation between species can concern the preferences, before concerning the geography (Popper 1974a: 141).

In an article published in 1982 (Popper 1982a), and in a lecture delivered in 1986 (Popper 2014), Popper revisits his proposal for improving Darwinism. Several points should be noted.

First, to make his proposal better understood, Popper introduces the distinction between "passive Darwinism" and "active Darwinism" (Popper 1982a: 37–42; 2014: 119–120). For the first, evolution can be explained by two things: the variability of the genome and the physical environment (including other organisms). In contrast, for the second, evolution depends on a third thing: the activity of living organisms. For Popper, this activity plays a fundamental role: "I shall claim that the *only* creative element in evolution is the activity of the living organisms" (Popper 2014: 119).

Second, Popper presents his proposal without entering into a discussion of the different types of genes (p-genes, s-genes, and a-genes) and the order of modification of these genes. One can therefore assume that this element is not essential. What is essential is the idea that the adoption of new preferences by organisms directs the selective pressure they undergo (Popper 1982a: 40–42; 2014: 119).

Finally, Popper emphasizes *the teleological dimension* of the activity of living beings. In 1974, he hesitates to affirm that living beings act according to an aim, preferring to affirm that they seek to solve problems (1974a: 142; 1974b: 272). But the distinction seems thin, if not null. In 1982, and again in 1986, he frankly uses the notion of aim[12]:

> If we assume that such active and explanatory and mindlike (and presumably partly conscious) behavior is the result of the very early evolution brought about by passively Darwinian forces, then we can also assume that, once this stage has been reached, evolution is no longer the passive result of heredity and a hostile environment. Mindlike properties of the organism will begin to play an increasingly important part in evolution: aims, such as preference for certain kinds of mates, preferences for certain locations for breeding, of for certain types of food; but also curiosity and explanatory behavior and, by contrast, conservative behavior; or change from one kind of behavior to another. (Popper 1982a: 40)

> Activity is a movement with an aim. And I think that we have to attribute some sort of activity to even the lowest organism... There cannot be adaptation without any aim. They cannot be knowledge without any aim. So, an aim is there. And we have to attribute all this to the lowest organisms. It is part of the distinction between living organisms and not-living crystals, let us say, liquid crystals. Crystals have many similarities with organisms, but they do not show activity in the sense in which organisms do. They do not show trial and error movements, which we find in the lowest organisms that we know. (Popper 2014: 125)

> The strange thing is that teleology enters the world with adaptation. Organisms are problem-solvers. Organisms seek better conditions. All of these are thoroughly teleological terms. (Popper 2014: 124)

[12] Obviously, this aim is not necessarily elaborated, nor conscious (Popper 1974a: 143; 1974b: 272; 2014: 125).

Table 1 Metaphysics of life and MRP for biology. Comparison between Popper and the Darwinists that he criticizes

	Some Darwinists	Popper
Metaphysical position on the living	*Reducibility of the living to the inert*: The apparently teleological activities of the living (including conscious activities) are in fact physical mechanisms	*Irreducibility of the living to the inert*: Living beings act in a teleological way. They have aims (conscious or not, elaborated or not)
Consequence	↓ Underestimation of the innovative capacity of the living	↓ Idea that the living has a capacity for innovation
MRP for biology	↓ *Passive Darwinism*: Evolution is explained by two things: the variability of the genome and natural selection	↓ *Active Darwinism*: Evolution is explained by three things: the activity of organisms, the variability of the genome, and natural selection

This last point is important, because Popper makes a logical link between the modified Darwinism he proposes and his conception of the irreducibility of the living to the inert (1974a: 141–143; 1982a: 40–41; 2014: 123–126). This link can be summarized as follows: thinking that the living is irreducible to the inert, because of its teleological dimension, leads to thinking that the living is a source of behavioral innovations, and this leads to thinking that the activity of the living plays a central role in evolution.

Table 1 highlights this logical connection between metaphysics of life and MRP for biology, comparing Popper to "passive Darwinism."

To conclude this part 2 of the chapter, let us remember that Popper has two reasons for wanting to modify Darwinism. The first is that he considers that Darwinism does not explain certain things well. The second is that he thinks that Darwinism is inspired by an erroneous metaphysical conception of life.

Popper then proposed what he considered an improved Darwinism: "active Darwinism." On the one hand, for him, this new MRP remains Darwinian in the sense that natural selection still plays a central role in evolution. On the other hand, this new MRP breaks with the mechanistic conception of life. In part 3, we shall see that there may be a tension between these two points.

3 Discussion of Popper's Position

The aim of this part is to discuss Popper's position on Darwinism, taking into account the criticisms that have been addressed to him. For maximum clarity, I will follow the order of part 2, discussing successively Popper's three assertions:

first, on the testability of Darwinism; then on the criterion of demarcation between science and nonscience, notably on its application to biology; and finally, on his proposal to improve Darwinism.

3.1 Discussion on the Testability of Darwinism

On this issue, several authors are opposed to Popper: Ruse (1977, 1981), Watkins (1995), Stamos (1996), Hull (1999), and Bradie (2016). Other authors, fewer in number, defend Popper's position: Platnick and Rosen (1987) and Vecchi and Baravalle (2015). But let us add that those who defend Popper's position refer to some philosophers of biology whose analysis on the testability of Darwinism echoes Popper's, notably Brady (1979) and Brandon (1980).

A first criticism, developed by some commentators, says that Popper contradicts himself by affirming, on the one hand, that Darwinism is not testable, and on the other hand, that Darwinism predicts gradualness (Ruse 1977: 649–650; Watkins 1995: 194). However, as I pointed out in 2.1, for Popper this prediction is not sufficiently discriminative to be a good test: Darwinism implies gradualness, but the reverse is not true, and many non-Darwinian theories also predict gradualness—Lamarckism, most of the theories of orthogenesis (W. Haacke, H. F. Osborn, etc.),[13] or the Bergsonian theory of the *élan vital*. Another important element is that gradualness admits of degrees. Therefore, if the prediction of gradualness is not *quantified* in some way, it is impossible to say whether or not it matches the facts, including in the fossil record.

A second and more often developed criticism is to reject Popper's central argument, claiming that Darwinism can make predictions on evolution and that these predictions can serve as tests. We will discuss some of these proposals in detail. But I would like to point out at the outset that these proposals still have two weaknesses. First, they concern predictions that are not precise enough, which makes it difficult to know whether the prediction is true in a particular situation. Second, they concern predictions that are not specific to Darwinism and that do not specifically test that theory.

The proposal made by Ruse in 1977 deserves a detailed examination, because Ruse discusses at length an example taken by Popper, while modifying his conclusion. In the previous section, we saw that, according to Popper, Darwinism does not *necessarily* predict strong evolutionary divergence. To justify this idea, he argues that the discovery of only three species of bacteria on Mars (or any other planet) would not contradict the Darwinian theory, because this theory could assert, for example, that the conditions for greater divergence were not present.

[13] For the various theories of orthogenesis in the nineteenth and twentieth centuries, see in particular Levit and Olsson (2006). For Osborn's theory in particular, see Ceccarelli (2021). Osborn's non-Darwinian theory includes the idea of a certain gradualness of evolution, and a certain adaptationism (Ceccarelli 2021: 175–177).

According to Ruse, Popper's mistake is that he does not specify certain conditions. The idea is that if certain conditions are better determined, it is possible to make a prediction that can serve as a test for Darwinism. Ruse mentions the following three conditions: different ecological niches, relatively isolated from each other; some genetic variability in existing organisms; and a sufficiently long time (Ruse 1977: 643–645). His conclusion then is as follows:

> Had one reason to believe that life on the planet was fairly old (e.g., through the fossil record or general complexity of structure), yet were one to find that absolutely no speciation at all had occurred, then I suggest that, *contra* Popper, modern evolutionists would be worried. Their theory, parts of it at least, would have been falsified. The claims that they make about speciation would seem not to hold. (Ruse 1977: 645)[14]

Ruse is right to specify certain conditions left out by Popper. However, this does not fundamentally change things, because we find the two weaknesses I mentioned. First, none of the conditions given by Ruse are quantified, so that if speciation is not observed, or only very little, it is possible to say that the conditions are not *sufficiently* fulfilled: the ecological niches might not be *sufficiently* different from each other or/and not *sufficiently* isolated (or too much) from each other; or/and the genome of the populations might not be *sufficiently* variable; or/and the time needed for speciation might not be *sufficiently* long. Another problem is that the expected result is not quantified either, in the sense that the number of expected species of bacteria, in order of magnitude, is not indicated: 10, 1000, 1 million, more? How then can we know if the prediction matches the fact? Second, this test proposed by Ruse is not discriminating enough because most non-Darwinian conceptions of evolution admit, or can admit, a certain adaptationism. For example, in Bergson's theory, evolution is conceived as being directed by a force immanent to life (the *élan vital*) toward the appearance of organisms endowed with reflexive consciousness. But this force is finite, so that evolution must adapt to given physical and historical circumstances. Evolution thus follows no plan, and its success is not guaranteed (Bergson 1998: 261, 265–267). In short, a certain teleology and a certain adaptationism are logically compatible.

The proposals made by Hull, in 1999, also deserve a detailed examination, because Hull refers directly to Darwin in trying to show that the latter already proposed various ways of testing his theory. Hull mentions the following Darwinian proposals:

> If it could be demonstrated that any complex organ existed, which could not possibly have been formed by numerous, successive, slight modifications, my theory would absolutely break down. (Darwin 1859: 189)

[14] Watkins (1995: 194) proposes a similar test.

> [Some naturalists] believe that very many structures have been created for beauty in the eyes of man, or for mere variety. This doctrine, if true, would be absolutely fatal to my theory. (Darwin 1859: 199)

> If it could be proved that any part of the structure of any one species had been formed for the exclusive good of another species, it would annihilate my theory, for such could not have been produced through natural selection. (Darwin 1859: 201)

> The abrupt manner in which whole groups of species suddenly appear in certain formations, has been urged by several paleontologists ... as a fatal objection to the belief in the transmutation of species. If numerous species, belonging to the same genera or families, have really started into life all at once, the fact would be fatal to the theory of descent with slow modification through natural selection. (Darwin 1859: 302)

> He who rejects [the imperfection of the geological record] will rightly reject my whole theory. (Darwin 1859: 342).[15]

These quotations show that Darwin was concerned with testing his theory. On this point, his sincerity is therefore obvious. But the problem is to know if what he proposes can play as a real test for his theory.

Proposal 1 amounts to the following: if the existence of a complex organ cannot be explained by the mechanism of natural selection, then the theory is refuted. But, as Popper notes, the mechanism of natural selection explains things "in principle," without making any precise predictions (Popper 1974a: 138). It is therefore difficult to see how this mechanism could not apply, "in principle," to every complex organ. Again, without introducing a quantification (of complexity, and of the time needed to produce a certain level of complexity by natural selection), how can we obtain a true test?

Proposals 2 and 3 amount to saying this: if it is proved that certain traits do not constitute a selective advantage for the individuals which possess them, then the theory is refuted. But here we find the two difficulties. First, given the multiple possible effects of a trait in a given environment, it may be difficult to know whether that trait constitutes a selective advantage or not (I will come back to this point shortly after). Second, Darwinism implies adaptationism, but the reverse is not true. Therefore, testing adaptationism is not specifically testing Darwinism.

Finally, proposals 4 and 5 concern gradualness. Here Darwin asserts that his theory would be refuted if it were shown that some species "have really started into life all at once," which is equivalent to speaking of a total absence of gradualness. But, again, without quantification, the notion of gradualness is imprecise. Moreover, many non-Darwinian theories also predict some gradualness. Therefore, this test is not discriminative enough.

Reading the responses that Ruse, Hull, and others make to Popper, one gets the impression that, for them, testing Darwinism amounts to showing its explanatory superiority over theories that deny evolution—in short, over the classical forms of

[15] All these passages are quoted by Hull (1999: 487–488).

creationism. But the problem does not end there. Historically and logically, there are several non-Darwinian conceptions of evolution, and some of these conceptions support a certain adaptationism and a certain gradualness.

In contrast, some biologists or philosophers of biology agree with Popper on the question of the testability of Darwinism: Platnick and Rosen (1987: 12–13)[16] and Vecchi and Baravalle (2015).

In their article, Platnick and Rosen refer to Brady (1979) who shows that the principle of natural selection is very difficult to falsify. This principle leads to explaining differential reproduction by fitness, but it is very difficult to establish an empirical criterion of fitness that is independent of differential reproduction. For Brady, this is because fitness depends on *an indefinite number of parameters*, and, in practice, it is impossible to take into account and quantify all these parameters (Brady 1979: 604–606). In other words, it is impossible to reduce the interactions between an organism and its environment to a determinate system:

> Neither Darwinism nor, in its present form, neo-Darwinism, contains a theoretical reduction of the organism to a determinate system, and thus neither contains a way of determining the contributions of various parameters, or even the number of parameters. In practice, of course, predictions are made on the basis of individual traits, but whenever anything goes wrong the resulting foray into ad hoc speculations invokes the notion that the summation of effects is really indeterminable—any parameter could be interfering in who knows how many ways. (Brady 1979: 615–616)

Brady then concludes that the principle of natural selection has not yet been seriously tested:

> If other parts of neo-Darwinism have been well scrutinized, natural selection has escaped any serious testing and will continue to do so until we know a great deal more about organism-environment interactions. We are so far from a science that can reduce these interactions to a determinate system that some critics have wondered whether it makes any sense to attempt to operationalize the principle at all. (Brady 1979: 616)

In the same sense, Vecchi and Baravalle (2015) refer to Brandon (1980) for whom the principle of natural selection is not an empirical principle, but an "organizing principle":

> The principle of natural selection is an organizing principle, or to put it another way, a schematic law. As a general schema, it is without empirical biological content, but it does serve to structure particular biological explanations of differential reproduction. (Brandon 1980: 432)

[16]In another article, Platnick seems to oppose Popper (Platnick and Gaffney 1978). In fact, in this article, Platnick and Gaffney try to show that Darwinism is not, in principle, untestable. But they recognize that it is hardly testable, which is in fact Popper's position.

The tone of the text is not as critical as in Popper, but the idea is not fundamentally different: the principle of natural selection has no empirical (testable) content; it is *a heuristic principle*.

In conclusion, even if Popper's position on the testability of Darwinism has often been criticized, this position retains a certain solidity, and it finds an echo among certain philosophers of biology.

3.2 Discussion on the Popperian Criterion of Demarcation Between Science and Nonscience

Popper's position on Darwinism can be discussed at another level, that of the criterion of demarcation between science and nonscience. According to Popper, this criterion of demarcation is testability, conceived as falsifiability. But by disputing this point, one could reject Popper's conclusion, that is, admit that Darwinism is hardly testable, while maintaining that it is a scientific theory.

The Popperian criterion of demarcation between science and nonscience has been much discussed in philosophy, without thinkers reaching a consensus (Hansson 2021). In this section, without being able to take into account the entire literature on the subject, I will try to briefly address the following two questions:

1. Is the problem of the demarcation between science and nonscience really relevant?
2. Is the Popperian criterion of demarcation relevant to biology?

On the first question, Larry Laudan points out that the search for a criterion of demarcation between science and nonscience presupposes that there are certain invariants of what is commonly considered as science. But this is not obvious. In fact, the opposite seems to be true:

> But we can go further than this, for we have learned enough about what passes for science in our culture to be able to say quite confidently that it is not all cut from the same epistemic cloth. Some scientific theories are well tested; some are not. Some branches of science are presently showing high rates of growth; others are not. Some scientific theories have made a host of successful predictions of surprising phenomena; some have made few if any such predictions. Some scientific hypotheses are *ad hoc;* others are not. Some have achieved a "consilience of inductions"; others have not. (Similar remarks could be made about several non-scientific theories and disciplines.) *The evident epistemic heterogeneity of the activities and beliefs customarily regarded as scientific should alert us to the probable futility of seeking an epistemic version of a demarcation criterion.* (Laudan 1983: 124)

For Laudan, the relevant philosophical question is therefore not: "what makes a belief scientific?" Rather, it is "what makes a belief well founded (or heuristically fertile)?" (Laudan 1983: 125).

Without claiming to decide between Popper and Laudan, I will first try to relativize the criticism developed by Laudan for the question that interests us, that of Darwinism. Then I will try to develop an argument in favor of Popper's position.

Let us assume that Laudan is right, i.e., that the problem of the demarcation between science and nonscience is not relevant in philosophy; what does this mean for Popper's thesis on Darwinism? First, following academic usage, one must continue to assert that Darwinism is a scientific theory. But second, using the notions considered relevant by Laudan (Laudan 1983: 124–125), one can affirm that, for the moment, Darwinism has not managed to make predictions that are sufficiently precise and specific to be tested. In this sense, it belongs to the set of *non-predictive* scientific theories, as opposed to *predictive* scientific theories, and its value is essentially heuristic. And consequently, it resembles a certain number of theories traditionally considered as metaphysical. Conclusion: in Laudan's terms, Darwinism can be considered a scientific theory whose specific characteristics make it resemble some metaphysical theories. Are we very far from Popper's position?

I now turn to an argument in favor of Popper. As Laudan notes, the scientific world presents epistemologically diverse theories: some are clearly predictive and well-tested, others not, whose value is then only heuristic. However, Popper could point out that, in science, testability remains central in the sense that it is either given or hoped for: *given* by testable theories and *hoped for* by theories whose value is only heuristic. For, in science, the goal of a theory whose value is heuristic is to produce empirical knowledge, i.e., testable knowledge. For example, some physicists claim that, for the moment, string theory is not testable (Smolin 2006). But it is clear that one of the main hopes of this theory is to become testable and corroborated by tests. This also applies to Darwinism. Now, how to name these theories, in science, whose value is only heuristic? On the one hand, we understand Laudan's logic: since these theories are developed by scientists, and since their goal is empirical knowledge, they should be called "scientific." On the other hand, one can also understand Popper's logic: since these theories do not yet constitute empirical knowledge, it is exaggerated and misleading to call them as the theories which constitute empirical knowledge.

Let us now consider the second question: is the Popperian criterion of demarcation between science and nonscience relevant to biology? For some philosophers of biology (Stamos 2007) but also for some biologists (Holliday 1999, Fincham 2000, Gallant 2000),[17] the answer is negative. For them, Popper relies only on physics to develop his epistemology, which makes that his conclusions are not applicable to other sciences, in particular to biology.

Among biologists, there are more precisely two criticisms. The first criticism is that, in contrast to what Popperian epistemology says, in biology certain hypotheses *can be verified*, in the sense that one can obtain complete certainty. To illustrate this

[17]Between 1999 and 2000, in the Correspondence pages of the journal *BioEssays*, several biologists took positions for or against Popper concerning method in science. I draw on these exchanges for the analysis which follows.

idea, Holliday mentions the example of Harvey's discovery of blood circulation. According to him, the hypothesis that a circulation of this type exists in all vertebrates is verified, i.e., certain. It is therefore useless to try to falsify this hypothesis (Holliday 1999: 890). A second criticism is that, in contrast to what Popperian epistemology says, in biology some hypotheses *cannot be falsified.* On this point, here is what Holliday writes:

> Thus, we do have certainty in genetics, but at the same time there are many exceptions to Mendelian inheritance in various biological adaptations to specific situations, where an alternative form of inheritance is beneficial. No matter how many such cases are documented, they in no way undermine the truth of Mendelian inheritance. How could this truth possibly be falsified? (Holliday 1999: 890)

In a few words, Holliday alludes to the fact that in biology, most natural regularities have exceptions. For him, the consequence is that in biology, a negative test is not a good reason to reject a theory. In other words, the consequence is that Popper's falsificationist method cannot work in biology, and therefore cannot define scientificity.

With respect to the first criticism, note that it does not challenge the idea that scientificity requires falsifiability. Thus, in the example taken by Holliday, we can continue to think that Harvey's hypothesis is scientific in the sense that it allows us to make falsifiable predictions. Therefore, this criticism does not change the problem encountered by Darwinism. Moreover, without claiming to settle the question of a possible "verification" in biology (in certain cases only), it should be noted that the Popperian critique of verification is purely logical: it only leads to assert that a statement cannot be absolutely certain if it concerns *a nonfinite class of objects.* Thus, in the example taken by Holliday, it only leads to assert that Harvey's hypothesis cannot be absolutely certain for past, present, and future terrestrial mammals (and even for possible extraterrestrial mammals). It does not claim that it is appropriate to devote one's efforts to trying to falsify any hypothesis.

The second criticism is more interesting for our reflection, because it directly concerns the Popperian criterion of demarcation between science and nonscience. However, it appears that this criticism is based on a misunderstanding of falsificationism. When a biologist formulates a theoretical hypothesis, he/she necessarily defines *a field of application* for this hypothesis, at least as a conjecture. For example, they may assert that their hypothesis is valid for all mammals, all animals, or all living organisms. And from this hypothesis, they may deduce a set of falsifiable predictions. If a test now gives a negative result, rejecting the original hypothesis may simply lead to redefining its scope, incorporating exceptions, for example, all mammals except mammals A, B, and C. From a logical point of view, the original hypothesis is falsified, because to speak of a scientific hypothesis without *a defined field of application* does not make sense. But the formulation of a new hypothesis is easy because it is only a matter of integrating certain exceptions. Popper's falsificationism does not exclude this possibility at all. It can therefore answer the "problem of exceptions" in biology.

This second criticism is sometimes expressed in another way. Thus, Stamos explains that Popper's falsificationism cannot be applied to biology because this falsificationism is thought to apply to a science that formulates laws, whereas it is difficult to think that biology formulates laws (Stamos 2007: 366–371). According to Stamos, the notion of law implies that of "physical necessity" (Stamos 2007: 368), whereas biological regularities are physically contingent. And this contingency would be the fundamental reason for the multiple exceptions observed for various regularities. However, Stamos introduces here obscure and unnecessary metaphysical considerations, absent from Popper's analysis. It is true that Popper sometimes asserts that the goal of theoretical sciences (as opposed to historical sciences) is to discover the "laws of nature." But one must keep in mind that his approach to science is logical, not metaphysical. Thus, for him, a "law of nature" is nothing more than a causal regularity whose knowledge allows us to make testable predictions (see, e.g., Popper 1959/1992: 38–39, 105–106). From this point of view, a "law of nature" does not necessarily have a cosmic scope, in the sense that it does not necessarily concern the most fundamental regularities of the universe, with the idea that these regularities would be invariable (when Stamos speaks of "necessary" laws, he seems to be talking about invariable laws). It may, for example, concern only a few species, or even a single species, i.e., something that live only on Earth, and for a given time—even if, in fact, the word "law" is rarely used in this case, probably because this word has a metaphysical connotation.

Finally, we come back to the same idea: for Popper, theoretical science is essentially characterized by its capacity to predict phenomena, because the latter present regularities. What he calls "law of nature" is thus defined only in relation to this capacity.[18] Consequently, for Popper, biology can *logically* proceed like physics, by producing predictive theories. And this does not imply that the living would be reducible to the inert (Popper thinks the contrary), or that it would not present certain specificities, for example, the absence of invariable regularities.[19]

As an alternative to the Popperian criterion of demarcation, Stamos argues that scientificity depends on the *more or less complete* satisfaction of a set of epistemic values and practices.[20] In particular, Stamos thinks of the following values: "simplicity, predictive accuracy, fertility, coherence with other things we know, unification, and of course testability." And he thinks of the following practices: "prohibitions against the invention of data, the encouragement of 'innovation and

[18] Obviously, the expression "law of nature" can designate something else in philosophy, in particular in metaphysics. But this is not the case in Popper. For such a minimalist conception of laws in science, see in particular Mitchell (2000).

[19] Evolution is not specific to the living world since, according to contemporary physics, it also concerns the universe. However, the theoretical framework favored by physics is that the universe evolves by obeying fundamental laws that *do not evolve*. That said, this framework is only a hypothesis, and some physicists do not hesitate to discuss it (see for example Smolin 2013: chap. 11).

[20] Stamos draws on several philosophers of science to defend this idea, including Kuhn (1983) and McMullin (1983).

deviance', a non-hierarchical structure, and the encouragement of honesty and reliability" (Stamos 2007: 373). But apart from predictive accuracy and testability (which, for Popper, are one and the same), metaphysics[21] can satisfy all these values and practices. Stamos's proposal therefore amounts to integrating metaphysics into science.

To conclude, if, like Laudan, one thinks that the problem of demarcation between science and nonscience is not relevant, then, as far as Darwinism is concerned, one can still formulate a conclusion close to that of Popper: Darwinism is a scientific theory whose specific characteristics make it resemble a certain number of theories traditionally considered as metaphysical. Moreover, the demarcation criterion proposed by Popper is not specifically designed for physics: it can just as easily be applied to biology and thus to Darwinism.

3.3 Discussion on the Popperian Proposal to Improve Darwinism

In this section, I discuss two criticisms of the Popperian proposal to improve Darwinism and then develop a third criticism.

A first criticism, essentially developed by Ruse, consists in contesting the reasons that Popper puts forward to justify his attempt to improve Darwinism. Let us recall that Popper puts forward three reasons: Darwinism (in its passive form) does not explain orthogenetic trends, evolution toward "higher" forms of life, and the separation of species during speciation. Overall, according to Ruse, Popper ignores the elements developed within Darwinism to explain these things (Ruse 1977: 651–652). On the first point, in particular, Ruse considers that Popper confuses natural selection and genetic drift: while the latter makes evolution a random walk, the former can produce trends (Ruse 1977: 653).

From one point of view, Ruse's remarks are justified because Popper develops his criticisms of Darwinism very little, probably because he thinks that these criticisms are classical, and therefore widely developed in the literature. From another point of view, however, Ruse neglects to put these criticisms in context. For Popper, Darwinism is not testable. It is therefore not possible to refute it scientifically. It is only possible to consider its explanations as insufficient. Thus, Popper obviously knows that natural selection can, *in principle*, explain orthogenetic trends. But he doubts that natural selection is, *in fact*, sufficient to explain these trends.

A second criticism, developed by many authors, is that, in its broad outlines, Popper's proposal is not original because it repeats an idea already proposed by several thinkers, in particular by Baldwin (Ruse 1977: 654; Watkins 1995: 191–192; Hull 1999: 493–494; Vecchi and Baravalle 2015: 537–538; Bradie 2016: 161). Some of these authors add that, in detail, about the different types of genes (p-genes, s-genes, and a-genes) and the order in which they are modified, Popper's

[21] In Popper's sense, and not in the positivist sense of the term

proposal is biologically unsound and unnecessarily complex (Ruse 1977: 655–656; Hull 1999: 494–495; Vecchi and Baravalle 2015: 538).

As I pointed out in Sect. 2.3, in 1982 and 1986, Popper presents his proposal without entering into a discussion of the different types of genes and the order in which they are modified. One can therefore think that this element is not essential.[22]

On originality, Popper's commentators are right to point out that his proposal repeats an idea already formulated. However, they underestimate an important element. According to Popper, the activity of living beings plays *a central role in evolution*, so much so that, for him, there are three fundamental factors in evolution: the activity of organisms, the variability of the genome, and natural selection (see Table 1 in Sect. 2.3). In contrast, for the architects of the Modern Synthesis, the "Baldwin effect"[23] is not central at all: for Simpson and Mayr, this effect is doubtful, if not marginal in evolution (Simpson 1953: 115; Mayr 1963: 610–612), and for Huxley, even if the Baldwin effect is an element to be taken into account, this element is not central in evolution (Huxley 1942: 17, 114, 304, 523–524). Moreover, even today, although there is a renewed interest in the Baldwin effect in biology, we are far from the Popperian proposal. This effect is invoked to explain certain aspects of evolution, notably for the human lineage (Durham 1991; Dennett 1991: 190; Deacon 1997). But it is not considered a central dimension of the whole of evolution. It is therefore difficult to avoid the conclusion that Popper's proposal has some originality.

In reality, this originality is explained by the fact that Popper bases his proposal on two ideas not shared by the defenders of Darwinism. The first is that modern Darwinism *is not enough to explain evolution*, and therefore one or more other major factors must be discovered. The second is that the activity of living beings has *an irreducible teleological dimension*: an irreducible capacity to invent and evaluate new behaviors, at least through sensations of pleasure and pain (Popper 1982a: 39–40). Popper's proposal therefore consists in hypothesizing that this teleological dimension is a major factor in evolution. It thus generalizes the proposal developed by Alister Hardy for the animal world, in *The Living Stream* (1965).[24] In contrast, the defenders of the modern synthesis leave aside, even contest, the teleological

[22] In 1974, Popper probably thinks that genetic mutations affecting behavior (preferences and skills) are less likely to be lethal than those affecting anatomy. This is why he imagines that genetic modification in a population tends to follow a certain order: first the p-genes, then the s-genes, and finally the a-genes. But his proposal can be maintained without this order.

[23] This expression was coined by Simpson (1953: 112).

[24] In this book, Hardy makes a critique of the Modern Synthesis (1965: 153–154, 209). Moreover, he explicitly mentions the "psychic side" of animal life (1965: 262–264). According to him, the teleological activity of animals plays a major role in their evolution (1965: 154, 262). Popper explicitly refers to Hardy (Popper 1982a).

Table 2 On the place of behavioral innovation in evolution. Comparison between Popper, Hardy, and modern Darwinism

	Modern Darwinism	Hardy	Popper
Position on Darwinism + **Position on teleology in the living world**	Defense of Darwinism + Hypothesis of a reduction of the teleological to the mechanical	Criticism of Darwinism + Hypothesis that the activity of animals is irreducibly teleological	Criticism of Darwinism + Hypothesis that the activity of all living beings is irreducibly teleological
Place of behavioral innovation in evolution	↓ Marginal place in evolution	↓ Central place in the evolution of the animal world	↓ Central place in the whole evolution

dimension of the activity of the living.[25] And among some contemporary defenders of the Baldwin effect, we find the same distrust of teleology.[26]

Table 2 summarizes the basic elements that show how Popper's proposal is original.

A third criticism, this time not inspired by Darwinism, can be addressed to Popper. This criticism concerns the question of *the emergence of the teleological* in the living world. Let us recall that, for Popper, the main merit of Darwinism is to propose a reductionist approach to the mechanisms of evolution, i.e., not to introduce any ad hoc hypothesis—as opposed to theistic, panpsychist, vitalist, or even Lamarckian approaches to evolution. However, in order to explain the existence of irreducibly teleological activities in the living world, Popper makes the hypothesis that this type of activity can somehow emerge from purely mechanical processes. This hypothesis is not supported by any independent reason: it is purely ad hoc. In this respect, the Popperian approach to evolution resembles other non-reductionist approaches.

To defend his emergentist hypothesis, Popper can only claim that this metaphysical position is the one most compatible with the scientific knowledge of our time. Against panpsychism in particular, defended by some evolutionary theorists,[27] Popper asserts that, for contemporary physics, elementary particles do not manifest any memory capacity (Popper 1977: 182).

[25] In 2004, Mayr explains that he admits that certain animal behaviors have a teleological dimension (Mayr 2004: 57). But he adds that this teleological dimension is reducible to mechanisms (Mayr 2004: 61).

[26] For some authors, the Baldwin effect is only acceptable if it is conceived as reducible to mechanisms, without anything fundamentally teleological (Downes 2003: 43, 48; Dennett 2003: 71–73).

[27] Among the advocates of a form of panpsychism, Popper mentions two biologists: Rensch (1900–1990) and Waddington (1905–1975) (Popper 1977: 179). Wright (1889–1988) and Haldane (1892–1964) should also be mentioned.

4 Conclusion

Popper's first assertion is that Darwinism (reduced here to the theory of natural selection) is difficult to test, and this assertion seems sound. The tests proposed by some defenders of Darwinism always have two weaknesses: they are based on unquantified predictions, which makes it difficult to know whether or not a particular fact (real or hypothetical) corresponds to the prediction, and they are insufficiently specific because they do not allow one to decide between Darwinism and some other conceptions of evolution.

From this first assertion, Popper deduces that Darwinism is less a scientific theory than a metaphysical research program for science. This second assertion is obviously based on Popper's conception of the demarcation between science and nonscience, a conception debated in philosophy. Without going into the details of this debate, I have tried to show that Popper's demarcation criterion can be applied to biology as well as physics. Moreover, the rejection of a clear demarcation between science and nonscience only leads to a reformulation of Popper's position as follows: Darwinism is a scientific theory whose specific characteristics make it resemble a certain number of theories traditionally considered as metaphysical.

Finally, considering that Darwinism hardly explains some aspects of evolution, Popper proposes to replace "passive Darwinism" by "active Darwinism." For the former, evolution is essentially explained by two things: the variability of the genome and natural selection. For the second, there is another factor: the activity of organisms. The originality of Popper's proposal is, if one may say so, to think that the Baldwin effect applies to the whole of the living world, including the humblest organisms. According to him, this proposal is justified by the fact that the activity of organisms is irreducibly teleological.

Regarding the deconstruction of Darwinism, the main interest of Popper's analysis is to highlight a major weakness of Darwinism, while explaining its success in biology. This weakness is its inability to make testable predictions, both qualitatively and quantitatively. Despite this, Darwinism plays a major role in biology, as a research program, because it aims to explain evolution in a mechanistic way, without invoking any special force. As Popper points out, "a successful reduction is, perhaps, the most successful form conceivable of all scientific explanations, since it achieves what Meyerson (1908, 1930) stressed: an identification of the unknown with the known" (Popper 1974b: 259–260). Now, to return to the questions asked in the introduction, this amounts to saying that, from Popper's point of view, Darwinism is more philosophical than scientific. Moreover, since no theory of evolution does better than Darwinism when it comes to testable predictions, this implies that the study of evolutionary mechanisms is on the whole more philosophical than scientific, and that the history of evolutionary biology is in part the history of a philosophical debate. This conclusion does not, however, imply that knowledge of the mechanisms of evolution can never be scientific. As Popper repeatedly points out, many ideas first belonged to philosophy before belonging to science.

Another interesting aspect of Popper's analysis is his proposal to improve Darwinism. Admittedly, it is not at all obvious that "active Darwinism" is the answer to the question of evolutionary mechanisms. However, this proposal show that Darwinism is flexible enough to include a significant part of teleology. In other words, it shows that Darwinism can be thought outside the metaphysical framework of radical mechanism.

References

Baldwin JM (1896) A new factor in evolution. Am Nat 30(354):441–451

Baldwin JM (1897) Organic selection. Science 5(121):634–636

Bergson H (1998) Creative evolution. Translated by Arthur Mitchell, Dover, New York

Bradie M (2016) Karl Popper's evolutionary philosophy. In: Shearmur J, Stokes G (eds) The Cambridge companion to Popper. Cambridge University Press, New York, pp 143–169

Brady R (1979) Natural selection and the criteria by which a theory is judged. Syst Zool 28:600–621

Brandon R (1980) A structural description of evolutionary theory. Proceedings of the biennial meeting of the Philosophy of Science Association 2:427–439

Ceccarelli D (2021) Recasting natural selection: Osborn and the pluralistic view of life. In: Delisle RG (ed) Natural selection: revisiting its explanatory role in evolutionary biology. Springer, Cham, pp 171–194

Darwin C (1859) On the origin of species: a facsimile of the first edition (1966). Harvard University Press, Cambridge

Deacon T (1997) The symbolic species. Norton, New York

Dennett D (1991) Consciousness explained. Little, Brown, Boston

Dennett D (2003) The Baldwin effect: a crane, not a skyhook. In: Weber BH, Depew DJ (eds) Evolution and learning: the Baldwin effect reconsidered. MIT Press, Cambridge, pp 69–79

Delisle RG (2021) Natural selection as a mere auxiliary hypothesis (sensu stricto I. Lakatos) in Charles Darwin's *origin of species*. In: Delisle RG (ed) Natural selection: revisiting its explanatory role in evolutionary biology. Springer, Cham, pp 73–104

Downes S (2003) Baldwin effects and the expansion of the explanatory repertoire in evolutionary biology. In: Weber BH, Depew DJ (eds) Evolution and learning: the Baldwin effect reconsidered. MIT Press, Cambridge, pp 33–51

Durham WH (1991) Coevolution: genes, culture, and human diversity. Stanford University Press, Stanford

Fincham J (2000) More on Popper and biology: the utility of induction. BioEssays 22(7):684

Gallant J (2000) More on Popper and biology: the utility of induction. BioEssays 22(7):684

Hansson SO (2021) Science and pseudo-science. In: Zalta EN (ed) The Stanford encyclopedia of philosophy, https://plato.stanford.edu/archives/fall2021/entries/pseudo-science/

Hardy A (1965) The living stream. Collins, London

Holliday R (1999) The incompatibility of Popper's philosophy of science with genetics and molecular biology. BioEssays 21(10):890–891

Hull DL (1999) The use and abuse of Sir Karl Popper. Biol Philos 14:481–504

Huxley J (1942) Evolution: the modern synthesis. Allen and Unwin, London

Kuhn TS (1983) Rationality and theory choice. J Philos 80:563–570

Lakatos I (1999) Lectures on scientific method. In: Motterlini M (ed) For and against method. University of Chicago Press, Chicago, pp 19–112

Laudan L (1983) The demise of the demarcation problem. In: Cohen RS, Laudan L (eds) Physics, philosophy and psychoanalysis: essays in honor of Adolf Grünbaum. Reidel, Dordrecht, pp 111–127

Levit GS, Olsson L (2006) "Evolution on rails": mechanisms and levels of orthogenesis. Ann Hist Phil Biol 11:97–136

Mayr E (1963) Animal species and evolution. Harvard University Press, Cambridge

Mayr E (2002) What evolution is, Phoenix, London
Mayr E (2004) What makes biology unique? Cambridge University Press, Cambridge
McMullin E (1983) Values in science. Philos Sci 50(Suppl):3–28
Mitchell SD (2000) Dimensions of scientific law. Philos Sci 67(2):242–265
Platnick N, Gaffney E (1978) Review of Popper's idea on evolutionary biology. Syst Zool 26:138–141
Platnick N, Rosen D (1987) Popper and evolutionary novelties. Hist Philos Life Sci 9:5–16
Popper KR (1957) The poverty of historicism. Routledge & Kegan Paul, London
Popper KR (1959/1992) The logic of scientific discovery. Hutchinson, London. Republished in 1992 by Routledge, London. All references are to the 1992 edition
Popper KR (1963) Conjectures and refutations. Routledge & Kegan Paul, London
Popper KR (1972) Objective knowledge. Oxford University Press, Oxford
Popper KR (1974a) Intellectual autobiography. In: Schilpp PA (ed) The philosophy of Karl Popper. Open Court, La Salle, pp 1–181. Republished with modifications from 1976 to 1992. See Popper KR (1992)
Popper KR (1974b) Scientific reduction and the essential incompleteness of all science. In: Ayala FJ, Dobzhansky T (eds) Studies in the philosophy of biology. University of California Press, Berkeley, Los Angeles, pp 259–284
Popper KR (1977) Some remarks on panpsychism and epiphenomenalism. Dialectica 31:177–186
Popper KR (1978) Natural selection and the emergence of mind. Dialectica 32:339–355
Popper KR (1980) Evolution. New Scientist 87:611
Popper KR (1982a) The place of mind in nature. In: Elvee RQ (ed) Mind in nature. Harper and Row, San Francisco, pp 31–59
Popper KR (1982b) The postscript to the logic of scientific discovery III. Quantum theory and the schism in physics, Hutchinson, London
Popper KR (1983) The postscript to the logic of scientific discovery I. Realism and the aim of science, Hutchinson, London
Popper KR (1992) Unended quest: an intellectual autobiography. Routledge, London
Popper KR (1994) The myth of the framework. Routledge, London
Popper KR (2014) A new interpretation of Darwinism. In: Niemann HJ (ed) Karl Popper and the two secrets of life. Mohr Siebeck, Tübingen, pp 115–129
Ruse M (1977) Karl Popper's philosophy of biology. Philos Sci 44:638–661
Ruse M (1981) Is science sexist? Reidel, Dordrecht
Settle T (1996) Six things Popper would like biologists not to ignore: in memoriam, Karl Raimund Popper, 1902–1994. Biol Philos 11:141–159
Simpson GG (1953) The Baldwin effect. Evolution 7:110–117
Smolin L (2006) The trouble with physics. Houghton Mifflin, New York
Smolin L (2013) Time reborn. Houghton Mifflin, New York
Stamos D (1996) Popper, falsifiability, and evolutionary biology. Biol Philos 11:161–191
Stamos D (2007) Popper, laws, and the exclusion of biology from genuine science. Acta Biotheor 55:357–375
Vecchi D, Baravalle L (2015) A soul of truth in things erroneous: Popper's "amateurish" evolutionary philosophy in light of contemporary biology. Hist Philos Life Sci 36:525–545
Watkins J (1995) Popper and Darwinism. In: O'Hear A (ed) Karl Popper: philosophy and problems, vol 39. Cambridge University Press, Cambridge, pp 191–206

Part III

Around and Beyond the Synthesis

Typology/Population Distinction and Its Role in the Marginalization of Nineteenth-Century Non-Darwinian Theories in Modern Historiography

Michał Jakub Wagner

Abstract

This article aims to deconstruct the division between typological and population thinking used by modern Darwinists, with a focus on examining the methodology employed to establish a connection between essentialism and typology. Specifically, the study analyzes the methods of two major proponents of this thesis: Ernst Mayr and David Hull. The typology/population dichotomy, initially proposed by Mayr, has significantly influenced the historical narrative surrounding the rise of non-Darwinian theories in the nineteenth century. However, this narrative oversimplifies the concept of species in non-Darwinian theories and reinforces the perceived link between Darwinism and the modern synthesis. By scrutinizing the methodology used by Hull and Mayr, it is crucial to assess the soundness of the methods employed to identify the influence of essentialism on nineteenth-century biology. This chapter's aim is to examine the rationale behind their claims and the adequacy of the analysis built upon these assumptions. The goal is to determine whether the argument presented, supporting the division between "accurate" populational theories of evolution and "erroneous" typological ones, is sufficiently supported by evidence and justified by the methods employed by Mayr and Hull in their historiography. The deconstruction of this classic interpretation of non-Darwinian theories inherently challenges a specific interpretation of Darwinism that was popularized by Mayr and Hull. More specifically, it challenges the vision of Darwinism as the representation of "good" science, differing significantly from other evolutionary theories influenced by ancient essentialism.

M. J. Wagner (✉)
Institute of Philosophy, Cardinal Stefan Wyszyński University in Warsaw, Warsaw, Poland
e-mail: m.wagner@uksw.edu.pl

R. G. Delisle et al. (eds.), *Unity and Disunity in Evolutionary Biology*,
https://doi.org/10.1007/978-3-031-42629-2_10

Keywords

Mutationism · Intertheoretical · Essentialism · Typology · Transmutationism

Ernst Mayr (1963, 673; 1969, 414) defines typology as follows: "A concept in which variability is ignored, and members of a population are regarded as replicas of a 'type,' the Platonic eidos." Mayr (1997, xvi) equates typology with ancient essentialist philosophy and contrasts it with modern population thinking, where one finds not classes but aggregates of unique individuals, that is, populations. Mayr was not the first scholar to write about typology; earlier, the existence of such a way of thinking about species had been pointed out by George G. Simpson and Theodore Dobzhansky. Undoubtedly, however, it was Mayr who contributed to the popularization of the term, extending and modifying its original meaning. In his predecessors' work, typology referred to a specific method of interpreting taxa, but Mayr transformed it into a negative philosophical factor derived from ancient essentialism.[1] For Mayr, typology became part of the narrative he was creating, which he used to reinterpret the history of biology (Witteveen 2015a, 21; Witteveen 2015b, 8). According to this narrative, essentialist philosophy dominated the natural sciences up to the nineteenth century and was only overcome by Darwin (Mayr 1959, 2; Mayr 2002, 74–75). This dominance led to the spread of the typological concept of species and, at the same time, delayed the acceptance of the idea of evolution.[2] Although Mayr (1997, 335–336) links typology to the idea of the immutability of species, he does not reject the possibility of evolutionary theories accepting this concept. According to Mayr (1963, 517; 1997, 348–349), post-Darwinian biology did not immediately adopt a population-based understanding of species, and relying on typology led to the creation of erroneous saltationist theories such as Hugo DeVries' mutationism (Mayr 1949, 514; 1957, 5; 1997, 308).

Thus modified, Mayr's version of typology entered the language of philosophy and the history of biology as a permanent feature in discussions about species concepts in non-Darwinian biology. Mayr later supported his "essentialist story" with the work of Arthur J. Cain, who, in his influential article on Carl Linnaeus, pointed out his connection to Aristotle's essentialism, and David Hull, who built on Popper's understanding of essentialism and highlighted its negative impact on taxonomy (Winsor 2006, 152–165). Today, historians dismiss Mayr's "essentialist story" as an overly simplistic interpretation of how species were understood before the rise of the modern synthesis. The critique of Mayr's notion of typology can be summarized in three points:

[1]The terms ancient essentialism or metaphysical essentialism, which I use in this chapter, will refer to the concepts of Aristotle and Plato, namely, those that, according to Ernst Mayr and David Hull, are said to have initiated typological thinking about species.

[2]Using Mary Winsor's formulation, I will refer to this historical narrative as the "essentialist story" in the subsequent part of the text.

(a) There is no historical evidence indicating the complete dominance of essentialist philosophy in pre-Darwinian biology (Honenberger 2015, 15; Winsor 2006, 155; Witteveen 2015b, 9; Amundson 2005, 13–16; Richards 2010, 17–20; Wilkins 2009, 232–233).
(b) The concept of type, if present in taxonomy, was not connected to metaphysical theses but emerged from the practice of classifying organisms (Faber 1976, 117–119; Amundson 2005, 24–25, 31–33, 43–47, 81; Wilkins 2009, 70–74, 90, 136–139, 232).
(c) Without substantiating historical evidence for the validation of the "essentialist story," it can be argued that it represents an oversimplification of the history of biology. This oversimplification arises from the whiggish nature of Mayr's historiography, as its primary objective was to fortify the association between the modern synthesis and Charles Darwin's theory while discrediting alternative evolutionary theories (Winsor 2006, 155–159; Richards 2010, 76; Wilkins 2009, 232–233; Chung 2003, 294–295; Wilkins 2010, 145; Levit and Meister 2006, 298–304).

Building upon the critique presented in the previous section, it is possible to characterize the "essentialist story" as modern biology's "invented tradition," employing the terminology coined by Eric Hobsbawm and Terence Ranger. According to Hobsbawm (2000b, 263–268), these traditions are symbolic actions that create the impression of being historical while actually being newly created. Invented traditions primarily serve to foster a sense of belonging in a community and legitimize the power of certain groups and institutions (Hobsbawm 2000a, 1–9). Critics argue that the "essentialist story" serves a similar function. They see it as an artificial construct, lacking historical evidence, designed to legitimize the modern synthesis (by discrediting "typological" theories) and to unite the community of biologists under a single theory of evolution. In this case, an invented tradition was created to promote a new interpretation of Darwinism as an example of a correct theory of evolution, unaffected by the detrimental influences of ancient metaphysics. The "essentialist story" played a pivotal role in establishing this interpretation among philosophers and historians of biology, making Darwinism an exception in nineteenth- and twentieth-century science that was to be dominated by incorrect and irrational theories of evolution. Consequently, it provided architects of the synthesis with another argument to claim its superiority over other alternatives.

This chapter addresses whether the methods used to identify the influence of ancient essentialism in non-Darwinian theories are valid. In addition to deconstructing the methodology of the proponents of the "essentialist story," it will be also examined whether the analysis built upon it presents a sufficient argument to accept the possibility of a historical division between "accurate" populational biological theories and "erroneous" typological ones. In other words, this chapter will deconstruct the methods used to perpetuate the "essentialist story" and, by doing so, will call into question the rationale behind the vision of Darwinism as the sole example of a "good" theory of evolution (i.e., not influenced by metaphysics).

1 The Question of Intertheoretical and Historical Continuity

As John S. Wilkins (2010, 142) and Mary P. Winsor (2006, 162–163) point out, the "essentialist story" likely originated from John Dewey's famous 1909 paper, "The Influence of Darwin on Philosophy." In this text, Dewey discusses the negative impact of ancient essentialism on the natural sciences, which has been counteracted by the efforts of individuals like Darwin (Dewey 1910, 4–8). He emphasizes the need for further reform in science to eliminate remnants of this philosophy (Dewey 1910, 18–19). Wilkins argues that Dewey makes a significant mistake by assuming that biological taxonomy follows the same principles as scholastic logic. The modern concept of species as a specific taxonomic category emerged through the work of Andrea Casalpino, John Ray, and Linnaeus (Wilkins 2009, 94–98). Prior to the eighteenth century, species primarily functioned as a logical category applicable across multiple levels of analysis, not solely in the realm of living beings (Richards 2010, 46–56). While scholastics understood species in essentialist terms, naturalists embraced a "generative conception of species"[3] (Wilkins 2010, 143–144). According to this view, species were understood as lineages that pass on certain sets of traits from generation to generation, so kinship primarily determined membership in a particular species (Wilkins 2009, 10, 229–230; Richards 2010, 56–67). "Therefore, Dewey misled generations of thinkers by identifying what the scholastic logicians, from whom we get the logical notions of *genus* and *species*, meant by these terms with what naturalists meant, despite his plea for sympathy with them"(Wilkins 2010, 143).

The validity of the "essentialist story" relies heavily on the assumption of continuity between ancient essentialist philosophies and later conceptions of species. Without evidence supporting such continuity, there may be insufficient grounds to accept its existence. The presence of the "generative concept of species" inherently suggests a lack of continuity, as this conception assumes the existence of interspecific variability, directly challenging Mayr's narrative that pre-Darwinian biologists were guided by typological thinking that disregarded such variability. Considering this, one might question whether the "generative" concept ultimately undermines the "essentialist story." According to Mayr (1957, 7–8), the "generative" concept can still be considered a form of typology, as it overlooks the relationships between individuals within a population and instead emphasizes the relationships between successive generations. As population thinking perceives individuals primarily as temporal manifestations of the gene pool, the "generative" concept can be seen as essentialist, as morphological similarity ultimately holds greater significance within it (1957, 11–14). The classification of "generative" concepts as part of typological thinking presents an evident challenge: the inherent ambiguity surrounding the category of typology itself. Typology appears to be a vague term encompassing various ways of conceptualizing species that both acknowledge and disregard variability. Moreover, a typologist could be either a proponent of creationism or

[3] I use Wilkins' (2009, x) term here; Richards (67) uses the term "historical concept of species."

an evolutionist who accepts saltationism. Such a broad understanding of typology is, on the one hand, elucidated by Mayr himself, who asserts that this concept underpinned numerous biological debates (1959, 2–3). On the other hand, it raises the question of whether it is truly valid to assert that these diverse ideas were all influenced by the same philosophical concept. This dilemma pertains to the establishment of connections between different theories and ideas in the history of biology and whether they can be definitively proven. As I will further explore, it becomes evident that the connections between essentialism and the theories identified by Mayr as representative of it hold validity only when we overlook the absence of historical and intertheoretical continuity between ancient essentialism and various pre- and post-Darwinian concepts of species.

The issue of continuity holds significant importance, especially in relation to non-Darwinian evolutionary theories that arose during the period known as the "eclipse of Darwinism." Assuming that Mayr created the "essentialist story" solely for propagandistic purposes, his main objective would have been the depreciation of evolutionary theories competing with Darwinism. Let us now examine the alleged typology found among non-Darwinian evolutionists of the "eclipse" period. Mayr identifies DeVries and William Bateson as prominent typologists who played significant roles in propagating "essentialism" in that period. Both were early proponents of Mendelian genetics and outspoken critics of Darwinism. Additionally, they both rejected gradualism in favor of saltationism. DeVries put forward the mutation theory, suggesting that new traits emerged through extensive transformations in hereditary material. Mayr (1997, 341–342) described DeVries' views, stating that while he made significant contributions to scientific progress with his work on Darwin's theory of pangenesis, his shift to a new theory of evolution had unequivocally negative consequences, notably impeding research on the phenomenon of mutation (1997, 311). Historians of science recognize that DeVries' perspectives on mutation were largely shaped by his study of *Oenothea lamarckiana*, which erroneously led him to believe in the existence of large-scale progressive mutations. The choice of this unfavorable subject of study may explain DeVries' mistake as a case of empirical research misleading an unsuspecting researcher. However, Mayr disagrees with this explanation. According to him, it was not that DeVries was deceived by his research; rather, he deliberately chose a research subject that aligned with his typological intuitions in order to substantiate them (1997, 346). So, which aspects of DeVries' theory would support this notion?

What supports an essentialist interpretation of mutationism is DeVries' utilization of Adolphe Queletet's statistics. According to DeVries, intraspecific variability was expected to follow a Gaussian curve, fluctuating around a mean value (DeVries 1906, 725–726). This average value could be perceived as representing the essence of the species, reminiscent of what Eliot Sober referred to as "the natural state model" (1980, 353). This model reflects essentialist thinking about variability and suggests inherent limits to species variation. It is believed to have originated, in part, from the philosophy of Aristotle (Sober 1980, 362), and in nineteenth-century science, it was specifically promoted by Quéletet (Sober 1980, 366). According to this model, complete transformation is deemed impossible as species variability always oscillates around an unchangeable mean value. DeVries aligns with a similar line of thought, as he suggests that non-mutational transformations can only be regressive, relying on the recombination of existing traits. True evolution, according to DeVries, can only occur through the emergence of new factors via mutation (DeVries 1906, 13–17, 16–18). Bateson's early work also exhibits similarities to the "natural state model." He recognizes the existence of limits to variability, determined by the structural constraints of organisms while rejecting the idea of their unlimited plasticity, as (he believed was) proposed by Darwinism (1984, 69–80). In his later work, this concept takes the shape of a theory suggesting that evolution entails the alteration of internal structure through chemical changes, which can involve the addition or removal of its constitutive factors (1913, 93–97).

Both DeVries and Bateson's theories share a common thread: the insistence on the existence of a natural state of the species, which, when disrupted, leads to evolution. However, the question arises whether saltationism alone, along with ideas resembling the "natural state model," qualifies these theories as manifestations of typological thinking. It is important to remember that typology, as used by Mayr and others, encompasses a broad range of ideas, not limited to just these two. In both DeVries and Bateson, for example, we do not find a denial of variability. On the contrary, as early Mendelians, they make variability their primary focus of study. They also criticize Linnaeus' systematics, which Mayr identifies as one of the prime examples of typological thinking (1982, 176–177). DeVries (1906, 10–12) dismisses Linnaeus' taxonomy as an artificial system incapable of reflecting inter-specific variability, while Bateson (1913, 6–9) criticizes it as a factor contributing to the popularity of creationism. Regarding the Mendelians, there is also no clear link to essentialist philosophy. One could consider Quéletet as a potential mediator, although the question remains whether he even was a Platonist or an Aristotelian. We know that Queletet read Aristotle's *Ethiká Nikomácheia*, which later influenced his criminology (Beirne 1987, 1159). Nevertheless, the question remains whether this alone is sufficient to regard his views as a continuation of Aristotelian essentialism. It is important to note that mere similarity of ideas does not necessarily imply a shared source, just as the presence of analogous organs in different species does not imply their close evolutionary kinship in biology. Therefore, without evidence indicating that saltationist theories originate from ancient essentialism, it would be unjustified to claim that they share the same ontological assumptions. Even if one were to argue that the presence of saltationism and the "natural state model" in these theories provide adequate evidence for such a relationship, their concept of essentialism should be viewed as fundamentally different from its ancient counterpart. For instance, they do not assume the existence of a great chain of being and acknowledge the possibility of evolutionary changes within species. As a result, the essentialism found in these theories can be considered radically different than ancient essentialism. Considering the disparity between Mayr's notion of typology and what we find in the theories of DeVries and Bateson, it is, at the very least, problematic to categorize them as representatives of the same way of thinking about species.

The key question arising from the above analyses is as follows: If we do not find similarities between what Mayr referred to as "typology" and the way naturalists like DeVries and Bateson wrote about species, what is the basis for claiming that they were part of the same style of thinking about species that originated with Plato and Aristotle? This problem can be resolved in two ways. One approach involves an in-depth analysis of both ancient metaphysical essentialism (e.g., Platonism and Hylomorphism) and the non-Darwinian concept of species (as seen in Neo-Lamarckism, German idealism, and theories of orthogenesis) to determine if there is any truth to the "essentialist story." It's important to note that there are articles and books that delve into this research and demonstrate how different the essentialism of the ancients was from the supposed essentialism of non-Darwinian naturalists (e.g., Amundson 2005; Wilkins 2009). Therefore, there is no need to replicate those works here. I would also argue that this is the wrong way to approach this problem. The "essentialist story," as the received view, should be primarily analyzed from the perspective of how it was originally conceived and formulated. The burden of proof should not be on those who are doubtful about the validity of the "essentialist story" but rather on those who proposed it. In other words, the second way to address the question above is to analyze the methodology underlying the philosophical analysis that validates the thesis of conceptual affinity between non-Darwinian evolutionary theories and ancient metaphysics to see if it has any merit. As I will demonstrate in the subsequent sections, using the works of Mayr and Hull as examples, there is no profound philosophical analysis supporting the connections they make when advocating for the "essentialist story." The categories they employ are often used in a purely arbitrary manner.

2 Ernst Mayr's' "Problematic History" Versus His "Essentialist Story"

As mentioned earlier, the basis of the "essentialist story" rests on establishing a link between ancient metaphysics and historical scientific theories. While critics rightfully highlight the lack of direct evidence supporting this assertion, there is an additional aspect to consider concerning the methodology employed to construct it. It is plausible that Mayr's approach may have unveiled insights that would be elusive using a different methodology. However, if it transpires that Mayr's philosophical framework is arbitrary enough to permit various interpretations of the relationships between philosophical theories and ideas, then it can be assumed that the "essentialist story" has not been sufficiently proven and has, thus far, been accepted a priori. Let us now delve into Mayr's historical methodology, which he himself labeled as "problematic history."

Problematic history aims to approach the historiography of sciences by focusing on the analysis of problems within the discipline and how they were addressed. The historian's main task is to study scientific problems and understand their correct solutions. Without such understanding, the historian would be unable to recognize instances of scientific progress or the hindrance caused by flawed solutions or improperly formulated problems (Mayr 1982, 6–8). Therefore, according to Mayr (1982, 10–11), to practice problematic history, one must possess a comprehensive understanding of the current state of knowledge. By its very definition, this historiography embraces a presentist perspective, as a contemporary viewpoint is essential for recognizing the value of past problem-solving approaches.

Mayr's historiography rests upon two crucial pillars that significantly shape his argumentation for the "essentialist story." When explaining delays in problem-solving or the rejection of certain solutions, Mayr (1982, 834–835) relies on Bernard Barber's concept of "cultural blindfolds." His historiography primarily adopts an internalist approach, except when considering theories that are deemed "wrong" from a contemporary perspective—their emergence and popularity are explained by external factors. These external factors, referred to as "cultural blindfolds," encompass philosophical, religious, and cultural concepts that either hinder scientists from accepting certain theories or give rise to pseudoscientific ones. Thus, Mayr's vision of science leans toward positivism, where metaphysics is viewed as the primary obstacle to scientific progress, interfering with proper scientific methodology (Junkier 1997, 53–55). The primary distinction between positivism and Mayr's philosophy lies in his emphasis on the notion of "concepts." According to Mayr (1982, 23–24, 856), scientific advancement primarily occurs through the development of "concepts" that enable the interpretation of empirical data in increasingly better ways. The recognition of "concepts" throughout the history of science leads to the second significant pillar of Mayr's historiography—Lovejoyanism. Influenced by the work of Arthur O. Lovejoy, Mayr (1982, 18–19) argues that historical research should focus on reconstructing how specific "concepts" (equivalent to Lovejoy's "ideas") influenced the development of particular fields. Therefore, the historian's task is to reconstruct the genealogy of ideas (Junkier 1997, 38; Hull 1994, 380).

Lovejoy's approach brings several significant presuppositions that profoundly impact Mayr's historiography. These presuppositions primarily revolve around the ontological status of ideas and their relationship to historical actors. Lovejoy (1948, 3–5) posited that ideas possess inherent qualities that manifest themselves at different times, evoking distinct intellectual and emotional responses among those expressing them. Since ideas have an independent existence, the role of historical actors is reduced to being passive carriers of them. Lovejoy asserted that ideas have an internal logical structure, of which their human carriers may not be fully aware, and this structure influences how they articulate the ideas through their works. Ideas can gradually alter beliefs, values, and tastes by introducing additional meanings or suggesting implications that may not have been clearly discerned by those employing them (Lovejoy 1948, 14). In his critique of Lovejoyanism, Thomas Bredsdorff (1975, 13–18) argues that such a characterization of ideas can lead to a situation where historians feel justified in "reading the mind" of the historical actors they study. If "ideas are caused by ideas, and they, in turn, cause deeds" (Bredsdorff 1975, 22), then historians, armed with knowledge of the history of a particular idea, may discover subtexts in the works they examine that the authors themselves were unaware of. Mayr (1982, 850–852) adopts this presupposition in a peculiar way by differentiating between the explicitly expressed but often untrue motivations of scientists and the implicitly contained true determinants of their work. Following Imre Lakatos (1971, 94), it can be argued that Mayr here incorporates another characteristic of positivism: the supposition of the existence of a "false consciousness" of scientists that prevents them from seeing when they are working under the influence of irrational factors. Based on this premise, Mayr (1982, 493) is able to conclude, for instance, that while Darwin acknowledged being inspired by Thomas Malthus, the actual determinant for the development of the theory of natural selection was Darwin's observations during his voyage on the Beagle. Darwin's admission of being influenced by *An Essay on the Principle of Population* would be considered an example of "false consciousness." These three elements—presentism, Lovejoyanism, and cultural blindfolds—form the methodological foundation upon which Mayr justifies the connections he establishes between non-Darwinian theories and essentialism. As we will see later, historiography built upon such a methodological basis falls into the trap of *petitio principii*, where its assumptions are constituted by its conclusions.

Unlike Lovejoy, who began his study of the history of an idea from its origins to trace the way in which it changed, Mayr adopts a different approach. He starts with the first and most significant expression of an idea and examines its history in reverse. This is due to the presentist nature of "problematic history." If new theories are correct, the past science should be examined from the perspective of the contemporary state of scientific knowledge. Thus, when Mayr explores the history of the idea of evolution, he examines it through the prism of the first theory that, according to him, best expressed it, namely, Darwin's (cf. Mayr 1982, 392–394, 852–853; Mayr 1988, 182–183). Past theories of evolution will therefore be judged in terms of how similar they were to Darwin's theory, and this comparison will demarcate between true evolutionary theories and pseudo-evolutionary ones. Since, according to Mayr (1982, 352), a key element of Darwinism is its anti-essentialist character, any theory that has essentialist assumptions will be judged not as evolutionist but as "transmutationist." Essentialism thus serves as the criterion that distinguishes anti-essentialist evolutionism from essentialist pseudo-evolutionary "transmutationism." Given that the focus is on the comparative analysis of ideas themselves, the cultural and social context surrounding the theories expressing them is ignored. It is irrelevant to Mayr whether transmutationist theories were valid in their historical context. His primary concern lies in assessing whether these theories approach the explication of evolution in a manner similar to Darwinism. Consequently, when comparing the influence of Lamarck and Charles Lyell on the development of biology, Mayr concludes that the creationist Lyell had limited influence on the emergence of Darwinism. Mayr argues that Lamarck, who expressed ideas akin to Darwin's such as uniformitarianism and gradualism, had a greater impact on the development of the idea of evolution than the "essentialist" Lyell (cf. Mayr 1982, 376–379). This assessment is of course, incorrect from the perspective of Darwin himself, who perceived his work as more aligned with the ideas presented in *Principles of Geology* than in the *Zoological Philosophy*. However, Mayr's research does not prioritize the scientist's personal sentiments; for him, they can be a part of "false consciousness." He asserts that the most crucial elements in theory development are often concealed within the unconsciousness of historical actors. In the end, what matters most is the similarity of ideas and their structure, which can stand in opposition to what a scientist thinks about what they have created or what has inspired them.

Similarly, Mayr explains the nature of Darwin's population thinking. According to Mayr (1982, 267–269), Darwin initially formulated his theory of evolution using population thinking but abandoned it under the influence of his correspondence with botanists. Consequently, in *On the Origin of Species*, Darwin adopted a nominalist perspective and expressed his theory using typological terminology (Mayr 1982, 268). Mayr does not view this presence of typology as indicative of Darwin's essentialism,[4] but rather as another manifestation of "false consciousness." In such a case, can the manifestations of essentialism in non-Darwinian theories also be attributed to the scientists "false consciousness"? Is it possible that the perceived resemblance between DeVries' theory and essentialist philosophy is merely superficial, while its underlying motivation stems from populational ideas? One might expect that under Mayr's methodology, such a hypothesis is plausible. However, Mayr himself does not entertain this possibility. The influence of essentialism in non-Darwinian theories is not considered by him as a manifestation of "false consciousness" and does not receive the same level of analysis as Darwin's theory. The assertion that Darwinism has been minimally impacted by nonscientific factors is not the result of rigorous examination but rather a preconceived assumption that only theories deemed true contemporarily meet the criteria of truthfulness. Other theories, rejected by modern science, are seen as products of "cultural blindfolds" such as essentialism. As a result, the notion that non-Darwinian theories were directly influenced by essentialism is presented as a certainty rather than a hypothesis that should be further examined. Why is this the case? Because at its core, Mayr's historical methodology relies on circular reasoning: the creation of non-Darwinian theories is attributed to the influence of essentialism because they are not contemporaneously accepted and do not share assumptions (they do not explicate similar ideas) with Darwinism—thus, they are judged to be created through "cultural blindfolds." Conversely, Darwinism is not seen as emerging under the influence of essentialism because it is contemporaneously accepted and does not share assumptions with non-Darwinian theories—hence, it is deemed to represent true and rational science. In the end, a circular argument emerges: everything contemporaneously accepted is deemed "good" science and assumed to be free from the influence of cultural blindfolds, while everything not contemporaneously accepted is labeled as "bad" science and attributed to the impact of cultural blindfolds. Why are contemporaneously accepted theories considered "good" science? Because they

[4]It is worth noting that Mayr did not initially attribute the origins of the populational concept to Darwin. Establishing Darwin as the progenitor of population thinking (between 1956 and 1959) was also a strategic move, as it clarified the connection between his theory and the modern synthesis (Chung 2003, 287; Winsor 2006, 154–155).

are presumed to be unaffected by cultural factors. How do we establish that they are unaffected by cultural factors? Because they are contemporaneously accepted. Conversely, how do we conclude that contemporaneously rejected, non-Darwinian theories were influenced by cultural factors? Because they are contemporaneously rejected.

3 David Hull's Phylogenetics of Science

The influence of Lovejoy may help explain why Mayr did not address the issue of historical continuity between concepts. If ideas possess an inherent logical structure, then their expression, regardless of the context—scientific, cultural, or historical—will exhibit similarity. Ironically then, Mayr's approach to the classification of ideas/concepts can be characterized as typological. However, Mayr was not the sole proponent of the "essentialist story." David Hull also played a significant role in providing the historical and philosophical groundwork for distinguishing between anti-Darwinian typology and pro-Darwinian population thinking. It might appear that Hull was more attuned to the problem of typology in historical classification. According to Hull (1985, 776–778; 1988, 12–17), historians, akin to taxonomists, should adopt a population-based perspective when examining scientific theories. In historical research, population thinking would mean viewing theories as "historical entities"—i.e., understanding that scientific theories are relative to a particular historical moment and, thus, subject to variation (Hull 1988, 509–511). Thus while discussing the history of Darwinism, to understand it as a "historical entity" is to acknowledge that the concept of Darwinism has evolved over time to the extent that what Victorian Darwinists regarded as Darwinism will differ from today's understanding. Therefore, as the meaning of a theory changes, its examination should resemble the reconstruction of phyletic lines in biology (Hull 1988, 16–17).

One might expect that Hull, acknowledging the mutability of theories, would present a more sophisticated argument supporting the "essentialist story." However, Hull falls into a similar fallacy of *petitio principii* as Mayr. Hull (1988, 25) ultimately constructed a historiographical narrative similar to Mayr's. According to Hull, pre-Darwinian biology was also dominated by essentialism, which Darwin was the first to challenge. However, while Mayr focused on the lack of variability as the primary characteristic of typology, Hull emphasized two additional aspects: the claim that a finite number of types formed the basis of species (Hull 1988, 41–42), and the association with the concept of natural kinds, which influenced the methodologies employed by typologists (such as defining through enumeration of characteristics, employing the principle of exclusion, and the possibility of establishing deterministic scientific laws; Hull 1989, 37–38). Hull (1989, 39) states, "From Huxley to DeVries, Goldschmidt, Schidenwolf, and Goudge, they hoped that evolution might prove saltative. The motivation (...) was to (...) retain the essentialist mode of definition and all that it entailed." This statement suggests that concepts, terms, and names are intrinsically linked to broader ontological consequences. Therefore, the concept, much like in Mayr's approach, cannot be arbitrarily employed, as its internal logic dictates how it can be utilized within the context of scientific theory (cf. Hull 1989, 156). In other words, a scientist who embraces evolutionism but desires to maintain an essentialist mode of definition will be compelled to adopt saltationism as the sole concept compatible with both viewpoints.

This assumption was likely adopted by Hull from the works of Karl Popper. Before Hull started co-creating the "essentialist story" with Mayr, he gathered information about essentialism from three sources: Simpson, Cain, and specifically Popper (Winsor 2006, 166–167). In his first significant article on the subject, titled "The Effect of Essentialism on Taxonomy," Hull relies on Popper's characterization of methodological essentialism. He defines it as the view that originated from Plato and was adopted by Aristotle, which posits that the process of definition involves recognizing the essence of the *definiendum* (Hull 1965, 317). Popper, in his own work, considered methodological essentialism as one of the main hindrances to the development of European science (1994, 226). Following Popper's perspective, Hull argues that the essentialism observed in Plato and Aristotle's philosophy continued in the field of biology through the work of Linnaeus, leading to significant challenges in the advancement of biological systematics (cf. Hull 1965, 316–317). In Hull's later work, this narrative becomes integrated into the rhetoric of the phylogeny of scientific ideas, where scientists by communicating their ideas establish the lineage of a particular scientific theory. Consequently, essentialism, replicated through the works of different scholars, creates a unified lineage of typological thinking that stretches from Plato and Aristotle, through Linnaeus, and beyond. Since replication occurs through interpersonal contact, the sociological factor becomes crucial in reconstructing such a lineage, particularly within the scientific communities where a particular theory or idea was disseminated. Therefore, the task of the historian becomes identifying an "exemplar," a scholar who represents a specific segment of the history of the studied idea and sets the boundaries of the community within which it was transmitted (Hull 1988, 113–114). As a result, Hull concludes that two major streams of biological thought existed in the nineteenth century. The first was idealism, represented by Johann von Goethe in Germany and Étienne Geoffroy Saint-Hilaire in France, embodying all the characteristics of typological thinking. The second was evolutionism, represented by Darwin himself, signifying a departure from essentialism in favor of population thinking (Hull 1988, 40–50, 73).

Similarly to Mayr, the main problem of Hull's philosophy lies in the methodological tools he employs to determine the phylogenesis of examined ideas and concepts. Therefore, the principles of classifying scientists into particular conceptual lineages are quite loose here, which stems from the multitude of criteria that Hull allows for grouping. While the conceptual similarity is a fundamental classifier for him, Hull permits exceptions. For instance, although Mayr has different views from Darwin on geographical isolation, Hull still classifies him as a Darwinist (1989, 235). This is because theories like Darwinism, lacking an essence, can undergo constant modification. As a result, scientists with differing perspectives can be categorized within the same lineage (1989, 236). To establish their relationship, it becomes necessary to consider the social aspect of scientific work, such as who cited whom (1989, 233). Hull (1989, 198) emphasizes the significance of spatiotemporal unity and continuity in reconstructing the phylogenesis of ideas, as they are transmitted through direct or indirect contact. So, in that sense, the link between Darwin and Mayr seems obvious—Mayr read Darwin, cited Darwin, comes from a sequence of scientists working in the context of evolutionary biology, so there is continuity between them. However, as Hull acknowledges, sometimes this method needs to be relaxed, for example, when relationships are not so apparent. Here, Hull gives an example of the relationship between Linnaeus and Aristotle: "In order for Linnaeus to have been influenced by Aristotle's ideas on classification, he must have read Aristotle, or read someone who did, or talked to someone who read Aristotle, and so on. As tenuous as these connections may get, they serve as a justification for treating ideas as historical entities" (1989, 199). This approach raises concerns about delineating the "kinship" of theories. If these connections can be established through indirect or even potential contact, it becomes possible to create or manipulate associations between scientists according to one's preference or agenda. In the same way, for example, Patrick Matthew could be regarded as being in the same lineage as Darwin since there are pieces of "evidence"[5] that Darwin read *On Naval Timber and Arboriculture* before the publication of *On the Origin of Species*. Hull, of course, assumes that Darwin only read Matthew after the publication of his work (1989, 233; 1982, 493) and therefore excludes him from the conceptual genealogy of the idea of natural selection. The underlying problem here, however, is not whether Darwin did or did not read Matthew, but rather the manipulability of the classification principles proposed by Hull. For instance, Hull (1989, 234) argues against the kinship between Lamarck's and Darwin's ideas of evolution, despite the spatiotemporal continuity between them (we know that Darwin was familiar with Lamarck's theory before he published *On the Origin of Species*[6]), due to significant conceptual differences. In this case, conceptual differences matter, whereas they did not in Mayr's case. In his article "Exemplars and Scientific Change" (1982, 493–494), Hull treats Richard Goldschmidt's theory similarly, not considering him a Darwinist despite the fact that he certainly read about and used Darwin's theory, as Goldschmidt's theory of systemic mutations differs too much conceptually from Darwinism. Interestingly, within the same text and paragraph, Hull acknowledges a kinship with Darwinism of Stephen J. Gould's Goldschmidt-inspired theory of punctuated equilibrium.

In his phylogenetics of ideas, Hull frequently employs criteria of similarity and spatiotemporal continuity in an arbitrary and often inconsistent manner, disregarding historical data. This arbitrariness becomes evident when Hull distinguishes between

[5]I placed the word "evidence" in quotation marks because it mainly stems from the controversial claims made by Mike Sutton (Sutton 2015). However, this does not alter my main argument: if Hull allows for even indirect, potential communication between scientists, then one should consider even such controversial claims.

[6]He mentions Lamarck's laws of evolution, for example, in his essay from 1844 (Darwin 2006, 10).

those who are considered evolutionists and those who are not. One significant factor in this differentiation is the presence of typological thinking. For instance, Hull regards Lamarck as a pseudo-evolutionist due to his supposed acceptance of a limited number of taxa into which species could change during evolution (Hull 1988, 87). However, it's worth noting that Lamarck was a nominalist rather than a typologist (cf. Lamarck 1914, 20), which questions the validity of Hull's categorization. Moreover, Hull's definition of evolutionism seems limited. In *Science as a Process* (1988, 91), when discussing the debate between Cuvier and Saint-Hilaire, Hull characterizes it as a discussion between two typologists regarding the number of types. Saint-Hilaire is discredited as an evolutionist because his theory presupposed the existence of a single type from which all species were derived. Similarly, while acknowledging the presence of transmutationist elements in Linnaeus's thought, Hull concludes that Linnaeus's typology prevents him from being considered an evolutionist (1988, 85). In a footnote on the same page, Hull admits to using the term "evolution" in a rather restrictive sense, relying primarily on a modern understanding of the concept.[7] Thus, Hull (1988, 33, 72) openly acknowledges his presentism, considering it a necessary aspect of historical analysis. Consequently, when investigating the history of evolutionism, he approaches it from a Darwinian standpoint which provides him an insight into how to understand the concept of evolution (Hull 1988, 35), and guides his decision on whom to classify as evolutionist.

Presentism lies at the core of the problems of Hull's methodology. In principle, Hull (1988, 510; 1982, 489) allows for the selection of any "exemplar" without the requirement that it be chosen from the most significant or influential scientists. The crucial factor is that the chosen individual belongs to a specific school of thought at a particular time. This flexibility in selection is justified by the understanding that members of a given group may hold divergent views. For example, Darwin and Huxley belong to the same group but differ on the issue of gradualism. Ultimately, it is the scientist's personal identification with a community that holds importance. Hull (1982, 488) posited that there is consistently a certain level of conceptual similarity within a sociological group, which becomes particularly evident during discussions between distinct scientific communities, especially when members defend shared beliefs against critiques from outsiders. However, these broad criteria for selecting exemplars can be used to challenge Hull's historiography, i.e., to incorporate "typological" scientists into the conceptual lineage of evolutionism. For instance, consider the case of Ernst Haeckel being designated as an exemplar of Darwinism. Haeckel, though not the most prominent Darwinist (which should not exclude him as a potential exemplar according to Hull's methodology), belonged to the sociological group of Darwinists. As an exemplar, Haeckel would establish a developmental trajectory of the concept of evolution that encompasses both German idealism and Lamarckian theory. In the end, Haeckel (1906, 59) viewed Darwinism

[7] As an aside, it is worth adding that Hull was aware of the problems that are associated with the use of anachronisms in historiography (1988, 76).

as nothing more than an extension and continuation of Lamarckism, drawing profound inspiration from the works of Goethe (Bowler 1988, 72). Thus, we have a compelling example that aligns with Hull's philosophy but contradicts his historiography, which excludes idealism and Lamarckism as instances of anti-evolutionist typology. It is worth noting that within Hull's methodology, it is conceivable to generate similar "typological" genealogies of evolutionism by utilizing exemplars such as Herbert Spencer or Thomas Huxley.

However, Hull imposed certain limitations on the selection of "exemplars," specifically addressing the potential influence of the community's name on the categorization of its members. In his work "Exemplars...," Hull (1982, 488) argues against classifying Mendel as a Mendelist because the Mendelian school is defined by figures such as Bateson, who did not work in the same period as the Moravian geneticist. The specificity of Hull's reasoning is better exemplified by the example he provides, illustrating a potential mistake in treating George J. Romanes, the creator of the term "Neo-Darwinism," as a representative of this evolutionary school of thought. Hull explains that such an error would be quickly detected, as it is evident from Romanes' writings that he coined the term "Neo-Darwinism" to describe supporters of August Weismann's theory, from which he distanced himself. Moreover, Hull states, "...a rigid application of the type-sample method would require that the name 'neo-Darwinian' be applied to Romanes and his scientific circle, rather than to workers such as Thiselton-Dyer, E. R. Lankester, and E. B. Poulton" (1982, 490). The mistake would be rectified when it becomes apparent that the classification did not encompass the appropriate scientists. This brings us to the core of the problem: Hull's classification relies on the historian already knowing who should be assigned to a particular group and who should not. As shown in the example above, historians will know beforehand that Lankaster, Poulton, and Thiselton-Dyer were neo-Darwinists, while Romanes was not. A more in-depth analysis is not necessary here, as classification, in the end, depends on an existing consensus regarding whom to classify where. As a result, it becomes derivative of certain initial assumptions established by the historian. Criteria such as conceptual similarity and sociological and spatiotemporal relationships are freely employed to confirm preconceived connections between theories and scholars.

At the beginning of "Exemplars...," Hull addresses the criticism of the "essentialist story" by stating, "Numerous contemporary authors argue that no one from Aristotle to the present has ever adhered to the previous view of species. Certainly, no philosopher or naturalist held precisely the view described above, but views of this sort were just as surely common during the period discussed" (1982, 481). As an illustration, he cites the examples of William Whewell and John Stuart Mill. Their views, because superficially similar to Aristotle's essentialism, allow Hull to categorize them within the same intellectual current. Despite significant differences from Aristotle's metaphysics and their existence in different time periods, they are considered to represent the same typological way of thinking as Aristotle did (it is noteworthy that Hull considered the temporal factor in his decision not to classify Mendel as a Mendelist). Despite Whewell (1860, 25–27) arguing against using Aristotle's concept of species and genus in biological taxonomy, he is still recognized as a typologist. Similarly, Mill who claimed (1843, 126) that classification should be based on observed similarities among entities rather than metaphysical essences is also recognized as a typologist. These crucial distinctions are disregarded, when in other cases (as was in the case of, e.g., Lamarck, Matthew, and Goldschmidt) conceptual differences were considered important in determining classification. All these aspects are overlooked in favor of upholding the "essentialist story." In conclusion, Hull assumes a Darwin-centric classification, acknowledging its bias, and interprets historical data accordingly using various taxonomic criteria. Consequently, the "essentialist story" serves as both the premise and the outcome of his historical analysis. Without any additional evidence to support it, it can be concluded that Hull simply perpetuated this received view and utilized his philosophical framework to provide justification for it.

4 The Sociological Aspect of the "Essentialist Story"

The main conclusion we can draw from the previous considerations is that the "essentialist story," as presented by its main proponents, appears as a preconception rather than a conclusion of their philosophical and historical analyses. I am not arguing here against the existence of a type-based method in the nineteenth century or some superficial similarities between non-Darwinian theories of evolution and ancient essentialist concepts like the "natural state model." However, the crucial point lies in the superficiality of these similarities. Neither Mayr nor Hull have provided sufficient arguments to establish that we are dealing with the same concepts. Their analyses, which aimed to demonstrate a connection between nineteenth-century typology and ancient metaphysics, relied on a weak methodology that was fundamentally whiggish and presentist. Such a methodology cannot serve to prove the rather strong thesis about the connection between nineteenth-century non-Darwinian natural science and the legacy of Plato and Aristotle. Despite the weak methodological foundations and lack of compelling evidence, the "essentialist story" has gained widespread acceptance in the modern history and philosophy of biology. Historians have often employed Mayr's distinction between population and typological thinking when analyzing the history of nineteenth-century biology, even when challenging the Darwin-centric narrative (e.g., Bowler 1988, 108). However, it is important to remember that the similarity of concepts does not necessarily imply their affinity or compatibility.

Undoubtedly, the emergence of the "essentialist story" is closely tied to the development of a synthesocentric historical narrative, co-authored by Mayr, with the intention of establishing a connection between the modern synthesis and Darwin's original theory. Contemporary historians of science (e.g., Bowler 1992; Largent 2009; Delisle 2021) highlight the problematic nature of this narrative, which is revisionist in nature and aims to downplay the contributions of non-Darwinian naturalists to biology. The distinction between typology and population thinking played a significant role in achieving this, as it allowed opponents of Darwinism to be categorized as perpetrators of irrational ancient metaphysics. Carl Chung (2003,

292–295) indicates that Mayr employed this distinction specifically to discredit his intellectual adversaries. So, if the typology-population distinction initially served as a propaganda tool in internal discussions among biologists, why has it been (mostly) unquestioningly adopted by philosophers and historians of biology? Three factors appear to be at play here. Firstly, its acceptance was tied to the initial monopoly that biologists held in studying the history of their discipline (it is worth noting that this situation was not exclusive to the history of biology but extended to the history of science as a whole: Hiebert 1994, 89). Early studies on the history of biology focused on the question of why the synthesis emerged and what delayed its development (Browne 2010, 366–367). The history of biology was intended to serve an instrumental purpose, primarily aimed at uncovering the origins of the discipline rather than striving for an accurate account of its history. Mayr even directly criticized historians who focused on thorough historical analysis, labeling them as "priggish" (1990, 304–305). Later, nonbiologists interested in the history of evolutionism conducted their historical analyses within the framework and terminology established by biologist-historians, thereby replicating the research goals they had set (Smocovitis 1996, 52–59). Secondly, there were noticeable similarities between essentialist metaphysics and the views of non-Darwinists, which made it easier for nonbiologists to accept the thesis regarding the existence of typological thinking. However, the problem lies in the fact that the "essentialist story" not only suggests similarity but also proposes a stronger thesis of a direct connection between non-Darwinian theories and ancient metaphysics. What then led to the adoption of this stronger thesis? Here, a third factor comes into play: the authority of the proponents of the "essentialist story." Mayr, having had a tremendous influence not only on biology but also on the history and philosophy of biology, was able to present his theses *ex-cathedra* without providing historical evidence. As Winsor writes (2006, 154–156), Mayr was not required to provide proof, as his historical works primarily served a utilitarian purpose, mainly to reinforce the validity of the emerging paradigm of the modern synthesis.

The spread of the opposition between typological and population thinking can be better understood thanks to Saul Kripke and Hilary Putnam's theory of reference. According to Kripke (1996, 266) the linking of a name to a referent relies on the "initial baptism," which is the moment when the name is first associated with its intended referent.[8] The meaning of a name is fixed through historical continuity in its usage, with each successive user relying on the past usage. This was the case with "typology." Certain conceptual similarities observed across different theories were recognized and identified as instances of typology. The initial baptism in this context was conducted by the architects of the synthesis, such as Simpson, Dobzhansky, and Mayr. Subsequently, later researchers, including Hull, reinforced this usage, leading

[8] A referent should be understood as a specific entity or thing in the real world to which a word or expression refers. Initial baptism is, therefore, an explanation of how terms are linked to their referents. In other words, it is the first act of naming something that establishes the connection between the real world (existing independently from the individual) and the unit of language.

to a fixation on the term "typology" and its referents. The significance of those performing the baptism was not coincidental either. Putnam (1996, 287–288) observed that the use of language is fundamentally social, involving a division of linguistic labor within any community. There are language users who possess greater knowledge of how certain terms should be used. Since speakers do not have complete knowledge of a term's full extension, they rely on the opinions of specialists to help them navigate problematic situations (Putnam 1996, 291). Mayr, as a specialist within the community of historians and philosophers of biology, played such a role in shaping the extension of the term "typology." However, the issue arises when a specialist presents a false extension, as Mayr did by applying the concept of typology to both nineteenth-century natural science and ancient essentialist metaphysics, suggesting a connection between the two. In this context, Ludwik Fleck's concept of esoteric and exoteric circles of science can shed further light on the matter.

According to Fleck (1986, 138–139, 146–148), intellectual communities, referred to as "collectives" in his terminology, possess a hierarchical structure consisting of two parts: the esoteric and the exoteric. The esoteric part comprises active members who generate and critically analyze scientific knowledge. At its core are highly skilled professionals in a specific field, followed by scientists in related disciplines, and at the periphery are highly educated individuals who are not considered professional scientists. On the other hand, the exoteric part represents the domain of laypeople who gain access to scientific knowledge through representatives from the esoteric part. However, communication between these two spheres is not seamless. While theories, discoveries, and theses are constantly discussed within the esoteric circle, their transmission to the exoteric circle is simplified and presented as dogma. Fleck explains this as follows: "A discovery that has been made by an expert has to rework itself in stages through second and third discoveries for successive, increasingly popular popularizers to make it accessible to the layman or expert in another field" (Fleck 2007, 269). Applying Kripke's sociolinguistic theory to Fleck's framework of the scientific community reveals that specialists communicate a simplified and controversy-free representation of the truth regarding the extension of terms. Furthermore, if the "initial baptism" of a term's meaning took place within the esoteric circle, its subsequent adoption by nonspecialists may also rely on a simplified version that was initially presented by specialists to the laymen in the exoteric circle (see Fig. 1).

In the case of typology, we encounter a term whose extension has been extensively debated within esoteric circles. This is exemplified by the dispute among Mayr, Simpson, and Dobzhansky regarding its application to the views of Carlton Coon (Witteveen 2015b, 6–8). However, in Mayr's historical and philosophical works, the term is presented as free from such uncertainties. Through Mayr's conversations with other members of the esoteric circle, a fairly clear picture emerges of what he intended to communicate to the exoteric circle. John C. Greene, in his book *Science, Ideology, and World View*, included an excerpt from his correspondence with Mayr regarding the article "Darwinism as a World View." In the article, Greene stated that the concept of Darwinism was understood

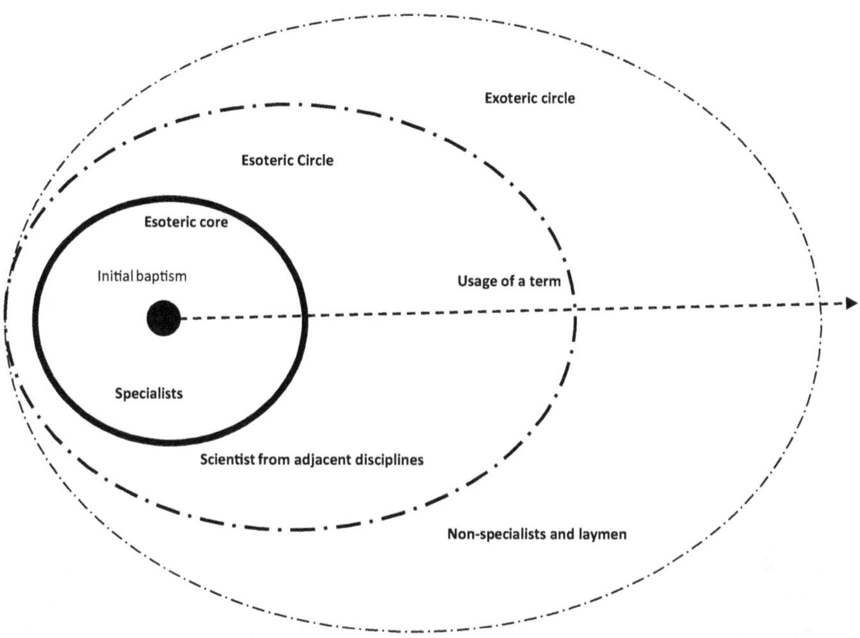

Fig. 1 Division of linguistic labor in the scientific community within esoteric and exoteric circles

differently in the nineteenth century, with a meaning distinct from its modern understanding. While Mayr agreed with this conclusion, he added, "...when it comes to using terms that are common coin in modern science, we cannot load them down with the uncertainties of history" (Greene 1981, 155). Another anecdote, shared by Joeri Witteveen, involves a philosopher questioning Mayr about authors who contributed to the development of population thinking. Mayr did not provide specific names and commented, "The trouble is that philosophers indulge too often in dichotomy. They still tend to follow the axiom of scholastic logic: 'Tertium non dat.' Actually, in most philosophical and scientific controversies, there are three or more alternatives" (quoted in Witteveen 2015b, 9). According to Mayr, the history of biology should be presented in a form devoid of uncertainties.

There are two conclusions that can be drawn from applying Fleck's model to the story of the typology/population distinction. Firstly, the critique of the "essentialist story" by historians and philosophers of biology arises from questioning the authority of Mayr's historical narrative. It was only when this authority was challenged that it became apparent how much the concept of typology oversimplified the historical development of the species concept. Even Winsor (2006, 169) acknowledged that she initially accepted Mayr's and Cain's work without question, only later in her career recognizing its problematic nature. Hence, historians had to penetrate the esoteric circle in which the "essentialist story" was constructed. Consequently, it was only when the history of biology established its own independent esoteric circle, separate from the influence of biologists themselves, that a reevaluation of the terminology used thus far became possible. This leads to the second conclusion: the relationship between biologists and humanists studying the history of their field was not one of equals, but rather an example of communication between the esoteric and exoteric circles. Mayr used history as a tool and treated his historical works as a simplified version for popularizing the development of evolutionary biology. In the case of Hull, we are not dealing with an individual who belonged to the esoteric circle, but rather someone who embraced a version of the history of biology communicated to the exoteric circle.[9] After all, his paper "The Effect of Essentialism on Taxonomy" relied on the authority of Simpson and Cain. Therefore, his philosophical analyses were based on the simplistic narrative of typology and population thinking, which Mayr and others presented to the exoteric circle of nonspecialists, including historians and philosophers of biology.

5 Conclusions

Typology and population distinction were created on a dubious philosophical basis, which calls into question the entire historiographical narrative that was based upon it. In the works of both Mayr and Hull—who are regarded as leading popularizers of the "essentialist story"—we can see how arbitrary and methodologically flawed their analyses were. In both cases, it can be observed that their rationale behind the categorization of certain theories as representatives of typology served a particular agenda, aimed at perpetuating a Darwin-centric vision of the history of biology and discrediting non-Darwinian theories of evolution. Their conclusion that non-Darwinian theories were linked to ancient essentialism was based on fallacious methodology, and in both cases, it was hindered by the fallacy of *petitio principii*, i.e., this link wasn't proven. The "essentialist story" should be regarded as a major simplification of the history of biology that was popularized by the architects of synthesis, especially Mayr, and later presented to nonbiologists as indisputable truth. Historians accepted it because of the authority of these architects and only later began to scrutinize whether the "essentialist story" was in any way correct. In this chapter, I've attempted to demonstrate that even without an in-depth analysis of the history of biology, we can see that the "essentialist story" should be approached with caution, as the argumentation behind it was methodologically flawed. Can we then completely reject the "essentialist story"? I would not argue for such a strong thesis. Instead, I would propose suspending the conclusions of Mayr and Hull. This is because the "essentialist story" persisted for so long in historiographical and philosophical discourse that it is highly unlikely that it was merely a fabrication.

[9]In a way, this is corroborated by Hull's (1985, 773–774) account of the American Association for the Advancement of Science conference, where he was surprised to find that biologists disagreed on the relevance of evolutionary theory, which contradicted the vision, promoted by the architects of the coherence of synthesis.

According to Putnam's sociolinguistic theory (1996, 291), the extension of a term is determined both sociologically, through its use in a specific community where specialists play a crucial role, and ontologically, by the existence of certain characteristics that allow entities to be recognized as designators of the term. This implies that even if the concept of typology was simplified for the general public by specialists, there must still be entities that laypeople could recognize as referents of "typological thinking." If it were a complete fabrication, historians and philosophers could not apply those terms in their work. In reference to Putnam, one could argue that there was an ontological determinant that influenced the development of the term "typology," which later "specialists" employed in constructing the "essentialist story." After all Darwin himself criticized the concept of species used by Linnaeus, noting that it presupposes an "unknown element of a distinct act of creation" (1859, 44). However, this does not mean that because there were some elements in past theories that were similar to essentialism, the conclusion that they represented a similar way of thinking about species or were part of the same ontology and metaphysics is valid. Let's take an example from evolutionary biology—we know that similar organs can evolve independently from each other, such as wings and eyes. So, in nature, there are similar organs that are structurally different. Based on this, biologists employing the principle of parsimony know that just because bats and birds have similar organs—wings—it doesn't imply that they are closely related. Returning to the case of essentialism, even if we observe some similarities, it doesn't necessarily mean we are dealing with the same concept or the same metaphysics. If we want to claim that, we need to prove it first. Without doing so, we are simply reiterating the received view, and one that was based on a flawed methodology. Observed similarities between non-Darwinian theories and ancient metaphysics can thus explain why the "essentialist story" was accepted for so long but cannot be used as the sole evidence for it. Until someone proves it, historians and philosophers of biology should also adhere to the principle of parsimony when discussing the affinity of past theories and philosophies.

As noted by Anthony Smith (2003, 129–130), for an "invented tradition" to gain acceptance within a community, it must be based on intersubjective truth about the past. Smith explains that the creation of traditions involves emphasizing certain events while marginalizing others from a community's history. Thus, the "essentialist story" can be seen as an "invented tradition" in Smith's sense. It drew on actual historical facts and real aspects of certain theories to paint a narrative of the history of evolutionism in which Darwin became the champion of a population-based approach to understanding species while competing theories were portrayed as continuations of ancient metaphysics. This "invented tradition" was promoted among biologists, historians, and philosophers to highlight the significance of the emergence of the modern synthesis and to foster unity within the scientific community under the banner of Darwinism. More precisely, this specific vision of Darwinism was envisioned as a truly scientific description of evolution that surpassed other alternatives, which were associated with unscientific essentialism. However, given the issues with the methodology that contributed to the construction of this vision of Darwinism, one should question whether the exceptionalism attributed to Darwin's theory of natural selection by the "essentialist story" was accurate. Consequently, it prompts us to consider whether its alternatives should still be regarded merely as extensions of ancient metaphysical concepts.

References

Amundson R (2005) The changing role of the embryo in evolutionary thought. Cambridge University Press, Cambridge, Roots of Evo-Devo

Bateson W (1913) Problems of genetics. Oxford University Press–Yale University Press, London–New Haven

Bateson W (1984) Materials for the study of variation treated with special regard to discontinuity in the origin of species. Macmillan, London–New York

Beirne P (1987) Adolphe Quetelet and the origins of positivist criminology. Am J Sociol 5:1140–1169

Bowler PJ (1988) The non-Darwinian revolution. Reinterpreting a historical myth. The Johns Hopkins University Press, Baltimore–London

Bowler PJ (1992) The eclipse of Darwinism. Anti-Darwinian evolution theories in the decades around 1900. The John Hopkins University Press, Baltimore–London

Bredsdorff T (1975) Lovejovianism—or the ideological mechanism. An enquiry into the principles of the history of ideas according to Arthur O. Lovejoy. Orbis Litterarum 30:1–27

Browne J (2010) Making Darwin: biography and the changing representations of Charles Darwin. J Interdiscip Hist 3:347–373

Chung C (2003) On the origin of the typological/population distinction in Ernst Mayr's changing views of species, 1942-1959. Stud Hist Phil Biol & Biomed Sci 34:277–296

Darwin C (1859) The origin of species by means of natural selection, in the preservation of favored races in the struggle for life. John Murray, Albemarle Street, London

Darwin C ([1842]2006) Essay of 1842. In: Darwin F (ed) The foundation of the origin of species. Two essays written in 1842 and 1844 by Charles Darwin. Cambridge University Press, New York, pp. 57–255

Delisle RG (2021) Introduction: in search of a new paradigm for the development of evolutionary biology. In: Delisle RG (ed) Natural selection. Revisiting its explanatory role in evolutionary biology. Springer, Cham, pp 1–8

DeVries H (1906) Species and varieties. Their origin by mutation. Lectures delivered at the University of California. The Open Court–Kegan Paul, Chicago–London

Dewey J (1910) The influence of Darwinism on philosophy. In: The influence of Darwin on philosophy and other essays in contemporary thought. Henry Holt and Company, New York, pp 1–19

Faber PL (1976) The type-concept in zoology during the first half of the nineteenth century. J Hist Biol 1:93–119

Fleck L (1986) Powstanie i rozwój faktu naukowego. Wprowadzenie do nauki o stylu myślowym i kolektywie myślowym. Wydawnictwo Lubelskie, Lublin

Fleck L ([1939]2007) Nauka a środowisko. In: Werner S., Zittl C., Schamtz F. (eds) Ludwik Fleck. Style myślowe i fakty. Artykuły i świadectwa. Wydawnictwo IFiS PAN, Warszawa

Greene JC (1981) Science, ideology, and world view. Essays in the history of evolutionary ideas. University of California Press, Berkeley, Los Angeles, London

Haeckel E (1906) Last words on evolution. A popular retrospect and summary. A. Owen & CO., London

Hiebert EN (1994) On demarcations between science in context and the context of science. In: Gavroglu K, Christiandis J, Nicolaidis E (eds) Trends in the historiography of science. Springer, Cham, pp 87–106

Hobsbawm E (2000a) Introducing: inventing traditions. In: Hobsbawm E, Ranger T (eds) The invention of tradition. Cambridge University Press, Cambridge, pp 1–14

Hobsbawm E (2000b) Mass-producing traditions: Europe, 1870-1914. In: Hobsbawm E, Ranger T (eds) The invention of tradition. Cambridge University Press, Cambridge, pp 263–307
Honenberger P (2015) Grene and Hull on types and typological thinking in biology. Stud Hist Phil Biol Biomed Sci 50:13–15
Hull DL (1965) The effect of essentialism on taxonomy–two thousand years of stasis (I). Br J Philos Sci 60:314–326
Hull DL (1982) Exemplars and scientific change. PSA: proceedings of the biennial meeting of the Philosophy of Science Association, Vol. 1982, Volume Two: Symposia and Invited Papers: 479-503
Hull DL (1985) Darwinism as a historical entity: a historiographic proposal. In: Kohn D (ed) The Darwinian heritage. Princeton University Press, Princeton, pp 773–812
Hull DL (1988) Science as a process. An evolutionary account of the social and conceptual development. The University Chicago Press, Chicago–London
Hull DL (1989) The metaphysics of evolution. State University of New York Press, New York
Hull DL (1994) Ernst Mayr's influence on the history and philosophy of biology: a personal memoirs. Biol Philos 9:375–386
Junkier T (1997) Factors shaping Ernst Mayr's concepts in the history of biology. J Hist Biol 1:29–77
Kripke S ([1972] 1996) Naming and necessity. In: Martinich A P (ed) The philosophy of language. Oxford University Press, New York, Oxford, pp. 255–270
Lakatos I (1971) History of science and its rational reconstructions. Boston Stud Philos Sci VIII:91–135
Lamarck JB ([1809]1914) Zoological philosophy. Macmillan, New York
Largent MA (2009) The so-called eclipse of Darwinism. In: Cain J, Ruse M (eds) Descended from Darwin insights into the history of evolutionary studies, 1900–1970. American Philosophical Society, Philadelphia, pp 3–21
Levit GS, Meister K (2006) The history of essentialism vs Ernst Mayr's "essentialism story": a case study of German idealistic morphology. Theory Biosci 124:281–307
Lovejoy AO (1948) The great chain of being. A study of the history of an idea. Harvard University Press, Cambridge-Massachusetts
Mayr E (1949) Speciation and selection. Proc Am Philos Soc 6:514–519
Mayr E (1957) Species concepts and definitions. In: Mayr E (ed) The species problem. American Association for the Advancement of Science, Washington, pp 1–22
Mayr E (1959) Darwin and the evolutionary theory in biology. In: Meggers BJ (ed) Evolution and anthropology. A centennial appraisal. The Anthropological Society of Washington, Washington, pp 1–10
Mayr E (1963) Animal species and evolution. The Belknap Press of Harvard University Press, Cambridge
Mayr E (1969) Principles of systematic zoology. McGraw-Hill Book Company, New York
Mayr E (1982) The growth of biological thought. Diversity, evolution, and inheritance. The Belknap Press of Harvard University Press, Cambridge–London
Mayr E (1988) Toward a new philosophy of biology. Observations of an evolutionist. Harvard University Press, Cambridge–London
Mayr E (1990) The myth of the non-Darwinian revolution. Biol Philos 1:85–92
Mayr E (1997) Evolution and diversity of life. The Belknap Press of Harvard University Press, Cambridge, London
Mayr E (2002) What evolution is. Weidenfeld & Nicolson, London
Mill JS (1843) The collected works of John Stuart Mill, volume VII–A system of logic part I. University of Toronto Press, Toronto
Popper K (1994) The open society and its enemies. New One-Volume Edition. Princeton University Press, Princeton and Oxford
Putnam H ([1973] 1996) Meaning and reference. In: Martinich A P (ed) The philosophy of language. Oxford University Press, New York, Oxford, pp. 284–291

Richards RA (2010) The species problem: a philosophical analysis. Cambridge University Press, Cambridge

Smith AD (2003) Nationalism and modernism. A critical survey of recent theories of nations ad nationalism. Routledge, London and New York

Smocovitis VB (1996) Unifying biology: the evolutionary synthesis and evolutionary biology. Princeton University Press, New Jersey

Sober E (1980) Evolution, population thinking, and essentialism. Philos Sci 3:350–383

Sutton M (2015) On knowledge contamination: new data challenges claims of Darwin's and Wallace's independent conceptions of Matthew's prior-published hypothesis. Philosophical Aspects of Origin 12:167–205

Whewell W (1860) On the philosophy of discovery, chapters historical and critical. John W. Parker and Son, London

Wilkins JS (2009) Species. A history of the idea. University of California Press, Berkely, Los Angeles, London

Wilkins JS (2010) What is a species? Essences and generation. Theory Biosci 129:141–148

Winsor MP (2006) The creation of the essentialism story: an exercise in metahistory. Hist Philos Life Sci 2:149–174

Witteveen J (2015a) "A temporary oversimplification": Mayr, Simpson, Dobzhansky and the origins of the typology/population dichotomy (part 1 of 2). Stud Hist Phil Biol Biomed Sci 54:20–33

Witteveen J (2015b) "A temporary oversimplification": Mayr, Simpson, Dobzhansky and the origins of the typology/population dichotomy (part 2 of 2). Stud Hist Phil Biol Biomed Sci xxx:1–10

Fisher, Wright, and Haldane: Three Philosophical Conceptions of Evolution

Joël Dolbeault

Abstract

Fisher, Wright, and Haldane are the main founders of population genetics, a central part of the Modern Synthesis. In addition to their scientific writings, each of them published *philosophical writings* presenting a certain conception of nature and evolution. However, a comparative study of these conceptions leads to a surprising result. First, Fisher, Wright, and Haldane do not share the same overall conception of evolution. What is more, Wright and Haldane's conceptions introduce elements that are foreign or opposed to Darwinism, such as the hypothesis of intentional forces, or that of a general directionality of evolution. In this sense, this study challenges the traditional narrative of unification held by some synthetists or historians. It shows that, among the synthetists, a wide variety of ideas on evolution continued to exist. First, I explain that, on the basis of an emergentist philosophy of nature, Fisher develops a philosophy of evolution consistent with Darwinism, in the sense that, for him, the main cause of evolution is natural selection. In a second step, I attempt to show that Wright's panpsychist philosophy leads him to hypothesize that the creativity at work in evolution depends essentially on purposeful forces intrinsic to organisms, at all levels of organization in nature. In a third step, I attempt to show that, on the basis of a philosophy that is also panpsychist, Haldane hypothesizes that some laws of organization of matter give evolution a general orientation toward more complexity and mind. At the end of the chapter, I discuss the interest of taking into account the philosophy of nature and evolution defended by these biologists.

J. Dolbeault (✉)
Department of Philosophy, University of Lille, Lille, France

R. G. Delisle et al. (eds.), *Unity and Disunity in Evolutionary Biology*,
https://doi.org/10.1007/978-3-031-42629-2_11

Keywords

Fisher · Wright · Haldane · Darwinism · Anti-reductionism · Emergentism · Panpsychism · Holism

1 Introduction

Fisher, Wright, and Haldane are the main founders of population genetics, a central part of the Modern Synthesis. However, in addition to their purely scientific work, these three biologists published writings that present a certain *philosophy of evolution*, in the sense of a global view of evolution, with a more or less speculative dimension. The reading of these writings leads us to two questions: do Fisher, Wright, and Haldane have the same global view of evolution? Moreover, is the global view of each of them perfectly consistent with the Modern Synthesis? The aim of this paper is to answer these two questions. The notion of "synthesis" could lead one to think that, for these two questions, the answer is affirmative. We shall see, however, that it is negative.

In Fisher, Wright, and Haldane, the global view of evolution is always based on a global view of nature, in particular on a certain conception of life and mind. For each of them, we shall therefore have to consider this view of nature, which is in fact a certain *philosophy of nature*. We shall see in particular that, even if Fisher, Wright, and Haldane do not hesitate to speculate on nature, their speculations are always guided by a concern for rationality. In other words, for them, it is not a question of expressing a personal preference on nature, inspired by a religious belief for example. It is a question of proposing a global view of nature and evolution in coherence with the whole of the observable facts.

This work is largely inspired by Delisle's work on the philosophy of the biologists who built the Modern Synthesis (Delisle 2009a, b). Like Delisle's work, it shows that, for many synthetists, Darwinism is accompanied by a philosophical reflection on nature in general. It also shows that this reflection produces results that are more or less consistent with Darwinism. Should we then give importance to these philosophical speculations, one might ask? But can we do otherwise, one might answer? Darwinism is strongly linked to a global view of nature. Thus, in one way or another, it is immediately on a philosophical ground.

Some publications have already been devoted to the philosophical ideas of Fisher and Wright on the one hand (Hodge 1992; Steffes 2007; Gayon 2008), of Haldane on the other (Sarkar 1992; Hammond 2009; Dronamraju 2017). However, a comparison of the three has never been done. Moreover, some important elements seem to have been left out by commentators, including the emergentism defended by Fisher, the panpsychism defended by Haldane, and his hypothesis of laws of organization of matter. Besides, another originality of the work presented here is to systematically ask the question of the coherence of these ideas with the Modern Synthesis.

The first three parts of this chapter will be devoted successively to Fisher, Wright, and Haldane. In each part, I will first explain the philosophy of nature developed by the author, then his philosophy of evolution, before asking the question of the relation of the latter to Darwinism. In the last part of the chapter, after a brief recapitulation of the previous parts, I will return to the interest of studying the philosophical ideas of the synthetists.

2 Fisher

Ronald Fisher (1890–1962) was a British biologist and statistician. His main contribution to evolutionary biology is the development of statistical methods in genetics. In addition to his scientific work, Fisher published two philosophical writings in which he developed certain ideas about nature and evolution:

- "Indeterminism and Natural Selection," *Philosophy of Science*, 1934
- *Creative Aspects of Natural Law* (transcription of a lecture), 1950

These two writings do not constitute a complete philosophy of nature and evolution. However, they allow Fisher to clarify his position on certain fundamental questions, in particular on the question of determinism. A comprehensive view of nature and evolution then emerges.

As these two writings present similar ideas, and as the 1950 one is more accomplished than the 1934 one, I will mainly refer to the 1950 writing. If necessary, I will indicate the possible differences between the two writings.

2.1 On Nature

The main focus of Fisher's philosophical writings is evolution. But what Fisher says about evolution is based on a certain conception of nature. We must therefore first consider this conception.

The main element of this conception is *indeterminism*, in the ontological sense. Here is how Fisher explains it:

> Perhaps I should begin by saying what I apprehend the word "creative" to mean. First, taking the word coldly and dryly, divesting it of emotional significance and moral associations, I take it to qualify effective causation; to imply that had the nature or intensity of the causal system been different, the effects which flow from it would also be different; merely different, not necessarily worse or better. This requires that the causal system itself might really have been different, and so have entailed different real consequences. Creative causation in this sense is thinkable in an indeterministic world, in which the causative system might indeed have been different. (Fisher 1950: 1)

In this passage, Fisher wants to explain that the word "creative" can be understood from the notion of indeterminism. What interests us here is the way Fisher explains this notion. For him, indeterminism does not call into question the principle "same causes, same effects." If some effects are not determined in advance, it is because there is some indeterminacy in the causes, as if the causes were not totally fixed. In other words, as if the causes could *vary by themselves* (without this variation being due to anything other than themselves) before producing their effects.

Now, according to Fisher, we have good reasons to think that indeterminism is true. A first reason is that, throughout its history, physics has never proved the truth of determinism at the level of individual atoms. Physics has even shown that the determinism observed at the macrophysical level can be understood as a statistical consequence of microphysical processes whose behavior could be indeterministic (Fisher 1950: 2). A second reason, stronger than the first, is that, for contemporary physics, the world seems to be indeterministic:

> The study of atoms, and sub-atomic particles and processes, can, it now appears, only be carried on by recognising indeterminism as inherent in their nature. It is particularly important, in this respect, that there is not a vague or uncertain degree of indeterminism, but a specific amount, calculable from one of the fundamental constants of physics. No appeal to subjectivity, or to human limitations, can explain it away. It is, for the present generation, one of its most important tasks to recognise what this change in our outlook requires, and to adjust our minds to its implications. (Fisher 1950: 2)

This passage obviously refers to the quantum theory, which, in 1950, was well established. Fisher emphasizes that, for this theory, indeterminism is not the consequence of a limitation of human knowledge. It is something inherent to physical processes, which can be quantified. This is why we can speak of *an ontological* (as opposed to epistemological) *indeterminism.*

In the 1934 article, Fisher develops two additional reasons, this time referring to psychology. First additional reason: “the existence of the power of effective choice, universally verified by subjective experience” (Fisher 1934: 107). This power of choice testifies to a form of indeterminism in the sense that consciousness appears as a cause *that varies by itself*, before producing its effect, namely an action. Second additional reason: while our consciousness has some access to past experience (thanks to memory), it has no access to future experience. For Fisher, this asymmetry is inexplicable within the framework of determinism because the latter implies that, given a certain present state, the future as well as the past is completely fixed (Fisher 1934: 107). In the 1950 article, Fisher still mentions the first additional reason (Fisher 1950: 15–16), but not the second one. On the whole, it may be thought that, for him, psychological reasons are less important than physical ones.

One question remains. By relying on physics and on psychology, Fisher actually introduces two kinds of ontological indeterminism. The first, based on physics, concerns processes *without* consciousness, whereas the second, based on psychology, concerns processes *with* consciousness. Does Fisher make a connection between these two indeterminisms? And if so, which one?

Fisher does not deal with this question in a systematic way, but in his two writings, he provides some elements of an answer.

First, it is clear that he rejects the idea of a reduction of indeterminism with consciousness to indeterminism without consciousness. This reduction would in fact amount to affirming that consciousness has no real influence on things. Fisher rejects this idea (Fisher 1934: 108; 1950: 15–17). In the 1934 article, he even develops a classic argument against it: if conscious choices were without any real effect on action, then they would represent no selective advantage, and their existence in some species would be purely fortuitous. Since this consequence is hardly acceptable from the point of view of the theory of evolution, the starting hypothesis must be rejected (Fisher 1934: 108).[1]

Moreover, Fisher does not propose the opposite reduction, i.e., the reduction of indeterminism without consciousness to indeterminism with consciousness. This reduction would correspond to panpsychism, since it would amount to affirming that physical processes have a minimal degree of consciousness. In fact, Fisher does not say anything on this point. We can therefore think that he does not defend this idea.

Besides, Fisher rejects the idea that the two indeterminisms are unrelated because, for him, consciousness originated in the evolution of inert matter (Fisher 1950: 5).

By deduction, we understand that Fisher defends *a form of emergentism* in the sense that, for him, a first form of indeterminism, without consciousness, has given rise to a second form of indeterminism, with consciousness.

For Fisher, the organization of matter is obviously the condition for the emergence of consciousness. On the other hand, if consciousness is not a simple epiphenomenon, it must have an action on the organism. In this sense, consciousness must be both *an effect* and *a cause* of organization. This double point of view, typical of emergentism, is expressed in the following formula: "If consciousness is, as it would seem, the symbol, or even the means, of unification in our being, this is the region to which creative activity could most fitly be traced (Fisher 1950: 17).

On life, as on consciousness, Fisher defends a form of emergentism. In this sense, he emphasizes that, throughout evolution, "new things" appear: growth, voluntary movement, striving and effort, joy and pain, etc. (Fisher 1950: 5).

In a passage from 1934, he stresses in particular the importance of the emergence of purposive (unconscious) actions, as precursors of conscious actions:

> An indeterministic world, then, is one in which the human qualities of aspiration, planning and foresight, are rationally possible and may be advantageous, and in which we can recognise the primitive precursors of these qualities, not as an epiphenomenon, but as having a real part to play in the survival or death of the organisms that evince them. Biologically it might be said that purposive action by the organism as a whole is the crowning stage of an evolutionary process by which relatively large masses of living matter have come to achieve that co-operation of parts and unity of structure which we call individuality. For, on statistical view of causation, spontaneity or creative causation is at its highest only when perfect unity is achieved. (Fisher 1934: 108)

The important point in this passage is the affirmation that purposive actions are not an epiphenomenon. For Fisher, these actions are *an effect* of the progressive

[1] This argument was first formulated by William James (1879).

organization of matter (thanks to evolution), but also *a certain cause* of organization, because of their role in the activity of organisms.

Commentators of Fisher's philosophy emphasize his indeterminism, but not his emergentism (Hodge 1992; Gayon 2008). Yet this point is fundamental. This may be because Fisher does not use the word "emergentism" to describe his conception, whereas he uses the word "indeterminism." However, this conception is both indeterministic and emergentist, for it asserts that mechanical processes have given rise to purposeful and conscious processes, and that the latter are irreducible to the former.

This interpretation of Fisher's writings is consistent with what he says about mechanism and vitalism. According to him, biology must go beyond these two positions:

> Biologists, perhaps, as this view [indeterminism] becomes more familiar, will be more inclined to elaborate independently concepts appropriate to their own field of study, and be less inclined to assume that all improvements in method must be those introduced from the "exact" sciences of physics and chemistry. They may also come to wonder what the controversy between "mechanism" and "vitalism" can have been all about. (Fisher 1934: 104; see also 1950: 6)

On the one hand, it is obvious that the living presents new types of behavior compared to inert matter. Therefore, mechanism (here, in the sense of the reduction of living processes to inert processes) is not defensible. On the other hand, these behaviors are made possible by the fundamental indeterminism of the physical world. They do not depend on a special force. Therefore, vitalism is not tenable either. The overcoming of these two positions implies a form of emergentism.

Finally, let us note that Fisher's philosophy of nature also implies *a form of holism*, defined as the idea that an organized whole is capable of types of behavior of which its parts are not capable. Thus, for him, some organisms are capable of purposive actions, sometimes conscious, while the parts of these organisms (their atoms, in particular) are not. Fisher explicitly assumes the use of the term "holism," while detaching himself from Smuts' holism (Fisher 1950: 16–17). We shall return to this. In fact, Fisher's holism is logically implied by his emergentist approach to nature because, for him, new types of behavior emerge through the progressive organization of matter.

2.2 On Evolution

On the basis of this philosophy of nature, Fisher proposes a certain philosophy of evolution. In the 1950 paper, in particular, his aim is to take a stand against Bergson (1859–1941) and Smuts (1870–1950).[2] Fisher wants to show that Bergson and

[2]In the 1934 article, Fisher does not mention these authors, but it is very likely that he has already read them and is already aiming to offer an alternative philosophical view.

Smuts are right to conceive of evolution as creative, but that, to prove this, they are wrong to introduce a special force or a special principle. In other words, Fisher wants to show that, by the mere mechanism of natural selection, and taking into account indeterminism, evolution is creative.

Fisher begins his demonstration by explaining what he means by the word "creative." According to him, this word has mainly two meanings. As we have already seen at the beginning of Sect. 2.1, the first meaning refers to the notion of indeterminism: a process is creative if it depends on causes capable of independent variation (Fisher 1950: 1).[3] The second meaning refers to the notions of importance and novelty: a process is creative if it produces something important and new, "not merely new in time like a new penny, but new in its nature and potentialities" (Fisher 1950: 3). In this way, the term is used to refer to intellectual, artistic, moral, or social work (1950: 4). This second meaning is emotionally charged, in the sense that the evaluation of the importance and novelty of something depends on a subjective appreciation.

These two meanings are obviously compatible. And according to Fisher, evolution is creative in both senses:

> Approaching the matter on this level, it is, therefore, almost axiomatic that the process by which living things, as we know them, have come gradually into existence, is, in the fullest sense, a creative process. It has created new things, pregnant with potentiality; it has produced among other things growth, voluntary movement, and appetite; striving and effort, joy and pain; consciousness, and, in Man at least, conscious self-criticism. (Fisher 1950: 5)

In this passage, "almost axiomatic" means almost logical. Because evolution is a natural process, it is an indeterministic process, which makes it creative in the first sense. And because evolution is also the continuous appearance of new things, it is creative in the second sense. Let us note that this passage is clearly emergentist because when Fisher speaks of the creation of "new things," he is speaking of things that had no existence in inert matter.

To better explain his position, Fisher develops a critique of the positions of Bergson and Smuts. We will consider successively these two criticisms.

In *L'évolution créatrice*, published in 1907, Bergson develops the hypothesis that evolution is explained primarily by a special force, psychic in nature (Bergson 1911: 59, 61, 96–97, 280). Metaphorically, he names this force the "*élan vital*." According to him, this force is immanent to life and, by acting on the germs in the course of generations (1911: 95–96), it aims at the development of consciousness in matter (1911: 274, 288). Fisher does not understand Bergson's position well, because he equates the *élan vital* with an agent distinct from living beings (Fisher 1950: 8–9). However, he does understand that the *élan vital* is a special force, psychic in nature, and his criticism is precisely on this point. A first criticism is that Bergson introduces

[3] In the 1934 article, Fisher goes so far as to think that the notion of cause is unthinkable without this idea of capacity for variation, which amounts to saying that determinism is incompatible with the notion of cause (Fisher 1934: 106).

a mythological entity whose action is incomprehensible (Fisher 1950: 9–10). A second criticism is that Bergson's hypothesis is not compatible with what science knows about genetic mutations: scientific observation shows that these mutations occur in a totally blind way (i.e., without the slightest intentionality), and that mutation rates are too low to be the main cause of evolution (Fisher 1950: 10).

In *Holism and Evolution*, published in 1926, Jan Smuts develops the hypothesis that evolution is explained mainly by the heredity of acquired characters (Smuts 1926: 204–205). Smuts knows Weismann's arguments against this type of heredity (1926: 191, 197–200), but according to him, this heredity is made possible by the fact that each organism, as a totality, possesses a force field whose type of action is organic. Thus, for Smuts, the cells of an organism can receive and accumulate information in their field, before this information is the cause of a variation directed toward the utility of the organism (1926: 191). Fisher does not discuss these ideas in detail. But he understands that Smuts defends a kind of Lamarckism by appealing to a special principle, "holism" (Fisher 1950: 11–12). Fisher addresses two criticisms to Smuts. The first is that genetics leads to the rejection of any form of heredity of acquired characters (Fisher 1950: 11). The second, also addressed to Bergson, is that mutation rates are too low to be the main cause of evolution (Fisher 1950: 12–13).

According to Fisher, the fundamental error of Bergson and Smuts is that they think that Darwinism implies determinism. To save human freedom, they therefore reject this theory by arbitrarily postulating an ad hoc force or principle (Fisher 1950: 15–16). But this is unnecessary because Darwinism is compatible with indeterminism. Moreover, indeterminism is scientifically sound, and it allows us to think that evolution is creative.

Starting from the idea that the main cause of evolution is natural selection, Fisher makes the following reasoning: evolution depends essentially on natural selection, which itself depends essentially on the activity of living beings, in the sense that this activity constitutes a selection pressure on each individual and a response to this pressure (Fisher 1950: 17–18); therefore, in evolution, creativity comes from *the activity of living beings*. In this sense, he writes:

> Just where does the theory of natural selection place the creative causes which shape evolutionary change? In the actual life of living things; in their contacts and conflicts with their environments, with the outer world as it is to them; in their unconscious efforts to grow, or their more conscious efforts to move. Especially, in the vital drama of the success or failure of each of their enterprises. (Fisher 1950: 16)

Fisher adds that, in conscious living beings, creativity comes largely from *their conscious activity*:

> The surface or limit separating the inner from the outer life of each living thing is also, in our experience, the true seat of our consciousness, the boundary of the objective and the subjective, where we experience, through our imperfect sense-organs, what comes to us from outside, and, with at least equal obscurity, that which rises into consciousness from within. If consciousness is, as it would seem, the symbol, or even the means, of unification in our being, this is the region to which creative activity could most fitly be traced. (Fisher 1950: 17)

Bergson and Smuts are therefore right to think that purposeful, even conscious, causes play a role in evolution. But they are wrong to think that these causes would act directly on the germs. In fact, these causes are part of the mechanism of natural selection proposed by Darwin. They lie in the activity of organisms.

Finally, and in line with the above, Fisher states that the evolution of life on Earth will depend more and more on *human activity*:

> For the future, so far as we can foresee it, it appears to be unquestionable that the activity of the human race will provide the major factor in the environment of almost every evolving organism. Whether they act consciously or unconsciously human initiative and human choice have become the major channels of creative activity on this planet. Inadequately prepared we unquestionably are for the new responsibilities, which with the rapid extension of human control over the productive resources of the world have been, as it were, suddenly thrust upon us. (Fisher 1950: 21)

2.3 Conclusion

In this conclusion, I will attempt to answer two questions: how can we situate Fisher's philosophical conception of evolution in relation to the Modern Synthesis? And what is the main problem possibly raised by this conception?

For the Modern Synthesis, evolution as a whole is explained mainly by natural selection. In this sense, Gayon writes:

> The synthetic theory, if it is a theory, is the monotonous claim that evolution as a process is mainly driven by natural selection acting at the genetic level. Or, in other words, Darwin's thesis of the "paramount power of natural selection" (Gayon 1997) reinterpreted in genetical terms. This thesis has been repeated thousands of times by all those who claim to adhere to the Modern Synthesis. It is a simple idea, but it has the merit of being clearly identifiable as a theoretical claim. It put an end to the quasi-metaphysical discussions of the nineteenth—and early twentieth—century evolutionists about the "factors of evolution". (Gayon and Huneman 2019: 532; see also Gayon 1997)

The conception of evolution developed by Fisher is fully compatible with this idea. Of course, for him, the indeterminism *without consciousness* of elementary matter also plays a certain role (described as "creative"). But this point does not contradict the Modern Synthesis, which can assume the idea of a fundamental randomness in matter, notably in genetic mutations. Certainly, for him, the indeterminism *with consciousness* of some animals also plays a certain role. But this role is minor in the directions taken by evolution. By comparison, we are very far from Popper's hypothesis that the purposeful activity of all organisms plays a major role in the directions taken by evolution (see Dolbeault, in this volume).

However, Fisher's conception of evolution is also *emergentist*, in the strong sense of the term, because it asserts that evolution creates new things that are irreducible to the old—in particular, purposeful behaviors that are *irreducible to blind mechanisms*. And it is clear that, since the Modern Synthesis, Darwinism is often accompanied by a metaphysical mechanism for which, in nature, all processes are reducible to blind mechanisms.[4] From this point of view, the conception developed by Fisher is original. It contradicts the philosophy that often accompanies Darwinism.[5]

The main problem with this conception is that it introduces two forms of indeterminism in nature, the first without consciousness (mainly in inert matter), the second with consciousness (in some living beings), and that it postulates that the first has caused the emergence of the second. In philosophy, a classical criticism of this conception is that it postulates something incomprehensible: how can consciousness emerge from a matter devoid of consciousness? And how can purposeful processes emerge from processes without purpose? To avoid this problem, many thinkers prefer to opt for a metaphysical mechanism—which probably explains the choice of such a metaphysics by some advocates of Darwinism. Other thinkers opt for ideas like panpsychism, Wright and Haldane in particular.

3 Wright

Sewall Wright (1889–1988) was an American biologist. His main contribution to evolutionary biology concerns population genetics, in particular the notion of genetic drift. In addition to his scientific work, Wright published a small number of philosophical writings in which he presents a global conception of nature, with some remarks on evolution. These are mainly three articles:

- "Gene and Organism," *American Naturalist*, 1953
- "Biology and the Philosophy of Science," *Monist*, 1964
- "Panpsychism and Science," in *Mind in Nature* (eds John Cobb, David Griffin), 1975

The presentation I propose will be based mainly on the 1975 article, which is the most complete and the most accomplished. If necessary, I will indicate the possible differences between the 1975 paper and the other two.

[4] In philosophy, metaphysical mechanism is the idea that, in nature, all processes are fundamentally explained by blind (i.e., non purposeful) forces, including the apparently purposeful behaviors of certain animals. This position is metaphysical, not epistemological, in the sense that it does not imply that science could ever show how apparently purposeful behaviors are reduced to blind mechanisms. Metaphysical mechanism differs from what is called "mechanism" today, in epistemology, which concerns the method of certain sciences, and not the metaphysical question of the reduction of purposeful behaviors (Craver and Tabery 2023, point 2).

[5] On the defense of metaphysical mechanism in Darwinism, see in particular: Monod (1970: 55), Dawkins (1986: 74), Provine (1988: 64–65), Dennett (1995: 80–81), Mayr (2004: 61, 91).

3.1 On Nature

Like Fisher, Wright defends a certain form of *indeterminism*. But Wright's indeterminism is associated with a panpsychist conception of nature.

Panpsychism is in fact Wright's answer to the following problem: what is mind in nature? In particular, what is mind in relation to the physical world? Wright is well aware that the essential work of scientists is not to address metaphysical questions. However, according to him, biologists are obliged to ask this question because their object appears problematic from the outset:

> The biologist is continually in trouble. The objects of his study range from a class of molecules that have the basic self-duplicating property of living things, through cells which suggest purely physical systems, through animals which give increasing evidence of having minds, to human beings in whom streams of consciousness seem to involve continual choices of action, at the opposite pole from control by impersonal laws of nature. The zoologist cannot escape the problem of the relation of matter and mind. (Wright 1975: 79)

While the physicist's object never manifests psychic properties, and the sociologist's object always manifests them (properties taken into account in sociological explanations), the biologist's object is diverse: sometimes it manifests psychic properties, sometimes not. Consequently, as far as his/her work as a scientist is concerned, which consists in explaining phenomena by looking for their causes, the biologist is obliged to wonder about the relation between mind and matter. The central question is obviously: can we explain certain biological phenomena by invoking psychic causes?[6]

Wright first rejects what he calls "complete materialism," the idea that mind is not a fundamental category of reality (Wright 1975: 81). According to him, this position is absurd because it is opposed to the experience of our own consciousness as something different from the physical world.

Wright also rejects what he calls "solipsism," which philosophers generally call idealism, or immaterialism, the idea that matter is not a fundamental category of reality (Wright 1975: 80–81). In the 1975 paper, Wright does not develop an argument against solipsism, even pointing out that it is impossible to prove its falsity. However, in his 1964 article, he develops the following classical argument: it appears that the sensations of which we are aware depend on something that is not our will, and which presents scientifically knowable regularities; this something can be called "the external world" (Wright 1964: 275).

Wright also rejects dualism, understood as the idea that mind could exist without matter. According to him, this conception is difficult to reconcile with natural sciences (Wright 1975: 81). Wright does not give any further details, but we can guess that he makes the following reasoning: from the point of view of evolution, the

[6]This problem is obvious when it comes to explaining the behavior of many animals.

mind as it appears in certain animals, including human beings, has a material origin; because of this origin, it cannot exist independently from matter.

Finally, Wright rejects the idea that mind could emerge from a matter devoid of any psychic property, that is to say, the idea defended by Fisher. In this sense, he writes: "Emergence of mind from no mind at all is sheer magic" (Wright 1975: 82). His argument is that such an emergence would be very different from that of a new organ, because mind does not appear as a new organ, but as a specific property, distinct from physical properties (Wright 1975: 81–82).[7]

Wright is thus led to defend a form of *panpsychism*,[8] by asserting that mind is as old as matter, and coextensive with it at any level of organization:

> We conclude that the evolution of mind must have been coextensive with the evolution of the body. Moreover, mind must already have been there when life arose and indeed must be a universal aspect of existence—still assuming that mind cannot arise from nothing. (Wright 1975: 82)

Wright adds that what is true for phylogeny is also true for ontogeny:

> The emergence of mind in the course of individual development from the fertilized egg presents a similar problem and one that is an everyday occurrence instead of a single event in the remote past. It would appear that the mind of a human being must develop from something of the nature of mind in the fertilized egg and, back of this, in the separate germ cells and in the nucleic acid molecules. (Wright 1975: 82)

More precisely, Wright defends a monistic panpsychism. In this sense, he affirms that "mind and matter are merely two aspects of the same reality: as it is to itself and as it seems to other minds with which it interacts" (Wright 1975: 82). Note that, according to this conception, mind has a certain priority over matter because it is reality "as it is to itself."

This panpsychism has a fundamental consequence: the rejection of *metaphysical mechanism*, that is, the rejection of the reduction of the teleological to mechanisms. For Wright, the opposite is true. What appears from the outside as devoid of teleology is in fact determined by a certain stream of consciousness. This applies to all levels of organization of matter:

> Under the dual-aspect view, the objective world of natural science with its hierarchies of physical and biological entities must exactly parallel an inner world of mind. Because of the flowing character of the latter, parallelism should be sought not in matter as mere mass and occupation of space, but rather in the incessant action which modern physics finds in its units; the wavelike properties of proteins, neutrons and electrons, photons and its other elementary entities, down to Planck's quanta as the ultimate known units of action. At a higher level is the incessant action within molecules, including nucleic acid and protein. The

[7] This anti-emergentist argument is classic among the defenders of panpsychism, particularly in the 19th and 20th centuries (voir Skrbina 2005: Chap. 5 and 6).

[8] Wright himself uses this term (Wright 1964: 278; 1975: 82).

> metabolic activities within a cell, and its activities in relation to other cells may be considered the external aspects of the cell's stream of consciousness. (Wright 1975: 83)

The question remains as to why many natural processes appear to us as mechanical. According to Wright, the fundamental reason is that there is no opposition between freedom and regularity: under identical conditions, an organism (atom, molecule, cell, etc.) tends to behave in an identical way. In other words, the same causes produce the same effects, including for systems endowed with consciousness (Wright 1975: 83). Besides, drawing on quantum physics, Wright points out that, fundamentally, natural regularities are only statistical (Wright 1975: 83).

Therefore, the "laws of nature" are not a fundamental reality that would determine all processes. They are merely statistical regularities that arise *from the choices of organisms*: "the laws of nature are ultimately statistical and no more preclude choice on the part of individual components than do statistical laws of human behavior" (Wright 1975: 83). Wright claims to owe this conception of the laws as statistical regularities to the mathematician Karl Pearson (1857–1936) (Wright 1975: 80). The panpsychist interpretation of this conception, however, is not found in Pearson.

Like Fisher, Wright thus has an *indeterministic* conception of nature. However, in contrast to Fisher, for him there is only one kind of indeterminism: at any level of organization of matter, indeterminism presupposes a certain degree of consciousness.

Like Fisher, Wright also defends *a kind of holism*. But this holism is linked to panpsychism. For Fisher, holism corresponds only to the idea that the organized whole has properties that its parts do not have (to any degree). For example, the property of acting in a purposeful way. For Wright, holism also corresponds to the idea that the whole is a fusion of the parts. Thus, Wright hypothesizes that the unity of the human mind is explained by the fusion of the streams of consciousness of the individual brain cells (Wright 1975: 83).

Finally, it should be noted that Wright makes a very clear distinction between science and philosophy. Here is how he summarizes his position on this subject:

> I fully agree with Pearson here that science is restricted to verifiable knowledge and thus must exclude the knowledge of our streams of consciousness, because it is unverifiable by anyone else. We must continue to accept a rigorous determinism as far as possible, and supplement this by probability distributions where necessary, even though we interpret the determinism philosophically as the external aspect of choices throughout the hierarchy of existence and make use of this philosophical interpretation in choosing topics for research. (Wright 1975: 86)

To assert philosophically that every organism is, internally, a certain stream of consciousness does not mean that science should seek to study that stream of consciousness. For Wright, science is concerned with observable regularities (fundamentally statistical), not with the ultimate causes of phenomena (which are psychic). Philosophy can only guide the scientist in the choice of the subjects he studies.[9]

3.2 On Evolution

In his philosophy writings, Wright says little about evolution. However, the few remarks he makes are sufficient to understand the general direction of his thinking. In his view, even if evolution is an object for science, something about evolution necessarily eludes science.

For Wright, the main role of science is to discover regularities. And evolution presents certain regularities. Think of the evolution of gene frequency in a population as a function of certain factors (selection, migrations, etc.). Therefore, evolution can be an object for science (Wright 1975: 86).

However, for Wright, something about evolution necessarily eludes science. Here is what he writes on this subject:

> Science is thus a limited venture. It must act as if all phenomena were deterministic at least in the sense of determinable probabilities. It cannot properly explain the behavior of an amoeba as due partly to surface and other physical forces and partly to what the amoeba wants to do, without danger of something like 100% duplication. It must stick to the former. It cannot introduce such principles as creative activity into its interpretation of evolution for similar reasons... In carrying out this program, the scientist should not, however, deceive himself or others into thinking that he is giving an account of all of reality. The unique inner creative aspect of every event necessarily escapes him. (Wright 1953: 17)

What escapes science is the creative activity of organisms. This activity escapes science because it is psychological in nature ("inner"), and because its physical results are unpredictable. Yet, according to Wright, this activity plays a role in evolution.

In a passage from 1931, republished in 1975, he says a little more:

> The present discussion has dealt with the problem of evolution as one depending wholly on mechanisms and chance. In recent years, there has been some tendency to revert to more or less mystical conceptions revolving about such phrases as "emergent evolution" and "creative evolution." The writer must confess to a certain sympathy with such viewpoints philosophically, but feels that they can have no place in an attempt of scientific analysis of the problem. One may recognize that the only reality directly experienced is that of mind, including choice; that mechanism is merely a term for regular behavior, and that there can be no ultimate explanation in terms of mechanism—merely analytic description... Whatever incompleteness is involved in scientific analysis, applies to the simplest problems of mechanics as well as to evolution. (Wright 1931: 154; 1975: 86)

[9] Let us note the similarity of this last idea (expressed at the end of the passage) with the claim developed by Popper that metaphysics can guide scientific research (see Dolbeault, in this volume).

In this passage, Wright again emphasizes that something about evolution eludes science, and that this something is essentially about the creative activity of organisms, which is psychological in nature. The sympathy expressed for Lloyd Morgan (who speaks of "emergent evolution")[10] and for Bergson (who speaks of "creative evolution")[11] supports this idea, even if Wright does not exactly defend Morgan's or Bergson's position on evolution.

Then Wright continues:

> As to evolution, its entities, species and ecologic systems, are much less closely knit than individual organisms. One may conceive of the process as involving freedom, most readily traceable in the factor called here individual adaptability. (Wright 1931: 154–155; 1975: 87)

Here, Wright argues that the role of the creativity of organisms in evolution could be detected, from the outside, at the level of *individual adaptability*. One thinks of course of the behavior of animals, plants, etc. However, in the 1931 paper (which contains this text), the notion of individual adaptability refers mainly to the activity of cells:

> The evolution of complex organisms rests on the attainment of gene combinations which determine a varied repertoire of adaptive cell responses in relation to external conditions. The older writers on evolution were often staggered by the seeming necessity of accounting for the evolution of fine details of an adaptive nature, for example, the fine structure of all of the bones. From the view that structure is never inherited as such, but merely types of adaptive cell behavior which lead to particular structures under particular conditions, the difficulty to a considerable extent disappears. The present difficulty is rather in tracing the inheritance of highly localized structural details to a more immediate inheritance of certain types of cell behavior. (Wright 1931: 143, 147)

In this passage, Wright argues that the phenotype of an organism depends not only on its genotype, but on how that genotype is used in response to the environment. Specifically, he argues that the cells of an organism have some autonomy in the use of the genome. Logically, then, he suggests that, in evolution, the creativity of organisms could be detected at the level of *this use of the genome by the cells*. In other words, he suggests that, with respect to this use, cells would be capable of innovation, and that this would play an important role in evolution.

Finally, let us note the connection between this philosophical idea of Wright and his scientific theory of the "shifting balance process." The aim of this theory is to explain how the evolution of a population can be accelerated by genetic drift. In summary, the mechanism is as follows: if a population is divided into partially isolated subpopulations, then these subpopulations tend to undergo genetic drift; the result is an increase in the genetic variability of the population, and thus in the probability of the appearance of adaptive novelties (Wright 1980: 832–840). For

[10]Wright reviewed a book by Morgan on evolution (Wright 1935). He knows this author well.

[11]Wright also knows Bergson well, mentioning him as one of his sources of inspiration in philosophy (Wright 1975: 80).

Wright, all evolution is a "creative process" (Wright 1980: 841). But in the case described by this theory, this is even more true: "Creativity is raised to the second power, however, if the field of variability is amplified by random differentiation among numerous partially isolated local populations..." (Wright 1980: 841). In scientific language, what is increased is the probability of new combinations of genes. In philosophical language, it is the probability of new innovations of cells with these genes.

3.3 Conclusion

In this conclusion, I take up the two questions I asked about Fisher: how can we situate Wright's philosophical conception of evolution in relation to the Modern Synthesis? And what is the main problem possibly raised by this conception?

First, it is clear that Wright's general philosophy is totally opposed to the philosophy that often accompanies Darwinism, in particular since the Modern Synthesis. For him, the ultimate causes of the behavior of organisms are psychic, at all levels of organization. In contrast, for the philosophy that often accompanies Darwinism, the ultimate causes of the behavior of organisms are mechanical: they are physical events obeying blind laws.[12]

Second, it is also clear that, from an epistemological point of view, Wright's conception of evolution is original. Whereas most biologists presuppose that science can provide a complete explanation of the causes of evolution, Wright asserts that something causal in evolution necessarily eludes science: *the creative activity of organisms*. Wright asserts this in all of his philosophical writings (Wright 1953: 17; 1964: 288; 1975: 86–87), and in some of his scientific papers (Wright 1931: 154–155; 1959: 965). For him, this point is therefore crucial.

Third, one aspect of Wright's conception of evolution is frankly opposed to the Modern Synthesis. Whereas for the latter, creativity in evolution depends essentially on natural selection, Wright asserts that it also depends on the purposeful activity of organisms, at all levels of organization. In this respect, one can compare what Wright argued about the cell in 1975 with what Shapiro argues today on the same subject:

> The metabolic activities within a cell, and its activities in relation to other cells may be considered the external aspects of the cell's stream of consciousness. (Wright 1975: 83)

> Living cells and organisms are cognitive (sentient) entities that act and interact purposefully to ensure survival, growth, and proliferation. They possess corresponding sensory, communication, information-processing, and decision-making capabilities. (Shapiro 2011: 143)

It should be stressed that this connection between Wright and Shapiro is philosophical, not scientific. Philosophically, both have a non-mechanistic conception of

[12] See notes 4 and 5.

life, of the cell in particular. But scientifically, only Shapiro has attempted to give precise meaning to this idea and to make it an important element in the explanation of evolution.[13]

Now, Wright's philosophical conception of evolution is part of a panpsychist conception of nature, and because of its originality, this conception raises a set of questions. Without going into a long discussion, I will mention what appears to be the main problem.

This problem is finally the symmetrical of Fisher's problem. Unlike Fisher, Wright postulates only one kind of indeterminism in nature: indeterminism *with consciousness*. This conception has the advantage of corresponding to a unified view of nature. But it implies an unintuitive idea, namely that a certain form or/and a certain degree of consciousness would exist at all levels of organization, including the first levels (atoms, molecules, cells). Without rejecting a priori this conception, one can expect from it other arguments than that of the unified view of nature.

By suggesting that conscious creative activity could be detectable at the level of the cell's behavior, Wright nonetheless proposes an idea of research.

4 Haldane

John Haldane (1892–1964) was a British biologist, specialized in genetics. His main contribution to evolutionary biology concerns population genetics, but Haldane worked on other questions, notably that of the origin of life.

Haldane develops philosophical ideas in many writings: in some scientific publications, as in the conclusion of *The Causes of Evolution* (1932); but also in some articles or books specially dedicated to this subject. The following presentation will therefore be based on many sources. However, I mention the main ones:

- *The Causes of Evolution*, 1932
- "Quantum Mechanics as a Basis for Philosophy," *Philosophy of Science*, 1934
- *The Marxist Philosophy and the Sciences*, 1938
- "Why I am a Materialist," *The Rationalist Annual*, 1940
- "Life and Mind as Physical Realities," *Penguin Science Survey*, 1963

4.1 On Nature

Haldane fixed his philosophical ideas about nature and evolution in the early 1930s and broadly maintained the same ideas for the rest of his life.

From the early 1930s on, Haldane advocated what he called "materialism." But, as we shall see, this materialism is very different from what we now call materialism, or physicalism. Although a materialist, Haldane considers psychic properties to be

[13]This work is based on knowledge of genetics and molecular biology developed well after Wright.

something fundamental and irreducible in nature. Like Wright, he is thus led to defend *a form of panpsychism* linked to *a holistic conception of organisms*.

From this point of view, the presentation of Haldane's philosophy that I propose is very different from that proposed by Sarkar in 1992. Strangely enough, Sarkar does not say a word about Haldane's panpsychism, which although appears in many texts between 1929 and 1963. Moreover, Sarkar suggests that between 1930 and 1937, Haldane may have moved closer to mechanistic materialism (Sarkar 1992: 385, 389, 395–396), while some texts indicate the opposite. In the 1934 article in particular (Haldane 1934), which Sarkar does not mention, Haldane clearly defends a holistic approach to organisms, linked to panpsychist ideas.

More generally, it should be noted that these ideas of Haldane are little discussed in the literature. Only one article seems to be devoted to his holistic approach to organization (Hammond 2009), and only a few works mention his panpsychism (Skrbina 2005: 189–190). Several general studies on Haldane do not mention it at all (Shapiro 1993; Dronamraju 2011, 2017; Subramanian 2020).[14] The fact that Haldane speaks of "materialism" (not panpsychism or holism) in connection with his philosophy perhaps explains this, as if commentators have difficulty in grasping the coherence of his ideas.

To clarify things, it is therefore important to begin with Haldane's materialism. As early as the end of the 1920s, Haldane expressed a certain sympathy for materialism, considering that an explanation of the origin of life and mind from matter alone was conceivable (Haldane 1929/1968: 10–11; 1932a: 154–155). Then, from 1933 on, he assumed this position more and more clearly (Haldane 1933: 14–16; 1934: 97). But what does Haldane mean by "materialism"? In a 1940 text, he answers this question:

> When I say that I am a materialist I mean that I believe in the following statements:
>
> 1. Events occur which are not perceived by any mind.
> 2. There were unperceived events before there were any minds.
>
> And I also believe, though this is not a necessary logical deduction from the former two, that:
>
> 3. When a man has died he is dead.
>
> Further, I think that it is desirable that other people should believe these statements. I do not mean that I believe that the universe is a machine, or that I am a machine; nor yet that consciousness does not exist, or has a lesser reality (whatever that means) than matter. (Haldane 1940/1968: 27)

Assertion 1 of the text says that there is a world (a set of events) independent of any mind, especially mine. In philosophy, this position is usually called "realism," as opposed to idealism. And assertion 2 says that the world is prior to the mind, which

[14] From a philosophical point of view, these studies evoke mainly Haldane's political or/and religious ideas. The metaphysical ideas are left aside.

leads to the view that the mind originates in the world. In philosophy, this position is usually called "naturalism," in the sense of the idea that the mind originates in nature. Haldane's materialism is therefore a metaphysical realism and a naturalism. More precisely, it is *a theory of the natural origin* of life and mind. But, as one can notice, this materialism is not *a theory of the reduction* of life and mind to something else. For Haldane, psychic properties are a fundamental and irreducible reality. And the same for organization. From this point of view, his materialism clearly differs from what is called today materialism, or physicalism.

To complete this explanation, let us add that Haldane is inspired by the materialism of Engels and Lenin, which he summarizes by the same two assertions: (1) The material world has an objective reality, (2) and it is prior to the mind (Haldane 1938: 24–26; 1941/1968: 30–31). Haldane also points out that the "dialectical materialism" of Engels and Lenin is in no way a commitment to the mechanistic view of nature (Haldane 1938: 26; 1941/1968: 31).

Fundamentally, then, Haldane defends *a non-mechanistic materialism*. This allows him to introduce ideas like panpsychism and holism. Let us start with panpsychism. Here are some passages where Haldane clearly expresses his sympathy for this conception:

> Materialism of course includes many forms far more subtle than the crude materialism of fifty years ago, and if you are willing to concede enough unexpected properties to so-called dead matter it becomes distinctly idealistic. (Haldane 1932a: 155)

> If we ever explain life and mind in terms of atoms, I think we shall have to attribute to the atoms the same nature as that of minds or constituents of mind such as sensations. (Haldane 1932a: 157–158)

> But if we regard the wave-like aspect of matter as more fundamental than the particle aspect, we can surmise that it is of the same general character as human experience, but less, or at any rate differently, organized." (Haldane 1934: 92)

> I suggest that the mind has physical properties analogous to those of a single electron or atom, properties which are lost in a group of trillions of atoms with the low degree of organization found in a stone, but which may be accentuated when the same number of atoms is organized in a mouse or a man. (Haldane 1938: 146–147)

In fact, as early as 1929 in "The Origin of Life," Haldane pointed out that there was a certain coherence between panpsychism and the idea that life originated in matter: if matter has psychic properties, then the separation between inert and living matter is less important than one would think (Haldane 1929/1968: 2). In 1932, in the conclusion of *The Causes of Evolution*, he expressed a clearer commitment to panpsychism, stating that elementary matter was probably endowed with sensation (see the first and second quoted texts). And in 1934, he devoted an entire article to this question, this time in an attempt to develop and justify this panpsychist conception of matter (see the third quoted text). I will come back to the details of this article later, when I will analyze Haldane's argumentation. Finally, it should be noted that Haldane maintained this metaphysical preference for panpsychism until the end of his life. One of the last articles he published again attempts to develop and justify this conception (Haldane 1963).

From this point of view, Haldane's position echoes that of Wright, even if the words used are not the same. Whereas Wright speaks of "mind" and "consciousness" for any level of material organization, Haldane seems to reserve these words for the biological level, and even for animals only (Haldane 1934: 94–95; 1963: 233–235). In 1963, however, Haldane uses the word "soul" to refer to certain properties of atoms and cells (Haldane 1963: 235).

Haldane's non-mechanistic materialism is also compatible with a holistic conception of organization. Here are some passages where he defends this conception:

> Life then would be a resonance phenomenon between molecules. One of its most obvious expressions is in morphological patterns which possess a certain stability. It gives the organism a certain bias in the direction of wholeness. (Haldane 1934: 95)

> On the contrary it is possible, without denying the validity of physical laws, to adapt biological theory to biological practice by saying that the organism is something more than a machine, in the sense that it has a unity of a type which the machine lacks. (Haldane 1938: 104)

> It is now clear that so called dead matter has a good deal more organization than it could have if it consisted of particles each moving under the attraction or repulsion of all the others. And the kind of organization which explains the stability of a molecule may very well explain the stability of a tree or a dog, which we call its life. A sensation in a human mind may or may not be a quantum transition, but it is a great deal more like a quantum transition than it is like a particle. (Haldane 1962, 1968: 162–163)

As early as 1932, Haldane expressed a certain skepticism about the ability to reduce "the unity of the organism" to physical processes (Haldane 1932b: 157). In the same way, in 1934, he underlined the difference between biological sciences and physical sciences: whereas the former deal with apparently teleological processes (self-repair, self-reproduction, etc.), the latter deal with apparently blind processes. Hence, among biologists, the permanent temptation of vitalism or animism (Haldane 1934: 80–81). But in the same article, in 1934, Haldane asserted that it was possible to explain organization without introducing a special force. To do this, it was necessary to think that an organized system formed *a united whole* endowed with specific behaviors. Taking as a model the notion of wave in quantum physics, he therefore made the hypothesis that, at each level of organization (atom, molecule, cell, etc.), the elements of a system could be in resonance and form a united whole (see the first quoted text). Finally, it should be noted that Haldane maintained this holistic conception of organisms until the end of his life (see the other quoted texts, but also Haldane 1947: 53, 56; 1958a: 62, 70). One of the last articles he published again attempts to develop and justify this conception (Haldane 1963).[15]

Haldane is well aware that his panpsychist and holistic conception of organization is speculative. But this speculation is not arbitrary. In fact, Haldane follows a rational path very similar to that of Wright. It can be summarized as follows:

(i) Life and mind have their origin in inert matter (Haldane's "materialism").
(ii) However, life and mind are irreducible to mechanisms (Haldane's anti-mechanism).
(iii) Therefore, it can be assumed that inert matter already has the fundamental properties of life and mind, namely organization and certain psychic properties.

Point (i) corresponds to Haldane's materialism. It is obviously based on evolutionary biology, since the latter leads one to think that life and mind have appeared progressively in the course of evolution. This materialism is a rejection of vitalism and dualism, understood as the introduction of special forces of unknown origin.

Point (ii) corresponds to Haldane's anti-mechanism. On mind, like Fisher and Wright, Haldane rejects epiphenomenalism: for him, the mind has a causal role, and this role is not reducible to blind mechanisms (Haldane 1932a: 161; 1932b: 157–158; 1934: 97; 1938: 142, 149; 1941/1968: 29). Haldane develops several arguments against epiphenomenalism. In particular, he takes up William James' argument based on evolution: if mind appeared and was preserved in the course of evolution, it is because it is functional (Haldane 1940/1968: 29).

Point (iii) corresponds to the solution proposed by Haldane, namely the panpsychist and holistic approach to organization. Like Wright, Haldane favors this view over emergentism, which he regards as a pseudo-explanation (1932a: 156–157). In other words, for Haldane, the fundamental properties of life and mind must already exist in the elements that constitute life and mind.

Haldane's argument for panpsychism and holism is thus similar to that developed by Wright. However, Haldane introduces another argument that is not found in Wright. This argument is that, since quantum physics presents *a holistic model of matter* at the microphysical level, this model of scientific explanation is conceivable for all other levels of organization, at least as a hypothesis. Haldane first expressed this idea in *The Causes of Evolution* (Haldane 1932a: 157–158). Later, he attempted to detail and justify it (Haldane 1933, 1934, 1938: 104–105, 142–148; 1941/1968: 33; 1963).

The quantum model of matter asserts that an elementary particle has the properties of a probability wave. Such a wave does not have an exact spatial and temporal localization (Haldane 1934: 78–79). Moreover, it makes it possible for several associated particles to lose their individuality, and thus form a kind of unity

[15]This analysis is consistent with Hammond's study (2009). In his article, Hammond makes it clear that throughout his life, before and after his commitment to materialism, Haldane advocated a holistic (as opposed to reductionist) approach to organisms.

(1934: 78–79). Finally, and if one may say so, it makes the particle act in a probabilistic way according to possible futures (1934: 88–89). Analogously, Haldane points out that the mind appears to be without exact spatial and temporal localization (1934: 86),[16] and that it acts according to possible futures. Drawing on the quantum model, one could therefore think of the mind as a wave whose resonators would be molecules or cells (1934: 87–88, 94–95). In a more general way, one could think that any organization is explained by laws *analogous* to those of quantum physics (1934: 83, 95).

Let us specify that, for Haldane, it is not at all a question of affirming that everything can be explained by the laws of quantum physics. On the contrary, it is a question of affirming that each level of organization obeys *specific laws*, analogous to those of quantum physics. In this sense, in 1963, Haldane speaks of "an extension" of the principles of quantum physics (Haldane 1963: 230). In short, this approach is clearly anti-reductionist.

As we shall see, this hypothesis of specific laws has consequences for the way of thinking about evolution.

4.2 On Evolution

Haldane's view of evolution as a whole differs from those of Fisher and Wright. Three points are worth noting.

The first point concerns natural selection. For Haldane, if it is true that natural selection plays an important role in evolution (Haldane 1932a: 20; 1958a: 12; 1958b: 103), it is not proven that it explains the whole of evolution. First, because some facts seem to oppose this idea: the existence of abrupt evolutionary changes, linked to hybridization in particular (Haldane 1932a: 138–139; 1959a: 711); the fact that evolutionary changes seem to depend on constraints linked to the development of organisms (Haldane 1932a: 139–141; 1958b: 101; 1959b: 147); or the fact that natural selection tends to favor the conservation of species rather than their evolution (Haldane 1958b: 98–99; 1959a: 712; 1959b: 124). In 1958, Haldane can thus write:

> I am tempted to think that the elementary theory of natural selection is now nearly complete. This does not mean that evolution has been explained. It does mean that we can see the next questions to ask. (Haldane 1958b: 101)[17]

Another reason is that, for Haldane, biology has not yet proved that Darwinism is sufficient to explain the complexity of organisms:

---

[16] For Haldane, the absence of exact temporal location of the mind corresponds to the "specious present" of consciousness (Haldane 1934: 86).

[17] For more details on this question, see in particular: Rao and Nanjundiah (2017).

> I have recently been reading Paley's *Evidences of Christianity*, as every good Rationalist should. Paley attempts, with very great skill, to prove the existence of a creator from the design of living organisms. Of course a good many of his arguments were met by Darwin. It is clear that, given the facts of heredity and variation, organisms tend to adapt themselves to their environment without any conscious planning by themselves or anyone else. But it is by no means proved that the whole course of evolution from single-celled organisms to oaks, daisies, ants, and men can be explained on these lines... The chemical organization of a cell is immensely complicated, and it is hard to see how an organism could work at all unless it were of extreme chemical complexity. If so, the very first steps in evolution are the hardest to explain. I think, therefore, that a reasonable man should be prepared to examine arguments which assume a measure of design in living creatures, even though I do not personally think that they are cogent. (Haldane 1944, 1968: 43)

In this 1944 article, Haldane develops a critique of the design argument. In his view, the idea that nature is explained by a divine plan is inconsistent with what nature shows, namely living beings whose activity consists, for the most part, in destroying others (Haldane 1944, 1968: 44–47). However, the article begins with the idea that the complexity of organisms remains a challenge for Darwinism (on this point, see also Haldane 1959a: 713).

Finally, Haldane considers that, from the point of view of the history of science, it is likely that Darwinism will be replaced, although it is also likely that the theory that will replace it will be different from pre-Darwinian theories:

> The history of science makes it almost certain that facts will be discovered which show that the theory of natural selection is not fully adequate to account for evolution. But the same history makes it extremely improbable that these facts will be in any way related to the criticisms at present made of it. The physics of Newton and Galileo have proved inadequate in several respects, and are being replaced by relativistic and quantum mechanics. These, however, are even further from the medieval physical theories than were the theories of Galileo and Newton. (Haldane 1959b: 148–149)

The second point concerns the role of the mind in evolution. According to Haldane, the evolution of animals depends largely on behavioral innovations whose ultimate cause is mental. In fact, Haldane assumes what has been called since Simpson (1953) the "Baldwin effect": in a given population, a behavioral innovation may have a selective advantage for the innovative individuals; if this is the case, natural selection will tend to favor traits (anatomical or otherwise) that facilitate the new behavior. But for Haldane, this effect has *a major role in animal evolution*, which contradicts the idea that evolution is a purely mechanical process:

> Many of the important steps in animal evolution must, I think, have depended on mental processes, particularly such fundamental steps as the origin of flight or climbing. Some animals in a species were slightly better pre-adapted for gliding than others. But not all which were so preadapted used their powers, any more than men use their powers fully. Those which did so, in certain cases, enjoyed a selective advantage... Critics of Darwinism have said that it reduces evolution to a mechanical process in which mind plays no part. This is a fair criticism of some vulgarizations of Darwinism, but not of the theory presented by Darwin or his more intelligent successors. (Haldane 1958b: 102)

The third point concerns the role of the laws of organization in evolution. As we saw in the previous section, Haldane hypothesizes that each level of organization of matter (atom, molecule, cell, etc.) depends on laws analogous to those of quantum physics: laws capable of creating a certain unity between several elements and new behaviors. Logically, this amounts to affirming that evolution does not only depend on natural selection, but also on laws of organization whose action directs matter *toward life and mind*. Of course, these laws can only come into action under given physical conditions. But without these laws, the same physical conditions would give nothing.

In *The Causes of Evolution*, Haldane already mentions these laws of organization of matter (Haldane 1932a: 157–158). Logically, he then hypothesizes that these laws make it possible the appearance of the mind:

> For such reasons as these I do not share the view that mind, as we know it, cannot be a product of evolution. An essential element of evolution is variation. Variation is at random in the sense that it may lead in many directions, mostly of no survival value, and that those which possess survival value for the individual may lead to degeneration and extinction of the species. But it follows chemical and biological laws, and only certain combinations will lead to mind. If we are to have mind at all, it must probably conform to certain laws. There is no need to suppose that these laws, any more than those of biochemistry, are products of natural selection. Selection no doubt accounts for certain details, but in all probability not for the general character of mind. (Haldane 1932a: 162)

A passage from 1959 echoes this text:

> Again it is urged that mind cannot be a product of evolution. This difficulty does not arise for a consistent idealist, nor for a consistent materialist, whether he adopts a materialism based on modern science, or on the Hindu notion of *prakriti*, which is the basis not only of matter but of individual mental phenomena, or on the Chinese notion that *li* (meaning something like substance) and chih (meaning something like organization) are omnipresent, the latter in various degrees. (Haldane 1959b: 148)

In this passage again, the question is to know if the evolution of matter can lead to mind as known in humans. And again, the answer is positive, provided we consider matter as endowed with organizational and psychic properties. Haldane evokes these properties by referring to Hindu and Chinese philosophical notions. But he also refers to "modern science," which here means quantum physics and its laws of organization of matter.

A question remains: if some laws of organization play an important role in the emergence of life and mind, should we think that these laws reveal some intention in nature? This question is asked in *The Causes of Evolution* (Haldane 1932a: 158). And it reappears indirectly when Haldane asks whether life and mind characterize *the universe as a whole*, if not the galaxies (1934: 91–92; 1962, 1968: 163–164; 1963: 237), for if the universe is a kind of mind, the laws could be conceived as its mode of action.[18] However, Haldane believes that, for the moment, we cannot answer this question.[19]

4.3 Conclusion

In this conclusion, I take up the two questions I asked about Fisher and Wright: how can we situate the philosophical conception of evolution proposed by Haldane in relation to the Modern Synthesis? And what is the main problem possibly raised by this conception?

First, it is clear that Haldane's general philosophy is opposed to the philosophy that often accompanies Darwinism. This is due to its *anti-reductionism*, based on the hypothesis of laws of organization of matter at different levels, and to its *panpsychism*, which amounts to affirming that, at different levels, certain causes are irreducibly purposeful. We measure the extent to which it is not possible to summarize this philosophy by the single term "materialism." From this point of view, and in contrast to what Sarkar asserts (Sarkar 1992), there is a certain continuity between the philosophy of J. B. S. Haldane and that of his father, the biologist J. S. Haldane (1860–1936).[20] J. B. S. Haldane explicitly suggests this at the end of his life (Haldane 1963: 229–230).

Besides, one aspect of Haldane's view of evolution is frankly opposed to the Modern Synthesis. While for the latter, the directions of evolution essentially depend on natural selection, Haldane makes this hypothesis that these directions also depend on the laws of organization of matter acting at different levels: atom, molecule, cell, etc. In other words, for Haldane, certain laws could frame evolution, giving it *a certain directionality*, toward more complexity and mind.

The main problem raised by this conception is the same as the one we mentioned for Wright: its speculative dimension. However, by introducing the hypothesis of laws of organization of matter, Haldane considers that his philosophical speculations on life and evolution can give rise to scientific research. With him, more than with Wright, philosophical speculation is never separated from a scientific objective.

[18] In 1932, Haldane suggested this idea (Haldane 1932a: 158). Moreover, Haldane is also interested in the idea that the laws of nature could evolve, as something living (Haldane 1938: 68–70; 1941/1968: 42).

[19] Note that, for Haldane, our inability to answer this question is not due to an impossibility of principle. According to him, this question depends largely on our knowledge of the universe on a large scale. Thus, Haldane conceives that an extremely precise and thorough observation of the universe (which could require millions of years) could allow us to know if it is alive and intelligent (Haldane 1962, 1968: 164–165).

[20] See in particular (Haldane 1935).

5 Summary and Discussion

Table 1 summarizes the analyses of the previous points and provides a global view:

Line 1 concerns the method followed by Fisher, Wright, and Haldane. All three respect the *naturalistic method* in philosophy, in the sense that none of them appeals to supernatural entities (God for example) to explain natural processes. Certainly, each of them postulates things whose existence is not recognized in science, for example, the phenomenon of strong emergence in Fisher, or the psychic properties of elementary matter in Wright and Haldane. But for them, these things are part of nature. Moreover, they are the result of an always scientifically informed speculation.

Lines 2 and 3 concern some facts considered as irreducible by the three thinkers: a certain indeterminism of natural processes, and the causal role of the mind in nature (i.e., the rejection of epiphenomenalism). There is obviously a connection between the two. One who thinks that the mind has a causal role in nature tends to think that the apparent indeterminism of conscious decision is made possible by a general indeterminism within nature. Fisher, Wright and Haldane make this connection.

Lines 4 and 5 concern the philosophical hypotheses developed by each of them about nature in general. Line 4 concerns the place of the mind in nature, in response to the crucial question of the origin of the mind as known in human. Line 5 concerns

Table 1 Fisher, Wright, and Haldane, on nature and evolution. Recapitulation and comparison

	Fisher	Wright	Haldane
1) Philosophical naturalism:	Yes	Yes	Yes
2) Indeterminism:	Yes	Yes	Yes
3) Causal role of the mind in nature:	Yes (rejection of epiphenomenalism)	Yes (rejection of epiphenomenalism)	Yes (rejection of epiphenomenalism)
4) On the place of mind in nature:	Emergentism	Panpsychism	Panpsychism
5) On organization:	Holism (the whole has emergent properties)	Holism (the whole is a fusion of the parts)	Holism (the whole is a fusion of the parts) + Laws of organization of matter
6) Elements explaining the directions of evolution:	↓ Natural selection	↓ Natural selection + The creativity of organisms, at all levels of the organization	↓ Natural selection + The creativity of animals + The laws of organization of matter (which give a certain directionality to evolution)

the living as opposed to the inert, and more generally the organized as opposed to the inorganized.

Note that, when it comes to *the origin of mind*, only Fisher defends a form of emergentism in the sense that, according to him, organisms endowed with a mind emerge from a matter which is devoid of it. In contrast, Wright and Haldane defend a form of panpsychism in the sense that, in their view, elementary matter already has psychic properties. Although surprising, this idea is not new to philosophy. Its history goes back to antiquity, with the argument that the dynamic properties of matter and living things can be explained by something psychic, analogous to our mind. However, the development of evolutionary biology in the nineteenth century renewed the panpsychist argument. For defenders of this conception, the idea is that if living beings originate in inert matter, then the fundamental properties of living beings, including the mind, must already exist in some form in inert matter. This argument can be found in E. Haeckel (biologist, 1834–1919), W. James (psychologist and philosopher, 1842–1910), W. Clifford (mathematician and philosopher, 1845–1979), F. Paulsen (philosopher, 1846–1908), M. Prince (neurologist, 1854–1929), J. Royce (philosopher, 1855–1916), C. Strong (psychologist and philosopher, 1862–1940), and others.[21] At the time of Fisher, Wright, and Haldane, panpsychism was therefore a well-known philosophical idea in the intellectual world. Under the influence of neo-positivism, this idea gradually lost its importance over the course of the twentieth century, before regaining interest today (Skrbina 2005).

With regard to *the living world in general*, Fisher also defends a strong emergentism, in the sense that, according to him, in the course of the evolution of living beings, new properties appear that are irreducible to the old ones: growth, voluntary movement, effort and pain, joy and pain, and so on. In contrast, Wright and Haldane reject this idea of strong emergence. According to them, the most elementary organized systems already have psychic properties, and more complex systems are formed by the fusion of elementary systems. The idea is obviously speculative. It is hotly debated today in the literature on panpsychism.

Finally, only Haldane introduces the idea that certain natural regularities can be explained by special laws. Fisher does not address this question. And for Wright, natural regularities are ultimately explained by the "choices" of organisms, at all levels of organization. This means that, for him, the ultimate causes of natural processes are psychic.

Line 6 concerns the basic elements that each introduces to explain the directions taken by evolution. While Fisher only introduces natural selection, Wright and Haldane also introduce non-Darwinian elements. For each, the black arrows indicate that the conception of evolution defended is largely derived from a certain conception of nature.

This summary immediately shows that Fisher, Wright, and Haldane have different *philosophical* conceptions of evolution. In other words, in addition to some

[21] For panpsychism in these thinkers and others, see Skrbina (2005: chap. 5 and 6).

scientific disagreements between them, there are more fundamental disagreements, so to speak, albeit speculative. Again, this relativizes the notion of "Modern Synthesis." This result echoes that obtained by Delisle about five other synthetists: Huxley, Dobzhansky, Rensch, Simpson, and Mayr (Delisle 2009b).

This summary also allows us to see that the conceptions of Wright and Haldane go beyond the Darwinian framework because, in order to explain the directions taken by evolution, they introduce elements that are foreign to this framework. Wright thus introduces the idea that evolution depends on the creativity of organisms, *at all levels of organization*. Specifically, as I explained above, he suggests that the cells of an organism are capable of learning to use their genome in a variety of ways and of transmitting this learning. Haldane, on the other hand, introduces the idea that the evolution of animals depends largely on their creativity. In this sense, he gives a major place to the Baldwin effect. Moreover, he hypothesizes that evolution as a whole depends on some laws of organization that make it possible the appearance of life and mind. This means that, for him, from a cosmic point of view, evolution has *a certain directionality*. This result is similar to the one highlighted by Delisle, concerning Huxley, Dobzhansky, and Rensch (Delisle 2009a: 120–126; 2009b: 379–380). For these three synthetists too, evolution is a cosmic phenomenon characterized by a certain directionality.

There is another point worth emphasizing. It is the question whether evolution is a purely mechanical process. In this regard, here is what Provine writes:

> The term "evolutionary constriction" helps us to understand that evolutionists after 1930 might disagree intensely with each other about effective population size, population structure, random genetic drift, levels of heterozygosity, mutation rates, migration rates, etc., but all could agree that these variables were or could be important in evolution in nature, and that purposive forces played no role at all. So the agreement was on the set of variables, and the disagreement concerned differences in evaluating relative influences of the agreed-upon variables. (Provine 1988: 61–62)

As we have seen, with respect to nature in general, Fisher, Wright, and Haldane unanimously reject radical mechanism (lines 3 and 4 of the table), and this already tends to undermine a purely mechanistic approach to evolution. But above all, as far as evolution is concerned, what Provine says does not apply at all to Wright and Haldane. For Wright, evolution depends crucially on the creativity of organisms, conceived as irreducibly purposeful. And for Haldane, the same is true of the evolution of animals, not to mention that, for him, the hypothesis of the laws of organization of matter leads one to the question of a more global intention in nature. In fact, only the philosophy of evolution proposed by Fisher fits into the framework defined by Provine. It cannot be said therefore that, among the synthetists, there was an agreement to reject the idea of purposeful forces in evolution.

Besides, note the proximity of Wright and Haldane's ideas to those of the "romantic" biologists evoked by Esposito (2017), who in the first half of the twentieth century, on the basis of a non-mechanistic conception of living things, attempted to develop a non-Darwinian approach to evolution. For Wright and Haldane, as for these biologists, evolution is a cosmic process, and its creativity depends crucially on the activity of organisms. The filiation between these biologists and J. B. S. Haldane is also obvious, since J. S. Haldane was a "romantic" biologist (Esposito 2017: 226–227).

It may be objected that, for the understanding of evolution, the philosophical ideas of the synthetists are of little interest, and that only their scientific work should count. But this is absurd because the philosophical ideas of the synthetists have a direct link to the science of evolution. First, these ideas say something about evolutionary biology, including its present or absolute ability to explain evolution. Thus, Wright repeatedly asserts that something fundamental about evolution eludes science. Besides, these ideas can also serve as a research program for biology. For example, what Wright says about the role of individual adaptability in evolution can encourage the scientist to study this adaptability at the cellular level.[22] In the same sense, Haldane's hypothesis about the laws of organization of matter is immediately a research program for science.

It may also be objected that the philosophical ideas defended by Fisher, Wright, and Haldane have little value in themselves, that they are only the expression of a personal taste—what Provine suggests about Wright (Provine 1986: 95–97), or/and of an outdated philosophical heritage. But both points are dubious. The first because, on reading the three synthetists, it is clear that each one attempts to ground his positions in a purely rational manner. The second because the questions of the reduction of life and mind to the inert are still very much alive in philosophy today, and ideas such as emergentism, panpsychism, and holism are still hotly debated.

6 Conclusion

Fisher, Wright, and Haldane are three synthetists from the same generation and the same field, population genetics. Yet, each of them defends an original view of evolution, different from that of the other two. Moreover, among these different philosophies of evolution, two introduce elements opposed to the Modern Synthesis.

The conclusion of this chapter is therefore that, *on the very broad level of philosophical ideas*, the debate on evolution has never stopped among synthetists. What's more, this debate has gone far beyond Darwinism. What is the living in relation to the inert? Does evolution obey a general directionality? Do certain purposeful forces play a role in evolution? For some architects of the Modern Synthesis, these questions have never stopped being on the table.

[22]Contrary to Provine's assertion (Provine 1986: 95–97), it is possible to make a connection between Wright's philosophical ideas and his own scientific research. For example, Steffes shows that Wright's philosophical ideas guided his choice to study certain questions scientifically, in particular the effect of gene interaction on phenotype (Steffes 2007).

References

Bergson H (1911) Creative evolution. Henri Holt, New York
Craver C, Tabery J (2023). Mechanisms in science. In: E Zalta, U Nodelman (eds) The Stanford encyclopedia of philosophy, URL. https://plato.stanford.edu/archives/fall2023/entries/science-mechanisms/
Dawkins R (1986) Sociobiology: the new storm in a teacup. In: Rose S, Appignanesi L (eds) Science and beyond. Blackwell, Oxford, pp 61–78
Delisle RG (2009a) The uncertain foundation of neo-Darwinism: metaphysical and epistemological pluralism in the evolutionary synthesis. Studies in History and Philosophy of Science Part C 40(2):119–132
Delisle RG (2009b) Les philosophies du néo-Darwinisme. Presses Universitaires de France, Paris
Dennett DC (1995) Darwin's dangerous idea: evolution and the meanings of life, vol 35. Simon and Schuster, New York, p 34
Dronamraju K (2011) Haldane, Mayr, and beanbag genetics. Oxford University Press, New York
Dronamraju K (2017) Popularizing science, the life an work of JBS Haldane. Oxford University Press, New York
Esposito M (2017) The organismal synthesis: holistic science and developmental evolution in the English-speaking world, 1915–1954. In: Delisle RG (ed) The Darwinian tradition in context: research programs in evolutionary biology. Springer, Cham, pp 219–242
Fisher RA (1934) Indeterminism and natural selection. Philos Sci 1:99–117
Fisher RA (1950) Creative aspects of natural law. Cambridge University Press, Cambridge
Gayon J (1997) The paramount power of selection: from Darwin to Kauffmann. In: Dalla Chiara ML, Doets K, Mundici D, van Benthem J (eds) Structures and norms in science, Volume Two of the tenth international congress of logic, methodology and philosophy of science. Kluwer, Dordrecht, pp. 265–282
Gayon J (2008) *L'évolution créatrice* lue par les fondateurs de la théorie synthétique de l'évolution. In: Fagot-Largeault A, Worms F (eds) Annales Bergsoniennes IV. Presses Universitaires de France, Paris, pp 59–84
Gayon J, Huneman P (2019) The modern synthesis: theoretical or institutional event? J Hist Biol 52: 519–535
Haldane JBS (1929/1968) The origin of life. The rationalist annual 1929. Republished in Haldane JBS (1968), pp 1–11
Haldane JBS (1932a) The causes of evolution. Harper and Brothers, London
Haldane JBS (1932b) Some consequences of materialism. In: Haldane JBS (ed) The inequality of man and other essays. Penguin Books, Harmondsworth, pp 157–170
Haldane JBS (1933) Modern physics and causality. The Rationalist Annual 1933:9–16
Haldane JBS (1934) Quantum mechanics as a basis for philosophy. Philos Sci 1(1):78–98
Haldane JBS (1938) The Marxist philosophy and the sciences. George Allen, London
Haldane JBS (1940/1968) Why I am a materialist. The rationalist annual 1940. Republished in Haldane JBS (1968), pp 27–35
Haldane JBS (1941/1968) The laws of nature. The rationalist annual 1941. Republished in Haldane JBS (1968), pp 36–42
Haldane JBS (1944/1968) The argument from design. The rationalist annual 1944. Republished in Haldane JBS (1968), pp 43–48
Haldane JBS (1947) What is life? Boni and Gaer, New York
Haldane JBS (1958a) The unity and diversity of life. Publications Division, Delhi
Haldane JBS (1958b) The present position of Darwinism. J Sci Ind Res 17A(3):97–103
Haldane JBS (1959a) The theory of natural selection today. Nature 183:710–713
Haldane JBS (1959b) Natural selection. In: Bell PR (ed) Darwin's biological work. Cambridge University Press, Cambridge, Some aspects reconsidered, pp 101–149
Haldane JBS (1962/1968) Beyond agnosticism. The rationalist annual 1962. Republished in Haldane JBS (1968), pp 157–166

Haldane JBS (1963) Life and mind as physical realities. In: Barnett SA, McLaren A (eds) Penguin science survey 1963 B. Penguin, Harmondsworth, pp 224–238
Haldane JBS (1968) Science and life. Pemberton Publishing, Islington
Haldane JS (1935) The philosophy of a biologist. Clarendon Press, Oxford
Hammond A (2009) J. B. S. Haldane, holism, and synthesis in evolution. In: Cain J, Ruse M (eds) Descended from Darwin. American Philosophical Society, Philadelphia, pp 49–70
Hodge J (1992) Biology and philosophy (including ideology): a study of Fisher and Wright. In: Sarkar S (ed) The founders of evolutionary genetics. Kluwer academic publishers, Dordrecht, pp 231–293
James W (1879) Are we automata? Mind 4(13):1–22
Mayr E (2004) What makes biology unique? Cambridge University Press, Cambridge
Monod J (1970) Le hasard et la nécessité. Seuil, Paris
Provine WB (1986) Sewall Wright and evolutionary biology. University of Chicago Press, Chicago
Provine WB (1988) Progress in evolution and meaning in life. In: Nitecki MH (ed) Evolutionary progress. University of Chicago Press, Chicago, pp 49–74
Rao V, Nanjundiah V (2017) Haldane's view of natural selection. J Genet 96(5):765–772
Sarkar S (1992) Science, philosophy, and politics in the work of J. B. S. Haldane, 1922–1937. Biol Philos 7(4):385–409
Shapiro AM (1993) Haldane, Marxism, and the conduct of research. Q Rev Biol 68:69–77
Shapiro JA (2011) Evolution: a view from the 21st century. FT Science Press, Upper Saddle River
Simpson GG (1953) The Baldwin effect. Evolution 7:110–117
Skrbina D (2005) Panpsychism in the west. MIT Press, Cambridge
Smuts J (1926) Holism and evolution. Macmillan, New York
Steffes D (2007) Panpsychic organicism: Sewall Wright's philosophy for understanding complex genetic systems. J Hist Biol 40(2):327–361
Subramanian S (2020) A dominant character: the radical science and restless politics of J.B.S. Haldane. Atlantic Books, London
Wright SG (1931) Evolution in Mendelian populations. Genetics 16:97–159
Wright SG (1935) The emergence of novelty. A review of Lloyd Morgan's 'emergent' theory of evolution. J Hered 26:369–373
Wright SG (1953) Gene and organism. Am Nat 87(832):5–18
Wright SG (1959) Genetics and the hierarchy of biological sciences. Science 130(3381):959–965
Wright SG (1964) Biology and the philosophy of science. Monist 48(2):265–290
Wright SG (1975) Panpsychism and science. In: Cobb J, Griffin D (eds) Mind in nature. University Press of America, Washington, pp 79–88
Wright SG (1980) Genic and organismic selection. Evolution 34(5):825–843

A Synthesis Without Darwin: Unification Attempts in Early Theoretical Biology

Jan Baedke, Alexander Böhm, Stefan Reiners-Selbach, and Vera Straetmanns

Abstract

In this chapter, we discuss the role of evolutionary theory—especially Darwinism—during the institutionalization of theoretical biology in the early twentieth century. During this time, theoretical biologists strived to establish a new theoretical framework for biology. Rather than unifying biological phenomena under Darwin's theory of natural selection, they aimed to establish a philosophical framework that could overcome the vitalism-mechanism divide and counteract disciplinary fragmentation and specialization. Their accounts ultimately turned out to be in line or in conflict with Darwin's theory, but discussing Darwinism was not the driving force to develop them. Our historical narrative is based on a twofold approach: First, we conduct a Digital Humanities analysis of central publications in the field; second, we discuss in detail how influential theoretical biologists, especially the zoologist Julius Schaxel, shaped "synthesis discourses." We show that theoretical biologists objected to evolutionary theories due to their underlying philosophies: mechanism and materialism (for Darwinism) and vitalism (for Lamarckism). Their new philosophical synthesis did not directly aim to reject Darwinian principles like natural selection nor to establish an evolutionary synthesis through population genetics. This makes it necessary to deconstruct established periodizations of past evolutionary theorizing, as such philosophically minded unification attempts neither directly contributed to the "eclipse of Darwinism" nor to the emergence of the Modern Synthesis. Finally, we argue that a stronger engagement with so-far "eclipsed" debates in early

J. Baedke (✉) · A. Böhm · V. Straetmanns
Department of Philosophy I, Ruhr University Bochum, Bochum, Germany
e-mail: jan.baedke@ruhr-uni-bochum.de

S. Reiners-Selbach
Faculty of Arts and Humanities, Heinrich Heine University Düsseldorf, Düsseldorf, Germany

R. G. Delisle et al. (eds.), *Unity and Disunity in Evolutionary Biology*,
https://doi.org/10.1007/978-3-031-42629-2_12

twentieth century theoretical biology can help casting a light on past and present developments in evolutionary theory.

Keywords

Theoretical biology · "Eclipse of Darwinism" · Modern Synthesis · Julius Schaxel · Philosophy of biology · Neo-Kantianism · Holism

1 Introduction

In the first decades of the twentieth century, theoretically interested biologists and philosophers, especially in German-speaking countries, were increasingly uncomfortable with the state of theory in biology. They sought a new conceptual foundation of biology that would counteract trends of the time that were leading toward disciplinary specialization and fragmentation. In addition, this foundation was expected to provide the means for unifying and systematizing biological knowledge with the aim of revealing theoretical connections between fundamental properties of life such as development, behavior, inheritance, and, ultimately, evolution. In this novel synthesis, evolutionary debates were not primarily driven by taking Darwinian or non-Darwinian standpoints. In fact, German-speaking theoretical biologists attempted to unify biological phenomena not by subsuming them under Darwin's theory. Instead, they started from a number of philosophical frameworks—neo-Kantian, holistic, dialectical—which structured their evolutionary reasoning and attempts to unify biological findings. This resulted in evolutionary theories that ultimately could be in line or in conflict with Darwin's theory, but for which discussing Darwinism was not the main motivation. What could be easily mistaken as an attempt to replace Darwinism by non-Darwinian theories, was, in fact, merely an epiphenomenon of what was really going on. It was not Darwin that was rejected, but rather a philosophical framework—in this case: mechanism—that, among others, Darwinians adopted at the time.

By drawing on bibliometrical analyses of central publications in the field and on case studies about influential theoretical biologists, like Julius Schaxel, Ludwig von Bertalanffy, and Emil Ungerer, and their philosophical frameworks (e.g., neo-Kantianism and holism), this chapter discusses different unification attempts that did not aim to reject Darwinism, nor help establish the Modern Synthesis. This analysis makes questionable established historiographies of past evolutionary theorizing, especially Peter Bowler's narrative spanning from the "eclipse of Darwinism" to the emergence of the Modern Synthesis.

We will proceed as follows: First, we will describe the standard "eclipse" to "unity" periodization in the history of early twentieth century evolutionary theory (Sect. 2). After giving a short introduction to the emergence, motivations, and aims of theoretical biology during this time (Sect. 3), we offer a macro-level bibliometrical analysis, which shows that Darwin and Darwinism were not widely discussed in bio-theoretical debates in the early twentieth century (Sect. 4). To explain why this was the case, we first look at the influential theoretical biologist Julius Schaxel: By drawing extensively on late nineteenth century neo-Kantian critiques of mechanistic versions of Darwinism among German-speaking biologists (Sect. 5), Schaxel rejected the underlying materialist philosophy of Darwinism (Sect. 6). However, against the common "eclipse" narrative, anti-Darwinian accounts did not fare much better among theoretical biologists. To show this, we study bio-theoretical criticism against vitalist philosophies underlying Lamarckism, put forward by German holists (Ganzheitsbiologie; Sect. 7). In the conclusion, we reflect on what this history of theorizing about biological phenomena without Darwin means for common periodizations of the history of evolutionary theory (Sect. 8). We also discuss how the common narrative of the "dark age of Darwinism" in the early twentieth century so-far prevented historians from better understanding several past and current developments in evolutionary theory, from the history of endosymbiotic theory to the recent "Extended Evolutionary Synthesis."

2 The Eclipse of Darwinism and Unification of Biology

In his books *The Eclipse of Darwinism* and *The Non-Darwinian Revolution*, Peter Bowler (1983, 1988, see also Bowler 2005) has developed a highly influential narrative of late-nineteenth and early-twentieth century evolutionary biology. According to Bowler, Darwin's view of evolution by natural selection (and that of advocates of Darwinism) was "eclipsed" by several anti-Darwinian lines of thought. These emerged out of older structuralist approaches to development and evolution and discussions of the effect of soft inheritance on phylogenetic change:

> From these phylogenetic studies emerged much of the support for *non-Darwinian mechanisms* such as Lamarckism, orthogenesis and saltationism. In their more extreme form, though, these ideas led to a *complete rejection of some of the key ideas of Darwinism*. The "eclipse" occurred as more and more biologists rejected the emphasis on function and adaptation, highlighting instead mechanisms that stressed the role of formal or structural constraints on development. (Bowler 2005: 24; emphasis added)

According to Bowler, Darwinism was rejected because the Darwinian mechanism of adaptive evolution by natural selection was replaced by mechanisms that highlighted organismal, developmental processes as primary drivers of inheritance and evolution. This "eclipse" narrative, and Bowler's general historiography of the transition from the nineteenth to the twentieth century, has gained substantial momentum since the 1980s. "I am reasonably confident," Bowler (2005: 23) states, "that much of what I argued for has now become taken for granted."

Ultimately, so the narrative goes, this theoretical fragmentation of evolutionary biology, with various rivaling evolutionary theories being tested, discussed, and continuously reworked, came to an end in the 1930–1940s. After that, we come to see a revival of selectionism and Darwinian principles that theoretically unified evolutionary biology. This unifying theory integrated Darwinian mechanisms, especially natural selection, and Mendelian genetics within the mathematical framework of population genetics. In short, evolutionary biology again found unity through the "Modern Evolutionary Synthesis."

This historiography of "eclipse-then-synthesis" by Bowler is, in fact, an older one. It was developed by the architects of the Modern Synthesis themselves, especially by Julian Huxley in his *Evolution: The Modern Synthesis* (1942).[1] Huxley described the state of evolutionary theory in the late-nineteenth and early-twentieth century as a "dark age" dominated by confused and contradictory approaches put forward especially by German-speaking scholars. In this phase, Mendelians fought against neo-Lamarckians, mutationalists against advocates of Weismannism, and geneticists against Darwinian zoologists. Against this situation of theoretical disunity and fragmentation, Huxley and others aimed at developing a "synthetic point of view" (Huxley 1942: 8). He expressed the desire not only to unify theoretical biology, but biology as a whole:

> Biology at the present time is embarking upon a phase of synthesis after a period in which new disciplines were taken up in turn and worked out in comparative isolation. Nowhere is this movement towards unification more likely to be valuable than in this many-sided topic of evolution; and already we are seeing the first-fruits in the re-animation of Darwinism. (Huxley 1942: 13)[2]

Huxley believed that the rebirth of Darwinism would bring new "light" to all of biology, both theoretically and methodologically:

> Biology in the last twenty years, after a period in which new disciplines were taken up in turn and worked out in comparative isolation, has become a more *unified science*. It has embarked upon a period of synthesis, until today it no longer presents the spectacle of a number of *semi-independent and largely contradictory sub-sciences*, but is coming to rival the unity of older sciences like physics, in which advance in any one branch leads almost at once to advance in all other fields, and theory and experiment march hand-in-hand. As one chief result, there has been a rebirth of Darwinism. (Huxley 1942: 26; emphasis added)

This argument that unification in evolutionary theory—through a new Darwinism—drove and enabled the unification of biology as a whole quickly became influential among both fellow scientists as well as historians. Theodosius Dobzhansky (1964, 1973) constructed evolution (governed by Darwin's mechanism of natural selection

---

[1]Even earlier, David Starr speaks in an unpublished manuscript of "The eclipse of Darwin" (ca. 1925), see Largent (2009).

[2]Huxley continues with a characterization of Darwinism: "By Darwinism I imply that blend of induction and deduction which Darwin was the first to apply to the study of evolution. He was concerned both to establish the fact of evolution and to discover the mechanism by which it operated [. . .]. On the one hand he amassed enormous quantities of facts from which inductions concerning the evolutionary process could be drawn; and on the other, starting from a few general principles, he deduced the further principle of natural selection." (Huxley 1942: 26). According to Huxley these facts are the tendency of organisms to increase progressively through reproduction, the tendency of a number of a given species to remain constant, and the fact that all nature shows variations. From these facts Darwin deduced the struggle for existence and the principle of natural selection.

and described by the Modern Synthesis) as an integrative principle of biology. He claimed that "nothing makes sense in biology except in the light of evolution" (1964: 449).[3] Dobzhansky argued that evolution theoretically unifies all biological knowledge: knowledge about the diversity of life-forms and unity of life (e.g., the universality of the genetic code) as well as about developmental and embryological patterns:

> Seen in the *light of evolution*, biology is, perhaps, intellectually the most satisfying and inspiring science. Without that light it becomes a pile of sundry facts—some of them interesting or curious but *making no meaningful picture as a whole* [. . .]. One of the great thinkers of our age, Pierre Teilhard de Chardin, wrote the following: "Is evolution a theory, a system, or a hypothesis? *It is much more—it is a general postulate to which all theories, all hypotheses, all systems must henceforward bow and which they must satisfy in order to be thinkable and true.* Evolution is a light which illuminates all facts, a trajectory which all lines of thought must follow—this is what evolution is." (Dobzhansky 1973: 129; emphasis added)

Consequently, so the story goes, the new Darwinian theory of evolution brought light to a dark age of biology. Other historians and scientists adopted this narrative by authors of the Modern Synthesis stating that, in the 1930–1940s, biology was unified under a new "Darwinian umbrella." Besides Bowler (1983), for example, contributions to the edited volume of Ernst Mayr and William B. Provine *The Evolutionary Synthesis: Perspectives on the Unification of Biology* (1980) as well as Betty Smocovitis' in-depth analysis of the "sad state of Darwinism at the turn of the century" (1996: 117) in her book *Unifying Biology* tie in with this narrative.

We are skeptical about the common historiography from "eclipse" to "unity." This is due to several reasons.[4] First, it presents a history that was largely written by the historical actors—synthesis authors—themselves. This threatens a kind of self-serving writing of history by those with the most to gain from the established narrative. Second, it favors a discontinuous historiography of evolutionary thought that tends to pejoratively discredit the relevant period as a "dark" time in which little scientific progress was made (see Largent 2009). Thus, this period can easily be misinterpreted by historians as lacking value and being of little interest. Nothing could be further from the truth. The early twentieth century, and especially the interwar period, was a lively and highly innovative phase in biology both theoretically and methodologically (see Harwood 1993, Baedke and Brandt 2022).

Rather than treating it as the dark age of Darwinism, here we wish to develop a different historiography. First, against the standard "eclipse" narrative, we aim to show that there were reasons for neglecting Darwin(ism) during this period other than questioning the importance of natural selection in evolution (as discussed, e.g., in Kellogg 1907). The reason some, especially German-speaking, scholars showed

[3]On differences between Dobzhansky's arguments in his 1964 and 1973 publications, see Giaimo (2023).

[4]For more reasons and an in-depth critical analysis of the eclipse narrative, see Meulendijks (2021).

less interest in Darwin was not so much because they actively tried to reject certain evolutionary mechanisms. Instead, first and foremost, they favored different underlying philosophical frameworks. That is, philosophical debates heavily channeled theoretical positions during this time, and led—as a side product—to the demise of, or rather disinterest in, Darwinism. Second, against the standard "unification" narrative, we hold that in the early twentieth century, various "syntheses" were developed that clearly differed from the Modern Synthesis. A number of theoretically minded scholars argued that biology, in fact, needed unification, but not by means of Darwinian principles, or evolutionary biology more generally. They claimed that what may unify biology is a shared and consistent philosophical framework that sets clear conceptual, explanatory, and methodological standards. In short, it is not evolution that unifies biology, but philosophy.

These two components of our new narrative from "eclipse" to "synthesis" can best be understood by looking at the field of theoretical biology, which emerged in the first decades of the twentieth century—first among German-speaking biologists and philosophers, then increasingly in an international network.

3 The Rise of Theoretical Biology

Somewhat in line with the historical reconstructions by authors of the Modern Synthesis, in the late-nineteenth and early-twentieth century biologists, indeed, faced a challenging situation. Advanced experimental techniques enabled the gathering of fascinating new data on topics like morphology, regeneration, inheritance, plasticity, and robustness (Driesch 1892; Woltereck 1909; Morgan 1910; Spemann and Mangold 1924; for discussion, see Baedke and Brandt 2022). As a result of this research, scientists observed a growing specialization and fragmentation within the field of biology (see Ringer 1969; Harwood 1993; Nyhart 1995). Viennese theoretical biologist Ludwig von Bertalanffy summarizes this situation as follows:

> It is a question of arriving at general theories for the phenomena of life, from which the individual phenomena can be derived in the same way as, for example, light, heat, magnetism, electricity etc. can be derived as special cases from the electromagnetic theory of light, or as the electron theory provides a common explanation for the phenomena of chemical affinity, the periodic system and the transformation of the elements, radioactivity etc. A Th. B. [theoretical biology] in this sense does not yet exist; nevertheless, only through it will biology be able to become a completed science, whereas today it represents only an *immense accumulation of more or less unrelated individual facts*. [...] We have enough facts that await theoretical processing—only the ideas are missing. When Kepler and Newton founded theoretical physics, they certainly had infinitely less experience than our biology has today. It is not the facts that we lack, but the Copernican action! (Bertalanffy 1929: 374, 376; emphasis added; German original)

Other biologists and philosophers shared Bertalanffy's call for a unifying theory in biology, similar to that provided for physics through theoretical physics. Philosopher and historian of biology Adolf Meyer-Abich (Meyer 1926[5]) stated:

> If one compares the various presentations of theoretical physics with each other, one will find small deviations in the manner of presentation and in the logical arrangement of the theories, but essentially the same doctrinal content [. . .]. The situation is quite different in biology. If one compares here the known general presentations of biological problems with each other, one could be tempted to believe in presentations of different sciences which to a certain extent only use the same terminology. (Meyer 1926: 6-7; German original)

Biology thus needed some clear theoretical framework that would counteract fragmentation trends and allow for consistent interpretation and comparison of data as well as clear methodological guidance. But what could provide such a framework? Many theoretical biologists increasingly believed that neither vitalism nor mechanism—the widely discussed opposing research frameworks of the time—could do so (see Allen 2005; Nicholson and Gawne 2015). While mechanistic accounts neglected the special characteristics of non-reducible, teleological phenomena of organisms, vitalist theories (in their metaphysical form) acknowledged these phenomena, but were not compatible with a materialist scientific worldview.

Rather than assuming—in line with Dobzhansky, Mayr and other Modern Synthesis authors—that Darwin's theory of evolution or a general theoretical emphasis on natural selection could come to the rescue and provide a unified framework of biological knowledge, theoretical biology turned to philosophy instead. Theoretical biologists believed what biology lacked was a consistent philosophical framework that could guide data collection and theorization in biology, but that was also superior to vitalism and mechanism. While agreeing on this general aim, they put different emphasis on what this new field should focus on. Some sought to establish the autonomy of biology from physics from a philosophical perspective (Driesch 1908), while others hoped that abstracting from details would counteract the fragmentation and specialization of the discipline (Reinke 1901). A significant group of scholars believed that the main duty of theoretical biology was to facilitate theoretical unification and conceptual clarification. This ranged from organizing theories through a consistent philosophical framework that establishes validity standards (Schaxel 1919a), to relating individual new empirical findings to nature as a whole (Uexküll 1928 [1920]), and to systematizing theoretical foundations of different biological disciplines to uncover the "logic of biology" (Meyer 1926). Given this diversity of aims, it is challenging to identify a *single* underlying philosophical framework that drove early theoretical biology. This is not least because there were at least four different philosophical traditions used as starting points for theoretical biology to unify biology: neo-Kantianism, dialectical materialism, holism, and organicism (see Baedke 2019; Baedke et al. 2024).

[5] In 1938, Adolf "Meyer" added his mother's family name "Abich."

To better understand the arguments central in this field, and how the history of these philosophical frameworks fares against standard periodizations, such as the eclipse of Darwinism and theoretical unification through the Modern Synthesis, let us take a twofold approach: first, we trace bibliometric patterns of Darwinian and evolutionary debates in theoretical biology; second, we look at influential theoretical biologists in detail and why their neo-Kantian and holistic unification attempts neither aimed to reject Darwinian mechanisms nor to establish an evolutionary synthesis in line with the Modern Synthesis.

4 Darwin's "Eclipse" in Theoretical Biology

A quantitative Digital Humanities approach can offer some first insights into how evolutionary debates in general, and discussions about Darwin's or Darwinian positions specifically affected early theoretical biology over time. Therefore, we created a corpus of 311 relevant documents (ca. 6.7 million words after preprocessing) published in the field of theoretical biology between 1900 and 1971. This includes three central book series, all international publications of the journal *Acta Biotheoretica* (founded by Adolf Meyer-Abich, C.J. van der Klaauw, and J.A.J. Barge in 1935), which was the first international journal on theoretical biology (Reydon et al. 2005), and relevant German-language and international monographies. The corpus takes into consideration that the field of theoretical biology first emerged in German-speaking countries, but then also thrived in the United Kingdom and United States (see Laubichler 2001; Esposito 2015; Nicholson and Gawne 2015; Peterson 2016; Baedke 2019). See Table 1 for a list of publications.

Table 1 Publications in theoretical biology included in the corpus

Publication format	Editor(s)	Title	Publication Time Span
Book series	Julius Schaxel	*Abhandlungen zur theoretischen Biologie* (Contributions to theoretical biology)	1919–1931
Book series	Adolf Meyer-Abich	*Bios: Abhandlungen zur theoretischen Biologie und ihrer Geschichte, sowie zur Philosophie der organischen Naturwissenschaften* (Bios: Contributions to theoretical biology and its history, as well as to the philosophy of the organic natural sciences)	1934–1947
Book series	Various	*Bibliotheca Biotheoretica*	1941–1971
Journal	Various	*Acta Biotheoretica*	1935–1971
Monographies		Various German-language publications (e.g., Reinke 1901; Schaxel 1919a; Ungerer 1919, 1922; Uexküll 1928 [1920]; Bertalanffy 1932, 1942; Ehrenberg 1923; Hartmann 1925), but also English ones (e.g., Woodger 1929; Haldane 1931; Bertalanffy 1933; Needham 1936)	1901–1971

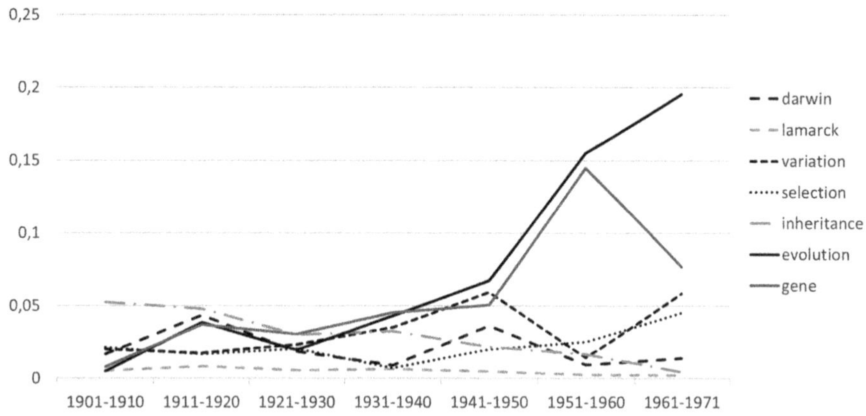

Fig. 1 Changes in the usage of keywords of evolutionary debates in early theoretical biology, 1901–1971. Depicted is the average frequency of keywords mentioned in 100 words per publication (0,1 means that 1 keyword is used every 1000 words); German or English keywords with wildcards (*): "darwin*", "lamarck*", "variation", selection (i.e., "natürliche selektion", "natural selection", "selektion" or "selection"), "inheritance"/"vererbung", evolution (i.e. "evolution*", "Artwandel", "Phylogenese" or "phylogenesis"), gene (i.e., "Gen", "gene", "genes", "genetisch", "genetic*" or "Genetik")

We scanned and pre-processed the selected sources, including OCR (Smith 2019). We searched these publications for keywords of evolutionary topics in early theoretical biology, linked to issues like evolution, Darwin and Darwinism, Lamarck and Lamarckism, natural selection, variation, inheritance, and genes. See Fig. 1 for results.[6]

For example, "darwin*" appears on average around 4 times per 10.000 words (per publication) in the 1910s, 2 times in the 1920s, once in the 1930s, and 3 times in the 1940s. In fact, all evolutionary keywords score relatively low in the whole corpus during the interwar period. Later, starting from the 1940s and 1950s, evolutionary topics increasingly gain relevance in theoretical biology, but seemingly less due to a discussion of classical Darwinian topics, ideas or principles like natural selection, but more linked to debates about genetics and the role of genes in evolution.

What can we learn—admittedly on a rather coarse-grained scale—from these results? First, evolutionary (including Darwinian) topics seemingly played a rather minor role in establishing the field of theoretical biology in the early twentieth century (see also Laubichler 2001). This might be due to the fact that many theoretical biologists were physiologists, embryologists or zoologists (like Julius Schaxel, Emil Ungerer, Jakob von Uexküll, Hans Leo Przibram, Max Hartmann, John Scott Haldane, Joseph Needham), or philosophers (like Ludwig von Bertalanffy, Adolf

[6]Due to the lower number of publications in the field until 1920, the results from 1920 onwards are more robust. Then the field started flourishing, with the publication of, e.g., Schaxel's *Abhandlungen* book series (since 1919) and Meyer-Abich's *Bios* (since 1934). The journal *Acta Biotheoretical* was founded in 1935.

Meyer-Abich, Joseph Woodger) not trained or working in evolutionary biology (but see, e.g., Friedrich Alverdes, Oskar Hertwig, Hans Böker, Conrad Hal Waddington). At the same time, this omission of evolution in the literature is surprising if we—following Dobzhansky and Mayr—assume that the theory of evolution should be a (or maybe the most) central target of any theoretical biology worth its name, as evolution is thought to theoretically unify all biological knowledge, including that of embryological, physiological, and morphological patterns. According to Dobzhansky, evolutionary ideas and concepts provide the ultimate theoretical framework to conceptualize and theorize about living beings. If it is correct that evolution theoretically unifies biology, why did early theoretical biologists care so little about evolution?

Second, our results also suggest that the vacuum of Darwinian debates was not filled by non-Darwinian positions, such as Lamarckian ones. The keyword "lamarck*" appears on average only around 0.5 times per 10,000 words from 1910 to the 1940s. This result is at odds with Huxley's and Bowler's "eclipse" historiography. Even if Bowler is right, and Darwin was eclipsed in other fields of biology, say evolutionary theory, during this time by theories like Lamarckism, one might wonder why such debates did not reverberate in discussions in theoretical biology. After all, a clash of central, seemingly opposed theories struggling for the supremacy when it comes to theoretically grasping and methodologically approaching phenomena of inheritance and evolution should be of high relevance for theoretical biology.

This leads us to our central questions: Why did theoretical biologists most widely ignore Darwin and evolutionary theory, and how did they aim at providing a theoretical synthesis for biology that did not go through (or against) Darwin? To address these points, let us put more detail to our coarse-grained analysis depicted in Fig. 1. First, let us identify those theoretical biologists in the corpus that, in fact, discuss Darwin and Darwinism. In the early years of theoretical biology during the interwar period, these entries are quite unevenly distributed, with most of the publications not discussing Darwin(ism) at all. Between 1911–1920, exceptions are Julius Schaxel's (1919a) influential *Grundzüge der Theoriebildung in der Biologie* (Foundations of Theory Construction in Biology), who used "darwin*" by far the most (16/10.000 words), and Schaxel (1919b), Kroner (1919) and Franz (1920), each with around 6/10.000 words. Between 1921 and 1930, Schmid (1921), Driesch (1928 [1908]) and Bertalanffy (1928) contain each 5/10.000 "Darwinian terms," and Rignano (1927), Alverdes (1921), and Ungerer (1922) 4/10.000 each. Between 1931 and 1940, Metallmann (1939) and Schindewolf (1937) used such terms the most with 17 and 13/10.000 words, respectively; Plate (1936) used 7/10.000 words, then Jordan (1935) and Ungerer (1936) with 4/10.000 each. But, again, these few discussions were rather the exception than the rule. For example, after the international journal *Acta Biotheoretica* was launched in 1935, nearly none of its bio-theoretical papers deemed it worth discussing or even mentioning Darwin (ism), Darwinian principles, Lamarck(ism), or evolution. Let us now address the reasons for this neglect by looking at some of the above listed authors in detail.

5 Julius Schaxel and Neo-Kantianism

Let us start with Julius Schaxel (1887–1943; Fig. 2).[7] Schaxel was a zoologist and theoretical biologist based at the University of Jena. He was one of the last personal students of the Darwinian prophet Ernst Haeckel. Schaxel obtained his doctorate through cytological analyses of oogenesis, which he carried out in Richard Hertwig's lab in Munich (supervised by geneticist Richard Goldschmidt). He then went on to study in detail the progressive determination of cells in development (Schaxel 1915). In 1916, he secured an "außerordentliche" professorship of zoology at the University of Jena. In Jena, he founded the small "Institute of Experimental Biology" in 1918, where he conducted embryological research, especially on axolotl development and regeneration. Schaxel increasingly faced tensions with members of the Zoological Institute in Jena, especially with race hygienist Ludwig Plate. This was not least because from the 1920s onward, Schaxel became a public spokesman for socialist politics, and actively attempted to interlink biology with dialectical materialism. After the National Socialists came to power, Schaxel was forced to emigrate to the Soviet Union in 1933, where he held, among others, a position at the Laboratory for Evolutionary Morphology at the Academy of Science in Leningrad. He was accused of espionage in 1938 and imprisoned for 2 months (see Fig. 2, right) and died in a sanatorium near Moscow in 1943 under unknown circumstances.

Besides his experimental work in embryology, Schaxel was a key figure in the institutionalization of theoretical biology. From 1919 to 1931, he edited the

Fig. 2 Julius Schaxel in 1919 (left) and in 1938 (right). The latter photo was taken (likely by soviet officials) around the time when he was imprisoned in Moscow, 1938 (Schaxel papers, Ernst Haeckel Haus, Jena; Box JS11)

[7]For discussions of the life and work of Schaxel, see Hopwood (1997) and Reiß (2007). For an autobiographical document by Schaxel, see Reiß et al. (2007).

influential book series *Abhandlungen zur theoretischen Biologie* (Contributions to Theoretical Biology; 31 monographs) with books by, for example, Paul Weiss, Ludwig von Bertalanffy, Hans Driesch, Emil Ungerer, and Friedrich Alverdes. This series was instrumental in building a German-speaking network of theoretically minded biologists. One of the founding works of the young field of theoretical biology was Schaxel's widely read 1919 book *Grundzüge der Theoriebildung in der Biologie* (Foundations of Theory Construction in Biology). In this work, in line with his peers, he argued that "[c]ontemporary biology is in a state of crisis" (Schaxel 1919a: 2; German original). Schaxel criticized:

> Due to internal reasons we do not have a theoretical biology equivalent to theoretical physics [. . .]. Problems and theorems successively and side by side ran wild [ins Kraut geschossen], without undergoing a strict test despite their mutual influence and conflict. [. . .] The abundance of material allures to, but does not justify this procedure. Instead of having a standard of validity through a theoretical biology as a science based on concepts, the whole biology constitutes of a collection of disorganized, nonequivalent theories, of which many are not even worthy of the name. (Schaxel 1919a: 4; German original)

He argued that in order to meet these problems a new field of "Theoretische Biologie" or "kritische Biologie" (see Schaxel 1917) should be established. It should provide a philosophically grounded framework that unambiguously guides data collection and theorization in biology through clear categorizations of empirical findings, rather than simply collecting more and more data. In the end of his 1919 book, Schaxel described which particular philosophy should underlie his "critical biology." This philosophy was most widely neo-Kantian. Schaxel (1919a: 201) suggested continuing especially Otto Liebmann's *Gedanken und Tatsachen* (1882–1904) and his *Zur Analysis der Wirklichkeit* (1911 [1876]), Eduard von Hartmann's *Das Problem des Lebens* (1906), Nicolai Hartmann's *Philosophische Grundfragen der Biologie* (1912), and (Heinrich Rickert's pupil) Richard Kroner's *Zweck und Gesetz in der Biologie* (1913).

Neo-Kantian discussions of biology and especially of Darwin's theory (and Darwinism) emerged in the second half of the nineteenth century as a reaction to the rise of materialism in biology (see Beiser 2014). Inspired by new developments in biophysics and physiology, for example through the work of Helmholtz, German materialists like Carl Vogt and Ludwig Büchner happily embraced Darwinism—even though Darwin himself was reluctant to adopt materialist views. By the 1860s, this new "marriage" between Darwinism and materialism developed into a radical position by arguing that all evolved life on earth (and even all consciousness) is nothing but matter in motion. By the time Ernst Haeckel became a spokesman for Darwin in the early 1960s, this materialist and mechanist form of Darwinism began a triumphal march in German science and the public. So much so that in his *Geschichte des Materialismus* (History of Materialism) neo-Kantian Friedrich Albert Lange (1875 [1866]) assessed that materialist debates have completely shifted into debates about Darwinism (see Teo 2002).

Darwinism gained influence among German-speaking scientists, exactly at the same time—in the 1860s–1870s—when neo-Kantianism emerged. At first, some neo-Kantians, like the naturalist Lange, were rather favorable toward Darwinism, while others, like Jürgen Bona Meyer remained skeptical (see Beiser 2014).[8] This balanced situation shifted at the end of the century toward clearly anti-Darwinian and anti-materialist positions. In addition, while neo-Kantians at first discussed (the empirical evidence for) Darwin's mechanisms and processes in detail (see, e.g., Meyer 1870) such as sources of variation, heredity, natural selection, and adaptation, Darwinisms' underlying philosophy increasingly became the center of attention. For example, Liebmann (1882–1904, 1911 [1876]) directly attacked the broader metaphysical consequences and underlying materialist-mechanistic philosophy of the works of German Darwinians like Vogt and Haeckel. For Liebmann, Darwin's theory did not fail on empirical grounds, but on philosophical grounds. He asked whether life could be explained mechanistically or only on the basis of concepts like teleology. Liebmann argued that a materialist worldview helps in understanding the mechanistic basis of species transformation, but it is a poor philosophy to address deeper questions, like the difference between living and non-living, the presence of form in living beings, and why there are living beings in the first place. With Kant, Liebmann defended a dualist position and claimed that materialism alone cannot answer such questions. Physical and chemical laws allow materialist biologists to merely explain blind actions, but not the purposive aspects of organisms (see Clarke 2021). According to Liebmann, concepts like *Lebenskraft* and *élan vital* could not address the latter issue either, since they violated the general laws of nature. Thus, Darwinism should be complemented by a transcendental philosophical approach that spells out overarching principles which determine the evolutionary process by drawing on teleological reasoning as a regulative tool.[9]

Within neo-Kantianism, this and similar lines of philosophical critique against Darwinism gained even more momentum with the publication of Haeckel's *Die Welträthsel* (1899). In this book, Haeckel not only forcefully linked his monist-materialist philosophy with Darwin's theory, but polemically attacked Kant and neo-Kantianism. This controversy would set the stage from which Schaxel developed his theoretical biology.

6 "Aphilosophical Darwinism"

Even though Julius Schaxel was a student of Haeckel, when it came to Darwinism he aligned himself not with Haeckel but with the neo-Kantians like Liebmann. Schaxel's critique of materialist versions of Darwinism sets the stage for a reception of Darwinism in theoretical biology that had two characteristics:

---

[8] For a detailed discussion of different lines of reception and critiques against Darwinism by neo-Kantians, see Clarke (2021).

[9] Besides Liebmann, many neo-Kantians were interested in discussing the role of teleology in biology. For teleological debates in the late 19th and early twentieth century biology, see Henning (2009) and Baedke et al. (2024).

First, it abstracted from much of the diversity among nineteenth-century materialist views defended by German Darwinists—from Vogt's or Büchner's materialist science that did not leave much room for philosophy to Haeckel's monist natural philosophy that endorsed materialism, but that, through pantheism, tried to avoid reducing the whole universe to dead matter. Second, this reception in theoretical biology (at least to some degree) narrowed down previous, more diverse neo-Kantian critiques of materialism and Darwinism—from Lange's critique of the metaphysical aspirations of materialists and Liebmann's description of the limitations of grounding historical explanations on materialism, to Rickert's argument that natural selection lacks normative or socio-political significance. Instead, Darwinism now primarily mattered as the—problematic or limited—theoretical paradigm located on one side of the vitalism-mechanism spectrum.

In his 1919 *Grundzüge*, Schaxel analyzed in detail existing theoretical frameworks that could be used as a starting point for a new theoretical foundation to biology, which would guide empirical research, systematize the use of concepts, and also reunite the fragmented biological fields. He was convinced that neither Wilhelm Roux's *Entwicklungsmechanik* nor the neovitalism of Hans Driesch could do so (see Hopwood 1997, Reiß 2007). Against Driesch he argued that neovitalism clearly offers an elaborate philosophical framework, but was built by deduction, and that the existence of vital forces was poorly supported by experimental evidence. In contrast, for Schaxel, Roux's mechanism merely qualifies as a loose research program, rather than a consistent biological theory. Mechanism was not a serious philosophy, which could address the most basic questions of life. He argued that, in the future, theoretical biology would have to develop a middle ground—a task that philosophically minded biologists and philosophers in movements like organicism and holism would happily accept from the 1920s onward (Nicholson and Gawne 2015, Peterson 2016, Baedke 2019).

Following the argumentative patterns developed before him by Liebmann and other anti-materialist neo-Kantians, Schaxel would forcefully attack the union of Darwinism and mechanism or materialism, respectively. While acknowledging the general virtues of mechanistic methodologies, Schaxel's critique against mechanistic Darwinism was harsh and polemic. He starts his 1919 book right in the first chapter with Darwinism. He acknowledges that "Darwin intervenes without tradition in the history of science. His work means a caesura." (1919a: 6; German original). But next to that he adds:

> We will see in a moment that already flowing currents of ideas carry Darwinism away to what it is. This unintentional and unnoticed assimilation of an idea is perhaps the most impressive example of uncritical theorizing that can be found in any science. Thus, it serves our purpose to explain the origin of the precarious situation of the scientific enterprise. (Schaxel 1919a: 6; German original)

For Schaxel, Darwinism was an exemplar for how public and popular science views infiltrated science since the middle of the nineteenth century. He argued that the generally poor theoretical situation in biology, with a lack of conceptual and methodological reflections, made scientists adopt ideas from popular movements like materialism. During this process, Darwinism changed from a scientific theory into a public opinion. Schaxel polemically criticized Haeckel for contributing to this problematic development—much to the disliking of Haeckel, who broke off his friendship with Schaxel.

According to Schaxel, the commonly endorsed version of Darwinism is synonymous with materialism, which is more of a worldview or ideology linked to religious, social, and political issues than an academic theory. This "extra-scientific, materialist propaganda" (1919a: 168), he stated, had contaminated evolutionary theory. It is overly empirically focused, anti-rational scientific and public actors are to blame for that:

> What actually helped Darwinism to become powerful are not so much the ideas flowing from embryology and cell theory as the spirit of biological content as much as *the spirit of anti-rationalism in general, which characterizes the empiricists*. Even stronger than in science itself this spirit works outside. In the political events of the middle of the [19th] century it finds its clear expression. The philosophical materialism, of whose leading names the biologists know especially L. Büchner, J. Moleschott and C. Vogt, has its strength not in deep thoughts." (Schaxel 1919a: 12; emphasis added; German original)

But not only materialists are to blame for this, Schaxel argued. This ominous "anti-rational" development in evolutionary theory started with Darwin himself:

> With the retreat of comprehensive ideas, individual knowledge accumulates [in the middle of the 19th century]. Less and less attention is paid to the origin and the interpretations given to the facts even by the most reserved researcher. This spirit also prevails in *Darwin's teaching*. The self-contemplation is neglected in biology in favor of *factual enrichment*. When the materialistic movement then takes over the leadership there is no opportunity to make up for what has been neglected before. Thus, in the last three-quarter of a century our science became richer in content, and known wider, broader, but not deeper. (Schaxel 1919a: 14; emphasis added; German original)

For Schaxel, Darwin was a collector, not a philosopher. Darwin and Darwinism neglected the central questions of life and, like Roux's *Entwicklungsmechanik*, emphasized individual observation over unifying theory. For example, Darwinism explains how organisms change and why they did not evolve differently, but not what they are.

Schaxel's résumé is crushing. He downgraded Darwin's theory of evolution to an inconsistent popular science ideology, not worth the name "theory." It fragments the organism into a set of properties or traits and, even worse, mixes atomistic and materialist views with historical and teleological thinking in unclear and conceptually inconsistent ways. A theoretical biology, he argued, needs to start exactly from these problems Darwinism could not solve. Biologists need to integrate fragmented empirical methodologies and data collection practices (focusing on different parts of organisms) as well as disunited theories (of mechanism and vitalism). A simple discussion of (the empirical support for) Darwinian principles of evolution, like natural selection, could not provide solutions for these deeper challenges. According to Schaxel, only a thorough reflection—through theoretical biology—of the underlying conceptual, theoretical, and methodological foundations of biology can do so. In short, biology needs more philosophy, not "aphilosophical Darwinism" (1919a: 51; German original).

The argumentation developed in Schaxel's pioneering *Grundzüge* drew on and reinforced older anti-materialist critiques against Darwinism by neo-Kantians.[10] It set the stage for debates especially among German-speaking theoretical biologists for the next three decades. Theoretical biologists most widely accepted Schaxel's neo-Kantian interpretation that Darwinism was supporting (or at least being the victim of) a simplistic materialist-mechanistic philosophy. For example, still 30 years later, holistic theoretical biologist Adolf Meyer-Abich, who started his career with a PhD thesis on Otto Liebmann (in Jena, 1916), stated[11]:

> Only when it would be possible to prove that an obvious regularity of all organic formations and processes was in truth only the result of completely goal- and purposeless forces and processes, the idealistic biology were hit in the core and inwardly finally overcome. To lead this proof was the final goal of Darwinism in its main and core piece, the selection theory. Only then the ground was prepared for the basic thesis of modern mechanistic biology, according to which all states and processes in the organisms do not differ in principle from those of the inorganic nature, that therefore biology represents only apparently an independent and autonomous science opposite to physics and chemistry, but in truth is nothing more than "not yet physics". (Meyer-Abich 1948: 135; German original)[12]

This and similar philosophical criticism made theoretical biologists most-widely neglect (or, if at all, discuss merely in passing) Darwinism or Darwin's principles of evolution (see Fig. 1). This omission was not because theoretical biologists directly opposed to the principles of evolution put forward by Darwin. They, more fundamentally, interpreted the materialistically biased and empirically focused framework of Darwinism an ill-starter for carving out a philosophically sound, unifying approach for biology.

7 Anti-Vitalism, Lamarckism, and Holism

Schaxel shut the door for Darwinism in theoretical biology. But what about those other evolutionary accounts, especially neo-Lamarckism, that historians and biologists like Bowler and Huxley, identified to have "eclipsed" Darwinism? Did theoretical biologists endorse those? Figure 1 above suggests they did not. Why is that?

Again, Schaxel gives us a hint. In his work, Lamarckism did not fare much better than Darwinism. He argued:

[10] For a critique on Schaxel's critique of Darwinism, see, e.g., Schultz (1921).

[11] On the life and work of Adolf-Meyer-Abich, see Amidon (2008) and Baedke et al. (2020).

[12] For a similar equation of Darwinism with mechanism, see, e.g., Ungerer (1919: 4–5, 247–248).

> In the framework of his not very unified philosophy, which encompasses the whole nature with inclusion of the human being, Lamarck teaches things that come close to Darwinian biology in several ways, especially because of the disorder [Ungebundenheit] of their principles." (Schaxel 1919a: 96; German original)

His main critique against (neo-)Lamarckism builds not on a discussion of Lamarckian processes like "soft inheritance," but on a philosophical argument about teleological reasoning.[13] Schaxel was worried that the active and purposive "self-adaptation" of organisms described by Lamarckism is often conceptualized in intentionalist or psychologistic ways. "What the observer knows from the experience of his actions, he transfers into the organism," Schaxel (1919a: 99; German original) criticizes. This "Psycholamarckismus," as he calls it, commonly adopts anthropomorphic views, or introduces spurious vital forces that should explain organisms' purposiveness. He concludes: "Psycholamarckism is the new vitalism" (Schaxel 1919a: 101); and vitalism, much like mechanism, does not provide the right conceptual framework for biology.

This philosophical skepticism against Lamarckism, through a fundamental critique of teleology and vitalism in biology, was enforced within early theoretical biology especially through holistic scholars. For example, Ludwig von Bertalanffy stated: "For we must not forget—as Schaxel and others have pointed out—that every Lamarckism already carries a trait towards vitalism in itself" (1927: 215; German original). He and plant biologist and philosopher of biology Emil Ungerer actively launched a campaign to replace discussions of purposiveness by descriptions of the dynamic equilibrium and holistic maintenance of the organisms' organization (see Baedke 2024, Baedke et al. 2024). Ungerer argued in his *Die Regulationen der Pflanzen: Ein System der teleologischen Begriffe in der Botanik* (The Regulations of Plants: A System of Teleological Concepts in Botany; Ungerer 1919, second expanded edition 1926) and in his *Die Teleologie Kants und ihre Bedeutung für die Logik der Biologie* (Ungerer 1922; Kant's Teleology and its Meaning for the Logic of Biology), that biologists should replace the idea of purposiveness by the concept of "wholeness" (Ganzheit). Only the latter "holistic view" ('Ganzheitsbetrachtung'; Ungerer 1919, IV) can overcome an anthropomorphic understanding of teleology. He follows Driesch's (1911) similar suggestion to replace "purposiveness" through "wholeness," but for Ungerer wholeness, against vitalism, should be accessible to empirical scientific investigation. To overcome the common "confusion between the holistic view and vitalism," Ungerer (1919, IV; German original) argued, we need to develop "Ganzheit" as a precise descriptive tool of science that can guide empirical research. Therefore, he conceptualizes wholeness as the intrinsic ability of living systems to maintain their organization, which can be observed especially in physiological processes. He studied physiological and developmental disturbances in plants in detail to show how organisms can reestablish their dynamic equilibrium and maintain themselves as organizational

[13] In his discussion of teleology, Schaxel parts ways with many devoted (neo-)Kantians, like Liebmann, that defended the general heuristic value of teleological reasoning in biology.

wholes (by restoring their form, their metabolic functions, and their coordinated behavioral activities).

This replacement of purposiveness through wholeness as a non-vitalist way to grasp the internal teleology of organism, had a large impact on the theoretical biology community, both on the German-speaking community and internationally (see Baedke et al. 2024). For example, Ludwig von Bertalanffy strongly builds on Ungerer, with whom he "especially agrees" when it comes to teleology (Bertlanaffy 1927: 230; German original). He stresses, against vitalism, that empirical investigations of teleological phenomena are indeed possible, if they focus on organisms' dynamically maintained organization (Bertalanffy 1927, 1928, 1929; see Baedke 2019). He states that living organisms are open systems far from thermodynamic equilibrium with metabolic self-organization: the "living organism [. . .] remains or establishes its state, by means of constant change of those substances and energies that build up the system as well as during external perturbations" (Bertalanffy 1932: 86; German original). In this view of teleology, living beings must maintain their homeostatic internal organization actively as open systems. This process is constrained by physical laws, especially non-equilibrium thermodynamics. Organisms must obey these laws—during development and evolution—to stay alive.

This and similar critiques against the conceptual alliance of purposiveness, intentional teleology, and vitalism made many holists, at least, skeptical of (neo-) Lamarckism.[14] For example, Ungerer (1919: 245) argued that in "psycholamarckism [. . .] teleology already completely coincides with the postulation of mental entities," and that "it was this lack of purity of the conceptual distinction [between purposiveness and intentionality] that brought teleology so much into disrepute" (German original). For many holists, Lamarckists had not solved a fundamental philosophical problem, namely developing a consistent theoretical framework (like a theory of goal-directedness through self-organization) that rescues organisms' internal teleology for biology. Instead, Lamarckists took refuge with vitalism.

8 The "Eclipse" of Theoretical Biology and Rise of the Modern Synthesis

We have come to see that in core theoretical debates in biology in the early twentieth century, Darwinism indeed played a minor role (see Fig. 1). But not due to the reasons assumed by the common linear historiography from "eclipse-to-unity," which biologists like Huxley and Mayr introduced and later historians like Bowler adopted. At least when it comes to the emerging field of theoretical biology,

[14]Please note that the anti-vitalist critique of Lamarckism was common among theoretical biologists, but there were other positions, too. Some scholars were (at least at some point) more open to Lamarckism, like Adolf Meyer-Abich, or rejected the mechanistic approaches adopted by some neo-Lamarckians.

Darwinism was not "eclipsed" by non- or anti-Darwinian theories of evolution, like neo-Lamarckism. The critique of theoretical biologists was not targeted against the evolutionary principles and mechanisms described by Darwinism. More fundamentally, they rejected its underlying materialist-mechanistic philosophy, at least the one defended by some nineteenth-century German-speaking Darwinians. At the same time, however, Lamarckism faced similar criticism. Its common vitalist treatment of teleological processes in organisms were also rejected on philosophical grounds. In short, they argued that neither mechanism (e.g., Darwinism) nor vitalism (e.g., Lamarckism) provided the right conceptual framework for biology.

For many theoretical biologists, Darwinism and Lamarckism were merely symptoms of a much deeper ailment, namely a lack of theoretical and conceptual unity in biology. The "synthesis" envisioned by theoretical biologists was not the Modern Synthesis, however. It did not aim to integrate principles of Mendelian inheritance and Darwinian evolution. Instead, its goal was to integrate mechanism and vitalism through a new philosophy of biology. While various international efforts were made by holists, dialectical materialists and organicists to reach this goal, this "philosophical synthesis" never materialized. This is largely due to sociopolitical and biographical reasons.

Theoretical biology witnessed a radical shift in the 1940s and 1950s, when philosophical discussions about the theoretical and conceptual foundations of biology were increasingly abandoned. Many older theoretical biologists died during this time (e.g., Schaxel, J.S. Haldane, Uexküll) or focused on different, less philosophical research. For many of the German-speaking theoretical biologists, the end of WWII was a break with the past. Many holistic advocates, like Meyer-Abich, Ungerer, and Bertalanffy, had at some point in their careers or personal lives endorsed National Socialist ideology (see Deichmann 1992, Pouvreau 2009, Dahn 2019, Baedke et al. 2024). Thus, after the war, holism—that formed the core of German-speaking theoretical biology in the 1930s—was considered highly suspicious. While some scholars, like Meyer-Abich and Ungerer, continued publishing on bio-theoretical issues until the 1960s, the original philosophically minded focus of the field was gone once and for all. Instead, theoretical biology shifted toward mathematical approaches, a focus it still has today. As a consequence, the philosophical "synthesis" envisioned by theoretical biologists in the early twentieth century never arrived. In fact, today, it is widely forgotten by biologists and historians of biology. It seems that it became "eclipsed" by a narrative about a different synthesis arriving around the same time—the Modern Synthesis.[15]

What should we learn from this? First, we should be skeptical about the long-accepted "eclipse" historiography and unilinear periodization of Darwinism in the early twentieth century. The history of theoretical biology tells us that what might look like an active attempt to replace Darwinism by non-Darwinian theories, might

[15] On the likely effects of the emergence of population genetics and the Modern Synthesis on topics in theoretical biology, see Fig. 1. In the 1950s and 1960s, evolutionary discussions gained relevance in theoretical biology, but, still, not discussions about Darwin and Darwinism.

be rather a side-effect of other discourses. It was not Darwin that was rejected, but rather a philosophical framework (mechanism) that, among others, some prominent Darwinians had adopted.[16] Second, we should learn that there were other lively unifying or synthesizing movements in the early twentieth century, apart from the one celebrated by Huxley and Bowler, which never directly contributed to the emergence of the Modern Synthesis.[17] In fact, we should come to see that, against Dobzhansky, evolutionary principles may not be the sole (or even main) unifying forces in biology. Early theoretical biologists argued that it was not evolution that unifies biology, but philosophy. Especially given the widespread critique of disciplinary and theoretical fragmentation in biology (what Schaxel labeled "crisis") in the first decades of the century, we should not assume that there was only one unifying movement that shaped biology during this time.

What is more, once we deconstruct the common "dark age of Darwinism" narrative, we start to see thus far overlooked patterns in past and present evolutionary biology. In fact, this is in line with Bowler's original aim to motivate historians to rediscover long forgotten traditions and alternatives that had been dismissed in the Modern Synthesis period. Bowler stated:

> When the author of this chapter first began to study the 'eclipse of Darwinism' in the 1980s, the triumph of the modern Darwinian theory made it easy to dismiss the alternatives as blind alleys into which scientists had been led temporarily. It was necessary to argue that, right or wrong, those theories had played so significant a role that historians who ignored them would not produce a balanced view of how evolutionism actually developed. Hindsight was not a valid reason for dismissing non-Darwinian theories as a trivial side issue. (Bowler 2017: 196)

Bowler was surely right about that. In fact, once we take a closer look at this period, we begin to uncover innovative contributions to evolutionary biology even in debates that did not much discuss Darwin or Darwinism. The early twentieth century was anything but a "dark" period in which little scientific progress in evolutionary biology was made. Let us just mention one of many examples: Starting in the early 1940s, theoretical biologist Adolf Meyer-Abich developed a comprehensive theory of evolutionary change through "holobiosis," which he derived in substantial part out of holistic philosophical ideas. While completely overlooked by historians of evolutionary biology, Meyer-Abich's theory had a surprisingly modern outlook, as it anticipated key elements of endosymbiotic theory, defended famously by Lynn Margulis from the late 1960s onward (see Baedke et al. 2020).

In addition, casting a light on this "dark" period helps us understanding current developments in evolutionary theory. In recent years, there have been several calls for integrating developmental and organismal phenomena into evolutionary theory.

[16] To be more precise, Schaxel, directly contributed to an "eclipse of Haeckelism" since Haeckel's materialism was his main target. However, other critique by theoretical biologists usually extended Schaxel's argument to a general critique against the mechanistic foundations of Darwinism.

[17] For a different take on synthesizing efforts within early twentieth century organicism and holism that differed from the Modern Synthesis, see Esposito (2017).

This so-called "Extended Evolutionary Synthesis" (EES, Laland et al. 2015) argues that evolutionary theory should not only (or primarily) explain evolutionary phenomena through changes in gene-frequencies in populations but through the ways in which organisms' development and behavior biases and drives evolutionary change. Both for advocates and critics of the EES, discussions about the historical status of this new theory usually adopt an overly narrow contrastive approach that compares it with the Modern Synthesis. Advocates of the EES argue it differs from and extends the Modern Synthesis, while critics argue that it does not. This historiography is limited and currently hinders fruitful theoretical debates about what is epistemologically entailed by the framework of the EES. Once we realize that—against what the "eclipse" narrative suggests—there are other "lighted syntheses" throughout the twentieth century, besides the Modern Synthesis, we can start exploring similarities between them and the EES. Then, we come to see that currently the EES unwittingly reintroduced certain explanatory roles granted to the organism in early twentieth century by holistic and organicist theoretical biologists, but which later became lost or marginalized in evolutionary biology, such as evolutionary effects of reciprocal organism-environment interaction and the role of agency in evolution (see Baedke & Fábregas-Tejeda 2023). The EES is more than a simple counter-draft to the Modern Synthesis. It ties in with strong, but long neglected theoretical traditions in biology.

These cases should exemplify that there is much in past and present biology, and especially in evolutionary biology, that, unfortunately, has been eclipsed until today. It is time that biologists and historians of biology cast a light on those "dark" periods.

Acknowledgments J.B. developed the argument of the chapter, collected the archive material, and wrote the text. A.B., V.S., and S.R.-S. created the corpus for the bibliometrical study and conducted the digital analysis (in Sect. 4). We thank Daniel Brooks for constructive comments on an earlier version of this chapter. J.B., A.B., and V.S. gratefully acknowledge the financial support from the German Research Foundation (DFG; BA 5808/2 – 1).

References

Allen GE (2005) Mechanism, vitalism and organicism in late nineteenth and twentieth-century biology: the importance of historical context. Stud Hist Phil Biol Biomed Sci 36:261–283

Alverdes F (1921) Rassen- und Artbildung. Borntraeger, Berlin

Amidon KS (2008) Adolf Meyer-Abich, holism, and the negotiation of theoretical biology. Biol Theory 3:357–370

Baedke J (2019) O organism, where art thou? Old and new challenges for organism-centered biology. J Hist Biol 52:293–324

Baedke J (2024) Plant agency: a short history from Kant to plant psychology, then to holism, and back again. In: Fábregas-Tejeda A, Baedke J, Prieto G.I., Radick G. (eds.) The riddle of organismal agency: new historical and philosophical reflections. Routledge, Milton Park, pp 61–76

Baedke J, Böhm A, Reiners-Selbach S (2024) From Kant to holism: the decline of neo-Kantianism and the rise of theoretical biology. In: Pulte H, Baedke J, Koenig D, Nickel G (eds) New perspectives on neo-Kantianism and the sciences. Routledge, Milton Park, pp 253–277

Baedke J, Brandt C (2022) Between the wars, facing a scientific crisis: the theoretical and methodological bottleneck of interwar biology. J Hist Biol 55:209–217

Baedke J, Fábregas-Tejeda A (2023) The organism in evolutionary explanation: from early twentieth century to the extended evolutionary synthesis. In: Dickins TE, Dickins BJA (eds) Evolutionary biology: contemporary and historical reflections upon core theory. Springer, Cham, pp 121–150

Baedke J, Fábregas-Tejeda A, Nieves Delgado A (2020) The holobiont concept before Margulis. J Exp Zool Part B 334:149–155

Beiser FC (2014) Encounter with Darwinism. In: Beiser FC (ed) The genesis of neo-Kantianism, 1796–1880. Oxford University Press, Oxford. https://doi.org/10.1093/acprof:oso/9780198722205.003.0012

Bowler PJ (1983) The eclipse of Darwinism: anti-Darwinian evolution theories in the decades around 1900. Johns Hopkins University Press, Baltimore, MD

Bowler PJ (1988) The non-Darwinian revolution: reinterpreting a historical myth. Johns Hopkins University Press, Baltimore, MD

Bowler PJ (2005) Revisiting the eclipse of Darwinism. J Hist Biol 38(1):19–32

Bowler PJ (2017) Alternatives to Darwinism in the early twentieth century. In: Delisle RG (ed) The Darwinian tradition in context: research programs in evolutionary biology. Springer, Cham, pp 195–217

Clarke E (2021) Neo-Kantianism, Darwinism, and the limits of historical explanation. Br J Hist Philos 29(4):590–613. https://doi.org/10.1080/09608788.2021.1912707

Deichmann U (1992) Biologen unter Hitler: Vertreibung, Karrieren, Forschungsförderung. Campus, Frankfurt

Dahn R (2019) Big science, nazified? Pascual Jordan, Adolf Meyer-Abich, and the abortive scientific journal *Physis*. Isis 110:68–90. https://doi.org/10.1086/701352

Dobzhansky T (1964) Biology, molecular and organismic. Integr Comp Biol 4:443–452. https://doi.org/10.1093/icb/4.4.443

Dobzhansky T (1973) Nothing in biology makes sense except in the light of evolution. Am Biol Teach 35:125–129

Driesch H (1892) Entwicklungsmechanische Studien. III. Die Verminderung des Furchungsmaterials und ihre Folgen (Weiteres über Theilbildungen). IV. Experimentelle Veränderungen des Typus der Furchung und ihre Folgen (Wirkungen von Wärmezufuhr und von Druck). V. Von der Furchung doppeltbefruchteter Eier. VI. Ueber einige allgemeine Fragen der theoretischen Morphologie. Z Wiss Zool 55:1–62

Driesch H (1908) The science and philosophy of the organism. Black, London

Driesch H (1911) Die Kategorie "Individualität" im Rahmen der Kategorienlehre Kants. Kant Studien 16:22–53

Driesch H (1928) Philosophie des Organischen. Original (1908). Quelle & Meyer, Leipzig

Ehrenberg R (1923) Theoretische Biologie. Springer, Berlin

Esposito M (2015) Romantic biology, 1890–1945. Routledge, New York

Esposito M (2017) The organismal synthesis: holistic science and developmental evolution in the English speaking world, 1915–1954. In: Delisle RG (ed) The Darwinian tradition in context. Springer, Cham, pp 219–241

Franz V (1920) Probiologie und Organisationsstufen: Eine Hypothese und ihre Anwendung auf die Morphologie. Borntraeger, Berlin

Giaimo S (2023) On citing Dobzhansky about the significance of evolution to biology. Integr Org Biol 5(1):obac047. https://doi.org/10.1093/iob/obac047

Haeckel E (1899) Die Welträthsel. Emil Strauß, Bonn

Haldane JS (1931) The philosophical basis of biology. Hodder and Stoughton, London

Hartmann M (1925) Allgemeine Biologie: Zelle, Statik, Dynamik und Stoffwechsel, vol 1. Fischer, Jena

Hartmann N (1912) Philosophische Grundfragen der Biologie. Vandenhoek und Ruprecht, Göttingen

Harwood J (1993) Styles of scientific thought: the German genetics community, 1900–1933. University of Chicago Press, Chicago

Henning W (2009) An annotated bibliography to Kant's teleology. Kant Yearbook 1:249–266
Hopwood N (1997) Biology between university and proletariat: the making of a red professor. Hist Sci 35:367–424
Huxley J (1942) Evolution: the modern synthesis. Allen & Unwin, London
Jordan HJ (1935) Das Problem der "Ganzheit" in der Biologie. Acta Biotheor 1:100–112
Kellogg VL (1907) Darwinism today. Henry Holt and Company, New York, NY
Kroner R (1913) Zweck und Gesetz in der Biologie. C.B, Mohr, Tübingen
Kroner R (1919) Das Problem der historischen Biologie. Borntraeger, Berlin
Laland KN, Uller T, Feldman MW, Sterelny K, Müller GB, Moczek A, Jablonka E, Odling-Smee J (2015) The extended evolutionary synthesis: its structure, assumptions and predictions. Proc R Soc B 282:20151019. https://doi.org/10.1098/rspb.2015.1019
Lange FA (1875) Geschichte des Materialismus und Kritik seiner Bedeutung in der Gegenwart, vol 2, 2nd edn. Original (1866). Baedeker, Iserlohn
Largent MA (2009) The so-called eclipse of Darwinism. Trans Am Philos Soc 99(1):3–21
Laubichler M (2001) Mit oder ohne Darwin? Die Bedeutung der darwinschen Selektionstheorie in der Konzeption der Theoretischen Biologie in Deutschland von 1900 bis zum Zweiten Weltkrieg. In: Hoßfeld U, Brömer R (eds) Darwinismus und/als Ideologie. VWB, Berlin, pp 229–262
Liebmann O (1882-1904) Gedanken und Tatsachen, 2 vols. K. J. Trübner, Straßburg
Liebmann O (1911) Zur Analysis der Wirklichkeit. Original (1876). K. J. Trübner, Straßburg
Mayr E, Provine WB (eds) (1980) The evolutionary synthesis: perspectives on the unification of biology. Harvard University Press, Cambridge, MA
Metallmann J (1939) Der Kampf um die Autonomie des Lebens. Acta Biotheor 5(1):1–10
Meulendijks M (2021) Eclipsing the eclipse? A neo-Darwinian historiography revisited. J Hist Biol 54(3):403–443. https://doi.org/10.1007/s10739-021-09650-9
Meyer A (1926) Logik der Morphologie im Rahmen einer Logik der gesamten Biologie. Springer, Berlin
Meyer JB (1870) Die Entstehung der Arten: Der Darwinismus. In: Meyer JB (ed) Philosophische Zeitfragen: Populäre Aufsätze. Adolph Marcus, Bonn, pp 39–103
Meyer-Abich A (1948) Naturphilosophie auf neuen Wegen. Hippokrates-Verlag Marquardt, Stuttgart
Morgan TH (1910) Sex limited inheritance in Drosophila. Science 32:120–122
Needham J (1936) Order and life. Yale University Press, New Haven
Nicholson DJ, Gawne R (2015) Neither logical empiricism nor vitalism, but organicism: what the philosophy of biology was. Hist Philos Life Sci 37:345–381. https://doi.org/10.1007/s40656-015-0085-7
Nyhart LK (1995) Biology takes form: animal morphology and the German universities, 1800–1900. University of Chicago Press, Chicago, IL
Peterson EL (2016) The life organic: the theoretical biology club and the roots of epigenetics. University of Pittsburgh Press, Pittsburgh
Plate L (1936) Hypothese einer variabelen Erbkraft bei polyallelen Genen und bei Radikalen, ein Weg zur Erklärung der Vererbung erworbener Eigenschaften. Acta Biotheor 2(2):93–124
Pouvreau D (2009) The dialectical tragedy of the concept of wholeness: Ludwig von Bertalanffy's biography revisited. ISCE, Litchfield Park
Reinke J (1901) Einleitung in die theoretische Biologie. Paetel, Berlin
Reiß C (2007) No evolution, no heredity, just development—Julius Schaxel and the end of the evo–devo agenda in Jena, 1906–1933: a case study. Theory Biosci 126(4):155–164
Reiß C, Springer S, Hossfeld U, Olsson L, Levit GS (2007) Introduction to the autobiography of Julius Schaxel. Theory Biosci 126(4):165–175. https://doi.org/10.1007/s12064-007-0015-7
Reydon TAC, Dullemeijer P, Hemerik L (2005) The history of Acta Biotheoretica and the nature of theoretical biology. In: Reydon TAC, Hemerik L (eds) Current themes in theoretical biology. Springer, Dordrecht, pp 1–8
Rignano E (1927) Das Leben in finaler Auffassung. Borntraeger, Berlin

Ringer FK (1969) The decline of the German mandarins: the German academic community, 1890–1933. Harvard University Press, Cambridge, MA
Schaxel J (1915) Die Leistungen der Zellen bei der Entwicklung der *Metazoen*. Fischer Verlag, Jena
Schaxel J (1917) Mechanismus, Vitalismus und kritische Biologie. Biol Centralbl 37:188–196
Schaxel J (1919a) Grundzüge der Theoriebildung in der Biologie. Fischer, Jena
Schaxel J (1919b) Über die Darstellung allgemeiner Biologie. Borntraeger, Berlin
Schindewolf OH (1937) Beobachtungen und Gedanken zur Deszendenzlehre. Acta Biotheor 3(3): 195–212
Schmid B (1921) Von den Aufgaben der Tierpsychologie. Borntraeger, Berlin
Schultz J (1921) Julius Schaxel: Grundzüge der Theorienbildung in der Biologie. Jena, Fischer, 1919. 221 S. Über die Darstellung allgemeiner Biologie. (Abhandlungen zur theoretischen Biologie, I). Berlin, Bornträger, 1919. 62 S. Annalen der Philosophie 2(2):293–299
Smith R (2019) Tesseract documentation: tesseract 4.1.1. https://tesseract-ocr.github.io/. Accessed 24 March 2023
Smocovitis VB (1996) Unifying biology: the evolutionary synthesis and evolutionary biology. Princeton University Press, Princeton, NJ
Spemann H, Mangold H (1924) Über Induktion von Embryonalanlagen durch Implantation artfremder Organisatoren. Archiv für mikroskopische Anatomie und Entwicklungsmechanik 100:599–638
Teo T (2002) Friedrich Albert Lange on neo-Kantianism, socialist Darwinism, and a psychology without a soul. J Hist Behav Sci 38(3):285–301. https://doi.org/10.1002/jhbs.10062
Ungerer E (1919) Die Regulationen der Pflanzen. Springer, Berlin
Ungerer E (1922) Die Teleologie Kants und ihre Bedeutung für die Logik der Biologie. Brontraeger, Berlin
Ungerer E (1936) Die Hypothese der Keimgangmutationen. Acta Biotheoretica 2(1):23–58
von Bertalanffy L (1927) Studien über theoretische Biologie I. Biologisches Zentralblatt 47:210–242
von Bertalanffy L (1928) Kritische Theorie der Formbildung. Borntraeger, Berlin
von Bertalanffy L (1929) Zum Problem einer theoretischen Biologie. Kant Studien 34:374–390
von Bertalanffy L (1932) Theoretische Biologie, vol 1. Borntraeger, Berlin
von Bertalanffy L (1933) Modern theories of development. Translated by J. H. Woodger. Oxford University Press, Oxford
von Bertalanffy L (1942) Theoretische Biologie, vol 2. Borntraeger, Berlin
von Hartmann E (1906) Das Problem des Lebens. Volksverband der Bücherfreunde Wegweiser, Berlin
von Uexküll J (1928) Theoretische Biologie. Original (1920). Springer, Berlin
Woltereck R (1909) Weitere experimentelle Untersuchungen über Artveränderung, speziell über das Wesen quantitativer Artunterschiede bei Daphniden. Verhandlungen der Deutschen Zoologischen Gesellschaft 19:110–173
Woodger JH (1929) Biological principles. Routledge, London

The Strange Story of Mosaic Evolution

Silvia Caianiello

Abstract

Mosaic evolution, whereby transitional forms display a mixture of primitive and derived features, is nowadays recognized as a ubiquitous pattern of macroevolutionary change. Less known is that this expression was coined by Gavin de Beer in 1954. I move from the philological symptom of the frequent oblivion of de Beer's authorship of this notion to investigate the history of its reception by G.G. Simpson and E. Mayr in the context of the "hardening" Modern Synthesis. I argue that the incorporation of mosaic evolution by the Modern Synthesis was the result of a process of selective appropriation, which was accomplished with the overshadowing of de Beer's original formulation by Mayr in 1963. In fact, mosaic evolution touched several sensitive issues at the junction between embryology and paleontology: the role of natural selection in the origin of complex adaptations, the origin of higher taxa, and the explanation of macroevolutionary trends. As a litmus test about the hardening of Modern Synthesis, mosaic evolution reveals several cracks and fissures in the consolidation of the synthetic framework and sheds a realistic light on the defensive function of the hardening itself, as well as on its uneven success. In fact, the dismissal of developmental explanations for mosaic evolution pushed both Simpson and Mayr to adopt Huxley's progressivist notion of grade to account for macroevolutionary trends, loosening the strict requirement that adaptation be driven by natural selection.

S. Caianiello (✉)
Istituto per la storia del pensiero filosofico e scientifico moderno (ISPF), Consiglio Nazionale delle Ricerche, Naples, Italy

Stazione Zoologica Anton Dohrn, Naples, Italy
e-mail: silvia.caianiello@ispf.cnr.it

R. G. Delisle et al. (eds.), *Unity and Disunity in Evolutionary Biology*,
https://doi.org/10.1007/978-3-031-42629-2_13

Keywords

Mosaic evolution · Modern Synthesis · Macroevolution · Modularity · Grade · Progressivism · Paleontology · Gavin de Beer · George Gaylor Simpson · Ernst Mayr

Mosaic evolution, the independent rate and time of change of different features within a lineage, is acknowledged as a widespread pattern of evolutionary change not only in paleontology but across many biological scales[1]. From the recent discovery of *Australopithecus sediba*, exhibiting a mosaic of primitive (*Australopithecus*) and derived (*Homo*) features (Kivell et al. 2011; Parravicini and Pievani 2019; Shubin 1998), to the different subregions of the genome (Bernardi 1993), mosaic evolution is «one of the most important principles of evolution» (Futuyma 2005: 54).

So, the strange story about mosaic evolution is by now not about the pattern itself, and the whimsical forms it may engender, but about this fortunate expression. In fact, its "philological" origin traces back to Gavin de Beer's 1954 paper "Archaeopteryx and evolution," where it was raised to the status of a principle "of general applicability (. . .) in the study of transitions from one type of animal to another" (de Beer 1954a: 165, 169). However, de Beer's authorship has been often neglected or misquoted. Only recently, and mostly in paleontology, it is being more systematically acknowledged.[2]

The evidence I started collecting about this frequent oblivion intrigued me. I could see no reason why de Beer's contribution didn't get firmly fixed in the

[1] Cf. Stebbins 1983; Zuckerkandl 1994; Purugganan 1998 as early examples of the breakthrough of the modular understanding of evolutionary change. With evo-devo, modularity will be hailed as an "ineluctable feature of biological order" (Raff 1996). For a general overview, cf. Carroll 2008; Caianiello 2011; for a contemporary assessment of the "modularity model" in the paleontological investigation about the origin of higher taxa, cf. Kemp 2016.

[2] Clean quantitative data in support of my statement are very difficult to collect, because the notion is overdetermined, so deeply entrenched in evolutionary parlance that it has branched in different disciplinary traditions. To keep at "philology," Purugganan 1998, Schlosser 2005, Prothero 2007 all refer to Gould as the source; Raff and Wray 1989 to Mayr. Equally significant in my view is the high frequency of misquotation of de Beer, either referring to the other 1954 publication (1954) where there is no mention of the expression "mosaic evolution" nor it is discussed as a principle, or to the third edition of *Embryos and ancestors* in 1958, where not even the term mosaic is ever mentioned (cf. for example Wray 1992; Witmer 2002; Wellnhofer 2010). More recently, and especially in paleontology where it probably never completely waned, the acknowledgment of de Beer's authorship—with the correct quotation—is becoming widespread: see Feduccia 2012; Felice and Goswami 2018; Parravicini and Pievani 2019. But a further difficulty is that even papers that correctly quote de Beer as a source, as Bock (1965, 2000), patently convey Mayr's adaptationist version of the notion. Maybe the best evidence of the oblivion of de Beer's authorship of this notion is that all major papers devoted to de Beer's work (except for the very first, Barrington 1973), that is Horder 2006 and Brigandt 2006, and even the recent Zou 2021, never mention his contribution to mosaic evolution.

historiography, had it even been a mere terminological innovation; the more as it belonged to a time when he was at the apex of his professional success. By 1954, in fact, de Beer was Director of the British Museum (Natural History)[3]; an outstanding comparative evolutionary embryologist and morphologist[4]; competent in experimental methods enough to author with his mentor Julian Huxley a still mainstream handbook on experimental embryology (Huxley and de Beer 1934)[5]; author in 1938 of a disruptive critique of genetic as well as developmental criteria for establishing homology (Horder 2006; Brigandt 2006); and apparently still so close to Julian Huxley and the inner circle of the Modern Synthesis that some, included Gould, have even listed him among its architects.[6]

To tackle this incongruity, I tried to reconstruct the early history of de Beer's paper in the context of the Modern Synthesis. The results of my research reveal a convoluted trajectory, which convinced me that the frequent oblivion of de Beer's name is not just an example of Proustian "intermittencies of the heart" of historical memory, but the result of a process of censorship and selective appropriation that involved different actors and steps.

To highlight these steps, I will follow a somewhat unorthodox methodology, that of a textual microhistory, to unveil at least parts of the network of mutual references that are woven into the texture of any scientific publication, in some cases even down to the detail of how these references are exploited. While this level of accuracy may prove fastidious, it may allow at times a sensitive recording of the tiny shifts, oscillations, and uncertainties of science in its making, and even of the subtle power games that may be missed in larger narratives but that for exactly this reason may challenge too monolithic crystallizations.

It may be useful to provide first of all a summary of the "philological" facts that I will try to put into context in the following paragraphs. De Beer's paper was initially well received and much discussed up to mid-1960s, even by scientists near to the

[3]The Museum parted way from British Museum in 1963 and was renamed in 1992 "Natural History Museum."

[4]De Beer 1937. His major work, *Embryos and Ancestor* had 3 editions (1940, 1951, 1958), and was itself a revised edition of *Embryology and evolution* of 1930. The third, de Beer 1958, was republished again in 1962 and 1971. A comparative study of these editions hasn't yet been accomplished. On the relevance of comparative evolutionary embryology in the "intellectual lineage" of evo-devo, see Raff & Love 2004.

[5]This handbook, republished in 1963, represented according to Churchill 1992: 109 an early attempt at an "embryological synthesis." De Beer was trained in experimental techniques at the Strangeways Laboratory in Cambridge and briefly with Spemann in Berlin, sent there by Julian Huxley, his mentor together with Edwin Goodrich. However, he soon abandoned experimental practice to pursue his work in comparative evolutionary embryology and morphology; cf. Ridley 1986; Hall 2000; Horder 2001.

[6]Cf. Gould and Eldredge 1972; Bowler 1996; Futuyma 2015. Nuño de la Rosa and Müller (2021: 7) list de Beer among the "outsiders" of Modern Synthesis. On the name game about the architects, see Gayon and Huneman 2019. De Beer contributed to both the most relevant British-based collections listed by Provine in 1980 as "major works of the synthesis" (1980: 400): Huxley 1940 and Huxley, Hardy and Ford 1954.

Modern Synthesis movement. George Gaylord Simpson's reception was more contentious, but equally overt. However, as Mayr in the early 1960s officially incorporated mosaic evolution into the Modern Synthesis framework, he did so by intentionally omitting de Beer's contribution and by reframing the notion in adaptationist terms. The neglect of de Beer's role, however, was not due only to Mayr's impressive historiographical influence, but also to Gould.[7] Last but not least, this oblivion was partly due to de Beer himself, who did not introduce the "principle" in the revised edition of *Embryos and Ancestors* in 1958. In fact, although he republished the 1954 paper in a collection in 1969 (De Beer 1969), his later references to mosaic evolution were never accompanied by a claim of authorship, and mostly limited to a mention of the 1954 paper hidden in a note. So, if there was some sort of a ban, apparently de Beer complied with it, a behavior that reveals something of the "tact and good sense" of his general attitude (Barrington 1973: 78), and even more about his relationship with the Modern Synthesis at this time. In fact, the date of publication of the paper is of importance: 1954, the year after Simpson's *Major features of evolution*, a landmark in the hardening of Modern Synthesis according to Gould (1983; cf. Delisle 2017: 160).

My aim is not to overstate the importance or modernity of de Beer's version of mosaic evolution, although some features of his "modular" interpretation of evolutionary mosaics corroborate the current image of de Beer as a precursor of evo-devo (Brigandt 2006; Zou 2021; cf. Sect. 1.4). The point I wish to emphasize is that in the subsequent crystallization of the expression the twist of his argument got largely lost. Thus, at least in the perspective of the present paper, its significance is related to the uneasiness his argument stirred in some of the major architects. I will use therefore this historical episode as a litmus test for detecting several, and not always overlapping, cracks and fissures in the synthetic conceptual framework at the time of its hardening.

This chapter is structured as follows: In Sect. 1, I will introduce de Beer's paper and highlight the sensitive topics that he—in my view intentionally—touched upon. In Sect. 2, I will turn to the early mostly positive reactions to his paper and contrast them with the one by George Gaylord Simpson, more complex and ambivalent, but nonetheless overt. In Sect. 3, I will deal with Ernst Mayr's appropriation and reframing of the notion and speculate on the reasons for his deliberate omission of de Beer's paper. In the conclusion, I will argue that the selective appropriation of de Beer's mosaic evolution corroborates the thesis of the "hardening" of Modern Synthesis. However, it also highlights how much this hardening was a defensive strategy against the different and not overlapping objections that these architects had

[7]Cf. below Sect. 3.2. My greatest surprise, considering his role in reassessing de Beer's legacy (Gould 1977: 221–222), was in fact the discovery that Stephen Jay Gould did—consciously, because he surely knew the paper (Gould 1970)—perpetrate this oblivion as he elaborated on the concept of mosaic evolution ever since *Ontogeny and Phylogeny*, tracing it back directly to Dollo's *chevauchement des specialisations*. In the present paper I will deal only with the fortune of this expression within Modern Synthesis, even if Gould contributed equally to the oblivion of de Beer's authorship.

to face each in his own disciplinary field. The conceptual cost of this strategy with respect to mosaic evolution was, however, significant, as it forced both Mayr and Simpson to embrace Julian Huxley's progressivist notion of grade to account for macroevolutionary trends not amenable to environmental change and ultimately to weaken exactly the selectionist approach of the hardened Synthesis.

1 Between the Lines of de Beer's *Archaeopteryx and Evolution* (1954a)

The first fossil of Archaeopteryx, the most celebrated "bones of contentions" (Chambers 2002), was discovered in 1861, 2 years after the publication of *On the Origin of Species* and was acquired by the then British Museum (Natural History). The much sought-after evidence of an "intermediate" form between reptiles and birds corroborated the Darwinian view of evolution as descent with modification and endorsed its fundamental gradualist tenet (Wellnhofer 2010). Nonetheless, Thomas H. Huxley did not confer Archaeopteryx the full status of a "missing link," but rather of an "intercalary type": evidence of descent, but unable to "indicate the nature of the steps by which the transition from one group to the other was effected" (Huxley 1877).[8]

In 1954, de Beer, thence Director of the British Museum (Natural History), published two papers concerning Archaeopteryx. The first, *Archaeopteryx lithographica: a Study Based upon the British Museum Specimen*, was a "splendidly illustrated monograph" (Barrington 1973: 76), still valued as the standard description of the London specimen (de Beer 1954b; cf. Witmer 2002; Wellnhofer 2010). The term "mosaic" is here only employed descriptively. The aim of the second paper, "Archaeopteryx and Evolution," the Presidential address to the Section of Zoology of the British Association for the Advancement of Science, was instead to provide "evidence on which *I hope to demonstrate a principle of evolution which deserves further attention*" (de Beer 1954a: 160, my italics). The principle is mosaic evolution: animals evolve "one piece at a time," "with different pieces evolving separately, and some of them too fast." Such a principle "may appear to be so well known" in paleontology "as to make this paper redundant" (1954a: 169), yet de Beer states that it still is in need to be "demonstrated."

To "demonstrate" mosaic evolution, de Beer selects a peculiar gallery of forms, which are intermediate between different classes of vertebrates. The explicit criterion is that none of them is an "ideal transitional form" (165): that is, in Thomas Huxley's terms, they are not "linear" but rather "intercalary types." In fact, they all exhibit

> one or more characters that rule them out from the direct ancestry of the descendants, either because they have already lost some structure which the early members of the descendant

[8] Thomas Huxley as first postulated a genealogical relationship between dinosaurs and birds and created the taxon Sauropsidae to unite them. For an overview of the still current controversy about the phylogenetic position of Archaeopteryx see Chatterjee 2015, Chap. 4.

> group still have, or because they possess a structure which the early members of the group in question have not yet evolved (1954a: 169).

De Beer lists for each of these intermediates the features that belong respectively to the ancestral and to the upcoming class (reptilian vs avian for Archaeopteryx[9]; fish vs amphibian for Ichthyostega; amphibian vs reptilian for Seymouria; reptilian vs mammal-like in Ictodosauria and in Ornithorinchus). The five shelves—fishes, amphibia, reptiles, birds, and mammals—correspond to levels of organization in vertebrate evolution and match the progressivist approach that will lead Julian Huxley 3 years later to uncouple "grades" ("general organizational advance," Huxley 1957) from "clades" (cf. Sect, 2.3). The independently evolving "pieces" de Beer focuses on are the "fundamental structural and functional requirements" (1954a: 163), which underlie the attainment of the upper grade.

Precisely because they "spoil" the picture of missing links, these "animals in transition" allow for de Beer to gain an insight into what he calls the "hard work of evolution," avoiding the fallacy of evaluating them at the touchstone of the "next shelf."

In particular, they allow the unbiased framing of a critical question: "what precisely is meant by the statement that an animal is intermediate between two classes" (1954a: 162)—that is, how we understand gradualism.

The next shelf, in fact, is composed of animals whose parts are all "stamped with the identity of their class." The danger of a teleological perspective to the "next shelf" is to support the illusion of a "gradual and simultaneous conversion of all the parts of the animal," whereby its parts "would be intermediate in structure between the two classes." The evoked "animals in transition" show that this is not the case. Archaeopteryx's features are either "perfectly reptilian" or "no less perfectly avian (…) the modifications which it has undergone have affected some structures to produce their *complete* transformation while other structures have not yet been affected at all" (1954a: 162). Furthermore, the enlisted diagnostic features of the next shelf are no simple single characters but, as Schaeffer (1956; see Sect. 2.2) will correctly point out, "functionally integrated structural complexes." Thus, for de Beer, the integrated parts arise already "complete," by "discrete steps in change"—as in the case of Archaeopteryx' wings, whose "arrangement of feathers is identical with that of flying birds" (de Beer 1954b: 44).

Introducing mosaic evolution in this short paper, de Beer deliberately focuses on patterns, not processes. Unlike *Embryos and ancestors*, neither the relationship between development and evolution nor mechanistic explanation in terms of heterochrony driven by "rate genes" is brought up to support his modular interpretation of mosaic changes.[10] However, as Schindewolf (1957) correctly highlighted, his

[9]In Archaeopteryx, the reptilian ones are long tail, claws on forelimbs, simple brain and vertebrae, and the avian ones are feathers and their asymmetric arrangement in the wing, furcula, grasping feet.

[10]"Rate genes" were Mendelian genes supposed to control speed and onset of developmental processes, based on the experiments of Goldschmidt on the reaction velocities of different genes and later of Huxley and Ford on Gammarus. It was de Beer however that first established them as

terminological innovation is but a conceptual transfer from development to evolution. Although the transfer is largely metaphoric, venturing into the rich developmental subtext that underlies his notion of mosaic provides important clues for understanding the piecemeal mode of transformation purported by de Beer.[11]

In Huxley's and de Beer's *Elements of experimental embryology* (1934), "mosaic" refers to a stage in development occurring after neurulation. It is the phase in which the gradient system of the whole egg—originally established by the action of external factors and further depending on the inductive differentiation chain elicited by the organizer—breaks up into "a patchwork of independent chemo-differentiated regions," similar to a "multiple tissue-culture" (Huxley and de Beer 1934: 354, 57). The predelineated—but not prelocalized—"organ-fields" self-differentiate independently and progressively split into a secondary "mosaic of invisibly determined subregions" (1934: 225), more and more partial fields. All these pieces are individually endowed with a capacity of self-regulation, but at this stage, the embryo is no longer a global Drieschean, self-regulating "harmonious equipotential system" (1934: 352–353). On the contrary, "single portions can be removed or substituted by others without exerting any effect on the rest of the system." The mosaic stage is pre-functional and is followed by a stage of functional differentiation in which "biological integration" is reinstated, driven by the working of the by now developed nervous and vascular (humoral) systems (1934: 57–59).

It is beyond the scope of this chapter to deal with the overall historical significance of the theory of gradient fields as supported by de Beer and Huxley (cf. Churchill 1992; Horder 2001). I just wish to underline that the characterization of the mosaic stage in development is largely consistent with the idea of a piecemeal change of self-organizing not only "organ-fields" but also restricted subfields of organs, while the overall viability of the resulting organism is ensured by the subsequent stage of functional co-adaptation of the parts. So, it is not surprising that already in 1928 de Beer hinted at the dissociability of different functional structures in evolution: "the biological success of birds is probably due in no small measure to the fact that the adaptations for flight in no way interfere with or involve the hind pair of limbs" (de Beer 1928: 447; cf. Felice and Goswami 2018).

De Beer's selected intercalary animals are mosaics of ancestral features, which may remain, as in the case of Seymouria, in "a profoundly archaic condition," and of new features, which appear as "adaptation *in a broad way*" to a new environment

---

the mechanistic basis of heterochrony (Ridley 1986: 61–62). For an updated view of their historical significance in the understanding of gene action cf. Raff 1996: 15–18.

[11] There was at the time another well-established notion of mosaicism in genetics that might have indirectly contributed to the insight that that the development of different characters in an organism might enjoy a considerable independence, while bringing about utterly viable organisms. The experimental study of genetic mosaics (spontaneous, such as gynandromorphs, or induced) aimed at determining the chromosomal localization of genes responsible for phenotypic characters and was a founding technique for Drosophila genetics and more generally for the tracking of cell lineages. Genetic mosaicism was however more focused on the genetic determinants of phenotypic traits than on the developmental process of their making. Cf. Sturtevant 1932; Stern 1956; for a brief overview of a huge and still lively research field, see Spinner & Conlin 2014.

(1954a: 167; my italics). For de Beer, the real "hard work" of evolution takes place within each shelf. In-between the shelves there is a "very wide tolerance for the amount of change of size, shape, proportions, accretion or loss, which a strain of evolving animals may show" (1954a: 164).

De Beer provides two examples of the "hard work" of evolution. The first is adaptive radiations and the process of diversification by which "different lines become more and more closely adjusted to special environments." However, immediately after, he adds: "There are also the trends of evolution which Watson has shown to occur in each level, unrelated to any specific adaptation of the environment, but progressing in a secular manner with time."

At this point of the paper, it becomes clear why neither here, nor in the subsequent cursory references to mosaic evolution in later texts, de Beer delves into the long and complex history of mosaic evolution *avant la lettre* (cf. Rieppel 2020 for an overview). In fact, it is apparent that de Beer, himself no paleontologist, is mainly interested in setting the stage for a confrontation between two authorities in the paleontological field, Simpson and David Meredith Sears Watson.[12]

The reference to Simpson, in particular, may partly explain why in this chapter de Beer refrained from speculating about the possible developmental processes underlying the pattern of mosaic evolution. In 1953, Simpson had downsized the explanatory relevance of paedomorphosis in the origin of higher taxa with an argument that might appear to corroborate the dismissing stance toward embryology classically attributed to Modern Synthesis: "The real evolutionary process, the change in populations (. . .) could be studied about as well if nothing were known about the individual developmental steps intervening between mutation and observed phenotype" (1953: 97–98).[13] However, Simpson did not always disdain embryological explanations (Simpson 1953: 25ss; cf. Sect. 1.2), and it is much more probable that what he disliked was rather the resemblance of de Beer's paedomorphosis with the "saltatory" theories of Schindewolf that he goes on refuting in the same page.

This backstory may help understand de Beer's choice to stick to the level of patterns in the 1954 paper, as well as his convoluted strategy of playing two paleontological authorities, Watson and Simpson, against each other. In fact, the two major sources he quotes as evidence that the principle is so well-known in paleontology "as to make this paper redundant" are Simpson's principle of the independence of characters, applied in 1953 to "tooth-length and tooth-height in the evolution of horses," and Watson's 1919 paper on Seymouria, the extinct genus

[12] D.M.S. Watson (1921–1951) had been his colleague at the University College in London and de Beer thankfully mentioned him in the Preface of de Beer 1928; in 1958 he co-edited the Festschrift in his honor with Swinton (de Beer and Swinton 1958).

[13] In the same chapter, Simpson attempts, without really succeeding, to spare de Beer from this critique. Of course, de Beer was well aware of the lack of embryological information in the fossil record, the major argument against speculations from pattern to mechanisms, but proposed nonetheless an alliance between phylogeny and embryology: "since phylogeny is but the result of modified ontogeny, there is the possibility of a causal analytical study of present evolution in an experimental study of the variability and genetics of ontogenetic processes" (de Beer 1940: 98).

"built up of perfectly well developed amphibian characters and equally decisive reptilian features, those of intermediate type being very rare" (Watson 1919: 292; De Beer 1954a: 170). But this juxtaposition is no more innocent than the one between adaptive radiations and "trends (. . .) unrelated to any specific adaption of the environment" (De Beer 1954a: 164). In fact, in 1953, Simpson went a long way to refute exactly the argument of Watson about evolutionary trends, which involved a substantial uncoupling of macroevolution from microevolution, and challenged the selectionist explanation of trends that the architects came to endorse more forcefully at this stage of the Modern Synthesis development.

Evolutionary trends were directly involved in mosaic evolution, because the same advanced characters arising in a mosaic fashion—those enabling the passage to the "next shelf"—eventually appear in parallel lineages. Thus, with his paper, de Beer poked the stick in the hornet's nest of orthogenetic vs selectionist explanations of evolutionary trends. He did it, for instance, by stressing the case of the reptile to mammal transition, which Watson also emphasized (1.1). Unlike the intermediate forms between the other grades, characterized by "a clean-cut and sudden adaptation to a new medium (. . .) in the production of the mammals there was no conquest of a new medium." Their major diagnostic mammalian feature, "the squamoso-dentary articulation of the lower jaw, cannot be correlated with the solution of any major problem of adaptation"; it just enabled them "to chew and to hear more efficiently." "Mammals are only perfected reptiles, more efficient, better adapted, and possessed a greater degree of independence of their environment" (de Beer 1954a: 167–168).

To better understand de Beer's subtle association between the mosaic mode in major transition and evolutionary trends, a step backward is therefore necessary to the point of contention between Watson and Simpson.

1.1 Watson's Secular Trends

Watson (1886–1973) Jodrell Professor of Zoology and Comparative Anatomy at University College of London, was an outstanding British paleontologist, committed, as Osborn's pupil William King Gregory, to "inferring function from fossils" (Bowler 2007: 13).[14] The program of merging paleontology with functional morphology entailed a shift "from structure to function," from homologies to the "conditions under which the transformation took place" (Bowler 2007: 26) and therefore to adaptive explanations. Darwin Medal in 1942, Watson was fluent enough in contemporary genetics to reframe "internal factors" in genetic terms and speculated on the possibility that new diagnostic characters (pre-adaptations) "accumulated one by one as a result of a chance association in a single animal of a group of unrelated genes" (Watson 1951: 161). He recognized that "the theory of

[14]Together with Robert Broom, these "functional morphologists" contributed to the "fins to limbs transition" problem by establishing the descent of amphibia from Crosspterygians (teleosts) (Bowler 2007).

natural selection is the only explanation of the production of adaptations which is consonant with modern work on heredity" (Watson 1936: 45). Nonetheless, as many contemporary paleontologists, he did not consider natural selection as the only nor the foremost driving force of evolution (Bowler 2007; Ochoa 2021). His life-long work on amphibians convinced him that the diagnostic features of classes arise by parallel evolution and are independently acquired by diverse branches of the same class. Such *orthogenetic* trends—the "identical changes all in a uniform direction, though at very different rates, in all the numerous stocks of a class" (Watson 1940: 229)—are not amenable to natural selection. The flattening of the skull and the reduction of cartilage bone are unrelated to the multifarious changes in environment and habits amphibians underwent "over a prolonged period of thousands or even millions of generations" (Watson 1951: 194). Natural selection, at least in the Darwinian meaning of the "precise fitting of an individual to its environment" (Watson 1948: 80) has a role only in "secondary effects" of functional coordination between "primary" changes. In the 1940 paper, the proposed explanation of the genuinely parallel evolution of diagnostic characters is thus not framed in neo-Lamarckian, but in genetic terms: "a restriction of the available material of variations on which it <natural selection> has to work," corroborated by the "widespread occurrence of identical mutations in allied animals" (Watson 1940: 230).

It is to this 1940 paper that Simpson reacts briefly in 1944, in a mild attempt to deconstruct such "secular" trends as possible artifacts and suggesting that "the broader and more general trends that may persist regardless of changes in habit were simply adaptations effective in the widest range of circumstances rather than specific to one habitat or one habit" (Simpson 1944: 169)—a suggestion that Watson appears to have taken very seriously in later writings, radicalizing his argument in a manner that will elicit a much stronger and more worried reaction by the later Simpson of *The major features of evolution.*[15]

In 1948, Watson reframes his argument by distinguishing between two major meanings of adaptation. "Special adaptation," due to the action of natural selection, tunes the animals to a "special ecological circumstance." This kind of adaptation involves the "persistence of definite ecological conditions (. . .) and also of habit in the animals considered." However, the "main evolution, which brings about large structural differences" in a large group of animals along a prolonged time follows a different pathway. Here, the selected evidence is that of mammal evolution (the same case evoked by de Beer). The changes in their skeleton, which keep constant over millions of generations are to be considered the "true adaptation," or a kind of change that "increases the efficiency of the animals which display them, often not in relation to some very restricted habitat, but for *life in a variety of conditions* (. . .); they are adaptations to *life in general*" (Watson 1948: 77–78, 83; my italics; cf. de Beer's "adaptation in a broad way" above). For Watson, the "factors which have

[15] For a recent assessment of the long-standing question of how far—and why—Simpson changed his mind between 1944 and 1953, see Sepkoski 2019 and Ochoa 2021.

called them into existence are in fact not environmental ones (. . .) but arise with the animal itself," and, especially in the later book, he goes on speculating on "some feature of the genetical make-up" (Watson 1951: 199, 195).

Notwithstanding his professed pluralism about the modes of evolution, Watson draws in the 1948 paper a major consequence from his new argument:

> there may be considerable difference between the processes of micro- and macro-evolution, between the establishment of sub-species, and of species, and those more fundamental changes common to related stocks within a relatively large group of animals (. . .) and it is important to consider whether or not there be any *genetic changes of a still different quality* (Watson 1948: 82; my italics).

1.2 Simpson's Confrontation with Watson, 1953

> It is worth while to devote a little space to this <Watson's> argument, because the example is similar but more extreme and clear-cut and better worked out than a large number of others in which *nonadaptive trends* are claimed to exist (Simpson 1953: 278; my italics).

Simpson, the major bridge builder between paleontology and Modern Synthesis, championed the "measurement and interpretation" of "tempos," the rates of morphological change in evolution, to infer modes, that is processes (Gould 1995: 131; cf. Sepkoski 2019). In 1944, the mode of evolution at "the origin of taxonomic units of relatively high rank" is labelled "quantum evolution." Quantum evolution is the explanatory device to address the absence of forms transitional to higher taxa, as both the speed of the transition and the low number of organisms involved would explain their failure to fossilize (Futuyma 2015). The difference between such major transitions and the phyletic, "horotelic" evolution, issued from the accumulation of small adaptive changes, is that its "mode" implies "a form of discontinuity," "all-or-none reaction (. . .) most important and distinctive at relatively high levels" (Simpson 1944: 199). The quantum evolution model, inspired by Sewall Wright's shifting balance theory, involved the existence of an inadaptive phase, in which the small isolated subpopulation "explores" by "some trial and error mechanism (. . .) the region surrounding the small portion of the field which it occupies" (Wright 1932: 4). Only in the following, preadaptive phase the population starts climbing the "peak" under the action of a high selection pressure, and moves on to the adaptive phase, the attainment of a new population equilibrium (Simpson 1944: 207).

It has been taken as a symptom of the hardening of Modern Synthesis that Simpson in 1953 constrained quantum evolution into a continuum of adaptive evolution, gradual throughout the whole range from tachytelic, horotelic and bradytelic rates. As Eldredge and Cracraft emphasize, this shift required abandoning the "inadaptive phase" of quantum evolution and involved a "model of selection as a means, not just of climbing peaks, but of traversing the valleys as well" (Eldredge and Cracraft 1980: 252). This approach, promoted by Dobzhansky in 1937, was foreshadowed already in 1944, when Simpson attempted to minimize the autonomy of preadaptation:

> the difficulty disappears when it is recognized that preadaptation and postadaption are phases of a single process in which selection is a conditioning factor throughout (. . .). Preadaption is better conceived as adaptation to a specific environment which becomes preadaptive for another.

So, in principle any change in this process will have "a definite survival value" (Simpson 1944: 186–188; cf. 270).

Interestingly, Simpson's reaction to Watson's radicalized challenge does not directly concern the mechanistic continuity between micro- and macroevolution, but rather the enlarged meaning of adaptation as general functional efficiency, to "life in a variety of conditions," that is Watson's reframing of Simpson's own suggestion of 1944. At this later stage, in fact, Watson's notion of "true adaptation" is equated by Simpson to a "nonadaptive" explanation of trends. The only legitimate driver of genuine adaptation is natural selection: the apparently different environments encountered by labyrinthodonts in the course of evolution might share crucial features, so that "the occurrence of the trend in 'different' environments is not a valid argument against its being similarly adaptive in both." Such may be the case of the dorsoventral flattening of the skull, adaptive to the presence of a "barren substrate" across otherwise different environments. As to the reduction of the cartilage bone, Simpson evokes a possible developmental explanation, by which the cartilage is viewed, after Romer 1946, as an embryonic adaptation, whose retention in adults might be due to the "adaptive neoteny" of a caenogenetic character. De Beer himself is invoked here as an authority to invert Watson's argument about labyrinthodonts, an "unusually good example of primary trends adaptive to the environment and oriented by selection and of secondary trends similarly oriented and adaptive to the primary trends" (Simpson 1953: 279–280).

There are good reasons to wonder whether de Beer appreciated Simpson's 1953 interpretation of neoteny as "adaptive." Clarifying this point may help add a further piece to the puzzle of the peculiar tactical strategy de Beer deploys in this paper.

As we can read in all the editions of *Embryos and ancestors*, for de Beer "ontogeny is the result of the action of external factors in evoking responses from the internal factors," which are inherited by the organism. De Beer is actively involved in the endeavour, promoted by Walter Garstang, to extend the action of natural selection to the whole process of ontogeny. "Qualitative evolutionary novelties can and do appear at all stages in ontogeny" and are all ultimately originated by the interaction between genes and environment, but *their proximate cause is beyond the province of embryology*. In 1958, de Beer assigns rather to genetics the task of unravelling their cause in terms of "variation and selection" (1958: 174). Thus, the crucial point is that while innovations may well be, for de Beer, adaptive to a youthful condition, as in Garstang's example of the torsion in gastropods, or of cartilage in labirinthodonts, heterochrony (of which neoteny is but one of the modes) is not (De Beer 1958: 170–171, 133, 88).

Heterochronic shifts—changes in timing, rate or duration of a developmental process—are due for de Beer mostly to "internal factors," random mutations in "rate genes" controlling the speed and onset of developmental processes, which are not per se adaptive (de Beer 1958: 19).[16] His rather baroque casuistry of heterochronic modes (radically reframed by Gould 1977) is but a tool to disentangle the distinct evolutionary potential of different kinds of temporal shifts, and plot them against the magnitude of the adaptation that can eventually follow the "changing of the gears" (de Beer 1958: 38) that they effect.

The crucial task de Beer sets himself is in fact to relate developmental to phylogenetic changes according to the time of onset of heterochronic shifts. In his classification, only the combination of early developmental innovations (caenogenesis or deviation) with neoteny can have macroevolutionary impact (1958: 114–115, 97). As neoteny allows the retention of the early innovation, the adult phenotype may exhibit a big evolutionary change "in one stroke." Because neoteny does but amplify tiny developmental innovations, it may fully "explain gaps in the fossil record" without contravening Darwinian gradualism (1985: 58, 124). Furthermore, the retention of early innovations in adults endows the evolving animal with greater *phylogenetic* or *evolutionary plasticity* (in modern terms, evolvability, cf. Brigandt 2006), providing an "escape" from specialization and ultimately from phylogenetic senescence (cf. Swetlizt 1995).[17] Unsurprisingly, de Beer's equates paedomorphosis with Severtsov's aromorphosis, which is framed in ecological terms and corresponds to "large innovations which do not restrict the possible habitats in which the organism may live" (1958: 113. Cf. Adams 1980: 205). One may therefore assume that, even understanding the neoteny proposed by Simpson as an adaptation, it would rather endorse than refute Watson's "adaptation to life in a variety of conditions" that Simpson tries here to exorcise.

It is highly probable that the more urgent reason while Simpson goes all this trouble to dismantle the argument of broad adaptations driven by increased efficiency is that Watson applied it in the 1948 paper also to the evolution of horses, Simpson's flagship against classic orthogenetic explanations of macroevolution (cf. Ochoa 2021). Watson's extension of the notion of adaptation seems to have been perceived at this time by Simpson as the reassessment of a form of orthogenesis not amenable to the more acceptable orthoselection, which requires the constancy of environmental pressures. An explanation that Simpson himself had considered for the evolution of horses (1944: 163), but, significantly, dropped altogether in 1953.

Two major evolutionary trends were classically recognized in the evolution of Equidae: hypsodonty and increase in size. Furthermore, these two trends are positively correlated.

Hypsodonty made a wonderful example of preadaptation enabling horses to "traverse the valley" up to the next peak, from grazing to a diet based on the

[16] His argument is that external factors "can also operate in modifying the rates of action of the internal factors, but they will on the whole act more or less equally on all of the internal factors, and will not therefore be of such importance as the internal factors in producing heterochrony" (de Beer 1958: 36); as far as heterochrony is concerned, selection has no direct causal role.

[17] De Beer proposed a genetic explanation of the increased plasticity, the availability of a large set of "unemployed" genes after the structures they controlled are lost by neoteny (de Beer 1958: 116–117).

"abrasive, silica-rich grasses that came to dominate the Miocene" (Miller et al. 2022). As I will detail below (Sect. 1.3), Simpson related hypsodonty to the action of natural selection, the more as it stopped when it reached an "adaptive limit" (Simpson 1953: 257, 263). While acknowledging the independence between the two evolutionary trends, he rather hinted at the primacy of hypsodonty as «greater hypsodonty assists the survival of larger animals» (Simpson 1944: 8, 53, 11). So, for Simpson the driving trend is the one amenable to natural selection.

Watson agreed that such "molarization of pre-molars" might be explained by natural selection but contested that it didn't occur in Artiodactyls, the other major group of hoofed mammals, notwithstanding their similar way of life and food requirements. He speculated that "some feature of the genetical make-up," such as "a fundamental mechanism which cannot be seriously modified without wide ranging and serious effects," must have prevented the evolution of hypsodonty in this suborder. However, these two groups share the same trend to the increase in body size—the trend well known under the unproper name of Cope's rule, a bulwark of orthogenesis.[18] Consistently with his notion of "true adaptation," Watson reframes increase in size as adaptive because of the higher thermodynamical efficiency of larger animals. As it is shared with other orders, increase in size is the *primary* trend, while hypsodonty—the trend amenable to natural selection—becomes a *secondary* trend, "rendered necessary primarily by that increase in size" (Watson 1948: 79–80).

In 1953, Simpson reacted to the strength of this argument (cf. Erwin 2021: 7) by supporting further evidence of the independency between the two trends, but finally disposed of the matter by reframing in more general terms the tenet of a continuum of adaptation: "both are, in a truer sense, dependent on the whole pattern of adaptive change, which is in the last analysis the true and primary trend to which all others contribute and within which all others are intricately interdependent" (Simpson 1953: 280). The whole line of Simpson's argument in 1953 shows his difficulty in coping with Watson's version of orthogenesis. He starts by equating it to an internalist, nonadaptive explanation of trends, to be countered by a stricter notion of adaptation as "precise fitting" to a specific environment. But, in the end, he is forced to concede the same status to Watson's "true adaptations" and to those directly issued from the action of natural selection.

1.3 Whose Mosaic?

Simpson's hypsodonty is proposed by de Beer as an example of the "principle of independence of characters." However, hypsodonty is far from being a good example precisely of evolutionary independence of characters.

Actually, Simpson had emphasized the importance of "differences in rate of evolution between different structures" in the same phylum or related phyla as one

[18] Polly 1998 shows that the real Cope's rule is rather the "Doctrine of the Unspecialized," to which I will refer later, and that body mass never was a focus of Cope's thesis.

of "the most important factors in the differentiation of supra-specific groups" already in 1937. At this time, however, the purported explanation was that static and evolving structures were endowed with "different degrees of variability" to be explained eventually in genetic terms (Simpson 1937: 240); a view largely consistent with Watson's one.

In 1944 and 1953, the independence of characters becomes one of the "theorems concerning the rates of evolution" (Simpson 1944: 12; 1953: 16). In 1953, he also acknowledges, with Mayr, that the characters involved in the rise of higher categories are usually rather "blocks" than single units, provided that their peculiar combination "has adaptive value different from (. . .) that of the various separate characters" (Simpson 1953: 347). Hypsodonty is such a case, as it is "the resultant of two other and simpler characters," tooth-length and tooth-height.[19] However, these two simple characters evolve "with considerable, not complete, independence from each other"; their correlation cannot be in principle "material of evolution" as they are genetically independent (1953: 63; cf. Simpson 1944: 86–87). But it is precisely their independence that Simpson strives to curtail, in order to put their seemingly uncorrelated variation under the control of natural selection. The problem is circumvented because as a "block" the two characters form "unquestionably a unit in its interaction with the pressure of natural selection." Thus, although "morphologically it is the resultant of two other characters which are correlated in one way and not in another," hypsodonty makes a unit of selection. Even more, it is an example of how two characters, each "unitary from the descriptive, morphological point of view, may combine functionally to form another character of great importance for the evolution of the animals conceived" (1953: 10–11, 16, 20).

Thus, de Beer apparently evokes as an example of Simpson's principle of independence of characters—to be applied further also to the "evolution of the classes of the vertebrates" (de Beer 1954a: 169)—the most straightforward case of selectionist explanation of an evolutionary trend, where the independent evolution of the two characters plays no autonomous evolutionary role.

Yet, in the conclusion of the paper, de Beer seems to take side with Watson. The latter interpretation of Seymouria, "a magnificent example of the way in which the evolution of great groups must have taken place" leaves him no doubt that mosaic evolution deserves the name of "Watson's rule" (de Beer 1954a: 170).

However, even this association is not entirely accurate. In fact, it is only in 1919 that Watson emphasized the "mosaic" mode of evolution as internal integration and absence of intermediate formations. In his further writings examined here (and already in 1920), not only the notion of mosaic disappears, but Watson tends to emphasize rather the gradualism of change, to the point of inverting the logical order

[19]Hypsodonty is currently measured as an index of the ratio of "tooth length across the occlusal surface/unworn crown height" (Macfadden and Higgins 2004). In modern parlance, it is amenable to functional modularity, i.e. a module built by a functional, rather than genetic or developmental, association of traits.

of his 1919 argument about Seymouria.[20] Rather than mosaicism, he stresses how diagnostic characters are formed in time "by a series of minute modifications, each complete in the sense that the individual character is either amphibian or reptilian, *the sum of all showing a gradual transition* from one to the other structure." In his 1951 book, the emphasis rests just on the parallel evolution of the same advanced characters in allied stocks, "so widespread in the animal kingdom that it is the general rule" (Watson 1951: 115, 184), with no further reference to mosaicism.

Thus, both the juxtaposition of Watson and Simpson in de Beer's argument and the attribution of mosaic evolution to them sound somewhat specious, rich as they are of hidden and sometimes contrasting implications. The strict association of mosaic evolution with trends, whereby the new diagnostic "pieces" are born as already complete and "broad" adaptations, is of de Beer's own making.

1.4 From Waddington's Epigenotype to de Beer's Mosaics

There is a further reason why it would have been more appropriate to name mosaic evolution "de Beer's rule." It is the only hint de Beer gives to a possible causal explanation of mosaic evolution, embarking on a path that to my knowledge he didn't pursue further, that of a *modularization* of the epigenotype or developmental system.[21]

The key concept is "the epigenetic climate of a developing organism," generically referred to Waddington (de Beer 1954a: 169). Just the year before, Waddington published "Epigenetics and evolution," his first vocal critique to the foundation of Modern Synthesis on mathematical population genetics (Smocovitis 1996: 22; Provine 1980: 402). The main topic is the epigenotype, "the system of causal relationships by which the newly fertilized zygote (. . .) becomes realized as the fully developed adult," as the new framework needed to overcome the dualism between the genotype, "the province of the geneticist" and phenotype, the one of "the taxonomist." It is the working of this system, in fact, that dictates which genetic variation will find its way to the phenotype:

> Changes in genotypes only have ostensible effects in evolution if they bring about with them alterations in the epigenetic processes by which phenotypes come into being; and the kinds of change possible in the adult form of an animal are limited to the possible alterations in the epigenetic system by which it is produced (. . .) only certain types of alteration of the whole

[20] I could find no evidence of any reaction from Watson to de Beer's "Watson's rule," while he wrote a highly appreciative review of de Beer 1954b (Watson 1955).

[21] Raff has stressed that de Beer's view of heterochrony is "global" rather than modular, measured as it is on the time of onset of sexual maturation (Raff 1996: 261). Whether this necessarily points to a global, in the sense of an organismic view of heterochronic change, deserves a further discussion, the more as Raff's interpretation of de Beer is heavily influenced by Gould, the main target of Raff's criticism. It is nonetheless only in this 1954 paper that de Beer explicitly envisages a decentred modularization of the evolving organism.

> system can be brought about by any conceivable alteration of a single member of the gene complex (Waddington 1953: 190–191).

Waddington's uncoupling of genetic and phenotypic evolution pushed at an open door. Already in 1934 Huxley and de Beer prophesized that "the old concept of Mendelian *characters* will disappear," as soon as gene action will be finally framed in the context of developmental processes (Huxley and de Beer 1934: 413). De Beer's 1938 critique of genetic criteria to ascertain homology—which, interestingly, he incorporated only in the third edition of *Embryos and Ancestors*—reassessed from another perspective the uncoupling between genotype and phenotype (de Beer 1938: 65–67).

However, in Waddington's view of the epigenotype lingers an ambiguity concerning the viability of "piecemeal" changes in de Beer's sense.

In the 1940s, Waddington's view of development as "a set of alternative canalized paths" (1941: 564; 1957: 19) also accounts for the "existence of discontinuities between one set of intergrading types and the next" (1941: 109). In principle, not all developmental pathways need to be equally canalized. Homeotic alterations "have certainly played a large part in evolution *on the grand scale*, which gives rise to the major phyla" (1941: 15, my italics). "A switch mechanism," possibly operated by specific genes, may drive certain regions toward an abnormal path in development (1941: 564). Nonetheless, Waddington's major emphasis rests on the *equilibrating* power of the developmental system as a whole. Thus, in his critical revision of the mechanism underlying Goldschmidt's "unbridgeable gaps," he states that apparently "systemic" mutations are better explained as the process in which "a developmental path has to be equilibrated (. . .). If, by an early-acting gene, a whole set of paths are thrown out of their old equilibria, a very considerable modification of the genetic background will be called for" (1941: 15; cf. 1957: 79; Woolf and Markow 2003). The resulting change is therefore deemed to be global.

This aspect will be further emphasized since the 1950s, as he increasingly embraces the cybernetic imagery (cf. Sect. 3). "The complexly interwoven nexus of cybernetic interactions" (Waddington 1953) within the epigenetic system doesn't seem to allow alternatives to either a homeostatic "resistance to change" by developmental buffering (Waddington 1955), or a global restructuring of the "interconnected pegs" (representing the genes) that support the "epigenotype" from beneath (Waddington 1957: 3). In the 1950s, while still acknowledging the theoretical possibility of a modular change in the phenotype (1953: 120; 1957: 37–38, 215), his emphasis rests nonetheless on the property of the tightly integrated epigenotype to constrain genetic variation (Waddington 1957: 115; 1953: 191). Thus, the least one can say is that uncorrelated variation isn't at the heart of Waddington's concerns.

De Beer's mosaic evolution "one piece at a time" attempts in my view a possible third way between the constraining function of the developmental system and a radical reshaping of its structure.

By applying the developmental notion of mosaic to macroevolutionary changes, de Beer hints at a modularization of the epigenotype, that allows different "organ fields" to evolve and self-organize independently in fully integrated parts. This implies that heterochrony (i.e. the "rate genes" whose mutations de Beer envisaged as responsible for heterochronic changes) may act locally, and single organ-specific developmental pathways may be "thrown out of gear" (Needham 1933) with respect to other regions. The more restricted the action of heterochrony (i.e. rate genes) is, the less these local changes will disrupt the integration of the resulting phenotype.[22] In this framework, even the more disrupting heterochronic changes, those affecting early development, may be locally constrained, and thus viable and relatively neutral with respect to natural selection. De Beer's main argument in support of this hypothesis is namely evolutionary viability:

> It is legitimate to think that changes restricted to single organ-systems are more likely to have occurred than progressive changes affecting the entire organism. It is further probable that changes restricted to single organ-systems at a time were more viable and less likely to succumb to selection (. . .) Selection may therefore be expected to have acted with greater rigour against organisms varying in more than one direction at a time, *unless the directions were correlated* (de Beer 1954a: 169; my italics).

This latter case is exemplified by the evolutionary correlation between hypsodonty and limbs, for Simpson strictly dependent on the habitat (Simpson 1953: 158). Thus, de Beer does embrace the selectionist explanation of apparently uncorrelated trends, when the different 'key characters' involved can be shown to be functionally related (as in hypsodonty). However, he maintains that the mosaic, "piecemeal" mode of change—which is, as he finally states, "the way in which the evolution of great groups may have taken place" (de Beer 1954b: 170)—is neutral with respect to natural selection. In fact, it is neutral in the same sense as heterochronic shifts are.

This is made explicit in a passage of the later edition of *Embryos and Ancestors*. Commenting on Simpson's and Mayr's "key characters," the "evolutionary novelties which enabled their possessors to spread into new ecological zones," he overtly associates them to paedomorphosis, the more as this latter endows animals with enough "phylogenetic plasticity" as to support adaptive radiations, the capacity to "begin as species and to rise from that rank" (1958: 115–117).

The focus on evolution within the shelves, seen at the benchmark of "intercalary" types, reveals in my view de Beer's commitment to Watson's view that "pre-adaptation does indeed occur, but does so accidentally without reference to the possibilities of utilisation" (Watson 1948: 83). Without neoteny, early novelties would remain adaptations to the youthful stage with no influence on the adult phenotype. But, unlike early developmental novelty, neoteny is for de Beer, as all

[22] Reframed in terms of changes in regulatory genes, whose action is not pleiotropic but modularly restricted, this thesis is the workhorse of early evo-devo, although supported by a completely new understanding of genetic regulation. See Schlosser & Wagner 2004; Callebaut and Rasskin-Gutman 2005. Modules are group of traits within an organism which are integrated with each other but relatively independent of other sets of traits, an architecture that is thought to facilitate adaptive evolution because changes in parts would not affect the whole organism.

heterochronies, a substantially random process. By stressing the randomness in the appearance of a new structure, de Beer uncouples the origin of trends and of evolutionary novelty from ecological opportunity. His stance reminds of the contemporary conundrum recently labelled *macroevolutionary lag*, the time that elapses between the inception of a novelty and its taxonomic or phenotypic diversification (Jablonski and Bottjer 1990; Erwin 2015; Jablonski 2017).

The broad adaptations early heterochronic shifts may engender are but "prophetic," an orthogenetically laden term that de Beer doesn't shy away from using in subsequent papers, among which the one he writes with Swinton in the Festschrift for Watson that they will edit together in 1958 (de Beer and Swinton 1958).[23] Besides, the genetic changes that underly such broad adaptations also are, as foreshadowed by Watson, genetically specific, in so far as they are triggered by mutations in rate genes working in early development.

But, at the same time, de Beer can stay true to synthetic gradualism: as all heterochronic shifts are genetically small, "there is no difference in principle between 'small' and 'large' evolution" (de Beer 1958: 115), no challenge to the gradualism of genetic evolution. Thus, in 1954, de Beer proposes a different, "mosaic" understanding of gradualism, which in his view is fully compatible with Darwinian gradualism. The "quantity of evolutionary change" necessary to accomplish the passage to the next shelf can be thought of as "moderate," and not involving any "upheaval or sensational reconstruction of the body" (de Beer 1954a: 164). This aspect will not be lost to the early appreciative readers of his paper.

2 The German Reception: Mosaic Evolution Versus Typostrophism

In 1956 Walter Gross, "one of Germany's best vertebrate palaeontologist" (Reif 1986: 120), wrote a paper entitled "Über die 'Watsonsche Regel'" welcoming de Beer's mosaic evolution as fresh ammunition against Schindewolf's typostrophism, to which Schindewolf himself reacted forcefully (Schindewolf 1957).[24] Schindewolf's theory—the major target, together with Goldschmidt, of the Modern Synthesis' attack to saltationism—postulated the sudden rise of higher taxa due to macromutations in early ontogeny, that set in "a new evolutionary cycle with increased intensity" (Schindewolf 1957: 52). According to this view, "the first bird hatched from a reptile egg," as the emerging key character marks already the appearance of the new type, whereby Archaeopteryx became the first representative

[23] Agassiz introduced the notion of "prophetic types" in a paper in which he also foreshadowed the notion of mosaic evolution (1850: 435); Wedekind, a supporter of orthogenesis and Schindewolf's teacher (Reif 1986), introduced in 1918 the term "prophetic characters." (Wedekind 1918).

[24] Gross had been one the first German (actually Latvian-born) palaeontologists to criticise typostrophism in 1943 (Reif 1986). It deserves to be mentioned that, notwithstanding this open contrast, Schindewolf helped Gross substantially to build a career in Western Germany as the latter emigrated from East Berlin in 1961, after the Berlin Wall was built.

of birds rather than a transitional form (Schindewolf 1950: 277; echoed by Goldschmidt 1940: 395).

In his accurate and sympathetic discussion of de Beer's argument, Gross defines the "Mosaikmodus der Evolution" as "heterochrononen Merkmalsverschiebung," heterochronic displacement of characters, although heterochrony is employed in the descriptive sense that "organs evolve (...) at different times with different tempo (...) largely independent of one another." Ancestral forms are "mosaic-like mixture of characters of ancestors and descendants," presenting *no "fabric of strict correlation"* (my italics). The "unbridgeable" differences between types "do not arise by leaps and bounds, in clean-cut separation and simultaneous formation of all characters, but stepwise, in an irregular and mostly rapid fashion" (*ungleichmässig und meist rasch*). A single character—such as the Archaeopteryx feathers, which appear before the capacity to fly—can change "without major perturbation of the body plan," and does not by itself mark the appearance of a new type. The "important paleontological rule" formulated by de Beer, by which "the different parts of organisms are capable of independent evolution, proceeding at different rates," is to the paleontologist already "in the air," "hinted at by many authors (...) but its clear formulation is de Beer's achievement" (Gross 1956: 32, 39, 35, 20).

The mosaic mode of Gross and de Beer will be soon incorporated in the "additive theory of the genesis of types" in the second edition of *Die Evolution der Organismen* (Heberer 1957), considered "the most comprehensive account of the evolutionary Synthesis in Germany before 1945" (Juncker and Hossfeld 2002: 228). In this book, in fact, Heberer welcomes Simpson's demise of the notion of megaevolution in 1953, as "there are no qualitative difference nor between micro- and macrophylogeny nor between macro- and megaphylogeny, as far as evolutionary mechanisms are concerned" (Heberer 1957: 859). The additive theory postulates the "gradual addition (*Aufsummung*) of small distances (*Abstände*)" and supports, against Schindewolf's theory of the "genesis of types in early ontogeny" (*frühontogenetische Typenentstehung*), the concept of a "phyletic continuity without saltation" (1957: 861, 891, 884; cf. Amrein 2010: 52).

It is in this context that de Beer's mosaic evolution is valued against Schindewolf: "single characteristics possess their own slow or accelerated speed of evolution, that is they evolve heterochronically" (Heberer 1957: 873). It is called in support of Heberer's own "additive microevolutionary" explanation of adaptive radiations: "Each radiation began below the species level (*infraspezifisch*), but the evolving characters shifted upwards in time in the taxonomic levels" (1957: 888). However, unlike Gross, Heberer perceives the problem of "synorganisation" (1957: 892, 895), or how types as functionally integrated adaptive systems come into being if not by macromutations. The theory of the *additive Typogenese* supports the idea that "a sum is composed in a mosaic fashion from the addition of single building blocks one after the other." Thus, the "additive" mode may well explain the origin of novelty within a solid mechanistic framework, without invoking unspecified "Ganzheitsfaktoren." As with the addition of some atoms a new molecule may arise out of the endless possible atomic combinations, so a new "idiotype" may arise that is more of the sum of its constituents (Heberer 1957: 865–866).

Simpson reviewed positively Heberer's "additive microevolutionary" approach (Simpson 1958),[25] without taking note of mosaic evolution. Much more influential on his later overt acknowledgment, in 1959, of this notion was another paper, by his "lifelong" friend the American paleontologist Bobb Schaeffer (Hull 1988: 70), published in 1956, together with the progressivist redefinition of the notion of grade that Julian Huxley will push forward since 1957.[26]

2.1 Mosaic Evolution According to Schaeffer

Last offshoot of the school of William King Gregory and one of the early disciples of Simpson and Mayr, and later one of Gould's teachers, the American paleontologist Bobb Schaeffer welcomed de Beer's concept as an important theoretical innovation (1956).

Schaeffer applies de Beer's notion to the case of subholosteans, extinct Late Permian ray-finned fishes that were important contributors to the Triassic radiation of Actinopterygians (Friedman 2022). Subholosteans are in fact transitional forms (Brough 1936) between the primitive palaeoniscoid and the advanced holostean (neopterygian) conditions, understood by Schaeffer as two successive paraphyletic "broad evolutionary grades."

The 11 diagnostic characters that make according to Schaeffer the major differences between the palaeniscoid and the holostean grades revolve around locomotory and feeding mechanisms. These mechanisms involve no "single organ-system" but "functionally integrated structural complexes" which may even comprise elements of other organ systems. The characters that remain unaltered (in the primitive condition) or change (in the advanced condition) in a correlated fashion build each a single complex. In this case their association is "true"; they are possibly ruled by a single mechanism (which for Schaeffer may be as well genetic as ontogenetic) and their correlated change is acted upon by selection, as in the case of the complexes belonging to the advanced condition that are favoured in the break-through to a new adaptive zone. Nonetheless, these "piecemeal" changes occur in a mosaic mode: that is, each change is progressive (trend-like), and occurs not only at different rates and times in different genera, but also under widely diverse selective pressures in the changing environments from Permian to Triassic. The problem Schaeffer is confronted with is whether a unified selective explanation holds for

[25]Neither Heberer (a zoologist and anthropologist actively involved in the Nazi *Rassenforschung*, Juncker and Hossfeld 2002) nor Bernhard Rensch, the two main supporters of synthetic gradualism in Germany, were paleontologists. Thus, only Walter Gross stood up to Schindewolf in his own field in Germany. On the differences between Gross' and Heberer's criticism to Schindewolf, cf. Amrein 2010: 52–63.

[26]For a historical overview, cf. Tassy and Fischer 2021. The notion of grade is still in use in paleontology, although devoid of progressivist implications, as "an informal grouping united by a level of morphological or physiological complexity supposedly reflecting adaptation, without the strict phylogenetic requirements implied by 'clades'" (Serrelli and Gonthier 2015: 8). Heavily discredited after the triumph of cladistics, it still has some supporters, cf. Willner et al. 2014.

the transition from paleoniscoids to holosteans feeding and locomotory systems. In this case, the progressive changes in both systems might be explained as a single trend driven by natural selection throughout. Along this line, Stebbins (1951) had successfully confirmed "the guidance of natural selection" in angiosperms, showing the "true" association between apparently independent trends, those in floral and vegetative characters.

To assess whether the associations between characters are true or spurious, Schaeffer applies the chi-square method, a statistical procedure for quantifying the frequencies and co-occurrence of characters and calculating standard deviations. He applies his test to the association among different skull characters (involved in feeding), to the one among different fin characters (involved in locomotion), and to the association of skull with fin characters.

However, his results do not match those by Stebbins. An "unexpectedly high" percentage of complexes which evolve at different times in different parts of the body prove to be unrelated to one another (Schaeffer 1956: 201, 209). This is especially the case of the skull with fin associations, which turn out to be spurious.

> It seems probable that the inconceivably complicated association network forming the integrated organism must include many structures so separated morphologically and functionally, as well as in rate of evolutionary change, that they are almost entirely independent of each other (1956: 207).

Schaeffer is therefore unable to robustly extend Stebbins conclusion to the feeding and locomotor systems of subholosteans, notwithstanding both sets of characters have an immediate adaptive value (1956: 209). He tries therefore a different path, that is to frame the problem in terms of evolutionary *experimentation*:

> It is probable that the transitional populations 'try out' all the combinations of primitive and advancing characters that are genetically, ontogenetically, and functionally possible" (Schaeffer 1956: 209).

A considerable degree of chance was already intrinsic in de Beer's view of the "hard work" of evolution, whose "wide tolerance" implied it to be largely indifferent for natural selection. Experimentation, however, restores the autonomy of the "inadaptive" phase that Simpson had just tried to downplay.

Schaeffer's compromise solution is to tame the autonomy of the experimentation process by adopting the enlarged meaning of adaptation.[27] The new characters that have made it to the "new adaptive level" are unrelated to a specific environment "or mode of life" but "were clearly adaptive since they increased the efficiency of the food-gathering and the swimming mechanisms." Ultimately, his strategy consists in placing broad and special adaptations into a continuous range, blunting the divide between general increase in efficiency and the strict "precise fitting of an individual

[27]In 1965, the forces taming the autonomy of experimentation will be even more: ancestry, selection "and particularly the canalization of morphogenesis" (Schaeffer 1965: 327), consistently with Mayr's reframing of mosaic evolution, see Sect. 3.2.

to its environment." This continuum encompasses in fact both cases in which "selection favored both the trend and the association net of an integrated complex" and cases—such as the association of skull and fin characters in subholosteans—in which the association is "spurious, but on the other hand, there may be *broad functional integration* between them, implying at least closely correlated trends, or perhaps even a single trend, and true association" (Schaeffer 1956: 210; my italics).

As remarked by Olson, Schaeffer's problem, not unlike Heberer's, is to reconcile the mosaic pattern of acquisition of new characters with the cohesiveness of the taxon as an adaptive whole. In the new synthetic framework, the absence of "strict correlation" in transitional forms can no longer be seamlessly acknowledged as in Gross. The problem of "the integrative control of the evolutionary mosaic" (Olson 1965: 340) is solved by Schaeffer by appealing to an enlarged notion of adaptation that, although unrelated to local environments, is nonetheless effective in driving the "broad functional integration" among distinct trends.

Schaeffer dismisses Watson's distinction of "true" and "secondary" adaptations by blurring them into a unique continuum. In this way, the strict selectionist interpretation of macroevolutionary trends becomes anchored to a progressivist interpretation of evolution, proceeding, as in the classes of ray-finned fishes, through the stepwise achievement of "broad evolutionary grades" (Schaeffer 1956: 302). Schaeffer's interpretation of de Beer's mosaic evolution highlights how the difficulty in explaining evolutionary trends "on the grand scale" by natural selection leads the synthetic movement to embrace more fully Huxley's progressivist view of evolution. This process will be facilitated by the reassessment of the notion of grade by Julian Huxley, just a year after, which will foster the incorporation of the notion of grade into the synthetic framework.

2.2 Grades of Improvement

To capture the history of this convergence it may be useful to start from the end. In 1986, Ernst Mayr defends the notion of biological progress as upheld by the Modern Synthesis under the banner of Darwin from the attack of the American historian of science John C. Greene. In this context, he states that the original view expounded by Julian Huxley in 1943 "although upholding the main thesis of Darwin, were not typical of those of the architects of the synthesis." In particular, "his claim that evolution at the species level was ultimately irrelevant as far as evolutionary trends are concerned (. . .) were criticized by Simpson, Dobzhansky, Mayr" (Mayr 1986: 231; cf. Shanahan 2004).

Nonetheless, it rather appears that it was exactly for explaining evolutionary trends above the species level that Simpson incorporated Huxley's progressivist notion of grade.[28]

[28] Cf. Eldredge 2001, for whom Simpson "never embraced (. . .) the notion of species and speciation" of Dobzhansky and Mayr. "To Simpson, speciation is trivial—and the real stuff of evolutionary change lies always in the origin of supraspecific taxa" (343–344).

Ever since 1936, Huxley's 'objective' measures for biological progress are increase in "control over the environment and independence of it." In so far as these two features give "advantages" to their owners that are related to the conditions of life, progress may be safely explained as an outcome of natural selection (Huxley 1936: 604; cf. Shanahan 2004). In 1942, however, Huxley uncouples more sharply progress from specialization. They are both *improvements*, but the first is "all-round," "in efficiency of living in general," while the latter is "one-sided," "in efficiency of *adaptation* for a particular mode of life" (my italics). Adaptation by natural selection is therefore restricted to specialization, whose outcome may not always be progressive and beneficial. Furthermore, he openly identifies the "all-round" improvement, as a "raising of the upper level of biological efficiency," as the genuine precondition of adaptive radiations. "All-round" improvement brings about broad, and "relatively unspecialized" traits, corroborating Cope's rule of the unspecialized,[29] while the subsequent adaptive radiation remains "a product of selection (...) relative to environmental conditions" (Huxley 1942: 562, 564, 496).

In 1957, improvement covers all three types of evolutionary processes that Huxley resumes under the headings of *cladogenesis*, *anagenesis*[30] and *stasigenesis*. The first two definitions are inspired by Rensch and match the former distinction between "one-sided" and "all-round" efficiency. Rensch in fact shared with Huxley both the adherence to Cope's rule and the tenet that only anagenesis brings about, with its "inventions" of "general perfectioning" (*Vervollkommnung*), the "real higher evolution" (*Höherentwicklung*; Rensch 1947: 303, 323).

For Huxley (1957) there is rather a division of labour among the three kinds of evolutionary processes, all sharing the fundamental Darwinian tenet of improvement with respect to the conditions of life: cladogenesis—the only monophyletic mode—brings about divergence and variety; anagenesis, a "general organizational advance" or *grade*; and stasigenesis, as the process of stabilization of "adaptively differentiated (...) delimitable and persistent anagenetic units," the persistence of types.

It is apparent that, under these assumptions, for Huxley it is unproblematic to conceive of grades as improvements unrelated to the local environment (Huxley 1942: 568). In this way, however, the stricter notion of adaptation by natural selection loses centrality, becoming only one of the possible modes of improvement. Several interpreters have emphasized Huxley's divergence from the course of "hardening" embraced by the other architects (Olby 1992: 70) the more as he never committed himself to extrapolationism (Adams 2021: 210; Schwartz 2021: 338). In his defence of grades, Gould positively evaluated Huxley's contribution as a

[29]Huxley 1942: 562: « there is no certain case on record of a line showing a high degree of specialization giving rise to a new type. All new types which themselves are capable of adaptive radiation seem to have been produced by relatively unspecialized ancestral lines». Cf. Swetlizt 1995. Also Rensch adhered to Cope's rule (Rensch 1947: 305). The debate about specialization and extinction is far from dead, see Raia et al. 2016.

[30]The meaning of the term anagenesis has changed significantly from this time, cf. Vaux et al. 2016: 167.

"frank display of a bias contrary to orthodoxy" (Gould 1976: 120; cf.). But apparently, it is precisely in the "panadaptationist" phase of Modern Synthesis (Gould 2002: 572) that a looser notion of adaptation without *fitting to a specific environment* gains traction and merges with Huxley's grade. In fact, it is to Huxley's grades that Simpson turns in 1959 to reframe the issue of evolutionary trends (cf. Ereshefsky 2001: 59), and it is in this new framework that mosaic evolution finds its place. The new connection between grade and mosaic evolution established by Simpson in this paper lays the ground for the *qualitative* argument that the evolutionary taxonomy of Modern Synthesis will oppose to cladistics (Sect. 3.2).

2.3 Simpson and Mosaic Evolution

In 1959, in "The nature and origin of supraspecific taxa," in a section of the Cold Spring Harbor Symposia devoted to "Trends of evolution," Simpson adopts Huxley's notion of grade to account for the mosaic mode of evolution of higher taxa.

To the weary repetition that "the basic processes are the same at all levels of evolution," he now adds "*although the circumstances leading to higher levels are special*" (1959: 255, my italics). What is special about higher taxa is that they "typically arise with—if not by—important basic changes in adaptation" (1959: 264) or a "basic general adaptive complex." In most cases examined, this complex turns out to correspond to "a general improvement that broadened adaptation as a whole rather than narrowing adaptation to a particular climate or vegetation complex" (1959: 257).

In principle, Simpson rejects, as Schaeffer, any divide between broad and special adaptations (1959: 266). Already in 1944, he warned that the notion of specialization remains "both complex and vague" if not measured in biogeographic terms, as "inverse width of the zone of tolerance" (Simpson 1944: 143).[31] But, as a matter of fact, a new divide lingers in the 1959 paper between broad and special adaptation that cannot be solved by the post-hoc criterion of the "breadth," the number of ecological sites in which the new adaptation provides a selective advantage (Simpson 1961: 223). In fact, Simpson recognizes that the preadaptation that commonly elicits adaptive radiation is the "broad" one, that *follows* the "achievement of a new grade of organization" (1959: 271), which, under Huxley's definition, is generally paraphyletic. Thus, even Simpson's attempt to fit "grades into ecological spaces"[32] doesn't solve the riddle of how the new grade of organization arises in the first place.

This difficulty becomes explicit when Simpson deals with the varied phenomenology of the "prospective" adaptation at the origin of higher taxa. At one extreme, there is the case of the transition from reptile to mammal, in which, as stressed by de Beer (and Watson), no entrance to a new adaptive zone is at stake. Simpson invokes

[31] This passage belongs to a paragraph entitled "The survival of the unspecialized."

[32] As Gould's comments about the "hardening" of Dobzhansky in 1951 (Gould 1982: XXXVII).

here "a general adaptive change or improvement in numerous ways or in virtually the whole organization" (Simpson 1959: 270). What is at stake in fact is not a single basic adaptive complex, but rather "a whole series of general structural and physiological improvements that made mammals *more efficient* within some of the same ecological situations" (Simpson 1959: 266; my italics). Moreover, these improvements come together in a mosaic fashion, at independent times and rates in different lineages. So, it is no question of a direct or integrating action of selective pressures, but at most of a complex interplay between "general adaptations" evolving in parallel. Thus, it rather appears that the grade already achieved sets the condition of possibility of the mosaic gate of improvement.

At the other extreme, Simpson sets de Beer's "piecemeal change," or mosaic evolution proper. It is the case when a leading new faculty—such as flight—allows the entrance in a new adaptive zone. Apparently, it should be the new advantageous key character that drives the post-adaptive process which eventually establishes the harmony of the new grade. However, the origin of the new "basic adaptive complex" proves to be an equally thorny issue. In fact, the "flying apparatus is (. . .) complex in itself" and involves a series of concomitant structural and functional changes that are acquired at different times and rates. In *Principles of animal taxonomy* of 1961 Simpson details the problem of the origin of the single *piece*, whereby parts of the same adaptive complex "evolve at different times and rates within one lineage." The only solution Simpson seems to find is to invoke, again, grades. "The slow attainment of a complex basic adaptation" is in principle similar to cases where no basic adaptation is at the origin of the group but "only progressive trends, or improvements in Huxley's sense" (Simpson 1961: 225). Thus, to elaborate a consistent alternative to de Beer's heterochronic explanation of the origin of the new "adaptive complex," Simpson turns again to grade as the force that drives piecemeal change.

Schaeffer and Hecht correctly summarized Simpson's 1959 paper as the statement that "Huxley's grades reflect general improvements that evolve in a mosaic fashion in a number of different related phyletic lines" (1965: 247). I think that Simpson was uncomfortable even with this solution.

Besides the coming together of a new "basic adaptive complex," Simpson is clearly concerned with the challenge that uncorrelated variation raises to the continuity of organismal adaptedness. He expresses his uneasiness with the notion of mosaic, a misleading term "if taken to imply quite independent evolution of different avian characteristics or lack of adaptive integration at any point in the process." Even the "clumsy" Archaeopteryx was "undoubtedly well-integrated," although "on a different basis from that of later birds" (Simson 1959: 266, my italics). As he will almost confess much later, his major difficulty was in accepting that "rates of evolution vary greatly not only within and between lineages but also among different characteristics, whether functional or structural, of any one lineage" (Simpson 1983: 181; see below), short the lack of "integrative control of the evolutionary mosaic."

In all commentaries about de Beer's mosaic evolution until 1961 there is no patent sign of annoyance. Only in 1971, in an obituary of the American paleontologist William King Gregory in 1971, he adds a rather peevish remark whose target is de Beer, although he is not explicitly mentioned. Discussing Gregory's "palimpsest theory," Simpson writes namely:

> It may contribute to a somewhat cynical view of the history of science that the same principle (it is not quite a theory) was proclaimed as new under the name "mosaic evolution" by another biologist years later and now generally goes under the latter name without reference to Gregory (Simpson 1971: 170).

The preference for the alleged Gregory's version of mosaic evolution shows how little Simpson was comforted by Heberer's additive strategy, by Schaeffer's inadaptive experimentation, and even by his own adhesion to a larger notion of adaptation as "organizational advance" driving the functional integration of the independent parts.

Gregory's palimpsest theory regarded the mixed nature of "any animal type at a given geologic period" (Gregory 1947: 8) as a complex of *heritage* and *habitus*. Both encompass "the totality of habits and anatomical characteristics" related to a specific environment, but the earlier is *palaeotelic*, relative to the characteristics "inherited from earlier life habits" while the latter is *caenotelic*, relative to those "assumed during the latest life habits of the type" (Gregory 1947: 8; Gregory 1910).

The new *habitus* features wipe out the *heritage* features "somewhat as the later writing on a palimpsest hides the partly erased image of the earlier writing." This overwriting concerns preferentially the parts "which come more directly and simply into relation with (. . .) special environments" (Gregory 1910: 111). Heritage features may be refunctionalized in order to adapt to the demands of the new environment, but, for Gregory's aim, they lose in this way any informative value for phylogeny (Gregory 1936: 430; cf. Rainger 1991).

In brief, for Gregory heritage embodies the ghost of past environments: "the present habitus of any given animal represents, as it were, the telescoping and net residue of all earlier ancestral habitus beneath the surface layer of later habitus characters" (1947: 45). Both heritage and habitus are adaptations in the tighter sense of *fitting to a specific environment.* It is always specializations that are inherited or cancelled along the course of evolution. Past and new specializations build together a layered continuum of adaptation. Clearly, in this perspective, no broad, generically progressive adaptation is needed; Simpson can stick to his former view that a preadaptation is but an "adaptation to a specific environment which becomes preadaptive for another" (Simpson 1944: 188). It is no wonder that Simpson preferred this seamless adaptationist approach to mosaic evolution to de Beer's version, although he didn't succeed in supporting it in 1959.

Simpson's uneasiness with de Beer's mosaic evolution is further confirmed by a retrospective remark appeared in the above mentioned 1983 primer *Fossils and the history of life*, which in the German version brought mosaics even in the subtitle ("Mosaiksteine zur Geschichte des Lebens"). He mentions a hesitation—clearly, his own—in the reception of the principle of mosaic evolution, due to its supposed resemblance to a missing link imagery. This remark, although patently incorrect with respect to de Beer's text, may reveal Simpson's discomfort at the manifest attempt by de Beer to reframe the problem of transitional forms in terms of "intercalary types," which challenged their claimed absence in the fossil record.[33] The uneasiness becomes overt annoyance when Simpson can't help adding that the principle was proposed by the "famous embryologist" Gavin de Beer "in his only cursory excursion into paleontology" (Simpson 1983: 181)—quite a blow below the belt, although delivered long after the death of its target.

By this time, however, much had changed. As already highlighted, in the 1950s Simpson had sensed in de Beer's theory of the origin of higher taxa by paedomorphosis the ghost of saltationism. This might explain why he refrained from recruiting de Beer's mosaic evolution against Schindewolf, as Gross and Heberer had done. If this is the case, I think that his interpretation of de Beer's piecemeal mode of mosaic evolution was more accurate than Gross' and Heberer's. The new "basic adaptive complex" doesn't arise for de Beer gradually, but "in one stroke," by an early developmental novelty plus neoteny. Simpson may have been reminded of this fact by the only critique that the tactful de Beer had inserted in his review of *The major transitions*—that of neglecting the role of clandestine evolution "in the origin of groups leading to higher categories" (de Beer 1954c: 1108). Clandestine evolution is in fact the case when a "qualitative novelty," adaptive to early ontogeny but not affecting the adult state, causes after neoteny an "unexpectedly abrupt modification" in phylogeny (de Beer 1958: 30–31). By 1983, however, Simpson was in a very different mood about saltationism, and rather claiming a substantial continuity between his quantum evolution and the theory of punctuated equilibria of Eldredge and Gould (Simpson 1984).

As to the problem of the emergence of complex adaptations, this other ghost had been exorcised in the meantime by Mayr. By updating the classical Darwinian explanation, Mayr incorporated mosaic evolution in the Modern Synthesis framework, purging it of the most controversial aspects of de Beer's version and erasing his very name.

3 Beyond Beanbag Genetics

Huxley's introduction of stasigenesis in 1957 hints at a change in the intellectual climate (Swetlizt 1995) which can be loosely related to the rising cybernetic imagery. The new emphasis on stasis (Swetlitz 1995; Sepkoski 2019: 697) shifted the inquiry on the role of stabilizing "homeostatic" mechanisms in evolution. This shift was significantly assisted in the Modern Synthesis by the (selective) appropriation of

[33] Cf. Prothero 2007: 82: "there are many transitional fossils *between larger groups* (...) If a series of stable fossil species shows an overall trend linking one major group to another, each of those fossil species is a 'transitional form', even though we rarely get all the fossils of the transitions *between* those species".

Schmalhausen's theory of stabilizing selection, in particular by Dobzhansky, and of Waddington's notion of the epigenotype by Mayr.[34]

The change was publicized by Mayr in 1959 as the opposition of "the newer population genetics" to what he called, with one of his typical "sweeping categorical statements" (Mayr 1982: 9; Borges 2005), "beanbag genetics" (Mayr 1959).[35] This shift has been variously described in the literature as from genetic selectionism and reductionism to gene-interactionism, and even "population organicism" (De Winter 1997; Dreuil 1997; Hagen 2021).

Beanbag genetics—or more soberly "the classical hypothesis"—envisaged evolution as the "gradual substitution and eventual fixation of the more favorable (. . .) gene alleles and chromosomal structures" (Dobzhansky 1955: 3). Fitness being assigned to independently segregating genes, natural selection would be mostly "purifying": favouring one allele over others, it would spread homozygosis at most loci and eliminate mutations as overwhelmingly deleterious. This hypothesis entailed the assumption of low genetic variation in populations, which would have hampered their traversing the valley toward new "peaks." Although at the time no direct evidence about genetic variation in natural populations was still available (Lewontin 1987), especially the Russian school of population genetics since Chetverikov had shown experimentally a high amount of standing genetic variation in sexually crossing species. For Chetverikov, recessive mutations were rather "absorbed" and hidden by heterozygosis (Depew 2011). The alternative or "balance" hypothesis assumed therefore that natural selection doesn't act only as a "sieve", but as a creative power, in that it favours more adaptable and "generalist" genotypes, whose norm of reaction would be, because of genetic variation in alleles at various loci, more plastic and fit in fluctuating environments (Hagen 2021: 162–167). By orchestrating beneficial combinations among random gene variants, natural selection could both direct and initiate evolutionary change (Beatty 2019: 719). "A genetic good mixer becomes superior to a genetic rugged individualist" in the "organismic system" of the gene pool (species), bound together by heritage and gene flow. It is therefore the gene pool, "the collective or corporate genotype," and not the gene the unit of evolution and the bearer of fitness (Dobzhanski 1955: 3, 12).

By promoting "a certain level of obligate heterozygosity" in fluctuating environments, natural selection fosters the evolution of a well-integrated and coadapted gene complex, endowed with "genetic homeostasis" (Dobzhansky 1955: 3; Lerner 1954). This property refers to the ability of populations to self-regulate their own genetic composition in order to maintain the balance between the maximization of average fitness and the maintenance of enough genetic diversity to

---

[34] On Schmalhausen and cybernetics, cf. Schmalhausen 1960; Levit et al. 2006; Vorob'eva 2010. On the reasons of Dobzhansky's preference for Schmalhausen over Waddington, cf. Gilbert 1994. On Mayr and Waddington see also Hagen 2021: 156–157.

[35] The attempt to shrug off the early phase of mathematical population genetics by lumping together Fisher, Haldane and Wright, elicited the legitimate grievances especially of this latter, see Wright 1960; Haldane 1964. Cf. De Winter 1997 for an overview. However, the direct target was rather Hermann Muller, cf. Lewontin 1987.

adapt to changing environments and/or explore suboptimal niches. In the common condition of ever-changing environments, of course, the balance between "adaptedness" and "adaptability" is never really achieved. Evolutionary change is always underway, but it is no longer a matter of only allelic substitution: it "will alter the whole genetic system and promote a repatterning of the gene pool" (Dobzhansky 1955: 11, 3–4).

Schmalhausen's stabilizing selection, driving the production of "new form of ontogenesis which are more protected against the disturbing influence of the environment" (Schmalhausen 1949: VI-VII), provided for Dobzhansky the "missing link" (Dobzhansky 1949: IX) between embryology and Modern Synthesis. In 1955, Dobzhansky incorporated the argument of Lerner (1954) of a more direct causal connection between "genetic" and "developmental homeostasis," that is Waddington's canalization, as both would be due to the prevalence of heterozygosis. Stabilizing selection would therefore promote through the same mechanism both the genetic variability needed for further evolution and "canalized" phenotypes capable of buffering genetic and environmental changes. The purported continuity between genetic and developmental homeostasis—although later partially revised by Lerner (Wolff and Markow 2003)—helped extend seamlessly the property of homeostasis from gene pools to genotypes to phenotypes (Plutynski 2008; cf. Newman 2003: 171).

In an otherwise appreciative review of Lerner, Waddington remained unconvinced about the role of overdominance as the cause of developmental homeostasis (Waddington 1955). Furthermore, in 1957 he openly contested the conflation of homeostasis at the individual and population level, arguing that the evolution of canalization might be rather antagonistic to genetic homeostasis (Waddington 1957: 42; cf. Debat and David 2001; Hall 2005). Nonetheless, he shared and helped promoting the anti-atomistic turn of the "newer" population genetics, and the tenet that cohesiveness and self-regulation both in the epigenotype and the genotype are the result of a constructive—and therefore creative—action of natural selection (Waddington 1957: 189).

3.1 Mayr's Recruitment of Waddington's Epigenotype

Mayr was much less impressed than Dobzhansky by Schmalhausen's stabilizing selection. In his landmark book *Animal Species and Evolution*, this "elusive concept" just reinstates the obvious fact that extreme phenotypes, as typically arise in the subpopulations peripherical with respect to the optimum peak, will be counterselected (Mayr 1963: 282). It is rather to Waddington that he turns to resume the "great conceptual progress" achieved in the view of the phenotype no longer as a sum of "gene-controlled characters" but as a "complex interacting system, the total epigenotype" (1963: 6).

Apparently, Mayr doesn't conflate genetic and developmental homeostasis. The earlier is rather a "brake on all forces that attempt to change the contents of the gene pool," while developmental homeostasis prepares the ground for upcoming genetic revolutions. In "normal conditions"—those nearest to the optimum peak of the adaptive landscape—it is developmental homeostasis that, while preserving the "constancy of the phenotype" against genetic and environmental perturbations, allows the maintenance and accumulation of genetic variability shielded from natural selection. Nonetheless, for Mayr the working of developmental homeostasis—although "not fully understood"—"is, no doubt, one aspect of the general phenomenon of the unity of the genotype (Mayr 1963: 264, 282).

Mayr's transubstantiation of the genotype in an epigenetic system,[36] both expressions of the "closely knit functional integration of the genotype" (1963: 264), has many interesting implications.

The first is the strategy of selective appropriation of developmental explanations that is enacted at this time of the Synthesis (Amundson 2005: 194). Development is no longer black-boxed as in Simpson's *The major features* (1953: 97–98). Rather, the developmental system is recruited as an "immensely conservative force" in much the same way as genetic homeostasis. In fact, both kinds of homeostasis must be broken for evolutionary change to take place.

> Although minor gene substitutions may be frequent, the well-buffered system of developmental canalizations shields the phenotype against major changes. There is opportunity for speciation, but *a major alteration is impossible as long as the epigenotype is intact* (Mayr 1963: 614, my italics).

The conditions for such a double break are the same of the peripatric mode of speciation, or the peak shifting by genetic revolution: a small founder population, geographically isolated, in which balanced heterosis is disrupted by the genetic drift and homozygotic "soloists" are favored in the drastically changed genetic environment (1963: 533), while the intensive selection pressure by the new niche will rapidly lead to a completely new homeostatic equilibrium of coadapted "good mixers" around a new peak.

But, secondly, the recruitment of the epigenotype has major consequences on the macroevolutionary scale. This becomes apparent in the chapter "Historicity, Temporalities and Causality: A Confusion at the Heart of Debates on Darwinism," devoted to *Species and transpecific evolution*. Its overall aim is to refute all "misconceptions," short all instantiations of essentialism or archetypal thinking represented by saltationism and orthogenesis, to reinstate the fundamental tenet that transpecific evolution (Rensch's 1947 revised term for macroevolution) is but the "extrapolation and magnification" of microevolutionary processes (Mayr 1963: 586).

Orthogenesis, the first target of Mayr's attack to the "misconceptions" about macroevolution, is quickly disposed of as the result of the constraining action of the epigenotype.

---

[36]Mayr 1963: 613–614: "Since most actions of genes are interactions during development, the genotype is an epigenetic system, or, as Waddington called it, an epigenotype"; the epigenetic system, in turn, is responsible of "the 'wholeness' of the genotype."

> The "wholeness" of the genotype, owing to its epigenetic system, explains many other phenomena that are difficult to interpret in terms of beanbag genetics, for instance, "tendencies" in certain families and orders that are absent in others. So-called orthogenetic trends are due to the fact that the evolutionary changes of the phenotype owing to natural selection are limited by the possible amplitude of response of the epigenotype (1963: 613-614).

After that Waddington had disparaged Simpson's attempt to "minimize" the reality of evolutionary trends (Waddington 1957: 64), Mayr is ready to reframe the issue in Waddingtonian terms. This quite radical revision assigns a greater weight to internal factors, now conflated in genotype + epigenotype. This move brings nontrivial implications. Mayr acknowledges in fact—at least at this time—both the constraining (albeit not directional) role of the epigenotype on evolutionary change and its *heritability*. This basic step will be of importance, as I will show below, for Mayr's early approach to the notion of grade.

3.2 Mayr's Mosaic Evolution

Mosaic evolution finds its place in the refutation of the second "misconception," saltationism, here represented by three related assumptions: typostrophism, reduced to the thesis that "a new type comes suddenly into existence and evolves harmoniously from that point on" (1963: 596); the generalization of Cope's rule of the unspecialized; and Goldschmidt's thesis that "the big categories exist first, and that in time they split in the form of the genealogical tree into lower and still lower categories" (Goldschmidt, quoted in Mayr 1963: 600).

In 1958, an early acknowledgment that "most evolution is of the 'mosaic' type" mentioned de Beer's authorship (Mayr 1958). It is therefore the more significant that in 1963 Mayr introduces the notion—"implicit in the work of outstanding paleontologists for nearly 100 years" (Mayr 1963: 598)—with a reference to Heberer, patently omitting his original source, Gavin de Beer.

Mayr reiterates Heberer's arguments against typostrophism. "Every evolutionary type," which is but a local population raised to the status of a "new higher category," "is a mosaic" resulting from the "unequal evolution" of its components, some key structures rushing ahead while others stagnate. There are no "missing links (. . .) half way between the types which they connect" but only a stepwise acquisition of diagnostic characters (Mayr 1963: 596).

But besides safeguarding Heberer's gradualist tenet about the origin of types, mosaic evolution has a new function: to challenge Cope's law of the unspecialized, a remnant of archetypal thinking. This outmoded phylogenetic principle is responsible for the "vain search for" or the fanciful invention of "unspecialized archetypes" in phylogenies (Mayr 1963: 598; 1965: 169).

Mosaic evolution is evidence to the fact that "nearly all major evolutionary breakthroughs were made by specially preadapted and hence *highly specialized forms*" (Mayr 1963: 596, my italics). Its most frequent case is in fact the one in which "the origin of a new group is very closely correlated with, and even sometimes preceded by, an evolutionary invention." Unlike Simpson, who reluctantly classified the "basic adaptive complex" as a "broad adaptation," Mayr stresses its nature of "specialized preadaptation" (1963: 596).

How does the species, the veritable "evolutionary pioneer," become "founder of a new (. . .) higher category" (1963: 588)? Sewall Wright, in the Symposium that became a landmark in the difficult dialogue between paleontology and Modern Synthesis (Smocovitis 1996: 157; 2023: 46), had recognized "a large measure of truth" in Goldschmidt's descending sequence from "big categories" to species (Wright 1949: 387). But he upheld that this pattern could be explained in terms of population dynamics. He sketched out two possible ways by which a species could move from a "single" peak to a higher "system of peaks." The first is a major ecological opportunity, the entrance in an empty niche. The selected example is Lack's studies on the speciation of Darwin's finches, which accounts for a rapid adaptive radiation, but hardly for the establishment of higher categories. The second way is the emergence of a feature of "general rather than special significance," possibly by a change of function. If the adaptive success of the new feature exceeds a given threshold, the improvement can be very rapid and is limited only by the "store of variability." The elimination of competition changes the whole surface of selective values in the isolated population, and the "divergent selective pressures" of the new system of niches foster an extensive "tachytelic" adaptive radiation. Further, secondary and tertiary radiations come to establish increasingly lower categories (Wright 1949: 387).

Thus, after Simpson's emphasis on the peculiarity of higher evolution as entailing the emergence of a basic adaptive complex, and Wright's acknowledgment that the qualitative nature of the adaptation has an intrinsic impact on the possibility for a species to move from a single peak to a "new system of peaks," it is easier to understand why in 1960 Mayr felt the need to address the "neglected problem" of the emergence of evolutionary novelties and refute the older saltationist meaning of "preadaptation" (Mayr 1960: 364).[37]

It is in this respect that emerges the possibly major reason for the censorship of de Beer's mosaic evolution in 1963. In 1960, in fact, the dismissal of saltationist explanations of evolutionary novelties is accompanied by the damnation of "all sorts of fancy terms like 'pedogenesis', 'palingenesis', 'proterogenesis' and whatnot," their "stultifying effect on the analysis" being so great that "the less said about this type of literature, the better" (Mayr 1960: 351). The association with Schindewolf's "proterogenesis" completes the picture, leaving no doubt about where Mayr would have placed de Beer's paedomorphosis as well. Even in 1963,

[37] The reference is to Cuenot. The evo-devo literature has considered Mayr's seminal 1960 paper as the only genuine attempt to develop a theory of the origin of complex adaptations, the long-standing problem of Darwinism since its inception (Müller and Wagner 1991: 230–231).

where Mayr confronts—much more respectfully—the theory of the origin of chordates by neoteny supported by Romer (1942),[38] the prominent American paleontologist who participated to the Synthesis debate (Mayr 1980: 42; cf. Romer 1942; Grodwhol 2019; cf. Sect. 1.2), he proposes an utterly selectionist explanation of neoteny as due to the pressure of natural selection, exerted by the new niche, to prolong the larval stage. However, even this "adaptive" neoteny is not acknowledged as a mode of macroevolutionary change. "These authors forget that the juvenile phenotype and the adult phenotype are expressions of the same genotype" (Mayr 1963: 607). For a prolongation of the larval stage by a selective pressure to be possible, genetic and developmental homeostasis must already have loosened, and the genetic revolution already triggered. Neoteny is therefore adaptive, but it may only follow the release of a new genetic potential, rather than initiate change.

Mayr's approach to novelty is the same as Heberer's: to account for the transition from a "quantitative to a qualitative" change in a "structure or property" (Mayr 1960: 351). However, in 1960 Mayr cautiously avoids tackling the problem of evolutionary novelties in relation to the origin of higher categories, notwithstanding the fact that most of the examples he presents stem from macroevolution (Mayr 1960: 351). Not so in 1963, where evolutionary novelties are explicitly considered for their function in reaching a "new system of peaks."

The classical problem of how a new structure can "be gradually acquired when the incipient structure has no selective advantage until it has reached considerable size and complexity?" (1963: 602) is equally classically answered, in 1963, with continuity of function and/or *exaptation* of already functional structures to a new purpose, the whole process assisted by an "ever present pressure of selection" (1960: 375). As prominent evo-devo scientists have commented, such an approach did but bypass the "generative problem," largely preventing further analysis about the origin of highly complex structures (Müller and Wagner 1991: 230–231). It is significant in this respect that Mayr now omits the most sophisticated case of novelty that he had examined in 1960, ironically the one that would have better matched the case of complex structures,[39] such as Archaeopteryx's wing, with its novel asymmetrical arrangement of feathers. In 1963, Mayr is adamant that the simpler case of change in function is "the most important cause of the origin of new structures."

In the most frequent case of mosaic evolution, that of piecemeal change, the "ever present pressure" is exerted differentially on different structures. Consequently, only

[38]De Beer also had proposed a neotenic explanation of the loss of flight in Ratites (de Beer 1956).

[39]This mode—presented under the cryptic heading "A change in the structural environment"—entails the former merging of preexisting building blocks, but requires that a "certain degree of perfection" be already attained before natural selection can exploit it opportunistically (372–375). Thus, much room was left to experimentation in the combination of different features. However, the description of this mode was quite convoluted, and Mayr seems to oscillate between an opportunistic and a creative role for natural selection. He might have been aware of this difficulty, the more if this mode was applied to the production of complex structures in mosaic evolution, where even the single building blocks should have evolved 'at different times and different rates' and spontaneously merged each time in the same way.

some specially preadapted "structure or structural complexes" evolve rapidly after the entrance in a new adaptive zone (1963: 598). That change is not "steady and harmonious," or the lack of integration of the transitional form, are no longer a concern, as, in Mayr's framework, any such change is only possible if the former genetic and developmental cohesiveness is already broken, and will only be restored with reference to a whole new surface of selective values.

However, the case of specialized advanced characters running ahead because of the heightened selective pressures of a new niche doesn't cover all instances of mosaic evolution of new types. This mode matches the case of Archaeopteryx, which remains a reptile "in nearly all its characters." But it doesn't match the evolution of mammals (again) from the therapsid, which required the independent and successive acquisition of four diagnostic characters in as much as nine different lines. In this case, Mayr is aware that differential selective pressures on specific organs are no longer a sufficient explanatory device (1963: 608).

The case of mammals, a new group arising as "the result of a broad and varied adaptation," is traced back by Mayr to a "cryptic" change, initially not reflected in diversified features (Mayr 1963: 601). Such change follows from the release of a new "evolutionary potential," a new "predisposition" that must have been common to the "same group of therapsid reptiles," following which they could evolve in a mosaic fashion toward the mammalian *grade*. Thus, the advent of a new "store of variability" (Wright 1949) by a genetic revolution comes to substantiate the notion of grade.

Here, the conflation between genotype and epigenotype is of essence to identify the new shared evolutionary potential with Huxley's "felicitous term 'grade.'" The grade in fact is made equivalent to "a gene complex and epigenetic system," which together make the "common heritage" of the evolving group. The incorporation of the epigenetic system (implicitly assumed as heritable) within the grade allows to explain the "true" parallelism occurring in mosaic evolution.

However, even for this case, Mayr still holds that the group endowed with the new "evolutionary potential" must be subject to "similar selection pressure" to evolve the same advanced traits (1963: 607–609); that is exactly the circumstance that Watson had radically challenged. While Simpson—probably more sensitive to the objections from Watson the paleontologist—had acknowledged the relative independence of mosaic evolution from local selective pressures, Mayr sticks in 1963 to a selectionist explanation of evolutionary parallelism in mosaic changes, untouched by the paleontological evidence of radically changing selective pressures across time. Nonetheless, Mayr's selectionist reframing of mosaic evolution will exert a long-lasting influence over Modern Synthesis (Schaeffer 1965; Stebbins 1983).

However, when cladistics becomes in the 1970s an accomplished opponent to Modern Synthesis' evolutionary systematics, the role of natural selection in mosaic evolution is overshadowed, while the tight connection of mosaic evolution and grade become integral part of the synthetic argument against cladistics.[40] In 1974, Mayr argues that mosaic evolution is evidence to the fact that common ancestry is not an effective criterion to select similarities. In mosaic evolution, the acquisition of the same apomorphies in parallel by different lineages at different times and rates is primarily due to the attainment of a new grade. It manifests itself with a new "well-integrated adaptive complex" that precedes the entrance into a new adaptive zone. The following abundant adaptive radiation results from "bountiful speciation and various modifications in the basic adaptive theme of the grade," but no mention is made of the role of natural selection in the repeated evolution of the same traits. Furthermore, now the grade is no longer referred to "a gene complex and epigenetic system," but only to the "percentage of shared genes." Nonetheless, it is still the grade that provides a criterion—which cladistic lacks—to discriminate between parallelism and convergence (Mayr 1974: 435, 439, 443, 445, 449–451; cf. Ereshefsky 2001: 59). So, the grade becomes also for Mayr the evolutionary force that drives mosaic evolution and presides over the stepwise stabilization of the new "type."

4 Conclusion

The strange story of mosaic evolution I have tried to reconstruct can be understood as a history of selective appropriation, accomplished with the overshadowing of de Beer's original formulation by Mayr (and perpetrated by Gould).

In my view, this story doesn't contradict the thesis of a hardening of Modern Synthesis, in the general sense of a reduced tolerance to non-adaptive or pluralist explanations. Such a shift is corroborated for instance by Simpson's changed attitude about the independent rate of the evolution of characters in 1937 and 1953 (Sect. 1.3), the first hinting at differential variational properties (evolvability) in traits, the second rather focusing on the guidance of natural selection on genetically independent, but functionally related, characters. But the case of mosaic evolution highlights also some remarkable oscillations in the way single architects strived to comply with the "hardening." In their shared programmatic effort to establish a new vehicular language for evolutionary biology based on population genetics and evolutionary externalism, in fact, each architect had to cope with the older and newer challenges raised by scientists working in his own disciplinary field.

In the 1950s the synthetic movement was engaged on various fronts. The turn toward "population organicism" (Dreuil 1997) and "organismic" biology (Dobzhansky 1964), was as much a reaction against the disruptive breakthrough of molecular biology as the purported divide between proximate and ultimate causation

[40] Rieppel 2020, an excellent source for the history of mosaic evolution *avant la lettre*, has highlighted the role of mosaic evolution in cladistics, especially through Thaktajan's version of heterobathmy, neglecting however the fact that mosaic evolution was employed by Mayr and other architects precisely against cladistics.

(cf. Borges 2005; Hagen 2021: 25). But the "organicist turn" of Modern Synthesis brought again to the fore the evolutionary role of developmental mechanisms, and their possible "homeostatic" function as vindicated by Schmalhausen and Waddington in the wake of the new cybernetic imagery. To manage this shift was the more difficult as the whole synthetic movement had been primarily concerned with replacing those former "evolutionary syntheses (. . .) that forged evolution with embryology" (Gilbert 1994: 144). In many respects, de Beer was a representative of these older traditions, particularly of comparative evolutionary embryology and morphology, that were inextricably linked, ever since von Baer and Kovalevsky, with phylogeny and paleontology. In this changing context, de Beer may have become an increasingly uncanny figure, notwithstanding his support of the Modern Synthesis and Darwinian gradualism (and his prestige as editor of Darwin's works). In particular, de Beer's only "excursion" in paleontology might have been particularly unwelcomed, the more as his geneticization of heterochronic changes hinted at a possible developmental explanation of the mosaic structure of transitional forms.

More generally, de Beer's account of mosaic evolution represented a challenge to the selectionist explanation of macroevolutionary trends. It highlighted how ecological opportunity alone was insufficient to explain "identical changes all in a uniform direction, though at very different rates, in all the numerous stocks of a class" (Watson 1940: 229).

De Beer's cautious proposal was to explain the sudden and irregular emergence of advanced "novelties" by modular changes in the developmental system, small at the genetic level and "moderate" at the phenotypic one. In fact, changes restricted to single "organ-fields" would not disrupt organismal integration, which would be ensured by the functional coadaptation of parts in the later stages of development. Consequently, ontogeny would bring about "abrupt modification in phylogeny" without the precondition postulated by Mayr, of a relaxing of stabilizing selection.

The random emergence of such a modular change might give rise to an adaptation *in a broad way*, much in the modern evo-devo sense of opening up "a new character space (. . .) that potentially allows the occupation of more niches" (Galis 2001: 584). To push a bit further—with a tiny bit of wiggish modernization—the implications of de Beer's principle, the opening of a new character space may establish a new evolutionary trend and account for the parallel evolution of the same trait in related taxa. If the trend is "progressive" it will be favored by natural selection and lead to the next "shelf" (or grade), where subsequent adaptive adjustments optimize the coadaptation of parts for fitting to the new local environment.

As the modular change wouldn't affect organismal integration, it may remain invisible to selection, and even "prophetic" for an extended time, consistently with the modern notion of macroevolutionary lag. This modernized version of de Beer's argument is coherent with his tenet that "phylogeny is the result of modified ontogeny" (de Beer 1940: 98; cf. Brigandt 2006), though this tenet didn't shatter in my view de Beer's faith in natural selection as the only veritable directional force of evolution (de Beer 1958: 33; de Beer 1966).

Thus, de Beer's reassessment on new grounds of the causal role in evolution of early developmental changes rehabilitated both crucial features of the "older syntheses" between embryology and paleontology and more proximate challenges from physiological genetics, concerning not only "saltationism" but also the randomness in the emergence of new complex structures. Both became more menacing as the issue of evolutionary novelties came to the fore of the macroevolutionary question (Sect. 3.1).

Both Simpson and Mayr rejected developmental explanations for the rise of new advanced characters in mosaic evolution, although paedomorphosis continued to be evoked even by paleontologists of Simpson's circle (cf. Hecht 1965). Both Simpson and Mayr, confronted with mosaic evolution, attempted to understand preadaptation (the advanced diagnostic features) as a local specialization liable to lead a population toward a whole new "system of peaks." However, this view was difficult to reconcile with the fact that the same advanced, and inherently complex, characters eventually evolved in parallel in different lineages in quite different selective environments.

The episode of his confrontation with Watson shows how Simpson faced a greater pressure from the paleontological community, a pressure that, I think, de Beer artfully exploited in the 1954 paper. In my view, Simpson's 1959 paper is evidence to the fact that he honestly, albeit reluctantly, came to recognize the impossibility of a rigorous selectionist explanation of macroevolutionary trends, and even to reinstate some asymmetry between micro- and macroevolution (Sect. 2.3). Simpson's unease with the notion of mosaic evolution may be seen as a further example of the well-known difficulty the Modern Synthesis encountered in incorporating paleontology (Sepkoski 2019). As Forey commented: "From a paleontological perspective, it is remarkable that this synthesis happened at all. Neo-Darwinism emphasized the environmental selection of small nondirectional mutations. Paleontology had been demonstrating directional change with occasional abrupt appearance of species" (Forey 2005: 166).

On the contrary, Mayr reacted to the developmentalist challenge by swiftly appropriating selected aspects of Waddington's epigenotype. Artfully blurring, in 1963, the genetic and the epigenetic systems (while purging genetic assimilation[41]), Mayr could embrace development as a largely conservative force that warrants the constancy of the phenotype, and a useful conceptual device for dismantling orthogenetic explanations of trends. Unlike de Beer's, Mayr's epigenotype would break up, given the appropriate population conditions, but in no way "modularize"; it is rather the action of natural selection that brings together the same adaptive mosaic changes. Overall, Mayr's omission of de Beer's argument about mosaic evolution appears the more intentional as this latter challenged precisely the conservative view of development that Mayr championed, with the unintended complicity of Waddington.

However, Mayr's selectionist explanation of mosaic evolution of 1963 will not hold in his later confrontation with cladistic. The role of the attained grade will be more prominent, as a force driving adaptive radiation, while natural selection will be

[41] De Beer on the contrary was positively impressed by Waddington's genetic assimilation (1958: 20).

reduced to a mechanism acting post-hoc to ensure the gradual optimization of the "basic adaptive theme" to different environments. So, both Simpson and Mayr came to endorse, with *grades*, a non-selectionist notion of adaptation at the macroevolutionary level, tainting the picture of the hardening of the Modern Synthesis as straightforward "panselectionism" (Gould 1983).

More generally, the strange story of mosaic evolution offers a glimpse at Modern Synthesis in its making, and at the different compromises that the architects made in time to cope with the specific challenges from their own disciplinary community. It corroborates the view of Modern Synthesis as a *process* (Serrelli 2015: 115), rather than as an axiomatic system that stands the proof of time, and of its hardening as the phase in which its strategies assumed an increasingly defensive slant.

References

Adams MB (1980) Severtsov and Schmalhausen: Russian morphology and evolutionary synthesis. In: Mayr E, Provine WB (eds) The evolutionary synthesis: perspectives on the unification of biology. Harvard University Press, Cambridge, MA, pp 193–122
Adams MB (2021) Little evolution, BIG evolution: rethinking the history of Darwinism, population genetics, and the "synthesis". In: Delisle RG (ed) Natural selection. Revisiting its explanatory role in evolutionary biology, Springer, Chaim, pp 195–230
Agassiz L (1850) On the difference between progressive, embryonic and prophetic types in the succession of organized beings. Proc Am Assoc Adv Sci 2:432–438
Amrein M (2010) Der Vogel aus dem Reptilienei. Eine Untersuchung von Otto H. dewolfs Evolutionstheorie. Grasshof, Bern
Amundson R (2005) The changing role of the embryo in evolutionary thought. Roots of evo-devo. Cambridge University Press, Cambridge
Barrington EJW (1973) Gavin Rylands de Beer 1899-1972. Biogr Mem Fellows R Soc 19:65–93
Beatty J (2019) The Creativity of Natural Selection? Part II: The Synthesis and Since. J Hist Biol 52:705–731
Bernardi G (1993) The vertebrate genome: isochores and evolution. Mol Biol Evol 10:186–204
Bock WJ (1965) The role of adaptive mechanism in the origin of higher levels of organization. Syst Biol 14:272–287
Bock WJ (2000) Explanatory history of the origin of feathers. Amer Zool 40:478–485
Borges RM (2005) Ernst Mayr and evolutionary biology: polemics and synthesis. Curr Sci 89:947–954
Bowler PJ (1996) Life's splendid drama. The University of Chicago Press, Chicago
Bowler PJ (2007) Fins and limbs and fins into limbs: the historical context, 1840–1940. In: Hall BK. Fins into limbs. The University of Chicago Press, Chicago and London, pp 7–14
Brigandt I (2006) Homology and heterochrony: the evolutionary embryologist Gavin Rylands de Beer (1899–1972). J Exp Zool B 306:317–328
Brough J (1936) On the evolution of bony fishes during the Triassic period. Biol Rev 11:385–405
Caianiello S (2011) Les modules de la variation. L'évo-dévo ou la nouvelle genèse des formes. Critique 67:130–142
Callebaut W, Rasskin-Gutman (2005) Modularity. The MIT Press, Cambridge, Ma/London
Carroll SB (2008) Evo-devo and an expanding evolutionary synthesis: a genetic theory of morphological evolution. Perspective 134:25–36
Chambers P (2002) Bones of contention: the fossil that shook science. John Murray, London
Chatterjee S (2015) 225 million years evolution. The rise of birds, 2nd edn. John Hopkins University Press, Baltimore

Churchill FB (1992) The elements of experimental embryology: a synthesis for animal development. In: Waters CK, Van Helden A (eds) Julian Huxley. Rice University Press, Houston, Biologist and statesman of science, pp 107–126
Dobzhansky T (1949). Foreword. In Schmalhausen II (1949) Factors of evolution: the theory of stabilizing selection. Blackiston, Philadelphia, pp IX–XI
De Beer G (1954b) Archaeopteryx lithographica: a study based upon the British museum specimen. In: British museum (natural history), London
De Beer G (1966) Genetics: the centre of science. P Roy Soc B-Biol Sci 164:154–166
De Beer GR (1928) Vertebrate zoology. Macmillan, New York
De Beer GR (1937) The development of the vertebrate skull. Clarendon Press, Oxford
De Beer GR (1938) Embryology and evolution. In: de Beer GR (ed) Evolution. Essays on aspects of evolutionary biology presented to prof. E.S. Goodrich. Clarendon Press, Oxford, pp 57–78
De Beer GR (1940) Embryos and ancestors. Clarendon Press, Oxford
De Beer GR (1954a) Archaeopteryx and evolution. Adv Sci 11:160–170
De Beer GR (1954c) A modern view of evolution (BR of Simpson 1953). Nature 4415:1107–1109
De Beer GR (1956) The evolution of ratites. Bull Br Mus (Natural History) 4:59–70
De Beer GR (1958) Embryos and ancestors, 3rd edn. Clarendon Press, Oxford
De Beer GR (1969) Streams of culture. Lippincott, Philadelphia
De Beer GR, Swinton WE (1958) Prophetic fossils. In: de Beer GR, Swinton WE (eds) Studies on fossil vertebrates, presented to DMS Watson. The Athlone Press, London
De Winter W (1997) The beanbag genetics controversy: towards a synthesis of opposing views of natural selection. Biol Philos 12:140–184
Debat V, David P (2001) Mapping phenotypes: canalization, plasticity and developmental stability. Trends Ecol 16:555–561
Delisle RG (2017) Introduction: Darwinism or a kaleidoscope of research programs and ideas. In: Delisle RG (ed) The Darwinian tradition in context. Springer, Chaim, pp 1–11
Depew DJ (2011) Adaptation as a process: the future of Darwnism and the legacy of Theodosius Dobzhansky. Stud Hist Phil Biol Biomed Sci 42:89–98
Dobzhansky T (1955) A review of some fundamental concepts and problems of population genetics. In concept and problems of population genetics. Cold Spring Harbour Symp Quant Biol 20:1–15
Dobzhansky T (1964) Biology, molecular and organismic. Am Zool 4:443–452
Dreuil D (1997) Organicisme populationnel. Le modèle de l'organisme dans le neodarwinisme contemporain. In: Tort P (ed) Pour Darwin. PUF, Paris, pp 327–342
Eldredge N (2001) The nature and origin of supraspecific taxa revisited–with special reference to Trilobita. In: Adrain JM, Edgecombe GD, Lieberman BS (eds) Fossils, phylogeny and form: An analytical approach. Springer, Chaim, pp 341–375
Eldredge N, Cracraft J (1980) Phylogenetic patterns and evolutionary process. Columbia University Press, New York
Ereshefsky M (2001) The poverty of Linnean hierarchy: a philosophical study of biological taxonomy. Cambridge University Press, Cambridge/New York
Erwin DH (2015) Novelty and innovation in the history of life. Curr 25:R930–R940
Erwin DH (2021) A conceptual framework of evolutionary novelty and innovation. Biol Rev 96:1–15
Feduccia A (2012) Riddle of the feathered dragons. Yale University Press, New Haven/London
Felice RN, Goswami A (2018) Developmental origins of mosaic evolution in the avian cranium. PNAS 115:555–560
Forey PL (2005) Systematics and paleontology. In: Williams DM, Forey PL (eds) Milestones in systematics. CRC Press, Boca Raton, pp 149–180
Friedman M (2022) The macroevolutionary history of bony fishes: a paleontological view. Annu Rev Ecol Evol Syst 53:353–377
Futuyma DJ (2005) Evolution. Sinauer Associates, Sunderland, MA

Futuyma DJ (2015) Can modern evolutionary theory explain macroevolution? In: Serrelli E, Gontier N (eds) Macroevolution. Explanation, interpretation, and evidence. Springer, Cham, pp 29–86
Galis F (2001) Key innovations and radiations. In: Wagner GP (ed) The character concept in evolutionary biology. Academic Press, San Diego, pp 581–605
Gayon J, Huneman P (2019) The modern synthesis: theoretical or institutional event? J Hist Biol 52: 519–535
Gilbert SF (1994) Dobzhansky, Waddington, Schmalhausen: embryology and the modern synthesis. In: Adams MB (ed) The evolution of Theodosius Dobzhansky. Princeton University Press, Princeton, pp 143–154
Goldschmidt R (1940) The material basis of evolution. Yale University Press, New Haven/London
Gould SJ (1970) Seeking confluence (BR of de Beer 1969). Science 168:717
Gould SJ, Eldredge N (1972) Punctuated Equilibria: An Alternative to Phyletic Gradualism. In: Schopf TGM. Models in Paleontology, New York, pp 82–115
Gould SJ (1976) Grades and clades revisited. In: Jerison HJ, Masterton RB, Hodos W (eds) Evolution, brain, and behavior. Taylor & Francis, Milton Park, pp 115–122
Gould SJ (1977) Ontogeny and phylogeny. Harvard University Press, Cambridge, MA
Gould SJ (1982) Introduction to Simpson, genetics and the origin of species, 3rd edn. Columbia University Press, New York, pp XVII–XXXIX
Gould SJ (1983) The hardening of the modern synthesis. In: Grene M (ed) Dimensions of Darwinism. Cambridge University Press, Cambridge, pp 71–96
Gould SJ (1995) Tempo and mode in the macroevolutionary reconstruction of Darwinism. In: Fitch WM, Ayala FJ (eds) Tempo and mode in evolution. Genetics and paleontology 50 years after Simpson, pp 125–144
Gould SJ (2002) The structure of evolutionary theory. Harvard University Press, Cambridge
Gregory WK (1910) The orders of mammals. B Am Mus Nat Hist 28:3–524
Gregory WK (1936) Habitus factors in the skeleton of fossil and recent mammals. Proc Am Philos Soc 36:429–444
Gregory WK (1947) The monotremes and the palimpsest theory. B Am Mus Nat Hist 88:1–52
Grodwhol J-B (2019) Animal behavior, population biology and the modern synthesis (1955-1985). J Hist Biol 52:597–633
Gross WR (1956) Über die 'Watsonsche Regel'. Paläont Z 30:30–40
Hagen JB (2021) Life out of balance. Homeostasis and adaptation in a Darwinian world. The University of Alabama Press, Tuscaloosa
Haldane JBS (1964) A defense of beanbag genetics. Perspect Biol Med 7:343–359
Hall BK (2000) Balfour, Garstang and de Beer: the first century of evolutionary embryology. Amer Zool 40:718–728
Hall BK (2005) Fifty years later: I. Michael Lerner's genetic homeostasis (1954)—a valiant attempt to integrate genes, organisms and environment. J Exp Zool B 394:187–197
Heberer G (1957) Theorie der additiven Typogenese. In: Heberer G, Dingler H (eds) Die Evolution der Organismen; Ergebnisse und Probleme der Abstammungslehre, 2nd edn. Fisher, Jena, pp 857–914
Hecht MK (1965) The role of natural selection and evolutionary rates in the origin of higher levels of organization. Sys Zool 14:301–317
Horder TJ (2001) The organizer concept and modern embryology: Anglo-American perspectives. Int J Dev Biol 45:97–132
Horder TJ (2006) Gavin Rylands de Beer: how embryology foreshadowed the dilemmas of the genome. Nature 892:892–898
Hull DL (1988) Science as a process. University of Chicago Press, Chicago
Huxley J, de Beer G (1934) The elements of experimental embryology. Cambridge University Press, Cambridge
Huxley JS (1936) Natural selection and evolutionary progress. Nature 138:603–605
Huxley JS (1942) Evolution. The modern modern synthesis, 3rd edn. Allen & Unwin, London

Huxley JS (1957) The three types of evolutionary process. Nature 180:454–455
Huxley JS (ed) (1940) The new systematic. Clarendon Press, Oxford
Huxley JS, Hardy AC, Ford EB (1954) Evolution as a process. Allen & Unwin, London
Huxley TH (1877) The three hypotheses respecting the history of nature. In collected essays IV; downloaded from the web at http://aleph0.clarku.edu/huxley/CE4/LecEvol.html
Jablonski D (2017) Approaches to macroevolution: 1. General concepts and origin of variation. Evol Biol 44:427–450
Jablonski D, Bottjer DJ (1990) The origin and diversification of major groups: environmental patterns and macroevolutionary lags. Clarendon Press, Oxford
Juncker T, Hossfeld U (2002) The architects of the evolutionary synthesis in national socialist Germany: science and politics. Biol Philos 17:223–249
Kemp TS (2016) The origin of higher taxa. Oxford University Press/The University of Chicago Press, Chicago/London
Kivell T et al (2011) Australopithecus Sediba hand demonstrates mosaic evolution of locomotor and manipulative abilities. Science 333:1411
Lerner IM (1954) Genetic homeostasis. Wiley, New York
Levit GS, Hossefeld U, Olsson L (2006) From the "modern synthesis" to cybernetics: Ivan Ivanovich Schmalhausen (1884–1963) and his research program for a synthesis of evolutionary and developmental biology. J Exp Zool B 306:89–106
Lewontin RC (1987) Polymorphism and heterosis: old wine in new bottles and vice versa. J Hist Biol 20:337–349
MacFadden BJ, Higgins P (2004) Ancient ecology of 15-million-year-old browsing mammals within C3 plant communities from Panama. Oecologia 140:169–182
Mayr E (1958) The sequence of the songbird families. Condor 60:194–195
Mayr E (1959) Where are we? Cold Spring Harb Symp Quant Biol 24:1–14
Mayr E (1960) The emergence of evolutionary novelties. In: Tax S (ed) Evolution after Darwin: the University of Chicago centennial. The University of Chicago Press, Chicago, pp 349–379
Mayr E (1963) Animal species and evolution. The Belknap Press
Mayr E (1974) Cladistic analysis or cladistic classification? In: Mayr E (ed) Evolution and the diversity of life, vol 1976. The Belknap Press, Cambridge, Massachussets/London, pp 433–476
Mayr E (1980) Prologue. In: Mayr E, Provine WB (eds) The evolutionary synthesis. Harvard University Press, Cambridge/London, pp 1–48
Mayr E (1986) The death of Darwin? In RSyn 4:229–236
Mayr R (1982) The growth of biological thought. Belknap Press, Cambridge, Massachusetts
Miller AH, Stroud JT, Losos JB (2022) The ecology and evolution of key innovations. Trends Ecol 38:122–131
Müller GB, Wagner GP (1991) Novelty in evolution: restructuring the concept. Annu Rev Ecol Evol Syst 22:229–256
Needham J (1933) On the dissociability of the fundamental processes of ontogenesis. Biol Rev 8: 180–223
Newman SA (2003). Hierarchy. In Hall BK, Olson WM (eds), Keywords and Concepts in Evolutionary Developmental Biology, pp 169–174
Nuño de la Rosa L, Müller GB (2021) A reference guide to evo-devo. In: Müller GB (ed) Nuño de la Rosa L. Evolutionary developmental biology. A reference guide. Springer, Cham, pp 3–12
Ochoa C (2021) Inertia, trend, and momentum reconsidered: G.G. Simpson—an Orthogeneticist? In: Delisle R (ed) Natural selection. Revisiting its explanatory role in evolutionary biology. Springer, Chaim, pp 261–190
Olby R (1992) Huxley's place in XX century biology. In: Waters CK, Van Helden A (eds) Julian Huxley. Rice University Press, Houston, Biologist and statesman of science, pp 53–78
Olson E (1965) Summary and comment. Syst Zool 14:337–342
Parravicini A, Pievani T (2019) Mosaic evolution in hominin phylogeny: meanings, implications, and explanations. J Anthropol Sci 97:1–24
Plutynski A (2008) The rise and fall of adaptive landscape? Biol Philos 2008(23):605–623

Polly PD (1998) Cope's rule. Science 282:50–51
Prothero DR (2007) Evolution: what the fossils say and why it matters. Columbia University Press, New York/Chichester
Provine WB (1980) Epilogue. In: Mayr E, Provine WB. The evolutionary synthesis. Harvard University Press, Cambridge/London, pp 399–411
Purugganan MD (1998) The molecular evolution of development. BioEssays 20:700–711
Raff RA, Wray JA (1989) Heterochrony: Developmental mechanisms and evolutionary results. J Evol Biol 2:409–434
Raff RA (1996) The shape of life. The University of Chicago Press, Chicago/London
Raff RA, Love AC (2004) Kowalevsky, comparative evolutionary embryology, and the intellectual lineage of evo-devo. J Exp Zool B 302:19–34
Raia P et al (2016) Progress to extinction: increased specialisation causes the demise of animal clades. Sci Rep 6:30965
Rainger R (1991) An agenda for antiquity. The University of Alabama Press, Tuscaloosa/London
Reif W-E (1986) The search for a macroevolutionary theory in German paleontology. J Hist Biol 19:79–130
Rensch B (1947) Neuere Probleme der Abstammungslehre. Die transspezifische Evolution, 2nd ed. Enke, Stuttgart
Ridley M (1986) Embryology and classical zoology in Great Britain. In: Horder TJ et al (eds) A history of embryology. Cambridge University Press, Cambridge, pp 34–65
Rieppel O (2020) Morphology and phylogeny. J Hist Biol 53:217–230
Romer AS (1942) Cartilage an embryonic adaptation. Am Nat 76:394–404
Romer AS (1946) The Early Evolution of Fishes. Q Rev Biol 21:33–69
Schaeffer B (1956) Evolution in subholostean fishes. Evol 10:201–212
Schaeffer B (1965) The role of experimentation in the origin of higher levels of organization. Syst Zool 14:318–336
Schaeffer B, Hecht MK (1965) Introduction and historical resume. Sys Zool 14:245–248
Schindewolf OH (1950) Grundfragen der Paläontologie. Schweizerbart'sche Verlagsbuchhandlung, Stuttgart
Schindewolf OH (1957) Über Mosaikentwicklung. Neues Jahrb Geol Palaontol Ab:49–52
Schlosser G (2005) Amphibian variations: the role of modules in mosaic evolution. In: Callebaut W, Rasskin-Gutman (eds) Modularity. The MIT Press, Cambridge, Ma/London, pp. 143–180
Schlosser G, Wagner GP (eds) (2004) Modularity in development and evolution. The University of Chicago Press, Chicago—London
Schmalhausen II (1949) Factors of evolution: the theory of stabilizing selection. Blackiston, Philadelphia, pp IX–XI
Schwartz JH (2021) What's natural about natural selection? In: Delisle RG (ed) Natural selection. Revisiting its explanatory role in evolutionary biology. Springer, Chaim, pp 329–390
Sepkoski D (2019) The unfinished synthesis? Paleontology and evolutionary biology in the 20th century. J Hist Biol 52:687–703
Serrelli E, Gontier N (2015) Macroevolutionary issues and approaches in evolutionary biology. In: Serrelli E, Gontier N (eds) Macroevolution. Explanation, interpretation, and evidence. Springer, Cham, pp 1–28
Shanahan T (2004) The evolution of Darwinism. Selection, adaptation, and progress in evolutionary biology. Cambridge University Press, Cambridge/New York
Shubin N (1998) Evolutionary cut and paste. Nature 394:12–13
Simpson GG (1937) Supra-specific variation in nature and in classification from the view- point of paleontology. Am Nat 71:236–267
Simpson GG (1944) Tempo and mode of evolution. Columbia University Press, New York
Simpson GG (1953) The major features of evolution. Columbia University Press, New York
Simpson GG (1958) BR of G Heberer, Die Evolution der Organismen. Q Rev Biol 33:148–149

Simpson GG (1959) The nature and origin of supraspecific taxa. Cold Spring Harb Symp Quant Biol 24:255–271
Simpson GG (1961) Principles of animal taxonomy. Columbia University Press, New York
Simpson GG (1971) William king Gregory (1876-1970). Am J Phys Antrhop 35:155–174
Simpson GG (1983) Fossils and the history of life. Freeman & Co, New York/San Francisco (German edition: Fossilien. Mosaiksteine zur Geschichte des Lebens. Spektrum der Wissenschaft, Heidelberg 1983)
Simpson GG (1984) Introduction: forty years later. In: Simpson GG (ed) Tempo and mode in evolution (facsimile of 1944 edition). Columbia University Press, New York, pp xiii–xxx
Smocovitis VB (1996) Unifying biology: the evolutionary synthesis and evolutionary biology. Princeton University Press, Princeton
Smocovitis VB (2023) Every evolutionist their own historian: the importance of history, context, and the extended evolutionary synthesis. In: Dickins TE, Dickins BJA (eds) Evolutionary biology: contemporary and historical reflections upon core theory. Springer, Cham, pp 25–52
Spinner NB, Conlin LK (2014) Mosaicism and clinical genetics. Am J Med Genet Part C Semin Med Genet 9999:1–9
Stebbins GL (1951) Natural selection and the differentiation of angiosperm families. Evolution 5: 299–324
Stebbins GL (1983) Mosaic evolution: an integrating principle for the modern synthesis. Experientia 39:823–834
Stern C (1956) The genetic control of developmental competence and morphogenetic tissue interactions in genetic mosaics. Roux Arch Dev Biol 149:1–25
Sturtevant AH (1932) The use of mosaic in the study of the developmental effects of genes. In: Jones DF (ed) Proceedings of the VI international congress of genetics. Brooklyn Botanic Garden, Menasha, Wis, pp 304–306
Swetlizt M (1995) Julian Huxley and the end of evolution. J Hist Biol 28:181–217
Tassy P, Fisher MS (2021) "Cladus" and clade: a taxonomic odyssey. Theory Biosci 140:77–85
Vaux F, Trewick SA, Morgan-Richards M (2016) Lineages, splits and divergence challenge whether the terms anagenesis and cladogenesis are necessary. Biol J Linn 117:165–176
Vorob'eva EI (2010) Evo-devo and the I.I. Schmalhausen concept of the evolution of ontogeny. Biol Bull 37:106–1131
Waddington CH (1941) Evolution of developmental systems. Nature 3717:108–110
Waddington CH (1953) Epigenetics and evolution. Symp Soc Exp Biol 7:90–94
Waddington CH (1955) The resistance to evolutionary change (BR Lerner 1954). Nature 4445:51–522
Waddington CH (1957) The strategy of genes. Allen & Unwin, London
Watson DMS (1919) On Seymouria, the most primitive known reptile. Proc Zool Society Lond:267–301
Watson DMS (1920) The structure, evolution and origin of the amphibia. The 'orders' Rachitomi and Stereospondyli. Philos Trans R Soc Lond 209:1–73
Watson DMS (1936) A discussion on the present state of the theory of natural selection. In: Watson DMS et al (eds) A discussion on the present state of the theory of natural selection, vol 121. Proc Royal Soc B, pp 43–73
Watson DMS (1940) The origin of frogs. Trans Roy Soc Edin 60:195–231
Watson DMS (1948) The mechanism of evolution (the Hooker lecture). Proc Linn Soc Lond 160: 75–84
Watson DMS (1951) Palaeontology and modern biology. Yale University Press, New Haven
Watson DMS (1955) Book review of G de Beer, Archaeopteryx lithographica. Nature 4462:785–786
Wedekind R (1918) 1918. Beiträge zur progressiven Entwicklung der Organisation der Organismen. Ges Beförderung d ges Naturwiss Marburg, Sitzungsberichte 2:25–26

Wellnhofer P (2010) A short history of research on Archaeopteryx and its relationship with dinosaurs. In: Moody RTJ et al (eds) Dinosaurs and other extinct saurian: a historical perspective. The Geological Society, London, pp 237–250
Willner W, Hülbner K, Fischer MA (2014) Return of grades: towards objectivity in evolutionary classification. Preslia 85:233–243
Witmer LM (2002) The debate on avian ancestry. In: Chiappe LM, Witmer LM (eds) Mesozoic birds. University of California Press, Berkeley
Woolf CM, Markow TA (2003) Genetic models for developmental homeostasis: historical perspectives. In: Polak M (ed) Developmental instability: causes and consequences. Oxford University Press, Oxford, pp 99–114
Wray GA (1992) The evolution of larval morphology during the post-Paleozoic radiation of echinoids. Paleobiology18: 258-287
Wright S (1932) The roles of mutation, inbreeding, crossbreeding and selection in populations. Proceedings of the XI International Congress of Genetics 8:209–222
Wright S (1949) Adaptation and selection. In: Jepsen GL, Mayr E, Simpson GG (eds) Genetics, paleontology, and evolution. Princeton University Press, Princeton, pp 365–389
Wright S (1960) Genetics and twentieth century Darwinism. A review and discussion. Am J Hum Gen 12:365–372
Zou Y (2021) Gavin de Beer (1899–1972). In: Muller GB (ed) Nuno de la Rosa L. Evolutionary developmental biology. A reference guide. Springer, Cham pp, pp 289–298
Zuckerkandl E (1994) Molecular pathways to parallel evolution: I. Gene nexuses and their morphological correlates. J Mol Evol 39:661–678

Deconstructing the Extended Evolutionary Synthesis: Do We Need a New Theory of Evolution?

Carlos Ochoa

Abstract

This chapter deconstructs the framework of the so-called Extended Evolutionary Synthesis—in the sense of separating and analyzing its fundamental assumptions—to highlight its shortcomings. In so doing, I promote new representations of evolution not characterized by "extending" or "expanding" the Modern Synthesis framework. To emphasize the shortcomings of the Extended Evolutionary Synthesis, I propose a hypothetical new theory of evolution—which I call the "Unmodern Synthesis"—whose structure is incompatible with the "core" assumptions of the Modern Synthesis. Historically, this model builds upon anti-Darwinian models and is based on the epistemological data of contemporary Evo-Devo. By considering the Unmodern Synthesis, we come to realize that integrating some Evo-Devo notions into the Modern Synthesis framework is not possible. This discrepancy suggests a complete replacement of such a model rather than adding to its content. Likewise, this chapter shows that the Extended Evolutionary Synthesis does not imply a significant change to the theory of evolution, since it is not much different from the Modern Synthesis in its most fundamental parts: it supports natural selection as a creative force, the gene-centered perspective (through the notion of "inheritance"), and the extrapolationist argument. A true challenge to fundamentally modify the theory of evolution involves rejecting these tenets, as the Unmodern Synthesis does. Overall, this work aims to encourage new interpretations beyond the framework of the Modern Synthesis, which could make a fruitful contribution not only to historical and philosophical studies of evolution but also to contemporary evolutionary theory.

C. Ochoa (✉)
Department of Evolutionary Biology, Faculty of Science, Universidad Nacional Autónoma de México, Coyoacán, Mexico City, Mexico
e-mail: carlos008a@ciencias.unam.mx

R. G. Delisle et al. (eds.), *Unity and Disunity in Evolutionary Biology*,
https://doi.org/10.1007/978-3-031-42629-2_14

Keywords

Extended Evolutionary Synthesis · Modern Synthesis · Darwinism · Unmodern Synthesis · Anti-Darwinism · New theory of evolution · H-Evo-Devo Theory · Evo-Devo · Developmental constraints · Natural selection as a creative force · Organism-centered perspective · Gene-centered perspective · Heredity · Modular thinking · Saltational evolution · Macroevolution

1 Introduction

Frequently, when biologists refer to the contemporary theory of evolution, most of them have in mind the Modern Synthesis. The term "Modern Synthesis" (MS) comes from Julian Huxley's book, *Evolution: The Modern Synthesis* (1942), and although there have been several discussions attempting to understand its meaning (Mayr 1980, Provine 1980, Smocovitis 1996; Gayon 1998; Reif et al. 2000; Delisle 2011; Dickins 2021), most agree that the MS, as a process, took place between the 1930s and 1940s and refers to the integration of different disciplines—e.g., genetics, systematics, and paleontology—to support the idea that the mechanisms of population genetics—mutation, genetic drift, natural selection—are sufficient to explain evolution at all levels—i.e., macroevolution is nothing more than accumulated microevolution (Ochoa 2017). An interesting point of this evolutionary model is that it relegated the organism and its structural properties from the evolutionary scope, giving genes the central role in the processes of evolution (Wagner 2015; Walsh 2015). As a result, the MS encountered difficulties in integrating those disciplines that were relevant for understanding macroevolutionary phenomena—e.g., morphology (Coleman 1980), embryology (Hamburger 1980), and even paleontology (Gould 1980), despite the latter discipline apparently being part of this movement (see Ochoa 2021a).

Despite these difficulties, the framework formed by the MS was ultimately accepted, and the knowledge of the theory of evolution continued to expand (Gould 2002; Kutschera and Niklas 2004). However, some scholars currently appeal to a new evolutionary model known as the "Extended Evolutionary Synthesis" (EES). This model builds upon the MS framework by incorporating topics such as inclusive inheritance, phenotypic plasticity, niche construction, and developmental biology (Pigliucci 2007, 2009; Pigliucci and Müller 2010; Laland et al. 2014, 2015; Müller 2014, 2017). In recent years, the EES model has gained considerable popularity, resulting in numerous publications and meetings aimed at either explaining, supporting, or challenging its conceptual framework (see Chiu 2022). Perhaps the most well-known publication of this movement is the book *Evolution: The Extended Synthesis*, edited by Massimo Pigliucci and Gerd B Müller (2010), which emerged from an influential meeting held at the Institute for Evolution and Cognition Research in Altenberg, Austria, in July 2008.

The main problem with this proposal is that its advocates do not provide a clear idea about the meaning of this model, e.g., they present such diverse views it becomes ambiguous to know what is modified or preserved from the previous theory (Huneman 2017; Lewens 2019; Buskell 2020; dos Reis and Araujo 2020). For example, according to its supporters, the EES does not contradict the main assumptions of the MS but rather adds new theoretical and empirical data to the original model (Pigliucci 2007, Müller 2017). However, sometimes they refer to these changes as "revolutionary," in the sense that the EES framework radically modifies the structure of the theory of evolution (e.g., Laland et al. 2014, 2015; Jablonka and Lamb 2020; Müller 2021). On the other hand, another group of scientists criticizes this stance, arguing that the phenomena suggested as inclusive—e.g., developmental bias, phenotypic plasticity, and niche construction—do not significantly alter the theory; instead, they are only considered as complementary within a traditional central argument (e.g., Wray et al. 2014; Charlesworth et al. 2017; Futuyma 2017). Pievani (2016b) categorizes the first group as "reformists" and the second as "conservatives."[1] In any case, Love (2017) comments that the idea of expanding or extending a theory means increasing its content, rather than reorganizing or replacing the previous framework. This lack of clarity is particularly evident in Gerd Müller's perspective through the inclusion of Evolutionary Developmental Biology:

> [T]he EES should not be misunderstood as an *extension of* the MS, appending merely peripheral add-ons to an established core framework. "Extended" is used in the sense of "broad," "comprehensive," or "encompassing." Although the EES incorporates standard parts of the MS, its resulting logical structure and the derived predictions are radically different. (Müller 2021, p. 1136; his emphasis)

Furthermore, in the British magazine *The Guardian*, Stephen Buranyi (2022), by recounting the discontent of the conservatives about the discourse of the reformists—particularly referring to the congress' invitation organized by the Royal Society held in London in 2015—argues that the problem of the so-called "Standard Evolutionary Theory"—i.e., the one that maintains the central assumptions of the MS—is that it continues to describe evolution as gene changes in populations:

> The theory's ideas are still deeply embedded in the field, yet no formal reckoning with its failures or schisms has occurred. To its critics, the modern synthesis occupies a position akin to a president reneging on a campaign promise—it failed to satisfy its entire coalition, but remains in office, hands on the levers of power, despite its diminished offer.

That is, although today everyone criticizes the MS, so far there is no theoretical framework to replace it. For example, while reformists advocate for expanding the MS by incorporating new processes and phenomena, this approach does not

[1] This difference in opinions can be appreciated in a divided paper published in *Nature* magazine whereby the two groups attempt to answer the question "Does evolutionary theory need a rethinking?". On the one hand, Laland et al. (2014) respond: "Yes, urgently," and on the other hand, Wray et al. (2014) answer: "No, all is well."

represent a true challenge to the MS: “They are reformers, not revolutionaries.” Interestingly, Buranyi notes that during the Q&A session at a 2017 Massimo Pigliucci conference, an audience member remarked that the conflict between reformists and conservatives appears to be more of a cultural war rather than a scientific debate. And according to another attendee: “Pigliucci basically said: ‘Sure, it’s a culture war, and we’re going to win it,’ and half the room burst out cheering.” In short, the EES contains numerous contradictions and shortcomings, necessitating a thorough examination of its theoretical foundations.

According to Love (2010), there is more than one way to represent the theory of evolution. Considering this, my strategy to deconstruct the EES framework—in the sense of separating and analyzing its fundamental assumptions to expose its inconsistencies—consists of proposing a hypothetical new theory of evolution with a structure that fundamentally clashes with the core assumptions of the MS. The construction of such an evolutionary theory—which I will call the Unmodern Synthesis (US)—is sustained by the historical record and contemporary data from Evolutionary Developmental Biology (Evo-Devo). Regarding its historical aspect, this model is built upon anti-Darwinian theories such as neo-Lamarckism, orthogenesis, and saltationism, thus diverging from the framework of traditional Darwinism. Regarding contemporary Evo-Devo, it asserts that certain key concepts in this discipline challenge the core assumptions of the MS. So, since the EES aims not to replace the MS but rather to expand its content with certain notions from Evo-Devo, the US will allow us to identify points of contradictions within the EES about such an integration. Taking this into consideration, we will see that the EES does not significantly deviate from the original MS regarding the role of natural selection (as a creative force), the conception of heredity (evolutionary entities are replicators), and the extrapolationist premise (macroevolution results from the accumulation of microevolutionary events). Unlike the EES, the US would offer a reorganization of evolutionary theory, supporting that natural selection is not creative, units of evolution are cohesive entities, and macroevolution is completely decoupled from microevolution. Hence, we would conclude that the EES does not entail any substantial changes to the evolutionary theory, even though most reformists believe so.

In Sect. 2, I will critique Pigliucci’s (2009) Venn diagram—which is used to illustrate the EES conceptual framework. Here I will use it as a model to outline the structure of the theoretical frameworks of Darwinian Theory (DT) and the MS. Consecutively, I will demonstrate their incompatibility with some key Evo-Devo concepts. In Sect. 3, we will explore the foundations of a prospective new theory of evolution that I will call the “Unmodern Synthesis” (US). This model is supported by historical and epistemological data related to Evo-Devo, which suggests a complete shift in perspective. In Sect. 4, I will compare the underlying assumptions of the MS and the EES with those of the US. There, I will show the incongruence of the structure of the EES, revealing that many of its assumptions are not so different from the MS framework.

In this regard, I do not mean that there are insurmountable obstacles in the theory of evolution. Rather, this work serves as an invitation to explore other ways of thinking about evolutionary theory that goes beyond the narrow notion commonly associated with the term "Darwinism." Mainly, I emphasize that the models that pretend to extend or expand the MS are not the only approaches through which we can frame the theory of evolution. In so doing, my proposal has the potential to promote new interpretations of evolution—not characterized by "extending" or "expanding" the MS framework—which could make a fruitful contribution not only to historical and philosophical studies of evolution but also to contemporary evolutionary theory.

2 Traditional Frameworks

One of the most well-known representations of the EES framework is Pigliucci's Venn diagram (2009, see Fig. 1a), which displays a set of inclusive ellipses that, as a whole, are supposed to represent the development of evolutionary theory. The description of the figure is as follows: "Conceptual representation of the continuous expansion of evolutionary theory in terms of ideas, phenomena studied, and fields of inquiry." At the center, we observe the least inclusive ellipse, representing Darwin's contribution to the development of two concepts: *common descent* and *natural selection*. The next level of inclusion adds other main concepts arising from the achievements of the MS—e.g., *Mendelism*, the *mathematical theory of population and statistical genetics*—that result in the unification of disciplines such as *natural history*, *genetics*, and *paleontology*. Moving on to the most inclusive ellipse, we appreciate the extension of this previous knowledge by incorporating new ideas—e.g., *epigenetic inheritance*, *phenotypic plasticity*, *multilevel selection theory*, and *niche construction*—which are established in other unified disciplines such as *Evo-Devo*, *genomics*, *ecology*, etc. This final contribution formed the so-called EES.

According to Pigliucci (2009, see also 2007), this diagram has a historical structure, beginning with Darwin's ideas that challenged the notion of the fixity of species and Paley's idea of intelligent design. Although, initially, Darwin's ideas conflicted with Lamarckism due to his adherence to the notion of blending inheritance. Other problems related to heredity were orthogenesis and saltationism, particularly the latter, which was linked to Mendelism, making it difficult to maintain the gradualist view of Darwinism. The conflict was eventually resolved through the synthesis of Mendelism with Darwinism thanks to the contribution of Fisher, Haldane, and Wright during the development of population genetics. As a result, neo-Lamarckism, orthogenesis, and saltationism were rejected. Later, Dobzhansky, Mayr, and Simpson corroborated the use of population genetics within disciplines such as experimental genetics, natural history, and paleontology, respectively. Yet, the inclusion of the concepts from the EES in this scheme is interpreted as what was missing from the MS. For Pigliucci, this progress of scientific knowledge can be explained by Thomas Kuhn's proposal of scientific revolutions. For example, under this view, the paradigm shift was only attained when Darwin postulated the theory of natural selection, leaving aside intelligent design ideas. Thus, scientific knowledge has continuously progressed, thereby no new paradigm shift has occurred since

Darwin's contribution, and the EES represents only the latest phase of this scientific revolution.

Interestingly, according to Love (2013, 2017), the idea that there is a single evolutionary theory with central concepts contributes to fostering a spatial metaphor that serves as a basis for popularizing the EES framework. However, some critics began to challenge such a model of central and unalterable concepts (Callebaut 2010; Love 2013; Fábregas-Tejada and Vergara-Silva 2018; Dickins 2021). For instance, to avoid both a schema of core concepts and a perspective of linear scientific progress, Fábregas-Tejada and Vergara-Silva (2018) opted for a graphical representation in the form of a spiral (Fig. 1b). Still, in opposition to the typical linear

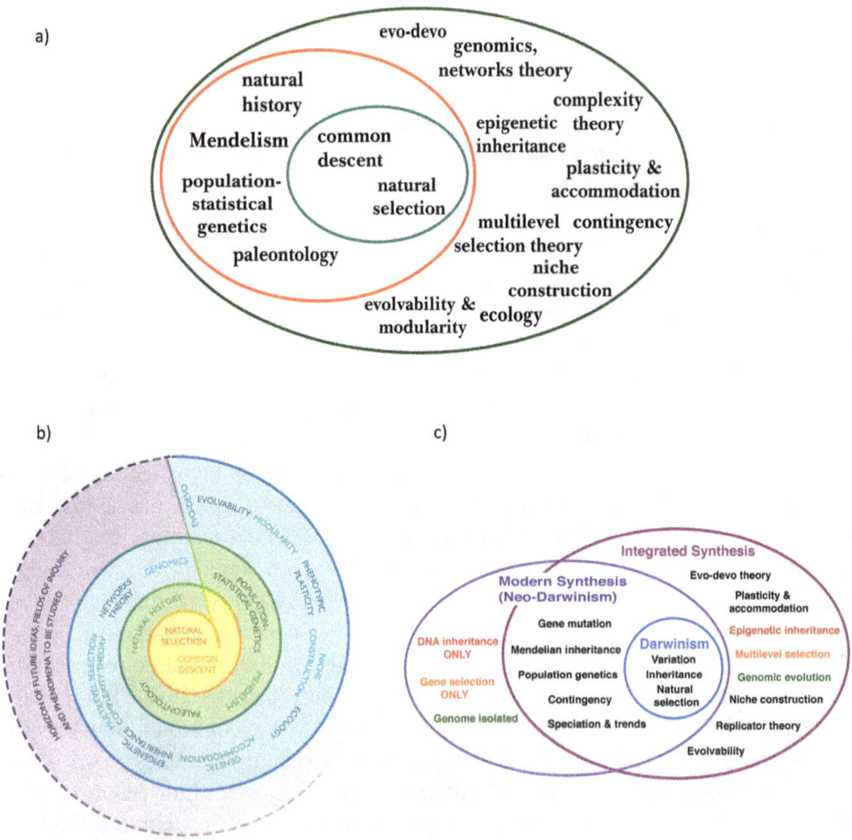

Fig. 1 Diagrams of spatial metaphors used to represent the development of evolutionary theory. (**a**) Pigliucci's (2009) Venn diagram representing the framework of the EES. See text for more details. (**b**) Fábregas-Tejada and Vergara-Silva's (2018) diagram represents the EES framework as a spiral. Unlike the previous one, this diagram avoids a linear representation of the progress of evolutionary biology. (**c**) Noble's (2015) diagram that represents the framework of the "Integrated Synthesis." Here, unlike the previous ones, the new concepts are not added to the MS. For this reason, more than an extension, this model represents a replacement of the previous theory

view of evolutionary thought, other authors argue that these new fields of inquiry contribute to a plural panorama of interpretations whereby they could be inconsistent with the assumptions of the previous conceptual framework (Love 2013, 2017, 2020; Huneman 2017; dos Reis and Araujo 2020). As Love notes: (2013 p. 328): "the spatial container metaphor [of the EES] with its amorphous interior structure might be blind to the nature of these types of challenges to standard evolutionary theory." In this regard, Noble (2015, see Fig. 1c) presents a diagram in which we observed that many of the present-day concepts, particularly those related to inheritance, do not overlap in the MS ellipse. Therefore, to Noble, more than an extension, these notions represent a replacement of the MS framework.

For the purposes of this work—and to support the previous argument—I will show the progress of evolutionary thought as envisioned by Pigliucci; but instead of assuming "a continuous line of ideas, phenomena studied, and fields of inquiry," I will indicate traditional frameworks represented in less inclusive ellipses and some key Evo-Devo concepts shown in a more inclusive ellipse (Fig. 2a). The point is that the MS comprises Darwinian Theory (DT), but the Evo-Devo ideas are so opposed to such theoretical frameworks that they should not be viewed as an extension of them. We must consider that many of the concepts developed by Evo-Devo do not come directly from the achievements of those less inclusive ellipses (i.e., from Darwinism), but rather they stem from anti-Darwinian theories (Fig. 2b). Below, I will describe each of these theoretical frameworks in order to highlight their implications for evolutionary theory.

2.1 Darwinian Theory (DT) Framework

To identify this least inclusive ellipse, it is not necessary to delve into the study of Charles Darwin's thought.[2] We will use the expression "Darwinian Theory" (DT) to refer to a scientific tradition that admits a certain range of stances where it is proposed natural selection as a creative force (Gayon 1998; Gould 2002; Beatty 2016). But what do we mean by "creative force"? Though there may be ambiguity about this notion (see Welch 2023), we can say that natural selection is creative concerning some properties of variation (Gould 2002, pp. 141–146). First, natural selection should preferably act on continuous or "small" non-extensive variations because if it acted on complex single-step adaptive traits, then, it would lose its creative role. Second, natural selection works on random variations because if variation was driven by itself in a particular direction, then natural selection would lose its creative role as well.

[2]Given that many of Darwin's assumptions were innovative in this time regarding evolution as a fact, those ideas might correspond to any evolutionary theory, e.g., see the assumptions of the "Darwinian Core" according to Vidya et al. 2023. Besides that, since there are many interpretations of Darwin's thought, some historians recommend revisiting his most fundamental ideas (see Delisle 2019, 2021).

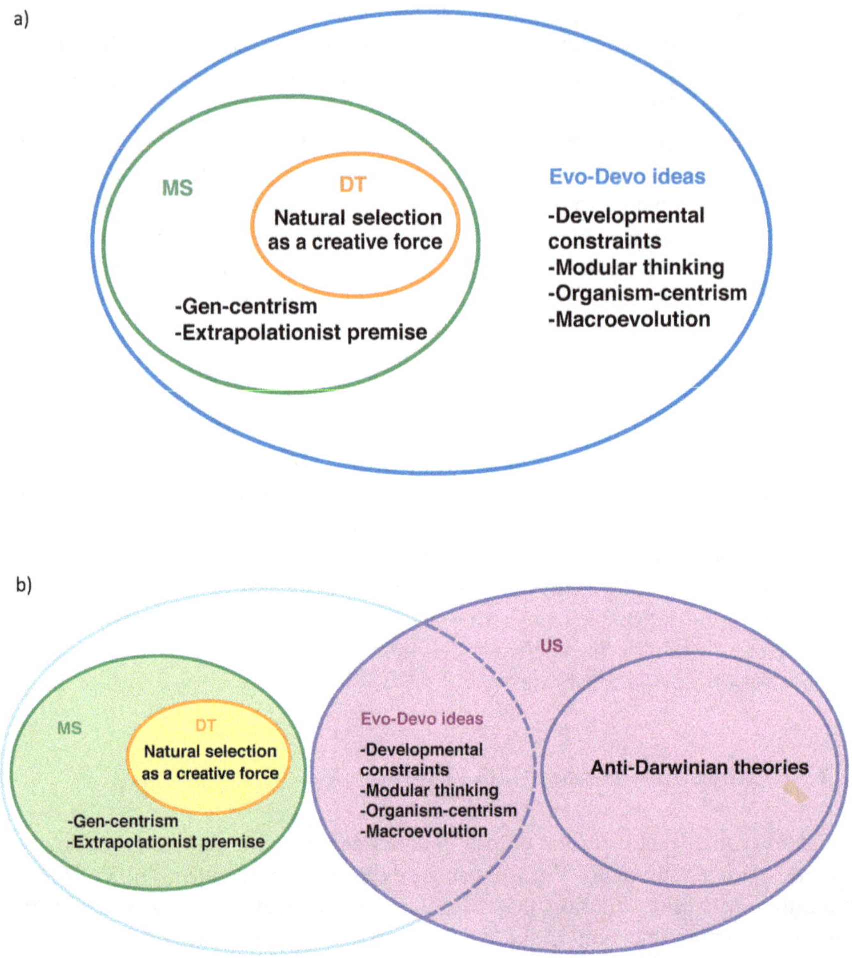

Fig. 2 Representation of the progress of evolutionary knowledge following the spatial metaphor. (**a**) Venn diagram that emulates Pigliucci's (seen in Fig. 1a). However, here the less inclusive ellipses show the DT and the MS frameworks, while the more inclusive ellipse shows some key Evo-Devo concepts. This image illustrates that part of the knowledge of Evo-Devo is opposed to the fundamental principles of DT and the MS since their roots lie in anti-Darwinian theories (**b**). The colored ellipses show that the MS and the US are different theoretical frameworks that do not overlap

Now, these criteria support the theoretical framework of a view that we can call "Darwinism," which implies the following: (1) *Gradualism*. The origin of the units of evolution (e.g., species) occurs through intermediate steps between one form and another. Since natural selection works by accumulating slight variations, individuals within a population would slowly adapt to different environments. Over millions of years, this process would account for the origin of taxonomically recognized structures (evolutionary uniformitarianism). (2) *Random variation*. Individuals

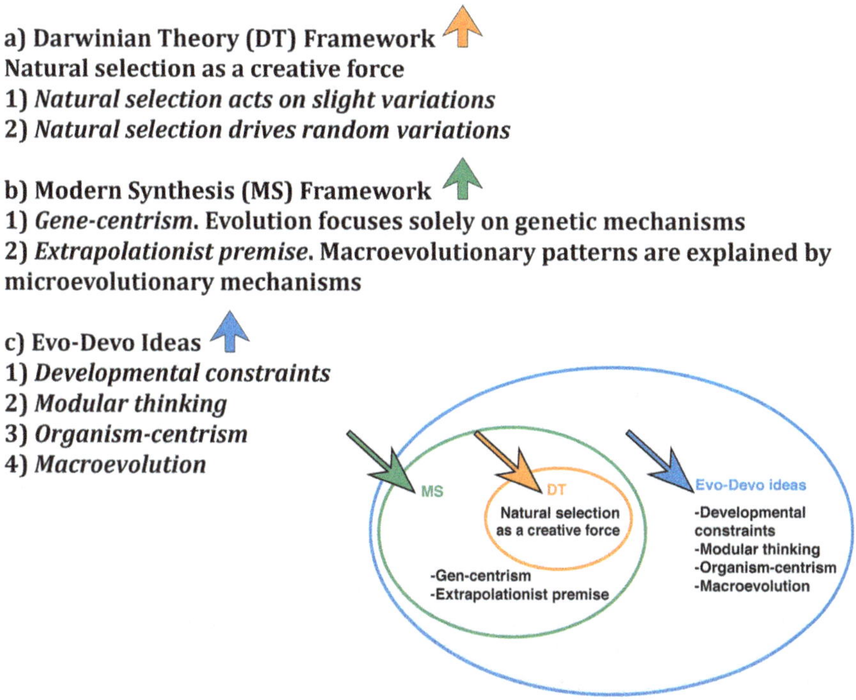

Fig. 3 Description of the Venn diagram shown in Fig. 2a, indicating: (**a**) DT framework. (**b**) MS framework. (**c**) Evo-Devo ideas

vary in multiple directions. Natural selection controls such variations, leading populations toward the best adaptive pathway. In a nutshell, the DT framework, which involves the belief in natural selection as a creative force, is built from two primary tenets: (1) *Natural selection acts on slight variations*. (2) *Natural selection drives random variations* (Fig. 3a).

2.2 Modern Synthesis (MS) Framework

While some scholars correctly point out that the MS framework lacks core elements (Callebaut 2010; Love 2013; Fábregas-Tejada and Vergara-Silva 2018)—e.g., the founders differed in many aspects making it difficult to establish a common framework among all of them (see Gayon 1998; Delisle 2011; Ochoa 2021a; Smocovitis 2023)—nevertheless, the architects of population genetics shared the common goal of constructing a statistical model to quantify genetic variation within a population in a limited number of mechanisms (natural selection, migration, genetic drift, etc.) (Gayon 1998). In accordance with this agenda, some architects—such as Dobzhansky, Mayr, and Simpson—interpreted the origin and maintenance of variation in terms of genetic changes. When they endorsed continuous variation, they argued that genetic changes might only transcend in populations when they had continuous effects (Ochoa 2017). And if the phenotypic changes were continuous—even observable in the laboratory—there was no other way but to rely on those microevolutionary mechanisms to explain the patterns of macroevolution (Adams 2021). In this regard, we can say that if we want to build a theoretical framework that describes the MS, it has at least two fundamental tenets: (1) *Gene-centrism*. Evolution focuses solely on genetic mechanisms. (2) *Extrapolationist premise*. Macroevolutionary patterns are explained by microevolutionary mechanisms[3] (Fig. 3b).

2.3 Evo-Devo Ideas, Instead of the Extended Evolutionary Synthesis (EES) Framework

The EES framework is traditionally structured from what was missing in the MS. Early drafts of the EES included at least three core research topics that were intended to supplement—or significantly modify—the MS framework: (1) niche construction (Odling-Smee et al. 2003), (2) inclusive inheritance (Jablonka and Lamb 2005), and (3) Evo-Devo (Müller 2007). However, in this modified diagram (Fig. 2a), there is no explicit indication of the EES framework; instead, I show a set of key Evo-Devo ideas (Fig. 3c) that the advocates of the EES might use as elements to be added; however, these concepts may conflict with the previous conceptual frameworks (Fig. 2b). I provide a brief description of these concepts below:

1. *Developmental constraints*. Given that developmental interactions can constrain the process of morphological change during evolution (Alberch 1982), this concept has two crucial implications. On the one hand, it reflects the stability of developmental systems, maintaining a specific steady state within the morphospace. On the other hand, it manifests the channeling of phenotypic variation to specific positions within the morphospace (Gould 2002 pp. 1025–1061; Ochoa forthcoming). This concept is important to our discussion because, as we will see later, it challenges the creativity of natural selection.
2. *Modular thinking*. Multicellular organisms are composed of subunits that are dynamic, discrete, and autonomous (Raff 1996; Bolker 2000; West-Eberhard

[3] Note that there are points of intersection between DT and the MS, particularly regarding continuity, randomness, and evolutionary uniformitarianism. But a crucial difference is that the interpretation of inheritance in the MS is accompanied by a theory of genetics (see Vidya et al. 2023). For example, in the introduction of one of the most influential books about the contemporary conception of the MS, *The Evolutionary Synthesis*, Mayr (1980, p. 1) asserts that general agreement of the MS rested on two fundamental conclusions: "gradual evolution can be explained in terms of small genetic changes ("mutations") and recombination, and the ordering of this genetic variation by natural selection; and the observed evolutionary phenomena, particularly macroevolutionary processes and speciation, can be explained in a manner that is consistent with the known genetic mechanisms."

2003). These subunits, or morphological modules, are defined as parts of an organism that are highly integrated in developmental terms but loosely connected to other parts of the same organism (Schlosser 2002, 2022; Kuratani 2009; Ochoa and Rasskin-Gutman 2015). Since these modules possess the property of maintaining themselves in a steady state over time (DiFrisco et al. 2020; Ochoa forthcoming), this fact challenges our notion of heredity (inheritance thinking).
3. *Organism-centrism.* A theory that sustains the "organism" as the center of study considers that the organism's development is the outcome of patterns of hierarchical interactions within the developmental process. In this schema, the phenotype is not only a result of the genes but rather it is constructed by different levels of organization—genetic regulatory networks, proteins, cells, tissues—including the organism's interaction with the environment. Accordingly, it is important to note that, in addition to natural selection, processes such as niche construction (Odling-Smee et al. 2003), phenotypic plasticity (West-Eberhard 2003), and behavioral choice (Diogo 2017) are relevant to this conception (Ochoa forthcoming). In any case, this idea challenges the gen-centered perspective of the MS.
4. *Macroevolution.* Although there are differences in the use of the concept of macroevolution (Ochoa forthcoming), it was originally defined in relation to the morphological discontinuity that occurred in lineages at the time of divergence (e.g., Goldschmidt 1940). However, for those biologists who placed their trust in gradual evolution, macroevolution meant only the accumulation of long-term microevolutionary processes (e.g., Dobzhansky 1937). Consequently, during the 1970s, though paleontologists recognized macroevolutionary patterns—such as the sudden origin of groups seen in the fossil record—they continued to argue in favor of microevolutionary mechanisms to explain these patterns (e.g., Eldredge and Gould 1972). That is, a true macroevolutionary theory entails saltational evolution, and simple recognition of macroevolutionary patterns is not enough (Løvtrop 1981). Thus, saltational evolution challenges the traditional extrapolationist premise (i.e., macroevolution is nothing more than extended microevolution).

In conclusion, although the EES attempts to incorporate some concepts from Evo-Devo, this does not necessarily mean that these are compatible with the theoretical frameworks of either DT or the MS. Therefore, the development of an alternative theory of evolution is indeed necessary.

3 Evolution: The Unmodern Synthesis

According to Love (2010), there is more than one way to represent the evolutionary theory, and while there is probably no single correct way to interpret it, some of these approaches may have greater benefits to satisfy certain research goals. Likewise, he claims that the structure of evolutionary theory may need a reorganization of its content rather than a mere reevaluation and addition of existing content. Regarding these observations, it is not unreasonable to structure a theory contrary to DT and the

MS as long as one has the appropriate elements. Interestingly, some scholars have pointed out the existence of two evolutionary theories, one orthodox based on upholding natural selection as the primary mechanism and another that focuses on the generation and limitation of variation (West-Eberhard 2003, p. 474). While one focuses on the continuous change and adaptive traits seen in microevolution, the other focuses on discontinuous modification and non-adaptive characters as relevant to explain macroevolution (Grene 1958; Ochoa 2017). While one pretends to develop a synthesis between genetics and evolution, the other calls for a synthesis between development and evolution (Gilbert et al. 1996; Amundson 2005; Esposito 2013; Ochoa 2017). Another relevant point is that these two theories stemmed from distinct styles of reasoning, which have unfolded in debates such as formalism vs. functionalism (Russell 1916; Asma 1996; Gould 2002; Amundson 2005; Ochoa and Barahona 2014), internalism vs. externalism (Alberch 1989; Ochoa 2017; Tamborini 2022), Evo-Devo vs. genetics populations (Gilbert et al. 1996; Amundson 2005; Esposito 2013; Huneman 2017; Ochoa 2017, 2021b; Love 2020), etc.

Regarding this, Welch (2017) notes that those statements affirming that Darwinian theory is either wrong or incomplete are not recent. In fact, he mentions that the contentions of Darwinism's critics in the past are very similar to those of the present, hence he questions whether this insistence is a symptom of the lack of ambition of some scientists to take other directions in evolutionary theory and he doubts it. Of course, this issue is not minor, since a well-structures theory of evolution that opposes the assumptions of the MS would replace any current model of evolution (Noble 2015; Walsh 2015; Huneman 2017) unless a Darwinian-like theory had sufficient elements to stay afloat (Depew and Weber 2013).

Taking this into account, my goal at this point is to use all these elements to develop a hypothetical evolutionary theory contrary to the tenets of DT and the MS in a way that I can show the shortcomings of the EES. I will call this theory the "Unmodern synthesis" (US) in consideration of Gilbert, Opitz, and Raff's proposal (1996) in their paper titled "Resynthesizing Evolutionary and Developmental Biology." For them, according to the MS framework, the only mechanisms available to account for variation were those belonging to genetics. Therefore, if there was a Moder Synthesis between genetics and evolution, then there must have been a relegated "Unmodern Synthesis" that would have linked embryology with evolution. So, as a consequence of the new advances in developmental genetics as well as the reinterpretation of the fossil record in contemporary science, Gilbert, Opitz, and Raff (1996) announced that:

> A new and more robust evolutionary synthesis is emerging that attempts to explain macroevolution as well as microevolutionary events. This new synthesis emphasizes three morphological areas of biology that had been marginalized by the Modern Synthesis of genetics and evolution: embryology, macroevolution, and homology. (Gilbert et al. 1996, p. 357)

In so doing, the US proposes that the causes of evolution are the outcome of the development, and not of genes by themselves. One of the most interesting consequences of his model is that: "[t]he role of natural selection in evolution... is seen to play less an important role," and population genetics would be "destined to change if it is not to become as irrelevant to evolution as Newtonian mechanics is to contemporary physics" (Gilbert et al. 1996, p. 368). So, if this is one of those ways to interpret the theory of evolution, perhaps the time has come to take it into consideration. In order to structure and justify this model, I will use historical and epistemological data related to Evo-Devo since, as Love (2010, 2017) has suggested, any alternative to the structure of evolutionary theory must be based on how we integrate the ideas of development with evolution.

3.1 History of Evolutionary Thought

Lewens (2019) and Smocovitis (2023) assert that the EES debate is also a historical one, and as Delisle (2011, p. 58) points out: "current calls for an expanded and extended Darwinism need to avoid erecting a straw man when considering its history." As we saw above, Pigliucci (2007, 2009) endorses a very traditional historical narrative that highlights the achievements of the MS to explain the EES. According to Lewens (2019, p. 709), the advocates of the EES are committed to this narrative, believing that the core of contemporary evolutionary theory developed beginning in the 1930s and 1940s. The supporters of the EES describe this core in terms of "tenets" or "assumptions" to build a dominant conceptual framework. Interestingly, Lewens comments that "conservatives" accept the historical narrative crafted by "reformists"; that is, they agree that the fundamental assumptions of the MS remain valid to the modern structure of evolutionary theory, despite the advances of contemporary science. In other words, conservatives say that there is nothing to significantly change that conceptual framework since the theory has been corrected and expanded over time (see Vidya et al. 2023). While reformists claim that these headways have significantly altered the original synthesis, thus, it should no longer be considered the same theory (see Laland et al. 2015, p. 2). Regarding this, there are two pivotal historical problems to consider.

First, reformists and conservatives develop a linear narrative in which it is assumed that there are core assumptions of the MS that arose during the 1930s–1940s, which continue to hold as the basis of current evolutionary science. Smocovitis (2023) points out that the picture of the MS painted by advocates of the EES is accompanied by a "standard historical narrative." However, she stresses that such a historical narrative does not exist since there was no common agreement between the actors of the MS and the historians who dedicated themselves to reconstructing what happened during the development of the MS. Furthermore, the story of the MS recreated by the participants of the EES in their writings is utterly riddled with inconsistencies, and there were not professional historians within the EES to corroborate such a narrative:

> [A]lthough history is being evoked and used to justify the need for the EES, it is incomplete, and so rudimentary, that it resembles a caricature, not just of evolutionary theory itself, but of

> its historical emergence as well as its subsequent evolution, or lack thereof. At best, it is a kind of "scientist's history" because it is deployed against a kind of convenient caricature, literally a textbook account, to make the case for EES; at worst, it is no real history at all; indeed, at times it is weirdly *ahistorical*, especially for an argument that rests on a foundation of history. (Smocovitis 2023, p. 43; her emphasis)

On the other hand, Esposito (2021) states that there are two distinct lines of knowledge: one synchronous that represents contemporary science and another diachronic that symbolizes the historical development of scientific ideas. The point is that these two lines of knowledge do not necessarily overlap. For example, although many researchers today consider natural selection as central to evolutionary theory, this does not mean that researchers of the past held the same view. According to Esposito, this historical fact has sometimes been overlooked by many scholars with a scientific background—as opposed to a background in history—since they generally tend to develop historiographies that support their own lines of inquiry: "[They] used history for underpinning particular scientific options. . . and is based on a linear narrative emphasizing great figures, core concepts and foundations" (p. 13). Esposito suggests that rather than a linear representation, the structure of evolutionary theory should resemble that of mycelium (fungi), with a great plurality of ideas and concepts that are no longer central.

Second, in recent years, the traditional historical narrative of the MS has been challenged by some workers who seek to reframe the events that occurred during the development of the MS (e.g., Amundson 2005; Delisle 2009, 2011, 2019; Stoltzfus and Cable 2014; Ochoa 2017, 2021a, b; Adams 2021; Schwartz 2021; van der Meer 2021; Dickins 2021; Smocovitis 2023). This challenge is important because the traditional narrative of the synthesis was primarily told by the architects of the synthesis themselves as well as Darwinian historians who continued to uphold that account. This historiographic fact made it challenging to truly appreciate various important contributions to the science of evolution, particularly those centered on the progress of morphology and embryology (Amundson 2005; Ochoa 2017; Tamborini 2022). For example, the framework of the MS set aside anti-Darwinian theories such as neo-Lamarckism, orthogenesis, and saltationism—which focused on the link between embryology and evolution—because the architects could not find genetic mechanisms to explain their causes. Consequently, their contributions were often dismissed or distorted in the conventional narrative because historians assumed that the architects of the MS somehow had scientific and reasonable arguments to have rejected them (see, for example, Provine 1971; Bowler 1983). But let's examine this point in a bit more detail.

At the end of the nineteenth century and the beginning of the twentieth century, there were some theories that opposed traditional Darwinism: neo-Lamarckism, orthogenesis, and saltationism. Neo-Lamarckism refers to the idea that organisms acquire new variations through their direct interaction with the environment, which stimulates them to change favorably during their lifetime; subsequently, after a few generations, offspring inherit these new traits from birth without environmental stimulus. Orthogenesis is a model that suggests that variations are constrained or channeled to follow the same pathways, irrespective of adaptation. Finally, saltationism is the theory that indicates that species and higher taxa emerge through abrupt change in one or a few generations, opposing gradual change. It is important to note that these theories contradict the DT framework regarding the creativity of natural selection. For instance, neo-Lamarckism and orthogenesis challenge the assumption that "natural selection drives random variation," while saltationism disagrees with the idea that "natural selection acts preferably on slight variations."

Thus, as I state in another work (Ochoa forthcoming), the issue is that these theories showed phenomena of the variation that could only be explained through an organism-centered perspective, which emphasized the structural properties of the organism in development, as well as the active interaction of the organism with the environment. That is, such theories could not be explained by the gene-centered perspective, which confined the behavior of variation within the parameters of what is strictly "hereditary" (limited to change in genes). However, neo-Lamarckism stressed that the organism is an active agent capable of favorably responding to its environment. Orthogenesis showed that "developmental constraints" were the most likely causes of evolutionary trends. Saltationism stressed that units of evolution are embryologically cohesive entities, suggesting that evolutionary change must occur discontinuously. Taking this into account, if we realize the need for a reinterpretation of such anti-Darwinian theories in light of Evo-Devo (Amundson 2005; Bowler 2017; Ochoa 2017), then, we can reasonably challenge the event of the synthesis between Mendelism and Darwinism (Stoltzfus and Cable 2014; Ochoa 2017, 2021b, forthcoming), which has been upheld for many years as a crucial step in scientific progress concerning the theory of evolution.

Given the points detailed above, adhering to a misrepresented and linear narrative centered around natural selection as a creative force and that presents the MS framework as the necessary foundation for contemporary knowledge can hinder our understanding of current evolutionary thinking: "conforming our understanding of evolutionary biology to fit the Synthesis narrative not only distorts history but also distorts *science*" (Stoltzfus 2017, p. 16; his emphasis). This perspective has led the advocates of the EES to believe that there is a fixed theoretical framework given by the achievements of the MS, leaving little room for alternative viewpoints beyond a set of established assumptions. And as Stoltzfus (2017, p. 12) clearly emphasizes:

> Though evolutionary biology is not governed by any master theory, and incorporates views excluded from the OMS [Original Modern Synthesis], the recognition of these changes has been hindered by woolly conceptions of theories, and by historical accounts, common in the evolutionary literature, that misrepresent the disputes that defined the OMS.

It is not a coincidence that the architects of the MS rejected those elements that today are intended to be added, just because they were "missing" pieces from their discourse. In fact, the MS could not incorporate the evolutionary theories that were based on embryology at the time since they were basically anti-Darwinian (De Renzi 1999; Amundson 2005; Ochoa 2017). While one of the goals of the EES was to introduce some elements of Evo-Devo into the MS framework, they ran into a final structure full of contradictions. For example, Ochoa and Barahona (2018) showed that some of the assumptions of the EES are related in one way or another to those anti-Darwinian theories (neo-Lamarckism, orthogenesis, and saltationism). In this regard, we can emphasize that contemporary evolutionary thought does not have to be guided by certain "assumptions" established within the MS framework because as Stoltzfus (2017, p. 16) points out: "For each of the past five decades, evolutionary thinking has diverged further from the OMS, both into completely new territory, and into territory previously considered non-Darwinian."

So, if we disregard the linear narrative, we will understand that these anti-Darwinian lines of reasoning lead to many of the key ideas and concepts of contemporary Evo-Devo (see Fig. 2b); e.g., the developmental constraints are linked to either orthogenesis or saltationism; the active role of the organism in its environment is connected to neo-Lamarckism, the notion that macroevolution is decoupled from microevolution is associated to saltationism, etc. If this is so, we are clearly not talking about an extension of a previous synthesis but rather a potential new evolutionary theory (see below).

With this, I do not mean that we need a better historiography of the EES, rather we can establish a new historiography that shows us a different way from the progress depicted by the framework of the MS. Naturally, the US is not structured from central assumptions; instead, its history is outlined through certain research traditions that are curiously opposed to the assumptions of the MS. In fact, my goal here will be to emphasize this contrariety.

3.2 Evo-Devo

As I have touched upon, all those scientists who accepted the MS framework by prioritizing genes while relegating the organism and its structural properties from the evolutionary scope, consequently, had difficulties integrating aspects relating to morphology and embryology.[4] But during the 1980s and 1990s, there was such uncanny progress in developmental biology research that it became hard to ignore. Clearly, it is untrue that it was not until the arrival of the EES that greater attention was paid to embryonic development since many Evo-Devo studies in tune with its key concepts already existed (see Hall 1999; Hall and Olson 2003). In fact, such studies have always challenged orthodox views (e.g., Raff 1996). Nevertheless, the announcement of a "new synthesis" has attracted attention. On the one hand, socially because this collective endeavor has financed books, papers, and meetings. And, on the other hand, epistemologically because the discourse of the supporters of the EES sounds quite conciliatory. For example, Müller (2019, 2021), one of the main

[4]Despite this, some conservatives such as Futuyma (2017, 2023) claim this is a false accusation. On the other hand, Dickins (2021, pp. 229–230) states the architects of the MS did not ignore development; rather, they assumed that an embryological explanation should align with their perspective on genetics and evolution. In so doing, they anticipated that the details could be effectively addressed by embryologists.

proponents of the EES, stresses that the theoretical consequences of Evo-Devo do not contradict "all" the assumptions of the MS, rather they only challenge its conceptual framework concerning the differences in predictions of phenotypic variation. In short, according to Müller, Evo-Devo does not invalidate the MS framework, instead, it includes another level of explanation. But despite this rhetorical and conciliatory argument, in my opinion, Müller's (2014, 2017, 2021, and personal communication) claims in favor of Evo-Devo can sometimes seem quite challenging to the MS framework.

Alternatively, when Laland et al. (2015) mention developmental processes, they emphasize what they call "constructive development" (the ability of an organism to shape its own embryonic trajectory) and "reciprocal causation" (the development of the organism is also the cause of its evolution) mainly to highlight the phenomena of phenotypic plasticity and niche construction. But Fábregas-Tejada and Vergara-Silva (2018) correctly assert that those topics are not all that the Evo-Devo agenda covers. Indeed, they state that: "Evo-Devo ideas discussed in Laland et al. (2015) represent a small fraction of the themes and concepts that, throughout recent history, have been deemed important or fundamental for the discipline" (p. 179). For example, the advocates of the EES employ some key concepts such as "developmental bias" and "facilitated variation," and briefly mention concepts such as "modularity" and "evolvability," without giving any details of their contribution to the EES framework. However, "The potential for including more themes of evolutionary developmental biology and developmental evolution to the EES is enormous" (Fábregas-Tejada and Vergara-Silva 2018, p. 180).

But although there is a unifying persistence through these criteria, the truth is that the MS model proposed since the 1930s–1940s regarding the explanation of the origin of variation is obsolete. The embryonic development of a phenotypic character implies a large series of interactions at different hierarchical levels. Starting from a complex system of gene interaction, which provokes both molecular and cellular interactions at other levels, and where the interaction between the system with the environment is considered as well. Of course, this hierarchical model goes beyond the simple scheme of MS that establishes that genetic mutations cause small-scale phenotypic effects (Amundson 2005; Laubichler and Renn 2015; Gawne et al. 2017; Huneman 2017; Love 2020).

But, according to the developmental process, "if recognized, the limitation to variation will always and everywhere affect evolution, and this means that evolution is a movement on rails, instead of wandering through the vast space of adaptation" (Popov 2009, p. 216). And if we accept that natural selection acts as "merely a filter for unsuccessful morphologies generated by development" (Gilbert et al. 1996), then we must admit that such views do not belong to the Darwinian paradigm (Popov 2009). Given these contradictions, some have proposed that the solution could be found in a difference of levels; that is, at the microevolutionary level we would apply the Darwinian view, while at the macroevolutionary level, we would need another type of explanation (see Verschuuren 1986, pp. 66–67). Or, since Darwinism and Evo-Devo work independently in their own research programs, the answer would not lie in considering a synthesis between both agendas, but in accepting a heterogeneity of interpretations in evolutionary theory (Craig 2015). However, Amundson (2005 p. 256) states that these solutions only disguise the problem, and to build a new synthesis one of the two (or both) must go. But we should keep in mind that the developmental approach has a greater explanatory power compared to the orthodox Darwinian view because, as Love (2017 p. 175) notes: "An appropriate structure of ideas with general explanatory power for evolutionary theory has development and morphology at the center, not fitness and evolutionary genetics or the theoretical principles associated with them."

In a nutshell, while there is a significant desire among some members of the scientific community to synthesize Evo-Devo with the MS from various perspectives (e.g., Minelli 2010; Craig 2015; Futuyma 2017, 2023; Müller 2019; Vidya et al. 2023), it is undeniable that Evo-Devo represents a major challenge to the traditional MS framework (Laubichler 2010; Depew and Weber 2011; Bateson 2014; Depew 2017; Huneman 2017; Love 2017; Folguera and Lavagnino 2018). As Laubichler (2010 p. 199; his emphasis) put it clearly: "any claim that evolutionary developmental biology (evo-devo) offers a significant *conceptual* challenge to the neo-Darwinian paradigm in evolutionary biology has, to a certain degree, revolutionary implications." Indeed, the US is built upon the Evo-Devo insights and thus fundamentally challenges the purportedly well-structured MS framework. In other words, the US does not represent an extension or expansion of the MS but rather a complete replacement.

4 Assumptions

One of the most influential works representing the conceptual framework of the EES is a paper published by Laland et al. (2015) titled "The Extended Evolutionary Synthesis: its structure, assumptions and predictions." The authors, who are mostly the main protagonists of this movement, argue that to explain the phenomena of evolution, a conceptual framework must be formulated based on a set of assumptions. They further claim that conceptual frameworks change as a result of theoretical innovations and new empirical findings and that this is what has led to the development of the EES.

Their strategy involves identifying six core assumptions of the MS framework and comparing them with six new assumptions of the EES. In so doing, they aim to demonstrate that while some assumptions of the MS are still relevant, the model is enriched and reinterpreted with new assumptions based on scientific practices. According to them, evolutionary biology is transformed giving way to new and/or alternative conceptual frameworks, which encourage novel perspectives: "the EES entails not only new research directions but also new ways to think about, and interpret, new and familiar problems in evolutionary biology" (p. 6).

But despite this enthusiastic endeavor to modify evolutionary theory, we will see that the structure of the EES is not only full of inconsistencies, but it also does not represent a model significantly different from the traditional MS. My strategy is to deconstruct the conceptual framework of the EES through the assumptions that

Laland et al. (2015) propose for both the MS and the EES. I will show that the EES does not differ from the MS regarding its consideration of natural selection's role as a creative force, the notion of "inheritance," and the extrapolationist argument (because natural selection as a creative force is the "core" tenet of DT, and both the "notion of inheritance within gene-centrism" and the "extrapolationist argument" are tenets of MS). Of course, if there is no challenge to these premises, there would be no relevant change in evolutionary theory. Additionally, through the construction of an alternative conceptual framework that I call the "Unmodern Synthesis" (US), we shall see that the assumptions developed from the advancements of Evo-Devo suppose a view completely contrary to DT, the MS, and, therefore, to the EES. Therein the idea of a potential new theory of evolution. Below, we will see the differences between the assumptions defined by the three conceptual frameworks mentioned (see summary in Table 1).

(1) Natural Selection as a Creative Force

MS

(1) *Natural selection directs variation*. The major directing or creative influence in evolution is natural selection, which alone explains adaptation.

EES

(1) *Natural selection directs variation along with other processes*. Developmental processes, operating through developmental bias and niche construction, share some responsibility for the direction of evolution with natural selection.

For conservatives, despite recognizing that there are some important factors that influence the direction of variation, natural selection is the only process that increases fitness and produces adaptive traits (see Futuyma 2017, p. 9). In contrast, for reformers, processes such as developmental bias and niche construction are mechanisms that work in concert with natural selection. However, it should be noted that reformers do not challenge the creative role of natural selection, they only argue that there are other "creative" factors that contribute in the same way to the direction of variation and, therefore, to adaptation. Regardless, according to the US:

US

(1) *Natural selection has no role in the direction of variation*. Developmental bias is the only cause of the direction of variation. Then, natural selection is not creative.

For some early twentieth-century anti-Darwinians, like William Bateson, the creativity of evolution resides in variation itself, not in natural selection. The morphological characters appear discontinuous, already ordered, and well-

Table 1 A comparison of the assumptions between the MS, the EES, and the US

Issues	MS assumptions	EES assumptions	US assumptions
(1) Natural selection as a creative force	*Natural selection directs variation*. The major directing or creative influence in evolution is natural selection, which alone explains adaptation.	*Natural selection directs variation along with other processes*. Developmental processes, operating through developmental bias and niche construction, share with natural selection some responsibility for the direction of evolution.	*Natural selection has no role in the direction of variation*. Developmental bias is the only cause of the direction of variation. Then, natural selection is not creative.
(2) Heredity	*Genetic inheritance*. Genes constitute the only general inheritance system.	*Inclusive inheritance*. Heredity extends beyond genes to encompass epigenetic inheritance, physiological inheritance, ecological inheritance, cultural inheritance, etc.	*Modular thinking*. Genetic "inheritance" does not cause the evolving entities' stability. To mention "heredity" as "transmission of genetic information" does explain why a character emerges, nor why it is maintained over generations either. Units of evolution are not replicators.
(3) The direction of variation	*Random genetic variation*. The direction of genetic variation is random. And hence natural selection orients genetic variation at the phenotypic level.	*Non-random phenotypic variation*. The direction of phenotypic variation is non-random. Developmental bias means that some phenotypic variants are more likely than others.	*Non-random phenotypic variation*. The direction of phenotypic variation is non-random. But if variation does not occur randomly, natural selection is not creative.
(4) Rates in evolution	*Gradualism*. Evolution via mutation of large effects is unlikely. Phenotypic transitions typically occur through multiple small steps.	*Variable rate of changes*. Variants of large effect are possible, allowing rapid evolutionary change. Saltation can occur.	*Saltational evolution*. Variable rates of change imply gradual evolution, not saltational evolution. However, most evolutionary change occurs in jumps (giving rise to instantaneous morphological gaps

(continued)

Table 1 (continued)

Issues	MS assumptions	EES assumptions	US assumptions
			between entities). Thereby, natural selection is not creative.
(5) Entity-centered perspective	*Gene-centered perspective.* Evolution requires changes in gene frequencies. Populations evolve through changes in gene frequencies (by natural selection, genetic drift, mutation, and gene flow).	*Organism-centered perspective.* The developmental system can facilitate adaptive variation and modify selective environments.	*Organism-centered perspective.* But if we accept the organism-centered view, then, we do not need the MS framework.
(6) Macroevolution	*Macroevolution.* Macroevolutionary patterns are explained by microevolutionary processes (selection, genetic drift, mutation, and gene flow).	*Macroevolution.* Additional evolutionary processes, including developmental bias and ecological inheritance, help explain macroevolutionary patterns and contribute to evolvability.	*Real macroevolution.* If saltational evolution is the most common mode, then, it is necessary to decouple macroevolution from microevolution.

structured. The role of natural selection, then, is to preserve such characters in a particular environment, but not to create them: "Selection determines along which branch Evolution shall proceed, but it does not decide what novelties that branch shall bring forth" (Bateson 1909, p. 96).[5] Nowadays, Alberch's (1982) model of developmental constraints is very clear in this regard. According to Alberch, developmental constraints are the cause of the distribution of variation: the variation appears discontinuous as well as with specific orientations in the morphospace. As I stated above, developmental constraints are expressed in two ways: one in a negative sense, maintaining a specific form in morphospace, and the other in a positive sense, providing direction to the variation (see Ochoa forthcoming). It is in this latter sense the notion of "developmental bias" is highlighted, and it can be defined as the tendency of developmental systems to vary more easily in some ways than in others (Arthur 2004).

It is important to understand that developmental bias is the sole cause of variation's directionality since development channels variation towards some

[5] For more detail, see Ochoa 2021b.

pathways and impedes others. In other words, while developmental bias generates a certain spectrum of possibilities for variation, the function of natural selection is restricted to preserve some of these pre-oriented options to fit them within a specific environment: "the role of selection is basically stabilizing, being responsible for 'pruning out' the nonfunctional morphologies, and for determining the differential survival of morphological types" (Alberch 1982, pp. 318–319). Therefore, considering the dynamics of developmental constraints, natural selection is not creative. This fact, of course, challenges one of the core tenets of DT: ~~Natural selection drives random variations.~~[6]

(2) Heredity

MS

(2) *Genetic inheritance*. Genes constitute the only general inheritance system.

EES

(2) *Inclusive inheritance*. Heredity extends beyond genes to encompass epigenetic inheritance, physiological inheritance, ecological inheritance, cultural inheritance, etc.

Conservatives believe that one of the great successes of the MS was the recognition that the problems of heredity were solved with the science of genetics: heredity is the transmission of "genetic information." While they do not ignore the possibility of other non-genetic forms of inheritance, they question their relevance in explaining evolution as a whole: "Documenting the frequency and distribution of non-genetic inheritance in natural populations, and its importance for evolved differences among populations and species, will be a task for empirical research" (Futuyma 2017, p. 9). On the other hand, reformists argue that they are organism-centered rather than gene-centered, but apparently, they do not deny that genes are the main system of heredity. Rather, they merely include other entities that are "heritable": "Heredity extends beyond genes." However, the US proposes another alternative:

US

(2) Modular thinking. Genetic "inheritance" does not cause the evolving entities' stability. To mention "heredity" as "transmission of genetic information" does not explain why a character emerges, nor why it is maintained over generations either. Units of evolution are not replicators.

[6]Note that the strikethrough text symbolizes the rejection of assumptions from traditional frameworks. On the other hand, consider that the developmental constraints also challenge the tenet of natural selection acting on slight variations as we will see below.

Although the concept of "heredity" is essential in understanding evolution, it is important to note that the notion of "heredity" or "inheritance" predates evolutionary thinking. It stems from the old theories of "generation" from the ancient Greeks through the eighteenth century and continued with the debates carried out during the nineteenth century on medical pathologies, breeding practices, hybridization, learning and cultural transmission, and so on (Waller 2003; Cobb 2006; Ceccarelli 2021). That is, the inheritance of physical characters was only a small part of a wide range of phenomena that were associated with this term. It is noteworthy that many of these discussions, which formed the basis of the evolutionary concept of "heredity," were concerned with the similarity between parents and offspring. But just as the science of genetics was dawning, Danish scientist Wilhelm Johannsen criticized this concept of "heredity":

> BIOLOGY has evidently borrowed the terms "heredity" and "inheritance" from every-day language, in which the meaning of these words is the "*transmission*" of money or things, rights or duties—or even ideas and knowledge—from one person to another or to some others... The *transmission* of properties—these may be things owned or peculiar qualities—from parents to their children, or from more or less remote ancestors to their descendants, has been regarded as the essential point in the discussion of heredity in biology as in jurisprudence... The view of natural inheritance as realized by an act of transmission, viz., the transmission of the parent's (or ancestor's) *personal qualities* to the progeny, is the most naïve and oldest conception of heredity. (Johannsen 1911, p. 129; his emphasis)

Johannsen believed that the concept of "heredity" should align more closely with the phenomena of embryology. Nevertheless, according to Amundson (2005), the discourse of heredity during the establishment of the MS completely abandoned the embryonic development program. Heredity was viewed as the transmission of genetic information that could be traced from parents to children. This view endorsed the gradual evolution of characters. Consequently, Darwinian geneticists focused on investigating the correspondence between genes and adult characters, rather than explaining how those characters arose during embryonic development: "So evolution became *changes in genes*, rather than changes in developmental processes" (Amundson 2005, pp. 201–202; his emphasis).

Currently, this tradition that I call "inheritance thinking" is typical of many of the Darwinian discourses: Heredity is the transmission of genes through generations, and evolutionary change occurs when this information is altered. From this point of view, the organism and its structural properties are irrelevant since the organism is seen as nothing more than a vehicle that carries the genes (Wagner 2015). This view of heredity is undoubtedly characteristic of the gene-centered perspective developed from the MS approach, which evidently lacks integration of aspects of embryology. Another option in evolutionary thought is what I call "modular thinking." This type of reasoning is based on the idea of the modules of morphology, being parts of an organism that are highly integrated in developmental terms (Schlosser 2002, 2022; Kuratani 2009; Ochoa and Rasskin-Gutman 2015). These systems include a series of processes that interact at different levels, not just at one, as is generally represented in "inheritance thinking." Because such interactions are highly integrated, they would explain why characters are maintained over time (DiFrisco et al. 2020; Ochoa forthcoming).

It should be noted that these different ways of thinking differ in their interpretation of evolutionary discourse. For example, under "inheritance thinking," if genes are faithfully passed down to the next generation (i.e., they are replicators), then the traits may remain fixed for generations. However, this view does not explain why a character emerges or why it is maintained over time. In contrast, under "modular thinking," characters are not replicators since they develop from scratch in each generation (Sattler 1984; Wagner 1989; Brigandt 2002; DiFrisco 2023), and the phylogenetic stability of the characters is explained by either developmental constraints (Alberch 1982; Ochoa forthcoming) or Character Identity Mechanisms (DiFrisco et al. 2020). While "Inheritance thinking" tends to focus on specific and small traits that pass from parents to children, "modular thinking" focuses on explaining the stability of organs through millions of years of evolution. In conclusion, this assumption of the US challenges the skewed notion of "heredity" that arose from the gene-centered view developed by the MS framework.

(3) The Direction of Variation

MS

(3) *Random genetic variation*. The direction of genetic variation is random. And hence natural selection orients genetic variation at the phenotypic level.

EES

(3) *Non-random phenotypic variation*. The direction of phenotypic variation is non-random. Developmental bias means that some phenotypic variants are more likely than others.

In the view of the conservatives, genetic variation is random in the sense that it is not driven by some "need," it does not necessarily mean that all modifications are equally likely. The significance lies in the fact that most mutations that affect survival are deleterious (Futuyma 2017). For reformers, non-random variation does not precisely refer to the genetic level, but rather they correctly recognize that the direction of variation depends solely on the phenotypic level. Nonetheless, they should have to admit that this assumption challenges one of the main tenets of DT, as stated in assumption (1): ~~Natural selection drives random variations~~. Therefore:

US

(3). *Non-random phenotypic variation*. The direction of phenotypic variation is non-random. But if variation does not occur randomly, natural selection is not creative.

(4) Rates in Evolution

MS

(4) *Gradualism*. Evolution via mutations of large effects is unlikely. Phenotypic transitions typically occur through multiple small steps.

EES

(4) *Variable rates of changes*. Variants of large effect are possible, allowing rapid evolutionary change. Saltation can occur.

One of the most confusing aspects of the study of evolution is that of evolutionary tempos. But before proceeding with this debate a point of clarification is needed. Under the term "gradualism" it is understood that entities change continuously over time irrespective of the velocity at which this process occurs. For example, for an entity "A" to become an entity "E," it must have passed through a series of intermediate forms, in this case, "B," "C," and "D." On the contrary, "saltational evolution" is identified by lacking intermediate steps, for example, "A" becomes "E," and such a process occurs in only one or a few generations (see Ochoa forthcoming). In any case, we should consider that this gradualism "as intermediacy of transitional forms" is not the same notion as the "gradualism" (anagenesis) that the advocates of punctuated equilibrium criticize (see Eldredge and Gould 1972). Because, as I demonstrate in another work (Ochoa forthcoming), punctuation (or speciation) is as gradual as any other previous theory on the tempos in evolution. That is, gradualism as intermediacy of transitional forms can occur at different rates of change (for example, either rapid or slow).

Once this point is clarified, we can note that there is ambiguity when reformists attempt to separate gradual evolution from saltational evolution. For example, they sometimes refer to "discontinuous change" as saltational, but other times it seems to mean rapid gradual change (or tachytely, in the vocabulary of the tempos in evolution): "The dynamics of biological systems illuminates *the capacity of continuous selectional regimes* to produce the *nongradual phenotypic change* frequently observed in the paleontological record" (Pigliucci and Müller 2010, p. 13; my emphasis). Of course, by making this distinction clear, conservatives condemn reformists stressing that there is nothing relevant to this point since it has always been recognized that gradual evolution has different rates of change: "Of course, advocates of the ES [Evolutionary Synthesis], beginning with Simpson and Rensch, have recognized that rates of evolution are highly variable; evolutionary rates at both the phenotypic and genomic levels have been a focus of extensive research" (Futuyma 2017, p. 9). On the other side, they recognize that saltational evolution is possible, but only in rare cases such as polymorphism seen in plants; however, "there is still no evidence that single mutations are responsible for the multiple character differences that typify most genera or other higher taxa" (Futuyma 2017, p. 3). Despite that, the US advocates for real saltational evolution:

US

(4) *Saltational evolution*. Variable rates of change imply gradual evolution, not saltational evolution. However, most evolutionary change occurs in jumps (giving rise to instantaneous morphological gaps between entities). Thereby, natural selection is not creative.

As I mentioned above, the real challenge to the rates of change is saltational evolution, rather than rapid gradual evolution. Interestingly, most modern biologists take for granted that the rates of evolutionary change occur more frequently by gradual evolution. In so doing, they have dismissed the possibilities of saltational evolution. This is likely due to their early education, where their teachers emphasized the idea that saltational evolution is wrong. But, if we examine the nature of variation more closely, gradual evolution appears to be more problematic than saltational evolution.

It is widely recognized in biology that, in the *Origin of Species*, Darwin challenged his own theory by claiming that: "If it could be demonstrated that any complex organ existed, which could not possibly have been formed by numerous, successive, slight modifications, my theory would absolutely break down" (Darwin 1859, p. 189). However, this Darwinian challenge had already been overcome by William Bateson in the early twentieth century, when he showed that most changes observed in characters occurred discontinuously. Furthermore, his research on developmental genetics demonstrated that transitional forms lacked genetic persistence (e.g., most of the continuous characters were due to the organism's reaction to the environment, see Ochoa 2021b). Finally, it is worth noting that the architects of the MS grasped the idea that continuous genetic changes (mutations) were somehow related to the continuous phenotypic changes of simple characters—e.g., color, proportion, size, etc. But they did not focus on the embryological origin of such phenotypic variations because they assumed that those were caused simply by either genetic mutations or recombination.

Nowadays, any proposal on the rates of evolutionary change must be consistent with the possibilities of change permitted by developmental systems (Anderson 1987). It should also be kept in mind that phenotypes are generated from complex non-linear interactions at different levels of organization (Laubichler and Renn 2015; Gawne et al. 2017; DiFrisco et al. 2023). Such interactions yield complex characters—i.e., those of body plan recognition—which are distinguished by being highly cohesive. This characteristic of the developmental systems implies discontinuous change, rather than gradual evolution (Balon 2002; Salazar-Ciudad and Jernvall 2005; Theißen 2009; Frazzetta 2012; Jaeger et al. 2012; Linde-Medina and Newman 2014; Peterson and Müller 2016). While saltational evolution has been controversial, recent research suggests that it is not impossible, given our contemporary knowledge of the origin of variation. However, we lack models to accurately understand its dynamics (Laubichler and Renn 2015). Despite these difficulties, some progress has been made in this regard, particularly concerning the old proposal made by Goldschmidt (1940) about hopeful monsters (e.g., Bateman and DiMichele 2002; Theißen 2009; Diogo 2017). Even my own work has taken this route, with which I attempt to show that saltational evolution is quite plausible (see Ochoa forthcoming). So, if we admit that rates of evolution occur by discontinuous steps rather than by the accumulation of slight variations through geological time, then, the pivotal tenet of DT would be completely disregarded: ~~Natural selection acts on slight variations.~~

(5) Entity-Centered Perspective

MS

(5) Gene-centered perspective. Evolution requires changes in gene frequencies. Populations evolve through changes in gene frequencies (by natural selection, genetic drift, mutation, and gene flow).

EES

(5) Organism-centered perspective. The developmental system can facilitate adaptive variation and modify selective environments.

For conservatives, if organisms evolve due to their genetic changes, then there is no way to conceive evolutionary change without considering a gene-based hereditary system. Or as Futuyma (2023, p. 534) asks in a recent publication: "Can anyone, however dedicated to the importance of development or niche construction, imagine a theory of evolutionary change without genetics?" Yet, as we saw above in assumption (4), it is not that we completely ignore the advances in genetics, but rather that it is not possible to recognize the function of genes without considering the great complexity of interactions that take place in embryonic development, including those that come from outside the organism. For this reason, I consider the position of the reformists to be correct when they say that there are many factors that act upon the organism that contribute to its development. The problem with the EES is that its supporters have not admitted that if we choose an organism-centered perspective, then we do not need the gene-centered perspective of the MS framework. As Walsh (2015, p. 229; his emphasis) correctly comments:

> Organisms as agents were omitted from the Modern Synthesis, but not because someone forgot to put them in. They were left out because the Modern Synthesis is an object theory, it has no way to accommodate agents. One cannot simply graft the idea of organisms as agents onto Modern Synthesis evolutionary biology, while leaving the rest of the theoretical apparatus intact. Consequently, an evolutionary theory that deals in agents, purpose and affordances is no mere extension of the Modern Synthesis; it is "a different *kind* of theory."

Regarding this, the US would be as follows:

US

(5) *Organism-centered perspective.* But if we accept the organism-centered perspective, then, we do not need the MS framework.

A new evolutionary theory would need to set aside the reductionist perspective of the MS, which has upheld genes as the main actors in evolutionary change. This new model would include natural selection, phenotypic plasticity, niche construction, and behavioral choice as mechanisms involving organism-environment interaction. In addition, it would re-evaluate the concept of species (along with a new interpretation of taxonomy) as well as the speciation processes that lack proper integration of embryology within the MS (see Ochoa forthcoming). In short, if we opt for an organism-centered perspective, we do not extend the framework of the MS but rather replace it completely: ~~Gene-centrism. Evolution focuses solely on genetic mechanisms~~.

(6) Macroevolution

MS

(6) *Macroevolution.* Macroevolutionary patterns are explained by microevolutionary processes (selection, genetic drift, mutation, and gene flow).

EES

(6) *Macroevolution.* Additional evolutionary processes, including developmental bias and ecological inheritance, help explain macroevolutionary patterns and contribute to evolvability.

For a long time, much attention has been devoted to the question of whether macroevolution (e.g., the origin of species) is nothing more than the sum of long-term microevolutionary processes, this is the extrapolationist premise (see Adams 2021). Intriguingly, most evolutionary biologists place too much confidence in this premise, even those who support theories such as punctuated equilibrium and species selection. For example, as I noted in another work (see Ochoa forthcoming), the problem with punctuated equilibrium is that, although it rightly points to macroevolutionary patterns in species at the geological scale (i.e., punctuation, stasis, and extinction), this model explains those patterns through microevolutionary processes provided by the MS—e.g., allopatric speciation is the cause of punctuation. When it comes to species selection, it is even worse since it might not even be a real phenomenon considering our contemporary knowledge of biology (see Ochoa forthcoming). Therefore, a truly macroevolutionary theory must be able to efficiently distinguish macroevolution from microevolution, either by identifying distinct evolutionary mechanisms or by the kind of variation being studied (e.g., simple or complex characters).

The main issue with the EES is that it does not offer a macroevolutionary perspective distinct from that of the MS as it assumes the extrapolationist argument, i.e., the EES supporters accept that macroevolutionary patterns are explained through microevolutionary mechanisms—such as genetic drift, mutation, selection, and so on—and only incorporate other "microevolutionary" mechanisms like developmental bias, ecological inheritance, among other considerations. Moreover, the EES does endorse some proposals with "macroevolutionary" implications such as multilevel selection theory (see Pigliucci 2009; Laland et al. 2015; Müller 2017). However, this multilevel selection theory, in its traditional sense, is very questionable (DiFrisco 2021; Eronen and Ramsey 2022; Ochoa forthcoming). This issue regarding macroevolution is significant because as Pievani (2016a) has pointed out, the EES lacks a hierarchical view of evolution. So, the US proposes something different since it completely challenges the extrapolationist premise:

US

(6) *Real macroevolution*. If saltational evolution is the most common mode, then, it is necessary to decouple macroevolution from microevolution.

Once again, we can see that adopting a saltational view of evolution offers an opportunity to move away from the extrapolationist argument. In fact, as I have previously noted, such a view may be our best option for a new evolutionary theory (see Ochoa forthcoming). Contrary to what many biologists might believe, given our current understanding of the origin and maintenance of variation, the possibility of saltational evolution seems more plausible than the gradualist perspective. In any case, if we can establish a model that allows for this type of explanation, we would not only reject the gradualist paradigm of DT but also the second most important premise of the MS: ~~Extrapolationist premise. Macroevolutionary patterns are explained by microevolutionary mechanisms~~. Therefore, we do not need an extension of the previous theory, but one that replaces it entirely.

5 Conclusions: Yes, We Need a New Theory of Evolution

As far as I am concerned, the EES has many inconsistencies in its conceptual theoretical framework. It is so disorganized that we should reconsider whether it is necessary to take it into account to represent the structure of contemporary evolutionary theory. For example, it outlines its history from a traditional linear narrative that ignores other lines of reasoning, e.g., those focused on morphology and embryology. The EES supporters believe that evolutionary theory stems from a set of assumptions developed by the MS during the 1930s and 1940s. However, as we have seen, we should consider that there was much more progress in the evolutionary theory of the anti-Darwinian line of reasoning—which gave rise to many of the key Evo-Devo notions—than what was provided by the MS. This creates a problem with integrating Evo-Devo into the MS framework. Even though the EES's discourse is to extend the previous knowledge of the MS with this field of study, many Evo-Devo notions oppose most of the interpretations made by the original advocates of the MS to explain evolutionary phenomena. For example, developmental constraints challenge the argument of natural selection as a creative force, modularity and the organism-centered perspective challenge the gene-centered perspective, and the acceptance of saltational evolution as a frequent mode of evolution challenges the extrapolationist premise, etc. The EES does not differ from the MS on these points, except for the consideration of the organism-centered perspective. However, it seems that the advocates of the EES still support the gene-centered perspective when the concept of "inheritance" is mentioned.

Keeping these problems in mind, I recommend abandoning not only the EES framework but also the MS framework—even though it may pain conservatives. As I commented above, this does not mean throwing away everything that the theory of evolution can offer, rather this is an invitation to explore other ways of thinking in evolutionary theory. Now, the original aim of this chapter was to propose a hypothetical new theory of evolution in order to highlight the shortcomings of the EES. But this point inevitably leads us to ask: Is it possible to construct an Unmodern Synthesis? Structuring a new theory of evolution is worthwhile if it has pragmatic research benefits (Love 2010). Of course, we do not need a "standard" evolutionary framework with core assumptions because as Love (2010) has stated, evolutionary theory has a wide range of research programs with contributions from different disciplines. However, just because evolutionary theory is pluralistic, it does not mean that we should consider all the assumptions of the MS as valid. Many evolutionary phenomena explained via the MS framework can be accounted for by other types of explanations.

As I pointed out above, despite the many criticisms faced by the MS, modern biologists offer no other model than an extension or expansion of the MS. To date, few alternatives to the EES have been proposed. One of these is the so-called *Hierarchy Theory of Evolution* (H-TE) by Niles Eldredge and Stanley Salthe (1984). But, as noted by one of its supporters (Pievani 2016a), the H-TE is another version of an "extended synthesis," "a multilevel extended synthesis," but "still strongly Darwinian in its core." In general, the problem with the H-TE is that it upholds the gene-centered perspective of the MS, and while it is presented as a hierarchy theory of evolution, it also falls into the extrapolationist premise.[7] On the other hand, an interesting alternative is *Organic Nonoptimal Constrained Evolution* (ONCE) proposed by Rui Diogo. According to Diogo (2017), ONCE "is intended to be an integrative, multidisciplinary view of biological evolution as a whole." Indeed, while Diogo presents ONCE as a kind of Extended Synthesis,[8] it differs from the EES by giving greater importance to behavioral choice, developmental constraints, and historical contingencies as key factors of "organic evolution."

Another different option is my own evolutionary model, which I call the *Hierarchical Evolutionary-Developmental Theory* (H-Evo-Devo Theory). It can be considered as a kind of US because, unlike other models, it does not integrate any contributions from the MS. This model proposes that the organism's internal properties reflect highly integrated entities—from the developmental point of

[7]For a more detailed critique of the H-TE see Ochoa forthcoming. On the other hand, Fábregas-Tejada and Vergara-Silva (2017) have examined the potential points of convergence between the H-TE and the EES. Nevertheless, as emphasized before, both frameworks are rooted in the MS, which is problematic.

[8]This means that it does not exclude some concepts developed by the MS such as genetic drift, gene flow, the notion of heredity, and so on.

view—that evolve at different scales. Developmental constraints operate at different levels, producing distinct states of morphological stability that are reflected as subspecies, species, higher taxa, and homologues, which are examples of units of evolution. While I do not intend to provide a detailed exposition of this model here,[9] I want to highlight some of its approaches that deviate significantly from the MS:

1. The mechanism of developmental constraints (in the negative sense) is essential in this representation because it accounts for the stability and recognition of evolving entities. Unlike the traditional view that mainly focuses on the mechanism of natural selection.
2. The model is structured around the idea of modules (homologues) as quasi-autonomous elements. Given that they are highly integrated systems in developmental terms, both their character identity and character states account for the phylogenetic significance of units of evolution. Whereas the MS states that genes are solely responsible for the stability of evolutionary entities over time.
3. The model is completely organism-centered. For example, in order to explain adaptations that occur at the subspecies level, it considers different organism-environment interaction factors such as phenotypic plasticity, niche construction, behavioral choice, and natural selection. This differs from the MS, which states that the organism is a passive agent guided only by genetic changes and natural selection.
4. Under this model, macroevolution is totally decoupled from microevolution. For instance, microevolution is explained by the dispersal model, while macroevolution is explained by the hopeful monster model. This is one of the reasons whereby we can distinguish microevolution from macroevolution regarding the origin of their units.[10] And furthermore, the H-Evo-Devo Theory also contemplates an additional level, *megaevolution*, which refers to the origin, stability, and extinction of higher taxa as active and real entities. In contrast, the MS considers that macroevolution is only the sum of microevolutionary processes: the taxonomic categories are no more than the historical record of the divergence process that occurs at a single level.

My goal here is not to promote my own evolutionary model but rather to exhibit the H-Evo-Devo Theory as an example of how Evo-Devo insights can be integrated into a new framework without relying on traditional MS notions. In any case, other proposals that organize the Evo-Devo ideas into a new framework are also welcome.

---

[9] See Ochoa forthcoming.

[10] According to the H-Evo-Devo Theory, subspecies evolution is related to microevolution, while species evolution is associated with macroevolution. Subspecies are those groups that are separated from each other by small morphological gaps in simple characters (such as color, pattern, size, proportion, etc.), they reflect the phenomenon of diversification and adaptation of the species to different environments, protecting them from extinction. While species are the least inclusive body plan within a hierarchy of body plans (under this view, higher taxa are more inclusive body plans than the species body plan).

My main point is that there are many ways to represent evolutionary theory (Love 2010), and some of these pave the way to discard the framework of the MS. At the end of the day, this intellectual effort could be very worthwhile because, if a potential new evolutionary theory works well in practice—or works better than the previous one— (Buskell 2020, Love 2013), then, undoubtedly, we will have a completely new representation with revolutionary implications within evolutionary theory.

Acknowledgments Many thanks to the editors for their invitation to contribute to this volume, especially Richard Delisle. I also want to thank Sandra Naitzé Martínez Byer for the English editing. I am very grateful to the reviewers for their suggestions and corrections that improve this work. Finally, I thank to Conahcyt for the financial support provided under my appointment within the National System of Researchers (SNI). To the memory of our colleague Mark B. Adams, who found in our group a place where evolutionary orthodoxy does not prevail.

References

Adams MB (2021) Little evolution, BIG evolution: rethinking the history of Darwinism, population genetics, and the "synthesis". In: Delisle RG (ed) Natural selection: revisiting its explanatory role in evolutionary biology. Springer, Cham, pp 195–229

Alberch P (1982) Developmental constraints in evolutionary processes. In: Bonner JT (ed) Evolution and development. Springer, Berlin, Heidelberg, pp 313–332

Alberch P (1989) The logic of monsters: evidence for internal constraint in development and evolution. Geobios 22:21–57

Amundson R (2005) The changing role of the embryo in evolutionary thought: roots of Evo-devo. Cambridge University Press, Cambridge

Anderson DT (1987) Developmental pathways and evolutionary rates. In: Campbell KSW, Day MF (eds) Rates of evolution. Allen and Unwin, London, pp 143–155

Arthur W (2004) Biased embryos and evolution. Cambridge University Press, Cambridge

Asma ST (1996) Following form and function: a philosophical archaeology of life science. Northwestern University Press, Evanston, Illinois

Balon EK (2002) Epigenetic processes, when *natura non facit saltum* becomes a myth, and alternative ontogenies a mechanism of evolution. Environ Biol Fish 65:1–35

Bateman RM, DiMichele WA (2002) Generating and filtering major phenotypic novelties: neoGoldschmidtian saltation revisited. In: Cronk QCB, Bateman RM, Hawkins JA (eds) Developmental genetics and plant evolution. Taylor and Francis, London, pp 109–159

Bateson P (2014) New thinking about biological evolution. Biol J Linn Soc 112:268–275

Bateson W (1909) Heredity and variation in modern lights. In: Seward AC (ed) Darwin and modern science: essays in commemoration of the centenary of the birth of Charles Darwin and of the fiftieth anniversary of the publication of the origin of species. Cambridge University Press, Cambridge, pp 85–101

Beatty J (2016) The creativity of natural selection? Part I: Darwin, Darwinism, and the mutationists. J Hist Biol 49:659–684

Bolker JA (2000) Modularity in development and why it matters to Evo-devo. Am Zool 40:770–776

Bowler PJ (1983) The eclipse of Darwinism: anti-Darwinian evolutionary theories in the decades around 1900. The Johns Hopkins University Press, Baltimore

Bowler PJ (2017) Alternatives to Darwinism in the early twentieth century. In: Delisle RG (ed) The Darwinian tradition in context: research programs in evolutionary biology. Springer, Cham, pp 195–217

Brigandt I (2002) Homology and the origin of correspondence. Biol Philos 17:389–407

Buranyi S (2022) Do we need a new theory of evolution? https://www.theguardian.com/science/2022/jun/28/do-we-need-a-new-theory-of-evolution

Buskell A (2020) Synthesising arguments and the extended evolutionary synthesis. Stud Hist Philos Sci C 80:101244. https://doi.org/10.1016/j.shpsc.2019.101244

Callebaut W (2010) The dialectics of dis/unity in the evolutionary synthesis and its extensions. In: Pigliucci M, Müller GB (eds) Evolution: the extended synthesis. MIT Press, Cambridge MA, pp 443–481

Ceccarelli D (2021) History of inheritance. In: Shackelford TK, Weekes-Shackelford VA (eds) Encyclopedia of evolutionary psychological science. Springer, Cham, pp 3732–3740

Charlesworth D, Barton NH, Charlesworth B (2017) The sources of adaptive variation. Proc R Soc B 284:20162864. https://doi.org/10.1098/rspb.2016.2864

Chiu L (2022) Extended evolutionary synthesis: a review of the latest scientific research. John Templeton Foundation, Pennsylvania

Cobb M (2006) Heredity before genetics: a history. Nat Rev Genet 7:953–958

Coleman W (1980) Morphology in the evolutionary synthesis. In: Mayr E, Provine WB (eds) The evolutionary synthesis: perspectives on the unification of biology. Harvard University Press, Cambridge, pp 174–180

Craig LR (2015) Neo-Darwinism and Evo-devo: an argument for theoretical pluralism in evolutionary biology. Perspect Sci 23:243–279

Darwin C (1859) On the origin of species by means of natural selection, or preservation of favored races in the struggle for life. John Murray, London

De Renzi M (1999) El Neodarwinismo y las ciencias de la forma: el porqué de una alianza inicial imposible. Revista Española de Paleontología 14:115–120

Delisle RG (2009) Les philosophies du néo-darwinisme: conceptions divergentes sur l'homme et le sens de l'évolution. PUF, Paris

Delisle RG (2011) What was really synthesized during the evolutionary synthesis? A historiographic proposal. Stud Hist Phil Biol Biomed Sci 42:50–59

Delisle RG (2019) Charles Darwin's incomplete revolution: the *Origin of Species* and the static worldview. Springer, Cham

Delisle RG (2021) Natural selection as a mere auxiliary hypothesis (sensu stricto I. Lakatos) in Charles Darwin's *Origin of Species*. In: Delisle RG (ed) Natural selection: revisiting its explanatory role in evolutionary biology. Springer, Cham, pp 73–104

Depew DJ (2017) Natural selection, adaptation, and the recovery of development. In: Huneman P, Walsh D (eds) Challenging the modern synthesis: adaptation, development, and inheritance. Oxford University Press, New York, pp 37–67

Depew DJ, Weber BH (2011) The fate of Darwinism: evolution after the modern synthesis. Biol Theory 6:89–102

Depew DJ, Weber BH (2013) Challenging Darwinism: expanding, extending, replacing. In: Ruse M (ed) The Cambridge encyclopedia of Darwin and evolutionary thought. Cambridge University Press, Cambridge, pp 405–411

Dickins TE (2021) The modern synthesis: evolution and the organization of information. Springer, Cham

DiFrisco J, Love AC, Wagner GP (2020) Character identity mechanisms: a conceptual model for comparative-mechanistic biology. Biol Philos 35. https://doi.org/10.1007/s10539-020-09762-2

DiFrisco J (2021) Integrating composition and process in levels of developmental evolution. In: Brooks DS, DiFrisco J, Wimsatt WC (eds) Levels of organization in the biological sciences. The MIT Press, Cambridge MA, pp 111–134

DiFrisco J (2023) Toward a theory of homology: development and the de-coupling of morphological and molecular evolution. BJPS 74:771–810

DiFrisco J, Wagner GP, Love AC (2023) Reframing research on evolutionary novelty and co-option: character identity mechanisms versus deep homology. Semin Cell Biol 145:3–12

Diogo R (2017) Evolution driven by organismal behavior: a unifying view of life, function, form, mismatches, and trends. Springer, Cham

Dobzhansky T (1937) Genetics and the origin of species. Columbia University Press, New York
dos Reis CRM, Araújo LAL (2020) Extended evolutionary synthesis: neither synthesis nor extension. Biol Theory 15:57–60
Eldredge N, Gould SJ (1972) Punctuated equilibria: an alternative to phyletic gradualism. In: Schopf TJM (ed) Models in paleobiology. Freeman, Cooper and Co, San Francisco California, pp 82–115
Eldredge N, Salthe SN (1984) Hierarchy and evolution. In: Dawkins R, Ridley M (eds) Oxford surveys in evolutionary biology, vol I. Oxford University Press, Oxford, pp 182–206
Eronen MI, Ramsey G (2022) What are the 'levels' in levels of selection? BJPS. https://doi.org/10.1086/718990
Esposito M (2013) Heredity, development and evolution: the unmodern synthesis of E.S. Russell. Theory Biosci 132:165–180
Esposito M (2021) Cathedrals, corals and mycelia: three analogies for the history of evolutionary biology. In: Delisle RG (ed) Natural selection: revisiting its explanatory role in evolutionary biology. Springer, Cham, pp 11–35
Fábregas-Tejada A, Vergara-Silva F (2017) Hierarchy theory of evolution and the extended evolutionary synthesis: some epistemic bridges, some conceptual rifts. Evol Biol 45:127–139
Fábregas-Tejada A, Vergara-Silva F (2018) The emerging structure of the extended evolutionary synthesis: where does Evo-devo fit in? Theory Biosci 137:169–184
Folguera G, Lavagnino NJ (2018) The dual role of Evo-devo mechanisms and the extended evolutionary synthesis. Principia 22:251–275
Frazzetta TH (2012) Flatfishes, turtles, and bolyerine snakes: evolution by small steps or large, or both? Evol Biol 39:30–60
Futuyma DJ (2017) Evolutionary biology today and the call for an extended synthesis. Interface Focus 7:20160145. https://doi.org/10.1098/rsfs.2016.0145
Futuyma DJ (2023) The evolving evolutionary synthesis. In: Dickins TE, Dickins BJA (eds) Evolutionary biology: contemporary and historical reflections upon core theory. Springer, Cham, pp 517–541
Gawne R, McKenna KZ, Nijhout F (2017) Unmodern synthesis: developmental hierarchies and the origin of phenotypes. Bio Essays 40. https://doi.org/10.1002/bies.201600265
Gayon J (1998) Darwinism's struggle for survival: heredity and the hypothesis of natural selection. Cambridge University Press, Cambridge
Gilbert SF, Opitz JM, Raff RA (1996) Resynthesizing evolutionary and developmental biology. Dev Biol 173:357–372
Goldschmidt RB (1940) The material basis of evolution. Yale University Press, New Haven
Gould SJ (1980) G. G. Simpson, paleontology, and the modern synthesis. In: Mayr E, Provine WB (eds) The evolutionary synthesis: perspectives on the unification of biology. Harvard University Press, Cambridge, pp 153–172
Gould SJ (2002) The structure of evolutionary theory. The Belknap Press of Harvard University Press, Cambridge MA
Grene M (1958) Two evolutionary theories (I). Br J Philos Sci 9:110–127
Hall BK (1999) Evolutionary developmental biology. 2nd ed. Kluwer Academic Publishers, Dordrecht
Hall BK, Olson WM (eds) (2003) Keywords and concepts in evolutionary developmental biology. Harvard University Press, Cambridge
Hamburger V (1980) Embryology and the modern synthesis in evolutionary theory. In: Mayr E, Provine WB (eds) The evolutionary synthesis: perspectives on the unification of biology. Harvard University Press, Cambridge, pp 97–112
Huneman P (2017) Why would we call for a new evolutionary synthesis?: the variation issue and the explanatory alternatives. In: Huneman P, Walsh D (eds) Challenging the modern synthesis: adaptation, development, and inheritance. Oxford University Press, New York, pp 68–110
Huxley J (1942) Evolution: the modern synthesis. Allen and Unwin, London

Jablonka E, Lamb MJ (2005) Evolution in four dimensions: genetic, epigenetic, behavioral, and symbolic variation in the history of life. MIT Press, Cambridge MA
Jablonka E, Lamb MJ (2020) Inheritance systems and the extended evolutionary synthesis. Cambridge University Press, Cambridge
Jaeger J, Irons D, Monk N (2012) The inheritance of process: a dynamical systems approach. J Exp Zool (Mol Dev Evol) 318B:591–612
Johannsen W (1911) The genotype conception of heredity. Am Nat 45:129–159
Kuratani S (2009) Modularity, comparative embryology and Evo-devo: developmental dissection of evolving body plans. Dev Biol 332:61–69
Kutschera U, Niklas (2004) The modern theory of biological evolution: an expanded synthesis. Naturwissenschaften 91:255–276
Laland KN, Uller T, Feldman M, Sterelny K, Müller GB, Moczek A, Jablonka E, Odling-Smee J (2014) Does evolutionary theory need a rethink? Yes, urgently. Nature 514:161–164
Laland KN, Uller T, Feldman M, Sterelny K, Müller GB, Moczek A, Jablonka E, Odling-Smee J (2015) The extended evolutionary synthesis: its structure, assumptions and predictions. Proc R Soc B 282:20151019. https://doi.org/10.1098/rspb.2015.1019
Laubichler MD (2010) Evolutionary developmental biology offers a significant challenge to the neo-Darwinian paradigm. In: Ayala FJ, Arp R (eds) Contemporary debates in philosophy of biology. Wiley-Blackwell, Malden MA, pp 199–212
Laubichler MD, Renn J (2015) Extended evolution: a conceptual framework for integrating regulatory networks and niche construction. J Exp Zool (Mol Dev Evol) 324B:565–577
Lewens T (2019) The extended evolutionary synthesis: what is the debate about, and what might success for the extenders look like? Biol J Linn Soc 127:707–721
Linde-Medina M, Newman SA (2014) Limb, tooth, beak: three modes of development and evolutionary innovation of form. J Biosci 39:211–223
Love AC (2010) Rethinking the structure of evolutionary theory for an extended synthesis. In: Pigliucci M, Müller GB (eds) Evolution: the extended synthesis. MIT Press, Cambridge MA, pp 403–441
Love AC (2013) Theory is as theory does: scientific practice and theory structure in biology. Biol Theory 7:325–337
Love AC (2017) Evo-devo and the structure(s) of evolutionary theory: a different kind of challenge. In: Huneman P, Walsh D (eds) Challenging the modern synthesis: adaptation, development, and inheritance. Oxford University Press, New York, pp 159–187
Love AC (2020) Situating evolutionary developmental biology in evolutionary theory. In: Scheiner SM, Mindell DP (eds) The theory of evolution: principles, concepts, and assumptions. The University of Chicago Press, Chicago, pp 144–168
Løvtrop S (1981) Macroevolution and punctuated equilibria. Syst Zool 30:498–500
Mayr E (1980) Prologue: some thoughts on the history of the evolutionary synthesis. In: Mayr E, Provine WB (eds) The evolutionary synthesis: perspectives on the unification of biology. Harvard University Press, Cambridge, pp 1–48
Minelli A (2010) Evolutionary developmental biology does not offer a significant challenge to the neo-Darwinian paradigm. In: Ayala FJ, Arp R (eds) Contemporary debates in philosophy of biology. Wiley-Blackwell, Malden MA, pp 213–226
Müller GB (2007) Evo-devo: extending the evolutionary synthesis. Nat Rev Genet 8:943–949
Müller GB (2014) EvoDevo shapes the extended synthesis. Biol Theory 9:119–121
Müller GB (2017) Correction to 'why an extended evolutionary synthesis is necessary'. Interface Focus 7:20170065. https://doi.org/10.1098/rsfs.2017.0065
Müller GB (2019) Evo-Devo's challenges to the modern synthesis. In: Fusco G (ed) Perspectives on evolutionary and developmental biology: essays for Alessandro Minelli. Padova University Press, Italy, pp 29–39
Müller GB (2021) Evo-devo's contributions to the extended evolutionary synthesis. In: Nuño de la Rosa L, Müller GB (eds) Evolutionary developmental biology. Springer, Cham, pp. 1127–1138

Noble D (2015) Evolution beyond neo-Darwinism: a new conceptual framework. J Exp Biol 218:7–13

Ochoa C (2017) El eclipse del antidarwinismo: la historia detrás de la síntesis moderna. Centro de Estudios Filosóficos, Políticos y Sociales Vicente Lombardo Toledano, México

Ochoa C (2021a) Inertia, trend, and momentum reconsidered: G. G. Simpson—an orthogeneticist? In: Delisle RG (ed) Natural selection: revisiting its explanatory role in evolutionary biology. Springer, Cham, pp 261–290

Ochoa C (2021b) The origins of theoretical developmental genetics: reinterpreting William Bateson's role in the history of evolutionary thought. In: Delisle RG (ed) Natural selection: revisiting its explanatory role in evolutionary biology. Springer, Cham, pp 137–170

Ochoa C (forthcoming) Hierarchical evolutionary-developmental theory. Springer

Ochoa C, Barahona A (2014) El Jano de la morfología: de la homología a la homoplasia, historia, debates y evolución. UNAM; Centro de Estudios Filosóficos, Políticos y Sociales Vicente Lombardo Toledano, México

Ochoa C, Barahona A (2018) La revolución no-darwiniana: ¿es darwiniana la extensión de la Síntesis Moderna? In: Muñoz-Rubio J (ed) La biología evolutiva contemporánea: ¿una revolución más en la ciencia? Universidad Nacional Autónoma de México, México, pp 71–94

Ochoa C, Rasskin-Gutman D (2015) Evo-devo mechanisms underlying the continuum between homology and homoplasy. J Exp Zool B Mol Dev Evol 324:91–103

Odling-Smee FJ, Laland KN, Feldman MW (2003) Niche construction: the neglected process in evolution. Princeton University Press, Princeton

Pievani T (2016a) Hierarchy theory and the extended synthesis debate. In: Eldredge N, Pievani T, Serrelli E, Tëmkin I (eds) Evolutionary theory: a hierarchical perspective. The University of Chicago Press, Chicago, pp 351–364

Pievani T (2016b) How to rethink evolutionary theory: a plurality of evolutionary patterns. Evol Biol 43:446–455

Pigliucci M (2007) Do we need an extended evolutionary synthesis? Evolution 61:2743–2749

Pigliucci M (2009) An extended synthesis for evolutionary biology. Ann N Y Acad Sci 1168:218–228

Pigliucci M, Müller GB (2010) Elements of an extended evolutionary synthesis. In: Pigliucci M, Müller GB (eds) Evolution: the extended synthesis. MIT Press, Cambridge MA, pp 3–17

Peterson T, Müller GB (2016) Phenotypic novelty in EvoDevo: the distinction between continuous and discontinuous variation and its importance in evolutionary theory. Evol Biol 43:314–335

Popov I (2009) The problem of constraints on variation, from Darwin to the present. Ludus Vitalis 17:201–220

Provine WB (1971) The origins of theoretical population genetics. University of Chicago Press, Chicago

Provine WB (1980) Epilogue. In: Mayr E, Provine WB (eds) The evolutionary synthesis: perspectives on the unification of biology. Harvard University Press, Cambridge, pp 399–411

Raff RA (1996) The shape of life: genes, development, and the evolution of animal form. The University of Chicago Press, Chicago

Reif WE, Junker T, Hoßfeld U (2000) The synthetic theory of evolution: general problems and the German contribution to the synthesis. Theory in Biosci 119:41–91

Russell ES (1916) Form and function: a contribution to the history of animal morphology. John Murray, London

Salazar-Ciudad I, Jernvall J (2005) Graduality and innovation in the evolution of complex phenotypes: insights from development. J Exp Zool (Mol Dev Evo) 304B:619–631

Sattler R (1984) Homology: a continuing challenge. Syst Bot 9:382–394

Schlosser (2002) Modularity and the units of evolution. Theory Biosci 121:1–80

Schlosser G (2022) Rebuilding ships while at sea—Character individuality, homology, and evolutionary innovation. https://doi.org/10.1002/jmor.21522

Schwartz JH (2021) What's natural about natural selection? In: Delisle RG (ed) Natural selection: revisiting its explanatory role in evolutionary biology. Springer, Cham, pp 329–389

Smocovitis VB (1996) Unifying biology: the evolutionary synthesis and evolutionary biology. Princeton University Press, Princeton

Smocovitis VB (2023) Every evolutionist their own historian: the importance of history, context, and the extended evolutionary synthesis. In: Dickins TE, Dickins BJA (eds) Evolutionary biology: contemporary and historical reflections upon core theory. Springer, Cham, pp 25–54

Stoltzfus A (2017) Why we don't want another "synthesis". Biol Direct 12:23. https://doi.org/10.1186/s13062-017-0194-1

Stoltzfus A, Cable K (2014) Mendelian-mutationism: the forgotten evolutionary synthesis. J Hist Biol 47:501–546

Tamborini M (2022) The architecture of evolution: the science of form in twentieth-century evolutionary biology. University of Pittsburgh Press, Pittsburgh

Theißen G (2009) Saltational evolution: hopeful monsters are here to stay. Theory Biosci 128:43–51

van der Meer JM (2021) The concept of natural selection in Theodosius Dobzhansky: its development and interpretation. In: Delisle RG (ed) Natural selection: revisiting its explanatory role in evolutionary biology. Springer, Cham, pp 291–326

Verschuuren GM (1986) Investigating the life sciences: an introduction to the philosophy of science. Pergamon Press, Oxford

Vidya TNC, Dey S, Prasad NG, Joshi A (2023) The Darwinian core of evolutionary theory and the extended evolutionary synthesis: similarities and differences. In: Dickins TE, Dickins BJA (eds) Evolutionary biology: contemporary and historical reflections upon core theory. Springer, Cham, pp 271–328

Wagner GP (1989) The origin of morphological characters and the biological basis of homology. Evolution 43:1157–1171

Wagner GP (2015) Reinventing the organism: evolvability and homology in post-Dahlem evolutionary biology. In: Love AC (ed) Conceptual change in biology: scientific and philosophical perspectives on evolution and development. Springer, Dordrecht, pp 327–342

Waller JC (2003) Parents and children: ideas of heredity in the 19th century. Endeavour 27:51–56

Walsh DM (2015) Organisms, agency, and evolution. Cambridge University Press, Cambridge

Welch JJ (2017) What's wrong with evolutionary biology? Biol Philos 32:263–279

Welch JJ (2023) The creativity of natural selection and the creativity of organisms: their roles in traditional evolutionary theory and some proposed extensions. In: Dickins TE, Dickins BJA (eds) Evolutionary biology: contemporary and historical reflections upon core theory. Springer, Cham, pp 65–107

West-Eberhard MJ (2003) Developmental plasticity and evolution. Oxford University Press, New York

Wray GA, Hoekstra HE, Futuyma DJ, Lenski RE, Mackay TFC, Schluter D, Strassmann JE (2014) Does evolutionary theory need a rethink? No, all is well. Nature 514:161–164

Part IV

Deconstructing the Historiography of Evolutionary Biology

Deconstructing and Reconstructing the History of Evolutionary Thought: An Agenda for a "Post-Darwinian" Historiography

Maurizio Esposito

Abstract

In this chapter, I argue in favor of a new historiography of evolutionary thought that reassesses the place of Darwin in it. I challenge two myths that have haunted many previous histories: the idea that, before Darwin, most scholars believed in a static world populated of "essential types" and the related idea that Darwin is the undiscussed protagonist of any general history of evolutionary thought. I also claim that, while many scholars have highlighted the antiquity of "evolutionism," they did so under the problematic categories of "precursors" and "forerunners." Those categories assume that history must be written in terms of "pre-Darwinian" anticipations of Darwin. Instead, I argue that a post-Darwinian, historiography should move away from those categories altogether and consider the following questions: how did the modern concept of "evolution" become possible? What kind of conceptual ingredients made such an idea compelling? I explore the possibility that the very idea of "evolution" emerged by the confluence of ancient materialism and the burgeoning "diachronization" of nature as hinted by many different and influencial scholars such as Descartes, Leibniz, Burnet, Kant, Buffon, and Lamarck, among others. I suggest that modern materialism and the discovery of "deep time" in the seventeenth and eighteenth centuries, which prompted a progressive historcization of nature and naturalization of human and non-human history, constituited the two fundamental pillars sustaining and stimulating the development of an evolutionary worldview.

M. Esposito (✉)
Department of History and Philosophy of Sciences, Interuniversity Center for the History of Science and Technology, University of Lisbon, Lisbon, Portugal
e-mail: Mesposito@ciencias.ulisboa.pt

R. G. Delisle et al. (eds.), *Unity and Disunity in Evolutionary Biology*,
https://doi.org/10.1007/978-3-031-42629-2_15

Keywords

Evolutionism · Darwinism · Historiography of Evolutionary Biology · Materialism · Historicism

> To write history is so difficult that most historians are forced to make concessions to the technique of legend.
>
> Auerbach (2003, 20)

> If the old story is untrue, what are we supposed to teach our students now?
>
> M. Adams (2024), in this volume (Chapter "The Evolution of "Darwinism": Up Close and Personal")

1 Introduction

In 1988, Ernst Mayr declared that essentialism "...had dominated Western thinking for more than 2000 years after Plato." Essentialists, Mayr remembered, had believed that: "...the changing variety of things in nature is a reflection of a limited number of constant and sharply delimited underlying *eide*, or essences." And concluded: "...no genuine and testable theory of evolution could develop until the possibility was accepted that species have the capacity to change, to become transformed into new species, and to multiply into several species" (1988, 172). Mayr believed that the hero who overcame the pernicious essentialist dogma of antiquity was Charles Darwin. The British naturalist was the leading man who ended those 2000 years of claustrophobic and tedious eternity by introducing us to a new world of restless change and transformation. In this sense, for Mayr, the history of evolutionary thought (HET onward) was mainly the history of emancipation from the ancient yoke of ideological stability. The historian's task was thus to grasp how such an emancipation happened and how Darwin succeeded in the feat despite the inertia of millenary prejudices. In this chapter, I argue that this often-repeated narrative is fundamentally misleading. It underpins a sort of "Darwin mania" that needs to be scrutinized, reassessed, and eventually questioned. By paraphrasing Mayr's statement above, I claim that such a "Darwin mania" has dominated Anglo-Saxon historical thinking for more than 100 years after John Dewey. In fact, in the same year as Darwin's celebration in Cambridge in 1909, Dewey delivered a lecture at Columbia University that was reprinted in the *Popular Science Monthly* as "The Influence of Darwin on Philosophy." The lecture offered a powerful and convincing version of "Darwin centric mania." Dewey observed that Darwin broke with 2000 years of platonic epistemology privileging the knowledge of the eternal and necessary things over the transient, historical, and contingent phenomena. Before Darwin, the protagonists of the scientific revolution, Galileo, Kepler, and Descartes, had partially questioned the static universe of Aristotle and his medieval peers and redirected their attention toward the physical properties of movement. Darwin fulfilled such a conceptual shift by applying the dynamic view of the universe to life:

> The conceptions that had reigned in the philosophy of nature and knowledge for two thousand years, the conceptions that had become the familiar furniture of the mind, rested on the assumption of the superiority of the fixed and final; they rested upon treating change and origin as signs of defect and unreality. In laying hands upon the sacred ark of absolute permanency, in treating the forms that had been regarded as types of fixity and perfection as originating and passing away, the "Origin of Species " introduced a mode of thinking that in the end was bound to transform the logic of knowledge, and hence the treatment of morals, politics, and religion (Dewey 1910, 2).

Dewey defended two related claims: (1) Before Darwin, most scholars believed that the cosmos and species were largely static, and (2) Darwin's *Origin of Species* introduced a new way of thinking and doing science and philosophy. Precisely because Darwin constituted a break between the past and the future, between the science obsessed with the permanent and science fascinated by the transient, the heart and soul of the history of evolutionary thought is, and always was, Darwin. Dewey's conclusions have been resumed several times throughout the twentieth century by scientists and historians alike (including Mayr and then S. J. Gould).[1] Yet, as I show in this chapter, his Darwin centric view was historically flawed and philosophically spurious as they were its subsequent avatars. In what follows, I argue that deconstructing the historiography of evolution coincides with challenging the two powerful self-sustaining myths that feed most Darwin-centered historiographies: i.e., the idea that, with a few exceptions, before Darwin almost everyone believed that the universe was a perfect and static system directly concocted by God and the related idea that the Darwin was the undiscussed protagonist of the HET, as the one who put the last nail in the coffin of ancient essentialism.

The task is far from simple. To challenge and eventually debunk those claims means to welcome a sort of Copernican revolution in the historiography of the life sciences. We have been led to believe that Darwin was the shining fixed "star" at the center of an intellectual system where many small and medium-sized planets (scholars, thinkers, etc.) rotated. The image is so entrenched in our historical

[1]Dewey might have inspired some scholars, but it is more probable that Dewey's historical framework has have been reinvented a few times throughout the twentieth century I think that the reason why such a Darwin-centered obsession" thrived and persisted throughout the twentieth century, despite its evident historiographic shortcomings, should be the object of a sociological essay. I will not deal with this issue here. In this volume, Mark Adams provides a few insights explaining why some historical accounts thrive and persist, despite of their evident shortcomings. For what concerns this chapter, it must be said that Dewey's historical viewpoint is like those philosophical myths of foundation, akin to Hobbes' *Homo homini lupus*, Rousseau's Noble Savage or Adam Smith's self-interested individuals. To be sure, Dewey's historical narrative served him as intellectual support for his instrumentalism or pragmatism, based on the dynamic and evolutionary relation between organisms and their environments.

subconscious that a special intellectual effort is required to overcome our cognitive inertia. And we would be naïve if we hope to break the inertia by merely reconsidering the precursors and forerunners of Darwin since antiquity. An important number of scholars have already brilliantly charted many "pre-Darwinian" ideas and theories.[2] But I contend that when we use the label "pre-Darwinian," we inevitably reinforce the traditional historiography that puts Darwin as the historical "unit of measurement" from which all evolutionary thought should be compared. We should move a step further and dispense with "pre-Darwinian" narratives as well as labels such as "precursors" and "forerunners." In this chapter, I claim that no "stars" illuminate the intellectual system we call evolutionary thought. There are not Victorian "geniuses," paradigmatic books, "dangerous ideas," or great uncompromising heroes, but a complex and assorted debate that has concerned many people in different places ever since antiquity. The debate addressed a persistent and challenging philosophical problem: the nature and origin of complex entities we usually call "living things." Accordingly, rather than wondering why and how Darwin and Darwinism succeeded and in what did the Darwinian revolution consist of, I suggest to tackle the following questions: how did the very idea of evolution become possible? What kind of conceptual ingredients made such an idea compelling? How did thinkers start to conceive nature historically? Of course, I can only draft a few possible answers in this chapter in the hope that other scholars might take this challenge seriously.

My proposal is both destructive and constructive. Throughout the chapter, I explore possible ways to restructure the historiography of evolution by revisiting two fundamental historical moments: the rediscovery of ancient materialism since the 15th century and the rediscovery of "history" since the 16th century, which introduced a new conception of "deep time" in both natural and human sciences.[3] My tentative argument is that the concept of "evolution" was made possible by the confluence of ancient materialism with burgeoning "historicism" as gradually shaped by figures as different as Descartes, Leibniz, Kant, Buffon, and Lamarck. In short, between the seventeenth and nineteenth centuries, we have a parallel movement consisting of naturalizing history and historizing nature. Darwin was dragged in the big intellectual wave that began in the seventeenth and eighteenth centuries and exploded in the nineteenth century. The structure of the long argument I defend in this chapter runs as follows: in Sect. 2, I challenge the "Darwin mania" and the two myths underpinning it: the idea that before Darwin, almost everyone hold that the world was static and the idea that Darwin must be at the center of the HET because he freed us from essentialism, opening the intellectual gates to an evolutionary worldview. In Sect. 3, I address the conceptual and historical implications of the rediscovery of ancient materialism in the early modern period. Finally, in Sects. 4 and 5, I show how modern materialist speculations might have

[2]From Zirkle (1941) to Sloan (2019)

[3]On the history of the idea of "deep time" in Geology, see Rudwick 2007.

created the conditions for the development of an evolutionary perspective of the physical and biological world.

2 Deconstructing the Myths

Is it true that before Darwin, mostly believed that the universe was a sort of static "machine" made of substantial forms created by a powerful Demiurge, an eternal prime mover or a benevolent God? Following Dewey, we might argue that Greek metaphysics and epistemology, which supposed that anything worth knowing was about the eternal, necessary, universal, and so forth, could not accept something like an evolutionary process. In a Platonic world of abstract forms or an Aristotelian universe peopled of eternal substances, the idea that complex entities might result from a sort of material and the contingent process was a sort of oxymoron. For an uncompromised Platonist or Aristotelian, the world we observe is already definitive and complete. We do not need to ask about its origins but inquire about the eternal laws making it possible and persisting. Anything transient could not be an object of science for the simple reason that what matters it revealing the stable and fundamental core making transitiveness possible. The very Aristotelian notion of *eidos* (species) was the consequence of a philosophy of knowledge obsessed with permanence. Second, along with Mayr, we might suppose that the fundamental stumbling block for having an evolutionary thought was not epistemology in general but the Aristotelian essentialism about species. The idea that all organic species held essential and immutable characters was utterly incompatible with an evolutionary view of life. The metaphysical distinction between essences and accidents made an evolutionary standpoint unthinkable. "For an essentialist—Mayr claimed—there can be no evolution, there can only be a sudden origin of anew essence by major mutation of saltation" (Mayr, 1988, 172). But whether we follow Dewey, who believed that the whole Greek metaphysics was the culprit, or Mayr, who blamed the notion of *Eidos*, both views lead to the same conclusion: The Greek worldview thwarted any attempt to imagine a dynamic and transformative conception of nature and life.

Today we know that Dewey and Mayr's views are deeply problematic for two reasons. First, the Aristotelian view did not coincide with the whole Greek view. Platonic and Aristotelian metaphysics was only an option among others, even in the ancient world. And second, the notion of *eidos* and, therefore, the supposed Aristotelian essentialism, were not what Mayr assumed it to be. Let us start by assessing the latter and then address the former. In the 1980s, David Balme persuasively argued that Aristotle's biology was not essentialist (Balme 1987). He observed that both notions of essence and form did not correspond to "species" in Aristotle. The essentialism Mayr decried for being an epistemic obstruction impeding an evolutionary view was an invention of the Middle age, where diverse separate concepts, essence, form, and species collapsed into each other.[4] Balme's insight was developed by Fran O'Rourke in 2004 and Mary Winsor in 2006). In a very long essay titled "Aristotle and the metaphysics of evolution", O'Rourke claimed that as Aristotle's biology could not be considered "essentialist", it could not have precluded the development of an evolutionary perspective (2004). Winsor argued that the Platonic and Aristotelian "essentialism" story was primarily manufactured by Mayr between 1953 and 1968 and then uncritically accepted by many other scholars (2006). After Balme, Mayr himself eventually recognized that his understanding of Aristotle's essentialism was somewhat spurious and distinguished the notion of Platonic *eidos* from the Aristotelian one:

> One of the reasons why Aristotle has been so consistently misunderstood is that he uses the term *eidos* for his form-giving principle, everybody took it for granted that he had something in mind similar to Plato's concept of *eidos*. Yet the context of Aristotle's discussions makes it abundantly clear that his *eidos* is something totally different from Plato's eidos (I myself did not understand this until recently). Aristotle saw with extraordinary clarity that it made no more sense to describe living organisms in terms of mere matter than to describe a house as a pile of bricks and mortar. Just as the blueprint used by the builder determines the form of a house, so does the *eidos* (in its Aristotelian definition) give the form to the developing organism, and this *eidos* reflects the terminal telos of the full-grown individual (Mayr 1988, 56).

Suppose Balme, O'Rourke, Winsor, and the above-mentioned view of Mayr are right. Then the notion of Aristotelian essentialism cannot be used to distinguish a pre-Darwinian from a post-Darwinian world. If Aristotle's biology was not essentialist and, even more, if it could eventually accommodate an evolutionary perspective, then the relation between Darwinism and the past must be entirely retooled (O'Rourke 2004). In short, the idea that Darwin introduced us to a new dynamic world requires further unpacking. The distinction between static and dynamic nature is neither the obvious nor the most appropriate frame for understanding the presumed novelty of Darwinism. Undoubtedly, Darwin can still be considered the protagonist behind the development of evolutionary thought, but not because he rejected Aristotelian essentialism or any other imaginary static world. One might highlight one or more concepts constituting the Darwinian worldview (whether common ancestor with modification or natural selection—something I will address in the last sections), but one cannot single out a fabricated version of "essentialism" to characterize a supposedly new anti-essentialism. While Mayr's conception of essentialism is non-Aristotelian, we can still agree with Dewey's stance that Greek

[4]As Balme wrote: "In *GA* as in *Metapk. Z*, neither essence nor form corresponds to 'species', whether in its Neoplatonist sense of an inner real form or in its modern sense of a class or population that has the same bodily structure and lifestyle... The extraordinary later misinterpretations of Aristotle, the magical entelechies and real specific forms, must be largely due to these imported concepts - Species, Essentia, Substantia - which presided like three witches over his rebirth in the Middle Ages, but should be banished to haunt the Neoplatonism from which they came" (Balme 1987, 306).

epistemology made evolutionary thinking impossible until Darwin. This option, though, supposes that before Darwin, the universe (and all living creatures in it) was conceived Platonically or Aristotelian by virtually everyone. Is that historically tenable? Were Plato and Aristotle's philosophies the only games in town? What about materialist or materialistic approaches? After all, Aristotle's metaphysics and epistemology were also a reaction against the atomists and materialists of his time. In what follows, we will see, both the essentialist story and Dewey's narrative share the same problem: a discrete silence about materialism.

Oddly enough, when we explore how historians of evolutionary thought have used the notion of "materialism," we realize that it is the proverbial elephant in the room. The idea is everywhere, but few really delved into it and when those few did it, "materialism" is very often portrayed in an overly schematic way. If we look at some articles and books that have become classics in the field, we realize that there is something odd about how "materialism" is used and conceived. For example, Dewey mentions the word "materialism" only once in his paper. And he did it because he wanted to emphasize that Darwin's theory was not "materialist" and, therefore, the theory could not be accused of being necessarily irreligious. In Mayr's monumental *Growth of Biological Thought* (1982), "materialism" is used with three different things: (1) Engels and Marx's dialectical materialism (pp. 9 and 863), (2) A sort of misplaced or dogmatic reductionism (p. 128), and (3) An irreligious or atheist approach as opposed to any form of mysticism (p. 535). Our short review might also consider Michael Ruse's *The Darwinian Revolution* (1979), where "materialism" is mentioned twice to characterize Lamarck's evolutionary stance based on the idea that there was no essential difference between organic and inorganic worlds (1979, pp. 7–8). We can follow with another monumental work, S. Jay Gould's *The Structure of Evolutionary Theory (2002),* where "materialism" is mentioned seven times in 1343 pages and mainly with meanings 2 and 3 mentioned above (2002, pp. 173, 261, 273, 274, 276, 329). Our short literary journey might end with Peter Bowler's *Evolution: The History of an Idea* (1989), where we find a more complete and compelling use of materialism. Yet, Bowler's use of the notion is particular and contextual. While discussing the views of preformationists in the seventeenth century, he observed that:

> The naturalists of the seventeenth century cannot altogether be blamed if their more primitive version of *materialism* forced them to conceive of preformation in a more literal sense. Inevitably, though, they understood the theory in the context of a worldview in which each species was fixed by a design established by its creator (1989, 47).

The history of knowledge is undoubtedly filled with opportunistic attempts to conciliate opposed ideas or contradictory traditions. While creationist materialists might have existed, yet it seems to me that the "primitive version of materialism" should instead be a "primitive version of mechanism" because materialists—those inspired by the encompassing view of Lucretius—could never accept the idea that "...each species was fixed by a design established by its creator" (1989, 47). True, and especially after the seventeenth century, we have different, and often incompatible, kinds of "materialisms." Historians have provided us with a wealth of different, sometimes opposed, definitions (Lenoir, 1982; Reill, 2006; Zammito, 2017). However, the ancient materialism of Epicurus, then rediscovered in northern Europe in the 15th century with Lucretius' *Rerum Natura*, was far less ambiguous than it has been commonly portrayed in the historiographies of biology. "Materialism" was not just a vague idea upholding a seemingly reductionist, irreligious, or mechanist perspective. It was a research program with a long history and deep philosophical contents. In the next section, I will go through some of these contents, which hardly touch some of those expressed by Mayr, Ruse, Gould, and Bowler. For now, it suffices to anticipate three things: (1) Materialism was a widely influential agenda in antiquity and it became fashionable again in modernity after being eclipsed for more than one thousand years. (2) Materialists emphasized the importance of movement, change, and transformation over stability and essentialism. (3) Most materialists believed that anything could be explained by material and efficient causes. No final *telos*, entelechies, mystical forces, or gods were needed to understand complexity (whether organic or inorganic).[5] If we concede these three claims, then again, Dewey and Mayr's "Darwin mania" is even more untenable. Before addressing some crucial elements of ancient and modern materialism, let us briefly address one of the main implications of the second myth: the idea that Darwin must be hailed as the protagonist of any general history of evolutionary thought.

If Dewey's story is incorrect and Mayr's conception of essentialism an historical "forgery," then Darwin cannot be hailed as the one who freed us from a static worldview. If Darwin did not emancipate us from natural stillness, in what sense he can still be considered as the founding father of evolutionary thought? Put differently, once we accept that Darwin did not introduce us to a new open and dynamic world or did not emancipate us from an imaginary essentialist tradition, why shall we consider him the heart and soul of history of evolutionary thought? I have seriously addressed these questions elsewhere and I will not repropose the analysis here. (Esposito 2021) For the sake of argument, it suffices to recall that: 1) Evolutionary

[5]Even a sophisticated intellectual historian such as John C. Greene did not seriously consider the materialist tradition as it unfolded since the 15th century. As he observed in a famous essay published in the late 1950s, in the eighteenth century: "...dominant sentiment of the stability and permanence of the great features of nature: the fixed stars, the everlasting hills, the eternal seas, the created species. Change was recognized as a real aspect of nature, but a superficial aspect" (1981, 12). Greene's conclusion might be right when applied to Newton and the British background (although I have my reservation), but in the continent, the materialist ideas of Lucretius were hotly debated for centuries. And, for Lucretian materialism, change was far from superficial (as we will see in the next sections). Much probably, if Greene had seriously considered the spread of materialism since antiquity, he would have realized that his "static" view of nature was not overwhelmingly dominant after all. Since the Greeks, cyclical and non-cyclical views of the universe were constantly debated. When we consider, for instance, the book 5 of Lucretius' *De Rerum Natura*, we can appreciate that cyclical and non-cyclical perspectives coexist.

biologists are entitled to put Darwin or whoever else they like at the center of the evolutionary theory (or theories), but historians have no obligation to follow scientists in their ideological (or even non-ideologically justified) bid to use Darwin as their founding father.[6] And 2) I am not interested here in reharsing the history of the Darwinian view of evolution by providing a few corrections to it. I am interested in something prior and more fundamental: I want to understand how the very concept of "evolution" evolved; i.e, how the idea that nature and living beings have a particular kind of history emerged and how the knowledge of this "history" became epistemically relevant to understand the structure and function of all organisms. In this sense, I am persuaded that HET cannot be grounded in the mind of one or few smart and inspired scholars but it lies in the much more confusing depths of conceptual history. The idea of "evolution" was a gradual achievement that took centuries of intense discussions addressing the nature and origin of complex beings. On this account, I argue that the materialist tradition, and the debates surrounding it, is the fundamental historical element we shall seriously reconsider.

3 Ancient and Modern Materialisms (and Why They Matter to the HET)

If we ask, "What is materialism?" We will easily end up with many inconsequential and contradictory definitions. Yet, if we ask: "From where and how did the materialist view emerge?" Our answer could be more specific and potentially illuminating. Historians of philosophy typically grant Leucippus and Democritus as the founders of the materialist tradition. Both are usually remembered as "atomists" active in the fifth century BC. They believed in a deterministic world constituted by the incessant swirling of indivisible particles they called atoms. The other important referent of the materialist school is the post-Socratic philosopher Epicurus. Unfortunately, not much is left of the ancient materialist tradition. As Catherine Wilson reminds us, until the fifteenth century, materialists were known mainly through their detractors. Little remains of Leucippus and Democritus, and almost nothing remains of the 37 books composing *Peri Physeos* of Epicurus. Yet, unlike Leucippus and

[6]Here we need to distinguish between two alternatives: is Darwin the founding father of "contemporary" evolutionary biology? Or, in alternative, is Darwin the main protagonist of the history of evolutionary thought? Elsewhere, I showed that the two questions inspire two very different agendas and should not be confused (Esposito, 2021). The history of evolutionary thought should not be identified with the actual content of evolutionary biology. I did argue that the task of thinking about the structure of evolutionary thought should not be confused with the different task of exploring the history of evolutionary thought. To overlap both tasks, namely, a synchronic systematization with a diachronic reconstruction of evolutionary thinking, leads quickly to pernicious forms of presentism. I.e., to say that Darwinism is at the heart of evolutionary theory (whatever it means) is not the same as to say that Darwinism is at the heart of the history of evolutionary thought. As simple as it might sound, such sort of fallacy has often been committed by many scientists turned historians (including Mayr and Gould). In this chapter I focus on the second question.

Democritus, Epicurus had the good fortune of having a disciple who managed to write a successful introduction to his thought: the Roman poet Titus Lucretius Carus (94 a.C.–c. 50 a.C). We do not know much about the biography of Lucretius, and we do not know how his *De Rerum Natura* managed to evade centuries of neglected and convoluted history. We do know that in 1416, a books hunter, Poggio Bracciolini, discovered a copy of the lost poem in a library in Germany, probably in the abbey of Fulda. The manuscript was composed of 6 untitled books. It portrays a universe that was nothing like that of Plato and Aristotle and nothing like medieval Platonic and Peripatetic reinterpretations. The history of the discovery, the editions, translations, and interpretations cannot be told here. As Wilson and Stephen Greenblatt have convincingly demonstrated, the book exerted an immense influence in the centuries after its discovery. Seminal thinkers as different as Bruno, Bacon, Spinoza, Leibniz, Gassendi, Galileo, Holbach, Diderot, and many others were deeply impressed, and the impact on modern science could hardly be overestimated. In fact, the dynamic universe that Dewey attributed to the sole inspiration of Galileo, Descartes, or Kepler was in part the outcome of the dissemination of materialist ideas. The corpuscular hypothesis, the difference between primary and secondary qualities; the exclusion of formal and final causes from the domain of scientific intelligibility; the idea that physical and organic things could be explained without recurring to any divine intervention; the infinity of the universe, the plurality of worlds and many other ideas were immortalized in the verses of Lucretius.

In his *The Swerve* (2012), Greenblatt uncovered the rediscovery of Lucretius's book and listed some central tenets that would characterize the materialist tradition in the centuries to come, which I summarize as follows:

- The atomic perspective assumes that matter is infinite and eternal.
- Everything that does exist results from the constant collision of atoms (seeds).
- Physical and biological order is not the result of a designer or creator.
- Novelty and change in the universe are due to pockets of indeterminism (clinamen, translated as swerve).
- The universe was not created for the existence of humans. There were no golden ages to be remembered.
- There is no intrinsic rationality in the world and universe. The world and universe are far from perfection.
- The order we perceive in nature is due to constant experiments, whereby more organized things persist, and less organized ones tend to disappear.

These points (and others I am not listing because they are not directly relevant in this discussion) constituted the materialist's essential intellectual toolbox. No doubt, in the centuries following Lucretius' rediscovery, many other combinations were tested. Especially from the seventeenth century onward, materialist options were coupled with non-materialist elements. Probably the most notable examples are Francis Bacon in the seventeenth century and the catholic priest Pierre de Gassendi, who was instrumental in resurrecting Epicurean materialism in the eighteenth century.[7] Of course, it would be impossible to do justice to all the materialist projects that proliferated before and after the nineteenth century. But it might be instructive to focus on one aspect of book 5 of Lucretius' *Rerum Natura*, where he directly addressed the generation of organisms. The book drew the attention of Leibniz and many other first-rate scholars between the seventeenth and eighteenth centuries. As I show in the next section, it stimulated a productive debate over the origins of complex things (including organisms).

Lucretius' Book 5 is extremely rich in terms of contents and hermeneutical possibilities. It addresses the constitution of our universe, the generation of living creatures on Earth, and the beginning of human civilization, language, and inequality. Particularly relevant for us is Lucretius' central claim that our world is far from perfect and could not be the product of a benevolent being. He recognized that there was an important connection between observable imperfection and the lack of an intelligent design. The universe and its physical elements (Earth, water, air, and fire) emerged out of atomic collisions. In this frame, the formation of our planet was itself the outcome of deterministic laws and chance. Lucretius speculated that, in the early ages of earth formation, many "monsters" (portenta) emerged. Those "monsters" were mostly deprived of appendages and organs, feet, hands or eyes. Most of them were functionally unfit for survival:

> Nature also tried to create many monsters simultaneously, which arose with fantastic appearance and limbs...Some were found bereft of feet, other in turn bereft of hands, also dumb without mouths and blind without faces, and bound by the adhesion of their limbs all along their bodies, so that they could not do anything, nor avoid evils, nor take up what was o use. She created other kinds of these monsters and portents, in vain since nature forbade their growth, nor were they able to reach the desired flower of age nor find food, nor be joined through the work of Venus. For we see that many things need to come together in creatures for them to be able to forge a race by breeding; firstly that there is food, then secondly that they have a way by which they generative seeds can trickle through the limbs when the members have relaxed, and thirdly so that females may be able to be joined to males, and they have the physical means whereby they can both exchange mutual joys between themselves. And it is inevitable that many species of living creatures become extinct at that time and that they were not able to found a race by breeding. For whatever you see grazing on the life-giving breezes, either cunning or strength or finally speed has kept that species safe since the origins of its birth... [835-865] (Campbell 2003, 32-33).

The passage is deep and needs some unpacking. Lucretius argues that, at the beginning of the Earth, there was a chaotic proliferation of forms. Some of those forms were viable, and some were not. The organic forms that persisted were not only viable, but they were also apt (today, we would say fit). In order to avoid extinction, organic forms require at least two elements: they had to be morphologically and physiologically viable and avoid intraspecific annihilation and extinction. Thus, Lucretius supposed two kinds of selection behind the origins of news creatures: selection of viable forms (forms that can survive and reproduce) and selection of fit forms (forms that can avoid being eaten or destroyed by other

[7]On Bacon and the materialist tradition, see Manzo 2001.

creatures). To put it shortly, nature first experiments and produces innumerable monsters. Then, environmental constraints eliminate unviable and unfit forms. Only viable and fit persist. And those viable and persisting forms remain unchanged until they eventually become extinct. Gordon Campbell, who commented in detail the book 5, argued convincingly that Lucretius cannot be deemed a proto-evolutionist (although he put forward a sort of evolutionist view about human development) for two fundamental reasons: not only in his view there is no hint of common ancestry, but, more crucially, he believed that those successful "monsters" persisted but did not transmute. Lucretius does not contemplate the existence of historical lineages or a unity of type. In his dynamic universe, organic forms only generate and reproduce, they never change.[8]

But Lucretius's importance does not lie in the fact that he might be deemed a seminal proto evolutionist. His fundamental importance lies in providing an alternative, non-theological, and non-teleological approach with aiming to explain the emergence of organic complexity. In the seventeenth century, this simple fact did not escape Leibniz's philosophical wit. In the dialogue titled *Conversation du Marquis de Pianese et du père Emery Eremite*, written around 1680, Leibniz described an imaginary discussion between a learned Hermit and an Epicurean libertine addressing whether "...variation and selection could produce order from chaos" (quoted in Wilson 2008, 102). The libertine tried to convince the former that the Lucretian view, according to which: "...badly made objects perish, and that the well-made ones are preserved ... thus although there are an infinity of badly-made objects, they cannot maintain themselves amongst the others" (2008, 102). When we observe the body of an organism, whereby all its parts fit together for the benefit of the whole, we should not necessarily conclude that a superior maker must have necessarily created it. It could result from a selective process; whereby more organized things persist over the less organized ones. Necessity and contingency generate order from below. The Hermit wryly replied that he had not observed something that was halfway complex. The natural world looked as if it were already perfectly constituted. The Hermit did not exclude the possibility that, given an infinite number of worlds, some worlds could be more perfect than others (and, given the endless collision of atoms, some functional organic bodies might emerge). Yet, he did emphasize that this possibility was infinitely remote. Living organization, as any other kind of organization and complexity, out of chaos is: "...as little credible as to suppose that a library forms itself one day by a fortuitous concourse of atoms" (2008, 102). Leibniz would concede that order without makers is possible but not probable. Yet, notwithstanding its improbability, it was a hypothesis he deemed worth to play with, and after Leibniz, many others seriously played with it.

No doubt, there were some fundamental shortcomings implicit in the materialist hypothesis. As a devoted Aristotelian, the Leibnizian Hermit recognized that the Epicurean (and Lucretian) kind of materialism was exposed to an assorted battery of

[8]The only moment we might glimpse a dim ray of evolutionism in Lucretius is when he reconstructs the history of humanity.

formidable objections. To grasp the full force of thoss objections, we need to take many steps back in the history of philosophy. We need to go back to the fourth century BC, when the materialist hypothesis explaining the origin of living beings was famously crushed by Aristotle in Book II of his *Physics*, in the *Alpha* of *Metaphysics,* and in Book 1 of *Parts of Animals*. The intensity and reiteration of those objections suggest that the materialist hypothesis was widespread well before Epicurus. But Aristotle's objections against materialism were so powerful and persuasive that it is surprising that materialists were still around centuries after Aristotle's death. The most general argument against materialism is probably in the book *Alpha*, where Aristotle noticed that Anaxagoras and Empedocles had argued that everything results from "combination and differentiation." "From this account—Aristotle observed—it might be supposed that the only cause is of the kind called *material.*" But this could be enough because: "It is surely not the substrate itself which causes itself to change. I mean, e.g., neither wood nor bronze is responsible for changing itself; wood does not make a bed, nor bronze a statue, but something else is the cause of the change. Now to investigate this is to investigate the second type of cause: the source of motion, as we should say" (1933, M 1, III, 9–13). Matter itself could not cause order and change, as wood cannot be the actual cause of the table or bronze, the cause of the statue. In short, matter itself cannot be an agent of change.

The objections against the materialist hypothesis get even more detailed and devastating in the second book of *Physics*, where Aristotle observed that *chance* cannot be considered a proper "cause" of something. Empedocles is once again the critical target because: ". . .he says in his Cosmogony—Aristotle commented—... that the parts of animals mostly came to be as the outcome of luck." (1933, II, 20, 32). Aristotle did not deny that contingency existed. He was skeptical about the possibility of using "chance" to explain something persitently regular occurring in nature. In the eighth chapter of his *Physics*, Aristotle wonders why we cannot consider "concurrence" (contingency) as the cause of order and complexity. "Why should not everything be like the rain? Zeus does not send the rain in order to make the corn grow: it comes of necessity" (1970, 8, 15–20, 39). The rain is not there for the sake of corn. The corn is there because it rains:

> What, then, is to stop parts in nature too from being like this—the front teeth of necessity growing sharp and suitable for biting, and the back teeth broad and serviceable for chewing the food, not coming to be for this, but by coincidence? And similarly with the other parts in which the 'for something' seems to be present. So, when all turned out just as if they had come to be for something, then the things, suitably constituted as an automatic outcome, survived; when not, they died, and die, as Empedocles says of the man-headed calves. (1970, II. 8, 20-30, 39)

Aristotle observed that the problem with "concurrence" is that it is precisely the opposite of order. *Chance* is the exception of a natural and persistent order and cannot be itself the cause of that order. In other words, chance exists precisely because nature is normally and consistently organized. And chaos and blind selection cannot explain order:

> The point is most obvious if you look at those animals other than men, which make things not by art, and without carrying out inquiries or deliberation. Spiders, ants, and the like have led people to wonder how they accomplish what they do, if not by mind. Descend a little further, and you will find things coming to be which conduce to an end even in plants, for instance leaves for the protection of fruit. If, then, the swallow's act in making its nest is both due to nature and for something, and the spider's in making its web, and the plant's in producing leaves for its fruit, and roots not up but down for nourishment, plainly this sort of cause is present in things which are and come to be due to nature. And since nature is twofold, nature as matter and nature as form, and the latter is an end, and everything else is for the end, the cause as that for which must be the latter (1970, II. 8, 40-41,)

Nature is constituted by matter and form; the latter is the cause for the organization of the former. The Aristotelian form (Eidos) is not primarily what Dewey and Mayr have thought to be—an essential or typological view of a species—but the *conditio* sine qua non for explaining order and complexity in nature in opposition to the materialist alternative.

In Book 1 of *Part of Animals*, Aristotle delivers the final blow to the materialist hypothesis. Here, the culprit is mainly Democritus, who believed that from the constitution of matter, he could derive an explanation of the constitution of all things, from stones to animals:

> What Democritus has said, then, is too unqualified, and is said in the same way as a carpenter might speak about a wooden hand. Indeed, this is also the way the natural philosophers speak of the generations and causes of configuration. Ask them by what potencies things were crafted. Well, no doubt the carpenter will say an axe or an auger, while the natural philosopher will say air and earth-albeit the carpenter's response is better; for it will be insufficient for him to say merely that when the tool fell this became a depression and that flat. Rather, he will state the cause, the reason why he made such a blow and for the sake of what, in order that it might then come to be this or that sort of shape. It is clear, then, that these natural philosophers speak incorrectly. . .Clearly, one should state that the animal is of such a kind, noting about each of its parts what it is and what sort of thing it is, just as one speaks of the form of the bed (2001, 640b, 25-35, 5,).

Aristotle argued that spontaneous events are infrequent, especially when we observe living beings. When we observe organs, limbs or any other animal part, we see purpose and order everywhere. To argue that matter and contingency could have produced such an order is, at best, intellectual laziness. For Aristotle, one could certainly play with the materialist hypothesis as a genuine, although desperate, attempt to provide a natural explanation of the origin of complex forms. Unfortunately, in the face of what we empirically observe, the hypothesis is less plausible than supposing the existence of a demiurge or, as Aristotle did, positing the eternity of the cosmos' order. Aristotle's objections persisted for centuries and worked as effective blueprints for many anti-materialist arguments. One notable example was the argument put forward by the Roman stateman and contemporary of Lucretius, Marcus Tullius Cicero.

In his *De natura deorum*, Cicero distinguished between two perspectives—the teleological and anti-teleological—by emphasizing two sorts of orders: order *ab infra* and order *de supra*. The former summarized the materialist idea that complex things were essentially the outcome of the movement of less complex things. The latter referred to the notion that order can only be imposed from the outside. Cicero argued that there could be no order without something more complex, bestowing its stamp on less articulated things. Cicero was undoubtedly much less sophisticated than Aristotle because the latter conceived *form* as something consubstantial with matter. Unlike Cicero, Aristotle did not believe that complex natural things were the direct outcome of an external maker. Like any other natural (biological) entity, organisms were the effects of internal, purposeful principles. What Aristotele and Cicero shared, however, was the idea that something less complex could not give rise to something more complex. To put it differently, while materialists wanted to explain complexity from below (i.e., self-organization), Aristotelians and Cicero replied that complexity could only be understood by supposing the existence of a higher (more complex) principle. For example, if we observe a complex entity like an organism, we must assume that its organization is not something that emerges spontaneously but is directed or shaped by some internal or external force. This was implicit in Aristotle's metaphysical view of the material world: if matter itself cannot be an agent of change—like bronze does not produce any statue for itself—then something more complex had to be invoked to organize it. If bronze alone cannot form statue, why shall we suppose that matter alone can generate organisms?.

When Lucretius was rediscovered in the fifteenth century, many of those ancient discussions were resurrected. In the beginning, the result of those debates was not significantly different from antiquity: the existence of complex forms screams loud the presence of a designer or some principle of organization. The possibility that something complex can emerge and persist out of sheer chance was still deemed hugely implausible. Anything organized supposes the existence of an organizer. As the Leibnizian Hermit pointed out: "If I found myself transported into a new region of the universe where I saw clocks, furniture, books, ramparts, I would venture everything I possessed that this was the work of a rational creature" (cited in Wilson 2008, 103). The example was famously rehearsed (or perhaps plagiarized) by William Paley in his 1802 *Natural Theology*. However, as I will show in the next section, something radically new happened between the late seventeenth and early nineteenth centuries. Something that made the materialist hypothesis finally plausible and the Aristotelian's formidable objections much less challenging. What made Paley's objections less convincing than those put forward by Aristotle, Cicero, and Leibniz's Hermit was the discovery (or perhaps the rediscovery) of "deep time."

4 The Rediscovery of "History"

While our historical conception of nature (and the world more generally) is mainly the upshot of modern discussions, the idea of a developing universe is much older. As the German historian Friedrich Meinecke noticed in the 1930s and Stephen

Toulmin reemphasized in the 1960s, the historical conception of nature took centuries to thrive (Toulmin 1965).[9] In antiquity, Milesian philosophers and historians had provided some incipient development theories of the cosmos, which included both nature and human society. As the historian Benjamin Farrington had noticed long ago: "The Ionian school of natural philosophers had offered a materialistic explanation of the evolution of the cosmos, they inculcated the ideal of positive science and the reign of universal law, they gave an account of the development of civilization in which man, thought his conquest of techniques, figured as the author of his own progress" (1953, 89). Toulmin was even more specific: the philosophers of Miletus put forward a developmental view of nature and humanity. Milesians conceived the whole universe as the result of a continuous creative process. The philosopher Anaximander (c. 610–c. 546 BC), for instance, envisaged a diachronic process of material creation whereby organic beings emerged through spontaneous generation and metamorphosis:

> The first animals were generated in the moisture, and were enclosed within spiny barks. As they grew older, they migrated onto the drier land; and, once they bark split an shed, they survived for a short of time in the new mode of existence. . .Man to begin with was generated from living things of another kind, since, whereas others can quickly hunt for their own food, men alone require prolonged nursing. . .Thus men were formed within these [fish-like creatures] and remained within them like embryos until they had reached maturity (Anaximander in Toulmin 1965, 36).

The wild speculative hypothesis of Anaximander about the origin of the cosmos and life was paralleled by the insights of the geographer and historian Hecataeus of Miletus (550–476 BC), who was one of the first scholars to introduce prose history (logography) as distinguished from Mythology in the reconstruction of human past events. The worldview of the Milesian was powerfully synthesized in Diodorus' world history, *Bibliotheca historica,* four centuries later. At the beginning of the colossal work, Diodorus sketched a cosmology that uncovered the universe's formation, the generation of the first animals, and the primordia of human civilization (see Toulmin 1965, 35). The examples show that the quest for reconstructing the whole history of the Earth, non-human and human animals, was considered a feat worth pursuing.

However, after the fifth century BC, the assumed continuity between nature and human history was shattered, Toulmin added. The Greek historians Herodotus and Thucydides did separate the history of nature from the history of human societies, considering the latter a much more interesting task to pursue. Then, by the fourth century BC, most philosophers moved away from speculations about the universe's origins and, eventually, processual nature. The two principal schools of thought of antiquity considered that either the universe was created in one stroke by a cunning and wise Demiurge (Plato) or it was itself caught in cyclical and repeating eternity (Aristotle). If someone had asked Aristotle what the universe's origin was and how

[9] See Meinecke 1972

the Earth was formed, he would have probably considered the question preposterous: the universe has no origins, and our planet and all animal forms were always there.[10] Precisely because a historical view of the cosmos was out of question, the central philosophical problems hinged on what was universal, eternal, and static in the natural or "unnatural" changes we observe. The Platonic abstract forms, or the most concrete Aristotelian forms (the notion of "Eidos" that concerned both Dewey and Mayr), were the essential tools that explained how an everchanging and passive matter could produce consistent and complex objects in this eternal, a-historical universe. The two great schools of antiquity that followed Aristotle, Stoicism and Epicureanism, could not reverse the conceptual discontinuity between nature and human society. While the Stoics conceived the cosmos as an instantiation of rational and divine principles, the epicureans -as mentioned in the previous section—saw it as the outcome of a specific configuration of matter. Yet, both regarded nature as the product of a cyclical and eternal processes of generations and corruptions, although materialists were significantly more ambiguous on this issue and left some space for diachrony. Lucretius' Book 5 of *Rerum Natura* exemplifies the tendency: nature is constituted by an infinite collection of swarming atoms that create and dissolve worlds constantly and cyclically, even though, at the end, the history of humanity was conceived in non-cyclical, historical terms.

This short historical review does not imply that no one after Milesians or Diodorus defended a progressive or transformist view of the cosmos. It would not be difficult to find exceptions to the trend.[11] Yet, the tendency to understand *nature* as stacked into a cyclical time and human societies as passing through a linear sequence of events was largely dominant. As Toulmin recalled, the four schools of antiquity did not give a significant place to "time" in their cosmologies. And with the spread of Christianity after the third century AD in the Roman Empire, things did not change radically. While Judeo-Christian thought supposed a linear and temporal history of creation that framed the human drama of fall and redemption, this did not spur any significant agenda for studying nature from a historical perspective. Most scholars of the Middle age—Albertus Magnus to St. Thomas—who synthesized the Christian *Genesis* and Greek philosophy, held that Creation itself was an allegory of God's will and were not bothered in understanding how such creation had actually happened. We must wait until the seventeenth century when historical considerations started to be applied to nature. The beginnings were allegedly ambiguous and fundamentally explorative. Francis Bacon, who was a fond follower of Democritus against the dogmatic Peripatetics, distinguished across three kinds of natural History: History of creatures (the course of natural phenomena), History of wonders (generation of natural irregularities) and history of arts (Nature as

[10]No doubt, there were some metaphysical reasons supporting the idea that the universe has origins. After all, if this is the case, what did happen before?

[11]One significant and interesting example is the Muslim peripatetic philosopher Ibn Sina (Avicenna). In his *Kitab Al-Shifa* (The Book of Healing), Ibn Sina suggested that the mountain ranges were the result of continuous upheavals (earthquakes) of the earth's crust plus the constant and slow action of water that produced valleys (Toulmin and Goodfield 1965)

transformed by humans). The first two were not genuinely historical in the sense that the latter was. Yet, Bacon's History of creatures and wonders was, in fact, a "cyclical" account of an essentially a-historical nature. Bacon's distinctions were entirely consistent with the materialist (Lucretian) separation between nature (as the realm of cyclical events) and human society (as the realm of linear events).

But soon after Bacon died in 1626, things started to change. In 1644, Descartes published *Principia philosophiae*. The book contained a speculative hypothesis about the mechanisms presiding over the formation of the Earth and other planets. Descartes' hypothesis purposedly proposed an historical, developmental, and linear view that replaced nature's most dominant cyclical views. Toulmin argued that Descartes' hypothesis was not yet properly historical insofar as it was an intellectual fiction that did not answer the question "how did really happen" but "how it could have happened?" For Descartes, Toulmin concluded: "…as for Plato earlier, the temporal flux of historical events lost all fundamental significance, acquiring a theoretical interest only by providing illustrations of the unchanging geometrical laws…" (Toulmin and Goodfield 1965, 80). Nevertheless, if we read the text very carefully, we might perceive a seminal, although timid, change in perspective. True, Descartes' conception of "nature" was still largely cyclical, however, he did argue that it was far better to understand natural events in terms of origins and genesis than in terms of simple and synchronic creation:

> …by taking into account the omnipotence of God, we must believe that everything He created was perfect in every way. But, nevertheless, just as for an understanding of the nature of plants or men, it is better by fare to consider how they can gradually grow from seeds than how they were created by God in the very beginning of the world; so, if we can devise some principles which are very simple and easy to know and by which we can demonstrate that the stars and the Earth, and indeed, everything which we perceive in this visible world, could have sprung forth as if from certain seeds (even though we known that things did not happen that way; we shall in that way explain their nature much better than if we were merely to describe them as they are now [or as we believe them to have been created] (Descartes 1983, 105-106)

The expression "could have sprung forth *as if*" is crucial. Descartes suggested that whereas we might forever ignore the actual steps leading to the world we perceive today, the intellectual effort to understand it in a linear and developmental way is more productive than understanding the world simply synchronically, even when our reconstructions are fictional. In short, Descartes argued not only that natural objects have a history but also that their history is relevant for achieving a scientific understanding of them. Perhaps we will not be able to recover the real history, but we can make sense of those objects by imagining a fictional history. The insight that we understand something better if we visualize its development should not be overlooked. The ancient distinction between cyclical and linear temporalities, one for nature and the other for humans, was challenged: nature has history as humanity itself, even though, given our epistemic limitations, the history we visualize is largely fictional. In short, the categorical difference between the time of nature and the time of humanity is denied and the cyclical time is gradually replaced by a rectilinear one.

In his seminal monograph, *The Dark Abyss of Time* (1987), the historian Paolo Rossi chronicled the explosion of physical theories based on "linear temporalities" between 1680 and 1780. While the outbreak reached its climax in the nineteenth century (usually considered the century of history), the leading intellectual transformation (or a paradigmatic shift) happened mainly between the seventeenth and eighteenth centuries.[12] Rossi regards the emergence of a distinct historical sense in the convergence between the history of the Earth and the history of the civil world. The debate concerning the nature of fossils and the formation of the Earth suggested that nature could have a history, especially a very long history. Then, the possibility of the antiquity of man in the backdrop of the new discoveries of other human communities overseas implied that biblical history could be flawed or needed to be reconciled with the findings of natural histories. The "abyss" of time seemed to swallow nature and human history into one all-embracing narrative. The linear temporality found in the Lucretian history of humanity was increasingly applied to the history of the Earth and the whole cosmos.[13] Undoubtedly, the first attempts at historicizing nature were rather clumsy. The British polymath and Robert Boyle's assistant Robert Hooke provided one of its first tentative pictures. In the series of lectures delivered at the Royal Society of London since 1667, "Discourse on Earthquakes," Hooke supposes that both the Earth and life could have history. Such a history did not necessarily disprove the biblical scripture. Both nature and the divine's temporality could converge. The significant transformations of the Earth could be squeezed into biblical times. Yet, the changes were inherently linear and referred to a past where the Earth (and the creatures living on it) might have been very different:

> In former times of the world [there may have been divers Species of creatures, that are now quite lost, no more of them surviving upon any part of the Earth. . .For since we find that there are some kinds of Animals and Vegetables peculiar to certain places, and not to be found elsewhere; if such a place have been swallowed up, 'tis not improbable but those Animal Beings may have been destroyed with them. . .There may have been divers new varieties generated of the same Species, and that by the change of the Soil on which it was

[12] On the idea of history and historicization of nature in the ancient and modern world, see Arendt 1958 and Foucault 1970

[13] Arendt's reflections on the difference between the circularity of biological life and the rectilinearity of human individual lives is particularly pertinent here. She noticed that most ancient Greeks distinguished between the immortality of nature, which in its eternal movements is an endless recurrence of events, and the mortality of humans, which follows a specific "life-story", a biography. As she writes: "This individual life is distinguished from all other things by the rectilinear course of its movement, which, so to speak, cuts through the circular movements of biological life. This is mortality: to move along a rectilinear line in a universe where everything, if it moves at all, moves in a cyclical order" (1958, 571). What I am arguing here is that, in the modern world, the rectilinear course of events that was traditionally attributed to human lives was increasingly applied to nature itself, so that nature, whether in the context of the history of the solar system, earth or life, was increasingly understood in terms of a "biographical" time. A proper concept of "evolution" could emerge only when the cyclical and eternal time of nature was replaced by the human rectilinear time.

> produced; for since we find that the alteration of the Climate, Soil, and Nourishment doth often produce a very great alteration in those Bodies that suffer it; 'tis not to be doubted but that alterations also of this nature may cause a very great change in the shape, and other accidents of an animated Body (Hooke quoted in Rossi 1984, 14-15)

Changes in living forms are connected with changes in the Earth's geology. The fossils we observe within the rocks are like "documents" that witness a past where other creatures existed. To see fossils as natural "documents" cemented the success of a new analogy that would feed many other speculations throughout the eighteenth century. Nature could be understood as we understand the human past: i.e., genealogically. Rossi explains: "The criteria of verification that are valid for human history can thus be applied to fossils as well since they are the 'Medals, Inscriptions, or Monuments of Nature's own stamping.' Even happy coincidences are not to be excluded: it is possible that some fables that contain 'the truth of History' conceal within them the account of events that the history of fossils might document" (Rossi 1984, 17).

It would be impossible to summarize here the complex and rich narrative put forward by Rossi in his book. I can only refer to a few crucial milestones that, among myriads of first and secondary figures, reveal the basic plot, which Rossi resumed as follows:

> Toward the middle of the seventeenth century, the problem of the *interpretatio naturae* tended to exceed the exclusively spatial and structural dimensions on which it had been based. It turned out to be connected with the dimension of time. Analyzing and interpreting a substance came to mean more than simply breaking it down, reducing it to particle motion, and studying it in its geometrical aspects. Other questions began to acquire meaning: How has a natural object come to be formed through time? How can we read it and interpret it from this point of view? How has nature produced a particular object during the course of time? (1987, 17)

History is not only a mere collection of political events or royal pedigrees. It becomes a fundamental epistemic tool for understanding nature, life, and humanity as they are today. All synchronicity requires the study of preceding diachronicity. All natural entity is fully comprehensible if its geneaology is taken into account. Of course, in the beginning, those new "natural historians" could not entirely emancipate from the Bible and other largely a-historical accounts. In his *Telluris Theoria Sacra,* published in 1680, Thomas Burnet tried hard to conciliate the time of nature and the time of the Bible. Drawing on Descartes' hypothesis on the origin of the Earth, Burnet believed that his book was, at the same time, a contribution to natural history and theology (Nicolson 1997). Explaining materialistically and mechanically (and especially historically) the origin and development of the Earth was, to him, entirely compatible with the words of the scriptures. Burnet's attempt was largely unsuccessful as a theology, although it was widely influential as a natural history. Newton himself was deeply impressed by Burnet's *Telluris,* but he could not accept the ungodly implications of the system. Newton perceived that a conception of a "historical" or "developmental" universe—where matter and laws were the cause of order—left God at the margins. Along with Newton, Robert Boyle, in a text published in 1665,[14] lamented that Epicureans and Lucretians did not distinguish between the origins of things and the development of nature. While the origins are inscrutable to us, we can only track down the visible history of the universe. We could observe and admire the universe's development as a material emanation of God while God's original prospect will be forever hidden from our investigations (Rossi 1984). In short, an historical view could only be endorsed once the origins of the universe were left in God's mind.

First-rate philosophers and naturalists expressed many hesitations and difficulties in accepting a truly historical and materialist view of nature. Leibniz's example is particularly instructive. In his *Protogea*, written in the early 1690s and published almost six decades later, he pictured a historical world that was itself the emanation and execution of a divine agenda. Leibniz toyed with the hypothesis that other organisms might have existed in the past (as the fossils constantly showed). He even toyed with the "dangerous" insight that organic variability over time could have been a possibility:

> ...it is possible that in some time and in some place in the universe the animal species are, have been, or will be more subject to change than they are among us at the present and several animals that have something of the cat, like the lion, the tiger, and the lynx, may have been of one single race and may now be like new subdivisions of the former species of cats (Quoted in Rossi 1984, 63)

While geology and life might have changed, those transformations could not be the results of chance. No Epicurean *chaos* could be invoked to explain order. Leibniz could not accept that such history could be seriously conceived as independent from God's creation. Unlike Lucretians, Leibniz was open to conceive nature as something that changes in a linear and perhaps progressive form. However, he could not see any fundamental relation between the history of nature and the materialist hypothesis. This is not surprising because an historical view of nature could be easily combined with a theistic account over the origin of the universe. To have a properly materialistic, evolutionary perspective, a further step was sorely needed: history itself needed to acquire autonomy from God, transcendent principles, or pre-established plans. In short, ancient materialism had to be *historized* and history of nature de-spiritualized and secularized.

5 Re-Historicizing Nature

John Zammito has recently observed that in the eighteenth century, natural history was historized as a consequence of four problems: (1) the emergence of regularities of the planets' orbits in the solar system; (2) the riddle of geological change; (3) The origin of life forms and their diversification and (4) the problem organic generation

[14] *About the excellency and Grounds of the Mechanical Hypothesis*

(Zammito 2017, 172). In what follows, I focus my attention on 2 and 3 and, primarily, on two figures who have been instrumental in providing a new, historized conception of nature: the French naturalist Georges-Louis Leclerc, Comte de Buffon, and the Prussian philosopher Immanuel Kant. Phillip R. Sloan has convincingly shown that Buffon was deeply indebted to the Leibnizian distinction between the abstract order of ideas and the concrete order of bodies. In fact, while debating with the Anglican theologian Samuel Clarke, Leibniz claimed that the reality of "time" and "space" had to do with the concrete succession of bodies. Without existing things, "time" and "space" would only be in God's mind. As Leibniz himself observed in the fourth letter to Clarke: "...The author contends that space does not depend on the situation of bodies. I answer it is true, it does not depend on such or such a situation of bodies, but it is that order which renders bodies capable of being situated, and by which they have a situation among themselves when they exist together, as time is that order concerning their successive position. But if there were no creatures, space and time would be only in the ideas of God" (Leibniz in Ariew 2000, 34). Leibniz's insight, Sloan suggested, led to a daring conclusion that many enlightened naturalists, including Buffon, exploited. "If time and space—Sloan claimed—have their ontological grounding only in the empirical relations of ordered succession and contemporaneous relationship of bodies, the history of the world cannot be separated from the very reality of time itself." (1979, 113). The German philosopher Christian Wolff (1679–1754) developed Leibniz's intuition and made it even more concrete. He emphasized, Sloan concluded, that: "...the 'existence' of time was in the ordered succession of concrete beings" (Sloan 1979, 114). Buffon was aware of Wolff's interpretation of Leibniz's distinction and made it a crucial premise of all his following works (Sloan 1979, 116). Thus, Buffon's historicization of nature was based on the fundamental metaphysical argument that things and time cannot be severed.

While the notion of "history" in Buffon's *Natural History* (1749–1788) is not what we would mean today (Hoquet 2010), it is difficult to deny that the French naturalist upheld a sort of historicist perspective of nature. True, in his "Natural History" the term "history" was still used according to the ancient meaning: i.e., the systematized inquiry into something. Yet, Buffon did put forward a seminal—although we might consider it as widely idiosyncratic—*History of Nature* (Roger 1989), where the term "history" acquired a genealogical flavor. Some scholars have insisted that Buffon's *history* of nature was not genuinely historical. John Eddy, for example, pointed out that there was no historical image in Buffon's *Natural History*. To think something historical, Eddy argued, is "...is to believe that things change incrementally and, in the end, fundamentally" (p. 651). In Buffon, Eddy remarks, we cannot find anything similar. We cannot identify a process whereby small and constant transformations lead to big qualitative changes. One of Eddy's main targets is Buffon's concept of "species." For Sloan, Buffon did introduce a historical view of "species" insofar as they are conceived as reified temporalities based on reproduction. In other words, "species" refers to an ordered succession of related individuals. Members of a "species" are similar because they share a common history (and eventually an ancestor) and not the contrary (see Sloan 1987). Eddy replied that

Buffon's concept of "species," and his theory of degeneration was essentially a-historical because we have no fundamental change due to a historical series but minor alterations due to a slow process of degeneration. In Buffon's "historical series," we have no proper novelties but small episodes of "involution." As Eddy concluded, "...contrary to what some think, he [Buffon] was no more a historicist than he was an evolutionist" (Eddy 1994, 661).

But is "historicism" a straightforward concept after all? Perhaps there are many ways to historicize nature. To be a *historicist* in the 17th and eighteenth century could mean different things. Some believed that the history of the cosmos and the Earth was a deployment of a divine plan (i.e., Leibniz). Others argued that it was the historical outcome of universal and eternal laws (the Telliamed of Maillet). Leibniz and Maillet could be considered "historicist" in a sense hinted by Descartes in the seventeenth century: to understand something by visualizing its genesis, whether the process is driven by God or by any natural force. We shall not expect "historicism" to necessarily be a post-Darwinian version like the radical contigentism of Stephen Jay Gould. If someone believed that "historical successions" were epistemically crucial for understanding the nature of any product (whether artificial or natural), then he or she might be considered as an "historicist" of some kind. I suggest that the crucial difference we shall consider here is not between historical and a-historical views of nature but between perspectives that propound for cyclical or linear temporalities of nature. The cyclical stance supposes that organized beings are the sole outcome of chance and natural laws while the linear stance assumes that chance and natural laws are not enough to explain them: organized entities are also the result of incremental and continuous formative processes (whether contigent, directed or a mixture of both). Thus, while we cannot consider Buffon, and most of his contemporaries, as "historicists" in a Gouldian or Eddyian sense, Buffon did provide a vision of nature based on a linear or rectilinear temporality.

As an Epicurean (and Lucretian), Buffon believed that the Earth was an aggregate of atoms and that living forms were generated at the beginning of the consolidation of the planet.[15] Like Lucretius, Buffon claimed that the first organic forms were generated spontaneously in the primordia of the Earth when the environment was particularly propitious for the formation of complex aggregates (See Caponi 2009). From the second volume of his *Natural History*, Buffon held that nature is intrinsically productive and creative. Living forms were not the result of highly improbable combinations of matter but the outcome of a widely common process of organic formation, which was justified by the intrinsic properties of the organic molecules. Unlike Lucretius, Buffon posited the existence of two kinds of particles: organic and brute. In the early ages of the Earth, organic molecules combined and formed complex organic aggregates. Those aggregates were generated according to a force alike to Newton's gravitation: what Buffon called *moule interieure*. The force was not directly visible but indirectly discernable through its effects on the organic molecules and aggregates. Following Lucretius, Buffon maintained that viable

[15] For a thorough discussion of Buffon's transformist views, see Gregory 2008.

aggregates persisted and reproduced. They could, therefore, spread over the Earth. Those first viable forms represent the "common ancestors" of the forms and "species" we can observe today. Buffon theory of degeneration was an ingenious complement to ancient materialism. In fact, once a new organic form turned out to be viable and persisting, it "degenerated" according to the different climates and food it could find and ingest.

The Buffonian "common ancestor" was a Lucretian surviving "monster" that reproduced and spread all over the Earth. The Lucretian monster was, for Buffon, the "founder" of a species, and species were historical collections of individuals that changed over time due to the changing conditions of the environment. Perhaps the most audacious description of his theory of degeneration appeared in the article Buffon dedicated to the "Ass" in his *Natural History*. He first recognized that the Ass appeared as a degenerated horse because it had a very similar structure with small differences, which: "...may be attributed to the long-continued influence of climate and food, and to a fortuitous succession of many generations of small wild horses, who, by gradually degenerating, at last, produced a new and permanent species, or rather a race of similar individuals, all marked with the same defects, and differing so widely from the genuine horse, as to be regarded as constituting a new species" (Buffon 1830, 112). But Buffon expanded the consideration further. He wondered whether two species could have a common ancestor in general: would it be possible that different animals spring from an original stock? Buffon's answer deserves to be quoted in full length:

> Philosophers will perceive the extent, the difficulties, and the importance of this question, which we shall here discuss, only because it for the first time occurs. It relates to the production of beings, and, for its illustration, requires that we should consider nature under a new point of view. If, from the immense number of animated beings which people the universe, we select a single animal, or even the human body, as a standard, and compare all other organized beings with it, we shall find that each enjoys an independent existence, and that the whole are distinguished by an almost infinite variety of gradations. There exists, at the same time, a primitive and general design, which may be traced to a great distance, and whose degradations are still slower than those of figure or other external relations: For, not to mention the organs of digestion, of circulation, or of generation, without which animals could neither subsist nor reproduce, there is, even among the parts that contribute most to variety in external form, such an amazing resemblance as necessarily conveys the idea of an original plan upon which the whole has been conceived and executed. When, for example, the parts constituting the body of a horse, which seems to differ so widely from that of man, are compared in detail with the human frame, instead of being struck with the difference, we are astonished at the singular and almost perfect resemblance. In a word, take the skeleton of a man, incline the bones of the pelvis, shorten those of the thighs, legs, and arms, lengthen the bones of the feet and hands, join the phalanges of the fingers and toes, lengthen the jaws by shortening the frontal bone, and, lastly, extend the spine of the back: This skeleton would no longer represent that of a man, but would be the skeleton of a horse... (Buffon 1830, 113-114).

And the conclusions are audacious:

> Let us next consider, that the foot of a horse, so seemingly different from the hand of a man, is, however, composed of the same bones, and that, at the extremity of each finger, we have the same small bone, resembling a horse-shoe, which bounds the foot of that animal. From these facts we may judge whether this hidden resemblance is not more wonderful than the apparent differences; whether this constant uniformity of design, to be traced from men to quadrupeds, from quadrupeds to the cetaceous animals, from the cretaceous animals to birds, from birds to reptiles, from reptiles to fishes, &c. in which the essential parts, as the heart, the intestines, the spine, the senses, &c. are always included, does not indicate, that the Supreme Being, in creating animals, employed only one idea, and, at the same time, diversified it in every possible manner, to give men an opportunity of admiring equally the magnificence of the execution and the simplicity of the design? In this view, not only the horse and ass, but man, monkeys, quadrupeds, and every species of animal, may be considered as one family. (Buffon 1830, 114).

All animals might belong to the same great "family." Perhaps all organic forms we see today are transformed organisms that derived from some old ancestor. But then, surprisingly, Buffon pulled the hypothesis back. If every species belongs to one family and one derives from one individual, then we might suppose that Nature "...have been able, in the course of time, to produce, from a single individual, all the organized bodies in the universe" (Buffon 1830, 115). But that goes against revelation and cannot be true: "...this is by no means a proper representation of nature. We are assured by the authority of revelation, that all animals have participated equally of the favors of creation; that the two first of each species were formed by the hands of the Almighty; and we ought to believe that they were then nearly what their descendants are at present" (1830, 115). Buffon played with the hypothesis of a common descent seriously. Yet, he rejected it for two reasons. The first reason was authority: it was contrary to revelation. The second reason was eminently empirical. As he explains:

> ...if one species could be produced by the degeneration of another, if the ass actually originated from the horse, this metamorphosis could only have been effected by a long succession of almost imperceptible degrees. Between the horse and ass, there must have been many intermediate animals, the first of which would gradually recede from the nature and qualities of the horse, and the last would make equal advances to those of the ass. What is become of these intermediate beings? Why are their representatives and descendants now extinguished? Why should the two extremes alone exist? (1830, 119).

Degeneration is not transmutation. We cannot deduce animal transformation from a degenerative process. Degeneration only accounts for minor variations, not for species transformation. But once Buffon complemented the materialist hypothesis of the spontaneous generation of organic forms (Lucretius' hypothesis) with the historical (and linear) view of nature, he realized that organic transformation could be a possibility (although, to him, a highly improbable one). Be as it may, Buffon synthesized Lucretian materialism with a new diachronic view of the world. He applied to nature the historical perspective that Lucretius had attributed to humans in the book V. In so doing, he explored the prospect of an organic historical transmutation that turned out to be too audacious for the French naturalist (and probably for many of Buffon's contemporaries).

Buffon was widely read and usually considered an international authority in natural History (Farber 1972). Outside France, he was appreciated by German naturalists and philosophers (Sloan 1979, Wilson 2016). One particular figure who was fond of Buffon's natural History was Immanuel Kant. In his first book, *Allgemeine Naturgeschichte und Theorie des Himmels* (Universal History of Nature and Theory of the Heavens), written around 1755, the presence of Buffon is particularly evident (Sloan 1979). Kant drew on Buffon also for his course in Physical Geography, which started as *Privatdocent* in 1757 (which he taught for 40 years). As Sloan recalls, Kant's fundamental distinction between *Naturgeschichte* (Genetic History of nature) and *Naturbeschreibung* (description of nature) was based mainly on Buffon. Even more, Kant seemed to prioritize the former over the latter. In his *Allgemeine Naturgeschichte* a diachronic conception of nature contextualizes the synchronic descriptions of natural phenomena while the whole scheme was soaked in materialism (although moderated). Kant was deeply fond of Lucretius, who studied as pupil at school and was his favorite author before Newton (Wilson 2016). In the *Allgemeine Naturgeschichte,* Kant made it clear from the beginning of his book:

> I will therefore not deny that Lucretius' theory or that of his predecessors, Epicurus, Leucippus, and Democritus, has much in common with mine. Like those philosophers, I posit a first state of nature as a universal dispersion of the original material of all world-bodies, or atoms as they call them. Epicurus posited a heaviness that caused these elementary particles to fall and this does not seem to be very different to Newtonian attraction, which I accept; he also accorded them a certain deviation from the straight linear motion of their fall, even though he had absurd notions of their causes and effects: This deviation to some extent corresponds to the change in the straight fall that we attribute to the repulsive force of the particles; finally, the whirlpools that arose out of the perturbed motion of the atoms were a centerpiece of the theories of Leucippus and Democritus, and they will also be found in ours (2012, 198)

He immediately clarified that the atheist implications of those materialist hypotheses did not apply to his own. Ancient materialists believed that chance could produce a well-ordered universe: "All of them together took this nonsense to the point that they made this blind coincidence the origin of all living creatures and really derived reason from the lack of reason" (Kant 2012, 198). Instead, the order we observe in the cosmos is connected to universal laws. Matter organizes itself through those laws. No chance or accident could be the cause of anything organized. Kant's materialism is a theologically committed materialism, where God plays the role of a law's trustee. Yet, this materialism without chance had another essential element: History. Kant felt the need to emphasize the concept a few times:

> Perhaps a number of millions of years has passed before the sphere of formed nature in which we find ourselves has grown to the perfection that now attends it; and perhaps an equally long period will elapse before nature takes an equally large step in the chaos: but the sphere of formed nature is incessantly occupied in spreading itself. Creation is not the work of one moment. After it has made a beginning with the production of an infinity of substances and matter, it is effective throughout the entire sequence of eternity with ever

> increasing degrees of fruitfulness. Millions and whole mountain ranges of millions of centuries will pass within which ever new worlds and world-orders will form and attain completion one after another in the remote distances from the centre point of nature...(2012, 266)

Creation is not an act of will but a developmental process that displays itself in many forms throughout centuries. God did not create the world in one stroke or 6 days. God set the rules for its indefinite expansion: "Creation is never complete. It is true that it began once, but it will never stop. It is always occupied with bringing forth more phenomena of nature, new things and worlds" (Kant 2012, 267). The idea of "flow," "succession," and "period" constituted an unbridgeable gap with the cosmologies put forward by ancient materialists. Kant did not refer to an indefinite succession of worlds but to one unique history of one universe: ours. He supposed that all elements of the visible universe, stars, comets, planets, and moons, could be explained by positing one unique process governed by universal law and principle: Newton's law of gravitation and the mechanical principle of repulsive and attractive force. There is one beginning and one linear development where new planetary systems are constantly formed. The idea of linear "time" in Kant's nebular hypothesis—a hypothesis set to explain the solar system's origin—is pervasive. The primordial cloud of particles, which constituted the disaggregated matter of an incipient system in formation, started to attract and collide mutually. They formed amalgams increasingly larger. The collision of small aggregates generated larger masses that cooled down and formed "planets." By introducing his theory of vortices, Descartes formulated a mathematical hypothesis (or fiction) accounting for the formation of the solar system. Kant, instead, put forward a historical hypothesis that embraced life on Earth (or perhaps on other planets...). The process goes on indefinitely and produces different material properties for each planet. Kant did not speculate on the origin of life on Earth, but he assured that:

> Countless animals and plants are destroyed daily and are victims of transience, but through an unexhausted generative capacity nature brings forth no less again in other places and fills the void. Considerable areas of the Earth that we inhabit are buried again in the sea from which a favourable period had dragged them; but in other places, nature replaces the deficiency and brings forth other regions that had been concealed in the depths of the water to spread new riches of its fruitfulness over them (2012, 269)

Change is everywhere, and nature is immensely fertile for producing order and abundance of living beings. The "evolutionary" view of the cosmos that Kant put forward witnessed a significant intellectual shift: from a cyclical time to a linear time of nature. Toulmin called it the "discovery of time." We can call it, more technically, the application of human rectilinear time to nature, which replaced the cyclical view of antiquity with a developing world. The speculations that followed Kant's "evolutionary" cosmos were corollaries of the same pattern while adding more and more details.

As is well known, Jean-Baptiste Lamarck embraced the hypothesis that Buffon and Kant had considered too audacious to accept. As a committed materialist,

Lamarck exploited the new intellectual possibilities that historized materialism had opened. His great transformist view, which outlined a history going from infusoria to vertebrates, synthesized the whole eighteenth-century debate over how to understand the natural world diachronically and, especially, how to account for organic complexity without recurring to divine, mystical, or unintelligible forces. The union of ancient a-historical materialism with genealogical views made the linear and processual view of the cosmos (and its living creature) possible. It was one of the most remarkable intuitions of the eighteenth century that only exploded, with all its potentialities, in the nineteenth century.

6 Toward a Tentative Conclusion

The traditional historiography of evolutionary through (HET) stands on two persisting myths: the idea that before Darwin, most people believed that the world was static, and that idea that Darwin is the undiscussed protagonist of any general HET. I suggested that deconstructing the traditional HET means to question and, eventually, to eradicate those myths. Only then will we be able to outline a new, more open, diversified, and pluralistic HET. I have argued that a new HET should reconsider the history of the materialist tradition and the rediscovery of the historical time in the seventeenth and eighteenth centuries. With the diffusion of Lucretius', *The Nature of Things,* scholars started again to play with the audacious hypothesis that organic forms have an origin and, especially, a natural and material one. However, those who, like Leibniz, explored the daring hypothesis were bounced back to the same Aristotelian (and Cicero's) conclusions: to explain organic complexity as the result of chance and necessity is preposterous: we cannot seriously believe that "...a library forms itself one day by a fortuitous concourse of atoms" (Leibniz quoted in Wilson 2008, 102). As long as scholars focused their debates on whether order and complexity might (or might not) emerge from matter in a static or cyclical universe, the more plausible answer remained the Aristotelian one: Order and complexity presupposed someone (God) or something (form) that organizes matter according to specific shapes and functionalities. If naturalists wanted to demolish Aristotle's arguments, they needed new, more powerful counterarguments. I suggested that the historicization of nature was the key element that made a new form of materialism plausible. Order and complexity did not emerge out of chaos, they emerged slowly and gradually through a diachronic process. The history of nature was the key to rejecting Aristotle's philosophical spell. The first attempts were, of course, a failure. Neither Leibniz nor Buffon and Kant could clearly grasp the relation between the history of nature, transformation, and the origin of complex forms. While they played with the hypothesis, they dismissed it as improbable or implausible. However, their cogitations were fundamental for establishing one essential insight: the history of nature is crucial for understanding nature. To comprehend nature we need to grasp its geneaology. The grand synthesis of Lamarck in his *Philosophie Zoologique* represented the crowning achievement of such an insight. Eighteenth-century scholars had put all the necessary ingredients on the table so that nineteenth-century scholars could finally constitute a veritable evolutionary view of nature and life.

It should be clear that this story sketched here does not admit precursors. Lucretius was not a precursor of Descartes, while Descartes was not a precursor of Leibniz or Buffon. Similarly, Buffon and Kant were not precursors of Lamarck, and accordingly, Lamarck could not be the precursor of Darwin. This history has no precursors because it is not a sequence of failed attempts leading to Darwin. Instead of precursors and forerunners, we have a collective history of many minds addressing some of the most challenging, thorny, and fascinating questions ever formulated: How can we understand organic complexity from a natural and material viewpoint? How did organic forms originate in a material universe? How could something as passive as matter generate the wondrous biodiversity surrounding us? In the eighteenth century, the fertile insight that emerged by addressing these questions was "History"; i.e., that the natural world (including living beings) has itself a "biography". No doubt, an Aristotelian could still reply that the combination of matter and chance cannot be the source of complexity, even when history is considered as an important variable in the equation. But the modern materialist and historicist could at least reply that organic complexity is not something formed spontaneously and immediately. Complexity is something generated over billions of years of small, continuous, and cumulative changes. In other words, the improbability of complexity could be distributed over a long stretch of time.

I am aware that this proposal will not convince many (if anyone). Biologists, still intoxicated by the HET two dogmas, would likely find it not interesting. We all like mythical sagas based on great heroes making extraordinary feats despite the tenacious resistance of less gifted people. Moreover, for many of what I previously called "scientist historians," history is often nothing more than a legitimation of some current scientific tradition (Esposito 2021). Historians, instead, will likely find my proposal too simplistic, suspiciously philosophical or extremely broad in the sense that I fail to focus on one specific historical space and moment (whether Lamarck's France, Darwin's England or Haeckel's Prussia). My reply to this justified objection is that the reassessment and reconsideration of HET is itself a philosophical enterprise (after all, the old historiography was itself the outcome of a philosophical stance informed by some historical data). The choice to focus on one period, country, institution or figure is far from neutral and is taken in the context of an implicitly or explicitly accepted framework. I can certainly be proven wrong but I think that we cannot understand how the concept of "evolution" evolved by focusing on one or few privileged periods or figures. The scale of our analysis must be much larger and inclusive. No doubt, the historian can reasonably reply that she is not interested in sweeping analyses. Yet the prize to pay for the unphilosophical historian is to continue to write histories under the spell of the traditional historiographic paradigm, which continues to exert its invisible influence, even in more nuanced and articulated narratives.[16] In fact, it is not enough to accuse the old paradigm of oversimplification. It is not enough to propose more complex histories by multiplying forgotten or

[16]What Mark Adams calls "just-so story" or "ever since Darwin" historical accounts in this volume.

overlooked figures. In a word, it is not enough to celebrate complexity per se. It is easy to forget that historical complexity is a cheap asset because complexity is virtually everywhere. A new, post-Darwinian historiography should certainly be complex, but it should not refrain from generalizations and encompassing reconstructions, even at risk of committing unintended oversights. Mistakes are the price we must eventually pay if we aim to deconstruct and replace an engrained tradition that does not hold water anymore. After all, let the historian who is without "epistemic sin" cast the first stone!

References

Adams MB (2024) The evolution of 'Darwinism': up close and personal. In: Delisle RG, Esposito M, Ceccarelli D (eds) Unity and disunity in evolutionary biology: deconstructing Darwinism. Springer, Cham, pp 83–121

Arendt H (1958) The modern concept of history. Rev Polit 20(4):570–590

Aristotle (2001) On the parts of animals, I-IV., Ed. J. Lennox. Clarendon Press, Oxford

Aristotle (1970) Physics. Books I and II., Ed. W. Charlton. Clarendon Press, Oxford

Aristotle (1933) The metaphysics. Books I-IX, Ed. H. Tredennick

Auerbach E (2003) Mimesis: the representation of reality in Western literature. Princeton University Press, Princeton

Balme DM (1987) Aristotle's biology was not essentialist. In: Gotthelf A, Lennox JG (eds) Philosophical issues in aristotle's biology. Cambridge University Press, pp 291–312

Bowler P (1989) Evolution: the history of an idea. University of California Press

Buffon GL (1830) The natural history of quadrupeds, vol 1. Thomas Nelson and Peter Brown, Edinburgh

Campbell G (2003) Lucretius on creation and evolution: a commentary on De rerum natura book 5 lines 772–1104. Oxford University Press, Oxford

Caponi G (2009) La miseria de la degeneración: el materialismo de Buffon y las 'limitaciones' de su transformismo. História, Ciências, Saúde-Manguinhos 16(3):683–703

Descartes R (1983) In: Miller VR (ed) Principle of philosophy. D. Reidel, Dordrecht

Dewey J (1910) The influence of Darwin on philosophy, and other essays in contemporary thought. H. Holt, New York

Eddy JH (1994) Buffon's histoire naturelle: history? A critique of recent interpretations. ISIS 85(4): 644–661

Esposito M (2021) Cathedrals, corals and mycelia: three analogies for the history of evolutionary biology. In: Delisle R (ed) Natural selection: revisiting its explanatory role in evolutionary biology. Springer International Publishing

Farrington B (1953) Greek science: its meaning for us. Penguin, London

Farber PL (1972) Buffon and the concept of species. J Hist Biol 5:259

Foucault M (1970) The order of things: an archaeology of the human sciences. Tavistock Publication, UK

Gould SJ (2002) The structure of evolutionary theory. Belknap Press

Greenblatt S (2012) The swerve: how the world became modern. W. W. Norton & Company

Greene JC (1981) Science, ideology, and worldview: essays in the history of evolutionary ideas. University of California Press, Berkeley

Gregory ME (2008) Evolutionism in eighteenth-century French thought. Peter Lang, Bern

Hoquet T (2010) History without time: Buffon's natural history as a nonmathematical physique. Isis 101(1):30–61

Kant I (2012) Universal natural history and theory of the heavens. In: Watkins E (ed) Kant's natural sciences. Cambridge University Press, Cambridge
Leibniz GW (2000) In: Ariew R (ed) Leibniz and Clarke: correspondence. Hackett Publishing Company
Lenoir T (1982) The strategy of life: teleology and mechanics in nineteenth-century German biology. Chicago University Press, Chicago
Lucretius (1924) On the nature of things. Harvard University Press
Manzo S (2001) Francis bacon and atomism: a reappraisal. In: Lüthy C, Murdoch JE, Newman WR (eds) Late medieval and early modern corpuscular matter theories. Leiden, Brill, pp 209–243
Mayr E (1982) The growth of biological thought: diversity, evolution, and inheritance. Belknap Press
Mayr E (1988) Toward a new philosophy of biology: observations of an evolutionist. Harvard University Press, Cambridge
Meinecke F (1972) Historism: the rise of a new historical outlook. Herder & Herder, New York
Nicolson M (1997) Mountain gloom and mountain glory: the development of the aesthetics of the infinite. University of Washington Press
O'Rourke F (2004) Aristotle and the metaphysics of evolution. Rev Metaphys 58(1):3–59
Reill PH (2006) Vitalizing nature in the enlightenment. University of California Press, Berkeley
Roger J (1989) Buffon. Fayard, Paris
Rossi P (1984) The dark abyss of time: the history of the earth and the history of nations from Hooke to Vico. Chicago University Press, Chicago
Rudwick MJS (2007) Bursting the limits of time: the reconstruction of geohistory in the age of revolution. Chicago University Press, Chicago
Ruse M (1979) The Darwinian revolution: science red in tooth and claw. Chicago University Press, Chicago
Sloan PR (1979) Buffon, German biology, and the historical interpretation of biological species. Br J Hist Sci 12(2):109–153
Sloan PR (1987) From logical universals to historical individuals: Buffon's idea of biological species. In: *Histoire du concept d'espece dans les sciences de la vie*. Gauthier-Villars, Paris
Sloan PR (2019) Evolutionary thought before Darwin. In: Zalta EN, Nodelman U (eds) *The Stanford encyclopedia of philosophy*
Toulmin S, Goodfield J (1965) The discovery of time. Chicago University Press, Chicago
Wilson C (2008) Epicureanism at the origins of modernity. Clarendon, Oxford
Wilson C (2016) The presence of Lucretius in eighteenth-century French and German philosophy. In: Lezra J, Blake L (eds) Lucretius and modernity: epicurean encounters across time and disciplines. Palgrave Macmillan, New York
Winsor M (2006) The creation of the essentialism story: an exercise in metahistory. Hist Philos Life Sci 28(2):149–174
Zammito JH (2017) The gestation of German biology: philosophy and physiology from Stahl to Schelling. Chicago University Press, Chicago
Zirkle C (1941) Natural selection before the 'origin of species'. Proc Am Philos Assoc 84:71–123

What if Darwin Had Published His 1844 Essay?

Derek Partridge

Abstract

There is a rigid backbone running through, and in effect pinning down, much of the standard discourse addressing the history of Darwin's development of his *Origin* theory of evolution. The core assumption is of a relatively continuous and smooth process that began in his 1830s notebooks, traversed two drafts, and culminated (to a first approximation) in the 1859 publication of *On the Origin of Species*. To sustain this well-established modern viewpoint, a complex network of supportive evidence has been constructed. Historiographic labels like "Darwin's delay", which presses for a positive answer to the counterfactual title, and the "origins of the *Origin*" in his 1830s Notebooks emerge as legitimate puzzles within the standard discourse. But without this spine of continuity, the standard puzzles and consequent necessity for explanations disappear or change form, and alternatives materialize. Under the aegis of the "disenchantment-based" discourse, to be developed, the 1859 *Origin*, for example was not a delayed publication, quite the opposite. Under pressure, in just 8 months Darwin rushed out an unreferenced abstract of his latest evolution theorizing. Similarly, in the absence of the chordate belief in continuity, what significance can be attached to finding *Origin* ideas in the Notebooks, and so on. As will be demonstrated, there is a consensus of new evidence that emerges as significant when the standard discourse is deconstructed, partly by means of evidence miss-interpreted and partly through evidence completely overlooked. The titular counterfactual

"All historians need to remember first that those in the past did not know what was going to happen, while we do, but, on the other hand, that those in the past thought they knew what might happen, including what did not". Jeremy Black, *What If? Counterfactualism and the problem of history*, 2008, pp. 2–3.

D. Partridge (✉)
University of Exeter, Exeter, UK
e-mail: d.partridge@exeter.ac.uk

R. G. Delisle et al. (eds.), *Unity and Disunity in Evolutionary Biology*,
https://doi.org/10.1007/978-3-031-42629-2_16

aimed directly at this rigid path through the standard discourse is the motivation to dig deeper under the too readily accepted explanatory surface. It is posed to stimulate thinking of what options were open at the time. It is definitely not a lead in to the transformation of "what if" to "if only"—the usual reading of the lament of "Darwin's delay".

Keywords

Darwin's delay · The 1844 Essay · The 1859 Origin · The disenchantment hypothesis · The barnacle diversion

1 An Outline of the Standard Discourse

Although the standard discourse centred on Darwin's publication of *On the Origin of Species*, admits a broad spectrum of interpretations and spans centuries, there is a surprisingly firm spine of belief running through the vast majority. This is a widely agreed acceptance that the two decades of intensive development that led up to his celebrated book, although puzzlingly elongated, was nevertheless a relatively continuous and smooth process that can be summarized as follows.

In 1842, Darwin began to draw together the pertinent threads of his Notebook entries—thousands of thoughts, observations, speculations, and deductions that he had amassed in the latter years of the 1830s (Barrett et al. 1987). The result was a terse rough sketch of his[1] first theory of evolution by means of natural selection, known as the 1842 Sketch, and composed of just 30-some pages. Two years later, he produced a much-expanded version, a complete statement of his evolution theory, the 1844 Essay, which occupied approximately 200 pages. Neither document was published by Darwin, but his son Francis published both in 1909 (Darwin ed. 1909). This version of his complete theory was, after some further development and a considerable delay, eventually published as *On the Origin of Species*, in 1859. There was "one major conceptual adjustment made" subsequent to the 1844 version; it was the inclusion of "a principle of divergence" (Browne 2002: 39). This "key-stone of my arch", as Darwin described the addition in an April 1859 letter to his publisher, was simply slotted in to consolidate his earlier outline theory.[2]

Standard discourse statements put it thus: Darwin's "theorizing about natural selection, first fully expounded in his manuscript 'Sketch of 1842' and 'Essay of 1844,' will be reformed, but not fundamentally altered, . . . when Darwin finally publishes the *Origin* in 1859" (Hodge 2013: 71). "Darwin, in the mid-1830s –

[1]For reasons that will become clear, we distinguish between Darwin's first two complete articulations of theories of evolution by means of natural selection—one in 1844, the other in 1859.

[2]Letter from CD to John Murray, 5 April [1859], dcp-lett-2447, accessed 03/09/2020.

always remember that it was in this decade that Darwin did his creative work, not the future decade of the 1850s when he finally published" (Ruse 2013: 4).

Historians have worked long and hard to iron out the puzzles that, nevertheless, arise. For example, why did Darwin wait so long to go public? The enigma of "Darwin's delay" features prominently in most versions of the standard narrative, and repeatedly provokes new explanations. Roderick Buchanan and James Bradley (2017: abstract) noted that "The suggestion that Darwin delayed publishing his species theory has long occupied a central part of his biographical storyline. The notion of a fretful delay reached a melodramatic apogee in Adrian Desmond and James Moore's best-selling 1991 biography". In Desmond and Moore's (1991: 317) summary: for Darwin, "publishing would be suicidal". Buchanan and Bradley favoured the idea that Darwin switched to the full barnacle study, which ate up nearly a decade, as a credentialing exercise. He needed to build his competence as a species man before going public with his evolution theory. John van Wyhe (2007) forcefully rejects the fear-based delay argument, substituting instead a "gap" caused by Darwin's plans to first complete the processing of all his *Beagle*-voyage specimens. In Peter Bowler's interpretation, "After putting together the outline of a naturalistic theory of species transformation, Darwin spent the next twenty years exploring its details and expanding its explanatory powers" (Bowler 1984: 164). Howard Gruber (1974: *xiv*) ascribes "Darwin's long delay in publication" to "fear of persecution and ridicule". In an essay entitled, *Darwin's Delay*, Stephen Jay Gould weighed in with his analysis of why Darwin waited "for more than twenty years to publish his theory" (Gould 1977: 22). He reasoned that "So complex an issue as the motivation for Darwin's delay has no simple resolution, but . . . the negative effect of fear must have played" a significant role (Gould 1977: 23). Robert Richards (1987: 152) headed a section *Mind, Instinct, and Darwin's Delay*, and also suggested that the delay was not due to "one factor alone", but "a complex intellectual, cultural, social and psychological environment" (Richards 1987: 156). There is not, it should be noted, anywhere within these diverse explanations any suggestion of technical difficulties significantly obstructing, if not quite blocking, theory publication.[3]

The many, widely differing explanations of the delay or gap constitute one area of finely structured alternatives within the standard discourse. Running through this web of differing secondary possibilities there is a sturdy limb of belief that, for one reason or another, soon after finishing the 1844 Essay, Darwin's called a halt to the immediate development of his evolution theorizing. It was not until late in 1854 that he picked up the Essay thread, refocused on evolution-theory development, and eventually published *On the Origin of Species* in November 1859. Why this long break? Notwithstanding "the bits and pieces of new information that come into Darwin's thinking between 1844 and 1859" (Ruse 2008: 6), Darwin had missed the opportunity to publish his celebrated theory many years earlier. Prior publication

[3]Richards does raise Darwin's perceived "fatal" problem with the instincts of neuter insects, but maintains that "It would be a mistake, however, to give this conceptual problem the full weight of delaying Darwin"(Richards 1987: 155).

would have avoided the unseemly rush caused by the arrival of a paper from Alfred Russel Wallace in 1858.

1.1 The Standard Discourse Summarized

The main hypothesis is that Darwin's theory of evolution by means of natural selection is the outcome of a relatively continuous and smooth process that began, and was more or less completed, in his 1830s notebooks. It then traversed two drafts and culminated in 1859 with belated publication of *On the Origin of Species*. Hence, most of the essential core of the *Origin* can be located in the Notebooks and is developed through his draft theories to its final blossoming in the *Origin*. An unnecessary delay of more than a decade, perhaps two, requires an explanation.

The titular counterfactual question, which is broadly a rephrasing of Darwin's delay, is posed to provoke consideration of what possible outcomes were open to Darwin at the time of this mid-step in his journey. It can be, and is, expanded, both forwards and backwards in time, to deliver a plausible deconstruction of a central element of the standard discourse. When the period, latter 1830s to November 1859, is subjected to a fresh examination, significant new meanings cascade out to elevate elements of the trivial to the status of historiographically noteworthy. Quite apart from delivering a non-standard response to the counterfactual question, the device proves useful because it drives systematic reconsideration of the relative importance of a range of factors in determining a particular historical path. As Jeremy Black (2008: 10) advocates the merits of a counterfactual approach to history: "It does so by requiring … [us] to consider how the presence/absence of a particular factor might have altered both the balance of events and the path-dependent interplay of those factors".

2 The Standard Discourse Deconstructed

Clear evidential misconstructions open fissures in the standard narrative—in particular, supposedly supportive facts that are patently wrong. Less clear-cut, there are standard interpretations that do not look so secure once elements of the standard discourse are questioned and reframed by the introduction of new factors. Unsurprisingly, the claimed misconstructions tend to be aimed in the direction of furthering the standard discourse. There is, of course, some chicken-and-egg circularity behind this observation. Other rejections of this standard narrative also exist; one such detailed questioning within a much broader conception begins with the assertion that "the problem of the origin of species or the species question" is "a, historically, surprisingly fluid and pluriform scientific issue" (Tanghe 2017: abstract). "The historiographical illusion that Darwin's Origin theory … persisted from at least his 1844 Essay outline," is called an "illusion of knowledge" and an "obstacle to discovery" (Tanghe 2017: 927). Although this is entirely in line with the basis of the present study, Koen Tanghe's very wide-ranging treatment is not, and probably could not be (given its vast scope), focused on the detailed science. A primary concern of this sweeping historiographic study of evolutionary thought is the way the "phrase 'origin of species' (or similar variants) has been, or still is, used in many different ways and contexts" (Tanghe 2017: 976). The *Conclusion* proposes that "the (pre)history of evolutionary thought is currently clearly under-recognized", partly because insufficient attention is paid to "particular problems which prominent theories in the past were designed to solve" (Tanghe 2017: 978). Whilst the present study largely eschews consideration of any prominent theories of the past, this is entirely fitting because it is fundamentally a comparison of two scientific theories, each of which was explicitly laid out in a document by Darwin. Although justified by the details of "how" each theory operates, the concern is centred on the overall evolutionary implications of the differences. Such an investigation barely rests on historical provenance, nor on what initially drove the details of Darwin's evolution theorizing. In both cases, these factors enter into consideration only insofar as they help elucidate a contemporary interpretation of the science expressed. The focus is on a comparison between two specific, mid-nineteenth-century structures separated by two decades. In this limited context, the very long (pre)history of evolutionary thought has minimal impact.

It is, though, instructive to begin by delving a short way back into history to reveal the sleight of hand that permeates discussions at all levels. Although it might appear to be harmless, this misconception constitutes a first fuzziness with the potential to seriously blur, and even block, paths of reasoning, and hence obfuscate the validity of the conclusions generated. Several of the unhelpful examples will emerge within the quotations presented below.

2.1 A Fundamental Ambiguity, and More

At the outset, it needs to be acknowledged that natural selection is *not* a theory of evolution. Beyond mere acknowledgement, some details of the role of natural selection in Darwin's evolution theorizing must be recognized and appreciated.

Following in less than a century after Darwin articulated his evolution theories, Ronald Fisher, in his classic study of *The Genetical Theory of Natural Selection*, felt impelled to emphasize this point in the first sentences of his *Preface*.

> Natural selection is not evolution. Yet, ever since the two words have been in common use, the theory of Natural Selection has been employed as a convenient abbreviation for the theory of evolution by means of Natural Selection. (Fisher 1930: *vii*)

At the very least, the latent ambiguity of this common two-word contraction is unhelpful. A reader is left to wonder whether a writer is referring simply to Darwin's preferred mechanism of organic change—variation and selection by means of competitive survival—or is it short-hand for Darwin's *Origin* theory of evolution by means of natural selection? Or is the reference to something less specific? The ubiquitous unqualified use of this two-word phrase leaves much commentary unsatisfactorily imprecise. An important knock-on effect is that popular casual usage as a short-hand reference clouds any intent to reference natural selection as no more than a mechanism of organic change. The insidious further development that stems from the popularity of short-hand usage is that awareness of the contraction aspect atrophies. It is then all too easy to slip into a mind-set in which "natural selection" is much more than a trope for "Darwin's theory of evolution". In effect, the ambiguity disappears and a loose equivalence between "natural selection" and "Darwin's theory of evolution" becomes an established belief.

But this equivalence is not even nearly true. Even Stephen Jay Gould, when he is striving to lay out natural selection as no more than a component of evolution, fails to disentangle it completely from the "worldview" within which it must function—these two main components are developed below. He comments on his specification of the "bare bones [of] natural selection". He writes that in reality, all that can be said is that natural "selection can work" (Gould 2002: 126, footnote). Whereas natural selection cannot "work" at all without a few necessary characteristics provided by its environment, the worldview—such as that organisms vary. Natural selection is a *component* of Darwin's theories of evolution that is quite separate from theorizing about the natural characteristics of the evolving organisms—how they vary; how inheritance works etc. This separation of concerns—mechanism of organic change and characteristics of the environment within which it is to operate—is vital to the disenchantment hypothesis to be developed below, for two primary reasons:

1. Clear identification of natural selection as a well-circumscribed *component* of an evolution theory opens up Darwin's evolution theorizing to encompass the possibility of unchanging natural selection alongside highly significant evolution-theory changes. We can neglect analogies, examples, justifications, stylistic options, and logical forms of the arguments, all the truly peripheral information.[4] It is the necessary "worldview component", as laid out below, that offers huge scope for change to Darwin's theorizing about evolution whilst, at the same time, preserving natural selection unchanged.
2. The worldview component is a generally unacknowledged, and generally *large* reservoir of evolution theory-change potential. Because it is quite separate from the theory of natural selection (or any other theory of organic change), it is possible to have multiple theories of evolution by means of natural selection that are radically different. The significant differences can all contained within the worldview component.

As introduced above, it is the frequent casual usage of the short-hand contraction for Darwin's *Origin* theory of evolution by means of natural selection that closes down the possible options for theory change. This is because it implies that the

[4]It is all peripheral in the sense of not contributing to the essence of the theory. Thus, any example can be omitted with no impact on the actual theory being presented. The impact, if any, is likely to be in terms of credibility support for the essence of the theory.

theory of natural selection is equivalent to its expansion, but for a few peripheral concerns. This may be seldom stated bluntly, but the invariable implication is that natural selection is in effect evolution. Then if natural selection is unchanged, where is the scope for evolution-theory change? It must be limited to minor concerns. It is only when natural selection is clearly and closely pinned down as no more than the name of a mechanism for a natural means of organic change that space for the extra theorizing necessary to build a theory of evolution is made manifest. It is within this "space" outside natural selection that a countless variety of different theories of evolution by means of natural selection can be formulated. Nevertheless, the role of natural selection in Darwin's evolution theorizing remains persistently misconstrued or misleadingly glossed over. It needs to be laid out explicitly.

2.2 The Place of Natural Selection in Evolution

Natural selection is a theory of *organic change*; another is based on the inheritance of acquired characteristics;[5] and yet another on the spontaneous generation of new species.[6] A theory of evolution requires both a theory of organic change and a *worldview*—i.e. a theory of how the organic beings and physical aspects of the world behave and interact. It is characteristics of the worldview—such as the frequency and causes of organic variation—that determine the operational details of the chosen organic change mechanism, or even whether it can operate at all. It is the interaction of these two component theories that dictate the patterns of organic change to be expected from the overall theory of evolution. On its own, natural selection theory delivers nothing more than a promise of the likelihood of unspecified organic change, and then only if the enclosing biosphere satisfies its preconditions. Even the loosely characterized organic change cautiously and carefully laid out by Gould—i.e. that "the average composition of the population will change in the direction of phenotypes favoured in the altered local environment" (Gould 2002: 126, footnote)—already implies some worldview characteristics,[7] as it must if any particular organic change patterns are to be identified.

In terms of the number of basic theoretical postulates necessary for a theory of evolution, laying out the organic change component is typically the smaller contribution. The bulk of an evolution theory is invariably occupied by all the relevant characteristics of the organisms, their potential environments, and how the two interact (once we neglect all the credibility supporting extras, such as, examples,

[5]Roughly, the organic change mechanism generally associated with Lamarck.

[6]The means of organic change proposed in *The Vestiges* (Anon 1844).

[7]Gould's attempt to define the actions of natural selection in the context of a minimal "geological and environmental stage" (Gould 2002: 161) nevertheless, implicitly assumes that organisms vary and that "local environments" alter. This, he admits in the same footnote, has long been one of his mistakes: over-stating the actions of "bare bones natural selection", when in reality all that can be said is that natural "selection can work" (Gould 2002: 126, footnote). Even though this requires basic worldview assumptions, such as, that organisms in a state of nature do vary.

justifications, and analogies that are not an integral part of either of the two component theories).

Furthermore, in order for an evolution theory to be viable the *preconditions* of the chosen organic change theory—i.e. those aspects of the environment that are necessary for the organic change mechanism to operate—must be satisfied by the worldview theory. In terms of Darwin's evolution theorizing: in order for natural selection to deliver the sort of organic change that could amount to species transmutation, the world within which it operates must satisfy all its preconditions. The four preconditions of natural selection are:

1. Organisms do vary—needed because variation in organisms is the raw material that kick-starts natural selection.
2. The variations can be inherited, otherwise, they cannot be passed down the generations.
3. Variations can confer a survival advantage, and these "beneficial" variations thus become preferentially established within the descendent population.
4. Individuals possessing a survival advantage, a beneficial variation, can become widely established, and will, over many generations, spread to eventually dominate the descendant population—a new adaptation can become the norm, a new variety or species.

Figure 1 presents an illustrative overview of the interlocking necessary for the components of an evolution theory based on natural selection to deliver a viable overall theory. Each of the four preconditions is illustrated with a geometric shape to represent a condition required by natural selection, and therefore one that must be satisfied by the worldview. The appropriate precondition number is attached to each

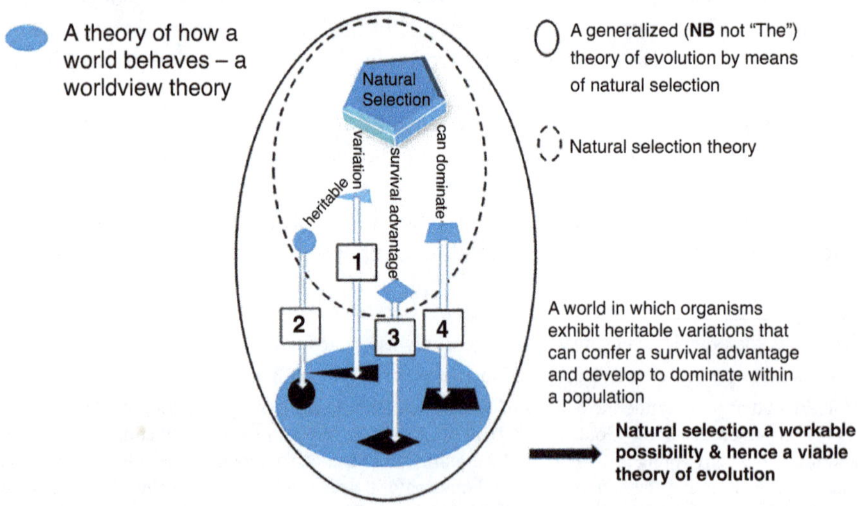

Fig. 1 The infrastructure of a general evolution theory based on natural selection

of the necessary interconnections. It should be emphasized that the worldview theory is illustrated as exhibiting no more than the characteristics necessary for natural selection to operate. Therefore, Fig. 1 does not illustrate a specific theory of evolution by means of natural selection. To specify a particular theory, such as Darwin's *Origin* theory, the details of precisely *how* the worldview satisfies each precondition is required. For example, is the variation of organic beings a localized rarity primarily triggered by external changes, or is it a common ubiquitous consequence of reproduction? The choice made, which is a worldview option, can play a major role in the patterns of evolutionary change to be anticipated. With this basic structuring in mind to replace the typically hazy uses of "natural selection", many weaknesses and misconceptions of the standard discourse can be exposed more easily. From the standpoint of the standard discourse, worldview theorizing very conveniently plays the sleeping partner, but its low profile belies a big presence.

A subsidiary general point concerns demarcations of the various theories. Theories are not well-defined objects but often must be treated as such. In the particular case under consideration, rough boundaries are helpful, although undoubtedly challengeable. The consequent objectification is, however, useful. Therefore, Fig. 1 also includes a casual visualization of the proposed boundaries. This allows us to treat our two component theories as separate objects explicitly interlinked by the four preconditions.

2.3 How Natural Selection Works in 1844 and 1859?

With regard to Darwin's theories of evolution, the details of the evolutionary changes to be expected are the consequences of the interactions between worldview features and natural selection. These operational particulars can be summarized as the answers to a sequence of five questions: when, where, why, how in detail, and with what impacts on the organic world, does natural selection operate? The answers as given in each of the two documents have been fully expounded elsewhere,[8] and are summarized below.

When is natural selection operative?

- In 1844: "only intermittently" (Ospovat 1981: 76–77), it is initiated when certain species are "unsettled by an alteration in their circumstances" (Browne 1995: 514) and ceases when "balance" restored, the unsettled organisms have been adapted to the new circumstances.
- In 1859: always; no beginnings and no endings because the quest for optimal adaptation is in a dynamic context that tends to change faster than natural selection can produce better adaptations.

Where is natural selection operative?

[8] Partridge (2018).

- In 1844: only on the particular unsettled organisms.
- In 1859: on all organisms everywhere.

Why is natural selection operative?

- In 1844: to "restore the balance" (Desmond and Moore 1991: 316) between organic and inorganic worlds by means of specific local corrections—in effect, it is a corrective feedback loop.
- In 1859: to improve the adaptation of all organisms to their organic–inorganic contexts—in effect, a never-ending global and dynamic process of function optimization.[9]

How, in detail, does natural selection operate?

- In 1844: in isolated, small populations in order to minimize free-crossings; and reasons for multiple common variants were suggested.
- In 1859: isolation was not essential; reasons for analogous variations were posited; unpressurized nooks were always available for divergent variants.

What are the impacts of natural selection-produced changes on the organic world?

- In 1844: some new varieties or possibly allied species, modified descendants of the "unsettled" species.
- In 1859: success of divergent variants led to branching descent with modification of all organisms; natural classification was delivered with new varieties and new species.

Coupled with these specifics, there are the more nebulous overall worldviews of 1844 and 1859, which are fundamentally different, and each, of course, meshes comfortably with the appropriate specifics listed above:

- In 1844: God had set the world up harmoniously—perfectly adapted organisms that simply needed occasional local readjustment, by means of natural selection, whenever external changes locally disrupted the "balance" (Desmond and Moore 1991: 316) or the "more or less stable" world of "plant and animal forms" (Browne 1995: 514) or "perfect adaptation" as Ospovat (1981: 84–86) maintained. Moreover, it was a theory of descent limited to allied species as was repeatedly stated (Darwin ed. 1909: e.g. 151 and 174). Darwin saw "no

[9]The function being optimized is adaptation of the organic world, within itself, and to the physical world. Given that in many instances the organic–inorganic world changes much faster than natural selection can operate to improve adaptations, and that improved adaptations for some organisms will be at the cost of a worsening for others, a global optimum will never be attained.

evidence" of "an inherent tendency to change and development in all species" (Darwin ed. 1909: 146).
- In 1859: God may well have breathed life into just one primordial organism (Darwin 1859: 490), and ever since, natural selection has been working continually on improving the adaptation of all the varieties and species that have resulted.

Dov Ospovat summarized the differences between Darwin's two theories, at the evolutionary level. After clarifying his use of the "fundamental ambiguity", we have: "In 1844 [Darwin's theory of evolution by means of] natural selection was a not a theory of organic development, but simply a theory of organic response to environmental change" (Ospovat 1981: 210). For Ospovat there were two different theories of evolution: "organic development", which underlies the *Origin* theorizing, and "organic response to environmental change" promoted in the Essay. And, of course, he did not mean that the theory of natural selection had changed, just the way it was employed in the two documents, and summarized above.[10] According to Ospovat, "The most important difference [between the Essay and the *Origin* theories] is in his conception of adaptation. Whereas in 1844 he believed adaptation to be perfect, by 1859 he supposed it to be relative" (Ospovat 1981: 84). "The contrast is not merely between unimportant differences in expression. It reflects a fundamental shift in Darwin's conception of evolutionary change" (Ospovat 1981: 85). On the other hand, defence of the standard discourse is still going strong. For example, with no justification offered, it is stated that the two-theory proposal (Partridge 2018) "exaggerates the differences between 1844 and 1859" (White et al. 2021: 102, footnote 38).

Whatever their assessed standing, the theory differences, summarized above, demonstrably exist. Adding significantly to these points of difference, a further central awkwardness associated with the Essay being a usefully publishable outline of Darwin's *Origin* theory revolves tightly around the notion of "divergence"—the tendency of descendants to become diversified in structure, constitution, and habits. Within the standard discourse, its inclusion within the *Origin* and omission from the Essay is widely acknowledged as the primary difference between the two theories.

2.4 The Miss-characterization of Divergence

A potential dislocation of the smooth spine of the standard discourse is introduced by Darwin's post-Essay realization of a powerful new key insight, the so-called "principle of divergence".[11] Two of its diverse repercussions are mentioned in the 1859

[10] Indeed, were it not for Dov Ospovat's untimely death in 1980, his research, as manifest initially (and as it tragically happened, also finally) in his 1981 book would most likely have done much to quash the spine of standard belief that the present chapter is challenging.

[11] Darwin announced his "principle" in late 1857 within a letter to the US botanist Asa Gray. He had enclosed "the briefest abstract of my notions on the means by which nature makes her species"—a

answers to the last two operational questions in the above summary of theory differences. The inclusion of this "principle" in the *Origin* demands some special obfuscation within the standard treatment. The difficulty hinges on reconciling an important change with no significant impact on the overall theory. The usual solution is reliant on the unity of a principle acting as a keystone that simply consolidates the previous theorizing. This view permits glossing over the complexities of its manifold boosts to the operational power of natural selection. With reference back once more to the "fundamental ambiguity", it should be noted that this again is instrumental in the maintenance of the standard narrative. Downplaying divergence as a significant theory element is aided and abetted by the undeclared but casually accepted mistake of holding "natural selection" to be more or less equivalent to "Darwin's theory of evolution", as, for example in the Ospovat quotation given above. The basis of popular belief rests on the contraction branch of ambiguity, which elicits the usual explanation: if natural selection had not been amended, the theory of evolution could not have been significantly amended.

The disenchantment-based view (developed below), by way of contrast, completely rejects the contraction option as anything more than a convenient short-cut that has runaway to assume the unwarranted mantel of equivalence. It is, therefore, untroubled by the possibility of two significantly different evolution theories, both by means of natural selection, laid out by Darwin, one in 1844 and the other in 1859. Hence, important changes to the first version no longer have to be interpreted as having a non-significant impact on the second version, not at all. Indeed, significant impacts are to be expected if the changes have been driven by deep dissatisfaction with the first version. The change of predisposition from standard smooth development to disenchantment and consequent disruption—from one theory to two that are significantly different—transforms an awkwardness into an expectation.

It is undeniable that the general notion of divergence—that descendants from any one species become diversified in structure, constitution, and habits—captured a crucial, post-Essay, insight about the organisms of the world. Although the general idea of divergence easily comes across as a compact principle, in Darwin's 1859 theory it is manifest as a collection of individual, very different, elements of the theory. It was an amendment whose distinct ramifications had to be incorporated into the overall theory. The triumph of this revision of Darwin's worldview was to boost important weaknesses in the operation of natural selection. Divergence, once properly integrated, improved the chances of the overall evolution theory delivering organic changes that could amount to species transmutation. According to David Kohn (1985: 255): divergence explained the branching-tree structure of natural classification; it accounted for sympatric speciation by vigorous natural selection; and, it pointed to crowded small and uniform areas as the places where vigorous selection would prevail. This powerful multifaceted conception was completed with

six-point summary of his evolution theory, in which this "principle" was the sixth point (Letter from CD to Asa Gray, 5 September [1857], dcp-lett-2136, accessed 08/02/2020).

the ecological division of labour.[12] Rather than the straightforward insertion of a keystone in an arch, the introduction of divergence demanded a multithreaded process of integration. This was an action more akin to the demands of those plastic puzzles in which a variety of pieces must be held precariously together until the tricky key piece interlocks them all into a solid whole, perhaps a sphere, a cube, or a pyramid. Divergence was just such an intricate key; it was not a simple keystone. It locked together the loosely connected components of his theory to deliver a more solid case for the production of new varieties or species, but only after its various prongs had been carefully threaded around the appropriate elements of natural selection. Darwin took about 3 years to achieve a satisfactory amalgamation of the various implications of his "principle of divergence" and his theory of evolution. David Kohn (1985: 256) maintained that "the principle of divergence was structured in November 1854". However, the evidence of Darwin's "species-Book"[13] drafts reveals that a major reorganization of the roles of divergence in this immediate precursor of the *Origin* was undertaken in the spring of 1858; by 12 June, he had "devoted over forty new pages to this topic" (Stauffer ed. 1975: 213).

Notice also that the incorporation of divergence had no effect on the structure of the theory of natural selection itself. It was a set of purely worldview readjustments. It boosted the power of natural selection by increasing the likelihood that Precondition (4), in particular, would deliver a new variety or species. The widely accepted "key-stone" nomenclature was fostered by Darwin.[14] Ever since, it has facilitated the standard belief that divergence simply slotted in and consolidated the theory without significantly disrupting the contention that Darwin should have published his 1844 Essay theory. A principle simply slotted in may pass muster as a theory amendment that does not seriously challenge the smooth continuity of the central spine. It is much more problematic to skate over the intricate-key interpretation as a non-disruptive change. Ultimately, divergence is no more than difference "writ large", but the expansion cascaded out into a variety of evolution theory boosts.

3 Disenchantment with the Essay

The key factor to be introduced as a basis for the present deconstruction is Darwin's dissatisfaction with his Essay theory. Traditionally, the phrase "natural selection" is dominant within any discourse centred on Darwin's evolution theorizing (recall Fisher's admonition). Darwin was clearly committed to this idea throughout the

[12] Such as Darwin's example of diverse organisms in a rectangle of turf, three feet by four feet (Darwin 1859: 114).

[13] For which he had chosen the title *Natural Selection*.

[14] Letter from CD to John Murray, 5 April [1859], dcp-lett-2447, accessed 03/09/2020.

period of the present study.[15] It is an acknowledgement of the fundamental ambiguity underlying the phrase "natural selection" that throws the scope for Darwin's dissatisfaction with his Essay theory of evolution wide open. If "natural selection" is used properly, in its own right, as no more than a theory of organic change, it is no longer the only significant, element of the theory; a worldview theory must exist alongside it. It is cognizance of the role that natural selection plays in conjunction with a worldview that reveals the poverty of the standard view. The composition of evolution theories in general, and Darwin's in particular as illustrated in Fig. 1, facilitates establishment of a vital point: a confident fixation on natural selection is no bar at all to the possibility of overall discontent with the Essay theory. Furthermore, it is also not the case that the changing of worldview characteristics will be limited to a minor partition of the overall theory. The essential details of a worldview component will tend to comprise far more of the theorizing than the specification of the organic change mechanism known as natural selection. As it happened, Darwin's dissatisfactions were all located in the worldview component of his evolution theorizing. It is now well-documented that Darwin had become disenchanted with his Essay theory within a few years of its completion (Partridge 2021). Nevertheless, a brief review of the major points of the argument will be undertaken below, and the term "disenchantment" will then be adopted to label an alternative to the standard discourse—the particular alternative that will be pursued in the present chapter. The consequent new elemental interpretations, when interlocked to generate a complete and coherent narrative, will contribute significantly to the credibility of an alternative discourse. By analogy with the scientific dictum of refutation not proof as the best route to theory assessment, the historical discourse that embodies the least puzzles is to be preferred, other things being equal. It is proposed that such an objective means is used to replace an unordered set[16] of alternative discourses with a ranked list in which the ranking is in terms of relative plausibility driven by "puzzle counting". Such relative assessment will add further value to investigations of unconventional interpretations of history.

With natural selection a more or less fixed component of his evolution theorizing, it was within the essential characteristics of his worldview that Darwin's dissatisfaction lay. From the very beginning, Darwin was dubious about the potential of his evolution theory to deliver the organic change needed for the production of new species or varieties. In his very first attempt to begin to lay out a coherent theory, he was quick to express one such concern: in a world where variation of organisms in a state of nature was rare,[17] so would be the chances of both parents exhibiting the same beneficial variation—an assumed requirement for the offspring to inherit the

[15] In a move into consideration of the latter five editions of Darwin's *Origin*, maintenance of his commitment to natural selection as virtually the only theory of organic change becomes more debatable.

[16] Strictly, sets are *unordered* collections, so the qualification employed is formally redundant but probably informative.

[17] A worldview characteristic that Darwin stated several times in his 1842 draft (Darwin ed. 1909: 4 and 9).

variation.[18] At the opening of his first draft of a theory of evolution he noted that as a consequence of free crossing, any favoured variety "will be constantly demolished" (Darwin ed. 1909: 2). As a result, Preconditions (3) and (4) would not be satisfied, and this, of course, means that his 1842 evolution theory (although little more than integrated notes) would be unlikely to deliver species transmutation. From the outset, Darwin viewed free crossing as an awkwardness that threatened to derail his theorizing. It was, moreover, as Darwin saw it, always a difficulty with his worldview theory. Prior to 1859, he seems not to have doubted the natural selection component, although logically, the problem could have resided in either component or in both.

Moreover, this particular source of dissatisfaction was one that Darwin was never confident that he had eradicated, despite multiple attempts spanning many years.[19] The criticisms that Darwin faced after publication of the 1859 *Origin*, and the various amendments that he made to the subsequent editions offer compelling testimony in support of Darwin's long-term uncertainty. In an 1869 letter to Wallace, written when he was preparing the fifth edition of the *Origin*, Darwin revealed his recent recognition of the "paramount importance" of "individual differences" over "single variations"[20]—i.e. recurrent small-scale variations as opposed to unique changes of large scope and import, often called "sports". Desmond and Moore (1991: 590), for example, depict Darwin and his defenders in an unceasing war of attrition against his religious and scientific opponents, which saw him in later editions of the *Origin* "hedging and wriggling…downgrading natural selection in some cases, backpedalling to admit 'spontaneous variations in the right direction' or a growth from increased use in others". In a similar vein, the second volume of Janet Browne's biography presents the post-1859 Darwin as "rattled … He was aware that he was becoming more environmentalist, more Lamarckian…and producing a poor-spirited compromise" in the fifth edition (Browne 2002: 315).

However, prior to 1859, Darwin was concerned about the likely demolition of a survival-enhancing variant. It seemed to be an inevitable consequence of uncontrolled mating under the worldview rubric of blending inheritance. This, of course, ran totally contrary to the requirement that such a variation will, over time, spread through the descendant population to become a characteristic of a new dominant form—a new race or species.[21] Within his first complete articulation of his evolution

[18] Another worldview characteristic advocated by Darwin: "Every peculiarity, let it be remembered, tends, especially where both parents have it, to be inherited" (Darwin ed. 1909: 236).

[19] Although the present study terminates with publication of the first edition of the *Origin*, Darwin's conceptual struggles with his evolution theorizing ran right through the five subsequent editions with no clear sign of diminishing, arguably quite the opposite.

[20] Letter from Charles Darwin to Alfred R. Wallace, 22 January [1869], dcp-lett-6567, accessed 09/04/2023.

[21] Customarily, the innovative productions of Darwin's evolution theorizing are labelled as "new species", but within his first theories, 1842 and 1844, he invariably hedges this outcome as "new races/varieties or species". For Darwin in 1844, "race" and "variety" were synonyms (Darwin ed. 1909: 204).

theory, the 1844 Essay, this hedging phrase (new race or species) is used repeatedly.[22] Whenever he does commit to the possibility of a new species, the statement itself is explicitly hedged, which further speaks for his uncertainty at this early date about whether his theory will actually deliver species transmutation. Examples are:

> "the production … of exquisitely adapted species, is at least *possible*" (Darwin ed. 1909: 134, original emphasis); and
>
> "the production of new races, indistinguishable from true species, is probable." (Darwin ed. 1909: 242–243)

A series of events between 1844 and 1859 are examined below. They do not speak of mere dissatisfaction but indicate a degree of abandonment of the Essay theory, especially when it is borne in mind that this does not mean that natural selection was abandoned. It was not, but a host of important Essay worldview characteristics were discarded or replaced. The evidence for this general claim has already been presented with the above comparison of the operating characteristics of natural selection within his evolution theory as written out in the 1844 Essay and in his theory as published in the 1859 *Origin.*

3.1 A Sequence of Events Indicative of Essay Disenchantment

There is a series of inter-theory events that point towards Darwin's deep dissatisfaction, and hence disenchantment, with his Essay theory. They are: his decisive switch in 1847 from a study of one peculiar specimen to an investigative survey of all barnacles; his 1848 declaration of intent to compose a new version of his evolution theory as a "species-Book"; in 1850, he confirmed burial of the Essay; in 1854, a return to focus on evolution theorizing, but not to the Essay; in 1856, no mention was made of the Essay for a possible role within his endeavour to construct a pre-emptive sketch of his theory; in 1858, Darwin did not want an excerpt from Essay to be presented and published. The evidential value of these specific events is highly context sensitive. Within the standard discourse, they do not tend to be singled out, and certainly not as discourse-determining factors. The context of each is developed below to reveal how the interplay of particular factors can re-structure the balance of events and hence the pathways of interpretation within an alternative discourse. We begin with "Darwin's delay" which enfolds the first event to be re-examined.

3.1.1 Darwin's Delay?

This two-word label denotes a straightforward major puzzle that emerges within the standard discourse and has been tackled repeatedly. Although variously explained, at base, the puzzle is that within his late-1830s Notebooks the origins of the *Origin* are

[22] Explicitly, Darwin ed. (1909: 209, 241 and 242–243).

already evident. Just a few years later he had written out his complete theory, the 1844 Essay.[23] Why then had Darwin waited until 1859 to publish his theory?

Fear of the repercussions is the most popular explanation of the delay (forcefully advocated and thoroughly developed by Desmond and Moore, 1991). Other reasons have been suggested and have been presented earlier. However, all assume that it was Darwin's free choice to turn away temporarily from a focus on evolution-theory development: possibly to complete a prior plan (van Wyhe 2007), or perhaps to build-up his expertise in species identification (Buchanan and Bradley 2017). None propose that Darwin was lured into a redirection of his primary research activity because of a logjam in the development of his evolution theorizing. The disenchantment-based proposal is that a variety of flaws in the Essay theory emerged within a few years of its completion. Darwin saw no clear route through the gridlock and was therefore open to the possibility of switching to a more promising activity that had presented itself. This is the interpretation that has been thoroughly developed previously (Partridge 2021); it will be summarized below.

The present analysis focuses on two aspects of "Darwin's delay". The first is a central concern within the standard discourse: why did Darwin undertake a long diversionary study to sort out the chaotic taxonomy of every barnacle that had ever lived? The second is a concern that only emerges as a consequence of the disenchantment-based discourse: was this redirection of his primary research interest inevitably no more than a temporary diversion from development of his evolution theory? This two-part approach hinges on two decisions—one in late-1847 and the other in late-1854—and the latter nicely illustrates the historiographic merit of a counterfactual approach. Realizing one of Black's points, it provides an illustration of how consideration of particular new factors can alter both the balance of events and the path-dependent interplay of those factors (Black 2008: 10).

3.1.2 1847: A Diversion into Barnacle Taxonomy

The major element of the delay was the many years Darwin spent focused on sorting out the chaotic taxonomy of all the world's barnacles. He began the study of a single peculiar specimen in late-1846,[24] and did not return to focus on his evolution theorizing until late-1854. By which time he had published three classic monographs laying out a full taxonomy of the *Cirripedia*, both living and fossils.[25] In order to fit this long break away from a focus on evolution theory development into the continuity of the smooth progression demanded within the standard discourse, it is maintained that an initial small study simply got out of hand. By looking back down along the standard storyline, Darwin can be seen frequently casting about somewhat

[23] First published by Charles Darwin's son Francis (F. Darwin ed. 1909: 57–255).

[24] The significance of the two dates, 1846 and 1847, is that in 1846 Darwin began sorting out the taxonomic status of a single peculiar specimen that he had collected on the Beagle voyage, but he did not embark on the full-class study until late in 1847.

[25] A monograph of the subclass *Cirripedia*, 2 vols, 1851 and 1854; two monographs of fossil *Cirripedia*: *Lepadidae*, 1851, and *Balanidae and Verrucidae*, 1854.

wildly for new specimens. It is, therefore, quite possible to perceive a loss of control, and this perception supports the standard explanation: the barnacle taxonomy study just snowballed (Desmond and Moore 1991: 351); its full extent crept up unexpectedly (Browne 1995: 475); the initial single specimen study "evolved into" the full study (van Wyhe 2007: 192). So, it is widely accepted that quite by chance, classifying barnacles ate up many more years than originally anticipated. Alternatively, we might wonder whether this appeal to serendipity is simply an artefact of the necessity to preserve the spine of the standard discourse. Serendipity by its very nature requires no explanation; its occurrence obviates the need for a rational explanation of Darwin's long diversion—it just happened. Perhaps Darwin did look away from his evolution-theory development, and before he could stop himself he was tumbled into a decade-long study.

However, the evidence speaks differently. In late-1847, just about 1 year after starting work on the taxonomy of one peculiar barnacle specimen (final unfinished business from his *Beagle*-voyage collections), Darwin announced his intention to undertake a monograph of the entire class of *Cirripedia*, both the living and fossils.[26] On 18 December 1847,[27] Darwin declared this new, vastly expanded intention.[28] The enlargement from one peculiar specimen to all that had ever lived was nothing less than a major redirection of his primary research activity. A decisive switch from one sample to a study of the full class of Cirripedia is not descriptive of a process that snowballed out of control.[29] Quite the contrary, it is a gigantic step change, beyond which a taxonomic study of barnacles had nowhere further to go. As such, it was also a commitment to many years of work that would not be focused on evolution-theory development.[30] As a serendipitous loss of control, the years away from development of his evolution theory can be accommodated within the standard framework: it was a blip, a surprisingly long blip that merely showed Darwin was human. But as a conscious and decisive multi-year move away from evolution-theory development, it throws something of a spanner into the works of the standard discourse. This evidential dislocation within the standard narrative is exposed by the efforts to

[26]The editors of Darwin's *Correspondence* state that Darwin's late 1847 commitment to "prepare a monograph of the entire sub-class" followed a suggestion by J.E. Gray (see dcp-lett-1139) (Burkhardt and Smith 1988, vol. 4: 390).

[27]Letter from CD to J. E. Gray, 18 December 1847, dcp-lett-1139, enclosed in dcp-lett-1146, accessed 29/01/2019.

[28]From a small and limited stopgap study whilst paused with a view to edit his Essay into a publishable form, Darwin suddenly declared a many-year shift into plain science as his primary research activity.

[29]This decisive switch is fully justified in detail in Partridge (2021).

[30]This is not to say that his mammoth taxonomic study did not contribute significantly to the development of his theory of evolution. It certainly did, as Browne (1995: 473–510) explained at some length, but this after-the-study fact does not appear to have been a driving force for his decision stated in December 1847.

reconcile an undeniably decisive step change with an evolving expansion. How can "the extent of the barnacle project [have] crept up unsuspected", when "Just before Christmas [1846?] he decided to make a full study of the entire order of barnacles"(Browne 1995: 475 and 472, respectively)? And, how can Darwin have "moved to take up the whole group" before 1846 ended, and yet "The barnacle studies evolved gradually" (van Wyhe 2007: 191 and 194, respectively)? The oddity of these irreconcilable juxtapositions is further deepened because the year 1846, first implied and then explicitly stated, should be 1847.[31] Darwin did not announce the expansion shortly after taking up the peculiar South American specimen in September 1846. No, he worked on it for something like a full year as a single specimen study, no more than another element of his *Beagle* collections backlog to be cleared. He did not announce the dramatic expansion until just before Christmas 1847.

How then is Darwin's conscious decision to turn away from development of his evolution theory to be explained? It has been proposed that the decision was based on a weighing up of future likelihoods and opportunities.[32] Looking forward from the latter 1840s, Darwin's disenchantment with his 1844 Essay theory had caused him to tell Hooker that he had buried it. The departure from his somewhat blocked focus on evolution-theory development also held the promise, as Darwin observed, of delivering a "Monograph of the Cirripedia [which] is now a pressing desideratum in Zoology".[33] Darwin was quoting the eminent biologist Louis Agassiz, whose prediction Darwin did fully realized in the early 1850s with his full barnacle study. For Darwin in 1847, the move was from intensely demanding and faltering scientific creativity, which was possibly doomed, to plain science that was almost guaranteed to deliver results commensurate with the effort put in. And deliver it did, quite spectacularly.

With disenchantment on the table as a major new factor, a rational explanation of Darwin's switch to a full study of the barnacles, as a conscious choice, opens up. In a nutshell:

> What choice was this really: continue with the extreme creative demand of chipping away at a misconfigured, possibly fatally ill-conceived, theory in the hope that it all might come out right in the end; or commit to immediately productive science that would almost certainly deliver a valued 'desideratum'? (Partridge 2021)

The balance was further tipped in the direction of plain science by the loss of his closest friend and confidant, Hooker. He left the UK in November 1847 on a plant-

[31] It is quite possible to account for this start date disagreement by the fact that from 1846 to 1854 it was all barnacle studying. However, in 1846 Darwin was focused on the study of just one peculiar barnacle from his *Beagle*-voyage collections. But in late-1847 he explicitly committed to a study of the complete class of *cirripedia*.

[32] Partridge (2021).

[33] Letter from CD to J. E. Gray, 18 December 1847, dcp-lett-1139, enclosed in dcp-lett-1146, accessed 29/01/2019.

hunting expedition in the Himalayas; he did not return until 1851. Earlier in 1847, Darwin expressed his dismay:

> I congratulate you over your improved prospects about India, but at the same time most sincerely groan over it: I shall feel quite lost without you to discuss many points with, & to point out (ill-luck to you) difficulties & objections to my species hypotheses.[34]

Not obviously, but perhaps unsurprisingly, some months later, in late-1847, Darwin choose the plain science option, which had also been Hooker's advice.[35]

3.1.3 1854: The Potential for Evolution of Intent

Quite apart from why Darwin chose to initiate this diversion away from his evolution theorizing, there is the question of was it necessarily no more than an interlude away from his evolution theorizing? It is hindsight that reveals it as a lengthy, but temporary, breakaway, and not a complete abandonment of evolution-theory development. However, despite Darwin's previously declared intention to return to his evolution theorizing,[36] he might well not have done so. Recall from the opening quotation "that those in the past thought they knew what might happen, including what did not".[37] Could it have happened that in 1854 Darwin did not realize his 1845 intention to return to evolution-theory development? It was declared almost a decade earlier, and circumstances had changed drastically with the triumphs of his barnacle study. Surely, non-return must have been a possibility? But the question is never asked because what actually happened is also the standard expectation. Consequently, the evidence is never quizzed in this novel direction.

However, with Essay disenchantment as a factor now in play, the balance of events and the path-dependent interplay of those factors is altered. There is a plausible path that does not switch back automatically to evolution-theory development as soon as the full barnacle study has been completed. Within the standard discourse, there is no surprise about Darwin's return to his full-time evolution-theory work in September 1854, after his unexpectedly extended barnacle study had been completed. Indeed, the return was entirely to be expected. Presumably, he was expected to return to editing his Essay into a publishable form? The puzzle of Darwin's delay suggests that this too would be the standard expectation. Or alternatively, a return to work on his "species-Book" idea might have been anticipated, especially if it was not intended to be a new version but an update and transfer of the Essay theory. Perhaps the expectation was a return to begin the transferral of information to the "species-Book". The drafting of chapters for this book was not begun in earnest until mid-1856. Whatever the detailed expectation, a return to his

[34] Letter from CD to JDH, 7 Apr [1847], dcp-lett-1077, accessed 23/07/2020.

[35] Darwin wrote of "plain barnacle work" which he contrasted with "theoretic species work" when he reminded Hooker that he had recommended the former, probably in late summer 1847 (CD to JDH, 12 October 1849, dcp-lett-1260, accessed 26/10/2019).

[36] Letter from CD to JDH, [5 or 12 November 1845], dcp-lett-924, accessed 04/05/2020.

[37] Black (2008: 3).

species work was Darwin's declared intention shortly before he began to examine the anatomy of the peculiar barnacle he had collected.[38] With his long diversion into barnacle taxonomy tidily and very successfully completed, why did Darwin not edit his Essay into a publishable form, and publish it? He did return to the development of his evolution theory, but he never returned to edit his Essay. This minor sub-puzzle—Essay or "species-Book" as the basis for theory development in 1854—only occurs in the context of the standard narrative. This is because neglect of the Essay is a firm expectation of the disenchantment hypothesis. In 1854, Darwin did not return to edit the Essay for publication because he had long left it behind. By 1848, he had determined to compose a new version of evolution by means of natural selection. Yet was Darwin's return to evolution-theory development by whatever route an inevitability?

Previous intentions by no means always transform smoothly into subsequent actions, especially when the intervening period has been productive way beyond all initial expectations. The multiple volumes on *Cirripedia* taxonomy that Darwin published in the early 1850s were, and still are after more than a century and a half, held as classics.[39] In 1853, Darwin was awarded the Royal Society's Royal Medal for Natural Science primarily on the basis of the research and scholarship captured in these works.[40] In a letter to Darwin, Hooker described the voting: the proposal and seconding for Darwin was "then, followed such a shout of pæans for the Barnacles that you would have sunk[?] to hear".[41] The editors of Darwin's *Correspondence* summed up this unexpected recognition of the barnacle work by the scientific establishment as follows:

> The announcement of the award ... mentioned both *Coral Reefs* and *Living Cirripedia* (1851), but it was the latter work that received detailed attention. (Burkhardt and Smith 1988: 406)

In fact, there is no explicit mention of *Coral Reefs* in the citation.[42] It was Darwin's revelations and organization of the various amazing anatomical details and whole-life functions of barnacles that received exclusive fulsome exposure.

In addition, Darwin had a well-documented history of deep interest in marine invertebrates (Sloan 1985; Love 2002). This apparent locus of fascination might well be considered an additional powerful reason why the triumph of his *Cirripedia* studies might well have induced Darwin to pursue further similar activities. But it did not.

---

[38] Letter from CD to JDH, [5 or 12 November 1845], dcp-lett-924, accessed 04/05/2020.

[39] According to the editors of Darwin's *Correspondence*: "Darwin was able to provide a thorough taxonomic study that has remained a standard work in cirripede morphology and systematics" (Burkhardt and Smith 1988, vol. 4: 388).

[40] It was his *Living Cirripedia* published in 1851, that received detailed attention in the medal citation (Burkhardt and Smith 1988, vol. 4: 406–407).

[41] Letter from JDH to CD, [4 November 1853], dcp-lett-1539, accessed 16/03/2023.

[42] It seems that *Coral Reefs* was mention in the initial proposal but never made it into the citation.

However, on the side of original-intention fulfilment, it is well-documented that barnacle work did deliver a wealth of valuable insights about organic change. So, in possession of a trove of species-development insights, Darwin had an unexpected extra impetus to simply return to his evolution-theory development, and that is what he did. With Essay disenchantment entertained as a factor, in 1854 Darwin reverted to what can be construed as the less promising option at the time. He returned to tackle his stalled evolution theorizing rather than further pursuing his long-term interests in marine invertebrates and, at the same time, capitalizing on his recent outstanding success. Within this new framework, for Darwin in 1854 there was nothing obvious about the decision he made. His evolution theorizing might never have progressed significantly, although we now know that it paid off handsomely. It is thoroughly whiggish reasoning, buoyed by the spine of the standard discourse, that projects back to Darwin's evolution theorizing in 1854 as an inevitable triumph just waiting to be realized. It is only the combination of the runaway fiction and neglect of the context that accepts Darwin's 1845 intentions[43] as simply realized in 1854. To do so is to presume that in the interim, he had achieved nothing that was likely to have impacted significantly on his aims declared years earlier. However, the interplay of evolution-theory disenchantment, and his glorious success with plain science applied in the domain of one of his long-term comfort topics, could well have dissuaded Darwin from his original intention to return to the demanding creativity of developing a problematic theory. The evidence is that it did not, but quite possibly not because this was simply a foregone conclusion as the standard discourse maintains.

3.1.4 1848: The "species-Book" Announcement

Within a few years of Essay completion, after the receipt of "many hostile facts"[44] and a detailed negative review,[45] Darwin announced his intention to pull all the new information together as a "species-Book".[46] When his repeated negative comments—such as becoming confounded in the mud,[47] being overpowered by

[43] Letter from CD to JDH, [5 or 12 November 1845], dcp-lett-924, accessed 04/05/2020) in which Darwin stated his intention to get on with his species work just as soon as, according to his Correspondence editors (dcp-lett-924, footnote 9), he had described his "*Beagle* invertebrate collection"—an intention, "in which, according to every law of probability, I shall stick & be confounded in the mud".

[44] Darwin's reference back to Hooker's "many hostile facts" is in (Letter CD to JDH, [9 October 1853], dcp-lett-1529, accessed 28/10/2019).

[45] Although no details of Hooker's review are known, it was in delivered by Hooker in 1847. Darwin referred back to it in 1849 when he wrote: "your decided approval of my plain Barnacle work over theoretic species work, had very great influence in deciding me to go on with former" (CD to JDH, 12 October 1849, dcp-lett-1260, accessed 26/10/2019).

[46] Letter from CD to JDH, 10 May 1848, dcp-lett-1174, accessed 04/02/2020.

[47] Letter from CD to JDH, [5 or 12 November 1845], dcp-lett-924, accessed 04/05/2020.

the task,[48] and the possible explosion of his theory like an empty puffball[49]—are not all dismissed as empty hyperbole, we have further evidence in support of the disenchantment hypothesis. Clearly, he did not view "species-Book" composition, which he did not actually begin drafting until after these negative comments were written, as a simple reframing of his Essay. His plan was eventually to embark on the very demanding task of composing a new version of his theory, one that successfully dealt with all the hostile facts that had been levelled at his Essay. His plan was *not* to undertake a relatively straightforward process of transferral of his Essay theory with up-to-date amendments and more justifications.[50] This is, of course, the standard discourse view demanded by "Darwin's delay".

3.1.5 1850: From Truth to Interment

In 1844, when he had "just finished" the Essay, Darwin wrote a codicil to his Will; he addressed it to his wife, Emma.[51] Within the standard discourse, it comes across somewhat oddly: perhaps, Darwin had written "Emma the strangest letter of his life" (Browne 1995: 446)? Although it was perhaps not so strange when not viewed through the lens of the standard narrative and accepted as a codicil rather than a letter.

Had this hypochondriac, desperately keen to ensure that his "true" theory would still be published "in case of my [i.e., his] sudden death", simply penned a quick amendment to his Will by means of a memorandum addressed to his wife? It was not run of the mill. It was also not entirely strange for a man always plagued by illness who believed that he had just completed a true theory that solved the mystery of mysteries.[52] He was also clear that he was not rushing to publish before the Essay had been subjected to a substantial editorial makeover. This intention was quite clear. He stated: "no intention of publication in its present *form*".[53] Moreover, in the codicil he laid out a comprehensive overview of the editorial work necessary before publication. In this 1844 memorandum, Darwin wrote "I believe that my theory is true".[54] Even if the exact meaning of "true" in this context is not clear, it is clear that at this moment in time, Darwin believed he had achieved a momentous goal. He

[48] Letter from CD to W. D. Fox 27 March [1855], dcp-lett-1656, accessed 11/06/2020.

[49] Letter from CD to JDH, 26 March [1854], dcp-lett-1562, accessed 27/04/2020.

[50] The 1844 Essay was "inadequately supported by the evidence" (Browne 1995: 446). It is proposed that Darwin wanted an "iron-clad argument" for the Essay (Costa 2014: 175).

[51] Memorandum from CD to Emma Darwin, 5 July 1844, dcp-lett-761, accessed 9/8/2019.

[52] This grandiose description was not one that Darwin used in the codicil, but he did use it in the second edition of his Beagle voyage book, which was published shortly after (Darwin 1845: 378).

[53] My emphasis, but as the body of the codicil made clear, it was "form" not "content" that Darwin was unhappy about.

[54] A generally ignored, but important, point. The omission is still going strong, for example (van Wyhe 2013: 32). Although (Desmond and Moore, 1991: 316) state of the Essay that Darwin "saw its truth", curiously though they reference this assertion to page *xix* of (Darwin ed. 1909), which is concerned with the dating of the two drafts, and is a work that omits Darwin's "theory is true" statement. However, on the following page they do quote Darwin's 1844 belief in the Essay's truth.

believed that he had composed a completely correct important theory, but he wanted its "form" improved before publication. That Darwin believed he had reached this peak of scientific achievement in July 1844 is not registered as such within the standard narrative. This omission leads to failures to register subsequent events and actions that do not further the standard discourse. Events such as his repeated concerns about his ability to compose a new version of his evolution theory; his up-front commitment to the full barnacle study; and his failure to ever return to edit the Essay, are simply backgrounded or totally ignored.

As the codicil states, in mid-1844 Darwin was keen to ensure that his Essay would be published whatever happened. But then in mid-1850, he wrote to Hooker that the Essay had "already done good & ample service, & may lay its bones in the earth in peace".[55] This drastic downgrading of Darwin's estimation of his Essay theory—from the ultimate scientific achievement to a worn-out carcass ready for burial—is a major obstacle in the standard historical path. It wrecks any pretence to continuous smooth progress. Just another awkwardness that is seldom, if ever, flagged up. It is, however, wholly in line with the disenchantment hypothesis that after the long barnacle study, he never did return to work on the Essay. It also further reinforces the contention that the "species-Book" announcement of 1848 was the declaration of a plan to compose a new version of his evolution theory, and not just an updated rewrite of the Essay.

3.1.6 1856: An Unacknowledged Oddity

In the spring of 1856, Darwin accepted (without great enthusiasm) a suggestion from Sir Charles Lyell to prepare and publish a pre-emptive sketch of his evolution theory. A week or so later, Lyell wrote to Darwin confirming his suggestion: "so out with the theory & let it take date—& be cited—& understood".[56] However, over the course of several months of discussion with a variety of correspondents as to how best he might compose this sketch, Darwin eventually abandoned the task. In the second half of the year, he turned instead to begin drafting chapters for his proposed "species-Book". Darwin's failure to compose the sketch is a commonly noted element of the standard history, and it is easily explained. The editors of *The Correspondence of Charles Darwin*, for example offer a good reason for non-composition of the sketch; they proposed that:

> Darwin was reluctant to squeeze his expansive material into such a small compass and soon abandoned Lyell's idea in favour of a full-length work that would do greater justice to his views. (Burkhardt and Smith 1988, vol. 6: *xiv*)

This seems to be entirely plausible. However, this reasoning has nothing to say about *non-consideration* of the Essay as a possible basis for the sketch. In fact, it *implies* some prior consideration of the Essay. How else could he have generated a

[55]Letter from CD to JDH, 13 June [1850], dcp-lett-1339, accessed 9/1/2020.

[56]Letter from Charles Lyell to CD, 1–2 May 1856, dcp-lett-1862, accessed 12/06/2020.

reluctance "to squeeze his expansive material" down? It was a reluctance that he seems not to have mentioned with any specific reference to "his expansive material" to Hooker, or to anyone else. Surely "his expansive material" is the editors' reference to the Essay, probably together with new information that he had collected during the intervening 12 years. It is undeniable that non-consideration is a peculiar omission of Darwin's in relation to the standard narrative because under this umbrella the Essay is an outline of the *Origin*. More than this: it is the *only* comprehensive outline in existence. Although the standard discourse usually includes Darwin's failure to *compose* the sketch as an in-action worthy of minor note (playing in as it does as a factor contributing to "Darwin's delay"), it is projected as nothing more. Yet, his several months discussing the possibilities with various correspondents and his failure to mention any possible use of the Essay, gets passed over completely. Hooker, as usual, was a primary actor in these discussions with Darwin about how Lyell's suggestion might best be satisfied. Hooker was also the one and only participant in these correspondence-based discussions who was thoroughly conversant with the Essay theory. He had read it through in 1847 and written out a page of notes (which unfortunately has disappeared).[57] Hooker had also discussed the Essay in detail with Darwin on a number of occasions in 1847, and apparently delivered a negative review. In 1849, Darwin reminded Hooker of his "decided approval of my plain Barnacle work over theoretic species work".[58] In 1856, Hooker had discussed in a sequence of letters with Darwin how the sketch might be constructed, but he too made no mention of using the Essay—not even as a possible basis for the sketch. The smooth progression, the spine of the standard narrative, more or less requires the stance of completely ignoring *non-consideration* of the Essay. The standard discourse only makes complete sense when the awkward logic of "non-consideration" by both of the principal actors is dodged. Otherwise, non-consideration of the Essay, despite an extended discussion of how best to compose the sketch, must pose a severe challenge to the possibility of a smooth narrative. The only way around this obstacle would be if neither Darwin's nor Hooker's mentions of the potential value of the Essay, although articulated, did not happen to be preserved in the historical record. However, any travel along such a path of presumed unfortunate omission must lead to an explosion of alternative historical interpretations. The result will be countless possible pathways laying out what might have happened despite a total lack of evidence that any of them did.

In addition, as the following sub-section makes clear, Darwin's failure to compose a preliminary essay also defies rational interpretation as an element of the standard discourse. Over the course of several months and working with Hooker

[57] Letter from CD to JDH, [14 March 1847], dcp-lett-1071, accessed 27/08/2020.

[58] Darwin wrote to Hooker chiding him for an about-face with regard to his earlier advice for Darwin to pursue the barnacle work rather evolution-theory development—seemingly a direct consequence of Hooker's review (CD to JDH, 12 October 1849, dcp-lett-1260, accessed 26/10/2019). The timing of Hooker's advice in favour of barnacle work, although difficult to pin down, appears to have happened at about the time when Darwin was about to switch to the full barnacle study, and may therefore have been an important factor in Darwin's decision.

(and others), he was unable to construct a statement that could be used to establish his priority—should the need arise. Yet 2 years later when the need arose, Hooker, working alone for at most 2 days, extracted a small portion of the Essay which served admirably (as the standard discourse maintains) to save Darwin's claim to priority. It is true that in 1856, Darwin was under little pressure to compose a pre-emptive strike, whereas in 1858 the arrival of Wallace's paper had put his back against the wall. The circumstances were quite different but the available evidence for long-term priority was not.[59] Under the disenchantment-based discourse, it was all as expected: in 1856, Darwin had no theory complete enough to support abstraction. Whereas in 1858, Hooker had simply ignored Darwin's instructions, as the following sub-section reveals.

3.1.7 1858: The Famous Linnean Meeting

A persistent cluster of errors within the standard discourse concerns the two contributions from Darwin's writing that were read at the famous Linnean Society meeting of 1 July 1858. Although many have arisen within the more popular accounts and are rejected in the scholarly literature,[60] one remains stubbornly resistant to correction. It is that Darwin *chose* the primary reading, an excerpt (widely miss-described in the plural as extract<u>s</u>[61]) from his 1844 Essay. This makes good sense as an element of the standard narrative, except that Darwin did *not* choose it. In point of fact, quite the opposite: he instructed Hooker, who was composing the defence of his priority, that he had sent him the Essay "<u>*solely*</u> that you may see by your own handwriting that you did read it"[62] (Darwin's heavy emphasis). In other words: "do *not* extract an excerpt to be read". Darwin's purpose in sending his strongly bound complete Essay to Hooker was for it to serve "mainly to establish the important fact that Hooker had read it" (Browne 2002: 39). Hooker's assurance was that it exhibited his 1847 annotations and therefore licensed him to assert Darwin's priority. Why was Darwin adamant that he did not want an excerpt read to support his priority claim? This action of Darwin's does not sit at all well on the spine of the standard discourse, which holds the Essay to be a potentially publishable draft of the *Origin*. Seemingly, for Darwin in 1858, the Essay was not a statement of his evolution theorizing that he was prepared to have declared publicly. Although Darwin's reluctance is perfectly in accord with the disenchantment hypothesis, with

[59] It was the case that by 1858, Darwin had composed, and Hooker had used as part of Darwin's defence, an 1857 document that Darwin described as "the briefest abstract of my notions" (Letter from CD to Asa Gray, 5 September [1857], dcp-lett-2136, accessed 08/02/2020). Its primary contribution was the introduction of the "principle of divergence", but little more, and it could not, of course, contribute to a claim for long-term priority.

[60] Such as: Darwin hastily prepared an abstract for presentation (Poulton 1896: 62), (Slotten 2004: 156) or (Wilson 2000: 181).

[61] For example Desmond and Moore (1991: 470) and Browne (2002: 40). A seemingly trivial error that subtly supports the standard discourse with its implication of theory coverage in comparison to the single excerpt presented and published.

[62] Letter from CD to JDH, [29 June 1858], dcp-lett-2298, accessed 19/08/2019.

regard to the standard discourse it is an interpretation that just does not fit in—and is therefore not generally acknowledged. However, as Janet Browne (2002: 40) has noted: "Darwin had little idea of what actually went forward" at the meeting as his first letters to Hooker after 1 July indicate. They also point to Darwin's belief that the primary defence of his priority would be a letter from Hooker and Lyell testifying to their friend's priority.[63] After sending Hooker material on 29th June from which his friend would construct his defence of priority, Darwin dropped out. He left Hooker (backed up by Lyell) to sort out the details of what would be presented in his name. The evidence reveals that Hooker had ignored Darwin's injunction and, in addition to a letter of support, selected an excerpt from the Essay to be read as Darwin's primary contribution. As it happened, or because of Hooker's wise choice, the published extract presents natural selection quite comprehensively. It also includes many of the essential worldview characteristics that enable natural selection to function and generate organic changes. But the worldview details are, of course, Darwin's 1844 selection and not his 1859 revisions and replacements—and the evolutionary consequences of the two sets are quite different, as summarized above. Yet, the standard discourse, flying squarely in the face of the evidence, holds Darwin's contribution to be the first announcement of his *Origin* theory. Yet, it requires no more than inspection of the details of the Linnean-meeting publications in comparison to those of the *Origin* to reveal it was no such thing.[64]

4 Justifying an Alternative Discourse?

Unashamedly a view through the lens of scientific theorizing, this study from a strict science perspective is also focused tightly on Darwin's development of the first edition of his *Origin* theory. This was a watershed event that closed one era and opened another: prior to publication, Darwin had kept his theory development as a very personal and private enterprise; after publication, it was a discussion topic for all and sundry, both secular and profane.

This chapter has offered a rethinking of the periodization of the development of evolutionary biology. The reassessment is driven by a close re-examination of just a couple of decades, specifically the years during which Darwin was creating and composing his *Origin* theory. Eschewing the vast scope of Darwinism, the present study is limited to the interval, late-1830s to November 1859—arguably, a critical period in the history of evolutionary biology. The result is an Essay-theory period, roughly 1842 until about 1847, and an *Origin* theory period, 1854 until 1859 (to a first approximation that neglects the subsequent editions of the *Origin*). Largely because of the *sans pareil* status of the 1859 *Origin*, important elements of the historiography of evolutionary biology have inevitably been preponderantly

[63] Such a letter was indeed written, signed by both Hooker and Lyell, read out at the meeting, and subsequently published (Darwin and Wallace 1858: 45–46).

[64] Partridge (2022) presents the full details of this comparison.

whiggish, and certain aspects of the interval under study are no exception. A number of specific cases were presented: for example, the long barnacle study conveniently viewed backwards which makes it look like a serendipitous breakaway despite the clear contradictory evidence; and that Darwin's Linnean-meeting readings announced his *Origin* theory of evolution, despite it being primarily an outdated Essay excerpt, the publication of which Darwin attempted to block.

With regard to any leap from "how it looks" to "what explains it", the great philosopher Ludwig Wittgenstein demonstrated his firm grasp of this logical lacuna when he asked one of his promising students:

"Why do people think that the Sun revolves around the Earth?"

"It is because it looks like it does," Elizabeth Anscombe is reputed to have replied.

"And how would it look if it didn't?" Wittgenstein asked.

This present attempt to forward a counterfactual event as an effective tool for opening useful alternative narratives in the history of evolutionary biology should be distinguished from the composition of a counterfactual history. The present paper does not develop a counterfactual history of biological science, such as Peter Bowler's in which he tries to imagine what would have happened had Darwin accidentally died as early as 1832 (Bowler 2013). Instead, the present analysis is limited to the proposition of a counterfactual event solely in order to present the unfolding history of science from an angle which is not that of established history. The consequence is not a counterfactual narrative, subject to weaknesses centred on plausibility, and questions about the role that contingency plays in the roll-out of evolution. These vulnerabilities are discussed in detail by Luca Tambolo (2020). The intent of the present study is to expose a new, and yet fundamentally factual, narrative. Although such a narrative is not immune to the challenge of plausibility, it is a simpler plausibility, one that can be evaluated by recourse to empirical evidence. Quite distinct from a counterfactual narrative, the new discourse is open to direct assessment via the historical record as the following section illustrates. Overall, we would adopt Tambolo's concluding suggestion that "a merit of counterfactuals that has gone unappreciated by their very advocates lies in their ability to put the present in a different light", and apply it with equal force to the introduction of counterfactual events (Tambolo 2020: 5). This is because they similarly open an approach to established history from a counterfactual angle, and present-day science appears in a new perspective.

The alternative discourse developed, which was triggered by consideration of a counterfactual event is well supported by the examination and reconfiguration of the unconsidered options that were open at the time. An important example is Darwin's return to his evolution theorizing in 1854, after his spectacular success with a favourite topic—marine invertebrates. "Because he always intended to", is not a wholly satisfactory answer. It completely glosses over the general impacts of time and contextual details of the moments. This standard answer is based on blind acceptance of the fixed nature of a prior intentional stance which is seemingly confirmed by the eventual outcome as if his intervening spectacular successes with barnacle taxonomy were run-of-the-mill with no significant impact. Painting the full barnacle study as a runaway process, not a conscious and decisive redirection of his primary research, bolsters the belief that his 1854 return to evolution-theory development was an inevitability. The disenchantment-based discourse opens a new potential pathway through the evidence. It is not a trail of counterfactual supposition, but simply a new potential path. Although it was not taken by Darwin, as a plausible option for him at the time, it introduces a generally unrecognized factor into interpretations of his follow-on actions.

Although no complete definitive answer to the titular counterfactual question has been generated, the preceding analysis strongly suggests a non-standard one. Non-publication of the Essay was not a missed opportunity for Darwin, it was a well-judged cautionary move. Although his pause for feedback and reappraisal did not deliver the anticipated outcome, it served its purpose. Publishing the 1844 Essay would have been a mistake that Darwin, the astute scientist, was never going to make.

But more important than debates over publish or not, counterfactualism has been demonstrated to be an effective tool for enriching the narrative possibilities in the history of evolutionary biology. For example, it has revealed an important weakness at the core of the standard discourse: quite contrary to the customary lament of Darwin's delay, Darwin was wise not to have published the Essay. The customary view has been reversed. Within this deconstruction of the standard discourse, Darwin emerges as the shrewd scientist, acutely aware of the flaws in his first version, the Essay theory, just a few years after completion. Despite his initial belief "that my theory is true",[65] his pause for reflection and editorial polishing, combined with Hooker's negative review and "many hostile facts",[66] soon made Darwin aware that his Essay theory was fundamentally defective. By 1848, he declared his intent to compose a new version, and by mid-1858 it was coming together.[67] It was the surprise arrival of Wallace's draft paper that propelled an unreferenced abstract of an unfinished manuscript into print in November 1859. Quite curiously, this move of Darwin's overrode the moral stance he had declared just a few years previously:

> it is really dreadfully unphilosophical to give a resumé, without exact references, of an unpublished work.[68]

The disenchantment-driven narrative diverges widely from that within the standard discourse. The deconstruction reveals the supposed delay as a rushed job. The *Origin* was a creation brought forth before its time, although whether Darwin would ever have published without such pressure is a moot point—but not a central one. The *Origin* was a publication precipitated by the threat of being forestalled because

[65]Memorandum from CD to Emma Darwin, 5 July 1844, dcp-lett-761, accessed 9/8/2019.

[66]Letter from CD to JDH, [9 October 1853], dcp-lett-1529, accessed 28/10/2019.

[67]By 18th June 1858 Darwin had drafted "about two thirds of the topics later presented in the *Origin of Species*" (Stauffer ed. 1975: 10).

[68]Letter from CD to JDH, 9 May [1856], dcp-lett-1870, accessed 02/06/2020.

Wallace had surprised him with a paper that Darwin read as, in effect, his own theory. Exceptionally, David Kohn has offered a similar view, although his reasoning is based on the centrality of the principle of divergence. For Kohn it was "the centrepiece in Darwin's explanation of the origin of species" (Kohn 2008: 88) but it "emerged only once Darwin had effectively set to work writing the Origin"[69] (Kohn 2008: 89). However, despite this viewpoint, which is breaking away from the standard narrative, he does not run his ideas through to their obvious endpoint: the conclusion that this must fracture, if not break, the spine of the standard narrative. Failure to publish a theory without its "centrepiece" should suggest simple wisdom, not foot-dragging that requires explanation.

5 A "better" Discourse?

The general rule with regard to assessing the merit of scientific theories is based on refutation rather than proof. Despite the equivalent status of *true* and *false* in formal logic, demonstrating falseness is invariably easier and more certain than proving truth. So, the ranking of alternative scientific theories is seldom pursued by attempting to prove one of them correct. Instead, it is typically approached along a path of assessing the relative degrees of incorrectness (i.e. falseness) each exhibits, and the least "incorrect" is to be preferred. This diversion into scientific practice is in reality far from as clean and clear-cut as the above suggests. The point to be made, however, is that assessment of complex propositions is founded on demonstrating error, not on proving truth. It is proposed that a similar philosophy could be adopted in historiography: specifically, the discourse with the fewer known puzzles is to be preferred over an alternative that embodies demonstrably more, all other things being more or less equal. This method delivers a justification for suggesting that a discourse developed around Darwin's disenchantment with his Essay theory may be more than plausible. It may be a preferred, alternative to the standard version.

Despite the significant differences between Darwin's two theories of evolution—as written out in 1844 and as published in 1859—Essay-theory disenchantment itself is often subject to a priori rejection. This is because almost everyone agrees that Darwin was remarkably constant in his belief in natural selection, at least within the period under investigation. This basis for blunt dismissal rests on two related misconceptions:

1. Firstly, a failure to appreciate that natural selection is very far from a theory of evolution as the earlier "fundamental ambiguity" section made clear. A world-view is also required, and moreover, neglecting examples, analogies and

[69] Curiously, some years earlier Kohn (1985: 256) maintained that "the principle of divergence was structured in November 1854". How does this square with: this principle "emerged only once Darwin had effectively set to work writing the Origin?" (Kohn 2008: 89). Drafting the pre-*Origin* species-book chapters did not begin until mid-1856, and actually writing the *Origin* began in mid-1858.

justifications, it is the worldview that accounts for the bulk of the evolution theory.
2. Consequently, Darwin's steadfast commitment to natural selection does not seriously limit the possibilities for theory dissimilarities. There is ample scope for difference within the worldview component, which the earlier theory-difference summaries have laid out explicitly.

Once the disenchantment hypothesis is given serious consideration, many of the standard discourse puzzles disappear. Some fail to put in an appearance because the new discourse infrastructure no longer pits conflicting interpretations one directly against another. Others fail to intrude because one of the clashing components no longer exists. Examples are: "Darwin's delay" is no longer a puzzle because the Essay is no longer viewed as, or close to, a publishable outline of the *Origin*; reconciling divergence as both "the centrepiece" and a non-significant theory amendment is no longer necessary, for precisely the same reason; and the conundrum of a serendipitous many-year excursion through the worlds' barnacles, the full scope of which Darwin had declared up-front, has been relieved of its awkward happenstance characterization. By mid-1858, Darwin had composed much of a new version of his evolution theory—primarily in the form of draft chapters for his "species-Book".[70] Non-consideration of the Essay as a basis for the pre-emptive sketch suggested by Lyell in 1856 has transformed into an expectation, not a mystery to be explained (or more usually, an awkwardness just skated over). Essay disenchantment and the public presentation of an excerpt in 1858 is no longer puzzling because far from choosing it, the evidence proclaims that Darwin attempted to bar the use of an excerpt. The modern perception of Darwin's Linnean-meeting publications as an early announcement of his *Origin* theory only has some credence when the Essay is mistakenly viewed as a meaningful outline of the *Origin*.

This is not to claim that the disenchantment-based discourse is totally puzzle free. However, it does appear to be free of straightforward evidential conflicts, such as an assumed chance expansion of what was recorded as a decisive step change. There is the puzzle of Hooker's "hostile facts" and negative review followed by his efforts to ensure that Darwin was not forestalled. Friendship was surely a factor, but as Jim Endersby (2008) has fully articulated, Hooker as a Victorian scientist without a personal fortune had to tread carefully along a narrow path. Consequently, his actions were constrained and controlled in complex ways. He did not have the simple freedoms open to both Darwin and Lyell, two independently wealthy men under no threat of being viewed as scientific tradesmen. Minor new puzzles also arise: such as why is there no record of Darwin's surprise in July 1858 when he discovered that, contrary to instruction, Hooker had included an excerpt from the

[70] Some years after the 1848 declaration to compose a "species-Book", Darwin gave the name *Natural Selection* to this intended book. He never did published it, but his draft chapters, which he began writing in mid-1856, were the primary source of information for his 1859 *Origin* Stauffer (ed. 1975) presents a full, detailed, and annotated transcription of Darwin's drafts and their developments.

Essay for publication? For Darwin, mid-1858 was a fraught period. In addition to the near certainty that he would be forestalled by Wallace, deadly disease had struck in the Darwin household.[71] The consequent grief in conjunction with Darwin's gratitude, relief, and joy when he learned from Hooker that at the Linnean meeting "all had gone on prosperously[72] may well have been quite sufficient to swamp any concern that Hooker had directly disobeyed instructions. Another small puzzle is: Why did Darwin never record his "disenchantment" with the Essay theory? Although there appears to be no record that Darwin used this term, in 1850 he did write that he had buried the Essay, which if anything is a rather more drastic. The proposed disenchantment hypothesis assumes no more than the abandonment of initial worldview characteristics. Overall, a strong case can be made for the claim that the disenchantment-based discourse is a good deal less puzzle-ridden than the standard alternative and hence merits serious consideration as a plausible alternative, possibly the more plausible alternative.

References[73]

Anon (1844) The vestiges of the natural history of creation (authored by Robert Chambers). John Churchill, London

Barrett PH, Gautrey PJ, Herbert S, Kohn D, Smith S (1987) Charles Darwin's notebooks, 1836–1844. Cornell University Press, Ithaca, NY

Black J (2008) What If? Counterfactualism and the problem of history. The Social Affairs Unit, London

Bowler PJ (1984) Evolution, the history of an idea. University of California Press, California

Bowler PJ (2013) Darwin deleted: imagining a world without Darwin. University of Chicago Press, Chicago, IL

Browne J (1995) Charles Darwin, voyaging. Jonathan Cape, London

Browne J (2002) Charles Darwin, the power of place. Jonathan Cape, London

Buchanan RD, Bradley J (2017) "Darwin's delay": a reassessment of the evidence. Isis 108(3): 529–552

Burkhardt F, Smith S (1988) The correspondence of Charles Darwin, vol 6. Cambridge University Press, Cambridge

Costa JT (2014) Wallace, Darwin, and the origin of species. Harvard University Press, Cambridge, MA

Darwin C (1845) A naturalist's voyage round the world, 2nd edn. London, John Murray

Darwin C (1859) On the origin of species (facsimile edn. 1964. Harvard University Press: Cambridge, MA). John Murray, London

Darwin F (ed) (1909) The foundation of the origin of species. Cambridge University Press, Cambridge

Darwin C, Wallace AR (1858) On the tendency of species to form varieties; and on the perpetuation of varieties and species by natural means of selection. J Proc Linn Soc Lond Zool 3:45–62

Desmond A, Moore J (1991) Darwin. Penguin, London

[71]The Darwins' baby son Charles had died of Scarlet Fever on 28th June (Browne 2002: 36).

[72]Letter from CD to JDH, 5 July [1858], dcp-lett-2303, accessed 26/08/2019.

[73]Note: In the references to the Darwin correspondence, CD is Charles Darwin, JDH is Joseph Dalton Hooker and the Correspondence website is https://www.darwinproject.ac.uk/letters.

Endersby J (2008) Imperial nature, Joseph Hooker and the practice of Victorian science. University of Chicago Press, Chicago

Fisher RA (1930) The genetical theory of natural selection. Clarendon Press, Oxford

Gould SJ (1977) Ever since Darwin. New York, Norton & Co

Gould SJ (2002) The structure of evolutionary theory. Harvard, Belnap

Gruber HE (1974) Darwin on man. Transcribed and annotated by Paul H. Barrett. Wildwood House, London

Hodge J (2013) The origins of the *origin*: Darwin's first thoughts about the tree of life and natural selection, 1837–1839. In: Ruse M (ed) The Cambridge encyclopedia of Darwin and evolutionary thought. Cambridge University Press, Cambridge, pp 64–71

Kohn D (1985) Darwin's principle of divergence as internal dialogue. In: Kohn D (ed) The Darwinian heritage. Princeton University Press in Association with Nova Pacifica, Princeton, pp 245–257

Kohn D (2008) Darwin's keystone, the principle of divergence. In: Ruse M, Richards RJ (eds) The Cambridge companion to the 'Origin of Species'. Cambridge University Press, Cambridge, pp 87–108

Love AC (2002) Darwin and "Cirripedia" prior to 1846: exploring the origins of the barnacle research. J Hist Biol 35(2):251–289

Ospovat D (1981) The development of Darwin's theory. Cambridge University Press, Cambridge

Partridge D (2018) Darwin's two theories, 1844 and 1859. J Hist Biol 51:563–592. https://doi.org/10.1007/s10739-018-9509-z

Partridge D (2021) Mind the step: did Hooker's judgement clinch Darwin's disenchantment? Notes Rec R Soc 76:445–461. https://doi.org/10.1098/rsnr.2020.0055

Partridge D (2022) The famous Linnean Society meeting: from old errors to new insights. Biol J Linn Soc XX:1–12

Poulton EB (1896) Charles Darwin and the theory of natural selection. Cassell, London

Richards RJ (1987) Darwin and the emergence of evolutionary theories of mind and behaviour. University of Chicago Press, Chicago and London

Ruse M (2008) The origin of the Origin. In: Ruse M, Richards RJ (eds) The Cambridge companion to the "Origin of Species". Cambridge University Press, Cambridge, pp 1–13

Ruse M (2013) Introduction. In: Ruse M (ed) The Cambridge encyclopedia of Darwin and evolutionary thought. Cambridge University Press, Cambridge, pp 1–31

Sloan PR (1985) Darwin's invertebrate program, 1826–1836. In: Kohn D (ed) The Darwinian heritage. Princeton, Princeton University Press in Association with Nova Pacifica, pp 71–120

Slotten RA (2004) A heretic in Darwin's court. Colombia University Press, New York

Stauffer RC (ed) (1975) Charles Darwin's natural selection: being the second part of his big species book written from 1856 to 1858. Cambridge University Press, Cambridge

Tambolo L (2020) An unappreciated merit of counterfactual histories of science. Stud Hist Philos Biol Biomed Sci 81(101):183

Tanghe KB (2017) A historical taxonomy of origin of species problems and its relevance to the historiography of evolutionary thought. J Hist Biol 50:927–987

Van Wyhe J (2007) Mind the gap: did Darwin avoid publishing his theory for many years? Notes Rec R Soc 61:177–205

Van Wyhe J (2013) Dispelling the darkness. World Scientific, Singapore

White RM, Hodge MJS, Radick G (2021) Darwin's argument by analogy: from artificial to natural selection. Cambridge University Press, Cambridge. https://doi.org/10.1017/9781108769518

Wilson JG (2000) The forgotten naturalist: in search of Alfred Russel Wallace. Arcadia, Kew, Australia

Redrawing the Boundaries of Darwinism: Addressing Darwin's Endorsement of the Inheritance of Acquired Characteristics in Darwin's Celebrations, 1909–1959–2009

David Ceccarelli

Abstract

Since the publication of *On the Origin of Species by Means of Natural Selection* (1859), the concepts of "Darwinism" and "Darwinian evolution" have long been debated. Alongside the theoretical adjustments evolutionary biology has undergone over time through the integration of research traditions, tools of analysis and sets of data, the narratives and representations of Charles Darwin's legacy have significantly changed. Undoubtedly, the in-depth study of Darwin's notebooks, which only became known to historians in the mid-twentieth century, has contributed to better reconstructing the phases of his theoretical formulation as well as his constant consideration of issues raised by critics. However, the very meaning of the concept of "Darwinian evolution" has been disassembled and reassembled in light of the debates in evolutionary biology and thus seems to be the product of a long and complex interplay between historiographical and scientific needs. This appears clearly when considering the way evolutionary biologists addressed Darwin's views on theoretical issues that turned out to be objects of great debate in twentieth-century evolutionary biology. In particular, this chapter will examine how scholars commented on Darwin's endorsement of the inheritance of acquired characters on the occasions of the 1909, 1959, and 2009 celebrations. Not only did these years represent symbolic occasions for evolutionary biologists to take stock of the state of affairs within their research field but were also an opportunity to reassess Darwin's legacy. This analysis will help deconstruct the category of Darwinism, showing that the boundary between "Darwinian" and "non-Darwinian" theories of evolution has undergone several shifts over time and that evolutionary biologists have often been concerned with

D. Ceccarelli (✉)
Department of Philosophy, Communication and Performing Arts, Roma Tre University, Rome, Italy
e-mail: david.ceccarelli@uniroma3.it

R. G. Delisle et al. (eds.), *Unity and Disunity in Evolutionary Biology*,
https://doi.org/10.1007/978-3-031-42629-2_17

identifying what the Darwinian research agenda should include and exclude, erecting research traditions, and confining inconvenient ideas.

Keywords

Darwin celebrations · Darwinism · Lamarckism · Darwin Days · Inheritance of acquired characters

1 Introduction: Standing on the Shoulders of Giants

> Why, ladies and gentlemen, are we here today, gathered together in the same thought? Why would a single name be sorrowfully whispered in every country of the civilized world? Why would the telegraph wires of a small and dark village in England be not enough to receive all the condolence telegrams that come from all over the world? [. . .] A man has died, but his name was Charles Darwin. [. . .] Darwin had long been a beacon of light in the great ocean of the unknown: pupils looked up to him to be close to him, opponents looked for him to avoid him; one could fight that man, or hate him, but it was impossible not to talk about him.[1]

With these words, leading Italian anthropologist Paolo Mantegazza (1831–1910) celebrated Charles Darwin at the Royal Institute of Higher Studies in Florence on May 21st, 1882. His vibrant, rhetoric-filled address exemplifies the atmosphere that accompanied Darwin's memorials after his passing at the age of 74 years. For many scholars, the scientific world had stopped at the news of Darwin's death, with even his critics expressing respect and admiration for his intellectual enterprise. Protestants, Catholics, bishops, spiritualists and even the "science priests" mourned the loss of Darwin (Ibid., p. 171). For once unanimous, "writers of all shades of opinion" from France, Germany, Austro-Hungary, Italy, and the United States had willingly paid tribute to the "modern philosopher" who had devoted his entire life to "attack the problems of Nature," Thomas Henry Huxley wrote in *Nature* on April 27th, 1882 (Huxley 1894, pp. 245–246).

Darwin had already become a worldwide celebrity toward the end of his life, as evidenced by his vast iconography (Browne 2005, 2009). Such an iconic status, historian James Secord (2009) pointed out, reflected an important facet of the emergence of professional science and the concurrent development of the communication and entertainment industries at the end of the nineteenth century. As a matter of fact, the commemorations on the occasion of Darwin's death represented

[1] "Perché, signore e signori, siamo oggi qui, raccolti tutti in uno stesso pensiero? Perché mai da più d'un mese in ogni paese del mondo civile un solo nome si mormora dalle labbra compunte al dolore? Perché mai i fili telegrafici di un piccolo e oscuro villaggio dell'Inghilterra non bastano a ricevere tutti i telegrammi di condoglianza che vi giungono da ogni parte? [. . .] Un uomo è morto, ma quest'uomo si chiama Carlo Darwin. [. . .] Darwin era da molti e molti anni un faro acceso nel grande oceano dell'ignoto: a lui guardavano i discepoli per avvicinarlo, a lui guardavano gli avversari per evitarlo; si poteva combattere quell'uomo, si poteva odiarlo, ma tacerlo era impossibile" (Mantegazza 1882, pp. 169–170).

the first of many celebratory events that would contribute to constructing one of the greatest icons of Western scientific thought.

Throughout the decades, a vast array of audiences has celebrated Darwin, looking at his legacy through the lens of the "social structures" of the time (Browne 2005, p. 272). Notably, Darwin's celebrations represented exceptional occasions for science outreach and engagement, as well as for the discussion of the status and development of evolutionary studies (Kolchinsky 2010). Over the years, evolutionary biologists have indeed exploited Darwin's anniversaries and celebrations to retrace the steps of progress within their research fields and, in many cases, to reevaluate Darwin's intellectual legacy in view of new scientific findings and challenges. As Marsha Richmond (2006, p. 448) noted, Darwin celebrations were not only to honor a "hero," but were also to reassess the underpinnings of Darwinism "at some critical junctures," serving a variety of institutional and disciplinary functions. As such, these celebrations represent unique and multilayered objects of investigation from a historical, epistemological, and sociological perspective, all the more so because scholars used these highly symbolic moments to redefine polarizing categories and renegotiate their conceptual boundaries, possible uses, and implications. Furthermore, Darwin's celebrations often provided experts from several fields of biological research with the opportunity to reflect upon their place in the history of life sciences with regard to Darwin's legacy.

This chapter argues that: (a) the perception of Darwin has changed in 1909, 1959, and 2009; (b) what we understand as "Darwinism" has undergone significant changes, as the demarcation between "Darwinian" and "non-Darwinian" theories of evolution has shifted multiple times over the years; (b) evolutionary biologists have long been concerned to identify what belongs in the "Darwinian" research tradition and what does not; (c) The image of Darwin has attained a nearly unparalleled iconic status, emerging as a symbol of laicism and serving as a reference point for scientific agendas and the secularized culture; (d) "siding with Darwin" has often represented a key strategy for evolutionary biologists. Quite understandably, this last assertion could easily be regarded as trivial or even slightly opportunistic, as one may argue that the reason evolutionary biologists sided with Darwin is simply that Darwin's theory, or at least some of its main explanatory components, had proven correct. Discussing the validity of Darwin's theory of evolution in light of contemporary research is beyond the scope of this chapter, which rather aims to show how, in many cases, scientists strove to emphasize the continuities between their research agendas and Darwin's scientific legacy, thus assembling and disassembling the very categories of "Darwinian" and "non-Darwinian" evolution. This appears clearly when one considers how evolutionary biologists addressed Darwin's views on theoretical issues that would represent hotly debated topics in twentieth-century evolutionary biology.

This chapter will use as a case study the way scholars assessed Darwin's endorsement of the inheritance of acquired characters (hereafter IAC) on the occasion of the three highly iconic years in the history of evolutionary biology: 1909, 1959, and 2009. These years indeed corresponded to the anniversaries of Darwin's birth (1809) and of the publication of the *Origin of Species* (1859). Each of these years bears witness to different stages of the institutionalization of evolutionary biology, as well as diverse theoretical, methodological, and epistemological approaches to the study of biological inheritance. As David Depew remarked, the very concept and ontological status of hereditary units changed across the three celebrations:

> In 1909, on the occasion of the fiftieth anniversary of the Origin's publication, genes were nothing more than theoretical entities invoked to explain Mendelian ratios. By the centenary celebration in 1959, they were physical entities – stretches of DNA located on chromosomes in the nucleus of each cell – that explained phenotypic trait distributions in populations, presumably by instructed links between DNA, RNA, the amino acids they code for, and the proteins into which these amino acids fold up. By 2009, genes were physical entities that can be directly observed and manipulated. Rather than inferring their existence from phenotypic changes, geneticists are now knocking out bits of DNA sequence from model organisms to determine their phenotypic effects and functions. (Depew 2009, pp. 358–359)

The chapter will consist of three main sections, one for each centenary. By examining these case studies, I aim to show that profoundly influential historiographical labels such as "Darwinism," "Lamarckism," and "non-Darwinian evolution" are the product of a long and complex interplay between historiographical and scientific needs. These various negotiations demonstrate the challenges associated with consolidating complex and dynamic theoretical structures into singular conceptual lineages (see Hull 1989). The utilization of labels such as "Darwinism" and "Lamarckism" frequently involves extrapolating epistemic components that carry significance within the context of current research agendas, thereby preserving, upholding, and at times, constructing traditions of thought.

2 1909—Disharmony and Pluralism in the Shadow of the "Eclipse"

Darwin's first centenary came at a time of great transformation and ferment in evolutionary studies. Later on, this phase would be famously labeled as the "Eclipse of Darwinism" (Huxley 1942, Bowler 1983), a highly evocative expression that has, however, met increasing criticism in the last few years (Largent 2009; Meulendijks 2021). By the end of the nineteenth century, the rise of experimental approaches to the study of inheritance, cell physiology, and organism development had led to the emergence of increasingly diverse languages and methodologies propelled by heterogeneous research goals (Love 2013). Within a few years, several works would, indeed, outline new standards for life sciences and evolutionary studies, reframing long-standing issues, concepts, and even antinomies. In the field of developmental mechanics (*Entwicklungsmechanik*), for instance, the almost old-fashioned controversy between preformationism and epigenesis had reappeared "as a modern-day problem in biology" about cellular differentiation in ontogeny (Hertwig 1894; Whitman 1894). At the beginning of the twentieth century, paleontologists were attempting to reassess the old notions from comparative anatomy of homology, analogy, and parallel evolution in view of the emerging understanding of hereditary and developmental processes (Gould 2002; Ochoa and Barahona 2014). At the same time, the concepts of "genotype" and "phenotype" introduced by Johannsen (1911) were contributing to a significant redefinition of the relationship between hereditary factors, individual mutations, and organism development.

As Vernon Kellogg wrote in the introduction of his *Darwinism to-day* (1907), between the nineteenth and the twentieth centuries the "extraordinary activity" in biology seemed essentially two-phased:

> There is going on a most careful re-examination or scrutiny of the theories connected with organic evolution, resulting in much destructive criticism of certain long-cherished and widely held beliefs, and at the same time there are being developed and almost feverishly driven forward certain fascinating and fundamentally important new lines, employing new methods of biological investigation. (Kellogg 1907, pp. 1–2)

Within this context, and 27 years after the eulogies and memorials that followed Darwin's passing, the year 1909 represented an occasion for the scientific community not only to celebrate Darwin's centenary but also to discuss the state of the art in the field and, at the same time, assess its development in the intervening 50 years from the publication of the *Origin of the Species* (1859).

The 1909 celebrations reflected the multifarious theoretical framework of early twentieth-century evolutionary studies. Narrowing the field of analysis to the English-speaking world, the two major events held at Cambridge and Baltimore[2] were hardly perceived as harmonic and cohesive. In his review of the two celebrations published in *The American Naturalist*, Vernon Kellogg famously claimed that the memorial meetings had left "the definitive impression on the scientific world of having celebrated Darwin's greatness and Darwinism's weakness" (Kellogg 1907, p. 383). Especially in Baltimore,[3] almost every speaker followed the same argumentative strategy, first praising the invaluable legacy of "the immortal master," and then proceeding to illustrate the flaws in his theory, "unloading their personal baggage" of alternative factors of evolution.[4] As Marsha Richmond pointed out (2006, p. 463), the organizers and speakers of the celebrations indulged in remarking that theoretical inhomogeneity prevailed among evolutionists,

[2]The Cambridge celebration was held from June 22 to June 24, 1909, and hosted 235 scientists from 167 countries (Richmond 2006). The American celebration, entitled "Fifty Years of Darwinism," was of quite a different tenor. The meeting was held before the American Association for the Advancement of Science in Baltimore on Friday, January 1st, and consisted of 11 addresses, the majority of which were held by American naturalists, with the only exception being British biologist Edward Poulton.

[3]According to Kellogg, The Cambridge celebration turned out to be far more representative of different research traditions, perspectives and subjects, which ranged from embryology, cytology, cosmology, botany and even theology (Ibid.).

[4]The only exception, Kellogg remarked, was Poulton, who defended Darwinism against criticisms "with a glorious patriotism that surpasses that of the defender of imperialism" (Ibid.).

with media emphasizing that, behind the veil of praise for Darwin, the actual leitmotiv of the celebrations consisted of controversies and dissension.

At first glance, such historical accounts would contribute to substantiating the idea that, between the 1880s and the 1910s, evolutionists were steering away from Darwin's research agenda, eclipsing his legacy. Upon closer examination, however, Darwinism continued to be largely understood—and employed—as an "umbrella term" (Richmond 2006), with biologists often seizing the occasion to add grist to their mill, striving to find legitimacy through the reinterpretation of Darwin's work. This was even more manifest when scholars publicly tackled divisive scientific issues such as the inheritance of acquired characteristics.

The 1909 celebrations took place during a phase of lively discussion on the relationship between biological inheritance, evolution, and the environment. In the aftermath of the controversy between "neo-Darwinism" and "neo-Lamarckism" that had erupted after August Weismann's writings on the separation of germplasm from somatoplasm, and amid the "Mendelian enlightenment" (Morris 1994), in 1909 different views continued to coexist in the debate on the physical basis of inheritance.[5] Despite its many detractors, in the early twentieth century, the IAC still represented a major factor for those who strove to preserve the role of the environment in initiating variations (Gliboff 2011, p. 51). On this point, the fact that Darwin had advocated a view of biological inheritance that encompassed the IAC, and even proposed a mechanism (that of *pangenesis*) to account for the transmission of somatic modifications to the germ cells, became a bone of contention for many evolutionists.

Darwin ventured to propose the hypothesis of pangenesis in *The Variation of Animals and Plants Under Domestication* (1868), where he posited that somatic cells released minute particles that circulated throughout the body system and finally congregated in the gonads. As he wrote to Fritz Muller on June third, 1868, parental characters seemed to be "photographed on the child" by means of the "atoms derived from each cell in both parents" (Darwin 1903, p. 98). Darwin called these atoms "gemmules," which were "supposed to be thrown off by every cell or unit, not only during the adult state but during all the stages of development" and could be "transmitted from the parents to the offspring" (Darwin 1868, p. 448). On more than one occasion, Darwin made it clear that pangenesis was only "a provisional hypothesis" or "speculation" (Ibid., p. 429), which could however still be of service to scientific discussion. Most notably, the hypothesis offered a new source for variations that could potentially accelerate the rate of evolutionary change and thus served as a defensive thesis against the well-known objections raised by Fleeming Jenkin: (a) the discrepancy between the time for natural selection to accumulate beneficial variations and the relative shortness of geological time; (b) the argument from blending inheritance, according to which minute variations

[5] See, for instance, Bowler 1989, Gissis 2011, Müller-Wille and Rheinberger 2012, Churchill 2015, Ceccarelli 2018.

tend to vanish at each generation and could not persist long enough to undergo selection (for more details, see Morris 1994, p. 342).[6]

Darwin's hypothesis of pangenesis received several objections regarding lack of experimental evidence (Galton 1870) and its inherent complexity, both in terms of the number of gemmules implied and the allocation processes through circulation (Brooks 1883, p. 78; Cope 1889, pp. 1068–1069). It is little wonder that this discussion was soon incorporated into the wider debate between neo-Darwinism and neo-Lamarckism.

During the 1909 celebrations, scholars often engaged with Darwin's attitude toward the IAC and Lamarckism. University of Strasburg anatomist Gustav Albert Schwalbe, who contributed to the commemorative volume *Darwin and the Modern Science* (the series of essays edited by botanist Albert Charles Seward for the Cambridge Philosophical Society) with an essay on the descent of the human species, openly addressed Darwin's views on the "Lamarckian" factors, substantiating the already popular claim that would long affect historiography, according to which Darwin had become increasingly Lamarckian over the years.[7] At an early stage of his work, Schwalbe argued, Darwin had paid scant attention to Lamarck's views. By drawing on Darwin's correspondence with Joseph Hooker during the 1840s, Schwalbe pointed out how Darwin defined Lamarck's views as nonsensical and absurd,[8] but then "became more convinced of the influence of external conditions" (Schwalbe 1910, p. 125). To substantiate this thesis, Schwalbe quoted the well-known passage of the letter Darwin sent to Moritz Wagner on October 12th, 1876: "In my opinion the greatest error which I have committed, has been not allowing sufficient weight to the direct action of the environment, i.e. food, climate, etc. independently of natural selection."[9] Schwalbe then remarked that had Darwin been able to publish a new edition of *The Descent of Man*, he would have probably emphasized the role of external conditions in the formation of human races and dismissed the influence of sexual selection on the origin of bodily characteristics.

Schwalbe's analysis tended to conflate Darwin's understanding of the influence of external conditions on evolution with Lamarckism *tout court*. Moreover,

[6]Further examination of Darwin's decision to adopt the hypothesis of pangenesis can be found in Vorzimmer 1963, Olby 1963, Geison 1969, Burchfield 1974. For a broader discussion on the impact of the "swamping argument" on Darwin's theory of natural selection, see Bulmer 2004.

[7]On this point, Johnson noted that Darwin's correspondence contains seemingly conflicting statements. As far as the *Origin of the Species* is concerned, Johnson argues that Darwin repeated that the use-inheritance mechanism assisted in some modification, but had a relatively minor role "both in terms of how often it is mentioned" and its "relative importance next to other causes" (Johnson 2015, pp. 165, 173).

[8]See for instance, the famous letter Darwin sent to Hooker in September 1849, where he wrote: "Lamarck is the only exception, that I can think of, of an accurate describer of species at least in the Invertebrate Kingdom, who has disbelieved in permanent species, but he in his absurd though clever work has done the subject harm" (Darwin 1961, pp. 139–140).

[9]Darwin Correspondence Project, "Letter no. 10643," accessed on April 30th, 2023.

Schwalbe omitted the following passage (emphasized in italics) from Darwin's letter to Wagner:

> In my opinion the greatest error which I have committed, has been not allowing sufficient weight to the direct action of the environment, i.e. food, climate, &c., independently of natural selection. *Modifications thus caused, which are neither of advantage nor disadvantage to the modified organism,* would be especially favoured, as I can now see chiefly through your observations, by isolation in a small area, where only a few individuals lived under nearly uniform conditions. (Ibid., emphasis added)

Here, Darwin seemed to be referring to what he used to call "undefined variations" elicited by the environment, a concept that he had discussed in the fifth edition of *The Origin of Species* (1869). This concept differed greatly from Lamarck's theory of adaptive change (Johnson 2015, p. 180), which rather implied a linear causal relationship between organisms' needs and modifications (Corsi et al. 2005, Burkhardt 2013):

> Indefinite variability is a much more common result of changed conditions than definite variability and has probably played a more important part in the formation of our domestic races. We see indefinite variability in the endless slight peculiarities which distinguish the individuals of the same species, and which cannot be accounted for by inheritance from either parent or from some more remote ancestor. Even strongly marked differences occasionally appear in the young of the same litter, and in seedlings from the same seed capsules. At long intervals of time, out of millions of individuals reared in the same country and fed on nearly the same food, deviations of structure so strongly pronounced as to deserve to be called monstrosities arise; but monstrosities cannot be separated by any distinct line from slighter variations. All such changes of structure, whether extremely slight or strongly marked, which appear amongst many individuals living together, may be considered as the indefinite effects of the conditions of life on each individual organism, in nearly the same manner as a chill affects different men in an indefinite manner, according to their state of body or constitution, causing coughs or colds, rheumatism, or inflammations of various organs. (Darwin 1869, p. 9)

Despite this, Schwalbe regarded Darwin's stress on the action of the environment as a mark of his reevaluation of Lamarckism.

For scholars committed either to the neo-Darwinian or neo-Lamarckian cause, Darwin's changes of mind and reflections on the action of the environment often led to diametrically opposed interpretations. August Weismann tended to downgrade Darwin's endorsement of the "Lamarckian principle." In his contribution to the volume *Darwin and the Modern Science*, he stated that Darwin was "not fully convinced of the transmissibility of acquired characters" (Weismann 1910, p. 22). Ernst Haeckel openly contested this interpretation in 1909. Like many other nineteenth-century advocates of neo-Lamarckism, he thought Darwin had increased a role for use-inheritance throughout the years (Johnson 2020, p. 178) and strove to bring to the fore the conceptual and theoretical continuities between Lamarckism and Darwinism. On this point, Haeckel devoted his commemorative speech delivered at the Volkshause of Jena on February 12th, 1909, to both Darwin's and Lamarck's evolutionary views, examining their differences and their shared tenets.

As he wrote to biologist Wilhelm Breitenbach on November 19th, 1908, the event was meant to be a "Darwin-Lamarck celebration."[10] On that occasion, Haeckel affirmed that Darwin, though having different mental dispositions and following different methodological and epistemological paths, had drawn similar conclusions to those of Lamarck, which came to represent the cornerstone of the modern doctrine of evolution (Haeckel 1909, p. 10). The two "pioneering masters" indeed shared the same goal, that of demonstrating the "natural unity" of the world. Lamarck had outlined the "magnificent edifice" of the theory of evolution, a portentous scaffolding that allowed scientists to understand the monistic nature of natural phenomena. The halls of this edifice, however, were mostly empty. It was Darwin who filled the vast spaces of the "monistic museum" erected by Lamarck with "thousands of illustrative objects," meticulously collected and empowered his explanatory system with the theory of natural selection (Ibid., pp. 11–12).

Haeckel's contribution to the volume *Darwin and the Modern Science* turned out to be even more forthright about Weismann's allegation, indicating that Weismann's thesis on the separation between germplasm and somatoplasm was not backed up by valid evidence:

> The notion of the absolute and permanent independence of the germ-plasm, as distinguished from the soma-plasm, is purely speculative; as is also the theory of germinal selection. The determinants, ids, and idants, are purely hypothetical elements. The experiments that have been devised to demonstrate their existence really prove nothing. (Haeckel 1910, p. 140)

Haeckel then went on to remark that Darwin was just as convinced as Lamarck of the transmission of acquired characters ("transformative inheritance" in Haeckel's lexicon) and its importance in the scheme of evolution. Darwin himself, Haeckel pointed out, had made clear that the IAC was an essential component of his theory, which blatantly collided with Weismann's account. "I had the good fortune," Haeckel wrote, "to visit Darwin at Down three times and discuss with him the main principles of his system, and on each occasion we were fully agreed as to the incalculable importance of what I call transformative inheritance" (Haeckel 1910, p. 140). Darwin's endorsement of "transformative inheritance" totally undermined the very concept of "neo-Darwinism" advocated by Weismann, according to which sexual reproduction (*amphimixis*) was the only source of variability. By dismissing the transmission of acquired characters, the so-called neo-Darwinians were in "express contradiction to the fundamental principles of Darwin and Lamarck" (Ibid.). Neither the label "neo-Darwinism" nor "ultra-Darwinism" (Romanes 1888) had a sound historical and conceptual foundation since the belief in the "omnipotence of natural selection was not shared by Darwin himself" (Haeckel 1910, p. 140). Natural selection had to be taken in conjunction with Lamarck's transformism, with which it was "in complete harmony" (Ibid., p. 141).

[10]Ernst Haeckel Online Briefedition, https://haeckel-briefwechsel-projekt.uni-jena.de/de/document/b_41077, accessed on April 30th, 2023.

Showing the divergencies between Darwin's theory of evolution and "neo-Darwinism" in the mid-1880s became one of the main rhetorical strategies used by the critics of Weismann's theory of inheritance. As American neo-Lamarckian paleontologist Edward Drinker Cope highlighted, Darwin himself had, in his works on human evolution, collected "numerous instances of the inheritance of various tricks of muscular movement of the face, hands, and other parts of the body" (Cope 1889, p. 1058). To scholars like Haeckel, the IAC was part and parcel of Darwin's explanatory framework, which had encompassed—and significantly improved—many of Lamarck's insights. Neo-Darwinians were thus misappropriating Darwin's scientific legacy, fabricating research traditions to consolidate a new research agenda. To some extent, by emphasizing the historical inconsistency of the very label of neo-Darwinism, Haeckel was attempting to cut the cord between Darwin and his self-proclaimed pupils, while at the same time claiming that his view of evolution well preserved the fundamental principles exposed by Darwin.

To the critics of the IAC and the advocates of neo-Darwinism, Darwin's understanding of the mechanisms of inheritance deserved further critical questioning. In his contribution to the American celebration, British zoologist Edward Poulton remarked that Darwin's views on the effects of external conditions widely differed from standard transformism as well as from neo-Lamarckism. Darwin had died a few years before Weismann's theory of germplasm continuity, which Poulton considered "the greatest change in evolutionary thought since the publication of the *Origin*" (Poulton 1909, p. 34). Like many of his contemporaries, Darwin was inclined to consider the transmission of acquired characters as a rather standard mechanism of heredity. Arguably, Darwin was led to several mistakes by "following the Lamarckian theory" (Ibid., p. 35), not the least of which was his interpretation of modern humans' feelings and facial expression patterns as inherited habits originally acquired by our ancestors.[11] Despite these naïve views of inherited habits, Darwin had devised a hypothesis, that of pangenesis, which appeared a far more reasonable way to account for the IAC compared to the ideas circulating at the time. Poulton, indeed, highlighted that Darwin was never satisfied by the well-known analogy between biological inheritance and memory proposed by authors such as Ewald Hering, Samuel Butler, Richard Semon, Ernst Haeckel, and Edward Drinker Cope and that even Francis Darwin would advocate.[12] Weismann's work had marked a point of no return in the history of evolutionary studies, yet "should it hereafter be proved that acquired characters are transmitted," Poulton wrote, "I can not but think

[11] Poulton makes reference to the letter Darwin sent to Edmund Gurney on July 8th, 1876, where he affirmed that the feeling of awe and wonder "excited by a grand cathedral may have some connection with the vague feelings of terror and superstition in our savage ancestors, when they entered a great cavern or gloomy forest" (Ibid.).

[12] See, for instance, the Presidential address Francis Darwin delivered before the British Association meeting in Dublin on September 18th, 1908: "Many attempts have been made to connect in this way the phenomena of memory and inheritance, and I shall ask you to listen to one more such attempt, even though I am forced to appear as a champion of what some of you consider a lost cause-the doctrine of the inheritance of acquired characters" (Darwin 1908, p. 354).

that the interpretation will be on the lines of Charles Darwin's hypothesis of Pangenesis" (Poulton 1909, p. 40).

Poulton's reading of pangenesis well exemplifies an attitude that would long characterize the way scientists engaged with Darwin's legacy retrospectively, which consisted of finding merits and positive implications even in those ideas that would be rejected. To Poulton, Darwin's view of heredity wrongly assumed that the somatoplasm could affect the germplasm, yet it entailed a corpuscular understanding of inheritance and steered away from the obscure analogy between heredity and memory. As biologists William E. Castle and Edmund Wilson pointed out in their contributions to the volume published for the American celebration, the hypothesis was short-lived, and yet it showed that Darwin had given much thought to the subject of hereditary units.[13] The tentative hypothesis of pangenesis remained "the only intelligible hypothesis of the transmission of acquired characters" (Wilson 1909, p. 112). This narrative held out for decades and re-emerged during the 1959 Darwin celebrations. However, if Darwin's support for the IAC seemed non-heterodox to both neo-Lamarckians and neo-Darwinians in 1909, by 1959 biologists would have substantially revised their perspectives.

3 1959—Darwin's "Miscalculation" and the Modern Evolutionary Synthesis

Darwin's centennials celebrated in 1959 coincided with a phase in which the so-called Modern Evolutionary Synthesis was already at an advanced stage and, as Stephen Jay Gould (1983, 2002) remarked, had already "hardened" around adaptationist and panselectionist theses. In many ways, the unprecedented number of events organized on a global scale to celebrate Darwin and the centenary of the publication of *The Origin of Species* served as a major device to publicly acknowledge the centrality of Darwinism in the history of evolutionary studies. Moreover, they represented a fundamental junction in the establishment of the so-called "Darwinian Industry" (Greene 1975; see also Delisle 2017, p. 134).

Scholars have generally highlighted that 1959 celebrations reflected the alleged phase of theoretical sclerotization that affected evolutionary biology during the 1950s. According to Gould (1983), Julian Huxley's keynote address on the occasion of the Darwin Centennial Celebration in Chicago, which was by far the most iconic and largest event held in 1959 to honor Darwin,[14] was a pure paean to panselectionism. His early pluralist and nonadaptationist views could not be seen

[13] Castle (1909, p. 144) specified that the hypothesis exerted some influence on the early twentieth-century debates on the behavior of hereditary units. An example was De Vries' theory of "intracellular pangenesis" (De Vries 1889).

[14] The celebrations took place between November 24th and 28th and saw the participation of at least 2.500 registrants and approximately 250 delegates from 14 countries (for further details, see Smocovitis 1999, p. 278).

in his address, which even took some quotes from the *Origin* "slightly out of context" to emphasize Darwin's belief in adaptationism and the omnipotence of selection. As Betty Smocovitis noted, the invitation list and discussion panels in Chicago well show how "unorthodox" evolutionists were almost entirely absent, with the sole exception of Conrad H. Waddington and Everett C. Olson, who did not fail to emphasize the weaknesses of the evolutionary synthesis with respect to the integration of evidence from embryology and paleontology (Smocovitis 1999, p. 298). As in 1909, Darwin's figure, life, and intellectual legacy continued to hold symbolic meaning. To postwar evolutionary biologists, the 1959 centennial served a broad range of functions and, above all, came to represent "a well-timed opportunity to retell the life story of Charles Darwin, who was now reinvented as the "founding father of their discipline" (Ibid., 279). Displaying the near-unanimous consensus on the centrality of natural selection and adaptationism represented a significant rhetorical device for an emerging scientific and institutional community (see Roger 1985, p. 813), even though it came at the expense of the inherent complexity of the debate on evolutionary processes. Much like academic conferences and meetings, media and broadcasts contributed to reinforcing the mythic narrative of Darwin's legacy, emphasizing how evolution was "the explanatory framework that unified the field of biology," whose theoretical building had been erected on the foundations laid by Darwin (Hall 2021, p. 251).

Surely, the 1959 celebrations served to reinforce the idea of historical and theoretical discontinuity between pre- and post-synthetic evolutionary biology, which often resulted in publicly dismissing the so-called "non-Darwinian" evolutionary theories that had circulated up to the early 1930s. A key passage from one of Julian Huxley's addresses held in Chicago (Panel II—"Evolution of Life") epitomizes this:

> We all accept the fact of evolution. We all agree that some combination of mutation and recombination is the raw material for change and that natural selection is the main directive or directional principle and that natural selection is not conscious – it is the result of the differential survival of variants through the generations. This means that biologists no longer need – and no longer can – think in terms of Lamarckism, or of so-called orthogenetic evolution – some inner urge, some *elan vital* that makes organisms evolve as they do. Selection acts on populations rather than on organisms, and it acts through the phenotype of the population – in R. A. Fisher's epigrammatic phrase, "Natural selection is a mechanism for generating an exceedingly high degree of improbability." It produces branching; it produces increasing adaptation, improvement, progress, or whatever you like to call it; and it produces horizontal persistence of branches, or stabilization. As a result, selection operates not only between individuals or populations but also, in the long run, between major groups – classes or even phyla. This results in the succession of so-called dominant types. I think that would summarize most of our main points. (Huxley et al. 1960, p. 139)

Within this context, Darwin's endorsement of the IAC was often stigmatized and regarded as an embarrassing but forgivable mistake. Following on a well-established narrative, Huxley himself pointed out that Darwin, despite having harshly criticized Lamarck's views of purposive adaptation in the 1840s, had attached more importance to the effects of the conditions of life and use and disuse in the later editions of the *Origin of Species*. It was this "Lamarckian error" with which "present-day biologists most often reproach Darwin" (Huxley 1960, p. 14). There were, however, several extenuating circumstances to be considered. Above all, Darwin's mistake had to be seen against the background of nineteenth-century views of biological heredity, most notably the idea of blending inheritance. To Huxley, in a context where the fusion of parental characters was deemed to swamp out favorable individual variations to the extent of potentially jeopardizing the action of natural selection, Darwin ascribed greater importance to the IAC as he needed to provide for "sources of more abundant variation" (Ibid., 14–15).

On the occasion of the annual general meeting held at the American Philosophical Society in April 1959 for the centennial of the publication of the *Origin of Species*, several scholars bolstered this narrative. Geneticist Theodosius Dobzhansky affirmed that Darwin, like all of his contemporaries who had "overlooked the work of an obscure monk called Gregor Mendel," shared the belief that parental characters undergo fusion after mating, which represented a major theoretical obstacle to solving the riddle of individual variability. Darwin, Dobzhansky highlights, conceded that a solution to this problem eluded him, but he never ceased investigating the mechanisms underlying biological inheritance. This even led him to make quite a miscalculation, as in his later years "he grasped the reed of Lamarckian inheritance of acquired characters, which he had spurned earlier" (Dobzhansky 1959, p. 252).

At the American Philosophical Society, University of California geneticist Michael Lerner (1959, p. 173) defined Darwin's idea of natural selection as "the binding principle for a unified theory of historical change in the living world." In combination with Mendelian inheritance, natural selection "had rendered obsolete alternative theories of evolution" such as those assuming the existence of "extra-mechanical agencies" or which posited the direct adaptation of organisms to their environments. Alternative theories of evolution and other "pathetically artless conjectures" turned out to sin against Occam's razor, which did not even spare Darwin's hypothesis of pangenesis (Ibid., p. 176). Lerner stressed that Darwin's endorsement of the IAC, however, was a forced option since he was well aware of the limits of his hereditary model:

> He was fully aware of the deficiencies in his knowledge of the mechanism of heredity. They led him to formulate a speculative and somewhat improbable, even in the light of a century ago, hypothesis of blending inheritance. This principle in turn forced Darwin to espouse the postulate that acquired characters were transmissible, because otherwise continued variability in natural populations could not be accounted for. (Ibid.)

Similarly to what Poulton had remarked in Baltimore 50 years earlier, in 1959 several scholars depicted Darwin's view of inheritance as quite reasonable considering the late nineteenth-century knowledge of hereditary mechanisms. In his thorough examination of Darwin's *Variation Under Domestication*, Swedish geneticist Arne Müntzing affirmed that Darwin's belief in the transmissibility of modifications induced by the action of the environment was ascribable to the influence of Lamarck and his followers:

> Though Darwin had a keen critical sense, we can see that in some cases he was misled by erroneous information from workers less critical than himself. The influence of Lamarck and his school is very obvious and lies behind Darwin's inability to distinguish between purely environmental modifications and changes caused by alterations in the hereditary constitution. (Müntzing 1959, p. 191)

Müntzing, however, emphasized that Darwin's hypothesis of pangenesis was "ingenious" and significantly diverged from purely transformist theories of organic change in at least two key aspects. In the first place, for Darwin, environmental conditions elicited both "direct and heritable effects" and "infinite and fluctuating variability," which constituted the polymorphic material on which natural selection may work (Ibid., p. 193). Most notably, it was the constitution of hereditary material that played a major role in determining individual variability:

> Though Darwin wrongly believed that variability is predominantly caused by environmental influences, it should be emphasized that he stressed the point that the constitution of the organisms will largely decide what kind of change will be induced. Thus, the organisms were supposed to have special constitutions which, though easily affected and changed by environmental influences, have a certain limited degree of independence. (Ibid., p. 217)

Narratives and comments were, however, not entirely uniform. On the occasion of the 1959 celebrations, English geneticist Cyril Darlington published the much-discussed book *Darwin's Place in History*, which was met with controversial reception from both scientists and historians. The essay was described as a rather unnecessary work imbued with precursorism (Gruber 1962, p. 428), which tended to assume that the history of the idea of evolution is one of those "with a reasonably clear-cut origin" (Gould 1960, p. 259). To biologist Cyril Bibby (1960, p. 170), Darlington's argument consisted of three main points: Darwin's failure to recognize pre-existent evolutionary traditions and relate his ideas with that of other scholars; his late application of the evolutionary theory to humans; and the fact that, in his late years, he modified his theory to cope with criticism "smuggling" Lamarckian inheritance as an alternative to natural selection. As a matter of fact, Darlington had claimed that once Darwin had adopted a mixed theory of IAC-*plus*-selectionism, his view had become barely distinguishable from anybody's else theory of evolution (Darlington 1961, p. 40). Darwin never denied "alternative possibilities of directed change or of the inheritance of acquired characters" (Ibid., p. 24). To Darlington, it was Fleeming Jenkin's argument from bending heredity that pushed Darwin to devise a "new theory" of evolution which openly endorsed that variations could be directed by the action of the environment:[15]

> Steadily he began to shift his ground away from natural selection. Retreating in good order he fell back on the assumption that acquired characters were inherited. The action of the environment in changing heredity constituted the 'persisting' cause of variation. It not only caused the variation: it now also directed it. The environment pushed a soft heredity. And when it had produced the variation the result was hard enough for natural selection to pull it further and to preserve it. The change to this new theory was produced by imperceptible modifications of successive editions. Darwin's friends noticed the changes, however, and noticed that they made his theory very like what had been said fifty years earlier. Darwin himself must have felt uncomfortable for he now proceeded to do something Lamarck had never attempted. He set to, work to invent a mechanism for Lamarckian inheritance. He supposed that changes in the environment and use or disuse affected the parts of the body concerned. From these parts the products of these effects condensed in the germ cells. This process of condensation was hard to reconcile with the advance of biology in the last thirty years. (Darlington 1961, pp. 37–38)

Darlington's judgment of Darwin's work and attitude toward his critics was harsh. Darlington considered Darwin's strategy far too flexible, not least on the issue of hereditary mechanisms and the inheritance of acquired characters, to the extent that it appeared hardly reconcilable with "even average intellectual integrity." Darwin was "slippery," much more so than figures such as Wallace, Lyell, Hooker, Chambers, or Spencer (Ibid., p. 60). He was able to "wriggle" to find a way through criticisms, and "wriggled so successfully that, in the end, he did not himself know where he stood" (Ibid., p. 62).

Despite Darlington's attempts to deconstruct the uncritical devotion to Darwin—which, as Bibby (1960, p. 170) remarked, was however imbued with "unworthy interpretations"—his view of Darwin's late theory of IAC-*plus*-selectionism as undistinguishable from other contemporary theories of evolution met with significant criticism. Several years after the publication of *Darwin's Place in History*, Ernst Mayr would point out that many historians had endorsed Darlington's claim that Darwin fell back on the IAC as a result of Jenkin's argument of blending inheritance. He further adds that:

> Admittedly, Darwin in his later years conceded a little more influence to soft inheritance than he had in 1859, but it never became a major component of his interpretation. Whenever he

[15] Such interpretation well reflects historian Susan Morris' analysis of the reception of Jenkin's work in the early twentieth century: "With the question of geological time declining in historical importance, and interest in heredity increasing, Jenkin's essay, and its influence on Darwin, came to be treated within the framework of Mendelian enlightenment, casting the rest of Jenkin's review into the shadows. One has only to note how universally Jenkin's article is described as having only two arguments - the swamping from blending inheritance, and the limits to variation - to realize how narrowly his article has been assessed. In truth, the article makes five separate arguments against natural selection: on the limits to variation, the swamping of variations by large numbers, the shortness of geological time, the problems of classification, and the logical rigour of Darwin's theory" (Morris 1994, p. 316).

> compared the contribution to evolutionary change made by the inheritance of acquired characters to that made by natural selection, he always made it quite clear that he continued to consider selection as the decisively important factor. (Mayr 1982, pp. 692–693)

To other scholars, however, insisting on the IAC as a somewhat core issue in the analysis of the relationship between Darwin and Lamarck's legacy represented a common and misleading cliché, which reiterated old-fashioned problems that were no longer central in modern evolutionary biology. During the panel discussion "The Evolution of Mind" held in Chicago, psychologist Ernest Hilgard followed up on Charles Gillespie's paper "Lamarck and Darwin in the History of Science," published in December 1958 in the *American Scientist*, arguing that instead of distinguishing between Darwin and Lamarck based on the IAC, scholars should rather focus on the quantitative implications of their theories. Indeed, it was "the fact that Darwin dealt with a mass of concrete relationships" that "laid the ground for eventual quantification," thus making it easy for evolutionary theory "to assimilate Mendelianism and population dynamics" (Hilgard et al. 1960, p. 188).

Not even during the hardening phase of the Modern Synthesis do we find a fully uniform assessment of Darwin's endorsement of the IAC and its relevance in the broader discussion on Darwin's place in history. Darwin's statements continued to mean different things to different scholars. In many cases, evolutionary biologists seemed committed to reiterating what Poulton had already avowed in 1909: Darwin's endorsement of the IAC was simply a response to a pressing theoretical need in a phase when the idea that the environment could elicit hereditable modifications was nothing more than a commonsensical view. With all its explanatory and empirical limitations, the hypothesis of pangenesis was a forced option that showed yet again how Darwin was always a step ahead of other naturalists.

4 2009—Re-expanding the Boundaries of Darwinism in the Era of the Epigenetic Turn

Darwin's bicentenary celebrations in 2009 represented a worldwide cultural phenomenon. The record of events entered in the Darwin Online Project includes more than 300 conferences, exhibitions and science festivals, and yet it is fairly limited to the English-speaking world.[16] Perhaps it is still difficult to make a comprehensive assessment of its several facets and implications in contemporary scientific culture. When compared to the 1959 celebrations, however, the bicentenary stands out for the interdisciplinary perspective that characterized the analysis of Darwin's legacy, as well as for the wide participation of professional historians of science and philosophers. Darwinism had become much more than a matter of scientific interest.

[16] http://darwin-online.org.uk/2009.html, accessed on April 30th, 2023. According to the record published on Pikaia.it, a total of 148 meetings, conferences, and exhibitions were organized only in Italy, https://pikaia.eu/darwin-1809-2009-tutti-gli-appuntamenti-previsti-in-italia/ accessed on April 30th, 2023.

In many ways, the 2009 celebrations continued in the wake of a trend that had been established around 1982 with the centennial of Darwin's death and further materialized in the 1980s with the emergence of new scholarly work in Darwinian studies. As Smocovitis (2005, p. 47) remarked, historians were almost entirely absent from the 1959 Darwin centenary held in Chicago, yet the celebration initiated the momentum that led to a new era in the history and philosophy of evolutionary biology. Between the late 1970s and 1980s, several influential works marked the field, including but not limited to Thomas F. Glick's edited volume *The Comparative Reception of Darwinism* (1974); Michael Ruse's *The Darwinian Revolution* (1979); James Moore's *The Post-Darwinian Controversies* (1979); David Oldroyd's *Darwinian Impacts* (1980); Robert Richards' *Darwin and the Emergence of Evolutionary Theories of Mind and Behavior* (1987); Peter Bowler's *The Eclipse of Darwinism* (1983) and *The Non-Darwinian Revolution* (1988).[17] In all likelihood, the most influential and representative volume of the decade was *The Darwinian Heritage* (1985), edited by Drew University historian David Kohn, whose origin goes back to the 1982 celebrations. A key passage from Kohn's introduction to the volume fully captures how the study of Darwin and Darwinism had become a matter for historians of science:

> The 1959 and 1982 centenaries were celebrated in qualitatively different ways. In 1959 the principal subject of interest was not Darwin, but modern evolutionary biology. The celebrations were organized by scientists, not by historians. The major collection of papers, *Evolution after Darwin*, edited by Sol Tax (1960), was concerned with evolution and genetics. A comparable collection of historical papers would have been inconceivable. Indeed, we should remember that in 1959, far from being a passionate subject of historical research, Darwin's significance was largely discounted by historians. Darwin was seen as a second-rate thinker – and observer and compiler of facts, a scientist of no great philosophical sophistication. In short, Darwin was regarded as a sort of inductive innocent. In contrast, the implicit premise for the current community is that Charles Darwin was a thinker of profound intellect and influence. Indeed, what characterizes the present community is a belief in the importance of Darwin. This high regard for Darwin is its central tenet. In an important sense, it is the revaluation of Darwin that has fostered the outpouring of sophisticated historical research on Darwin that we have witnessed between 1959 and the present. (Kohn 1985, pp. 1–2)

Many of these tenets were shared by the community during Darwin's bicentenary celebrations. As envisaged by Kohn (Ibid., p. 4), the "Darwin community" had considerably grown by 2009, with historians carrying out invaluable research on Darwin's notebooks, pieces of correspondence, and marginalia (Richards 2009). At the same time, the establishment of a close dialogue among historians, epistemologists, and scientists within the HPS has facilitated the development of an interdisciplinary view of Darwin's legacy (Ceccarelli 2020). By 2009, Darwin

[17] In the same space of time, evolutionary biologists such as Ernst Mayr and Stephen Jay Gould had made historical-epistemological reflection a tool for "concept analysis" and "clarification" (Mayr 1982, p. 17) and contributed to the field with influential volumes such as *The Growth of Biological Thought* (1982) and *Ontogeny and Phylogeny* (1977).

had gained intellectual respect and was recognized as a major thinker of his time whose work, life, and writings had implications beyond evolutionary biology. As Hossfeld (2009) has indeed shown the bicentenary celebrations focused on rather diverse thematic areas, such as Darwin's biography and the relationship between Darwinism and the humanities.

Interestingly, in continuity with the 100th anniversary of Darwin's death, in 2009 scholars were less committed to identifying "the areas of failure" of Darwin's work, whereas this had represented a fairly standard component of the 1959 celebrations led by evolutionists (Smith 1989, p. 452). There may be several possible explanations for this phenomenon. The first is the increasing historicization of Darwin's figure that resulted from the establishment of Darwinian studies in historical research, which arguably may have discouraged scholars from judging Darwin's work through the lens of contemporary biology. On the other hand, the emergence of pluralistic approaches in evolutionary biology has partially reoriented scientists' reading of Darwin's manifold work, re-expanding the conceptual boundaries of categories such as Darwinism and Darwinian tradition. As Henry Gee and Rory Howlett pointed out in the thematic issue published in *Nature* on February 1st, 2009, Darwinism had to be seen as an evolving research agenda. Even though "Darwin's first ideas" seemed "dated," throughout the decades biologists have continued to build on "their clarity of purpose," which proved to be the right path to take. Darwin thus had his "(however modified) descendants" who belonged to his intellectual lineage, whereas advocates of obsolete theories, such as Spallanzani's preformationism, had simply undergone extinction (Gee and Howlett 2009).

Among the new research trajectories that emerged during the second half of the twentieth century, that of "epigenetics" was decisive in the reassessment of Darwin's views on the IAC. As known, the very notion of "epigenetics" indicates a vast array of biological phenomena (Müller and Olsson 2003). The term was first introduced by British embryologist Conrad Hal Waddington in 1942 to designate the branch of biology that focuses on the "causal interactions between the genes and their products, which bring the phenotype into being" (Waddington 1942). Its use was soon—and sometimes loosely—extended to phenomena such as phenotypic plasticity, developmental canalization and, most notably, chromatin structure modifications (i.e., DNA methylation) which can lead to transgenerational inheritance without changing DNA sequences (Jablonka and Lamb 2010). By 2009, several studies had documented the transmissibility of epigenetic markers and their impact on the phenotype (Cervera et al. 2002; Vaughn et al. 2007; Johannes et al. 2009). In their comprehensive review published in 2009, Jablonka and Raz (2009) concluded that evidence of transgenerational epigenetic inheritance is ubiquitous across major taxa, albeit most common in fungi and plants. Despite the several technical and methodological limitations afflicting epigenetic research, not least the fact that most of the evidence discussed in these studies was obtained in non-ecologically relevant circumstances and from model species (Richards and Pigliucci 2020), the field was highly promising and was soon incorporated in the early version of the Extended Evolutionary Synthesis (Pigliucci and Müller 2010).

Within this theoretical framework, to scholars engaged in the field of epigenetic research the year 2009 represented much more than the bicentenary of Darwin's birth, all the more so because it was also the 200th anniversary of Jean Baptiste Lamarck's *Philosophie Zoologique* (1809). Biologists seized the occasion to reinterpret Darwin's hypothesis of pangenesis in light of modern research in post-genomics (Liu et al. 2009) and were often ready to dismiss the old-fashioned narrative of Lamarckian-vs-Darwinian evolution by normalizing Darwin's much-discussed endorsement of soft inheritance, which now served to vindicate Darwin's long-forgotten pluralistic view of evolution against the gene-centered versions of Darwinism that had become dominant through the Modern Evolutionary Synthesis. The anthology of articles titled "Evolutionary Biology 150 years after the 'Origin': Is a post-modern synthesis in sight?" and published in *Biology Direct* during the 2009 celebration, which was edited by Eugene Koonin, encapsulates this narrative. According to Koonin, the defining characteristic of the Darwinian discourse in 2009 was "the plurality of evolutionary processes and patterns." The major central tenets of the so-called neo-Darwinian orthodoxy had crumbled, to be replaced by "a new and incomparably more complex vision of the key aspects of evolution," which led post-modern evolutionists to be wary of any straightforward generalizations (Koonin 2009, pp. 473–475). In particular, post-genomics challenged the isotropic understanding of individual variations (Gould 2002), according to which random hereditable variations are the only material for natural selection. Random variations were relevant yet extremely diverse;[18] moreover, Koonin highlights, "the wide spread of stress-induced mutagenesis and related phenomena suggests the possibility of quasi-Lamarckian variation," which constituted "a part of Darwin's concept purged by the Modern Synthesis" (Ibid.).

Koonin and Wolf further explored the status of Lamarckian ideas in contemporary evolutionary biology in their paper "Is evolution Darwinian or/and Lamarckian?" where they claimed the need to overcome the somewhat hypostatized opposition between Lamarckism and Darwinism. By reviving a well-established narrative, the authors remarked that Darwin himself had slowly recognized the importance of Lamarck's work and the IAC in his late years. Throughout most of the twentieth century, and especially after the parabola of Austrian biologist Paul Kammarer and the well-known case of Trofim Lysenko in the Soviet Union,[19] "Lamarckism" came to represent a derogatory term, if not a stigma, for evolutionary biologists. However, a growing number of phenomena that came to the fore in modern research seemed to fit "a more broadly interpreted (quasi)Lamarckian paradigm," or simply showed that there was no sharp distinction between "Lamarckian" and "Darwinian" evolutionary scenarios, as these represented different forms of

[18] On this point, Koonin mentions the following mechanisms: nucleotide substitution, insertion and deletion; duplication of genes, genome regions and whole genomes; loss of genes and, generally, genetic material HGT including massive gene flux after endosymbiosis; and invasion and transposition of mobile selfish elements and recruitment of sequences from these elements (Ibid.).

[19] On Kammarer's case, see Koestler (1971), Gliboff (2006), and Berz (2009). On Lysenko's affair, see Joravsky (1970).

the interaction between organisms and their environment involved in evolutionary change (Koonin and Wolf 2009).[20] In a theoretical landscape where epigenetics finally allowed "the peaceful co-existence of Darwinian and Lamarckian evolution," scientists could be confident that in 2059 scholars will be celebrating Darwin *and* Lamarck (Handel and Ramagopalan 2010; see also Koonin 2009).

The theme of the relationship between Darwinism, Lamarckism, and contemporary epigenetics was present across several celebratory activities held in 2009, although in some rare cases conferences explicitly focused on the anniversary of the *Philosophie Zoologique*, or even made mention of Lamarck's name.[21] The theme of the IAC was distinctively tackled in the meeting held in Berlin in late 2008 entitled "Kultur der Evolution: Rethinking Evolutionary Theory from the Perspective of Cultural Studies." As Keith Benson and Ohad Parnes explained in the dedicated issue of the journal *History and Philosophy of the Life Sciences* (Vol. 31, n. 2), the meeting served to clarify how the use of oppositional language when addressing Darwinian and Lamarckian ideas lost sight of the historical and cultural complexity of the evolutionary debate, also and above all because it tended to homogenize complex theoretical systems creating contrived antinomies (Benson and Parnes 2009, p. 153). During recent decades, scholarship had finally started to dismantle such thematizations, "showing that Lamarckian elements were part of Darwin's own theory." Darwin himself had "endorsed the inheritance of acquired characters as part of his system and it was only after the establishment of the evolutionary synthesis in the 1920s and 1930s that mutation became the only accepted mechanism for the generation of evolution" (Ibid., p. 157).

Without any doubt, Eva Jablonka and Marion Lamb have been the major proponents of such a profound redefinition of the conceptual boundaries of Darwinism. As early as the 1980s, and especially with the publication of the landmark books *Epigenetic Inheritance and Evolution: The Lamarckian Dimension* (1995) and *Evolution in Four Dimensions* (2005), Jablonka and Lamb strongly advocated the need to revalue the Lamarckian "problematics" in view of contemporary genomics.[22] This could, in turn, contribute to reassessing the very notion of Darwinism. As they expressed in 2005, the gene-centered interpretation of Darwinism, albeit providing a "tidy framework for evolutionary thinking," was not the unique and

[20] Koonin and Wolf distinguished between bona fide Lamarckian and Quasi-Lamarckian phenomena, the difference lying in whether environmentally induced changes are adaptive in themselves, as in the case of stressed-induced mutagenesis: "Adaptive evolution resulting from stress-induced mutagenesis is not exactly Lamarckian because the stress does not cause mutations directly and specifically in genes conferring stress resistance. Instead, organisms evolved mechanisms that in response to stress induce non-specific mutagenesis which, however, appears to be fine-tuned in such a way so to minimize the damage from deleterious mutations in those rare genomes that carry a beneficial mutation. This type of mechanism is best defined as quasi-lamarckian" (Ibid.).

[21] See, for instance, the symposium *Lamarck-Darwin: 1809–2009. Doscientos años de evolucionismo*, held in Veracuz (Mexico) between September 28th and October 2nd, 2009.

[22] Loison (2021) highlights how Jablonka and Lamb understood Lamarckism as a "broad and inclusive" concept, often favouring expressions such as "Lamarckian dimension" or "problematics."

most significantly correct interpretation of Darwin's theory, whose long-forgotten pluralistic approach had ultimately been rehabilitated thanks to the epigenetic turn in life sciences. Contemporary research on epigenetics has confirmed the existence of hereditary phenomena that could be understood within the far too execrated category of "Lamarckism," and that represented only one of the multiple channels of inheritance systems (i.e., genetic, epigenetic, behavioral, and symbolic transmission). Such a view of inheritance, which encompassed Lamarckian principles, made it possible to rethink the theoretical structure of traditional twentieth-century neo-Darwinian views. Darwinism was "due for another transformation" (Jablonka and Lamb 2005, p. 40).

Quite remarkably, in 2009 Eva Jablonka and Snait Gissis organized a workshop to commemorate the 200th anniversary of Lamarck's *Philosophie Zoologique*, which would later lead to the publication of the well-known volume *Transformation of Lamarckism. From Subtle Fluids to Molecular Biology* (2011). The memorable bicentenary of Darwin's birth and the 150th anniversary of *The Origin of Species*, as stated in the preface of the volume, coincided with the anniversary of another, though "less noted," achievement in the history of biology which could be "equally monumental": the publication of the *Philosophie zoologique*. The workshop's aim was thus twofold:

> First, to redress a critical lapse in historical memory (to our knowledge, ours was the only full-scale international meeting devoted to commemorating the publication of *Philosophie zoologique*), and second, to bring into sharper focus new developments in biology that make Lamarck's ideas relevant not only to modern empirical and theoretical research, but also to problems in the philosophy of biology and to the writing of the history and sociology of evolutionary theory. (Gissis and Jablonka 2011, p. xi)

Labels such as Darwinism and Lamarckism, the editors stressed, had to be seen beyond their long-standing dichotomic understanding, as they denoted evolving research agendas. From a historical, theoretical, and epistemological perspective, scholars could actually recognize the "little similarity between Lamarck's descriptions of the subtle fluids that lead to the transformations of organisms during the vast epochs of time and modern ideas" and contemporary research on "the inheritance of epigenetic variations and the evolution of self-organizing structures" (Ibid.). However, the authors were keen to point out that "a certain logic" implied by Lamarckian stances, notwithstanding their historical and theoretical specificities, that is to say, their stress on developmental variations as opposed to predetermined and non-directed variations (Ibid., p. xii).

This use of labels such as "Lamarckism" and "neo-Lamarckism" has often come under criticism. Richards and Pigliucci (2020, p. 467) have deemed these terms as "vague" and "unnecessarily controversial" and rather tend to consider epigenetic inheritance as one of the many other findings and conceptual advances that led biologists to question some of the basic tenets of the Modern Synthesis. Historian Lauren Loison (2018) criticized the very analogy between transgenerational epigenetic inheritance and standard Lamarckian accounts based on the IAC mechanism, suggesting that, "given present-day knowledge, epigenetic inheritance would be better conceptualized as something different from a modern instance of Lamarckism" (Loison, 2021). Despite this, the tendency to use and reuse labels has proved to be particularly resilient over time, affecting the structure of the evolutionary debate. Interestingly, Jablonka and Lamb made clear that it is hard to give up using such terminology. As they explain in *Evolution in Four Dimensions*, the very notion of Lamarckism, as well as Darwinism, is problematical from the standpoint of contemporary biologists, as none today believe in Darwin's Darwinism or Lamarck's Lamarckism. These terms have undergone several semantic adjustments and "acquired very general meanings." Scientists could decide to avoid resorting to such mutable and controversial words, but Jablonka and Lamb highlight with great openness that this would come at a price: "You detach yourself from a tradition of thought and from some general but important things that these vague general terms refer to" (Jablonka and Lamb 2005, p. 354). Sticking to labels is much more than a nominal matter in scientific practice.

5 Conclusions: What is There to Celebrate?

This excursus of the main Darwin celebrations helps shed light on two key aspects of the history of evolutionary biology and, more specifically, of Darwinism. The first is that Darwin, his life, his work, and his scientific legacy have progressively become culturally significant objects whose meaning transcends the field of evolutionary biology. One may regard this as the long-term effect of narratives that started circulating as early as the 1860s and 1870s and that accompanied the reception of the *Origin of Species*, contributing to reinforcing the mythical consideration of Darwin's figure as the personification of the major qualities that characterized the ideal scientist during the Victorian age (White 2003). Nevertheless, as Kohn observed in 1985, the process that led Darwin to be recognized as the architect of a cultural revolution (Mayr 1995, see also Padian 2008) has been complex and anything but linear. However it happened, Darwin had become a symbol, and his celebrations, which have become known as "Darwin Days," greatly contributed to this transformation.

It is worth asking what we celebrate when we celebrate Darwin. As Italian zoologist Ferdinando Boero pointed out in 2015, Darwin celebrations such as modern Darwin Days are rituals that probably have no equal in contemporary science. We celebrate Darwin as a symbol of secularized culture but also because, unlike Galileo or Copernicus, the debate on evolution and its mechanisms is still open and meaningful:

> Why not celebrate Galileo's birthday, then? Or Copernicus'? The reason is very simple: there is no need to do so, because what they proposed is deeply incorporated into our culture. Only few and marginal religious fanatics still believe that the Earth is the centre of the universe, and that it is the Sun that rotates around it. For evolution, the debate is still hot.

> There have been controversies about evolutionary mechanisms, as happened when punctuated equilibria were opposed to gradualism. (Boero 2015)

Thus understood, scientific celebrations seem to be part and parcel of the scientific practice that characterizes progressive and expanding research programs (in Imre Lakatos' terms). Science festivals and celebrations, however, are not only a matter of science but also of culture (Krauss 2008; Durant and Ibrahim 2011; Bultitude 2011). Darwin Days have been instrumental in promoting scientific culture, engaging non-specialized audiences with life sciences and, in many cases, contrasting anti-evolutionary movements.[23]

The comparative analysis of Darwin's celebrations also enables us to better assess the strict and mutually affecting relation between what we celebrate and the ongoing debates in evolutionary biology. Darwin's celebrations in 1909, 1959, and 2009 illustrate scientists' deep commitment toward the history of science. As we have seen, evolutionary biologists often used public events to emphasize the elements of continuity between Darwin's scientific legacy and their research agendas, which arguably resulted in centralizing the discourse in evolutionary biology around the figure of Charles Darwin. As Hossfeld and Olsson (2013) have shown, Darwin's celebrations represented key moments in the construction of the so-called "Darwinian Industry," during which the contribution of other major figures to modern evolutionism, for instance the co-author of the theory of natural selection Alfred Russel Wallace, was often underestimated.

Scientists' commitment to restoring and maintaining traditions of thought has somehow forced them to stretch and adjust categories and labels. During Darwin's celebrations, this implied redrawing the border between categories such as "Darwinian" and "non-Darwinian" theories of evolution, for instance, by recasting the role of the IAC in the explanatory framework of Darwin's theory. On this point, pluralities of interpretation prevailed across the three celebrations, although some general arguments seem to have persisted throughout the years. Scholars committed to defending a somewhat "hard" conception of Darwin's evolutionary theory and heredity, whether it was nineteenth-century "neo-Darwinism" or the twentieth century "neo-Darwinian orthodoxy," often deemed Darwin's endorsement of the IAC as a minor issue that did not challenge the primacy of natural selection as the main driving factor underlying organisms' adaptative change. As Edward Poulton, Julian Huxley, and Theodosius Dobzhansky affirmed in their reconstructions,

---

[23]On this point, Boero specifically refers to several recent episodes in Italy, from 2004, when the Minister of Education proposed to remove evolution from the basic school curricula, to 2009, when the vice president of the National Research Council organized a conference whose proceedings were titled *Evolution: the decline of a hypothesis*. The latter episode aroused great controversy, which got the attention of international journals (Holden 2009; Harmon, 2009). Focusing on Italy, the promotion of Darwin Days has primarily been spearheaded by research centers, university departments in both natural and human sciences, as well as museums, public bodies, and local institutions. Notably, some celebrations have been openly endorsed by secularist movements, including the UAAR association (Union of Atheists and Rationalist Agnostics). Since 2003, the UAAR has played a significant role in coordinating Darwin's celebrations in Italy.

Darwin acted within the limited framework of nineteenth-century pre-genetic theories of inheritance, in which the belief in the transmissibility of acquired modifications was common sense, and thus could hardly avoid the miscalculation of the IAC. However, the way Darwin implemented it in his explanatory system, as vague as it was, represented a reasonable and ingenious response to the outstanding issues of nineteenth-century knowledge of hereditary mechanisms.

Over the years, scholars have questioned this narrative for diverse reasons. Darwin's endorsement of the IAC has largely been instrumental in contrasting ultra-selectionist accounts of evolution, or simply deconstructing the received notion of Darwinism that was consolidated during the Modern Synthesis. Especially during the celebrations of 1909 and 2009, biologists highlighted the divergence between Darwin's theory of evolution and subsequent iterations of "Darwinism" to assert Darwin's original pluralism.

What we today broadly call "epigenetic conceptions," including those held by Darwin, have represented boundary theoretical objects that, depending on the scholar and historical context, have been considered Darwinian or non-Darwinian. Even the Darwin of pangenesis, Johnathan Hodge (2010) famously remarked, seems, and has generally been regarded as, "a different Darwin." This is probably primary because Darwin never explicitly discussed the implications of pangenesis with respect to the mechanism of natural selection, not even in the last edition of the *Origin of the Species* (1872). Moreover, the hypothesis is far from the received view of the Darwinian theory of evolution that has been taught over the years, and thus sounds inherently "non-Darwinian." What underlies the very distinction between a "canonical" and a "different" Darwin is a somewhat retrospective assessment of his legacy in light of what matters today in evolutionary biology. Darwin's work was extremely multifaceted and often contained ambiguous terminologies and metaphors, as well as hesitancies, contradictions, and pluralities of explanation (Kohn 1989). It covered about half a century and underwent several stages of development, during which he never ceased to deal with criticism and to revise his theory and devise counterarguments. Hypotheses that today have little or no space in contemporary evolutionary biology were, to Darwin and his contemporaries, perfectly consistent with the biological knowledge and epistemological culture of the time and were thus worth further investigation. This is why, Hodge (2010, p. 133) pointed out, "the Darwin of pangenesis fits uneasily in scientists' history of science, much more comfortably in historians' history of science."

The historical narratives evolutionary biologists provided on the occasion of Darwin's celebrations exemplify the concept of "science-history," which, starting from the 1960s, historians of science have often stigmatized as "Whiggish history" or "Presentism" (Debus 1971; Brush 1995). It is by emphasizing scientists' retrospective—or even instrumental—interpretation of the past that historians of science have indeed demanded the autonomy of their discipline, contributing to the establishment and professionalization of a new academic field. The dichotomy between Whiggism and contextualism has somehow weakened over the years, as it did in the historiography of evolutionary biology. Several historians, epistemologists and evolutionary biologists have highlighted that a history of evolutionism informed to present-day problems is often desirable (Hull 1979) and that retrospective valorization is impossible to eradicate, both for scientists and for historians. If a substantial difference between "science histories" and "humanist histories" exists, this can hardly be reduced to the presence or lack of "presentism" (Esposito 2021, p. 31). By the very fact of posing some questions and selecting certain materials and sources, historians make choices and thus can never be considered as detached observers (Bachelard 1968; Koyré 1973; Junker 1996). At the same time, there are different ways historians can look to the present. As Stephen G. Brush (1995, p. 200) noted, one thing is the "present-minded" historian, whose questions about the past are "inspired by concerns of the present." Another thing is "presentism," which leads historians to give answers that are "distorted by those concerns." Scientists' and historians' accounts follow different modes and are arguably informed to different questions and purposes, yet they seem profoundly intertwined. Between the 1970s and the 1980s, the development of a new scholarship on the history of evolutionary biology, which saw historians rethinking the contribution of non-orthodox evolutionary traditions (Bowler 2017, p. 214), went hand-in-hand with the emergence of a new theoretical scenario whereby biologists sought to expand the category of Darwinism. Although they may not pursue the same goals, scientists and historians both contribute to co-constructing a network of knowledge whose hubs are deeply interconnected.

References

Bachelard S (1968) Epistémologie et historie des sciences. Rev Synth III:49–52

Benson KR, Parnes O (2009) Commemorating Darwin: the 2009 edition. Hist Philos Life Sci 31: 147–159

Berz P (2009) The eyes of the olms. Hist Philos Life Sci 31:215–239

Bibby C (1960) Darwin's place in history. Eugen Rev 52:169–170

Boero F (2015) Evolution: is celebrating Darwin still worthwhile? Ital J Zool 82:1–2. https://doi.org/10.1080/11250003.2015.1023042

Bowler PJ (1983) The eclipse of Darwinism. Anti-Darwinian evolution theories in the decades around 1900. The Johns Hopkins University Press, Baltimore

Bowler PJ (1988) The non-Darwinian revolution. Johns Hopkins University Press, Baltimore

Bowler PJ (1989) The mendelian revolution. The emergence of hereditarian concepts in modern science and society. John Hopkins University Press, Baltimore

Bowler PJ (2017) Alternatives to Darwinism in the early twentieth century. In: Delisle RG (ed) The Darwinian tradition in context. Springer, Cham, pp 195–217

Brooks WK (1883) The law of heredity. A study of the cause of variation, and the origin of living organisms. J. Murphy, Baltimore

Browne J (2005) Presidential address: commemorating Darwin. Br J Hist Sci 38:251–274

Browne J (2009) Looking at Darwin: portraits and the making of an icon. Isis 100:542–570

Brush SG (1995) Scientists as historians. Osiris 10:214–231

Bulmer M (2004) Did Jenkin's swamping argument invalidate Darwin's theory of natural selection? Br J Hist Sci 37:281–297

Bultitude K (2011) The rise and rise of science festivals: an international review of organised events to celebrate science. Int J Sci Educ 1:165–188

Burchfield JD (1974) Darwin and the dilemma of geological time. Isis 65:301–321

Burkhardt RW (2013) Lamarck, evolution, and the inheritance of acquired characters. Genetics 194:793–805

Castle WE (1909) The behavior of unit characters in heredity. Fifty years of Darwinism. In: Fifty years of Darwinism: modern aspects of evolution. Henry Holt and Company, New York, pp 143–159

Ceccarelli D (2018) Between social and biological heredity: Cope and Baldwin on evolution, inheritance, and mind. J Hist Biol 52:161–194

Ceccarelli D (2020) Historical narratives in scientific research: the case of evolutionary biology. S&F 24:25–41

Cervera MT, Ruiz-Garcia L, Martinez-Zapater JM (2002) Analysis of DNA methylation in Arabidopsis thaliana based on methylation-sensitive AFLP Markers. Mol Gen Genomics 268: 543–552

Churchill FB (2015) August Weismann: development, heredity, and evolution. Harvard University Press, Cambridge, MA

Cope ED (1889) On inheritance in evolution. Am Nat 23:1058–1071

Corsi P, Gayon JG, Gohau G, Tirard S (eds) (2005) Lamarck, philosophe de la nature. Presses Universitaires de France, Paris

Darlington CD (1961) Darwin's place in history, 2nd edn. The Macmillan Company, New York

Darwin CR (1868) The variation of animals and plants under domestication, vol II. Orange Judd & Company, New York

Darwin CR (1869) On the origin of species by means of natural selection, or the preservation of favoured races in the struggle for life, 5th edn. John Murray, London

Darwin CR (1903) In: Darwin F (ed) More letters of Charles Darwin, vol II. John Murray, London

Darwin F (1908) The address of the President of the British Association for the Advancement of Science. Science 28:353–362

Darwin CR (1961) In: Darwin F (ed) Charles Darwin autobiography with his notes and letters depicting the growth of the Origin of Species. Collier Books, New York

De Vries H (1889) Intracellulare pangenesis. Verlag von Gustav Fischer, Jena

Debus AG (1971) The relationship of science-history to the history of science. J Chem Educ 48: 804–805

Delisle RG (2017) From Charles Darwin to the evolutionary synthesis: weak and diffused connections only. In: Delisle RG (ed) The Darwinian tradition in context: research programs in evolutionary biology. Springer, Cham, pp 133–168

Depew DJ (2009) Darwinian controversies: an historiographical recounting. Sci & Educ 19:323–366

Dobzhansky T (1959) Variation and evolution. Proc Am Philos Soc 103:252–263

Durant J, Ibrahim A (2011) Celebrating the culture of science. Science 331:1242

Esposito M (2021) Cathedrals, corals and mycelia: three analogies for the history of evolutionary biology. In: Delisle RG (ed) Natural selection: revisiting its explanatory role in evolutionary biology. Springer, Cham, pp 11–38

Galton F (1870) Experiments in pangenesis, by breeding from rabbits of a pure variety, into whose circulation blood taken from other varieties had previously been largely transfused. Proc R Soc Lond 19:393–410

Gee H, Howlett R (2009) Evolution. Nature 457:807. https://doi.org/10.1038/457807a

Geison GL (1969) Darwin and heredity: the evolution of his hypothesis of pangenesis. J Hist Med 24:375–411

Gillispie CC (1958) Lamarck and Darwin in the history of science. Am Sci 46:388–409

Gissis SB (2011) Introduction: Lamarckian problematics in historical perspective. In: Gissis SB, Jablonka E (eds) Transformations of Lamarckism: from subtle fluids to molecular biology. MIT Press, Cambridge, MA, pp 21–32

Gissis SB, Jablonka E (2011) Preface. In: Gissis SB, Jablonka E (eds) Transformations of Lamarckism. From subtle fluids to molecular biology. The MIT Press, Cambridge, pp XI–XIV

Gliboff S (2006) The case of Paul Kammerer: evolution and experimentation in the early twentieth century. J Hist Biol 39:525–563
Gliboff S (2011) The golden age of lamarckism, 1866–1926. In: Gissis SB, Jablonka E (eds) Transformations of Lamarckism: from subtle fluids to molecular biology. MIT Press, Cambridge, MA, pp 45–55
Glick TF (ed) (1974) The comparative reception of Darwinism. University of Chicago Press, Chicago
Gould RP (1960) Darwin's place in history. By C. D. Darlington. Br J Philos Sci 11:259–260
Gould SJ (1977) Ontogeny and phylogeny. Harvard University Press, Cambridge, MA
Gould SJ (1983) The hardening of the modern synthesis. In: Grene M (ed) Dimensions of Darwinism. Cambridge University Press, Cambridge, pp 71–93
Gould SJ (2002) The structure of evolutionary theory. Belknap, Cambridge, MA
Greene JC (1975) Reflections on the progress of Darwin studies. J Hist Biol 8:243–273
Gruber JW (1962) Darwin's place in history. C D Darlington. Isis 53:428–429
Haeckel E (1909) Das Weltbild von Darwin und Lamarck. Alfred Kröner Verlag, Leipzig
Haeckel E (1910) Charles Darwin as an anthropologist. In: Seward AC (ed) Darwin and modern science. Cambridge University Press, London, pp 137–151
Hall A (2021) Remembering or deifying? The Darwin anniversaries of 1959 and 2009. In: Hall A (ed) Evolution on British television and radio. Transmissions and transmutations. Palgrave Macmillan, Cham, pp 241–273
Handel A, Ramagopalan SV (2010) Is Lamarckian evolution relevant to medicine? BMC Med Genet 11:73. https://doi.org/10.1186/1471-2350-11-73
Harmon K (2009) Italy science council funds creationist book. Scientific American, December 10th
Hertwig O (1894) Zeit- und Streitfragen der Biologie. In: Präformation oder Epigenese? Grundzüge einer Entwicklungstheorie der Organismen, vol 1. Gustav Fisher, Jena
Hilgard E, Gerard RW et al (1960) Panel four: the evolution of mind. In: Tax S, Callender C (eds) Evolution after Darwin. The evolution of life, vol I. University of Chicago Press, Chicago, pp 175–206
Hodge J (2010) The Darwin of pangenesis. C R Biol 333:129–133
Holden C (2009) Creationism at Italian science agency. Science 326:1597
Hossfeld U (2009) Quo vadis "Darwin industry"? Tendenzen und Trends im Darwin-Jahr 2009. Archiv für das Studium der Neueren Sprachen und Literaturen 161:330–344
Hossfeld U, Olsson L (2013) The prominent absence of Alfred Russel Wallace at the Darwin anniversaries in Germany in 1909, 1959 and 2009. Theory Biosci 132:251–257
Hull DL (1979) In defence of presentism. History and Theory, pp 1–15
Hull DL (1989) The metaphysics of evolution. State University of New York Press, Albany
Huxley TH (1894) Darwiniana essays. D. Appleton and Company, New York
Huxley J (1942) Evolution. The modern synthesis. Allen and Unwin, London
Huxley J (1960) The emergence of Darwinism. In: Tax S, Callender C (eds) Evolution after Darwin. The evolution of life, vol I. University of Chicago Press, Chicago, pp 1–21
Huxley J, Emerson AE et al (1960) Panel two: the evolution of life. In: Tax S, Callender C (eds) Evolution after Darwin. Issues in evolution, vol III. University of Chicago Press, Chicago, pp 107–143
Jablonka E, Lamb M (2005) Evolution in four dimensions: genetic, epigenetic, behavioral, and symbolic variation in the history of life. MIT Press, Cambridge-London
Jablonka E, Lamb MJ (2010) Transgenerational epigenetic inheritance. In: Pigliucci M, Müller GB (eds) Evolution—the extended synthesis. The MIT Press, Cambridge-London, pp 137–174
Jablonka E, Raz G (2009) Transgenerational epigenetic inheritance: prevalence, mechanisms, and implications for the study of heredity and evolution. Q Rev Biol 84:131–176
Johannsen W (1911) The genotype conception of heredity. American Naturalist 45(531):129–159 https://doi.org/10.1086/279202
Johannes F et al (2009) Assessing the impact of transgenerational epigenetic variation on complex traits. PLoS Genet 5:e1000530

Johnson CN (2015) Darwin's dice. The idea of chance in the thought of Charles Darwin. Oxford University Press, New York
Johnson CN (2020) Darwin's "Historical Sketch." An examination of the "preface" to the origin of species. Oxford University Press, New York
Joravsky D (1970) The Lysenko affair. Harvard University Press, Cambridge
Junker T (1996) Factors shaping Ernst Mayr's concepts in the history of biology. J Hist Biol 29:29–77
Kellogg VL (1907) Darwinism today. G. Bell & Sons, London
Koestler A (1971) The case of the midwife toad. Hutchinson, London
Kohn D (1985) Introduction: a high regard for Darwin. In: Kohn D (ed) The Darwinian heritage. Princeton University Press, Princeton, pp 1–5
Kohn D (1989) Darwin's ambiguity: the secularization of biological meaning author(s). Br J Hist Sci 22:215–239
Kolchinsky EI (2010) The Charles Darwin anniversary, 2009. Paleontol J 44:1467–1470
Koonin EV (2009) The origin at 150: is a new evolutionary synthesis in sight? Biol Direct 4:473–475
Koonin EV, Wolf YI (2009) Is evolution Darwinian or/and Lamarckian? Biol Direct 4:1–14. https://doi.org/10.1186/1745-6150-4-42
Koyré A (1973) Perspectives sur l'histoire des sciences. Etudes d'histoire de la pensée scientifique:390–399
Krauss LM (2008) Celebrating science as culture. Science 321:643
Largent M (2009) The so-called eclipse of Darwinism. Trans Am Philos Soc (New Ser) 99(1):3–21
Lerner M (1959) The concept of natural selection: a centennial view. Proc Am Philos Soc 103:173–182
Liu Y-S et al (2009) Darwin's contributions to genetics. J Appl Genet 50:177–184
Loison L (2018) Lamarckism and epigenetic inheritance: a clarification. Biol Philos 33:1–17. https://doi.org/10.1007/s10539-018-9642-2
Loison L (2021) Epigenetic inheritance and evolution: a historian's perspective. Biol Sci 376: 20200120. https://doi.org/10.1098/rstb.2020.0120
Love AC (2013) Theory is as theory does: scientific practice and theory structure in biology. Biol Theory 7:325–337
Mantegazza P (1882) Commemorazione di Carlo Darwin celebrata nel Regio Istituto di Studi Superiori in Firenze. In: Barsanti G, Barbagli F (eds) L'Uomo e gli uomini. Antologia di scritti antropologici (2010). Edizioni Polistampa, Firenze
Mayr E (1982) The growth of biological thought: diversity, evolution and inheritance. Harvard University Press, Cambridge
Mayr E (1995) Darwin's impact on modern thought. Proc Am Philos Soc 139:317–325
Meulendijks M (2021) Eclipsing the eclipse?: A neo-Darwinian historiography revisited. J Hist Biol 54:403–443
Moore JR (1979) The post-Darwinian controversies: a study of the protestant struggle to come to terms with Darwin in Great Britain and America. Cambridge University Press, Cambridge, UK
Morris SW (1994) Fleeming Jenkin and "The Origin of Species": a reassessment. Br J Hist Sci 27: 313–343
Müller GB, Olsson L (2003) Epigenesis and epigenetics. In: Hall BK, Olson W (eds) Keywords and concepts in evolutionary developmental biology. Harvard University Press, Cambridge, pp 144–123
Müller-Wille S, Rheinberger HJ (2012) A cultural history of heredity. University of Chicago Press, Chicago
Müntzing A (1959) Darwin's views on variation under domestication in the light of present-day. Proc Am Philos Soc 103:190–220
Ochoa C, Barahona A (2014) El Jano de la morfología: de la homología a la homoplasia, historia, debates y evolución. UNAM; Centro de Estudios Filosóficos, Políticos y Sociales Vicente Lombardo Toledano, México

Olby RC (1963) Charles Darwin's manuscript of pangenesis. Br J Hist Sci 1:251–263
Oldroyd D (1980) Darwinian impacts. An introduction to the Darwinian revolution. The Open University Press, Milton Keynes, UK
Padian K (2008) Darwin's enduring legacy. Nature 451:632–634
Pigliucci M, Müller G (eds) (2010) Evolution: the extended synthesis. MIT Press, Cambridge
Poulton EB (1909) Fifty years of Darwinism. In: Fifty years of Darwinism: modern aspects of evolution. Henry Holt and Company, New York, pp 8–56
Richards RJ (1987) Darwin and the emergence of evolutionary theories of mind and behavior. The University of Chicago Press, Chicago
Richards RJ (2009) Darwin's place in the history of thought: A reevaluation. PNAS 106:10056–10060
Richards CL, Pigliucci M (2020) Epigenetic inheritance. A decade into the extended evolutionary synthesis, in Paradigmi 3:463–494
Richmond ML (2006) The 1909 Darwin celebration. Isis 97:447–484
Roger J (1985) Darwinism today (commentary). In: Kohn D (ed) The Darwinian heritage. Princeton University Press, Princeton, pp 813–823
Romanes GJ (1888) Lamarckism versus Darwinism. Nature 38:413
Ruse M (1979) The Darwinian revolution: science red in tooth and claw. University of Chicago Press, Chicago
Schwalbe G (1910) The descent of man. In: Seward AC (ed) Darwin and modern science. Cambridge University Press, London, pp 112–136
Secord JA (2009) Focus: Darwin as a cultural icon. Isis 100:537–541
Smith R (1989) David Kohn (ed.). The darwinian heritage. Br J Hist Sci 22:452–456
Smocovitis VB (1999) The 1959 Darwin centennial celebration in America. Osiris 14:274–323
Smocovitis VB (2005) "It Ain't Over 'til it's Over": Rethinking the Darwinian revolution. J Hist Biol 38:33–49
Vaughn MW et al (2007) Epigenetic natural variation in arabidopsis thaliana. PLoS Biol 5:1617–1629
Vorzimmer P (1963) Charles Darwin and blending inheritance. Isis 54:371–390
Waddington CH (1942) The epigenotype. Endeavour 1:18–20
Weismann A (1910) The selection theory. In: Seward AC (ed) Darwin and modern science. Cambridge University Press, London, pp 18–65
White P (2003) Thomas Huxley: making the "Man of Science". Cambridge University Press, Cambridge
Whitman CO (1894) Evolution and epigenesis. In: Biological lectures delivered at the marine biological laboratory of Wood's Holl. Ginn & Company, Boston, pp 205–223
Wilson EB (1909) The cell in relation to heredity and evolution. Fifty years of Darwinism. In: Fifty years of Darwinism: modern aspects of evolution. Henry Holt and Company, New York, pp 92–113

The "Darwinian Revolution" as a Presentist Discourse: Ideological Implications Beyond the Anglo-Saxon Context

Juan Manuel Rodríguez Caso

Abstract

The history of evolutionary theory usually focuses on discussing its development and impact in specific countries such as the United States, Britain, France, and Germany. Although traditions analyse the introduction of Darwin's ideas in diverse cultural contexts, the starting point is a continued reassertion of diffusionism. In addition, given that biologists have strongly driven the history of biology, presentism is an approach often used, especially to defend the logic of the "Darwinian Revolution" itself. This commentary aims to reflect on diffusionism and presentism's impact in the specific case of Mexico. Part of the argument will focus on contrasting the approaches of both biologists and historians regarding the history of biology. On the other hand, the consequences of the "Darwinian Revolution" on the conception of biology in Mexico will be discussed.

Keywords

Diffusionism · Doctrines · Liturgies · Plasmogeny · Cardenismo

1 Introduction

Today, more than ever, there are contexts in which the place of science in society is discussed with notorious passion. Scientific discourse is sometimes accused of being "ideologised", which serves to demonise scientific work. However, it is essential to consider how science is understood, as it is sometimes common to maintain a "pure" view of science. Today, more than ever, scientific work can be understood as part of

J. M. Rodríguez Caso (✉)
Grupo de Estudios Sociales de la Ciencia y la Tecnología, Facultad de Ciencias, Universidad Nacional Autónoma de México, Mexico City, Mexico
e-mail: juan_manuel_rodriguez_caso@ciencias.unam.mx

R. G. Delisle et al. (eds.), *Unity and Disunity in Evolutionary Biology*,
https://doi.org/10.1007/978-3-031-42629-2_18

society and culture, and it is there where the role played by ideologies, or the ways of understanding the world, of each individual must be recognised. In the words of Robert Young: "The concept of ideology refers to legitimation and to the intrusion of values into putative facts. At a deeper level, it refers to how frameworks get constituted and how criteria for acceptable conclusions get established on the basis of value systems or worldviews" (Young 1992). In line with this approach, "[i] deologies reflect social locations and serve established or aspiring powers" (Young 1992). However, it was in earlier work that Young (1971) recovered a classic definition of ideology: "when a particular definition of reality comes to be attached to a concrete power interest, it may be called an ideology". How does this relate to the "Darwinian revolution"? Because, without questioning the importance of Darwin's ideas, when we speak of a "revolution" inspired by Darwin, we can propose a narrative supported in historical facts, or we can generate a history that underpins this "profound change". In short, a narrative—although it includes historical facts, these are interpreted according to the biases of the proponent—has been created around Darwin, which has allowed the creation of a history around Darwin, which justifies ideological claims such as the conflict between science and religion.

However, we can consider the question, are non-ideological narratives possible? Yes, albeit with nuances. One can think of creating narratives that present facts without expressing any opinion or point of view, or else an author may be perfectly aware of his or her own ideological biases and explicitly seek to avoid promoting them. Ultimately, whether or not non-ideological narratives are possible is a matter of definition. If we define ideology as a set of beliefs or values used to justify a particular course of action, then it is clear that all narratives are ideological. However, if we narrowly define ideology as a set of beliefs or values explicitly promoted in a narrative, then it is possible to imagine narratives that are not ideological.

In practice, most narratives fall between being entirely ideological and completely non-ideological. Most narratives will reflect the author's or narrator's values and beliefs to some extent, but they may also be open to multiple interpretations. Ultimately, ideology can be understood as part of developing a discourse, a narrative. The inspiration for this paper comes from the recent work edited by Mary Morgan et al. (2022) on the role of narrative in science. Three functions are suggested for narrative: "narrative-making operates as a means of making sense of their puzzling phenomena; it provides a means of representing that scientific knowledge; and it provides resources for reasoning about those phenomena" (Morgan in Morgan et al. 2022). Moreover, they are interrelated functions that allow scientists to understand the world around them to generate narratives that allow them to represent what they think they know and, from there, to reason.

The question is to speak of a general narrative and place it in a specific context. One of the characteristics usually given to scientific discourse is that of universality. The scope and importance of science are beyond doubt, but the fact that scientific work is conducted anywhere in the world does not justify understanding it as the same regardless of the context. Here is not the intention to delegitimise the importance of scientific work but to be critical of the use of science as a means to justify agendas or ideologies.

However, why does the narrative of the "Darwinian revolution" continue despite the lack of historical evidence in many cases? There is no completely satisfactory answer. Amongst historians, there is a heated debate about the value, or otherwise, of presentism. An initial problem is its definition, although the majority agreement is the interpretation of the past through the eyes of the present. Nevertheless, for this commentary, the importance of presentism can be understood in terms of its use in discourse. In line with Hakob Barseghyan's (2022) approach, it depends on the application that a type of presentism can be selected, the one that shows the intentionality of an actor, by a latter effect, by selecting a problem, or by a combination of all three. When people talk about the "Darwinian revolution", they are defending—sometimes dogmatically—a worldview in which science plays a central role, and Darwin is the one who opens the door to "modernity". A modern vision that recovers the vision of history promoted by nineteenth-century positivism by affirming that the only way to progress is by overcoming the past and the antiquated, that is, "magical thinking" and religion.

This commentary presents some reflections on discourses' impact on scientific narratives. In this case, to assess how a dominant discourse, such as that coming from the Anglo-Saxon academic world, can influence local narratives, as in this case, the history of science in Mexico.

2 "Darwinian Revolution" and Biology

Regardless of geography, it is common to refer to an interpretation that places Darwin and his ideas at the centre when discussing the history of biology. A review of the histories written about biology, especially in the twentieth century, reveals such a narrative. Now, it must be acknowledged that the accounts we have today of the history of biology—as a discipline, as a tradition—maintain that narrative around the centrality of Darwin's theory.

Today's historical accounts have mostly privileged that view, regardless of geography. A quick review of histories of biology published in English—including translated works that were initially published in other languages—are: *The History of Biological Theories* (1930) by the Czech biologist and philosopher Emanuel Rádl (1873–1942) or *The History of Biology: A Survey* (1935) by the Finnish Arctic explorer and geologist Erik Nordenskiöld (1872–1933), followed several years later by *A History of Biology to About the Year 1900: A General Introduction to the Study of Living Things* (1989) by the British historian Charles Singer and *The Epic History of Biology* (1993) by the American historian Anthony Serafini. In general, these works have in common that: (a) they present long-term narratives, usually beginning in the "ancient age", in cultures such as Mesopotamia, Greece, and Rome; (b) they are teleological histories, in which the "progress" of science can be observed; (c) the hagiographic narrative is privileged, with Darwin as the main protagonist; (d) after that, at different times, the "Darwinian revolution" is highlighted.

Few works have explicitly criticised the role of Darwin or the "Darwinian revolution" in the history of biology. Throughout the twentieth century, authors such as the US anthropologist Loren Eiseley, through works such as *Darwin's Century: Evolution and the Men who Discovered It* (1958), or the Danish embryologist Søren Løvtrup with *Darwinism: The Refutation of a Myth* (1987), put forward critical reconstructions of the historical development of the theory of evolution by emphasising—sometimes somewhat exaggeratedly—the role of other authors, such as Jean Baptiste Lamarck or Alfred Russel Wallace. In a way, these works can be seen as a form of deconstruction. Eiseley and Løvtrup started by analysing the different historical elements of the theory of evolution since the nineteenth century to propose an alternative version of the theory of evolution in what can be considered a deconstruction.

From a historiographical point of view, Joseph Caron's work is vital to understanding the dominant narratives about the history of biology. It takes four geographical contexts as reference points: France, Germany, the United States, and Great Britain. Each case presents the different ways in which the history of biology has been conceived. From here, it would be more convenient to speak in the plural, i.e., histories of biology. Now, in the specific case of the role usually given to Darwin and his ideas as the linchpin of biology, Caron very concretely makes this clear:

> The theory of natural selection of Charles Darwin (1809–1882) is the focal point of a [...] historical tradition which claims to account for the history of biology. No historian of the life sciences would for a moment entertain the thought of denying the crucial influence of Darwinian theory in the constitution of the general theoretical framework of the modern life sciences. The full acceptance of that fact does not, though, for an instant constitute any argument in favour of the creation of the science of 'biology' on the basis of evolutionary theory. A proper reading of the relevant histories cannot but show the reader the immense confusion which exists therein between biology and evolutionism. (Caron 1988)

However debatable the role of interpretation of historical evidence may be, narratives define what can ultimately be considered "authentic history". In this case, a theory such as evolution is associated with institutionalising a discipline such as biology. However, as Caron again clarifies, "this association is made for the most part implicitly. This is possible, in most cases probably, because the origins and the constitution of 'biological' science are simply not considered problematic" (Caron 1988). In other words, a recurrent attitude is to avoid facing the complexity of the biological sciences, both historically and methodologically.

Despite this, the "Darwinian revolution" remains a dominant historiographical category in evolutionary discourse. Through numerous works, classical authors such as Ernst Mayr and Michael Ruse have reaffirmed Darwin's centrality in the "correct way" of understanding evolution. For example, Ruse recently states revolution as "that change of worldview wrought by Darwin, taking us from the miraculous instantaneous creation of animals and plants, to a natural world of evolutionary change, to an ever-unfurling tree of life, from the monad to the man" (Ruse 2019). Mayr, on the other hand, insisted that the change promoted by Darwin included the elimination of anthropocentrism, as well as affecting ethical and metaphysical concepts. An essential part of both arguments is the break between science and religion, sometimes simplistically equated with creationism (Mayr 1972). These two positions start from the assumption of Darwin's ideas' impact on philosophy and biology.

Furthermore, although they appeal to a historical reconstruction, they start from a presentist and sometimes simplistic view. The latter can be seen in a less-than-complex understanding of social or cultural aspects of Darwin's context, such as the abovementioned case of the equation between religion and creationism. However, Gertrude Himmelfarb's (1996 [1959]) is one of the most detailed but controversial accounts. From the very title of the book, the author acknowledges the importance of Darwin's ideas in understanding nature, but unlike other authors, the impact of the revolution is discussed from the complexity of the Victorian era. Although a before and after is recognised, Himmelfarb speaks of a "conservative revolution". In short, Darwin and his work—particularly *On the Origin of Species*—is not seen as the trigger for change in society but as the culmination of a process of change that had begun long before.

What is not often said or acknowledged is all involved in accepting such a proposal without further discussion. The "Darwinian revolution" is an account that reaffirms Eurocentrism—or rather Anglocentrism—and a diffusionist view of scientific knowledge. By placing Darwin as almost the sole axis of any explanation, it ends up as a hagiographic account in which other characters' roles are relegated to second or third place.

Somehow, a narrative is imposed about the proper way to do science, the way Darwin did it, and the impression is that if you do not follow that path, you are not doing science. Moreover, this is reflected in seeking to "adjust" the development of science in a country like Mexico to what is proposed in other contexts.

3 Darwin in Mexico: From French Translations to the Anglo-Saxon "Invasion"

One way the "Darwinian revolution" has been reflected is, for example studies that seek to place Darwin and his ideas in different geographical contexts. The introduction, reception, and appropriation of Darwin's ideas in Mexico have been analysed significantly. The aim has been to argue that the natural sciences in Mexico, although they may have had an older origin, it was not until the "arrival" of Darwin's ideas that they were able to achieve the status of "modernity".

A key point about these processes is how different it can be to understand them from different historiographical perspectives. In the sense of the history of ideas, emphasis has been placed on questions such as looking for the "first mention" or the "first publication", as if this marked a before and after—which also seemed to be immediate—in the conception of science at the time. Furthermore, this encompasses an individual and the whole "community". To speak of Darwin in Mexico or Latin America—even in Ibero-America—takes us back to a tradition that, since the 1970s, has focused on studying and reconstructing that influence. However, the perspective almost unanimously defended to date is a diffusionist historiographical approach. Not only that, but at the same time, it is an approach that has reaffirmed aspects such as scientific imperialism. For example, the seminal work of historian Roberto Moreno de los Arcos (1984), based upon the work that formed part of Thomas Glick's project (1988) in the 1970s on the introduction of Darwinism in the world, presents the Mexican case in diffusionist terms, ignoring the discussions that were taking place in nineteenth-century Mexico about the scope of natural history. Furthermore, the crucial point is to show the "continuous" conflict between evolutionary ideas and religion by showing the discussion generated in the newspapers in the last quarter of the nineteenth century between the principal positivist authors, such as the brothers Justo and Santiago Sierra, and the Catholic Church.

On the other hand, if we pay attention to a broad historiographical discourse, the conclusions are quite different. Historical evidence leads us to understand that Darwin's work and many other authors, such as Herbert Spencer, came to the country through works written in French (Ruiz et al. 2016). A pending task is to assess the impact of these French translations in the Mexican context. In the specific case of Darwin's work, works such as *Darwin contre Darwin* (2009), by Thierry Hoquet, remind us how relevant a translation can be in interpreting Darwin's ideas, to the point of differing radically from what was initially proposed in the original work, without claiming that there could be a "correct" way of interpreting Darwin, a future task of the history of science in Mexico is to recognise what local scientific communities appropriated from these readings.

Over the years, the influence of French culture in Mexico waned, and academic circles ended up favouring materials produced in English, especially from the United States. One recent example of the problems that have arisen over the years is the uncritical view of teaching biological evolution based on translations or original works of US textbooks. In well-known works such as *Evolution* (2005) by Douglas J. Futuyma, in addition to setting out essential elements on the mechanisms of evolution, discussion of key concepts such as variation or species are complemented by chapters that discuss the importance of science education in society, starting with a problem that until a few years ago was characteristic of the United States: creationism. Given the social and cultural implications already seen in the US context, this last point becomes dangerous. At the time, a well-known Mexican scientist such as Antonio Lazcano (2005) spoke of the danger of creationist ideas reaching Mexico due to the growing importance of Protestant churches, especially from the United States. In the end, if one does not have the right theoretical tools—such as historical and philosophical training—the uncritical presentation of these positions becomes a kind of apologia rather than a critique. To this day, in general terms, the possible religious influences on the teaching and dissemination of biology in Mexico arise from the conceptions and beliefs that groups of individuals defend, irrespective of their belief or non-belief. This situation tends to be a mixture of different interpretations of doctrines and liturgies, but always in an eclectic sense. There is no concrete evidence that the Catholic Church—not only the majority religion, but with a solid political and social influence—or the Protestant churches have lobbied for changes in curricula, either through publications or specific sayings

(Torrens et al. 2023). Possible criticisms of biological evolution—and consequently, of Darwin—can be found in the discourses traditionally defended by groups such as Jehovah's Witnesses through publications such as *Did Man Get Here by Evolution or by Creation?*—published since 1967 and republished in numerous languages by the official JW publisher, the Watchtower Bible and Tract Society—are a revival of creationist arguments that defend the Bible's literal reading, hence rejecting scientific argumentation. Another example of the visibility of Protestant groups in Mexico today, which has served in recent years as a source of anti-scientific propaganda, is the permissiveness by the current government with which the doors have been opened to evangelical groups to form political parties such as the so-called Partido Encuentro Social—despite the traditional separation of church and state in politics, the logic of which only seems to apply to Catholicism.

What implications does this have for the history of science or the science itself? Unless we turn to historiographical proposals such as the circulation of knowledge or transnational contexts, a presentist perspective is defended, which, although it may have the good intention of defending the place of science in contemporary society, ends up defending implicit values, such as imperialism and scientism.

4 The Other History of Biology in Mexico: The Importance of Local Traditions

However, the history of biology in Mexico goes far beyond the role that Darwin's ideas can play, as if this were the only way to understand them. Authors such as Ana Barahona and Ismael Ledesma (1999), Ismael Ledesma (2002) and Rafael Guevara (2002, 2014) have shown us that when it comes to the history of biology, beyond the obvious, Mexico has its history. As in other places, it was in the nineteenth century that it was formally institutionalised, first thanks to the work of authors such as the Mexican naturalist Alfonso Luis Herrera and the French-born Mexican zoologist and botanist Alfredo Dugès and later with the establishment of academic institutions such as the Institute of Biology at the recently formed National Autonomous University of Mexico (UNAM) in 1929, or the National School of Biological Sciences at the National Polytechnic Institute (IPN) in 1936. On the other hand, the institutions resulted from disputes, as in the case of the Institute of Biology, where the confrontation—not only with different ways of understanding biology but also with opposing political visions—between Alfonso Luis Herrera and Isaac Ochoterena was definitive. Cases such as Herrera's are fascinating because he was one of the first authors in the world to provide a concrete theory to study the origin and evolution of life, "plasmogeny". In Herrera's own words,

> Plasmogeny is an experimental science whose object is to study the origin of the protoplasm, just as Cosmogony, for example, investigates the origin of the Universe, and Pathogeny, the origin of diseases. [...]
>
> Space is filled with a kind of electric mist which had been called ether, and as Plasmogeny considers everything in its meditations and proofs, we can generalise the

> above definition by saying that Plasmogeny is the science of the universal protoplasm and its origin, meditations and proofs, we can generalise the above definition by saying that Plasmogeny is the science of the universal protoplasm and its origin. (Herrera 1932)

Herrera was a figure of nineteenth-century Mexican science, both for his theoretical contributions such as the plasmogeny theory mentioned above—which, in addition to proposing the chemical origin of life, was an evolutionary theory, which reclaims not only Darwin's ideas but also French transformism—but also for his work in consolidating biology as a new discipline, through the consolidation of institutions such as the Commission of Agricultural Parasitology (1900) and the Directorate of Biological Studies (1915), which allowed for the institutionalisation of biology in Mexico. Now, authors such as Herrera were either trained in Europe—particularly in France—or had deep connections with academics outside Mexico, which allowed them to be "up to date" and, in that sense, to know and transmit the theoretical discussions in vogue.

The case of the IPN also has its particularity: it was an institution formed under the precepts of the so-called "Cardenismo"—proposed by the then President of Mexico, General Lázaro Cárdenas—which, amongst other things, brought scientific work closer to a political agenda also known as Mexican socialism. This proposal is one of the best examples of how biology was practised locally in Mexico since, on the one hand, it recovered an ideological stance—which for various reasons has been very successful in the Mexican context—under which biology was understood more as a tool—especially disciplines that could be applied to issues such as health and agronomy—for social development. Darwin, for example was understood through historical materialist interpretation. However, the scope of this biology was minimal because once Cárdenas' presidential term was over, everything related to "Cardenismo" faded into the background.

It is worth taking up an example of a writing that, despite its success at the time in disseminating evolution, has been ignored by the Mexican academic community. That book, published in its first edition in 1966 in Spain, is *La evolución* (The Evolution, 1986), published by the Biblioteca de Autores Cristianos (BAC, in English, Christian Authors Library). It is considered a turning point in the development of Spanish biological sciences, a work edited by some of the most important Spanish scientists of the twentieth century, such as the palaeontologists Miguel Crusafont i Pairó (1910–1983), Bermudo Meléndez Meléndez (1912–1999), and Emiliano Aguirre Enríquez (1925–2021). The book, which has been updated through four editions, has important features:

- It is a reflection of a theoretical discussion within evolutionism between authors who openly defended a finalist evolution (many of them with an evident influence of the proposal of the paleoanthropologist and Jesuit Pierre Teilhard de Chardin) and others who adopted the postulates of the Synthetic Theory.
- It is a work that was published in the context of Francisco Franco's dictatorship in Spain and represented one of the high points of the ceding of control of science to the Catholic Church.

- Despite being published in Spain, it is a text that to this day is widely disseminated and quoted in Latin American countries, although it is true that its circulation has been restricted to Catholic institutions. There are notable exceptions, such as the case of the UNAM—an officially secular public university, although with a scientistic and anti-clerical character in scientific areas—where the book can be used as a textbook.
- The book is organised through essays, ranging from the origin of life to human evolution. A striking detail is the discussion of science's philosophical, historical, and methodological aspects.

The case of *La evolución* has been analysed in recent years in Spain by members of the Sociedad Española de Biología Evolutiva (Spanish Society of Evolutionary Biology). Beyond ideological aspects, its impact was decisive in consolidating the Spanish scientific community. In the case of Mexico, it is a work that has been available in national libraries since its first publication. In an eminently Catholic country, it is a book that can influence the understanding of evolution, but, indeed, the history surrounding this publication and its possible impact on scientific discussions remains to be analysed.

The examples given here are only a sample of the complexity of the historical processes surrounding the case of biology in Mexico. A common feature in all of them is that the role of Darwin's ideas, although present, was not necessarily the only ones discussed, or the ones that defined the discussions.

5 Conclusions

Now, all of the above serves to reach some conclusions, which, although they may be preliminary, are helpful for future reflections on the implications of how the history of biology is being presented and understood in Mexico, particularly the ideas of Darwin and Darwinism.

One of the consequences that can cause the most problems is the promotion of a conflict that historically in Mexico today has not had significant repercussions. The conflict between evolutionism and creationism has been a characteristic identified above all with the United States throughout the twentieth century. In the case of Mexico, the problem has arisen from uncritical teaching—or, to put it another way, dissemination—of the history of biology, in which it has been assumed that the consequences of Darwin's ideas are potentially the same in different geographical contexts. Those ideas, or at least those credited to Darwin, are: evolution is explained—almost entirely—by natural selection; the latter is a random, creative, cumulative, and non-directional process; the primary material of evolution is variation; progress is rejected; there is no room for alternative evolutionary explanations such as those associated with Jean Baptiste Lamarck. Moreover, all of the above is basic to sustaining a naturalistic explanation of the world—which authors such as Mayr understand as a worldview—in understanding nature as synonymous with material (Mayr 1991). The point to understand here is that Darwin's ideas were introduced and appropriated in regions such as Latin America, not necessarily in scientific terms but in social terms. In countries such as Mexico, nineteenth-century discussions of Darwin were concerned with his implications for the existence of God and not with possible criticisms of his methodology, as presented in the work mentioned above by Moreno de los Arcos (1984).

This "revolution" continues to promote the formation of stereotypes, which in the Mexican case has brought with it a profound irony: Mexico is historically a predominantly Catholic country, and it is well known that Catholicism has had no significant complications or conflicts with biological evolution; the case of human evolution, as an exception, does not call into question that even the human body is the product of natural processes—which may ultimately be guided by God—which forces a rethinking of how the place of science is understood in a country like Mexico where the religious plays a fundamental role for the majority of the population. A "compatibilist" proposal between evolutionary theory and Catholic thought prevalent amongst the most progressive sectors of the Church is that of Pierre Teilhard de Chardin. Despite the controversy surrounding it, in recent years, it is remarkable to find increasing examples of its influence on Catholic thinkers, such as Pope Francis himself, who recurrently—in the recent encyclical *Laudato Si'* and the apostolic exhortation Laudate Deum—cites the work of Teilhard de Chardin as a justification of scientific endeavour as seen from a Catholic perspective.

Another consequence is the historical confusion generated by assuming a diffusionist discourse—with strong presentist tendencies—that defends a Eurocentric notion of science, which in the case of Mexico is mostly Anglocentric, with some Francocentric exceptions. Beyond a quasi-fundamentalist defence of decolonial proposals or the so-called "epistemology of the South", it is crucial to recognise the local traditions that have generated the disciplines at the time. This importance can have consequences for the training of new generations of scientists.

There is a latent risk—which is growing nowadays—of appealing to national, but above all nationalistic, narratives to present the development of science. Local narratives are only part of the solution to understanding the impact diffusionism has had on understanding the history of biology in a place like Mexico. The challenge lies in understanding the processes of appropriation, translation, and circulation of knowledge based on historical facts, which bring us closer to each context's social and cultural complexity.

Even a central element of the "Darwinian revolution", such as the "conflict" between science and religion, deserves a more profound analysis than, in the diffusionist logic, assuming that the same situation has arisen in any geography as in the United States. A genuine process of deconstruction should start with a critique of the dominant vision to propose new narratives or even reaffirm existing ones.

References

Barseghyan H (2022) Selection, presentism, and pluralist history. Stud Hist Phil Sci 92:60–70

Caron JA (1988) 'Biology' in the life sciences: a historiographical contribution. Hist Sci 26:223–268

Crusafont i Pairó M, Meléndez B, Aguirre E (eds) (1986) La evolución, 4th edn. Biblioteca de Autores Cristianos, Madrid
Futuyma DJ (2005) Evolution. Sinauer
Glick TF (ed) (1988) The comparative reception of Darwinism. University of Chicago Press, Chicago and London
Guevara Fefer R (2002) Los últimos años de la historia natural y los primeros días de la biología en México: la práctica científica de Alfonso Herrera, Manuel María Villada y Mariano Bárcena. UNAM, México
Guevara Fefer R (2014) El uso de la historia en el quehacer científico. Una mirada a las obras históricas del biólogo Beltrán y del fisiólogo Izquierdo. UNAM, México
Herrera AL (1932) La plasmogenia. Nueva ciencia del origen de la vida. Redacción y Administración Luis Morote, Valencia
Himmelfarb G (1996[1959]) Darwin and the Darwinian revolution. I.R. Dee, Chicago
Hoquet T (2009) Darwin contre Darwin. Comment lire L'Origine des espèces? Editions du Seuil, Paris
Lazcano A (2005) Teaching evolution in Mexico: preaching to the choir. Science 310:787–789
Ledesma Mateos I (2002) La introducción de los paradigmas de la biología en México y la obra de Alfonso L. Herrera. Historia mexicana:201–240
Ledesma Mateos I, Barahona Echeverría A (1999) Alfonso Luis Herrera e Isaac Ochoterena: la institucionalización de la biología en México. Hist Mex 48:635–674
Mayr E (1972) The nature of the Darwinian revolution. Science 176:981–989
Mayr E (1991) One long argument: Charles Darwin and the genesis of modern evolutionary thought. Harvard University Press, Cambridge
Moreno de los Arcos R (1984) La polémica del darwinismo en México: siglo XIX: Testimonios. Universidad Nacional Autónoma de México, México
Morgan MS, Hajek KM, Berry DJ (eds) (2022) Narrative science: reasoning, representing and knowing since 1800, 1st edn. Cambridge University Press
Ruiz Gutiérrez R, Noguera Solano R, Rodríguez Caso JM (2016) The ideology of the "Survival of the Fittest" during the Porfiriato in Mexico. In: Lightman B (ed) Global spencerism: the communication and appropriation of a British Evolutionist. Brill, Leiden, pp 149–172
Ruse M (2019) The Darwinian revolution. Cambridge University Press
Torrens Rojas E, Rodríguez Caso JM, Barahona Echeverría AR (2023) Laicidad y evolución. In: Capdevielle P, Salazar Ugarte P (eds) El Estado laico mexicano a 30 años de la reforma constitucional, 1°. UNAM-IIJ, México, pp 171–183
Young RM (1971) Evolutionary biology and ideology: then and now. Sci Stud 1:177–206
Young RM (1992) Science, ideology and Donna Haraway. Sci Cult 3:165–207

Historicity, Temporalities, and Causality: A Confusion at the Heart of Debates on Darwinism

Mathilde Tahar

Abstract

If Charles Darwin's work opened up the possibility of a true natural *history*, the significance of time in evolutionary processes was left unresolved. This ambiguity has led to various interpretations of what evolutionary history is, some seeing it as the pure unfolding of processes, others as a flow marked by contingency and unpredictability. These interpretations reflect underlying differences in the perception of causality: mechanical and uniform on the one hand, transformative and multifaceted on the other. This tension affects not only our understanding of the Darwinian text, but also our contemporary conception of evolutionary theory. Although many challenges in evolutionary biology are linked to time (the historicity of evolution, path-dependent processes, biological rhythms, the contingency of certain evolutionary events etc.), the importance attributed to time in biology remains unspecified. The aim of this chapter is to study the contradictions regarding the conception of history and causality in Darwin's writings and to analyse how they have permeated the different interpretations of evolutionary theory from the turn of the twentieth century to the present day. This analysis of the tensions within the different interpretations of Darwinism in history will lead me to address both the conception of biological time(s) and the approach to causality in contemporary evolutionary biology. I will conclude this chapter by proposing some clarifications about the significance of time in contemporary evolutionary biology.

Keywords

Time · Philosophy of biology · Charles Darwin · Evolutionary theory · Modern synthesis · French evolutionists · Process biology

M. Tahar (✉)
Department of Anthropology, University College London, London, UK

R. G. Delisle et al. (eds.), *Unity and Disunity in Evolutionary Biology*,
https://doi.org/10.1007/978-3-031-42629-2_19

1 Introduction

It is often said that time, in biology, does not have the same meaning as in physics (Longo 2021). Usually, this claim aims to support the idea that evolutionary biology, contrary to most parts of physics, *cannot be* predictive because of the importance of history. Indeed, in Newtonian physics, time appears as the exploration of a space of possibilities that can be described on the basis of a limited number of propositions. This is why, even in quantum physics, where processes defy prediction, it is still possible to make probabilistic judgements about their outcomes. In contrast, biological phenomena, and especially evolutionary ones remain mostly unpredictable, and even improbable. Theorists such as Baquero (2005), Montévil (2020), and Kauffman and Roli (2023) suggest that the unpredictability of evolution stems from the fact that, in the course of evolution, it is the space of possibilities itself that changes. More fundamentally, biological entities are defined by their very history (their phylogeny), whereas physical entities are defined by fixed criteria. This conception of biological historicity is linked to the idea that causality in biology is different from physical causality. In biology, new processes emerge, whereas the laws of physics do not change. Thus, biological events can produce genuine novelties that could not have been foreseen before their emergence (Longo et al. 2012).

All these characteristics are linked to a certain conception of the role of time in biology, especially in evolutionary biology, which is rarely made explicit. The link between temporality and causality in evolution remains elusive, with the lack of historical background on this problem maintaining the vagueness. Indeed, the ambiguities surrounding the significance of historical time in biology were already harboured by Darwin's formulation of the theory of evolution and have been maintained problematically ever since. Some consider that the long time frame of the Darwinian hypothesis is uniform—allowing small variations to accumulate in a straight line. Others see it as historical time—not the simple unfolding of an unchanging causality, but a time marked by discontinuity and chance. The vagueness of the concept of biological time(s), in Darwin's writings, but also in the different formulations of Darwinian theories throughout history leads to tensions within the theory of evolution. Without clarifying the role of history, biological temporalities, and the underlying conception of causality in evolutionary theory, it seems difficult to understand the methodological and epistemological issues related to the so-called specificity of evolutionary biology in relation to other disciplines.

The aim of this chapter is twofold. The first is to retrace some debates in the history of evolutionary theory, in the light of one of its most problematic concepts: evolutionary 'time'. I want to show that some of the tensions between the different interpretations of Darwinian theory and causality in evolutionary biology highlight a lack of clarity about the historicity of evolution and the role of time in biology. The second is to provide conceptual clarifications to address issues that are generally overlooked, but which are crucial to understanding some of the epistemological challenges facing biology: How can we synthesise the different roles that time plays in our understanding of evolution? What does it mean to say that evolution is a history? Why does biological time seem different from physical time?

In order to resolve the vagueness surrounding evolutionary time(s), I begin by examining the ambiguities within the Darwinian text. I then analyse how these ambiguities have permeated the different interpretations of evolutionary theory from the turn of the twentieth century to the present day. My ambition is also to provide a clarification about what is intuitively understood when we talk about the specificity of biological time, which mainly refers to a specific conception of causality. In the last two sections, I offer some clarifications on the importance of taking different temporalities into account in evolutionary biology, and of some of the epistemological consequences.

2 History, Time, and Causality in *The Origin of Species* (1859)

Charles Darwin (1809–1882) did not propose any reflection on the concept of history, or on the meaning of temporality in biology. However, his theory is based on a number of presuppositions about the importance of history, the temporality of biological processes and the causality of evolution.

Indeed, Darwin's theory is based on gradualism: small individual variations, accumulated by natural selection over a very large number of generations, lead to different varieties and even different species. For the theory to explain the extraordinary diversity of biological forms, a long, even quasi-infinite time is required. But there is an ambiguity in Darwin's very theory about the conception of time.

2.1 A Creative History?

On the one hand, evolution is considered as a truly historical process: the flow of time is not the mere unfolding of the same processes, producing different states: it produces changes of forms. Darwin developed his theory from considerations about the age of Earth. At the time Darwin was writing, the age of the Earth was still unknown. However, Charles Lyell (1797–1875) proposed that it was at least several hundred million years old, and Darwin's theory relied on this hypothesis. "He who can read Sir Charles Lyell's grand work on the Principles of Geology [...] yet does not admit how incomprehensibly vast have been the past periods of time, may at once close this volume" (Darwin 1859, 282). Only the depth of geological time can explain the evolution of species from individual variations.

Indeed, for Darwin, the passage of time has an effect on biological entities: it can truly both constitute and transform species. Thus, the importance of historical time directly relates to a special conception of causality in evolution. Over the course of evolutionary history, the *same* mechanism (natural selection) produces *new* effects, incommensurable with the previous ones. New species emerge over the course of time, a novelty that is explained by the fact that the past is still effective in the present (past variations are still present in the organism). The consequence of such a theory is that new species also generate *new processes*. For example, the regulation of glucose flow by the liver is a process that only exists in species that have evolved to have a liver. The liver is an organ that has evolved historically, and its evolution has given rise to new biological processes. Thus, causality in evolution is different from causality in physics because most of the 'laws' are historically constituted (Gould 1970). In other words, the processes themselves are path-dependent (Szathmáry 2006) and naturally (though not necessarily) linked to a contingent irreversibility (Desjardin 2011).

This approach has ontological consequences: since species have evolved progressively over time, starting with small variations between individuals, they are not eternal essences, they are only *logical* categories that allow us to make differences in our classifications. Naturalists do consider that there is fecundity of mongrels and sterility of hybrids, and that species are different in kind. But, for Darwin, this is only statistical: 'it is most difficult to say where perfect fertility ends and sterility begins' (Darwin 1859, 248). There are cases where hybrids are not sterile and cases where mongrels are not fertile. Species, for Darwin, cannot be identified by fixed physico-chemical properties, since they are generated in the course of time: they are transitory forms emerging from the individual variability. Therefore, Darwin's conception of time and causality can be interpreted as relying on an ontological claim: variability is fundamental in the living (Soto et al. 2016). Indeed (1) biological entities are not specified by invariants, like in physics, but by their history; (2) the conditions of the environment are also constantly changing. This means that both observables and initial conditions always change in biology, and the combination of the two produces unpredictable processes. This would justify that, for Darwin, the *explanandum* is less evolution than the "*the preservation of favoured races in the struggle for life*" (the subtitle of the *Origin of Species*). Evolutionary time allows individual variability to be channelled and stabilised in forms that are always unpredictable. In this sense, time appears a crucial factor of evolution, enabling a very specific type of causality, different from the deterministic one present in physics.

2.2 Uniformitarianism

However, the evolution of species takes place *progressively*, through the operation over time of the *invariable* process of natural selection. Thus, it could be argued that the transformation of forms would happen *over the course of time* but would not necessarily imply the intervention of *new* processes generated over time (what appears as new processes would only be determined as a result of the action of natural selection). This view implies that the invariable mechanism of natural selection follows a classical causal model: it has specific effects, depending on the situation. It holds that the unpredictability of evolution does not stem from a special role of history, or from fundamental variability, but from our inability to take account of the diversity of biological situations. Evolutionary history merely unfolds invariable mechanisms.

Darwin's gradualism is indeed based on uniformitarianism, which was the key principle allowing Lyell to define the age of Earth: the causes currently at work are the same as those at work in the past. If there is a change, it must be explained by the incremental action of the processes that we currently observe operating. Thus, time is considered a uniform succession, during which changes unfold. Similarly, Darwin draws upon the brief time span of individual variations and extrapolates them into the historical time frame of species. By studying natural selection in the context of human selection, which operates on a shorter timescale, he is able to extrapolate his findings to the longer processes of nature, as if the breeders' short- and medium-term forecasting justified the uniformity of the processes (and possibly the long-term forecasting of future biological events). This view implies that each moment in biological processes is equivalent to another, and causality thus appears to be uniform. Time's efficacy does not come from a special quality, but from its quantity: the more time passes, the more transformation takes place. History, meanwhile, is reduced to a series of discrete moments, whose sum total allows for change.

Natural selection then appears as a quasi-deterministic mechanism, with little room for contingency, and time appears as a mere unfolding. This is made explicit in Darwin's early writings where he assumes that the future of species could be foreseen by a being quite close to Laplace's demon (as in Pierre-Simon de Laplace 1749–1827):

> Let us now suppose a Being with penetration sufficient to perceive differences in the outer and innermost organization quite imperceptible to man, and with forethought extending over future centuries to watch with unerring care and select for any object the offspring of an organism produced under the foregoing circumstances; I can see no conceivable reason why he could not form a new race (or several were he to separate the stock of the original organism and work on several islands) adapted to new ends. (Darwin 1909, 85)

Such a hypothesis assumes that the mechanism of natural selection leaves nothing to chance, that biological causality is as determined as physico-chemical causality, and that no new biological process can emerge in the course of natural history. So, the unpredictability of evolution is not due to any particular characteristic of living beings but to our ignorance. What seems to us to be contingency would only reveal our ignorance, the causal chain being strongly determined: "If we must marvel, let it be at our presumption in imagining for a moment that we understand the many complex contingencies, on which the existence of each species depends" (Darwin 1859, 322).

There is therefore an unresolved tension in Darwin's conception of evolutionary history and, consequently, of the causality of evolutionary processes. Darwin's theory relies on the idea of continuous variability, which can be channelled in unpredictable ways over time, generating a history in the strong sense. Emphasising this aspect of Darwinian theory, one could consider that time is not only the unfolding of invariable mechanisms but a retention of the past in the present that would contribute to the generation of true novelties. However, Darwin's conception also requires the uniformity of the laws of nature over time. Thus, in another sense, there is no real change; time appears as a mere container, with the biological processes remaining the same.

As the conception of time is ambiguous, the conception of causality remains unclear. Darwin does consider evolution to be unpredictable, as it is a "changing history" (Darwin 1859, 106), and not the mere succession of causes and effects:

> Looking to the future, we can predict that the groups of organic beings which are now large and triumphant, and which are least broken up, that is, which as yet have suffered least extinction, will for a long period continue to increase. But which groups will ultimately prevail, no man can predict; for we well know that many groups, formerly most extensively developed, have now become extinct. (Ibid., 126)

But, as mentioned above, most of the time, this unpredictability seems to reveal our ignorance of the causes at work, and not a specific sort of causality, proper to organic evolution (Ibid., 67, 79, 195 *et passim*).

Thus, the role that Darwin ascribes to time seems paradoxical: time appears to be both the mere 'container' of unchanging mechanisms and a historical time, marked by contingency and improbability, that generates unpredictable new processes.

And it is not clear that the tension can be resolved merely by relying on Darwinian texts.

3 Different Conceptions of Time, History, and Causality: Different Interpretations of Darwinism

The question of the specificity of biological causality and the unpredictability of evolution have consistently puzzled biologists. Some argued for more physicalist interpretations, whilst others assumed a special causality for biology. These debates, especially vivid at the turn of the twentieth century, not only relied on different conceptions of evolution but of the various temporalities, relevant to biological explanations. To demonstrate this, I will study two interpretative traditions: the French reception of Darwinism, and that proposed by some of the theorists of the Modern Synthesis.

3.1 Evolutionary Theory in France

3.1.1 Darwinism in France

In France, the reception of Darwinism was delayed as it faced strong competition from (neo)Lamarckism. The popularity of Jean-Baptiste de Lamarck (1744–1829), especially his concept of the inheritance of acquired characteristics, may be partly due to a preference for the national author, but it also reflected a certain vision of what biology should be.[1] In France, at the end of the nineteenth century, the

[1]For a detailed overview of (neo)Lamarckism(s) in France, see Loison (2010).

biological science *par excellence* was physiology, thanks mainly to the discoveries made by Claude Bernard (1813–1878). Indeed, physiology offered the example of an experimental biological science, whose methodology was close to that of physical science, and which allowed the discovery of stable causal relationships as necessary as in physics. Physiology allowed biology to rise to the status of a 'real' science, a science based on determinism and whose hypotheses can be validated by crucial experiments. One strength of physiology is that its method is mainly experimental, relying on what can be observed *in the present*, or over a short period of time. With this in mind, it can be argued that the French preference for the inheritance of acquired characteristics inspired by Lamarckism is explained less by chauvinism than by the epistemic structure of French biology at the turn of the twentieth century. The full realisation of biology as a science depended on achieving the same level of deterministic or at least probabilistic precision as physics, which could only be attained through experimentation. Therefore, to study evolution scientifically, the mechanism underlying mutations had to be exhibited, i.e. the physiology of variation had to be understood. In this context, the Lamarckian hypothesis was particularly attractive because it suggested that the changes that cause evolution occurred during the lifetime of an individual and could therefore be directly observed and evaluated. Likewise, mutationism, supported notably by Hugo de Vries (1848–1935), was also appealing because mutations occurred within a single generation and could be studied over a short period of time, allowing for the discovery of their physiological mechanisms. Darwinism was not considered to be fully scientific precisely because the theory is based on a depth of historical time on which one could only make hypotheses but not experiments (Loison 2022). This implied that the relevant time scale for science was the short term, which makes it possible to observe efficient causality, and which we are allowed to extrapolate to longer time scales. Long, historical time, contributed nothing to explanation. Science needed to tend towards determinism and deduce a large number of effects from a small number of invariable laws—a concept that did not allow for a historical approach to biology.

This criticism of Darwinism, as untestable because based on the transformation of forms over a long period of time, was explicitly expressed by several French commentators of Darwin. The physiologist Pierre Flourens (1794–1867) was one of the first in France to provide a detailed review of *The Origin of Species*. However, he strongly opposed Darwin's ideas, and he criticised him for being unable to provide true—i.e. according to him: observable—examples of the transformation of one species into another (Flourens 1864). The same sort of objections were made by the naturalist Armand de Quatrefages (1810–1892), albeit he was less hostile to transformism than Flourens. The Darwinian theory was considered weak because it was not based on positive facts, results of experiments, which could be tested in the present (Quatrefages, 1870). One of those who welcomed Darwin positively was Charles Martins (1806–1889), director of the botanical garden in Montpellier, who republished Lamarcks's *Philosophie zoologique* (1873). In the introduction to the book, in a notable misinterpretation, he stated that Darwin consolidated Lamarck's theory. And he added that there remained, however, a great scientific task to be carried out: the physiological method needed to be applied to the study of variation if we were to understand the mechanism of evolution.

In all these authors, there was the idea that we would discover the physiological origin of species in the mechanisms of the present time, i.e. on the scale of one generation. The reference model of science remained the experimentation of the physical and chemical sciences.

Thus, there was a form of presentism (what is discovered in the present explains all past and future processes), accompanied by a nomological conception of causality. The only kind of causality envisaged was that discovered by physics—deterministic or probabilistic causality. Like the laws of physics, biological processes were conceived as immutable, history not bringing about their transformation but the successive stages of their unfolding.

3.1.2 French Neolamarckism: A Physiological Evolution with No History

Physiological anchoring and determinism explain both the mechanistic interpretation of Darwin in France and the success of Lamarck's theory. This coordination was especially embodied by the evolutionist Félix Le Dantec (1869–1917), a mechanist who defended the continuity of physics and biology. He subscribed to the mechanistic interpretation of Darwinism proposed by Ernst Haeckel (1834–1919), whilst at the same time calling for a new reading of Lamarck and for the physiological elucidation of the mechanisms of variation (Le Dantec 1902).

It is true that Neolamarckism is not a unified theory (for a detailed analysis, see Loison 2010). On the one hand, the first generation of Neolamarckians, including Félix Le Dantec, were fervent determinists and mechanists, who advocated for an experimental method allowing the secrets of evolution over the centuries to be elucidated in present-day experience. This was the case, for example with Edmond Perrier (1844–1921) and Alfred Giard (1846–1908) (Perrier 1879; Giard 1904). On the other hand, the second generation of Neolamarckians were vitalists and/or spiritualists who emphasised finality. This was the case with Albert Vandel (1894–1980) and Pierre-Paul Grassé (1895–1985), who considered that the theory should be complemented by a reference to the final causes, as only finality could explain the directions taken by evolution (Vandel 1949; Grassé 1973).

However, in both cases, history appears as the mere unfolding of predetermined events, according to invariable processes.

3.2 The Case of Henri Bergson

The philosopher Henri Bergson (1859–1941) wrote *Creative Evolution* in 1907 in this context of uncertainty about the processes at work in evolution. Although he was an evolutionist, he wondered whether the various theories of evolution under debate, be they based on mechanism (natural selection, the inheritance of acquired characteristics) or on finality (as in the vitalist interpretation of Neolamarckism) were capable of truly explaining evolution (Bergson 1911). His distrust of the debated theories stemmed from the fact that both mechanistic and finalist theories failed to comprehend the efficacy (*efficace*) of time: in one case, everything results from the initial cause; in the other, everything is directed towards the final cause. But in both cases, the passage of time between what we perceive as cause and its effect are inconsequential; any unpredictability in evolution is only due to our ignorance. In any case: "everything is given" (Ibid., 41), and history brings nothing new to the table. For Bergson, the determinism of the evolutionary theories of his time, based on a uniform and unilinear conception of causality, was linked to an inability to understand the historical dimension of biological evolution. Bergson interpreted Darwinism as such a theory, incapable of understanding the creativity of time. This led him to denounce, in both Darwinism and Neolamarckism, the inability to take into account the change of processes brought about by time. In contrast to this vision, Bergson envisaged the efficacy of time as something active. He did not think of time as a simple variable but as truly *creative*: it brings something new, the unpredictability of which comes not from our ignorance but from the very nature of history. And for evolution, in particular, Bergson proposed to think of this creative time as an *élan vital*. The *élan vital* is a propulsive and irreversible movement whose direction considered afterwards (by a retrospective glance) is like a straight line (Ibid., 124), but whose productive power has a circular dimension (living beings are like eddies, Ibid.). The objective of this image is to link, without contradiction, the directionality of time and its unpredictability. Regarding directionality, evolutionary time exhibits both linear and cyclical aspects. On the one hand, it is characterised by accumulation, an irreversible progression. On the other hand, it also involves recapitulation or resumption of the past, in such a way that history *never repeats*.

Although Bergson proposed a metaphysical conception of time that was difficult to translate into biological terms, he did clearly see that the stumbling block of the various evolutionary theories under discussion was the tension between a conception of causality based on physics and the little thought given to the importance of history.

The significance of Bergson's theoretical proposition has often been overlooked in the historiography. Biologists have been deemed uninterested in philosophical questions about the nature of historical time, which is thought to have led to a lack of serious consideration of Bergsonian philosophy in the field. If this may be generally true from the point of view of French biologists, it is not quite so simple for those who have been labelled as Modern Synthesis theorists.

3.3 The "Modern Synthesis"

3.3.1 Biogeographical Approach in Population Genetics

Julian Huxley (1887–1975) coined the phrase "The Modern Synthesis" to describe the integration of genetics and Darwinian theory by biologists from various disciplines and backgrounds (Huxley 1942). This synthesis led to a model referred to as "biogeographical" by Georges Canguilhem (1904–1995) (Canguilhem 2003, 177). Canguilhem used his term to contrast Darwin's theory with Lamarckism, according to which the environment had a physiological meaning, and what was important was the relationship between the internal and external environments. Canguilhem interpreted Darwinian theory as attributing causality to the interplay between living beings, whereby the difference between individuals structured the struggle for existence and thus natural selection. In this view, causality is based on otherness, and unfolds in *space* (and not time). Coupled with the rediscovery of Mendelian laws, this approach founded population genetics, i.e. the study of changes in allelic frequencies within a population, over time. In this perspective, time is a variable (and the difference between two varieties or two parent species is assumed to be proportional to the time elapsed). It is a measure of the difference that accentuates the effect but does not change the cause.

3.3.2 Creative Time and Contingency

Yet, some theorists of the so-called modern synthesis, reflected on the nature of time regarded as *historical,* i.e. as having such an efficiency that it would be *creative.* The reason is that truly new processes emerge in evolutionary history, generating an essential unpredictability. We find such reflection in Ronald Aylmer Fisher (1890–1962), Theodosius Dobzhansky (1900–1975), and Sewall Wright (1889–1988) (amongst others: Fisher 1950; Dobzhansky 1960; Dobzhansky and Boesiger 1968, 145–165; Wright 1964). And they all referred to Henri Bergson on this subject (not without criticism) as being the one who coined the link between history and creative causality in evolution—a link which, in their view, biology, unlike philosophy, cannot and should not take into account. Fisher, Dobzhansky, and Wright considered that the unpredictability of evolutionary history was related to the idea that natural selection may not be sufficient to explain all evolutionary phenomena: (1) there is a part of chance (migration, genetic drift etc. must be considered); (2) the solutions offered by evolution to ecological problems are unpredictable because biological situations never repeat themselves. Thus all three, in certain texts, questioned the nature of time as possibly producing something more than can be predicted by calculation (Dobzhansky 1960; Dobzhansky 1974; Wright 1964). According to them, in evolutionary history, the same causes cannot produce the same effects, since the same causes never occur twice.

This questioning of evolutionary temporality as truly creative comes from the recognition of the unpredictability of evolution, which is not solely attributed to our lack of understanding, but also to the historical aspect of the process itself, allowing for a degree of contingency. This contingency emanates from the fact that, in evolution, there are always original combinations of factors, combinations that have never been produced before and will never be replicated, as the factors themselves are shaped by historical processes. This prompts a more comprehensive reflection on the possibility of novelty in evolution, where novelty refers to an occurrence not predetermined in a space of possibilities.

However, it is important to note that these evolutionists questioned the creativity of evolution in papers that were more philosophical than scientific in nature. These interrogations stemmed more from the authors' personal curiosity and metaphysical reflections than from their scientific practice. To them, the question of the nature of biological temporalities, the meaning of historicity, and creativity of biological processes were not relevant to a biological theory, and should not be resolved by science but by philosophy (Wright 1964). In the background remain the idea that, for science, time cannot be anything other than a variable: a concept of creative irreversibility would necessarily be outside science. Thus, whilst all three addressed the nature of time, they did not dispute the primacy of the natural selection explanation and the biogeographical view of population genetics.

What these articles do show, however, is that the nature of evolutionary causation, and the specific concept of historical time associated with it, has not been resolved, even after the establishment of Darwinism. A tension remains, which is still evident in most contemporary debates.

4 A Confusion That Runs Through the twentieth Century

The problematic role of the diverse temporalities of processes played an increasingly important role in the sciences during the twentieth century. In particular, the irreversibility of time in quantum physics and the processual dimension of causal chains proposed by the theory of complex systems underlined the need for a robust concept of temporality, not confined to metaphysical debates but grounded on science theories. Hence, the (discreet) enthusiasm of certain contemporary scientists for Bergson's proposals. According to Ilya Prigogine and Isabelle Stengers, Bergson defined "a program that is beginning to be implemented by the metamorphosis science is now undergoing" (Prigogine and Stengers 1984, 83).

4.1 Gould, Dawkins, and Evolutionary Time(s)

However, even today, biologists rarely include a real theorisation of causality, let alone of biological time in their theories, which sometimes leads to deep misunderstandings. By way of example, it is worth mentioning a well-known disagreement, reported in detail by Kim Sterelny (2001): the one between Richard Dawkins and Stephen Jay Gould on the importance being attached to natural selection in evolution. I propose that this divergence is grounded in a more fundamental difference of perspective which, in the end, points to a dissimilar conception of biological time.

Dawkins considers time as the *framework* in which the invariable process of natural selection takes place. Indeed, in *The Blind Watchmaker,* he speaks of natural selection as an algorithm, which tends towards optimisation (Dawkins 1996, 46–50). Taking up this idea, Daniel Dennett (Dawkins' 'ally', goes so far as to conceive that natural selection might explore a 'Design Space'. He proposes that this Space contains all possible phenotypes, both in the past and in the future, and that natural selection acts as a mechanism for navigating through this space by selecting the most suitable genes to respond to ecological challenges (Dennett 1995). In so doing, Dennett reveals the implicit conception of time and causality underlying the algorithmic understanding of natural selection: time merely unfolds an invariable causality whose effects are all contained within an unlimited, yet fixed, space of possibilities.

Gould, on the other hand, adopts a historian's point of view and seeks to understand how past events can continue to have an effect today and how their significance has changed over time. First, he considers the plurality of processes at work beyond natural selection: extinction, metapopulation dynamics, and structuring of morpho-space over the centuries. Second, he focuses not only on the process but also on the specific conditions that led to the particular phenomenon under study: how variations were made available by history, and how the structuring of metapopulations in the course of history led to current biodiversity. This prompts him to (almost metaphysically) question the very nature of biological time, as a historical reality, leading to a creative causality. He grapples with the seeming contradiction that, whilst historical events are unique and unpredictable, there are nevertheless underlying natural laws that explain the recurrence of, yet not identical events. Hence, Gould conducts a reflection aiming to articulate the irreversibility and unpredictability of time—whereby the past is gone and cannot be repeated—with its cyclical nature, wherein the past retains influence and natural laws produce similar patterns of events (Gould 1988).

Stephen Jay Gould's perspective differs from Richard Dawkins' as he regards time not only as an arrow, representing a sequence of selections, but also as a cycle that allows for a resumption of events. Concretely, this means that past structures constitute reserves of potentialities that can take on new *functions* (exaptations), unpredictable from the outset—and Gould refers to Nietzsche's philosophy on this subject (Gould 2002, 1218). In the end, Gould considers the interplay between time linearity and circularity as enabling both the understanding of *contingency* (which pertains to the occurrence of unique and unpredictable biological situations) and the accounting for the fact that they have *biological and rational causes* (encompassing natural selection, morphological laws, and structural constraints) (Gould 1988).

In addition to offering an original reflection on biological time, Gould puts forward a new model of causality for thinking about the unpredictability of evolution: no longer the linear and necessary cause/effect causality, but interactions of plural, historically constituted, biological constraints producing original outcomes (Gould 2002, 1027–1037; Tahar 2022). This mixture of irreversibility and patterns leads to evolution not tending towards an optimum.

4.2 The Obliteration of Time

In Darwin's texts, the meaning of historical time remains a problem around which several theoretical tensions crystallize. Some examples show the creative irreversibility of time, the way in which the past extends into the present, creating 'oddities' (see, for instance, the torsion of the ovary of *Malaxis paludosa* in Darwin 2016, 200). But there is no explicit theorisation of the problems of biological temporality, and whilst it may have been considered a problem by some of the theorists of the Modern Synthesis, contemporary biology is characterised by an obliteration of time against which Gould is one of the few to speak out. As we see with Dennett's or Dawkins' theories, the temporal specificity of biological processes remains mostly unthought of, and thus neutralised. In such theories, biological history has no specificity, and the model of causality remains the same as the physical one: monolithic and nomological—ultimately uniform.

The French philosopher André Pichot denounced this obliteration of time, which leads, according to him, to evolutionary biology obliterating itself as an autonomous discipline and retreating to functional biology—whose object is not historical (Pichot 1987). Evolutionary biology is supposed to articulate phylogenetic history and the laws of functional biology, by referring both to the individual history of the organism possessing the organ or behaviour under study. But according to Pichot, this synthesis has been achieved by a cancellation of time. The phylogenetic time of palaeontology becomes the 'genetic code', or the 'genetic programme'; and individual history is reduced to being the expression of this programme. This avoids a real consideration of the effectiveness of time, by relating all historicity to a rather vague informative principle, relying on a material support. It is a way of turning the past into a piece of space in the present, of turning time into matter. The pitfall is twofold. (1) This materialisation is a cancellation (time is made space). (2) The transformation of time into space is supposed to turn its causal power into an informative one. But as the information causation in biology is not well defined, the explanations risk turning out to be purely verbal.

For Pichot, the main philosophical problem of evolutionary biology lies in this unresolved tension about the nature of time which is also an unresolved problem about the nature of causality. The historicity of evolution finds the specificity of evolutionary biology, but it is precisely this historicity that evolutionary biology refuses to consider, choosing instead to take the explanations of the non-historical sciences as a model.

4.3 A New Consideration of Time in Biology?

However, recent decades have witnessed a renewed interest in questions of temporality in evolutionary biology. In the philosophy of biology, more and more researchers are proposing to study life through an approach of 'process-biology', a conception in which biological causality is considered time dependent. These studies have been the subject of a recent synthesis (Nicholson and Dupré 2018), but the reflection on time remains rather classical: the authors conceive a linear time, rather close to a mere succession despite the attempt to take into account continuity (as a potential solution, some have suggested revisiting Bergson's ideas, see: Meincke 2021; Tahar 2023).

Despite its limitations, this approach nonetheless reveals a growing emphasis on the importance of time-related issues in evolutionary biology, and especially on the way in which this temporality invites us to reconsider the causality of evolution. Symbiosis, natural selection, gene transfer, posthumous phenotype... all these phenomena require us to study the interaction of different time scales. This is why some researchers have attempted to consider biological complexity not only causally but also temporally. This is the case of the Reticulate evolution studies which challenge the linear dimension of time and encourage us to think of complexity in a temporal way (Gontier 2016).

In these innovative approaches, time is no longer treated as a mere variable. But it is not reduced to a vague notion, which would explain the unpredictability of evolution, without it being possible to elucidate the processes underlying this unpredictability either. Instead, it is engaged in its intricate structure, moving beyond a 'merely-metaphysical label' treatment to become an operative concept for thinking about biological processes. This leads to a pluralisation of the concept of time: there is the historicity of evolution, which implies irreversibility and path dependency (in a very different sense than in physics, see Longo 2018). But there is also the problem of the entanglement of processes that take place on different timescales, particularly in the case of niche construction (Pocheville 2019). Furthermore, this approach necessitates a consideration of the diverse biological rhythms and their interplay with physical time, offering insights into their coordination (Longo 2021; see also Vaughan 2012 for a 'Bergsonian' perspective).

The subsequent sections of my chapter aim to build upon these theoretical advancements. This includes an analysis of the interrelation between the historicity of evolution and biological temporalities. Additionally, I will seek to understand the implications of these temporal problems for our comprehension of the causality of evolution and, therefore, for our ways of producing explanations.

5 The Creative Historicity of Evolution and the Temporal Approach to Agency

The Darwinian theory of evolution relies on the principle that variability is inherent in living organisms, both amongst individuals and across generations. Therefore, what requires an explanation is the stability or preservation of species. This means that the object of evolutionary biology is not stable: contrary to what happens in physics, the relevant observables are not invariants. Even seemingly constant biological processes, like meiosis or mitosis, are not timeless absolutes; *they are historical phenomena, having emerged at specific junctures in evolutionary history.* And nothing guarantees that these processes will last forever, even if they strongly constrain biological evolution today. Even in the historical parts of other natural sciences, history can be traced through the consistent application of mostly invariant processes over vast stretches of time (the changes are changes of values within pregiven dimensions, see Longo 2018). Biology, however, grapples with a history where processes themselves are in flux. During the course of evolution, the processes transform: they do not follow immutable types or laws but change over time. Evolutionary change is not merely a transformation of forms under fixed laws; it involves a fundamental alteration of the very 'laws' (if they can still be termed as such) governing evolution. The consequence is that time does not have the same meaning for the evolutionist as it does for the physicist: this continuous transformation of biological entities *and processes* introduces a degree of unpredictability and novelty that is unparalleled in the physical sciences.

5.1 The Historicity of Evolution: Beyond Uniformitarianism

5.1.1 Evolution: A Historical Time Marked by Rare Events

Because in the course of evolution, causal relationships themselves are transformed, each moment of time is not equivalent to another, and causality cannot be considered as uniform. Biological history is made of ruptures, contingency, and radical changes in possibility spaces. Admittedly, Darwin's uniformitarianism has made it possible to highlight the constancy of an adaptation imperative (organisms that are poorly adapted to their environment are less likely to produce offspring), which makes it possible now, as it did in the past, to explain the appearance of adaptive traits. However, this view does not account for the diversity of life forms observed and the form taken by biodiversity, nor does it predict the evolutionary paths of lineages. It may be because uniformitarianism is less an actual property of evolutionary time than a heuristic research principle, useful for understanding adaptation but potentially misleading in accounting for the diversity of biological forms.

The vast diversity in the biological world can be attributed not only to the variability of environmental conditions but also to the ever-changing biological processes through which organisms strive for existence. Because of this perpetual change, each biological event is individually *rare*, and this rarity impedes predictability (Longo 2018). This is visible, for instance, when a previously useless cryptic genetic variation begins to generate heritable phenotypic variation in a typical ecological situation. It is also the case when the environment experiences an exceptional situation (e.g. a drought that lasts only one summer), leading to the exceptional extinction of a population that was previously perfectly adapted to its environment (for a theoretical example of this, see Gould 2002, 665–666). In these cases, the event is rare in the sense that it is retrospectively improbable: it presupposes the convergence of different factors and manifests a unique complexity, itself resulting from the intersection of several histories (of the cell, the environment, the organisms, etc.). And it *disrupts* the typical sequence of causal events. The evolutionary implications of these rare events reveal that evolutionary causality is not unilinear.

In the evolutionary narrative, mathematical rarity intertwines with rarity, in the classical sense, characterised by infrequency. Certain infrequent events, though disconnected from the selective causal chains (the organisms and their environment), can exert profound evolutionary effects. A notable example is natural disasters, whose evolutionary importance has been studied and defended by David Raup (Raup 1991). Raup's hypothesis is that a certain number of species only became extinct independently of their adaptation, as a result of catastrophic events. These events would have caused cascading chains, resulting in a significant change in biodiversity, inexplicable by natural selection alone, inexplicable even by the interactions of organisms with each other and with their environment. Raup takes the example of tropical reefs: after each major extinction, species very different from those that went extinct repopulated the area. This difference comes from a time rupture. Even if the abiotic environment did not undergo much change, the rare event disrupted the evolutionary trajectories of the species that populated that environment, and these trajectories are not reproducible. This supports the idea that uniformitarianism alone is insufficient for understanding evolutionary time, as present causes may not be equivalent to those in the past. Without subscribing to Raup's radical position that catastrophes are much more central to evolution than natural selection, natural disasters do reveal that evolution cannot be predicted. And this unpredictability does not come from our ignorance but may be a property of evolutionary time itself. In the time of evolution, some events redefine the biological parameters and processes, and thus no invariants can be found. If biologists cannot produce reliable probabilistic forecasts, it is not due to the limitations of our science, but because 'the biosphere [...] in its persistent evolution, is doing something literally incalculable, nonalgorithmic, and outside our capacity to predict, not due to quantum uncertainty alone, nor deterministic chaos alone, but for a different, equally, or more profound reason: Emergence and persistent creativity [...] is real' (Kauffman 2000, x).

5.1.2 A Historical Causality: Constraints and Creativity

However, we do observe regularities. Meiosis and mitosis may not be eternal and invariable laws—they are indeed historically constituted processes—yet, today, they are processes that constrain future evolution. Additionally, evolution does not go in all directions: the evolutionary path of a species is linear and irreversible; convergent evolution results in similar structures in different lineages; and certain phenotypes appear inaccessible to some species (Fodor 2007). To account for these regularities, some have proposed to talk about *biological constraints* (Gould 2002; Montévil and Mossio 2015). This concept aims to replace the idea of necessitating efficient causes. In biology, there are fewer causes entailing their effects according to invariable laws than *constraints* enabling a channelling of processes at the same time as they make their transformation possible (Longo et al. 2012). These constraints are historically constituted and thus manifest the retention of the past in the present, but they are also transitory: they change under the effect of the passage of time: the future leads to their potential abolition (see Tahar 2022 for further analysis).

In evolution, processes evolve, observables change over time, and even constraints are historically constituted. Therefore, contrary to what happens in physics, history is not an exploration of a space of possibilities. Instead, historical time generates the perpetual change of the very space of possibilities, in ways that cannot be predicted (Longo et al. 2012; Kauffman, Roli 2023).

5.2 Evolutionary Creativity and Time-Based Approach to Agency

The creativity of the evolutionary causality comes from the following two (and correlated) properties of the evolutionary time. One is based on the Darwinian principle of individual variability but was not made explicit in Darwin's writings. (1) The irreversibility of evolutionary history results in the particularity of each biological situation: each situation is unique because an organism, like the environmental conditions in which it evolves, can never reproduce identically. The second also pervades Darwin's writing without being clearly conceptualised: (2) the irreversibility of evolutionary history has a creative dimension, made possible by the retention of the past in the present. This retention is concretely actualised in the *memory* of the living: the phylogenetic memory of genetic inheritance, but also the individual memory of the organisms, registered both in their motor mechanisms and in their cognitive ones.

This individual recording is processed differently from one organism to another: it is processed at different *rates*. This diversity in processing rates confers a distinct temporality to each organism, underpinning the originality and unpredictability of its actions. To understand this individual temporality more clearly, we can consider it in terms of three distinct but interconnected characteristics. There is the (i) *longevity of the organism*, which may be more or less out of step with the speed of change in the environment. The relationship between the longevity of individuals and the speed of environmental change plays a role in evolutionary processes. Let's imagine a population of individuals whose lifespans are short relative to the rate of environmental change. It is likely that these organisms will not have time to adapt over their lifetimes (individuals will have low adaptive capacity). However the population will evolve rapidly over the generations, with each new generation providing an opportunity for selection (the population will exhibit high evolvability). There are also (ii) *biological rhythms* (like the circadian or metabolic ones), related to longevity. These rhythms contribute to adjusting the organism to physical frequencies but are not identical to those frequencies (Longo 2021). The importance of these biological rhythms becomes especially evident when they are disrupted, as seen in cases where pollinators miss their timing with flowering plants, leading to ecological mismatches (Memmott et al. 2007). Finally, these different rhythms lead organisms to different (iii) ways of remembering the past and anticipating the future (*cognitive or psychological rhythm*). The different psychological rhythms ultimately lead to varied behaviours.

These rhythms, forming the organism's unique temporality, intertwine with physical time whilst maintaining their distinct characteristics. Their tuning with physical time carries significant ecological and evolutionary implications, influencing how organisms uniquely interact with their environments. Thus, organisms, by virtue of their temporal specificity, actively contribute to the unpredictability of evolutionary processes. In this sense, the very temporality of organisms contributes to the historicity of evolution.

Therefore, the plurality of rhythms invites us to consider the agency of organisms in terms of temporality. Darwin, without using the word agency, implicitly acknowledged it by positioning natural selection within the struggle for existence, thereby ascribing an active role to organisms in the process of selection. This perspective hinges on the idea that for a variation to be selected, it must prove useful for the organism in its interactions with the environment.

Recently, the role of organisms in evolution has received renewed attention (Corning 2014, Diogo 2017), especially through the concept of ecological agency (Walsh 2015). The main idea of this new research is that organisms *experience* the conditions of the environment as *affordances*, to which organisms respond in behaviours that both reflect the meaning the environmental conditions entail for them and potentially transform these conditions. Such an approach leads to reassess some evolutionary processes, such as the Baldwin effect (Baldwin 1896) to highlight the contribution of organisms on an evolutionary scale.

However, I suggest this conception would gain more concrete grounding if supported by a temporal approach. Indeed, a real study of the *plurality of biological rhythms* could help us understand how organisms experience environmental conditions and actualise them in unpredictable behaviours. This would pave the way for a more comprehensive approach not only to the active role played by organisms in evolution but to the link between short-term time-based organisational dynamics and long-term historical evolutionary processes.

But this understanding of evolutionary temporalities also has other implications: it reveals the specificity of biology as a historical science. Moving beyond the paradigm of the turn of the twentieth century, which predominantly viewed experimental science as the sole model for scientific inquiry, the historicity of evolution challenges us to develop new models of scientific understanding and explanation, capable of integrating the causal specificities of these temporal processes.

6 Nature of Time and Nature of Explanations

The implication of the above is that at least some evolutionary explanations must be historical and have a narrative dimension. This does not exclude the need for 'processual' explanations, i.e. explanations of how a process or mechanism (e.g. natural selection) produces a particular effect. But in evolutionary biology, these explanations need to be complemented by historical explanations whose role is not only to describe the context of an explanatory mechanism or process but to contribute to the explanation itself.

6.1 Historical Explanations

In evolutionary biology, several reasons require complementing the 'traditional' processual explanations of the nomological sciences with historical ones:

(1) Because evolution is characterised by historicity, there are no invariable laws governing efficient causes, but rather historically situated constraints. Therefore, no predictions can be made: *evolutionary explanations are mostly retrospective* and presuppose knowledge of the past (at least of phylogeny).
(2) Constraints in evolution are not monolithic but derive from a plurality of processes. Natural selection, often considered the main explanatory principle

of evolutionary phenomena, is not the sole causal process at work in evolution. Other processes, such as mutation, genetic drift, and symbiosis, also play significant roles. This requires an explanatory model capable of supporting a diversity of processes. One of the objectives of historical explanation in evolutionary biology is to *show how these various processes are articulated* to constitute constraints for the phenomenon under study.

(3) Moreover, due to the idiosyncratic nature of biological situations, these processes never generate the same effects. Thus, in evolutionary biology, there are queries that cannot be resolved by recourse to a general law, as they involve the analysis of *tokens* rather than *types* of events (Graham 1983; Tucker 2009). This is the case for inquiries such as: why does biodiversity have this configuration and not another? Or: why did this trait appear in this lineage? Or: how can we understand the speciation of this group? Hence the need for a historical explanation. Historical explanations can *account for the generation of the circumstances that enabled the occurrence of the unique phenomena under study.*

(4) In biology, each situation is a highly unique and improbable case since the space of possibilities is not predetermined but constantly changing. Therefore, *evolutionary explanations must demonstrate how a rare event occurred despite its improbability* (they should comprehend how-possible explanations, Dray 1968).

(5) Finally, because of the significance of organisms' temporal agency in evolution, *evolutionary explanations should be able to say something about the experience of organisms*, as this unique experience partially determines the evolutionary significance of the biological constraints. Without reference to this experience, part of the understanding may be lost since the rarity of evolutionary events often stems from the incongruity between the behaviour produced by organisms and the given conditions. Such an understanding of the organism's experience requires a semantic dimension that only historical explanations can support.

Therefore, historical explanations in biology are necessary, because (1) they are retrospective explanations, (2) which integrate multiple processes into a single coherent narrative. (3) They can trace the evolutionary trajectory of the case under study, (4) identify the decisive factors that made the improbable trajectory effective, and (5) formulate hypotheses about the meaning of the events and constraints for the agents who experienced them.

However, time in biology is not *only* historical. If evolution is the history of *rare* events opening unpredictable possibilities, biology is also law-based.

6.2 The Complementarity of Processual and Historical Explanations

The challenge is to think about how historical and processual explanations complement each other, as, at all stages of the historical explanation, nomological causality applies. According to Reydon (2021), a historical explanation accounts for the conditions that made a particular evolutionary trajectory possible (by specifying the properties of the organism, the population structure, and environmental factors), whilst a processual explanation explains how the trajectory actually occurred (specifying processes such as selection and drift). The explanatory force would thus be found in both the processual and the historical explanation.

However, given the historicity of evolution and the temporal agency of organisms, it is arguably more accurate to suggest that historical explanation goes beyond merely specifying the conditions necessary for processes to occur and may instead have explanatory precedence.

The first reason is methodological. Explaining how an event occurred in evolution requires taking into account clues from other sciences: genetics, molecular biology, but also geology, palaeontology, and so on. This necessitates the organisation of data across different systems of understanding, a task achievable only through the creation of empirically and rationally based historical narratives.

Secondly, in evolutionary biology, it is crucial to understand the articulation of the different processes in unique situations. And only a historical narrative can explain why processes interact in the way they do at a given time because the explanation must hold together different histories, characterised by different rhythms, and which are played out on different time scales. Only historical explanations can weave these stories together into coherent narratives revealing the articulation of processes.

Finally, and even more decisively, the active role of organisms in shaping their own evolutionary trajectories is not simply a contextual element that makes a process possible. And only a historical explanation can take agents into account because only a historical explanation can integrate an account of their own experience, i.e. hypothesise on the meaning biological events or constraints have for these agents, with regard to their individual history. This leads me to argue that, if the coordination of both types of explanation (processual and historical) is necessary in evolutionary biology, it is indeed history that has the explanatory priority, at least to answer some strictly evolutionary questions.

In any case, just as the historicity of evolution and the temporalities of biological processes underpin the specificity of biological phenomena, such a reflection on the complementarity of historical and processual explanations underpins the specificity of biology in relation to the other natural sciences.

7 Conclusion

The distinction between biological and physical time, as well as the historical nature of biological causality, are frequently acknowledged but rarely comprehensively explained. This chapter has sought to unravel the complexities surrounding the concept of time in evolutionary biology. It began with an analysis of the ambiguous conception of history in Darwin's works and the subsequent impact on the conception of causality. My aim was to shed light on how Darwin's lack of precision led to contrasting receptions of Darwinism and contradictory interpretations of causality in evolutionary biology. Moreover, I have highlighted how these uncertainties regarding the historicity of evolution, the temporality of biological processes, and their relationship to evolutionary causality have persisted from the publication of *The Origin of Species* to the present day and continue to influence contemporary epistemological issues. In the final two sections, this chapter brought together philosophical and biological perspectives to clarify these concepts and explore their methodological implications. This analysis has revealed the relationship between the temporality of biological processes and entities and the historicity of evolution, which remained implicit and confused in Darwin, and in most evolutionary biologists to this day. It has also shown the importance of revising our concept of explanation in evolutionary biology, calling for an epistemological and methodological reform inspired by the historical approach.

Acknowledgements I want to thank the members of PhilLiSci at Bielefeld University for their valuable feedback. I am also deeply grateful to Emily Herring who helped to significantly improve the quality of the text.

References

Baldwin JM (1896) A new factor in evolution. Am Nat 30(441–451):536–553

Baquero F (2005) Evolution and the nature of time. Int Microbiol Official Journal of the Spanish Society for Microbiology 8(2):81–91. Format

Bergson H (1911) Creative evolution (trans. Mitchell A). Macmillan, London

Canguilhem G (2003) Le vivant et son milieu. In: La Connaissance de la vie. Vrin, Paris, pp 165–197

Corning PA (2014) Evolution 'on purpose': how behaviour has shaped the evolutionary process. Biol J Linn Soc 112:242–260

Darwin C (1859) On the origin of species by means of natural selection, or the preservation of favoured races in the struggle for life. John Murray, London

Darwin C (1909) The foundations of the origin of species. Two Essays written in 1842 and 1844 (ed. Darwin F). Cambridge University Press, Cambridge UK

Darwin C (2016) The various contrivances by which orchids are fertilised by insects. In: Barrett PH, Freeman RB (eds) The works of Charles Darwin, vol 17. Routledge, Abingdon-on-Thames

Dawkins R (1996) The blind watchmaker: why the evidence of evolution reveals a universe without design. W. W. Norton & Company Inc, New York

Dennett DC (1995) Darwin's dangerous idea. Evolution and the meanings of life. Penguin Books, London

Desjardin E (2011) Reflections on path dependence and irreversibility: lessons from evolutionary biology. Philos Sci 78(5):724–738

Diogo R (2017) Evolution driven by organismal behavior: a unifying view of life, function, form, mismatches and trends. Springer, Heidelberg/New York/Dordrecht

Dobzhansky T (1960) Evolution and environment. In: Tax S (ed) Evolution after Darwin, vol 3. The University of Chicago Press, Chicago, pp 403–428

Dobzhansky T (1974) Chance and creativity in evolution. In: Ayala FJ, Dobzhansky T (eds) Studies in the philosophy of biology. University of California Press, Berkeley, pp 307–338

Dobzhansky T, Boesiger E (1968) Essais sur l'Évolution. Masson & Cie, Paris

Dray WH (1968) On explaining how-possibly. Monist 52:290–407

Fisher RA (1950) Creative aspects of natural law. The Eddington Memorial Lecture. Cambridge University Press, Cambridge UK, pp 179–184

Flourens P (1864) Examen du livre de M. Darwin sur l'origine des espèces. Garnier Frères, Paris
Fodor J (2007) Why pigs don't have wings. London Review of Books 29(20):19–22
Giard A (1904) Controverses transformistes. C. Naud, Paris
Gontier N (2016) Time: the biggest pattern in natural history research. Evol Biol 4(43):604–637
Gould SJ (1970) Dollo on Dollo's Law: irreversibility and the status of evolutionary laws. J Hist Biol 3(2):189–212
Gould SJ (1988) Time's arrow, time's cycle: myth and metaphor in the discovery of geological time. Harvard University Press, Cambridge US
Gould SJ (2002) The structure of evolutionary theory. Harvard University Press, Cambridge US
Graham G (1983) Historical explanation reconsidered. Aberdeen University Press, Aberdeen
Grassé PP (1973) L'Évolution du vivant. Albin Michel, Paris
Huxley J (1942) Evolution: the modern synthesis. George Allen and Unwin Ltd, London
Kauffman S (2000) Investigations. Oxford University Press, Oxford
Kauffman S, Roli A (2023) A third transition in science? Interface Focus 13:20220063
Lamarck JB (1873) Philosophie zoologique (preceded by a biographical introduction by Martins C). Librairie F. Savy, Paris
Le Dantec F (1902) Lamarck. La Revue blanche 29:356–377
Loison L (2010) French roots of French Neo-Lamarckisms, 1879–1985. J Hist Biol 44(4):713–744
Loison L (2022) Néolamarckisme et matérialisme. Revue d'histoire des sciences 65(1):61–79
Longo G (2018) How future depends on past and rare events in systems of life. Found Sci 23(3): 443–474
Longo G (2021) Confusing biological rhythms and physical clocks. Today's ecological relevance of Bergson-Einstein debate on time. In: Campo A, Gozzano S (eds) Einstein vs Bergson. An enduring quarrel of time. De Gruyter, Berlin
Longo G, Montévil M, Kauffman S (2012) No entailing laws, but enablement in the evolution of the biosphere. GEECO'12: Proceedings of the 14th annual conference companion on Genetics and evolutionary computation: 1379–1392
Meincke AS (2021) Bergson and process philosophy of biology. In: Wolf Y, Sinclair M (eds) The Bergsonian mind. Routledge, Abington-on-Thames, pp 432–445
Memmott J, Craze PG, Waser NM, Price MV (2007) Global warming and the disruption of plant–pollinator interactions. Ecol Lett 10(8):710–717
Montévil M (2020) Historicity at the heart of biology. Theory Biosci 141(2):165–173
Montévil M, Mossio M (2015) Biological organisation as closure of constraints. J Theor Biol 372: 179–191
Nicholson DJ, Dupré J (eds) (2018) Everything flows: towards a processual philosophy of biology. Oxford University Press, Oxford
Perrier E (1879) Le transformisme et les sciences physiques. Revue scientifique XVII:890–895
Pichot A (1987) L'étrange objet de la biologie, Biologie, physico-chimie et histoire. Fundamenta Scientae 8(1)
Pocheville A (2019) A Darwinian dream: on time, levels, and processes in evolution. In: Uller T, Laland KN (eds) Evolutionary causation. Biological and philosophical reflections. The MIT Press, Cambridge US
Prigogine I, Stengers I (1984) Order out of chaos. Man's new dialogue with nature. Bantam Books, New York
Quatrefages A (1870) Charles Darwin et ses précurseurs français. Baillière, Paris
Raup D (1991) Extinction: bad genes or bad luck. WW Norton, New York
Reydon TAC (2021) The proper role of history in evolutionary explanations. Noûs 57(1):162–187. in press
Soto A, Longo G, Montévil M, Sonnenschein C (2016) The biological default state of cell proliferation with variation and motility, a fundamental principle for a theory of organisms. Prog Biophys Mol Biol 122(1):16–23
Sterelny K (2001) Dawkins vs. Gould: survival of the fittest. Icon Books, Cambridge UK

Szathmáry E (2006) Path dependence and historical contingency in biology. In: Wimmer A, Kössler R (eds) Understanding change: models, methodologies, and metaphors. Macmillan, New York, pp 140–157

Tahar M (2022) Biological constraints as norms in evolution. Hist Philos Life Sci 44(1):9

Tahar M (2023) A Bergsonian perspective on causality and evolution. In: Bianco G, Wolfe CT, Van de Vijver G (eds) Canguilhem and continental philosophy of biology. History, philosophy and theory of the life sciences, vol 31. Springer, Heidelberg/New York/Dordrecht, pp 251–267

Tucker A (2009) Our knowledge of the past. A philosophy of historiography. Cambridge University Press, Cambridge UK

Vandel A (1949) L'Homme et l'Évolution. Gallimard, Paris

Vaughan M (2012) The time of biology and the time of life: Bergson and chronobiology. In: Abiko S, Fujita H, Sugiyama N (eds) Disséminations de L'Évolution créatrice de Bergson. Hildesheim, Olms, pp 224–230

Walsh DM (2015) Organisms, agency, and evolution. Cambridge University Press, Cambridge UK

Wright S (1964) Biology and the philosophy of science. Monist 48(2):265–290

Shaking the Tree: Discussing an Evolutionary Icon

Erica Torrens Rojas, Juan Manuel Rodríguez Caso, and Ana Barahona

Abstract

The history of Darwinism has been reconstructed and interpreted by historians, philosophers, and biologists. This could be seen as a pluralistic view of history. However, in the field of biology, a version of Darwinism privileged from an anachronistic perspective, mainly based on works of authors such as Ernst Mayr, Stephen Jay Gould, and Richard Dawkins has been defended. Historically, this tradition has emphasised the "scientific" elements of Charles Darwin's proposal to the detriment of an authentic understanding of these ideas' development in the context of Victorian England. This vision has resulted in the philosophical and social implications of Victorian science being uncritically taken up by biologists in their continued defence of Darwin's role in contemporary biology.

Against this background, in recent years there has been an increased effort to revise the grand narratives that have guided the study of the history of evolutionary biology and the resulting historiographies. In this regard, our contribution focuses on the visual metaphor of the "tree" of life. It is in our interest to study the ontological and epistemological commitments that systematists have with this image, given the recent discussion on whether this figure is indeed the best representation of the evolutionary process.

In this essay we turn our attention, not to the current scientific, historical, and philosophical debates about the Tree of Life, but to the tree as a concept and tradition for having entailed distinct epistemic and ontological commitments in the different periodisations of evolutionary biology, such as the Darwinian

E. T. Rojas (✉) · J. M. R. Caso · A. Barahona
Grupo de Estudios Sociales de la Ciencia y la Tecnología, Facultad de Ciencias, Universidad Nacional Autónoma de México, Mexico City, Mexico
e-mail: torrens@ciencias.unam.mx; juan_manuel_rodriguez_caso@ciencias.unam.mx; ana.barahona@ciencias.unam.mx

R. G. Delisle et al. (eds.), *Unity and Disunity in Evolutionary Biology*,
https://doi.org/10.1007/978-3-031-42629-2_20

Revolution, the Modern Synthesis, and novel proposals, that need to be examined.

Keywords

Periodisations · Anachronistic · Postsynthetic · Institutionalisations · Professionalisation

1 Introduction

One of the recurring discussions when talking about Charles Darwin and his ideas is his relationship with religion. Beyond the conflict that has been widely discussed, there is a little-accepted proximity: the one that leads us to recognise the appropriation of a religious—or quasi-religious—discourse within the study of evolution. According to Michael Ruse (2017), it means that "evolution tries to speak to the nature of humans and their place in the scheme of things, [so] we have a religion, or if you want to speak a little more cautiously a 'secular religious perspective'.

This idea reflects a recurrent attitude in a scientific community and any social group, which is the cohesion resulting from sharing a series of ideas or values. To some extent, an anthropological approach, such as that set out in the classic *Laboratory Life* (1979) by Bruno Latour and Steve Woolgar, allows us to appreciate that a group of biologists needs to have something that unifies them, in the same way, that a group of individuals has a better chance of survival to the extent that it sticks together. In many cases, that something has been Darwin. Beyond a reductionist view of these ideas or values for survival, there is the idea of recognising the importance of defending—sometimes dogmatically—an idea that, in principle, has a high explanatory value.

A set of ideas can result in a tradition, giving a concrete form of explanation to the extent that it can be confirmed through scientific practice, and its relevance to the group can be reaffirmed. In the case of Darwin's ideas, the "Darwinian revolution" has served to maintain and privilege one view of life over other alternatives. Darwinism is more than a theory. It is a way of seeing the world, which has Darwin at the centre of its explanation and from which it derives a history supported by a religious language. Thus, the step towards a dogmatic defence of ideas is relatively short.

The case of the "tree of life" shows how a "tradition", in this case, based on the "Darwinian revolution", can become the norm of a narrative. However, it is important to remember there are different historiographical proposals that have emerged throughout the twentieth century, which we consider crucial for a necessary re-signification of biology. We can speak of an "adequate" narrative to the extent that it consolidates and imposes itself on other proposals, for the most diverse reasons; however, this hardly implies that this "adequate" narrative cannot be criticised, or better still, deconstructed.

In the aftermath of WWII, not only were cultural and social processes reconfigured, but changes also occurred in the area of science itself, where the global circulation of scientific instruments, workforce, and ideas had become increasingly focused. Although some authors have highlighted the absence of the historian of science in this discussion (Turchetti et al. 2012), a contribution that had a tremendous impact on the way the history of science is made and on its increasingly marked split from Science and Technology Studies (STS), was *The Structure of Scientific Revolutions* published in 1962 by Thomas Kuhn. According to Wilson (2017), Kuhn was the first to look closely at the way the history of science had been told until the 1960s when in Chap. 11 of his book—a chapter particularly indebted to Ludwik Fleck, a Polish physician, biologist and sociologist—developed the idea that science constructs a history for itself, a history that provides a picture of continuous cumulative development, thus distorting the scientists of earlier times as having worked on the same set of fixed problems and according to the same set of fixed canons that hindsight has made to appear scientific.

This idea aroused much interest in reviewing how the past scientific disciplines had been written about, especially because epistemic universalism "(which understands science as a transcendent, truth-finding activity that, in principle, should not be affected by national, class or ethnic difference)" was the norm (Turchetti et al. 2012: 323).

According to Daston (2009), the reaction from the History of Science to Kuhn's call for attention was to be more explicit on the reappropriation of methods and perspectives of history, and that teleology, which is the notion of constant progress towards some end, should be repudiated as a narrative form in the history of science. One implication was to strive not to be anachronistic and to abandon "the language of winners and losers" (such as Lavoisier vs Priestley or Darwin vs Lamarck) (p. 802) that had dominated narratives in the history of science of the Anglo-Saxon world since the 1920s and instead strives to understand the science of the past on its own terms.

On the other hand, the STS ("a battery of disciplinary perspectives turned upon science and technology: first and foremost sociology, but also anthropology, political science, philosophy, gender studies, and history" (p. 800), focused on the idea that the content of science and technology (scientific facts, technologies, objects, actors) is open to social analysis and therefore, that in order to understand the nature of science, the sociocultural context, and the interaction between science and society must be taken into account. This has made the STS an interdisciplinary field, but also, in Daston's words, an "undisciplined" one (p. 811).

On the subject of Darwin and "Darwinism", according to Richard Delisle—one of the main spokesmen on the need to review the historiographic productions related to the interpretations that have been derived from Darwin's work—evolutionary biology has been an extraordinarily dynamic area since its foundation, but its true complexity has been hidden in the last 60 years under a narrative that is assumed to be orthodox about Darwinism, dominated mainly by English-speaking authors and categories such as Darwinian Revolution, Eclipse of Darwinism, Evolutionary Synthesis, and Postsynthetic Developments. Recently, the goal for the historians of science and STSers is the study of contributions and characters that are crucial to a more rigorous historical reconstruction, such as a broad understanding of the Victorian period—which includes processes such as disciplinary institutionalisations and the recognition of proposals considered to be "alternative" to Darwin's, just to mention a few examples.

The account presented here assesses the need to broaden the accounts of the history of biology. The role that can be given to Darwin and his ideas is part of the debate between historians and biologists, and without minimising their importance, biology has a much broader and more diverse history that deserves to be acknowledged. Now, discussing explanations in science is not only a conceptual issue but also a question of images. Possibly the most potent image in evolutionary biology is the "Tree of Life", which is at the heart of Darwinism. On the one hand, this image defends an explanatory logic that reaffirms Darwin, natural selection, descent with modification, and gradualism as the central concepts of evolutionary explanation. On the other hand, it is one of the most effective ways to convey ideas such as common ancestry, speciation, nested hierarchies, and divergence, contributing to the spread of a particular view of evolution associated with Darwin's ideas.

To speak of deconstructing a concept, or a theory, does not necessarily imply a pejorative critique. On the contrary, scientific work should be nourished by discussions and reflections that allow us to evaluate and, if necessary, redirect scientific explanations. The role of the history of science should not be to legitimise—ideologically, even dogmatically—a particular theory, but to generate explanations that even serve to improve the understanding of nature.

This chapter is organised as follows: first, a discussion of the role of anachronism, particularly presentism, in the history of evolutionary biology. This is followed by an account of the history of the "tree of life" concerning the history of systematics and the development of modern phylogenies. Finally, we return to the need for the history of evolutionary biology to be brought closer to the new historiographical currents.

2 The Role of Anachronisms in the History of Evolutionary Biology

Perhaps no other figure in the history of science is subject to such diverse interpretations as Charles Darwin. Although it may not be obvious, it is essential to know who writes—and reconstructs—history. The account provided by a scientist—such as a biologist in the case of Darwin and his ideas—a philosopher, a sociologist, an anthropologist, or a historian, is not the same, despite their focus on the same subject. Stephen Brush (1995) wonders whether scientists should write about the history of science, but it is clear to him that despite even sharing knowledge or skills, they have different ways of writing the history of science. However, is it so different from what everyone can say from their point of view?

We can take up here the "styles of reasoning" proposed by Ian Hacking or Michel Foucault's "episteme" to establish the different frameworks from which individuals seek to explain the world, or in this case, the history of science. For example, Hacking takes up a classic dichotomy: the philosopher and the historian. He accepts that each style can be complementary yet asymmetrical: "The historian may conclude that the philosopher's use of the tool is bunk, irrelevant to understanding the past. Nevertheless, the philosopher needs the history, for if the tool does not provide a coherent and enlightening ordering of the record, then it has no more place in sound philosophy than any other fantasy" (Hacking 1992, 1994). On the other hand, Foucault, in *The Order of Things*, speaks of episteme as, in general terms, a way of thinking, which we can place in the different ways of obtaining knowledge, which in turn allows the development of ideas and concepts that at a given time can be considered as valid (Foucault 2005). The above leads us to consider each academic community's particularities when obtaining or generating knowledge. Whether a philosopher, a scientist, or a historian, we can appreciate that even about the same historical facts, sometimes contrary interpretations can be given.

In this section, we want to reflect on the continued presence of presentism in the historical discourse, especially in defence of a particular vision of Charles Darwin and his ideas. In any case, it is crucial to recognise that reconstructing the past is complex. Ultimately, we can all be presentists insofar as we see the past from our present temporal position.

Now, here we want to start from the understanding that any concept, or an image as in our case, has a meaning or an intention. Moreover, each has a relationship with the context that arises and develops. When people talk about Charles Darwin, concepts such as natural selection, or images such as the tree of life, it is often mentioned that their context is nineteenth-century Britain. However, do we know what that refers to? It is now commonplace for Darwin to be the focus of any discussion or course on evolutionary biology. It is insisted that "his" idea of natural selection changed our understanding of the natural world. Moreover, this seems to be intended as a defence of the importance of a particular theory within contemporary biology.

As discussed in other chapters in this volume, the history of evolutionary biology presents a sharp contrast between clearly presentist views and others that appeal to a more contextual view. One of the most influential proponents of presentism in biology is Ernst Mayr. Its impact on how Darwin is understood today is important to recognise, although it does not limit us in terms of criticism. For example, in one of his papers in which he explicitly discusses historiographical issues, he mentions:

> With Ruse I conclude that it is by no means wrong to look at the past on the basis of an understanding of the present… To be sure, the historian must avoid the well known faults of bias, chauvinism, falsifications of priority, and finalistic interpretations, but this is true for any kind of historiography, developmental or not. On the other hand, selectivity is a necessity in developmental historiography. Also, the historian must be permitted to make evaluations when writing intellectual history, as Lovejoy has demonstrated so beautifully. A history that does not evaluate but merely records facts and presents documents is anti-intellectual – it is prig-history, as Harrison has called it. (Mayr 1990: 308–309)

Within the group of philosophers of biology that we can call "classical", one of the most influential proponents of presentism was David Hull, who starts by defending the intimate relationship between history and the philosophy of science. In his opinion, the academic division does not contribute to the understanding of science, and beyond the limits that disciplines can impose, there is the defence of free inquiry. His defence of presentism starts from considering three types of "relativity": meaning, reasoning, and truth. To some extent, Hull shows by this that knowledge is in permanent motion, which shows that it is not a simple task to understand the past in its terms. However, Hull states that "[e]ven the most compulsive historian is likely to know more about his age than the one he is studying", so it seems inevitable that any interpretation or historical reconstruction will end up being done from the present (Hull 1979: 15).

It is clear that, as we will see below, there is a significant recognition of the role and place of science in today's society. Especially in the context of the continuing questioning of science, there seem to be calls to maintain the importance of science by linking it to its past, but above all. It seeks to emphasise the importance of a discourse in which presentism can be understood in different ways or at least has the possibility of connecting us with our audiences, which are those that exist today:

1. "Substantive presentism may be defined as the belief that the present is the key to the past, in the literal sense that the present is substantively like the past
2. [. . .] methodological presentism. Following the geologists, a methodological presentist would argue that if we wish to understand past events, we should study similar recent or current ones
3. I call this form of presentism "motivational presentism" and argue that it is a good thing because hiding our motivations hamstrings us intellectually and stylistically, isolates us from potential audiences, and undermines our ability to speak persuasively about the value of our work" (Oreskes 2013: 600–601).

As a historian, Oreskes reflects on how Whig history—in Herbert Butterfield's sense—leads to the assumption that the present is better than the past. Furthermore, whilst historians may concede that the present may be "better" than the past, history is not a linear, progressive account. On the contrary, history is much more complex. Moreover, this opens the option to consider that one can be a "good presentist". Oreskes says, provocatively: "What matters to us about the past has everything to do with who we are, where we live, and what we think is important – to us, here and now, in the present. Our motivations are inescapably presentist" (Oreskes 2013: 603). The ultimate goal is to connect the past with the present, as a way to elaborate an authentically historical approach, in the sense of being the historians who defend and sustain a vision of the past that justifies the present.

Presentism presents challenges, such as whether it is possible to conceive of a good way of being presentist or whether presentism in all its forms is a historical misinterpretation. One historiographical approach that, according to Laurent Loison, can be useful in explaining both the contingency and rationality of science is a form of historical epistemology, based above all on work developed in the French tradition, with authors such as Gaston Bachelard, Georges Canguilhem, and Michel Foucault. For this purpose, Loison starts with a typology in which he distinguishes four forms of presentism: empirical presentism, descriptive presentism, causal-narrative presentism, and normative presentism (Loison 2016). The first is "the use of present-day knowledge to help specify the characteristics of the empirical sub-stratum based on which past interpretations were made, regardless of the content of these interpretations" (Loison 2016: 30). The descriptive view combines comparison, transcription, and translation of an explanation of the past into terms that are understandable in the present. The narrative approach involves using a series of events to identify causal historical processes to give perspective to a particular narrative (Loison 2016: 31–32). Finally, there is normative presentism, when "the use of present-day concepts and explanations to identify and emphasise the obstacles faced by earlier concepts to understand the same class of phenomena as well as their intrinsic limitations and theoretical shortcomings" (Loison 2016: 32).

However, another reflection worth taking up again is from the philosophy of science. In particular, we take up here what Hasok Chang recently said, who also highlights the usefulness of presentism, which, beyond being inevitable, can be understood and used in different ways at the time of writing history. Chang proposes three types: whiggism, philosophical history, and emancipatory presentism. Whiggism is not a synonym for presentism but "a strong optimistic assumption of progress that conceives present science to be unequivocally better than past science" (Chang 2021). The second type emphasises how philosophers write history, that is, characterising the development and nature of scientific knowledge from historical facts that serve as evidence; however, they start from contemporary philosophical concepts when proposing their historical reconstruction. The third approach differs from the previous ones since, in Chang's words, the historian will use the past to liberate us from its consequences. In a nutshell, Chang invites us to look for a tool in presentism, "[f]or the benefit of the wider society, I think it is far more interesting and useful if we professional historians give attention to the valuable bits of past science that modern scientists consider unimportant, outdated, or simply wrong" (Chang 2021: 107).

In the specific case of the "Darwinian revolution", a call to overcome it comes from a student of Mayr's, the renowned historian of biology Jonathan Hodge, who reminds us that the history of science needs to start from specific considerations:

1. "Go to the literature on political theory and political history [. . .] and one finds no agreement about the nature, causes and consequences of revolutions
2. If a revolutionary change in science is a sudden, total and irreversible regime shift, then historical scholarship has found no convincing cases.
3. Historians in various fields are currently questioning "revolution" as an interpretative category." (Hodge 2005: 102)

In other words, one of the most critical challenges facing the history of science is coming closer to professional historians' work. Questioning categories such as "revolution"—whether scientific or Darwinian—does not go beyond recognising that historical facts are often not definitive in affirming such situations. In methodological terms, it is important to appreciate that a concept like "revolution" is widely used within contemporary scientific language, although it has been the subject of numerous discussions. One of the best-known interpretations, perhaps even beyond the academic sphere, is that proposed by Thomas Kuhn. Nevertheless, classical works such as those by Bernard Cohen (1985) and Steven Shapin (2008) have helped to redefine the idea of revolution in the history of science. Whether by questioning the role of the alleged actors of these significant changes or by looking at the details that emerge from the understanding of the social context of the time, a necessary reflection starts with how we can reconstruct the history of the past from the present.

Recently, we also have reflections such as that of James Secord (2023), who proposes an interpretation of history that forces us to rethink current historiographical approaches. If speaking of a scientific revolution, then referring to a narrative created in the twentieth century, from a context such as that of the United States in the interwar period, we consider it relevant to analyse the use given to the history of science to justify narratives—or agendas—that in principle explain contemporary science. Moreover, as Secord concludes, we can consider the Darwinian revolution as a twentieth-century narrative, which, based on nineteenth-century theoretical elements, was put forward as a form of justification—not only scientific but even political and social—for a particular view of evolutionary biology.

At the very least, it is important to recognise the difficulties surrounding criticisms of presentism, usually understood as a pejorative form of anachronism. For example, Nick Jardine (2000) reminds us of the complexity of something we can take for granted: interpretation. Sometimes there is no "bad intention" in interpreting a text from the past, but we must ask ourselves what is the right way to do it. Approaching the text—or the image—does not necessarily imply approaching the author or the period. Moreover, the author's complexity and context go beyond possible generalisations. Jardine (2004) himself also reminds us of the implications of the distinction between etics and emics, based on the work of the American anthropologist Kenneth Pike, i.e. "the application of our theories in analysing others' behaviour and institutions" in the former case, and "the interpretation of others' worlds as they appear to them" in the latter case.

From the above analysis, is it inevitable that the history of science is presented from some form of presentism? Yes, at least insofar as those who reconstruct history do so from their own present. The problem is not presentism per se, but the implications it may have for historical reconstruction and interpretation. In any case, it is important to be aware of the implications of a presentist perspective, especially for the understanding of primary sources and their contexts.

Now, how is the "Darwinian revolution" interpreted from presentism? James Moore (1991), for example has already argued that Darwinism cannot mean whatever one wants it to mean, and that in historical terms there can be as many meanings as there are interpretations of that "loaded term". The "Darwinian revolution" is based on what sociologists of science such as the Colombian Olga Restrepo (2002) call a normative view of "Darwinism". This notion suggests that there is a "right" or "correct" way of understanding Darwin's ideas, and that this view continues to this day. Regardless of which version of presentism is held, the central point is to argue for a change in worldview based on science.

In the case we want to analyse in this chapter, the so-called "tree of life", the pertinent question is why concepts and images from the nineteenth century are still being used today. It is clear that concepts, or in this case, the meanings and interpretations of images, can change depending on a specific context. However, it is also true that each concept or image retains interpretive elements that can be seen as essential over time.

3 "Darwinism" and the Tree of Life

Being the main theme of this manuscript, the pictorial metaphor of the "Tree of Life" and the reasons why certain versions of it are susceptible to deconstructing, we will neither dwell on an analysis of the epistemological and methodological role of Darwin 1859's diagram in his theory of evolution—although we will mention some of its implications for contemporary problems of Darwinism (e.g. the commitements to the main tenets of vertical descent)—nor present Darwin's images and speculations prior to the publication of *On the Origin of Species*, such as his image of Notebook B of 1837, which was popularised on the bicentenary of his birth in 2009.

As a twentieth-century narrative, we already stated that the "Darwinian revolution" was put forward as a form of justification for a particular view of evolutionary biology, corresponding to the Modern Synthesis. Since the 1930s and 1940s, the professionalisation of evolutionary biology was launched thanks to the mathematical models that boosted population genetics and to influential evolutionists such as Theodosius Dobzhansky, Ernst Mayr, Julian Huxley, and George Gaylord Simpson, who were major characters in the so-called "modern synthesis of evolution" also known as the Neo-Darwinian Synthesis. Their standard narrative goes on to argue that the "synthesis" resolved significant difficulties and confusion in the field of evolutionary biology of the time. It reconciled genetics with evolution, reinstated natural selection as the primary mechanism of evolutionary change and showed that the significant transformations observed in the fossil record (macroevolution) could be explained by changes in local populations (microevolution). Regarding the interest in the reconstruction of the evolutionary history and relationships amongst or within groups of organisms (phylogenetics), during the period of the Evolutionary Synthesis, the pursuit of methods to link classifications with phylogenies of organisms remained limited, but in the 1950s and 1960s intensified when "some of the issues that had not been resolved in the late 1800s began to surface again" (O'Hara 1994: 14).

It is well known that from the second half of the twentieth century, especially Mayr and Simpson made great efforts to incorporate the theoretical edifice of the Modern Synthesis in formal education or textbooks (Rieppel 2008), which resulted in its incorporation into museums, magazines, and popular culture. By the 1960s, the central tenets of the Modern Synthesis were so heavily incorporated that, to a large extent, Mayr's ideas represent what both biologists and non-specialists understood (and still understand) by evolution. To this day, most primary and secondary school curricula include the "traditional pillars" of evolutionary theory by natural selection: variation, inheritance, selection, time, and adaptation. The high school sometimes also includes the periodisations given by the main historiographical labels related to this view of evolutionary biologies, such as the "Darwinian Revolution" and Modern Synthesis. According to this "synthetical narrative", evolutionary theory is the core of biology, and the Tree of Life is its unifying metaphor (Ragan 2009). According to Joseph Caron, Mayr extended the notion that biology was founded on Darwinian evolutionism. However, no evidence exists beyond an association "made for the most part in an implicit fashion" between biology and evolution (Caron 1988: 240).

It might be worth remembering that Darwin's evolutionary diagram is the only image present in his *On the Origin of Species* which may alert us to the epistemic, ontological, and heuristic importance it had for him. At the time when *On the Origin* was released (1859), putting a single image in a book was unusual because "in the intellectual life of the Victorian era, the sciences and the arts were not 'two cultures' but one" and natural history books were profusely illustrated (Donald 2009: 1). On the other hand, the diagram was printed on a paper more extensive than the book and had to be unfolded for viewing, which involved expensive printing. For Darwin to have put a single illustration in the book means he must have considered it key for his argumentation (Voss 2010). Indeed, after many years of experimenting with images to capture and develop his ideas on the generation of species, Darwin knew that to visualise evolution in the form of a branching diagram meant to understand it. He used this influential figure to introduce his theory of the formation of lineages through geological time—"descent with modification"—through natural selection as the principal agent of change. Moreover, he used it as evidence of his theory (Brink-Roby 2009).

The epistemic and ontological commitments Darwin's abstract and hypothetical diagram had for him are: it does not represent real species, since Darwin did not conceive of a method for inferring evolutionary relationships between true lineages. It shows how natural selection works through the generations by accumulating advantageous variations and by eliminating the unfavourable ones which allow gradual change of species over time; and, how this divergence can eventually lead to the emergence of new species (Fig. 1). That all species are interrelated; thus, several species can, over time, emerge from a single ancestral one—though Darwin deliberately omitted a single point of origin in his diagram. Extinction is a natural phenomenon in the evolutionary process.

One peculiarity of Darwin's figure is that, unlike genealogical or philological charts, which can be crosslinked, it branches continuously, and its branches do not reconnect after their point of origin. This is important because it illustrates two key concepts of Darwin's proposal: a community of descent—a common ancestor can originate several descendant species, but any of these descendants have only one immediate common ancestor—and divergence. Also, Darwin managed to integrate hypothetically, on the one hand, species as a classification unit and, on the other, the

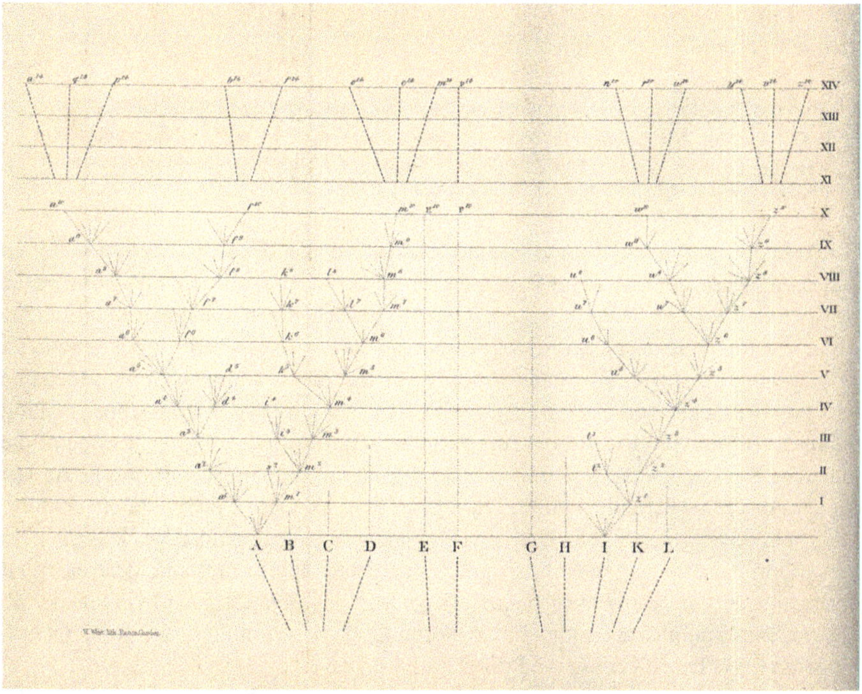

Fig. 1 Diagram in *On the Origin of Species*, 1859, pp. 116–117

ability of species to evolve. Therefore, the image of the branching lines is not only a symbol of the conceptualisation of history in nature but also a scheme in which time and taxonomy combine by linking synchronous elements—species on the horizontal axis—to diachronic elements—events in time on the vertical axis.

3.1 The Tree of Life as a Universal Phylogeny

In a nutshell, Darwin's commitments with the image that can explain the current need to "shake the tree" are gradualism, the notion of groups (species for him) as natural kinds (distinctive and discrete) or with well-defined boundaries, vertical inheritance (branches are constantly forking without merging) and divergence amongst the groups. How did Darwin's diagram turn *the* "Tree of life", which became synonymous with universal phylogeny? What are the implications?

Beginning with the semantic aspect, Darwin never used the word "tree" to refer his single figure in *On the Origin*, which he always referred to simply as a "diagram". Darwin employed his figure to show that natural selection was the cause of descent with modification through generations and that the natural system had to be genealogical in its arrangement. However, at the end of Chap. 4, Darwin felt he had to use a memorable metaphor to make his explanation more precise. It is then that he mentions evolution being like a Tree:

> The affinities of all the beings of the same class have sometimes been represented by a great tree. I believe this simile largely speaks the truth. The green and budding twigs may represent existing species; and those produced during each former year may represent the long succession of extinct species. (Darwin 1859: 129)

However, despite his elaborate and theoretically dense diagram, Darwin admitted that there were not any representations of genealogies of living beings since he considered it impossible to rebuild real lines of descent due to the imperfection of the fossil record. Thus, Darwin did not provide a formula or method for constructing a similar diagram for real species or taxa. After 1859, some naturalists became interested in achieving a diagram to represent evolutionary relationships, making the research of evolutionary relatedness amongst groups of living beings one of the practices that dominated much of the inquiry on the organic world around the last decades of the nineteenth century. During the first decades of the twentieth century, interest in developing a method of phylogenetic inference remained weak as the field was still in its early stages. Still, several fundamental approaches and ideas emerged during this period, laying the groundwork for later developments in phylogenetics. A few notable examples are comparative morphology, palaeontology and fossil record, biogeography, hybridisation, and polyploidy.

It was until the decades of the 1950s and 1960s that taxonomists attempted to be more explicit about their purpose and precise about their methods (Hull 1970). Three schools of biological systematics emerged and consolidated in the 1960s: evolutionary taxonomy (Mayr and Simpson), phenetics or numerical taxonomy (Sokal and Sneath), and phylogenetic systematics or cladistics (Hennig), which energetically debated each other. "The outcome of that debate was the victory of phylogenetic systematics/cladistics over the competing schools of thought" (Rieppel, 2008: 293), so that—the traditional narrative continues—a cladistics revolution took place that swept away evolutionary taxonomy as well as phenetics (Hull 1988). Since then, "Hennig's (1950, 1966) method of phylogeny reconstruction eventually became the generally accepted modus operandi in animal taxonomy" (Rieppel 2008: 296), burying alternative methods of character assessment that were based on background theories of inheritance, development, and function.

For all those involved in the "science wars between evolutionists, pheneticists and cladists" (Rieppel 2008: 294), under the paradigm of synthetic theory, one thing was clear: even if they disagreed on the method of reconstructing evolutionary history, it should be plotted in the form of a branching diagram, heir to Darwin's. Thus, the term evolutionary tree began to refer to phylogenies, and the idea of the possibility to come up with a tree for all the diversity of life on our planet became the "Tree of Life". "What Darwin gave us when he published *On the Origin* was a nontheistic reason for the discoverability and utility of such classifications. 'Community of descent is the hidden bond which naturalists have been unconsciously seeking, and not some unknown plan of creation,' he wrote (1859). There was to be an actual Tree of Life whose 'ramifying branches may well represent the classification of all extinct and living species in groups subordinate to groups'" (Doolittle and Brunet 2016: 1).

Besides, as many evolutionary biologists defended the plausibility and utility of the existence of a universal tree of life, a narrative which suggests that the reconstruction of the pattern of phylogenetic relationships is not only possible but epistemically accessible was pushed forward: "for there is, after all, one true tree of life, the unique pattern of evolutionary branchings that actually happened. It exists. It is, in principle, knowable. We do not know it all yet. By 2050 we should— or if we do not, we shall have been defeated only at terminal twigs, by the sheer number of species—. . ." (Dawkins 2003: 112).

However, even before phylogenetic systematics became the favoured method to reconstruct phylogenies, the equivalence of the graphic results of any of the schools with Darwin's diagram was already well established. In 1984, during the Biennial Meeting of the Philosophy of Science Association, Mayr wrote, "The theory of common descent of all organisms was nothing but a working hypothesis in Darwin's day. Nevertheless, the evidence for it is so overwhelming that it is no longer controversial except for cladistic detail. With the discovery that the genetic code is the same in all organisms, down to the prokaryotes, the descent of all existing organisms from a single origin of life is no longer questioned". And although Mayr recognised some exceptions to monophyly, he concluded that "on the whole our thinking about phylogeny today is very much the same as that of Darwin" (p. 148). A statement that obscures the fact that Mayr treated back in 1974 whether the interpretation of phylogeny given by evolutionary taxonomists or that offered by cladists corresponded more closely to Darwin's ideas on the subject.

Then, the 1960s and 1970s were marked by the birth and early advances of molecular phylogenetics (Koonin and Wolf 2009), which were expected to offer trees definable in terms of molecular data alone. Such trees could be compared with those based on the anatomical characters of organisms, and if they happened to agree, then real proof for macroevolution would be obtained. Most importantly, putting all prokaryotes and eukaryotes into a single universal Tree of Life (TOL) seemed possible.

3.2 Trees in the Centre of Discussions

"That LGT [Lateral Gene Transfer] might be a spoiler for the TOL was only a suspicion at the beginning of the era of molecular phylogenetics" (Doolittle and Brunet 2016: 2). Today, both the discovery of phenomena that can lead to genealogical discordance (such as hybridisation events, lateral or horizontal gene transfer, gene duplication, and incomplete lineage sorting) and the alternative interpretations of certain evolutionary phenomena given by different research areas (developmental bias and plasticity, inclusive inheritance and niche construction) has forced evolutionary biologists to question the feasibility of achieving the Tree of Life, making it a deeply contested representation. It seems that the process of evolution, especially of bacteria and plants, is better understood by complex webs of relationships rather than clean, dichotomously branching trees, thus revealing evolutionary history as a network, at least for these groups.

The conflict between trees and the presumed taxonomic placement of species and between trees derived from different genes has highlighted the complexity of evolution. This has turned many evolutionary biologists' attention to the role of networks as more accurate depictions of evolutionary history, and not only scientists but "philosophers have recently (and fruitfully!) turned their attention to horizontal gene transfer, especially in the context of microbiology, the concept of microbial species, and the Tree of Life (e.g. Franklin 2007; O'Malley and Dupré 2007; O'Malley et al. 2010; Bouchard 2010; Ereshefsky 2010; Franklin-Hall 2010; Velasco 2010; O'Malley 2014)" (Haber 2019: 4).

In short, during the last five decades, phylogenetic systematics has passed through substantial changes that have contributed to tackling important theoretical and methodological issues in systematics (Ebach et al. 2008), trying to "save" the predominant model of evolutionary connections amongst species as a bifurcating tree, just as Darwin depicted his figure in *On the Origin of Species*. However, "when the levels of data incongruence are great, researchers may resort to networks – often as a last option after all other tree-based methods have failed – to provide some way of making sense of the patterns within a dataset. Thus, the wide range of available network methods (Huson and Bryant 2006) has remained underutilised, likely because of the difficulties that arise when comparing trees and networks (such as matching tree branches to network edges)" (Schliep et al. 2016: 1212). We will explain below that there are powerful psychological and sociological reasons for scientists' clinging to the tree metaphor, which has made it persist, despite challenges to the "tree paradigm", for nearly 200 years, making Mayr's statement that "our thinking about phylogeny today is very much the same as that of Darwin" (Mayr 1984) uncontested in the mind of many specialists and non-specialists alike, idealising phylogenetic relationships.

3.3 Evolutionary Trees for the Public

Let us now turn to the public understanding of evolution. By the mid-twentieth century, the establishment in the collective imagination of the Western world that evolution is the organising principle of life, and trees are its embodiment began. Furthermore, the idea that trees follow directly from the central notions of Darwin's theory of evolution—since the conceptions of the non-fixation of species, common descent, and species diversification immediately lead to summarising life in the branching of modern forms from ancestral ones—was established. In short, the public was told that life has a historical narrative that can be conveniently represented by a tree, but not any tree. A *real* tree.

In Germany, the effort to build genealogies for groups of organisms experienced a significant boom at the end of the eighteenth century. Naturalist Ernst Haeckel read *On the Origin* in 1860 and again in 1861 (Richards 2008: 68). He was so impressed that he wrote to Darwin in 1864: "... Of all the books I have ever read, not a single one has come even close to making such an overpowering and lasting impression on me, as your theory of the evolution of species". Indeed, Haeckel became one of the most influential propagandists for evolution.

Haeckel, who in 1861 was 27 years old, had a doctorate in zoology and was newly appointed as a professor of comparative anatomy at the University of Jena, immediately accepted Darwin's two main ideas. That natural selection was the cause of descent with modification through generations and that the natural system had to be genealogical in its arrangement, as Darwin had shown with his branched diagram first and then with his metaphor of the "Tree of Life" described at the end of *On the Origin*'s Chap. 4 (Richards 2008).

Darwin's diagram is an abstract and hypothetical descent pattern consisting of straight lines, which rise from several points of origin at the base. Nevertheless, Haeckel devoted himself to building real organisms' genealogies based on the belief that the embryonic development of an organism—its ontogeny—revealed its evolutionary history—its phylogeny. He called this the "biogenetic law". This implied that if an individual repeated in its development the main stages of evolutionary change that their ancestors passed through during their long paleontological history (ontogeny recapitulating phylogeny), genealogical reconstructions could be made. It was Haeckel who coined the word "phylogeny" (Dayrat 2003).

In 1866, Haeckel published his *Generelle Morphologie des Organismen* (General Morphology of Organisms), which included eight genealogical trees or the so-called Stammbaum. These diagrams reflected his hypothesis on the origin of life and the genealogical relationships amongst groups of organisms. In 1868, Haeckel published *Natürliche Schöpfungsgeschichte* (The Natural History of Creation), containing the same type of material as *Generelle Morphologie* but designed for a broader audience. A curious thing regarding this book's genealogical trees is that they displayed Haeckel's new pictorial invention of stylised shrubs. Finally, in his book *Anthropogenie* (The Evolution of Man), published in 1874, Haeckel focused on embryonic development to explain the evolutionary process. In this book, he addressed the issue of human origin, its relationship with other primates and the nature of races. One can find perhaps the most emblematic tree image: his *Systematischer Stammbaum des Menschen* (Pedigree of Man), one of the few genealogies in *Anthropogenie*. This motif eventually became an archetypal image for popular representations of evolution (Fig. 2).

Before moving on, it is important to comment on Haeckel's diagrams. In constructing his phylogenies, Haeckel was thinking about the "development metaphor"—the analogy of evolutionary change with the growth and development of the individual—and never managed to build a complete genealogy or line of descent as Darwin had in mind. The biogenetic law does not reveal a series of hypothetical ancestors but their main morphological stages that are temporally rooted in a paleontological dimension. In addition, Darwin—who avoided conjecture and guesswork at all costs—drew an abstract diagram of the pattern of evolution, not daring (publicly at least) to build genealogies for actual groups. In Darwin's figure, the passage of time is represented merely as generations. Haeckel instead

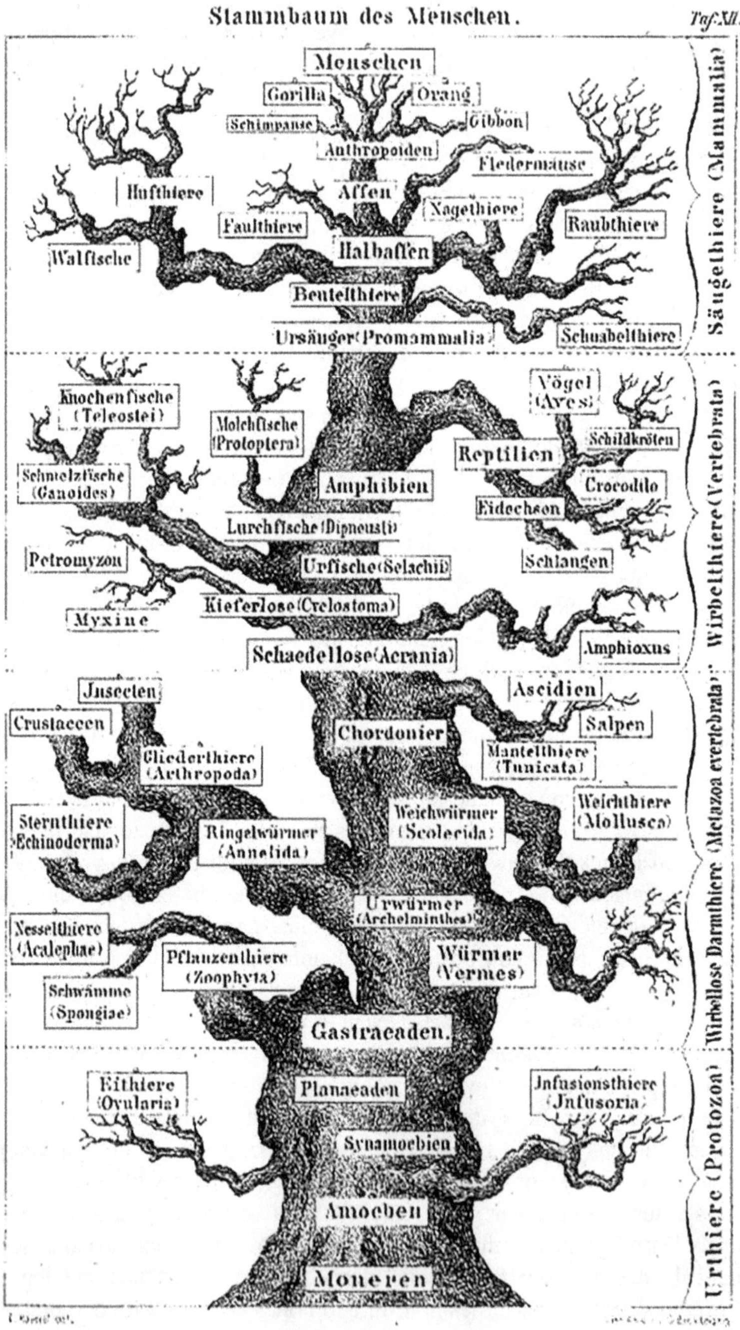

Fig. 2 Table XII Systematischer Stammbaum des Menschen (Genealogy or pedigree of Man), in Haeckel's *Anthropogenie*, 1874

decided to illustrate most of his genealogies as real trees or shrubs with labyrinthine branches connected to thick trunks with bark. He went as far as using his creativity where there was no scientific evidence, including modern forms called "living fossils" to represent older stages of some branches—knowing that each phylum continues to change after it arises—and tracing the origin of groups in geological time (Dayrat 2003).

In fact, after reading *Natürliche Schöpfungsgeschichte*, Darwin wrote to Haeckel on November 19, 1868:

> My dear Haeckel... your chapters on the affinities & genealogy of the animal kingdom strike as admirable and full of original thought. Your boldness, however, sometimes makes me tremble, but as Huxley remarks, someone must be bold enough to make a beginning in drawing up tables of descent [...] Although you fully admit the imperfection of the Geological record, yet Huxley agreed with me in thinking that you are sometimes rather rash in venturing to say at what periods the several groups first appeared. (Darwin's letter to Haeckel, dated November 19, 1868; Correspondence, Vol. 16 part II p. 850)

However, in chapter tenth of the fifth edition of *On the Origin* (1869), Darwin introduced his seeming approval of Haeckel's general employment of trees as diagrams of descent:

> Professor Haeckel, in his "Generelle Morphologie" and in several other works, has recently brought his great knowledge and abilities to bear on what he calls phylogeny, or the lines of descent of all organic beings. In drawing up several series he trusts chiefly to embryological characters, but draws aid from homologous and rudimentary organs, as well as from the successive periods at which the various forms of life first appeared in our geological formations. He has thus boldly made a great beginning and shows how classification will in the future be treated. (Darwin 1869: 676)

Thus, Darwin was ambivalent on whether Haeckel had gone too far in his experimental efforts to build stem trees to achieve a natural way of grouping organisms. Publicly, however, he never mentioned Haeckel having distorted his ideas. Something indeed interesting about this image is its repercussion in the posterior representation—and misconstruction—of evolution, particularly in the public realm.

Many scientists took Haeckel's *Stammbaum des Menschen* to represent a complete phylogeny of life, that is, a complete system of evolutionary descent amongst organisms, forgetting its justificatory conception of being an attempt to show the portion of the phylogeny corresponding to man, in an unbranched, lineal chain of forms.

By appreciating this image as a whole phylogeny, some fundamental notions, misconstrues—and prejudices—are aroused. The first was the assumption of Haeckel's linear, hierarchical and progressive view of the evolutionary process culminating—very similar to the chain of being—in humankind. The second was that Haeckel assumed that each side branch had retained the same level of complexity since it emerged. The third was Haeckel's conjecture that early apparition in time meant primitive but also numerically restricted.

An additional aspect of this image, which eventually contributed to misinterpreting other similar evolutionary trees, is that the thick trunk can be seen as a main stem of progress—the vertical axis involving two dimensions: time and level of complexity. By placing man at the centre, the logical conclusion was that "trees can uproot the most powerful and persistent of all ideological prejudices in the ordering of life, human domination by right of superiority" (Gould 1997: 31).

In the popular realm, Haeckel's tree, taken as representing a complete phylogeny of life, gave rise to multiple non-scientific images—eventually named "Trees of Life"—that spread a hierarchical and progressive view of evolution. These images soon became iconic (Fig. 3). However, it was not Haeckel's way of picturing evolution; the popularisers' eventual use of this tree caused the later confusion. In the following images that remained comparatively obscure, "one can barely detect Homo crouching on the upper right branch of the stem-tree of vertebrates, hardly represented as the teleological fulfilment of the evolutionary process" (Richards 2008: 142).

3.4 Shaking the Tree or the Need to Revise an Orthodoxy

We would like to begin wrapping up this text by explaining our choice of title. By shaking the tree, we meant paying close attention to the methodological, ontological, and epistemological commitments the "Tree of Life" has had in the different periodisations of evolutionary biology according to the "synthetic narrative" (such as the "Darwinian Revolution", the Modern Synthesis, and the Extended Synthesis), to give a historical and historiographical perspective to the current debate. By shaking up this perspective, we hope (perhaps not today or tomorrow, but when enough shocks come together) that those historiographical approaches, concepts, and historical labels that no longer hold up regarding the "Tree of Life", such as the knee-jerk defence of a tree that does not reticulate, the mythological narrative of winners centred on the idea of the "Darwinian Revolution" and the problem of presentism (or telling the story backwards to project our present expectations on tree diagrams onto the past), will fall away. Regarding this point, we refer to recent historiographies on the tree's history where tree diagrams from the late eighteenth and early nineteenth centuries are frequently styled as the "pre-evolutionary" or "pre-Darwinian" or even as the "roots" of later phylogenetic trees (Hellström 2016).

The main tenet of the "Tree of Life" is that all living organisms share a common ancestry and can be represented as branches on a tree. Whilst this metaphor has been widely used to illustrate the patterns of evolution, it has become increasingly contested in recent years due to several factors, such as:

- Horizontal Gene Transfer (HGT): The traditional tree metaphor assumes that genetic information is passed vertically from parent to offspring. However, scientists have discovered that genes can also be transferred horizontally between different species, especially microorganisms. This complicates the idea of a strictly bifurcating tree and suggests a more complex web of genetic relationships.

Fig. 3 A public domain evolutionary tree found on the web

- Hybridisation and Introgression: Hybridisation, the interbreeding of two distinct species, and introgression, the transfer of genetic material from one species to another, is more common than previously thought. These processes can blur the boundaries between species and result in the exchange of genetic material across branches of the tree, challenging the concept of a strictly hierarchical tree structure.
- Retention of ancestral traits: Some evolutionary lineages may retain ancestral traits for more extended periods of time, creating a mosaic pattern of genetic relationships that do not neatly fit into a tree structure. This can occur due to various factors, such as adaptation to specific environments or the persistence of genetic variation within populations.
- Incomplete or limited data: Our understanding of the tree of life is based on the available evidence, including fossil records and genetic data. However, our data is often incomplete, and our knowledge has many gaps and uncertainties. As more data becomes available, it may challenge or refine our current understanding of evolutionary relationships.
- Methodological challenges: Inferring evolutionary relationships from genetic data involves complex computational methods and assumptions. Different methods can produce different results, leading to debates and disagreements about the precise branching patterns in the tree of life.

In the popular sphere, the Tree of Life, as a metaphor for evolution, can sometimes lead to misunderstandings due to its simplicity and the complexities of evolutionary processes. Here are a few reasons why this metaphor can be misunderstood:

- The misconception of linear progression: The tree of life metaphor may give the impression that evolution is a linear process with a clear hierarchy, implying that some organisms are more advanced or superior. However, this is not the case. Evolution is a branching and diverging process, with species adapting to various environmental conditions. Each branch represents a distinct lineage, not necessarily a higher or more advanced life form.
- Limited representation of evolutionary dynamics: The tree of life metaphor provides a static snapshot of evolutionary relationships but does not capture evolution's dynamic nature. Evolutionary processes involve continuous adaptation, diversification, and extinction, resulting in a more intricate and complex pattern than a simple branching tree can convey.
- A simplistic view of species boundaries: The tree of life metaphor assumes clear boundaries between species, suggesting distinct branches on the tree. However, in reality, defining species can be a complex task. Hybridisation, introgression, and transitional forms challenge the notion of discrete species categories and can blur the boundaries between branches.
- Lack of emphasis on coevolution and ecological interactions: The tree of life metaphor focuses primarily on genetic relationships and evolutionary divergence. It may overlook the significance of coevolutionary dynamics and the interactions between species within ecosystems. Evolution is not solely about the branching of lineages but also the interplay between organisms and their environment.

Thus, in light of novel empirical discoveries and conceptual and theoretical advances, the tree figure is more likely to coexist with networks to depict vertical and horizontal inheritance (Blais and Archibald 2021). To many, there is a pressing need to achieve a more sophisticated understanding of reconstructing phylogenies through a concerted multidisciplinary effort. "This is a good thing, and it presents new avenues for advancement in phylogenetics while pressing us to reframe some of the central questions and commitments of phylogenetics" (Haber 2019: 17). However, "despite advances in both fields [tree and network inferences], the interface between trees and networks have rarely been bridged. The decision to use trees or networks is usually not based on arguments about the superiority of one approach over the other but rather on the evolutionary complexity of the group under investigation and the resulting data set. Nonetheless, tree-based methods remain the most common analytical choice" (Schliep et al. 2016: 1212).

3.5 Why an Orthodoxy?

Are there non-scientific reasons for this resistance to the use of networks? Taking up the ideas of the introduction about the appropriation of a religious—or quasi-religious—discourse within the study of evolution and mixing them with the ideas of Ludwik Fleck about "thought collective" (1981), we can see how the figure of the tree has tended to remain valid despite its recent explanatory ineffectiveness due to, at least in part, "Darwinism" as a shared system of opinion or thought within the scientific communities. Fleck argued that scientific knowledge is not solely the product of individual scientists but emerges from collective social processes. He proposed that scientific thought is shaped by a collective framework of shared beliefs, assumptions, and social interactions within a specific scientific community. This framework, which Fleck called the "thought collective", determines what is considered valid or meaningful within a particular scientific field. According to Fleck, the thought collective establishes the norms, concepts, and methods that guide scientific research. It shapes how scientists perceive and interpret data, constrains the range of acceptable ideas and interpretations, and influences the direction of scientific inquiry.

In this sense, the resistance to the use of networks, in line with Fleck's ideas, is the result of five factors: (1) the consideration that the presence of contradictions is practically impossible, (2) the contradictions present are not observed or (3) they are ignored, (4) great efforts are made to assimilate the contradictions, and (5) there must be many contradictions or a very deep one to eliminate the system (Fleck 1981: 73–80).

"When the levels of data incongruence are great, however, researchers may resort to networks–often as a last option after all other tree-based methods have failed–to provide some way of making sense of the patterns within a dataset. The only alternative is filtering 'rogue' taxa that are causing topological conflict and decreased branch support (Aberer et al. 2013) or analysing character subsets that each have an approximately tree-like history (e.g. non-recombining sequence blocks). Thus, the wide range of available network methods (Huson and Bryant 2006) has remained underutilised, likely because of the difficulties that arise when comparing trees and networks (such as matching tree branches to network edges)" (Schliep et al. 2016: 1212).

Thus, the "Darwinian revolution" has served to maintain and privilege a particular vision of life over other past and present alternatives (think of the inheritance of acquired traits, saltationism, Evo-Devo, and epigenetics, amongst others). Nevertheless, the thought collective is not static but evolves as new discoveries, technologies, and social changes impact the scientific community. Regarding the "Tree of Life", "students of animals and plants have long accepted that incomplete lineage sorting, introgression, and full-species hybridisation pose difficulties for the sorts of trees that Darwin might have had us draw. However, with their promiscuous willingness to exchange genes between widely separated branches of any "tree", microbes have most seriously jeopardised the neo-Darwinian synthesis in the oversimplified form that we have often presented to the public. More sophisticated understandings remain possible and should be debated in a more conceptually and science-historically self-aware context" (Doolittle and Brunet 2016: 5–6). Until then, systematists tend to accept that trees and networks both offer important evolutionary insights and are part of different research agendas. "While tree builders concern themselves with the inference of vertical descent, networkers are increasingly focused on community-based sources of genetic variation and their ecological implications. Moreover, they seek to incorporate aspects of biology traditionally excluded from phylogeny, such as co-evolution, symbiosis and environmental factors" (Blais and Archibald 2021: R311).

Finally, regarding the need to shake the tree to get rid of presentism, it is important to note that science historians need to adopt a more contextualised and nuanced approach. As previously mentioned, it involves understanding the historical context, appreciating the perspectives and constraints of historical actors, critically analysing primary sources, and recognising the limitations and uncertainties that existed at the time. By doing so, historians can provide a more accurate understanding of the development of scientific ideas and practices throughout history.

4 Conclusions

What is central to reconstructing the history of science? This question has historically led to the proposal of different approaches that, as if they were lenses, have different graduations. At one extreme microhistories have a narrow focus, and at the other, global histories have a much broader one. Microhistories focus on examining specific events, situations, or people in a wider historical context, with the difference from traditional history being that more detailed and lesser-known aspects of everyday life are analysed. This allows for a deeper understanding of historical reality, mainly because although the focus is narrow, microhistories are still placed in a broader historical context, which, in order to be studied, often requires interdisciplinary methods.

Concerning the question of what is central to reconstructing the history of science, the only simple answer would be that it depends. The last few decades have witnessed significant efforts to deconstruct (understood as a critical approach that questions traditional accounts and assumptions about scientific knowledge and its development) various scientific discourses. In evolutionary biology, deconstruction aims to question and dismantle the binary oppositions, hierarchies, and fixed meanings often present in interpreting scientific texts and discourses and the notion of a singular and stable meaning. It emphasises the fluid and contingent nature of knowledge production.

Taking into account the ideas we presented in the introduction on the formation of traditions, and taking "Darwinism" as a tradition of thought, it is understandable that its deconstruction is complicated and faces diverse resistances. Within "Darwinian thinking", a central idea is to understand the evolutionary process as a branching diagram, whose branches always diverge and never merge. This picture has several compromises, namely, that the process is gradual, that it is guided primarily by the mechanism of natural selection, that vertical inheritance is predominant and that the arrangement of the branches and nodes is determined by the presence or absence of specific traits or characteristics.

The fact that the evolutionary tree belongs to "Darwinism" as a way of seeing the world, implies a sometimes uncritical maintenance of this image, causing that at present tree representations coexist with those of networks and that the aspiration to achieve a theoretical unification has slowed down. We are not arguing that the image of the tree is not useful, since, as we have already explained, it has been incredibly fruitful empirically and epistemologically. Nevertheless, it may be dispensable.

Acknowledgements This work was possible by research grants PAPIIT IN401622: *Deconstruyendo el darwinismo: Hacia una nueva historiografía* of the General Direction of Academic Personnel Affairs (DGAPA) of the National Autonomous University of Mexico (UNAM), and Project CB-2017-2018-A1-S-8786-H-4180: *Ciencia Transnacional y Redes de Colaboración en el Estudio de las Ciencias de la Vida en México, 1960-1990* of the Nacional Council of Science and Technology, CONACyT.

References

Aberer AJ, Krompass D, Stamatakis A (2013) Pruning rogue taxa improves phylogenetic accuracy: an efficient algorithm and webservice. Syst Biol 62(1):162–166

Blais C, Archibald JM (2021) The past, present and future of the tree of life. Curr Biol 31(7):R314–R321

Bouchard F (2010) Symbiosis, lateral function transfer and the (many) saplings of life. Biol Phil 25(4):623–641

Brink-Roby H (2009) Natural representation: diagram and text in Darwin's on the origin of species. Vic Stud 51(2):247–273

Brush SG (1995) Scientists as historians. Osiris 10:214–231

Caron JA (1988) "Biology" in the life sciences: a historiographical contribution. Hist Sci 26:223–268

Chang H (2021) Presentist history for pluralist science. J Gen Philos Sci 52:97–114

Charles Darwin to Ernst Haeckel, dated November 19, 1868. Darwin Correspondence Project, "Letter no. 6466". https://www.darwinproject.ac.uk/letter/?docId=letters/DCP-LETT-6466.xml. Accessed 13 June 2023
Cohen IB (1985) Revolution in science. Harvard University Press, Cambridge and London
Darwin C (1859) On the origin of species by means of natural selection, or the preservation of favoured races in the struggle for life, 1st edn. John Murray, London
Darwin C (1869) The origin of species by Charles Darwin: a variorum text. University of Pennsylvania Press, Philadelphia
Daston L (2009) Science studies and the history of science. Crit Inq 35(4):798–813
Dawkins R (2003) The devils Chaplain: reflections on hope, lies, science, and love. Houghton Mifflin, Boston
Dayrat B (2003) The roots of phylogeny: how did Haeckel build his trees? Syst Biol 52(4):515–527
Donald D (2009) Introduction. In: Donald D, Munro J (eds) Endless forms: Charles Darwin, natural science and the visual arts. Yale University Press
Doolittle WF, Brunet TDP (2016) What is the tree of life? PLoS Genet 12:e1005912
Ebach MC, Morrone JJ, Williams DM (2008) A new cladistics of cladists. Biol Phil 23:153–156
Ereshefsky M (2010) Microbiology and the species problem. Biol Phil 25:553–568
Fleck L (1981) Genesis and development of a scientific fact. University of Chicago Press
Foucault M (2005) The order of things. Routledge, London and New York
Franklin LR (2007) Bacteria, sex, and systematics. Phil Sci 74(1):69–95
Franklin-Hall LR (2010) Trashing life's tree. Biol Phil 25(4):689–709
Gould SJ (1997) Redrafting the tree of life. Proc Am Phil Soc 141(1):30–54
Haber MH (2019) Species in the age of discordance. Philos Theory Pract Biol 11(21)
Hacking I (1992) 'Style' for historians and philosophers. Stud Hist Philos Sci Part A 23:1–20
Hacking I (1994) Styles of scientific thinking or reasoning: A new analytical tool for historians and philosophers of the sciences. In: Gavroglu K, Christianidis J, Nicolaidis E (eds) Trends in the historiography of science. Springer Science & Business Media, Dordrecht, pp 31–48
Haeckel, E. (1866). Generelle Morphologie der Organismen: Allgemeine Grundzüge der organischen Formen-Wissenschaft, mechanisch begründet durch die von Charles Darwin reformierte Descendenz-Theorie. Band 1: Allgemeine Anatomie. Band 2: Allgemeine Entwicklungsgeschichte. de Gruyter
Hellström P (2016) The great chain of ideas: The past and future of the history of ideas, or why we should not return to Lovejoy. Lychnos:179–188
Hennig W (1950) Grundzüge einer Theorie der phylogenetischen Systematik. Deutscher Zentralverlag, Berlin
Hennig W (1966) Phylogenetic Systematics. University of Illinois Press, Urbana, Chicago, London
Hodge J (2005) Against "Revolution" and "Evolution". J Hist Biol 38:101–121
Hull DL (1970) Contemporary systematic philosophies. Ann Rev Ecol Syst 1:19–54
Hull DL (1979) In defense of presentism. Hist Theory 18:1–15
Hull DL (1988) Science as a process: an evolutionary account of the social and conceptual development of science. The University of Chicago Press, Chicago
Huson DH, Bryant D (2006) Application of phylogenetic networks in evolutionary studies. Mol Biol Evol 23(2):254–267
Jardine N (2000) Uses and abuses of anachronism in the history of the sciences. Hist Sci 38:251–270
Jardine N (2004) Etics and emics (not to mention anemics and emetics) in the history of the sciences. Hist Sci 42:261–278
Koonin EV, Wolf YI (2009) Is evolution Darwinian or/and Lamarckian? Biol Direct 4:1–14
Latour B, Woolgar S (1979) Laboratory life: the social construction of scientific facts. Sage, Beverly Hills
Loison L (2016) Forms of presentism in the history of science. Rethinking the project of historical epistemology. Stud Hist Philos Sci Part A 60:29–37
Mayr E (1974) Cladistic analysis or cladistic classification? J Zool Syst Evol Res 12(1):94–128

Mayr E (1984) What is Darwinism today? PSA: Proceedings of the Biennial Meeting of the Philosophy of Science Association 2:145–156
Mayr E (1990) When is historiography Whiggish? J Hist Ideas 51:301–309
Moore J (1991) Deconstructing Darwinism: The politics of evolution in the 1860s. J Hist Biol 24: 353–408
O'Hara RJ (1994) Evolutionary history and the species problem. Am Zool 34(1):12–22
O'Malley MA (2014) Philosophy of microbiology. Cambridge University Press, Cambridge
O'Malley MA, Dupré J (2007) Size doesn't matter: towards a more inclusive philosophy of biology. Biol Phil 22(2):155–191
O'Malley MA, Martin W, Dupré J (2010) The tree of life: introduction to an evolutionary debate. Biol Phil 25:441453
Oreskes N (2013) Why I am a presentist. Sci Context 26:595–609
Ragan MA (2009) Trees and networks before and after Darwin. Biol Dire 4:1–38
Restrepo Forero O (2002) Leyendo historias sobre el Darwinismo. In: Puig-Samper MÁ, Ruiz R, Galera A (eds) Evolucionismo y cultura. Darwinismo en España e Iberoamérica. UNAM, Junta de Extremadura, Ediciones Doce Calles, Madrid, pp 21–46
Richards RJ (2008) The tragic sense of life: Ernst Haeckel and the struggle over evolutionary thought. University of Chicago Press, Chicago
Rieppel O (2008) Re-writing Popper's philosophy of science for systematics. Hist Philos Life Sci 30:293–316
Ruse M (2017) Darwinism as religion: what literature tells us about evolution. Oxford University Press, Oxford and New York
Schliep K, Potts AA, Morrison DA, Grimm GW (2016) Intertwining phylogenetic trees and networks (No. e2054v1). PeerJ Preprints
Secord JA (2023) Inventing the scientific revolution. Isis 1:5076
Shapin S (2008) The scientific revolution. University of Chicago Press, Chicago and London
Turchetti S, Herrán N, Boudia S (2012) Introduction: have we ever been «transnational»? Towards a history of science across and beyond borders. Br J Hist Sci 45(3):319–336
Velasco JD (2010) Species, genes, and the tree of life. Brit Jour Phil Sc 61(3):599–619
Voss J (2010) Darwin's pictures: views of evolutionary theory, 1837–1874. Yale University Press, New Haven
Wilson A (2017) Science's imagined pasts. Isis 108:814–826

www.ingramcontent.com/pod-product-compliance
Lightning Source LLC
Chambersburg PA
CBHW071812221225
37173CB00003B/11

* 9 7 8 3 0 3 1 4 2 6 2 8 5 *